MAGNETIC FIELDS THROUGHOUT STELLAR EVOLUTION

IAU SYMPOSIUM No. 302

COVER ILLUSTRATION:

Numerical dynamo in a rapidly-rotating spherical shell that models the magnetism of an active fully convective M dwarf. Due to the ordering influence of the dominating Coriolis force, convection develops as large scale convective columns that maintain the dynamo action. The surface topology of the magnetic field is dominated by its dipolar component. At depth, the magnetic field lines show a more intricate structure. The color of the field lines scale with the amplitude of the radial component of the magnetic field (red outward, blue inward) and the surface is made transparent to highlight the magnetic field structure at depth. Credits: Thomas Gastine

IAU SYMPOSIUM PROCEEDINGS SERIES

INTERNATIONAL ASTRONOMICAL UNION

UNION ASTRONOMIQUE INTERNATIONALE

MAGNETIC FIELDS THROUGHOUT STELLAR EVOLUTION

PROCEEDINGS OF THE 302nd SYMPOSIUM OF THE INTERNATIONAL ASTRONOMICAL UNION HELD IN BIARRITZ, FRANCE AUGUST 25–30, 2013

Edited by

PASCAL PETIT

Institut de Recherche en Astrophysique et Planétologie, Toulouse, France

MOIRA JARDINE

School of Physics & Astronomy, University of St Andrews, St Andrews, Scotland, UK

and

HENDRIK C. SPRUIT

Max-Planck-Institut für Astrophysik, Garching, Germany

CAMBRIDGE UNIVERSITY PRESS

CAMBRIDGE
UNIVERSITY PRESS

University Printing House, Cambridge CB2 8BS, United Kingdom

One Liberty Plaza, 20th Floor, New York, NY 10006, USA

477 Williamstown Road, Port Melbourne, VIC 3207, Australia

314-321, 3rd Floor, Plot 3, Splendor Forum, Jasola District Centre, New Delhi - 110025, India

79 Anson Road, #06-04/06, Singapore 079906

Cambridge University Press is part of the University of Cambridge.

It furthers the University's mission by disseminating knowledge in the pursuit of education, learning and research at the highest international levels of excellence.

www.cambridge.org
Information on this title: www.cambridge.org/9781107044982

First published 2014

A catalogue record for this publication is available from the British Library

Library of Congress Cataloging in Publication data

This journal issue has been printed on FSC-certified paper and cover board. FSC is an independent, non-governmental, not-for-profit organization established to promote the responsible management of the worlds forests. Please see www.fsc.org for information.

ISBN 978-1-107-04498-2 Hardback

Table of Contents

Foreword xii

Scientific Organizing Committee xiii

Conference photograph xiv

Participants xvi

General Introduction

Magnetic Fields in Stars: Origin and Impact 1
N. Langer

Session 1: Magnetized stellar formation

The Role of Magnetic Fields in Star Formation 10
R. E. Pudritz, M. Klassen, H. Kirk, D. Seifried & R. Banerjee

Magnetic field dispersion in the neighbourhood of Bok Globules 21
C. V. Rodrigues, V. de S. Magalhães, J. W. Vilas-Boas, G. Racca & A. Pereyra

The role of magnetic fields in pre-main sequence stars 25
G. A. J. Hussain & E. Alecian

Magnetic Fields in 7 Young Stellar Objects Observed with Nançay Radio Telescope 38
O. Bayandina, A. Alakoz & I. Val'tts

Can we predict the magnetic properties of PMS stars from their H-R diagram location? 40
S. G. Gregory, J.-F. Donati, J. Morin, G. A. J. Hussain, N. J. Mayne, L. A. Hillenbrand & M. Jardine

The magnetosphere of the close accreting PMS binary V4046 Sgr AB 44
S. G. Gregory, V. R. Holzwarth, J.-F. Donati, G. A. J. Hussain, T. Montmerle, E. Alecian, S. H. P. Alencar, C. Argiroffi, M. Audard, J. Bouvier, F. Damiani, M. Güdel, D. P. Huenemoerder, J. H. Kastner, A. Maggio, G. G. Sacco & G. A. Wade

V4046 Sgr: X-rays from accretion shock 46
C. Argiroffi, A. Maggio, T. Montmerle, D. Huenemoerder, E. Alecian, M. Audard, J. Bouvier, F. Damiani, J.-F. Donati, S. Gregory, M. Güdel, G. Hussain, J. Kastner & G. G. Sacco

X-rays from accretion shocks in classical T Tauri stars: 2D MHD modeling and the role of local absorption 48
C. Argiroffi, R. Bonito, S. Orlando, M. Miceli, F. Reale, G. Peres, T. Matsakos, C. Sthelé & L. Ibgui

Analysis of star-disk interaction in young stellar systems 50
N. N. J. Fonseca, S. H. P. Alencar & J. Bouvier

Magnetospheric Accretions and the Inner Winds of Classical T Tauri Stars 54
R. Kurosawa & M. M. Romanova

Building a numerical relativistic non-ideal magnetohydrodynamics code for astrophysical applications 64
S. Miranda Aranguren, M. A. Aloy & C. Aloy

3D YSO accretion shock simulations: a study of the magnetic, chromospheric and stochastic flow effects 66
T. Matsakos, J.-P. Chièze, C. Stehlé, M. González, L. Ibgui, L. de Sá, T. Lanz, S. Orlando, R. Bonito, C. Argiroffi, F. Reale & G. Peres

Magnetic higher-mass stars in the early stages of their evolution 70
J. H. Grunhut & E. Alecian

What do weak magnetic fields mean for magnetospheric accretion in Herbig AeBe star+disk systems? 80
A. N. Aarnio, J. D. Monnier, T. J. Harries & D. M. Acreman

Searches for the new magnetic intermediate-mass stars on various stages of MS evolution 84
E. A. Semenko

Chemical peculiarities in magnetic and non-magnetic pre-main sequence A and B stars 87
C. P. Folsom, S. Bagnulo, G. A. Wade, J. D. Landstreet & E. Alecian

Angular momentum evolution of young stars 91
S. P. Littlefair

The Effects of Magnetic Activity on Lithium-Inferred Ages of Stars 100
A. J. Juarez, P. A. Cargile, D. J. James & K. G. Stassun

Activity and Rotation in the Young Cluster h Per 102
C. Argiroffi, M. Caramazza, G. Micela, E. Moraux & J. Bouvier

X-ray emission regimes and rotation sequences in the M34 open cluster 106
P. Gondoin

The evolution of surface magnetic fields in young solar-type stars 110
C. P. Folsom, P. Petit, J. Bouvier, J.-F. Donati & J. Morin

Stellar models of rotating, pre-main sequence low-mass stars with magnetic fields 112
L. T. S. Mendes, N. R. Landin & L. P. R. Vaz

Session 2: Magnetic activity in the Sun and main-sequence stars with convective outer layers

Rotation and magnetism of solar-like stars: from scaling laws to spot-dynamos . 114
A. S. Brun

Probing the structure of local magnetic field of solar features with helioseismology 126
K. Daiffallah

Hanle and Zeeman effects: from solar to stellar diagnostics 130
A. L. Ariste

Coronal influence on dynamos 134
J. Warnecke & A. Brandenburg

A Bcool spectropolarimetric survey of over 150 solar-type stars 138
S. Marsden, P. Petit, S. Jeffers, J.-D. do Nascimento, B. Carter, C. Brown on behalf of the Bcool project team

High-resolution spectropolarimetry of κ Cet: A proxy for the young Sun. 142
J. D. do Nascimento, P. Petit, M. Castro, G. F. Porto de Mello, S. V. Jeffers, S. C. Marsden, I. Ribas, E. Guinan & the Bcool Collaboration

Theoretical evolution of Rossby number for solar analog stars 144
M. Castro, T. Duarte & J. D. do Nascimento Jr.

The large scale magnetic field of the G0 dwarf HD 206860 (HN Peg). 146
S. B. Saikia, S. V. Jeffers, P. Petit, S. Marsden, J. Morin, A. Reiners & the Bcool project

Starspots on Young Solar-Type Stars. 148
C. Brown, B. Carter, S. Marsden & I. Waite

Do Magnetic Fields Actually Inflate Low-Mass Stars? 150
G. A. Feiden & B. Chaboyer

A new spectropolarimeter for the San Pedro Martir National Observatory 154
E. Iñiguez-Garín, D. Hiriart, J. Ramirez-Velez, J. M. Núñez-Alfonso, J. Herrera & J. Castro-Chacón

Magnetic Fields in Low-Mass Stars: An Overview of Observational Biases 156
A. Reiners

On the spectropolarimetric signature of FeH in the laboratory and in sunspots . 164
P. Crozet, A. J. Ross, N. Alleq, A. L. Ariste, C. Le Men & B. Gelly

What controls the large-scale magnetic fields of M dwarfs? 166
T. Gastine, J. Morin, L. Duarte, A. Reiners, U. Christensen & J. Wicht

Magnetic fields in M-dwarfs from high-resolution infrared spectroscopy. 170
D. Shulyak, A. Reiners, U. Seemann, O. Kochukhov & N. Piskunov

Bridging planets and stars using scaling laws in anelastic spherical shell dynamos 174
R. K. Yadav, T. Gastine, U. R. Christensen & L. Duarte

Age, Activity and Rotation in Mid and Late-Type M Dwarfs from MEarth 176
A. A. West, K. L. Weisenburger, J. Irwin, D. Charbonneau, J. Dittmann & Z. K. Berta-Thompson

Magnetic fields of Sun-like stars 180
R. Fares

Stellar Magnetic Dynamos and Activity Cycles 190
N. J. Wright

Differential rotation and meridional flows in stellar convection zones 194
M. Küker & G. Rüdiger

Meridional flow velocities for solar-like stars with known activity cycles 196
D. Baklanova & S. Plachinda

On the reliability of measuring differential rotation of spotted stars........... 198
Z. Kővári, J. Bartus, L. Kriskovics, K. Vida & K. Oláh

Hyper X-ray Flares on Active Stars Detected with MAXI.................... 200
M. Higa, Y. Tsuboi, H. Negoro, S. Nakahira, H. Tomida, M. Matsuoka & The MAXI team

Modeling transiting exoplanet and spots For interferometric study............ 202
R. Ligi, D. Mourard, K. Perraut, P. Bério, L. Bigot, A. Chiavassa, A.-M. Lagrange & N. Nardetto

Stellar Magnetism in the Era of Space-Based Precision Photometry........... 206
L. M. Walkowicz

The new age of spotted star research using *Kepler* and CHARA 212
R. M. Roettenbacher, J. D. Monnier, R. O. Harmon & H. H. Korhonen

Rotation & differential rotation of the active Kepler stars.................... 216
T. Reinhold, A. Reiners & G. Basri

Starspots Magnetic field by transit mapping............................... 220
A. Válio & E. Spagiari

Investigating magnetic activity of F stars with the *Kepler* mission............ 222
S. Mathur, R. A. García, J. Ballot, T. Ceillier, D. Salabert, T. S. Metcalfe, C. Régulo, A. Jiménez & S. Bloemen

Detecting activity cycles of late-type dwarfs in Kepler data 224
K. Vida & K. Oláh

The effects of stellar winds and magnetic fields on exoplanets................ 228
A. A. Vidotto

Planetary protection in the extreme environments of low-mass stars 237
A. A. Vidotto, M. Jardine, J. Morin, J.-F. Donati, P. Lang & A. J. B. Russell

Planets spinning up their host stars: a twist on the age-activity relationship.... 239
K. Poppenhaeger & S. J. Wolk

Constraining Stellar Winds of Young Sun-like Stars......................... 243
C. P. Johnstone, T. Lüftinger, M. Güdel & B. Fichtinger

Bow shocks and winds around HD 189733b................................. 245
J. Llama, A. A. Vidotto, M. Jardine, K. Wood & R. Fares

Stellar Magnetism and starspots: the implications for exoplanets 247
C. Vilela, J. Southworth & C. del Burgo

On the effects of stellar winds on exoplanetary magnetospheres 251
V. See, M. Jardine, A. A. Vidotto, P. Petit, S. C. Marsden & S. V. Jeffers

Session 3: Origin and impact of magnetic fields in higher-mass stars with radiative outer layers

The nature and origin of magnetic fields in early-type stars 255
J. Braithwaite

The magnetic characteristics of Galactic OB stars from the MiMeS survey of magnetism in massive stars 265
G. A. Wade, J. Grunhut, E. Alecian, C. Neiner, M. Aurière, D. A. Bohlender, A. David-Uraz, C. Folsom, H. F. Henrichs, O. Kochukhov, S. Mathis, S. Owocki, V. Petit & the MiMeS Collaboration

Magnetic fields of OB stars 270
A. F. Kholtygin, S. Hubrig, N. A. Drake, N. Sudnik & V. Dushin

New observations of chemically peculiar stars with ESPaDOnS 272
V. Khalack, B. Yameogo, C. Thibeault & F. LeBlanc

The analysis of Li I 6708A line through the rotational period of HD166473 taking into account Paschen-Back magnetic splitting 274
A. V. Shavrina, V. Khalack, Y. Glagolevskij, D. Lyashko, J. Landstreet, F. Leone & M. Giarrusso

Bp stars in Orion OB1 association 276
I. I. Romanyuk & I. A. Yakunin

"Stellar Prominences" on OB stars to explain wind-line variability 280
H. F. Henrichs & N. P. Sudnik

Partial Paschen-Back splitting of Si II and Si III lines in magnetic CP stars 284
V. Khalack & J. Landstreet

The Dominion Astrophysical Observatory Magnetic Field Survey (DMFS) 288
D. A. Bohlender & D. Monin

Modeling surface magnetic fields in stars with radiative envelopes 290
O. Kochukhov

New Experiments with Zeeman Doppler Mapping 300
A. J. Martin, S. Bagnulo & M. J. Stift

Combining magnetic and seismic studies to constrain processes in massive stars 302
C. Neiner, P. Degroote, B. Coste, M. Briquet & S. Mathis

Magnetic fields of Ap stars from full Stokes vector spectropolarimetric observations 304
N. Rusomarov, O. Kochukhov & N. Piskunov

Magnetic Doppler Imaging of He-strong star HD 184927 306
I. Yakunin, G. Wade, D. Bohlender, O. Kochukhov, V. Tsymbal & MiMeS Collaborators

roAp stars: surface lithium abundance distribution and magnetic field configuration 309
N. Polosukhina, D. Shulyak, A. Shavrina, D. Lyashko, N. A. Drake, Yu. Glagolevski, D. Kudryavtsev & M. Smirnova

Roadmap on the theoretical work of BinaMIcS 311
S. Mathis, C. Neiner, E. Alecian, G. Wade & the BinaMIcS collaboration

Candidate Ap stars in close binary systems 313
C. P. Folsom, G. A. Wade, K. Likuski, O. Kochukhov, E. Alecian, D. Shulyak & N. M. Johnson

The unusual binary HD 83058 in the region of the Scorpius-Centaurus OB association 315
M. A. Pogodin, N. A. Drake, E. G. Jilinski & C. B. Pereira

Binary and multiple magnetic Ap/Bp stars 317
D. Rastegaev, Y. Balega, V. Dyachenko, A. Maksimov & E. Malogolovets

Wind channeling, magnetospheres, and spindown of magnetic massive stars 320
S. P. Owocki, A. ud-Doula, R. H. D. Townsend, V. Petit, J. O. Sundqvist & D. H. Cohen

X-rays from magnetic massive OB stars 330
V. Petit, D. H. Cohen, Y. Nazé, M. Gagné, R. H. D. Townsend, M. A. Leutenegger, A. ud-Doula, S. P. Owocki & G. A. Wade

Investigating the origin of cyclical spectral variations in hot, massive stars 334
A. David-Uraz, G. A. Wade, V. Petit & A. ud-Doula

The dichotomy between strong and ultra-weak magnetic fields among intermediate-mass stars 338
F. Lignières, P. Petit, M. Aurière, G. A. Wade & T. Böhm

UVMag: a UV and optical spectropolarimeter for stellar physics 348
C. Neiner, P. Petit, L. Parès & the UVMag consortium

Session 4: Magnetic fields in the ultimate stages of stellar evolution

Surface magnetism of cool giant and supergiant stars 350
H. Korhonen

Pollux: a stable weak dipolar magnetic field but no planet ? 359
M. Aurière, R. Konstantinova-Antova, O. Espagnet, P. Petit, T. Roudier, C. Charbonnel, J.-F. Donati & G. A. Wade

On dynamo action in the giant star Pollux : first results 363
A. Palacios & A. S. Brun

The Hertzsprung-gap giant 31 Comae in 2013: Magnetic field and activity indicators 365
A. P. Borisova, R. Konstantinova-Antova, M. Aurière, P. Petit & C. Charbonnel

Magnetic Field Structure and Activity of the He-burning Giant 37 Comae 367
S. Tsvetkova, P. Petit, R. Konstantinova-Antova, M. Aurière, G. A. Wade, C. Charbonnel & N. A. Drake

Strong variable linear polarization in the cool active star II Peg 369
L. Rosén, O. Kochukhov & G. A. Wade

Magnetic fields in single late-type giants in the Solar vicinity: How common is magnetic activity on the giant branches? 373
R. Konstantinova-Antova, M. Aurière, C. Charbonnel, N. Drake, G. Wade, S. Tsvetkova, P. Petit, K.-P. Schröder & A. Lèbre

Evolution of magnetic activity in intermediate-mass giants 377
P. Gondoin

Surface differential rotation of IL Hya from time-series Doppler images 379
Z. Kővári, L. Kriskovics, K. Oláh, K. Vida, J. Bartus, K. G. Strassmeier & M. Weber

Magnetic field of the classical Cepheid η Aql: new results 381
V. Butkovskaya, S. Plachinda, D. Baklanova & V. Butkovskyi

Activity on a Li-rich giant: DI Psc revisited................................ 383
L. Kriskovics, Z. Kővári, K. Vida & K. Oláh

Search for surface magnetic fields in Mira stars : first results on χ Cyg 385
A. Lèbre, M. Aurière, N. Fabas, D. Gillet, F. Herpin, P. Petit & R. Konstantinova-Antova

Magnetic fields around AGB stars and Planetary Nebulae 389
W. H. T. Vlemmings

Magnetic fields in Proto Planetary Nebulae 398
L. Sabin, Q. Zhang, A. A. Zijlstra, N. A. Patel, R. Vázquez, B. A. Zauderer, M. E. Contreras & P. F. Guillén

Polarimetry of R Aqr and PN M2-9.. 400
S. G. Navarro, L. Sabin, J. Ramírez & D. Hiriart

Measurements of the magnetic field in WD 1658+441 402
V. J. Ramírez, D. Hiriart, G. Valyavin, J. Valdez, F. Quiroz, B. Martínez, S. Plachinda& E. Iñiguez-Garín

Hydromagnetic Equilibria and their Evolution in Neutron Stars............. 404
A. Reisenegger

Hall Effect in Neutron Star Crusts.. 415
K. N. Gourgouliatos & A. Cumming

Magnetohydrodynamic equilibria in barotropic stars 419
C. Armaza, A. Reisenegger, J. A. Valdivia & P. Marchant

Magnetic field structures inside magnetars with strong toroidal field 423
K. Fujisawa

Axisymmetric and stationary magnetic field structures in neutron star crusts under various boundary conditions 427
K. Fujisawa & S. Kisaka

Magnetars: the explosive character of a small class of strongly magnetized neutron stars.. 429
N. Rea

NSMAXG: A new magnetic neutron star spectral model in XSPEC........... 435
W. C. G. Ho

Effects of strong magnetic fields in dense stellar matter...................... 439
A. Lavagno & F. Lingua

Search for Stable Magnetohydrodynamic Equilibria in Barotropic Stars........ 441
J. P. Mitchell, J. Braithwaite, N. Langer, A. Reisenegger & H. Spruit

Author index 445

Foreword

All phases of stellar evolution are influenced by the presence of magnetic fields in the interior and close environment of stars. Magnetic fields play a central role in the spindown of young stars, through magnetized outflows, star-disc interaction or magnetically-driven winds. They also impact the vertical settling of chemical species, leading to abnormal surface abundances observed in stars more massive than the Sun. In the advanced phases of stellar evolution, magnetic fields influence stellar evolution through their contribution to the mass-loss of cool giants and supergiants. Finally, extreme magnetic fields are observed in a small fraction of compact stellar remnants, powering X-ray and gamma ray emission.

Although most of these points have been identified decades ago, the ability to measure stellar magnetic fields and incorporate them in stellar models is relatively new. In this young and still growing research domain, the last few years have seen the dawn of a new era, with the advent of powerful tools strengthening both observational and modelling approaches to this field, rapidly changing our view of stellar magnetism throughout stellar evolution. The aim of this symposium was to bring together colleagues from all of these research areas. The topics covered spanned all phases of evolution, from the formation of stars and their early accreting years, through main sequence evolution for both low and high mass stars, and also the final stages of stellar evolution. Much of stellar astronomy now has relevance for the new field of exoplanets, and this brought another community to the symposium.

In addition to synthesizing the expertise of many research areas, the symposium also provided a forward look to the challenges and opportunities of the forthcoming decade. With an increasing number of present or future ground-based instruments in the visible and near infrared domains, stellar spectropolarimetry is now delivering direct magnetic field measurements throughout most of the Hertzprung-Russell diagram. Combined with tomographic modelling, spectropolarimetric data sets provide the surface distribution of the magnetic vector with increasing spatial and temporal resolution. Many indirect tracers of magnetic activity are also available from X-rays to sub-millimetric and radio wavelengths, providing us with observational clues on the effect of magnetic fields at various distances from the stellar surface (chromosphere, corona, accretion flows, winds, jets). Statistical studies based on huge samples are also obtained from space-borne observatories like KEPLER, offering a completely new view of stellar activity. They will soon be complemented by systematic activity measurements provided by the GAIA spacecraft. This wealth of observational material is progressively getting closer to the richness of solar observations, for which continuous monitoring is now available at extremely high spatial resolution and throughout most of the electromagnetic spectrum (e.g. HINODE, SDO). This symposium showed clearly that these tight observational constraints constitute a necessary guidance to numerical simulations of stellar magnetism, which now use the power of massively parallel supercomputers, enabling the investigation of stellar dynamos through global 3-D simulations of convective layers, as well as the evolution of magnetic fields in radiative zones. The future indeed promises to be a rich one for studies of stellar magnetism throughout stellar evolution.

Pascal Petit, Moira Jardine & Hendrik Spruit

SCIENTIFIC ORGANIZING COMMITTEE

- Gibor Basri (Univ. California, USA)
- Matthew Browning (Univ. Toronto,Canada)
- Corinne Charbonnel (Geneva Observatory, Switzerland)
- Jose-Dias do Nascimento (Univ. Natal, Brazil)
- Siraj Hasan (IIA, India)
- Moira Jardine (Univ. Saint Andrews, Scotland, co-chair)
- Oleg Kochukhov (Univ. Uppsala, Sweden)
- Renada Konstantinova-Antova (Bulgarian Academy of Sciences, Bulgaria)
- Hiroaki Isobe (Univ. Kyoto, Japan)
- Stephen Marsden (James Cook University, Australia)
- Pascal Petit (Univ. Toulouse, France, chair)
- Sami Solanki (MPS, Germany)
- Henk Spruit (MPA, Germany, co-chair)
- Klaus Strassmeier (AIP, Germany)
- Asif ud-Doula (Penn State, USA)
- Gregg Wade (RMC, Canada)

LOCAL ORGANIZING COMMITTEE

- Marie-Ange Albouy (UPS, Toulouse, France)
- Michel Aurière (IRAP, Tarbes, France)
- Jérôme Ballot (IRAP, Toulouse, France, co-chair)
- Boris Dintrans (IRAP, Toulouse, France)
- Maria-Eliana Escobar (IRAP, Toulouse, France)
- Dolorès Granat (IRAP, Toulouse, France)
- Fabrice Herpin (LAB, Bordeaux, France)
- Loïc Jahan (IRAP, Toulouse, France)
- Laurène Jouve (IRAP, Toulouse, France, co-chair)
- Nicole Le Gal (IRAP, Toulouse, France)
- Laura Léal (IRAP, Toulouse, France)
- Pascal Petit (IRAP, Toulouse, France)
- Pierre Vert (OMP, Toulouse, France)

EDITORS OF PROCEEDINGS

- Pascal Petit
- Moira Jardine
- Henk Spruit

CONFERENCE PHOTOGRAPH

Participants

a
Alicia Aarnio
Costanza Argiroffi
Cristóbal Armaza
Svitlana Artemenko
Michel Aurière

b
Dilyara Baklanova
Jèrôme Ballot
Fabienne Bastien
Olga Bayandina
Lionel Bigot
David Bohlender
Ana Borisova
Sudeshna Boro Saikia
Jonathan Braithwaite
Allan Sacha Brun
Varvara Butkovskaya

c
Matthieu Castro
Corinne Charbonnel
Patrick Crozet

d
Khalil Daiffallah
Alexandre David-Uraz
JoséDias do Nascimento
Stephanie Douglas
Natalia Drake

f
Rim Fares
marianne faurobert
Gregory Feiden
Colin Folsom
Nathalia Fonseca
Yori Fournier
Kotaro Fujisawa

g
Rafael Garcia
Thomas Gastine
Philippe Gondoin
Konstantinos Gourgouliatos
Scott Gregory
Jose Groh
Jason Grunhut

h
Elodie Hébrard
Huib Henrichs
Fabrice Herpin
Masaya Higa
Wynn Ho
Gaitee Hussain

i
Elisa Iñiguez Garin

j
Moira Jardine
Sandra Jeffers
Colin Johnstone
Laurène Jouve
Aaron Juarez

k
Viktor Khalack
Oleg Kochukhov
Renada Konstantinova-Antova
Heidi Korhonen
Zsolt Kovari
Levente Kriskovics
Dmitry Kudryavtsev
Manfred Küker
Andreas Künstler
Ryuichi Kurosawa

l
Norbert Langer
Andrea Lavagno
Agnès Lèbre
Jyri Lehtinen
Roxanne Ligi
François Lignières
Stuart Littlefair
Edward Liverts
Joe Llama
Arturo Lopez Ariste

m
Stephen Marsden
Alexander Martin
Stéphane Mathis
Titos Matsakos
Luiz Mendes
Sergio Miranda Aranguren
Joe Mitchell
David Montes

n
Silvana Navarro
Coralie Neiner

o
Stanley Owocki

p
Ana Palacios
Pascal Petit
Véronique Petit
Nikolai Piskunov
Mikhail Pogodin
Katja Poppenhaeger
Ralph Pudritz

r
Julio Ramirez
Denis Rastegaev
Nanda Rea
Ansgar Reiners
Timo Reinhold
Andreas Reisenegger
Claudia Rodrigues
Rachael Roettenbacher
Iosif Romanyuk
Lisa Rosén
thierry Roudier
Naum Rusomarov

s
Laurence Sabin
Victor See
Evgeny Semenko
Denis Shulyak
Aditi Sood
Hendrik Spruit
Deniss Stepanovs

t
Svetla Tsvetkova

v
Adriana Valio
Krisztián Vida
Aline Vidotto
Conrad Vilela-Lewandowski
Wouter Vlemmings

w
Gregg Wade
Lucianne Walkowicz
Joern Warnecke
Andrew West
Nicholas Wright

y
Rakesh Yadav
Ilya Yakunin

Magnetic Fields throughout Stellar Evolution
Proceedings IAU Symposium No. 302, 2013
P. Petit, M. Jardine & H. Spruit, eds.

doi:10.1017/S1743921314001628

Magnetic Fields in Stars: Origin and Impact

N. Langer

Argelander-Institut für Astronomie, Universität Bonn

Abstract. Various types of magnetic fields occur in stars: small scale fields, large scale fields, and internal toroidal fields. While the latter may be ubiquitous in stars due to differential rotation, small scale fields (spots) may be associated with envelop convection in all low and high mass stars. The stable large scale fields found in only about 10% of intermediate mass and massive stars may be understood as a consequence of dynamical binary interaction, e.g., the merging of two stars in a binary. We relate these ideas to magnetic fields in white dwarfs and neutron stars, and to their role in core-collapse and thermonuclear supernova explosions.

Keywords. Stars, magnetic fields, stellar evolution, supernovae

1. Introduction

Magnetic fields play a vital role in all stages of stellar evolution. This is already true during star formation. The magnetic support in collapsing molecular cloud cores is fundamentally affecting the fragmentation process (Price & Bate 2007). Later-on, during the accretion process, magnetic fields provide the required viscosity to bring in mass and to remove surplus angular momentum (Donati *et al.* 2007).

In this paper, we investigate the role of magnetic fields in stars once they are born. In order to do so, we distinguish various types of magnetic fields. First, it is useful to distinguish stable from dynamo fields. As stable fields we consider those which have a decay time of the order of the stellar life time or more, and which therefore do not need a dynamo action to continuously replenish them. Braithwaite & Spruit (2004) showed that combined toroidal-poloidal magnetic fields can survive in the radiative envelopes of stars for a long time, which they suggested to exist in magnetic A stars and in magnetic white dwarfs.

Other magnetic field geometries have so far been found unstable, e.g., such fields are expected to decay on their Alfvén time scale (e.g., Tayler 1973). However, inherently unstable field configurations may be present in stars over long time scales, if a dynamo process is continuously regenerating the field (Brandenburg & Subramanian). It may be expected that this regeneration process leads to some time variability. The prime example may be the Solar magnetic field, which is produced by a so called $\alpha\Omega$-dynamo, where the B-field is generated by an interplay between the differential rotation, which winds up poloidal field and generates toroidal field, and the α-effect, which generates a poloidal field from a toroidal one (Rüdiger *et al.* 2013).

While only stable and dynamo fields are long-lived and thus accessible to observations, there is some evidence for intermittent fields playing a role as well (Langer 2012). In particular during dynamical stellar merger events, which are suspected to lead to stable fields in the merger product (see below), the fields during the merger event itself are thought to be significantly stronger than thereafter. This strong intermittent component may be responsible for a removal of a large fraction of the angular momentum during the merging process.

From the observational perspective, it is also useful to distinguish various types of fields. For once, there are large scale fields, i.e. fields where the length scale over which local field maxima occur at the stellar surface is comparable to the size of the star itself. A classical example is a dipole field, i.e. a field which has only two points of maximum field strength at the stellar surface, which are located at different sides of the star. Dipole fields which have their magnetic axis inclined to the axis of rotation are in fact common amongst intermediate mass and high mass stars, although somewhat more complicated but still large scale field geometries occur as well.

In contrast to the large scale fields are small scale fields, for which the length scale of significant field variation is small compared to the stellar radius. An example for this are the Solar sunspots. The Sun also shows that stellar magnetic fields can have various components, as the small scale sunspots with field strengths of the order of 1000 G, and the global Solar dipole field with a strengths of about 1 G.

Finally, from the observational perspective, we want to distinguish toroidal magnetic fields as a third type, since toroidal fields are essentially hidden from direct observations. Still, they may strongly influence the evolution of stars, and may thus produce indirect evidence of their existence.

2. Toroidal fields: ubiquitous?

Spruit (2002) has suggested that a dynamo process can operate in differentially rotating radiative stellar envelopes. The main component of the produced magnetic field is toroidal, which is thought to counteract the differential rotation by producing a torque which transports angular momentum against the angular momentum gradient. While the model of Spruit has been criticized (Zahn, Brun & Mathis 2007), the main effect has been confirmed in simplified MHD models (Braithwaite 2006).

While the toroidal fields are not directly observable, there are currently two lines of observational evidence in their support. First, the nearly rigid rotation in the Sun beneath the Solar convection zone has been reproduced by Eggenberger *et al.* (2005) relying on the Spruit mechanism. Second, Mosser *et al.* (2012) found through oscillation measurements in a large sample of pulsating giants that the red giant cores rotate much slower than expected when only non-magnetic angular momentum transport processes are taken into account. While they claim that their results agree with the slow observed spins of white dwarfs, the evolutionary models of Suijs *et al.* (2007) which include angular momentum transport through the Spruit mechanism predict indeed white dwarf and neutron star spin periods which are close to the observed values (Fig. 1). While it can not be excluded that non-magnetic transport processes like gravity waves (Talon & Charbonnel 2008) could also reproduce the observational constraints, the results quoted above may speak in favor of the Spruit mechanism.

The only prerequisite of the Spruit mechanism is differential rotation. In case the Spruit mechanism works as expected, we may than conclude that toroidal magnetic fields are present in *all* stars, perhaps with the exception of stars with strong internal large scale fields, which perhaps rotate as rigid bodies, and of fully convective stars — where the Spruit mechanism may be overpowered by the predominance of convection.

3. Low mass stars

We define low mass stars as such stars which have convective envelopes during core hydrogen burning. The Sun is a low mass star. Magnetic activity in low mass stars is investigated since may decades, and many papers in these proceedings give the status of

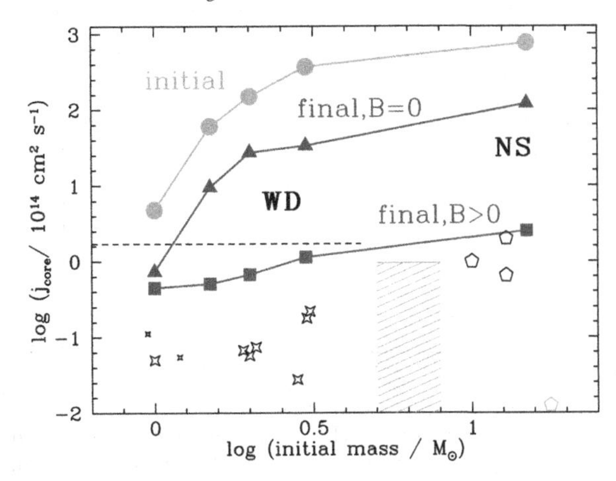

Figure 1. Average core specific angular momentum versus initial mass for the low and intermediate mass models of Suijs *et al.* (2008) and for the $15M_\odot$ model of Heger, Woosley & Spruit (2005) evolving from the zero age main sequence to their end stages (full drawn lines). The upper line corresponds to the initial models. Filled triangles mark the final models of the non-magnetic sequences, and filled squares the final models of the magnetic sequences. The dashed horizontal line indicates the spectroscopic upper limit on the white dwarf spins obtained by Berger et al. (2005). Star symbols represent asteroseismic measurements from ZZ Ceti stars, where smaller symbols correspond to less certain measurements. The green hatched area is populated by magnetic white dwarfs. The three black open pentagons correspond to the youngest galactic neutron stars, while the green pentagon is thought to roughly correspond to magnetars. See Suijs *et al.* (2008) for details.

the current research. We therefore restrict ourselves here to address the magnetic fields in low mass stars just for comparison to those in more massive stars.

The present conclusion is that *all* low mass stars show magnetic fields. I.e., the presence of a convective envelope is sufficient to develop a field. While rotation is a necessary ingredient to the $\alpha\Omega$-dynamo, and faster rotators tend to show higher magnetic activity, even slow rotators as our Sun possess an appreciable magnetic field. It is clear that these fields lead to an efficient angular momentum loss over the lifetime of these stars, to the extent that the their spin rate is a function of their age.

4. Intermediate mass stars

Intermediate mass stars prove mostly to be non-magnetic. Only about 10% of the core hydrogen burning stars in this mass range show a strong large scale field of more than a few hundred Gauss, while the remaining 90% appear to have fields which are weaker than about one Gauss. While the fields in the magnetic fraction of intermediate mass main sequence stars appear, partly, to have a rather complex morphology (Donati *et al.* 2006), their structure appears to be simple compared to the fields of low mass stars (Donati & Landstreet 2009).

The magnetic intermediate mass stars are generally slow rotators. I.e., while the field strengths of low mass stars are larger for faster rotators, the situation is almost the reverse for intermediate mass stars. While their cores are convective and could produce a magnetic field deep down (Brun *et al.* 2005), it appears unlikely that this field is transported deeply into the radiative envelope or even to the stellar surface (Charbonneau & MacGregor 2001, MacGregor & Cassinelli 2003, MacDonald & Mullan 2004). Since also their envelops are radiative, it appears not to be possible to explain their magnetic fields through a dynamo process.

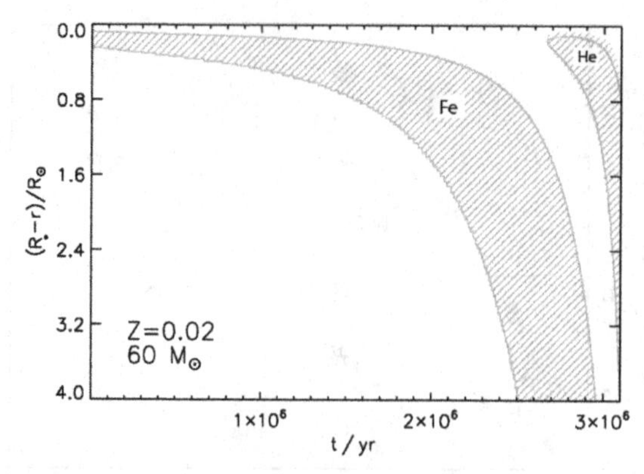

Figure 2. Evolution of the radial extent of the subsurface helium and iron convective regions (hatched) as function of time, from the zero age main sequence to roughly the end of core hydrogen burning, for a 60 $M_\odot$ star (Cantiello *et al.* 2009). The top of the plot represents the stellar surface. Only the upper 4 $R_\odot$ of the star are shown in the plot, while the stellar radius itself increases during the evolution. The star has a metallicity of $Z = 0.02$, and its effective temperature decreases from 48 000 K to 18 000 K during the main sequence phase.

As mentioned above, Braithwaite & Spruit (2004) found stable magnetic field configurations to be able to exist in these stars. However, while the suggested field geometries are quite compatible to those observed, this does not allow any conclusion on the origin of these fields. We return to this question on Sect. 6.

5. Massive stars

Evidence is accumulating that massive main sequence stars show both, the small scale fields produced by convective envelop dynamos, and the large scale stable fields just as they are observed in the intermediate mass stars.

5.1. *Small-scale fields*

Cantiello *et al.* (2009) pointed out that massive main sequence stars have convective envelops which may be capable to produce observable magnetic fields. While the convection zones occur beneath the stellar surface due to opacity peaks produced by iron and helium recombination, their distance to the surface is so small that magnetic flux tubes can buoyantly float to the surface in a short time. Their spatial extent is a significant fraction of the stellar radius (Fig. 2).

There is multiple observational evidence for the existence of these sub-surface convection zones. First, Cantiello *et al.* (2009) showed that the predicted dependence of the kinematic signature of these zones at the stellar surface on stellar mass, surface temperature and metallicity agrees with the observations of micro-turbulence as determined from spectroscopic measurements of a large number of O and early B stars in the Galaxy and the Magellanic Clouds. Secondly, stochastically excited pulsations have been measured in several massive main sequence stars (cf., Belkacem *et al.* 2009). And thirdly, the velocity field induced by the sub-surface convection may lead to a clumping of the hot star winds very near to their surface, in agreement with observations (Cantiello *et al.* 2009, Sundqvist & Owocki 2013).

Finally, and most relevant in the present context, magnetic spots at the surface of massive main sequence stars as predicted by Cantiello *et al.* (2009) may provide an explanation of the discrete absorption components (DACs) in their UV spectra lines. The DACs phenomenon appears to be best explained by assuming a disturbance of the

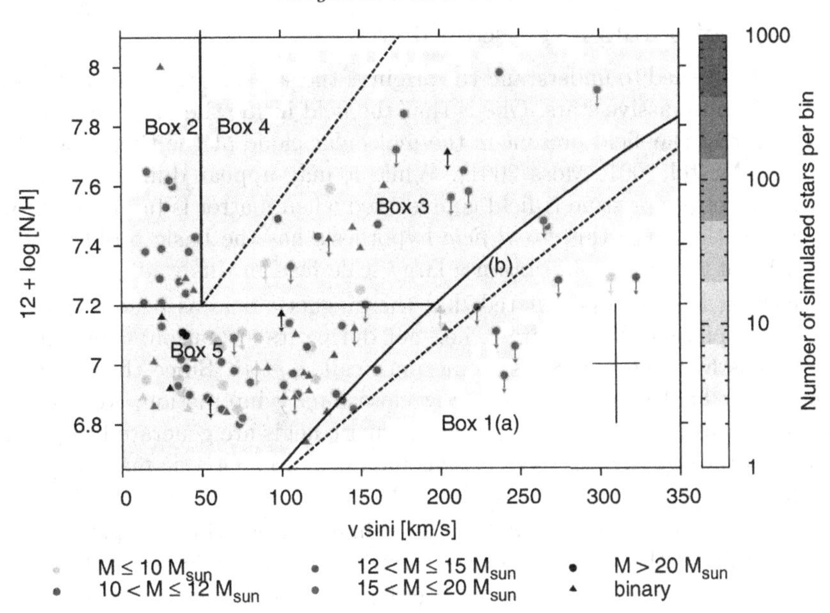

Figure 3. Hunter diagram for LMC early B type stars from the VLT-FLAMES Survey of Massive Stars (symbols), showing projected rotational velocity against their nitrogen surface abundance. Single stars are plotted as circles, radial velocity variables as triangles. A population synthesis simulation based on single star evolution models with rotational mixing (Brott *et al.* 2011a) is shown as a density plot in the background (Brott *et al.* 2011b). The color coding corresponds to the number of predicted stars per pixel. The cross in the lower right corner shows the typical error on the observations.

radiation driven wind of the star due to hot spots at its surface (Cranmer & Owocki 1996). As DACs are ubiquitous in O stars (Howarth & Prinja 1989), the tentative implication is that so are their small scale magnetic fields.

5.2. *Large-scale fields*

In recent years, the evidence has been growing that concerning large scale stable fields, the massive stars behave essentially as their intermediate mass counterparts. The magnetic fraction of OB stars has been determined to be of the order of 10% (Grunhut & Wade 2012), and the magnetic topologies are similar to those found in intermediate mass stars, with a predominance of highly inclined magnetic dipoles or low-order multipoles.

Indirect evidence for a magnetic fraction of massive stars comes from the Hunter diagram of LMC early B-stars (Fig. 3), which shows that 15% of them are nitrogen-rich slow rotators (Hunter *et al.* 2008, Brott *et al.* 2011b). While this group of stars is not reproduced from models of rotating single stars (Brott *et al.* 2011a), a magnetic field would allow to explain their slow rotation. Morel *et al.* (2008) identified a similar population in our Galaxy, and showed that a large fraction of these objects does indeed show a magnetic field. Further evidence is provided by Dufton *et al.* (2013), who showed that the velocity distribution of LMC early B-type stars is bimodal, with ~20% of them rotating with values below 100 km/s, very reminiscent of the situation in the A-type stars (Zorec & Royer 2012).

6. The origin of stable large-scale fields

Two ideas are pursued to understand the origin of the stable, large-scale magnetic fields in intermediate and massive stars. One is that the field inside the main sequence stars is a relic of the interstellar field present in the molecular cloud at the time when it formed the stars (e.g., Mestel 2001, Moss 2001). While it may appear difficult to understand how the memory of the pristine B-field is preserved when matter is funneled through the MRI-driven accretion disc, this *fossil field* hypothesis has the basic problem to explain why it is ~10% of the stars that obtain a large scale field in this way.

Alternatively, it has been postulated that the magnetic field as been acquired by the star in an earlier phase of its evolution, i.e., not during its formation. Confusingly, fields according to this hypothesis are also sometimes called *fossil*. Since the event with the largest appeal for field generation is strong close binary interaction, we speak here of *binary-induced fields*. The general idea that magnetic fields are generated through stellar mergers is supported by the dearth of close companions to magnetic main sequence stars at intermediate and high mass (Carrier *et al.* 2002).

Several people have suggested that strong binary interaction, in particular stellar merger, can result in the generation of strong, stable magnetic fields. Ferrario *et al.* (2009) and Tutukov & Fedorova (2010) suggested to explain the magnetic intermediate mass and massive stars through pre-main sequence mergers, while Tout *et al.* (2008) argued that the fields in magnetic white dwarfs may originate from white dwarf mergers.

However, there is evidence that a large fraction of the observed magnetic intermediate mass and massive stars are remnants of mergers between two main sequence stars. The latest determination of O star main sequence binary parameter distributions by Sana *et al.* (2012) implies that 8^{+9}_{-4}% of all Galactic O stars are indeed the product of a merger between two main sequence stars (de Mink *et al.* 2014). Furthermore, Glebbeek *et al.* (2013) predict a nitrogen enrichment in main sequence merger products which is well compatible with the values found in the magnetic early B-stars analyzed by Morel *et al.* (2008; cf. Sect. 5.2).

While indeed also a significant fraction of the intermediate mass Herbig stars is found to be magnetic (Hubrig *et al.* 2004, Wade *et al.* 2007), those might also be merger products on the pre-main sequence, possibly induced by circumstellar tides as proposed by Krontreff *et al.* (2012). The ratio of nitrogen-rich to nitrogen-normal magnetic massive main sequence stars may thus give an indication of the ratio of pre-main sequence to main sequence mergers.

A stellar merging process is a most drastic event which induces strong differential rotation on a timescale close to the dynamical timescale. Binary evolution may produce such a situation also in some cases where the final merger is avoided in the end, i.e. during common envelope evolution which leads to an *almost-merger*. Perhaps Plaskett's star, which is a very massive close binary just past its rapid mass transfer phase, and of which the mass gainer has been found to host a strong magnetic field (Grunhut *et al.* 2013), and the Polars, Cataclysmic Variables with magnetic white dwarf companions (Tout *et al.* 2008), belong to this category.

It appears in any case unlikely that thermal timescale mass transfer can lead to strong magnetic fields in the mass gainer, neither accretion during star formation, which occurs on a similar time scale. The point is that the products are not observed to be magnetic. Mass transfer in massive binaries is known to lead to Be stars, which may often be seen as single stars because their companion exploded in a supernova explosion (de Mink *et al.* 2013). While perhaps not all Be stars are binary products, *none* of the many analyzed Be stars has been found to be magnetic by Grunhut & Wade (2012). Analogously, if accretion during star formation would induce strong stable fields, all stars should possess such fields, which is obviously not the case.

7. Evolution

While in the sections above, we were mostly assessing main sequence star, one may wonder how the stable magnetic fields survive during the post main sequence evolution of intermediate mass and massive stars. Theoretical ideas for this are scarce, since after the main sequence, ever changing parts of the post main sequence stars become convective. In particular, most of these stars may evolve into red giants and supergiants, which have convective cores and deep convective envelopes and thus leave little room for a stable magnetic field. On the other hand, there is no stage where the whole star would become convective, and it may thus be possible that the stable field of the main sequence stars is preserved throughout their post-main sequence evolution deep inside the star.

It has indeed been postulated that the magnetic white dwarfs — again: about 10% of all white dwarfs — are the remnants of magnetic intermediate mass main sequence stars, as the magnetic fluxes of both are quite comparable. In the light of the previous section, it may be plausible to assume that the magnetic white dwarfs have perhaps two components, one evolving from magnetic main sequence stars, and the other from white dwarf mergers. The latter channel may lead to an on average larger mass of magnetic white dwarfs, while the former might stand out by a slower-than-average spin. If magnetic white dwarfs are really merger products, it may be interesting to note that they will only play a minor role in Type Ia supernovae, if any at all. However, magnetic fields could form in double degenerate mergers, which are thought to provide one channel towards Type Ia explosions.

While the fraction of neutron stars which have extreme fields, the magnetars, is not well established, it appears again compatible with an order of magnitude of $\sim 10\%$. Also the flux freezing argument could apply. If massive main sequence stars had a B-field of $\sim 10^4$ G in their core — which appears plausible as they have surface fields of up to $\sim 10^3$ G —, the resulting B-field in the neutron stars would be of the order of 10^{14} G, which is two orders of magnitude larger than typical neutron star magnetic field strengths (Ferrario & Wickramasinghe 2006). In contrast to the scenario by Duncan & Thompson (1992), where the magnetar field forms from an extremely rapidly rotating collapsing iron core, magnetars as successors of magnetic main sequence stars would form slowly rotating neutron stars. This appears not only to be more compatible with the young supernova remnants surrounding some magnetars (Vink 2008), but also with the dearth of progenitors which can produce rapidly rotating iron cores in a high metallicity environment as our Galaxy (Yoon *et al.* 2006), and would argue against magnetar-powered supernovae (Woosley 2010). As also proposed by Duncan & Thompson (1992), the proto-neutron stars with ordinary spin rates (corresponding to $j \simeq 10^{14}\,\mathrm{cm}^2/\mathrm{s}$; see Fig. 1) may well produce the 10^{12} G fields found in most neutron stars.

8. Outlook

While all low mass stars appear to have small scale, dynamo-produced surface fields, the stronger the faster they rotate, only a fraction of $\sim 10\%$ of the intermediate mass stars possess strong B-field, which are large scale and occur mostly in slowly rotating single stars. Both types of fields may be combined in the massive stars, the small scale ones in all of the, the large scale one again in a fraction of about 10%. In addition, all stars may contain internal toroidal magnetic fields induced by differential rotation, which couples their core and envelope spins.

The hypothesis for the formation of the large scale fields which is consistent with all currently known constraints is that of strong binary interaction, preferentially via stellar mergers. In contrast to the fossil field hypothesis, it makes several clear predictions. Due to the large observational efforts currently underway, we can expect that this topic will be settled within the next years.

The question of the influence and survival of the large scale fields during the post-main sequence evolution of intermediate mass and massive stars appears more difficult to answer. Since red supergiants will likely not allow to assess this question observationally due to their deep convective envelopes — which may produce its own field through a dynamo process —, it may be interesting to focus on blue supergiants and Wolf-Rayet stars. If descendants of magnetic main sequence stars evolve into long-lived blue supergiants, a fair fraction of them might show surface magnetic fields, although considerably weaker ones if the magnetic flux is conserved. Also some Wolf-Rayet stars may have magnetic main sequence stars as precursors, but a field detection in these objects appears difficult due to their strong winds (de la Chevrotière *et al.* 2013).

Whether the large scale fields survive even until the formation of the compact remnant remains an open question, although there may be more arguments in favor of this idea than against it (cf. Sect. 7). However, it remains a challenge to produce solid theoretical predictions about the survival of the field during the post-main sequence evolution, as well as to identify observational strategies which would allow to settle the case.

References

Belkacem, K., Samadi, R., Goupil, M. J., *et al.* 2009, *Science*, 324, 1540

Berger, L., Koester, D., Napiwotzki, R., Reid, I. N., & Zuckerman, B. 2005, *A&A*, 444, 565

Braithwaite, J. 2006, *A&A*, 449, 451

Braithwaite, J. & Spruit, H. C. 2004, *Nature*, 431, 819

Brandenburg, A., Subramanian, K. 2005, *Phys. Rep.*, 417, 1

Brott, I., de Mink, S. E., Cantiello, M., *et al.* 2011a, *A&A*, 530, A115

Brott, I., Evans, C. J., Hunter, I., *et al.* 2011b, *A&A*, 530, A116

Brun, A. S., Browning, M. K., & Toomre, J. 2005, *ApJ*, 629, 461

Cantiello, M., Langerm, N., Brott, I., *et al.* 2009, *A&A*, 499, 279

Carrier, F., North, P., Udry, S., & Babel, J. 2002, *A&A*, 394, 151

Charbonneau, P. & MacGregor, K. B. 2001, *ApJ*, 559, 1094

Cranmer, S. R. & Owocki, S. P. 1996, *ApJ*, 462, 469

de la Chevrotière, A., St-Louis, N., & Moffat, A. F. J. 2013, *ApJ*, 764, 171

de Mink, S. E., Langer, N., Izzard, R. G., Sana, H., & de Koter, A. 2013, *ApJ*, 764, 166

de Mink, S. E., Sana, H., Langer, N., Izzard, R. G., & Schneider, F. R. N. 2014, *ApJ*, in press

Donati, J.-F., Howarth, I. D., Jardine, M. M., *et al.* 2006, *MNRAS*, 370, 629

Donati, J.-F., Jardine, M. M., Gregory, S. G., *et al.* 2007, *MNRAS*, 380, 1297

Donati, J.-F. & Landstreet, J. D. 2009, *ARAA*, 47, 333

Dufton, P. L., Langer, N., Dunstall, P. R., *et al.* 2013, *A&A*, 550, A109

Duncan, R. C. & Thompson, C. 1992, *ApJL*, 392, L9

Heger, A., Woosley, S. E., & Spruit, H. C. 2005, *ApJ*, 626, 350

Hubrig, S., Schöller, M., & Yudin, R. V. 2004, *A&A*, 428,, L1

Hunter, I., Brott, I., Lennon, D. J., *et al.* 2008, *ApJL*, 676, L29

Eggenberger, P., Maeder, A., & Meynet, G. 2005, *A&A*, 440, L9

Evans, C. J., Taylor, W. D., Henault-Brunet, V., *et al.* 2011, *A&A*, 530, A108

Ferrario, L. & Wickramasinghe, D. 2006, *MNRAS*, 367, 1323

Ferrario, L., Pringle, J. E., Tout, C. A., & Wickramasinghe, D. T. 2009, *MNRAS*, 400, L71

Glebbeek, E., Gaburov, E., Portegies Zwart S., & Pols, O. R. 2013, *MNRAS*, 434, 3497
Grunhut, J. H. & Wade, G. A. 2012, *ASPC*, 465, 42
Grunhut, J. H., Wade, G. A., Leutenegger M., *et al.* 2013, *MNRAS*, 428, 1686
Howarth, I. D. & Prinja, R. K. 1989, *ApJS*, 69, 527
Korntreff, C., Kaczmarek, T., & Pfalzner, S. 2012, *A&A*, 543, A126
Langer N. 2012, *ARAA*, 50, 107
MacDonald, J. & Mullan, D. J. 2004, *MNRAS*, 348, 702
MacGregor, K. B. & Cassinelli J P. 2003, *ApJ*, 586, 480
Mestel L. 2001, *ASPC*, 248, 3
Morel, T., Hubrig S., & Briquet M. 2008, *A&A*, 481, 453
Moss D. 2001, *ASPC*, 248, 305
Mosser, B., Goupil, M. J., Belkacem, K., *et al.* 2012, *A&A*, 548, A10
Price, D. J. & Bate, M. R. 2007, *MNRAS*, 377, 77
Rüdiger, G., Kitchatinov, L. L., & Hollerbach R. 2013, *Magnetic Processes in Astrophysics*, Wiley-VCH, Weinheim
Sana, H., de Mink, S. E., de Koter, A., *et al.* 2012, *Science*, 337, 444
Spruit, H. C. 2002, *A&A*, 381, 923
Suijs, M. P. L., Langer N., Poelarends A-J, Yoon S-C, Heger, A., & Herwig F. 2008, *A&A*, 481, L87
Sundqvist, J. O. & Owocki S P. 2013, *MNRAS*, 428, 1837
Talon S. & Charbonnel, C. 2008, *A&A*, 482, 597
Tayler, R. J. 1973, *MNRAS*, 161, 365
Tout, C. A., Wickramasinghe, D. T., Liebert, J., Ferrario, L., & Pringle, J. E. 2008, *MNRAS*, 387, 897
Tutukov, A. V. & Fedorova, A. V. 2010, A.Rep, 54, 156
Vink, J. 2008, Advances in Space Research, 41, 503
Wade, G. A., Bagnulo S., Drouin, D., Landstreet, J. D., & Monin, D. 2007, *MNRAS*, 376, 1145
Woosley, S. E. 2010, *ApJL*, 719, L204
Yoon S.-C., Langer N., & Norman, C. 2006, *A&A*, 460, 199
Zahn J.-P., Brun A. S., & Mathis S. 2007, *A&A*, 474, 145
Zorec, J. & Royer F. 2012, *A&A*, 537, A120

Magnetic Fields throughout Stellar Evolution
Proceedings IAU Symposium No. 302, 2013
P. Petit, M. Jardine & H. Spruit, eds.

doi:10.1017/S174392131400163X

The Role of Magnetic Fields in Star Formation

Ralph E. Pudritz[1,2], Mikhail Klassen[1], Helen Kirk[1,3], Daniel Seifried[4] and Robi Banerjee[4]

[1]Department of Physics and Astronomy, McMaster University,
Hamilton, ON L8S 4M1, Canada

[2]Origins Institute, McMaster University, Hamilton, ON L8S 4M1, Canada
email: pudritz@mcmaster.ca

[3]Banting Fellow, McMaster University; now at Radio Astronomy Program, NRC, Canada

[4]Hamburg Sternwarte, University of Hamburg,
Gojenbergsweg 112 21209 Hamburg - Germany

Abstract. Stars are born in turbulent, magnetized filamentary molecular clouds, typically as members of star clusters. Several remarkable technical advances enable observations of magnetic structure and field strengths across many physical scales, from galactic scales on which giant molecular clouds (GMCs) are assembled, down to the surfaces of magnetized accreting young stars. These are shedding new light on the role of magnetic fields in star formation. Magnetic fields affect the gravitational fragmentation and formation of filamentary molecular clouds, the formation and fragmentation of magnetized disks, and finally to the shedding of excess angular momentum in jets and outflows from both the disks and young stars. Magnetic fields play a particularly important role in angular momentum transport on all of these scales. Numerical simulations have provided an important tool for tracking the complex process of the collapse and evolution of protostellar gas since several competing physical processes are at play - turbulence, gravity, MHD, and radiation fields. This paper focuses on the role of magnetic fields in three crucial regimes of star formation: the formation of star clusters emphasizing fragmentation, disk formation and the origin of early jets and outflows, to processes that control the spin evolution of young stars.

Keywords. magnetic fields, star formation, filaments, molecular clouds, protostellar disks, magnetic braking, outflows, protostellar rotation.

1. Introduction

The theory of stellar structure and evolution informs us that the mass, rotation, and composition of stars control their fates. Two of these, the distribution of initial stellar masses (the initial mass function, or IMF) and the distribution of initial protostellar rotation rates can in principle be directly affected by magnetic forces. While star formation involves the interplay of turbulence, gravity, radiation, and magnetic fields on a variety of scales, can magnetic fields significantly affect the nature of the distribution of stellar masses of newly formed stars? Can they control the rotation rates of stars as they arrive onto the zero-age main sequence? The short answer to both of these questions is yes. The subtlety lies in teasing out whether or not magnetic effects are significantly muted in the complex interplay of these physical processes.

The origin of the IMF is linked to the distribution of overdense regions in star forming clouds. This is clearly seen in Figure 1(a) where observations of the conditions leading to stellar birth in a low mass, star forming cloud (the Pipe Nebula) are shown. Dense knots of gas known as cores are distributed along a filamentary structure. The mass spectrum of these cores known as the (Dense) Core Mass Spectrum or CMF is compared with the IMF in Figure 1(b). The data shows that these distributions are similar in form, where masses can be well determined, except that the CMF is shifted to higher masses by a factor of roughly 3. (Alves *et al.* 2007) The results are similar to many other studies (Motte *et al.* 1997, Testi & Sargent, Johnstone *et al.* 2000). The CMF has a peak at around $1M_\odot$ (André *et al.* 2010) whereas the most abundant stars in the IMF occur at $\simeq 0.2M_\odot$, The shape of IMF can be well fit by a lognormal distribution with a power-law tail at high masses (Chabrier 2003). In another example of a more evolved, star forming cloud known as the Aquila Rift, 85% of the mass of the cloud can be fit by a lognormal distribution, while 15% falls inside the power-law tail (Andre *et al.* 2011, Schneider *et al.* 2013). Turbulence readily produces lognormal density distributions. A simple interpretation of the relation between the IMF and the CMF is that physical processes such as radiation from the central star, or hydromagnetic outflows manage to disperse typically 2/3 of the gas in the original cores.

A huge effort over the last decade has measured the distribution of spins of over 5000 young stars (review, Bouvier *et al.* 2013). A key finding is that young, low mass (0.1 - 1 $M_\odot$) pre main sequence (PMS) stars that are a few Myr old span a broad range of stellar spins with most ranging from periods of 1 - 10 days. More than half of the T Tauri stars are slow rotators; for typical TTS stars with masses $M_* = 0.5M_\odot$, radii $R_* = 2R_\odot$, and a period of 8 days, the ratio f of the observed spin period to the break up spin rate is $f < 0.1$. These systems undergo vigorous accretion which should push the rotation rates to near break up within a few 10^5 yr. Here too, magnetic forces in the form of spin-down torques may be playing a decisive role.

In this review, we descend from the large physical scales associated with magnetic fields in galaxies and molecular clouds, to accretion disks and outflows around young stars, down to the surfaces T-Tauri stars undergoing active magnetospheric accretion from their surrounding accretion disks. We trace some of the important roles that magnetic fields can play in star formation.

2. Magnetic fields and Filamentary Molecular Clouds in Galaxies

Magnetic fields in molecular clouds ultimately arise as a consequence of dynamo action in the large scale interstellar medium of galaxies. They can be traced by polarized

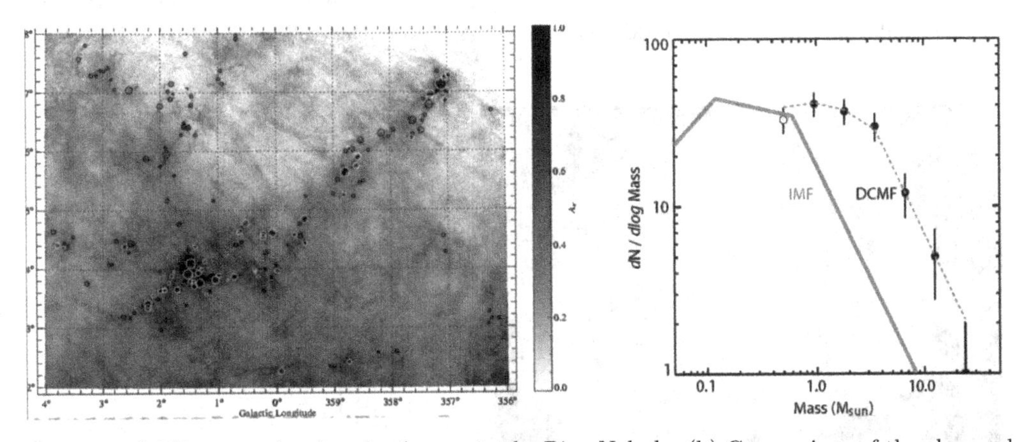

Figure 1. (a) Dense molecular cloud cores in the Pipe Nebula; (b) Comparison of the observed Core Mass Function (CMF) in the Pipe Nebular, describing the mass distribution of cores, with the Initial Mass Function which is the mass spectrum of newly formed stars. Adapted from Alves *et al.* (2007).

synchrotron emission mapped by multifrequency radio observations. This allows the strength of the field perpendicular to the line of sight to be measured. The parallel component of the field is obtained by measuring the Faraday rotation, or rotation measure (RM) (review, Beck, 2012). Ordered fields with spiral arm geometry can be traced a large variety of galaxies (Beck 2005) and are strongest in the inter-arm region between optical spiral arms, with a strength of 10-15 μG and are generated by galactic dynamos. In galaxies with strong spiral density waves, such as M51 shown in Figure 2(a), the field is compressed and sheared by the non-axisymmetric flows associated with the arm (Fletcher *et al.* 2011). The dominant dynamo mode is a quadrupolar type of field which includes a vertical component out of the galactic plane (Heeson *et al.* 2009). The distribution of field strengths shown in Figure 2(b) for several tens of disk galaxies is $B_{tot} \simeq 17 \pm 14\,\mu G$ with a mean field of $B_{mean} \simeq 5 \pm 3\,\mu G$. These measurements show that magnetic fields are in equipartition with energy in the cosmic rays and the interstellar medium on these scales.

These fields are subsequently incorporated into the formation of giant molecular clouds in galaxies. This may occur by several different mechanisms. Strong fields are buoyant as was first shown by Parker. In the self-gravitating gas of the galaxy, a combined Parker-Jeans instability operates (Elmegreen 1982) wherein the gravity is the strongest in the galactic plane and the magnetic buoyancy peaks far above the plane. The growth time of such an instability in a galaxy is about the Jeans timescale or $(4\pi G\rho_o)^{-1/2} \simeq 23n_o^{1/2}$ million years where n_o is the volume number density of the gas. Gas draining down the large scale buoyant magnetic loops rising through the galactic disk accumulates into magnetized 10^6 solar mass GMCs in 10 Myr in this scenario. Recent work has emphasized the possible role of the magneto-rotational instability (MRI) operating together with the Toomre gravitational instability for galactic disks, in creating magnetized molecular clouds (Sellwood & Balbus 1999, Kim *et al.* 2003). Since the orbital frequencies of gas in galaxies is a decreasing function of radius ($\Omega(r) \propto r^{-1}$), they will be subject to the MRI instability which creates vigorous turbulence. In sufficiently dense parts of the galaxy, the Toomre instability will act like a swing amplifier acting on the larger fluctuations, capable of building molecular clouds up to 10^7 solar masses. In either case, the magnetized gas in a galaxy is, together with gravity, a major participant in the creation of star forming GMCs.

Star formation is strongly linked to the structure of GMCs. It has been known for decades that some molecular clouds have obvious filamentary like structure. The recent

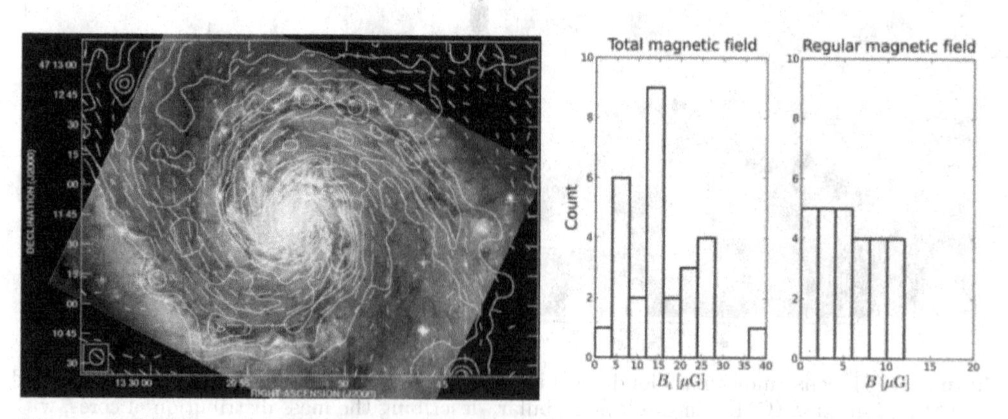

Figure 2. (a) Magnetic fields in the spiral galaxy M51. (b) Disitribution of the strengths of magnetic fields in spiral galaxies. Adapted from Fletcher *et al.* (2011).

Herschel space observatory observations made a major breakthrough however by showing that filaments are ubiquitous - they occur on 10 pc to less than 0.1 pc scales in all molecular clouds (review, André *et al.* 2013, André *et al.* 2010, Menschchikov *et al.* 2011, Henning *et al.* 2011). Filamentary networks are seen in clouds without much evidence for star formation (eg. Polaris Flare, Ward-Thompson *et al.* 2010). In the Orion cloud, more than 70 % of cores are strongly associated with filaments (Polychroni *et al.* 2013) indicating that the formation of filaments precedes star formation. Numerical simulations of supersonic turbulence in the interstellar medium have consistently shown that gas is compressed into filamentary networks (eg. Porter *et al.* 1994, Vasquez-Semadeni *et al.* 1994, Padoan *et al.* 2001). The addition of gravity in the simulations leads to the collapse of regions of denser gas to form stars (eg. Klessen & Burkert 2000, Bonnell *et al.* 2003, MacLow & Klessen 2004, Tilley & Pudritz 2004).

The observations also reveal that star-forming cores are found preferentially in filaments that are gravitationally unstable. Equilibrium calculations show that filaments whose mass per unit length exceeds a critical value of $m_{crit} = 2c_s^2/G$ in purely thermally supposed gas (c_s is the sound speed) are gravitationally unstable (Inutsuka & Miyama 1997). This condition is modified for magnetized filaments, or for turbulence where the condition becomes $m_{crit} = 2\sigma_{tot}^2/G$ for the total velocity dispersion including turbulence (Fiege & Pudritz 2000). Magnetic fields can both thread filaments (poloidal component) providing additional pressure support against gravity, as well as wrap around filaments (toroidal component) which tends to squeeze them. The orientation of the magnetic field with respect to a filament in the Serpens molecular cloud is shown in Figure 3a. The observations of infrared (H band) polarization of the cloud are overlaid on a column density map (Sugitani *et al.* 2011). The young Serpens South Cluster is at the intersection of a couple of filaments. The field strength is $\simeq 100\,\mu G$ and its direction is generally perpendicular to the filament axis. This orientation of the field is also seen in other molecular clouds. Submillimeter polarimetry observations of cores in the Orion molecular cloud also tend to show fields that are perpendicular to the filament axis, as well as a general large scale ordered field (H-B Li *et al.* 2009). Optical polarization studies of the prominent filament in the Taurus B211 cloud shows a similar pattern (Palmeirim *et al.* 2012). A possible emerging trend in the magnetic field observations is that fields tend to to be perpendicular to filaments that are dense and forming stars, whereas they are aligned parallel to more diffuse filaments or fine striations. The latter are often found perpendicular to major filaments, and it is possible that the accretion flows onto filaments from the surrounding gas could be magnetically channeled. More measurements are required.

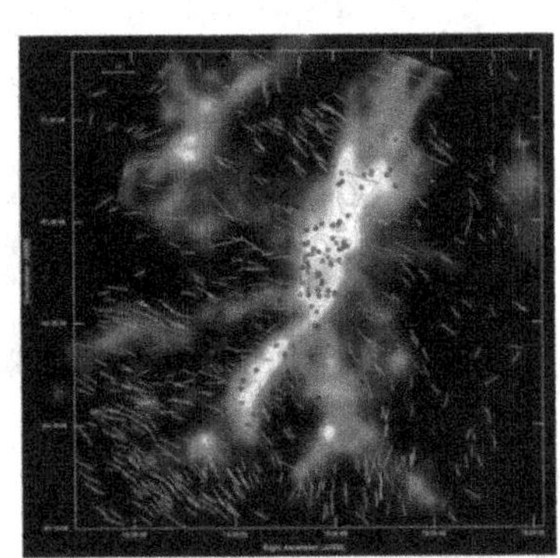

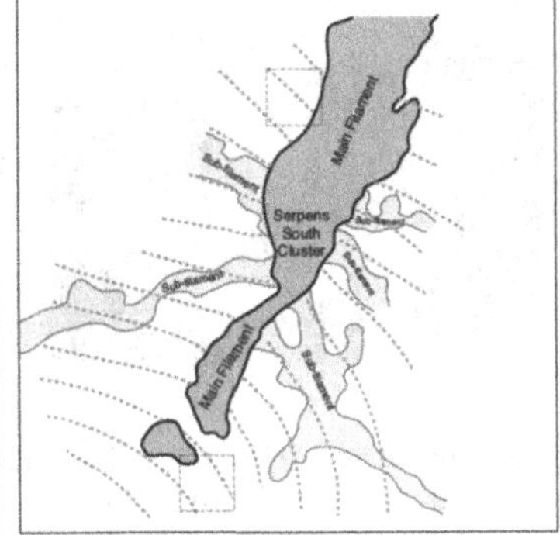

Figure 3. Orientation of the magnetic field with respect to the column density of gas in the filamentary cloud, Serpens-South in the Aquila complex. Adapted from Sugitani *et al.* (2011).

Magnetic field strengths in molecular cloud cores are detected by Zeeman measurements (Crutcher *et al.* 2010, Crutcher 2012). These observations show that the magnetic field strength is a constant in the low density ISM up to $n \simeq 500\,\mathrm{cm}^{-3}$, and increases with density as $B_{tot,max} = 10(n/300\,cm^{-3})^{0.65}\,\mu G$, at higher densities characteristic of molecular gas. In molecular gas the field can be compressed. A measure of the importance of gravity relative to the magnetic energy in clouds is the mass to flux ratio, $\lambda = 2\pi\sqrt{G}\Sigma/B$ where Σ is the column density of the gas and B is the field strength. The critical value, where gravitational density is comparable to the magnetic is $\lambda = 1$. The observations show that there is an upper envelope for the data of $\lambda \simeq 2 - 3$ (Crutcher, 2012). At this level, simulations show that magnetic fields suppress the fragmentation of filaments (eg. Tilley & Pudritz 2007).

We show, in Figure 4, the central 1 pc regions of two simulations of cluster forming clumps of mass $500 M_\odot$, early (150,000 yrs) in the evolution of a self-gravitating turbulent cloud. On the left is a purely hydrodynamic simulation and an MHD simulation with $\lambda \simeq 2.33$ is on the right. The hydro filaments are more compressed and more apt to fragment. The MHD run has more diffuse filaments (due to magnetic pressure support) and somewhat less fragmentation (Kirk *et al.* 2014). Turbulence is raising the critical line mass threshold above the thermal value (contours red, blue, and white at 5, 2, and 1 thermal critical line mass).

Simulations of supersonic turbulence have shown for nearly 2 decades that lognormal distributions of column density naturally arise. The reason is that shocks are a multiplicative process for building density fluctuations. In a medium that is repeatedly compressed by a random distribution of shock strengths, and in the limit of a large number of such events, a lognormal arises as a consequence of the central limit theorem (Padoan *et al.* 1997). Figure 5(a) shows that even a small number of shocks rapidly sets up a lognormal distribution in the density distribution (Kevlahan & Pudritz 2009). In Figure 5(b) high resolution simulations without gravity produce a lognormal to high accuracy. Once gravity is turned on in this medium, a power law tail develops as the denser gas filaments start to be pulled together and ultimately collapse (Kritsuk *et al.* 2011), which agrees well with the observations (Kainulainen *et al.* 2009). Adding MHD to this does not change the results very much - there is only a weak dependence on the Alfvénic Mach number of

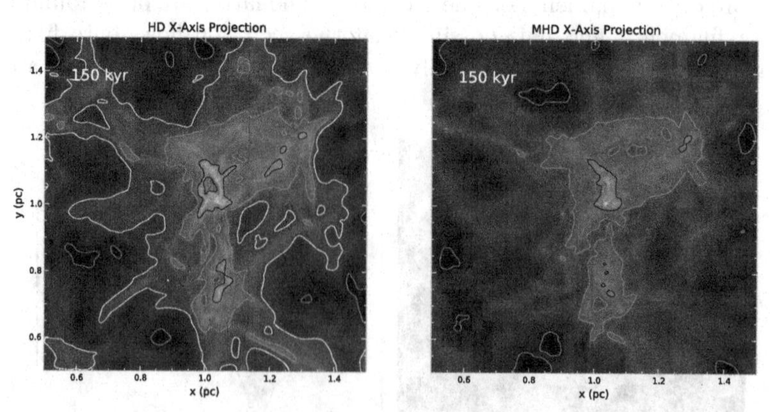

Figure 4. Filamentary structure: comparison of hydrodynamic and magnetohydrodynamic simulations of a turbulent, cluster forming clump. Column density maps for a 500 $M_\odot$ cloud: hydro (left) and MHD (right) with $\lambda = 2.33$. Red contour outlines 5 times m_{crit} for the *thermal* critical line mass, blue is 2, and 1 is critical thermal line mass. Adapted from Kirk, Klassen, Pillsworth, and Pudritz (in prep).

the hydromagnetic turbulence - the main effect being to broaden the effective half width of the lognormal (eg. Lemaster & Stone 2008, Collins *et al.* 2011).

In summary, magnetic fields generated in galactic dynamos can clearly control the formation of molecular clouds, the dynamics of their filamentary structure, and the extent to which fragmentation into star forming cores takes place. There is a significant effect therefore on suppressing the formation of low mass cores. We now descend to much smaller scales, of $\leqslant 0.04$ pc and examine the collapse of cores, and the formation of disks and outflows.

3. Magnetic Fields and the Origin of Protostellar Disks and Outflows

Protostellar disks are observed around protostars of all masses - from brown dwarfs to massive stars. The angular momentum that is associated with molecular cloud cores arises from the collision of oblique shocks that produce the filaments (eg. Jappsen & Klessen 2004, Banerjee *et al.* 2006). The specific angular momentum of cores on these 0.05 pc scales is of the order 10^{21} cm^2 s^{-1} which is 4 orders of magnitude greater than that of the young star that ultimately forms as a consequence of gravitational collapse. In purely hydrodynamic simulations, the collapse of such a region produces disks. The centrifugal radius that defines the outer edge of the disk under these conditions is typically 100 AU.

It has been known for some time that magnetic braking of rotating magnetized cores would easily occur (eg. Mouschovias & Paleogolou 1980, Basu & Mouschovias 1994). In the case of the gravitational collapse of a magnetized core, the faster rotating material deeper down in the collapsing region remains attached to the slower rotating material further out by the magnetic field, which therefore twists the field lines. The torsional Alfvén waves generated in this way carry off angular momentum, transferring it to the surrounding gas. What had not been anticipated was that in collapse calculations of cores threaded by ordered magnetic fields in clouds with $\lambda \leqslant 5$, the braking can be catastrophic, preventing the formation of rotationally supported disks (Mellon & Li 2008, Hennebelle & Fromang 2008, Duffin & Pudritz 2009, Seifried *et al.* 2012a, Santos-Lima *et al.* 2012). The effect is not seen for sufficiently weak fields ($\lambda > 10$) but as noted, this does not fit the data for many cores.

Several solutions to this "magnetic braking catastrophe" have been suggested (see review Li *et al.* 2013) involving processes such as magnetic reconnection or lowered braking

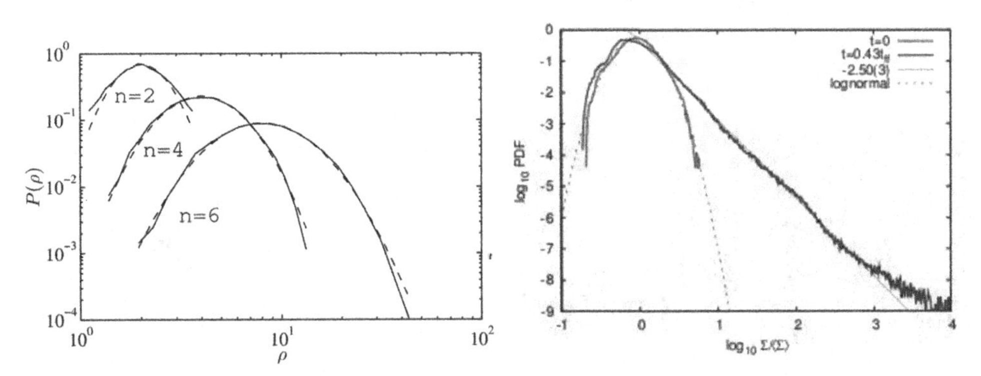

Figure 5. (a) Probability distribution function (PDF) of the density in a medium traversed by n shock waves. The dashed curves are best fit lognormal models. Adapted from Kevlahan & Pudritz 2009. (b) Simulations of highly resolved 3D turbulence that without gravity, creates a lognormal PDF (red curve). When gravity is turned on, a power-law tail is generated (blue curve). Adapted from Kritsuk *et al.* (2011).

efficiency by disks inclination to the mean field. Arguably the most robust solution to this problem is that turbulence facilitates the formation of rotationally supported disks (Seifried *et al.* 2012a - also Santos-Lima *et al.* 2012, Joos *et al.* 2013) . Figure 6(a) shows two simulation results for the formation of disks: the left figure shows that an ordered Keplerian disk does not form for ordered magnetic fields, while rotationally supported disks do form in the turbulent simulation (right). The rotation curves for disk-like structures in turbulent MHD collapse simulations are shown in Figure 6(b) for 4 different disks that form in an initially 100 $M_\odot$ cloud simulations with $\lambda = 2.6$ (Seifried *et al.* 2012a). In all cases, the rotation curves are Keplerian with much smaller radial inflow velocities.

The formation of magnetized accretion disks is accompanied by the launch of vigorous outflows. Outflows are among the earliest and most easily detectable signatures of star formation. This is well exemplified by the so-called Class 0 protostars in which much of the collapsing gas is still in the envelope and yet very energetic molecular outflows are present. Outflows can be driven by centrifugal acceleration (Blandford & Payne 1982, Pudritz & Norman 1983) or by the magnetic pressure gradient associated with the toroidal magnetic field, known as a tower flow (Lynden-Bell 1996, 2003). In a collapse, both types of outflow are observed in numerical simulations and there is now a vast literature on this subject featuring a wide range of numerical techniques (review, Pudritz *et al.* 2007, Li *et al.* 2013). Figure 7(a) shows the magnetic structure associated with both of these outflow types. The first is a slower tower flow that arises in the outer regions of forming disks and is likely associated with the formation of the first core (top frame). The second is the centrifugally driven wind from smaller disk radii (Banerjee & Pudritz 2006).

The slow component is a transient outflow that in the later stages of disk evolution is dominated by the centrifugal disk wind. This is shown in Figure 7(b) for two different distributions of magnetic fields in the underlying disk (Staff *et al.* 2010). One notes that the dense material in the jet is confined more towards the central outflow axis, which is collimated by a strong toroidal field component that provides the collimating pinch force for the jet. This kind of jet structure is robust and does not depend on the details of how magnetic field is distributed in the disk. By applying the conservation laws of axisymmetric outflows, such as the generalized Bernoulli theorem (Pelletier & Pudritz 1992), it is possible to show that both of these outflows are just different regimes of a single generalized acceleration condition that can nevertheless be successfully applied to dynamical outflow simulations (Seifried *et al.* 2012b).

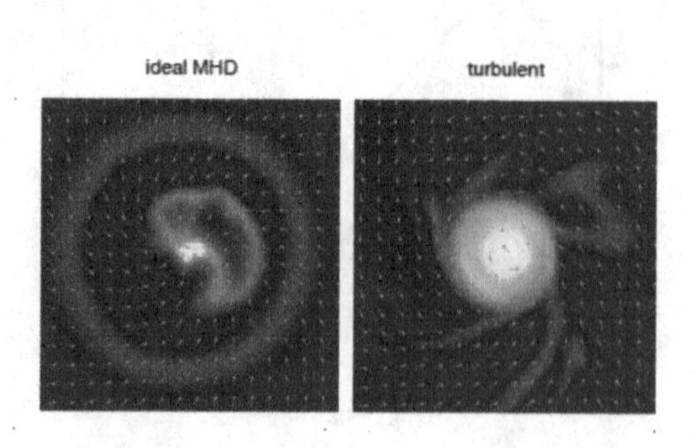

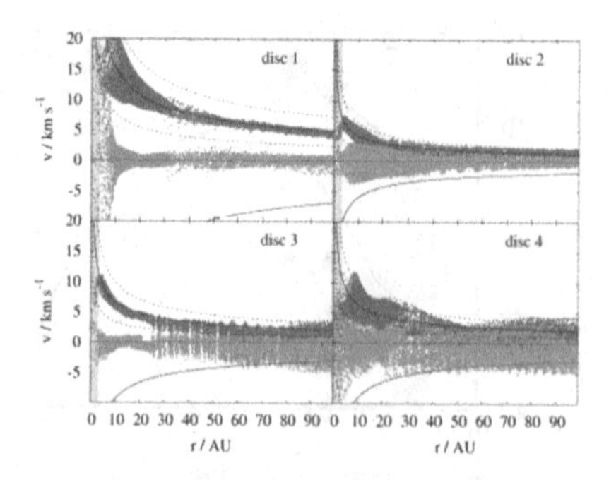

Figure 6. (a) Accretion of magnetized gas onto a central object in the absence (left) or presence (right) of turbulence. Adapted from Santos-Lima *et al.* (2012). (b) Rotation curves (red) and radial inflow speeds (green) through 4 different disks formed in 3D simulations of the collapse of magnetized turbulent, 100 $M_\odot$ cloud. Solid line traces Keplerian rotation. Adapted from Seifried *et al.* (2012a).

The importance of outflows for star formation is two fold. First, it is well known that MHD disk winds are highly efficient in extracting the angular momentum of disks - much more than viscous torques that arise through disk turbulence (Pelletier & Pudritz 1992). HST observations suggest that jets may carry at least 60 % of the angular momentum that must be shed by disks in order to enable accretion from disks onto their central stars (Baciotti *et al.* 2002, Anderson *et al.* 2005). The second important aspect of outflows is that they may be the agents that help disperse the collapsing gas in cores - giving rise to the conversion of the CMF to the IMF discussed in the introduction (Matzner & McKee 2000). In more massive star formation, radiation fields will also play a significant role in core dispersal.

4. Solving the Angular Momentum Problem for Young Stars

We have now arrived at the magnetosphere of an accreting protostar as it truncates the accretion disk. The magnetic stresses associated with the star's magnetic field are strong enough at this radius R_t to dominate the turbulent stress in the disk, so that matter free-falls along the field lines to the stellar surface. If the material at this last Keplerian orbital radius brings its angular momentum with it, the resulting spin-up accretion torque on the star is $\tau_a = \dot{J} = \dot{M}_a\sqrt{GM_* R_t}$, where $\dot{M}_a$ is the disk accretion rate. If unbalanced, this torque will spin up the star to break up speeds within 10^5 yr. Several different solutions have been proposed for this problem. In the disk-locking picture of Königl (1991), dipolar field lines penetrating the disk beyond the co-rotation radius, $R > R_{co} \simeq R_t$ would provide a spin-down torque. The difficulty with this idea is that the intense shear due to Kepler flow even a short distance from the co-rotation radius would create a very strong toroidal field, which would rapidly inflate and disconnect from the disk (Matt & Pudritz 2004). Simulations of the interaction of a rotating magnetosphere inside an accretion disk confirm that stellar field lines penetrating near to the co-rotation radius undergo episodic inflation and ejection, much like coronal mass ejections on the Sun (Zanni & Ferreira 2009).

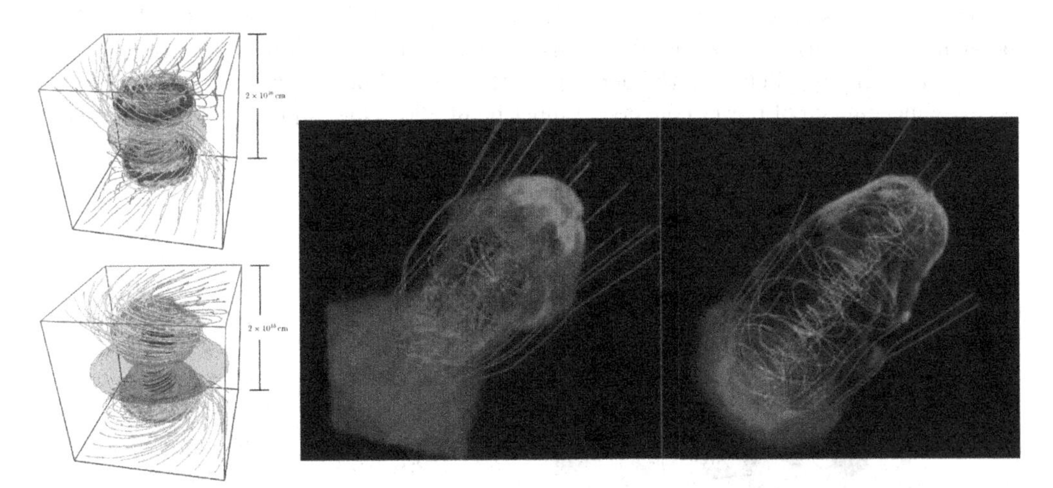

Figure 7. (a) Simulations of the collapse of a magnetized core; Top - magnetic field lines associated with a tower flow on 100 AU scales; Bottom - magnetic field lines associated with centrifugally driven disk wind from the inner region of the disk on 1 AU scalpes. Adapted from Banerjee & Pudritz (2006). (b) 3D simulations of jets from a Keplerian accretion disk for two different models for distribution of magnetic fields in the disk, including Blandford & Payne (1982) and Ouyed & Pudritz (1997). Adapted from Staff *et al.* (2010).

An alternative idea is that the excess angular momentum is not deposited back into the disk, but is instead carried away by a strong magnetized acretion-powered stellar wind (Matt & Pudritz 2005). The point is that the high accretion rates onto young stars liberate gravitational potential energy on impacting the photosphere. This powerful source of energy, if tapped at reasonable efficiencies, is available to drive a wind along the open magnetic field lines from the polar caps.

The four elements of a general theoretical framework involving both disk and stellar winds as well as magnetoshperic interaction are illustrated in Figure 8(a). At large disk radii beyond the interaction region is the disk wind. Magnetospheric accretion brings both gas and its angular momentum to the star where angular momentum is deposited and gravitational potential energy released. The third component is the accretion powered stellar wind carries off the accreted disk angular momentum. Finally, there is the interaction region between the stellar and disk winds. This interface between the disk and the magnetosphere is known from simulations to be highly time variable as accretion proceeds in bursts followed by the inflation of field lines (Goodson, 1999, Zanni & Ferreira 2013). The accretion powered stellar wind (APSW) exerts a pressure on the surrounding disk wind, which prevents it from collimating as much as it might in its absence.

The spin down torque upon the star due to the APSW is $\tau_w = \dot{M}_w \Omega_* r_A^2$ where r_A is the Alfvén radius of the stellar wind and $\dot{M}_w$ is the stellar wind mass loss rate. . In the theory of disk winds, the wind mass loss rate in disk wind theory is roughly a tenth of the observed accretion rate (Pudritz *et al.* 2007). The same kind of scaling is expected for APSW winds. Numerical calculations of 2D stellar magnetized winds assumed an initial dipole field and no disk or surrounding disk wind, for a variety of stellar magnetic field strengths, and found that for a 2kG field, a large Alfvén radius is obtained with $R_A \simeq 19.3 R_*$ (Matt & Pudritz 2008). The fraction of the accretion power that is converted into wind power (ϵ) is one of the main parameters of the APSW theory. Equilibrium solutions for the slow spins can be found for various objects and this fixes the efficiency parameter. For a strong stellar field of 2 kG, the efficiency can be as low as 17 % but this value must be higher for weaker dipole fields.

Our journey has now taken us to the accretion shock at the stellar photosphere. The question is what initiates the APSW? Thermal driving of the wind will not work because the cooling is highly efficient in the dense post-shock gas (Matt & Pudritz 2008). A much less lossy mechanism of power conversion must be in play. One important possibility that

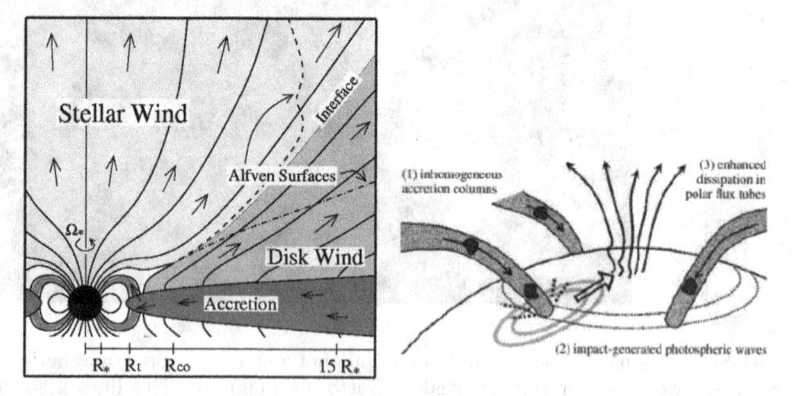

Figure 8. (a) A global view of disk winds, and accretion powered stellar winds associated with magnetized accretion disks and protostars. Adaped from Matt & Pudritz (2005). (b) Illustration of the excitation of wave driven outflow by lumpy magnetospheric accretion onto the protostar. Adapted from Cranmer (2009).

is the accretion power is converted to MHD wave driven winds (Matt & Pudritz 2008(b). Wave driven stellar winds are well established for late type stars (Hartmann & MacGregor 1980, DiCampli 1981). The shaking up of the stellar photosphere in the neighbourhood of the accretion column would launch waves propagating across the photosphere, as illustrated in Figure 8(b). Those regions with open magnetic field lines would be strongly excited leading to the launch of Alvén waves. The accretion column is likely to consist of lumpy gas. The energy released by each of these clumps in the inghomogenous accretion column is roughly 10% of the kinetic energy of the clump multiplied by the ratio of the clump density to the background flow. This scenario was worked out in some detail in two papers by Cranmer (2008, 2009) and found to be in rough agreement with observations although much more work remains to be done.

In conclusion, the role of magnetic fields in star formation is manifold and carries through many decades in physical scales. On the largest, magnetic fields can play a role in the formation and fragmentation of filamentary molecular clouds. The lognormal core of the IMF and CMF appears to be a direct consequence of turbulence, while gravity seems to be the dominant player in the creation of the high mass tail. Radiation and MHD together affect gravitational fragmentation and the CMF, with radiation dominating on smaller scales and MHD on larger. MHD turbulence, however, probably plays a dominant role in how disks are formed. Finally, angular momentum transport by magnetized winds and jets likely plays a central role in driving disk accretion and regulating the spins of young stars.

References

Alves, J., Lombardi, M., & Lada, C. J. 2007, *A&A*, 462, L17

Anderson, J. M., Li, Z. Y., Krasnopolsky, R., & Blandford, R. D. *ApJ*, 630, 945

André, Ph., Men'shchikov, A., Bontemps, S., *et al.* 2010, *A&A*, 518, L102

André, Ph., Men'shchikov, A., Könyves, V., & Arzoumanian, D. 2011, in Computational Star Formation, IAU Symp. 270, Eds. J. Alves *et al.*, p. 255

André Ph., Di Francesco J., Ward-Thompson D., Inutsuka, S-I., Pudritz, R. E., & Pineda, J. 2014, in: Protostars and Planets VI, (Eds. H. Beuther, R. Klessen, K. Dullemond, & T. Henning), in press.

Bacciotti, F., Ray, T. P., Mundt, R., *et al.* 2002, *ApJ*, 576, 222

Banerjee R. & Pudritz R. E. 2006, *ApJ*, 641, 949 1

Banerjee, R., Pudritz, R. E., & Anderson, D. W., 2006, *MNRAS*, 373, 1091

Basu, S. & Mouschovias, T. 1994, *ApJ*, 432, 720

Beck, R. 2005, in: Cosmic magnetic fields; Lecture Notes in Physics, 664, Eds. R. Wielebinski & R. Beck, p. 41.

Beck, R. 2012, *Space Sci Rev*, 166, 215.

Beck *et al.* 1999,

Blandford, R. D. & Payne, D. G. 1982, *MNRAS*, 199, 883

Bouvier, S. P. Matt, Mohanty, S., Scholz, A., Stassun, K. G., & Zanni, C. 2014, in: Protostars and Planets VI, (Eds: H. Beuther, R. Klessen, K. Dullemond, & T. Henning), in press.

Chabrier, G. 2005, *PASP*, 115, 763

Collins, D., Padoan, P., Norman, M. L., & Xu, H. 2011, *ApJ*, 731, 59

Cranmer, S. R. 2006, *ApJ*, 689, 316

Cranmer, S. R. 2009, *ApJ*, 706, 824

Crutcher, R. M. 2012, *ARA&A*, 50, 29

Crutcher, R. M., Wandelt, B., Heiles, C., Falgarone, E., & Troland, T. H. 2010, *ApJ, 725, 466*

Decampli, W. M. 1981, *ApJ*, 244, 124

Duffin, D. F. & Pudritz, R. E. 2009, *ApJL*, 706, L46

Elmegreen, B. G. 1982, *ApJ*, 253, 655

Fiege, J. D. & Pudritz, R. E. 2000, *MNRAS*, 311, 85

Fletcher, A. 2011, in The dynamic interstellar medium: a celebration of the Canadian Galactic Plane Survey, ASP Conference Series, 438, Eds. R. Kothes, T. L. Landecker & A. G. Willis, 438, 197.
Fletcher, A., Beck, R., Shukurov, A., Berkhuijsen, E. M., & Horellou, C. 2011, *MNRAS*, 412, 2396
Goodson, A. P. & Winglee, R. M. 1999, *ApJ*, 524, 159
Hartmann, L. & MacGregor, K. B. 1980, *ApJ*, 242, 260
Heesen, V., Krause, M., Beck, R., & Dettmar, R.-J. 2009, *A&A*, 506, 1123
Hennebelle, P. & Fromang, S.2008, *A&A*, 477, 9
Henning, Th., Linz, H., Krause, O., *et al.* 2010, *A&A*, 518, L95
Inutsuka, S. & Miyama, S. M. 1997, *ApJ*, 480, 681
Jappsen, A.-K. & Klessen, R. D. 2004, *A&A*, 423, 1
Johnstone, D., Wilson, C. D., Moriarty-Schieven, G., *et al.* 2000, *ApJ*, 545, 327
Joos, M., Hennebelle, P., & Ciardi, A. 2012, *A&A*, 543, A128
Kainulainen, J., Beuther, H., Henning, T., & Plume, R. 2009, *A&A*, 508, L35
Kevlahan, N. & Pudritz, R. E. 2009, *ApJ*, 702, 39
Kim, W.-T., Ostriker, E., & Stone, J. M. 2003, *ApJ*, 595, 574
Kirk, H., Klassen, M., Pillsworth, S., & Pudriz, R. E., 2014, in preparation.
Klessen, R. S. & Burkert, A. 2000, *ApJS*, 128, 287
Königl, A. 1991, *ApJ*, 370, L39
Kritsuk, A., Norman, M. L., & Wagner, R. 2011, *ApJL*, 727, L20
Lemaster, M. N. & Stone, J. M. 2008, *ApJ*, 688, 905
Li, H.-b., Dowell, C. D., Goodman, A., Hildebrand, R., & Novak, G., 2009, *ApJ*, 704, 891
Li Z-Y., Banerjee R., Pudritz R. E., Jorgensen J. K., Shang H., Krasnopolsky R., Maury A., 2014, in:Protostars and Planets VI, (edited by H. Beuther, R. Klessen, K. Dullemond, & T. Henning), in press.
Lynden-Bell, D. 1996, *MNRAS*, 279, 389
Lynden-Bell, D. 2003, *MNRAS*, 341, 1360
Matt, S. & Pudritz, R. E. 2004, *ApJL*, 607, L43
Matt, S. & Pudritz, R. E. 2005, *ApJL*, 632, L135
Matt, S. & Pudritz, R. E. 2008, *ApJ* 681, 391
Matt, S., & Pudritz, R. E. 2008, *ApJ*, 678, 1109
Matzner, C. D. & McKee, C. F. 2000, *ApJ*, 545, 364
Mellon, R. R. & Li, Z.-Y. 2008, *ApJ*, 681, 1356
Men'shchikov, A., André, Ph., Didelon, P., *et al.*2010, *A&A*, 518, L103
Motte, F., André, P., & Neri, R.1998, *A&A*, 365, 440
Mouschovias, T. C. & Paleologou, E. V. 1980, *ApJ*, 237, 877
Padoan, P., Nordlund, A., & Jones, B. J. T. 1997, *MNRAS*, 288, 145
Padoan, P., Juvela, M., Goodman, A. A., & Nordlund, A. 2001, *ApJ*, 553, 227
Pelletier, G. & Pudritz, R. E. 1992, *ApJ*, 394, 117
Pudritz, R. E., Ouyed, R., Fendt, C., & Brandenburg, A. 2007, in: Protostars and Planets V, (edited by B. Reipurth, D. Jewitt, and K. Keil), 277
Santos-Lima, R., de Gouveia Dal Pino, E. M., & Lazarian, A. 2012, *ApJ*, 747, 21
Schneider, N., André, Ph., Könyves, V., *et al.* 2013, *ApJL*, 766, L17
Seifried, D., Banerjee, R., Pudritz, R. E., & Klessen, R. S. 2012a, *MNRAS*, 423, L40
Seifried, D., Pudritz, R. E., Banerjee, R., Duffin, D., & Klessen, R. S. 2012b, *MNRAS*, 422, 347
Sellwood, J. A. & Balbus, S. A. 1999, *ApJ*, 511, 660
Staff, J. E., Niebergal, B. P., Ouyed, R., Pudritz, R. E., & Cai, K. 2010, *ApJ*, 722, 1325
Sugitani, K., *et al.* 2011, *ApJ* , 734, 63
Ward-Thompson, D., Kirk, J. M., André, P. *et al.* 2010, *A&A*, 518, L92
Tilley, D. A. & Pudritz, R. E., 2004, *MNRAS* , 353, 769
Tilley, D. A. & Pudritz, R. E., 2007, *MNRAS*, 382, 73
Zanni, C. & Ferreira, J. 2009, *A&A*, 508, 1117
Zanni, C. & Ferreira, J. 2013, *A&A*, 550, 99

Magnetic Fields throughout Stellar Evolution
Proceedings IAU Symposium No. 302, 2013
P. Petit, M. Jardine & H. Spruit, eds.

doi:10.1017/S1743921314001641

Magnetic field dispersion in the neighbourhood of Bok Globules

C. V. Rodrigues, V. de S. Magalhães, J. W. Vilas-Boas, G. Racca and A. Pereyra

Divisão de Astrofísica, Instituto Nacional de Pesquisas Espaciais
Av. dos Astronautas, 1758 – São José dos Campos – SP – Brazil
email: `claudiavr@das.inpe.br`

Abstract. We performed an observational study of the relation between the interstellar magnetic field alignment and star formation in twenty (20) sky regions containing Bok Globules. The presence of young stellar objects in the globules is verified by a search of infrared sources with spectral energy distribution compatible with a pre main-sequence star. The interstellar magnetic field direction is mapped using optical polarimetry. These maps are used to estimate the dispersion of the interstellar magnetic field direction in each region from a Gaussian fit, σ_B. In addition to the Gaussian dispersion, we propose a new parameter, η, to measure the magnetic field alignment that does not rely on any function fitting. Statistical tests show that the dispersion of the magnetic field direction is different in star forming globules relative to quiescent globules. Specifically, the less organised magnetic fields occur in regions having young stellar objects.

Keywords. magnetic field, star formation, Bok Globules, polarimetry

---

1. Introduction

The star formation results from the interplay between physical ingredients as the gravitational field, gas pressure, magnetic fields, and turbulence (e.g., Crutcher 2012). In particular, the importance of magnetic fields and turbulence is still an open issue. Other unanswered question is the origin of the interstellar turbulence.

There are some indicatives that star forming regions have less organised magnetic fields than quiescent regions. Pereyra (2000) studied the magnetic field direction along Musca Dark Cloud and found that there is an increase of the dispersion of the magnetic field direction (DMFD) near a young stellar object. The same happens in Pipe Nebula (Alves, Franco, & Girart 2008; Franco, Alves, & Girart 2010): the B59 region has the largest DMFD in this dark cloud. Additionally, the DMFD in the interstellar medium near Herbig-Haro objects is qualitatively consistent with that of the B59 region (Targon *et al.* 2011).

Bok Globules are among the simplest interstellar regions. Hence they are appropriate to study the role of different physical aspects on the star formation. Do the Bok Globule properties differ as a function of the presence of star formation? To help to answer this question, we performed an observational project to verify whether the magnetic field organisation differs as a function of the presence of star formation in a sample of Bok Globules. We used optical polarisation to map the magnetic field direction. This study is partially presented in Magalhães (2012).

Table 1. Estimates of the alignment degree of the magnetic field direction around our Bok globules sample. The table is separated in globules with and without star formation.

Star forming region		
Globule	η	σ_B
BHR 016	0.71	13.4
BHR 044	0.17	-
BHR 053	0.68	26.8
BHR 058	0.88	11.3
BHR 075	0.87	13.4
BHR 117	0.90	12.0
BHR 138	0.98	5.2
BHR 139	0.99	5.6
BHR 140	0.94	7.9
BHR 148-151	0.67	-
BHR 059	0.51	-
BHR 145	0.44	-
BHR 113	0.99	5.1

No star formation		
Globule	η	σ_B
BHR 034	0.92	13.8
BHR 074	0.91	10.6
BHR 121	0.97	7.1
BHR 126	0.93	11.2
BHR 133	0.91	13.8
BHR 144	0.73	15.1
BHR 111[1]	0.94	10.2

[1] BHR 111 is associated with a Very Low Luminosity Object (Maheswar, Lee, & Dib 2011).

2. The Bok Globules sample and their association with star formation

The Bok Globules sample used in this study is the same presented in Racca, Vilas-Boas, & de la Reza (2009) (see Tab. 1). These globules are not associated with bright nebulae or molecular complexes.

Racca *et al.* (2009) classified 10 globules as having star formation by the presence of an associated IRAS source. In the present study, we revise this classification based on new infrared surveys. Data from WISE (Wright *et al.* 2010), 2MASS (Skrutskie *et al.* 2006), AKARI (Murakami *et al.* 2007), and GLIMPSE (Benjamin *et al.* 2003; Churchwell *et al.* 2009) were used to obtain the spectral energy distribution (SED) for the identified sources. These SEDs were fitted using Robitaille *et al.* (2006, 2007)'s models to check if they are consistent with young stellar objects or reddened foreground/background objects. Table 1 presents the resulting classification.

3. Optical polarimetry and the alignment of the magnetic field

The interstellar magnetic field was mapped using optical polarimetry. The data were collected at the 0.6m telescope of the *Observatório do Pico dos Dias*†. We used a CCD polarimeter (Magalhães *et al.* 1996) and an I_C filter. This instrumental configuration provides polarimetry of point sources in a field-of-view of 11 x 11 $arcmin^2$ (e.g., Fig. 1.left).

Usually the alignment degree of the interstellar magnetic field direction is quantified by the Gaussian dispersion of the polarisation angle, σ_B (e.g., Fig. 1.right). However, a Gaussian fit is not always possible. An example is the BHR 059 field (Fig. 2). To let a quantitative analysis for all fields, we propose a new estimator of the magnetic field alignment, η:

$$\eta = \frac{\overline{P}}{\overline{|P|}}, \tag{3.1}$$

where $\overline{P}$ is the vectorial average of the polarisation vectors and $\overline{|P|}$ is the arithmetic average of the polarisation module of the field stars. If all objects have the same polarisation angle, $\eta = 1$; if the polarisation angles are randomly distributed, η tends to zero. Table 1

† Managed by *Laboratório Nacional de Astrofísica*/Brazil.

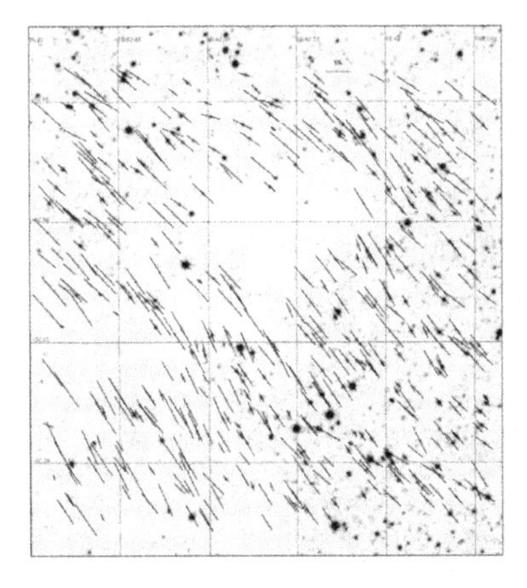

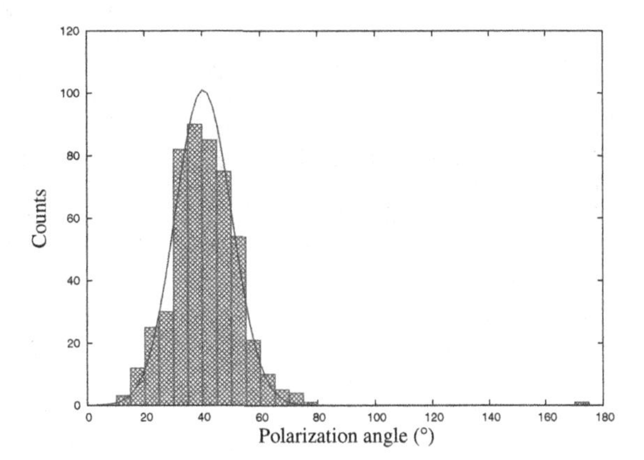

Figure 1. (Left) Optical polarimetry in the line of sight of BHR 111 in the I band superposed on a DSS2 red image. (Right) The corresponding histogram of the linear polarisation angle. The superposed line is a Gaussian fit.

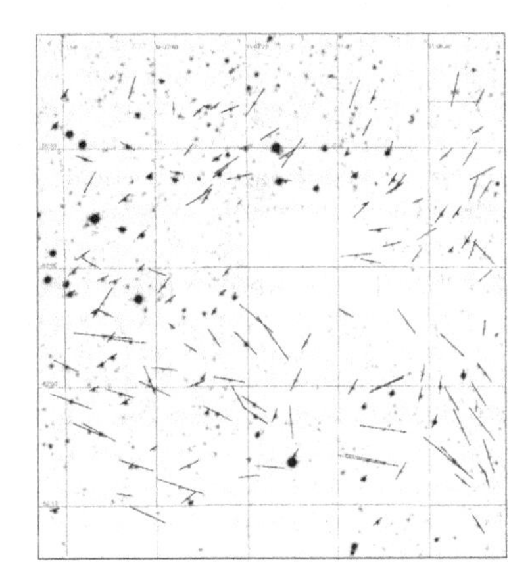

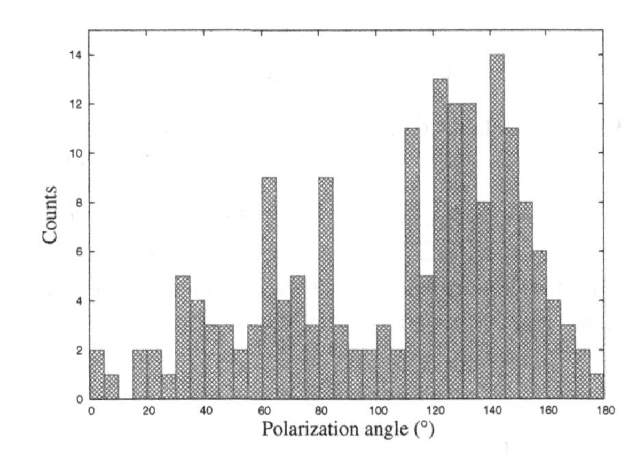

Figure 2. As Figure 1, for BHR 059.

presents η and σ_B for the observed fields. We could not estimate σ_B for BHR 059 and other fields presenting a non Gaussian distribution of polarisation vectors.

4. Results and discussion

Figure 3 shows graphically η and σ_B for our sample. The magnetic field organization decreases from the left upper region to the right lower corner. Organised magnetic fields are found in regions with or without star formation. However, the less organised fields are only found for regions presenting star formation.

Using the Kolmogorov-Smirnov test, we found that the probability that the globules with star formation have the same η (σ_B) distribution of quiescent globules is 8% (68%).

These results suggest that the alignment of the interstellar magnetic field is different in regions with star formation compared with quiescent regions, and less organised fields are found in star forming regions. A possible interpretation is that the star formation injects kinetic energy in the interstellar medium, increasing the turbulence and disorganising the magnetic field; the organisation of the magnetic field lines is related to the gas turbulence according to Chandrasekhar & Fermi (1953), considering energy equipartition. Another possibility is that the star formation is favoured in more turbulent regions.

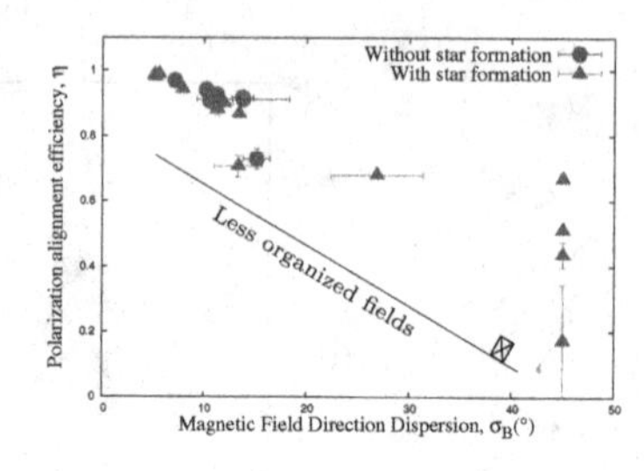

Figure 3. Polarisation alignment efficiency, η, and DMFD, σ_B, for star forming globules (red triangles) or quiescent globules (blue circles).

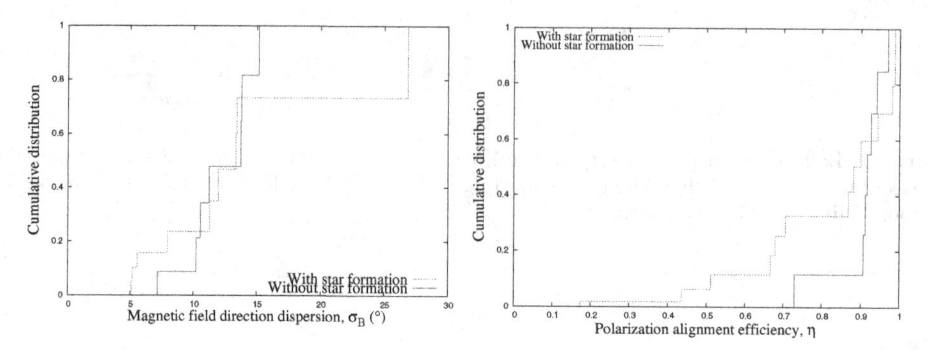

Figure 4. (Left) Cumulative distribution of σ_b for globules having star formation (red dotted line) or quiescent globules (blue solid line). (Right) The same for η.

Acknowledgements: CVR thanks Grant 2010/01584-8, São Paulo Research Foundation (FAPESP). This research makes use of: data products from WISE and 2MASS; the NASA/IPAC Infrared Science Archive; observations made with AKARI and the Spitzer Space Telescope; the SIMBAD database; the NASAs ADS Service; the NASA's SkyView facility; Aladin; and DSS.

References

Alves, F. O., Franco, G. A. P., & Girart, J. M. 2008, *A&A*, 486, L13
Benjamin, R. A., Churchwell, E., Babler, B. L., *et al.* 2003, *PASP*, 115, 953
Chandrasekhar, S. & Fermi, E. 1953, *ApJ*, 118, 113
Churchwell, E., Babler, B. L., Meade, M. R., *et al.* 2009, *PASP*, 121, 213
Crutcher, R. M. 2012, *ARA&A*, 50, 29
Franco, G. A. P., Alves, F. O., & Girart, J. M. 2010, *ApJ*, 723, 146
Magalhães, A. M., Rodrigues, C. V., Margoniner, V. E., Pereyra, A., & Heathcote, S. 1996, *ASPC*, 97, 118
Magalhães, V. S. 2012, *MSc Dissertation*, Instituto Nacional de Pesquisas Espaciais
Maheswar, G., Lee, C. W., & Dib, S. 2011, *A&A*, 536, A99
Murakami, H., *et al.* 2007, *PASJ*, 59, 369
Pereyra, A. 2000, *PhD Thesis*, Universidade de São Paulo
Racca, G. A., Vilas-Boas, J. W. S., & de la Reza, R. 2009, *ApJ*, 703, 1444
Robitaille, T. P., Whitney, B. A., Indebetouw, R., Wood, K., & Denzmore, P. 2006, *ApJS*, 167, 256
Robitaille, T. P., Whitney, B. A., Indebetouw, R., & Wood, K. 2007, *ApJS*, 169, 328
Skrutskie, M. F., *et al.* 2006, *AJ*, 131, 1163
Targon, C. G., Rodrigues, C. V., Cerqueira, A. H., & Hickel G. R. 2011, *ApJ*, 743, 54
Wright, E. L., Eisenhardt, P. R. M., Mainzer, A. K., *et al.* 2010, *AJ*, 140, 1868

Magnetic Fields throughout Stellar Evolution
Proceedings IAU Symposium No. 302, 2013
P. Petit, M. Jardine & H. Spruit, eds.

doi:10.1017/S1743921314001653

The role of magnetic fields in pre-main sequence stars

Gaitee A. J. Hussain[1] **and Evelyne Alecian**[2,3]

[1]ESO, Karl-Schwarzschild-Strasse 2, D-85748, Garching bei München
Germany
email: ghussain@eso.org

[2]UJF-Grenoble 1/CNRS-INSU, Institut de Planétologie et d'Astrophysique de Grenoble
(IPAG) UMR 5274, Grenoble, F-38041, France
email: evelyne.alecian@obs.ujf-grenoble.fr

[3]LESIA, UMR 8109 du CNRS, Observatoire de Paris, UPMC, Université Paris Diderot,
5 place Jules Janssen, F-92195 Meudon Cedex, France

Abstract. Strong, kilo-Gauss, magnetic fields are required to explain a range of observational properties in young, accreting pre-main sequence (PMS) systems. We review the techniques used to detect magnetic fields in PMS stars. Key results from a long running campaign aimed at characterising the large scale magnetic fields in accreting T Tauri stars are presented. Maps of surface magnetic flux in these systems can be used to build 3-D models exploring the role of magnetic fields and the efficiency with which magnetic fields can channel accretion from circumstellar disks on to young stars. Long-term variability in T Tauri star magnetic fields strongly point to a dynamo origin of the magnetic fields. Studies are underway to quantify how changes in magnetic fields affect their accretion properties. We also present the first results from a new programme that investigates the evolution of magnetic fields in intermediate mass (1.5–3 $M_{\odot}$) pre-main sequence stars as they evolve from being convective T Tauri stars to fully radiative Herbig AeBe stars.

Keywords. stars: activity, stars: accreting, stars: circumstellar matter, stars: magnetic fields

---

1. Background

Pre-main sequence (PMS) models show that all stars form from a fully convective low mass core. In Fig. 1, stars rise along the birth line (blue dashed line) until the strong accretion phase stops and then follow a quasi-static contraction along PMS tracks. We adopt Behrend & Maeder (2001) tracks here; different models predict slightly different positions for the PMS tracks (Siess *et al.* 2000, Tognelli *et al.* 2011) but they all show the same fundamental transitions to internal structure with mass. Higher mass stars naturally evolve much more quickly along these tracks compared to their lower mass counterparts. It is worth noting that A-type stars will undergo fundamental changes to their interior structures during their PMS evolution; starting as fully convective T Tauri stars, developing radiative cores; eventually becoming fully radiative Herbig stars (Grunhut *et al.*, *this volume*, Alecian *et al.* 2013). Finally, just before they reach the Zero Age Main Sequence they will develop a convective core. This review focusses on the magnetic properties of T Tauri stars, which either have fully convective interiors or outer convective envelopes. While these stars cover a relatively narrow range in spectral type (G to M), they encompass almost two orders of magnitude in stellar luminosity.

It has long been a requirement for T Tauri stars to host kilo-Gauss magnetic fields with strong dipoles in order to explain several key observational characteristics. Classical T Tauri stars, T Tauri stars that are accreting, have long rotation periods (typically

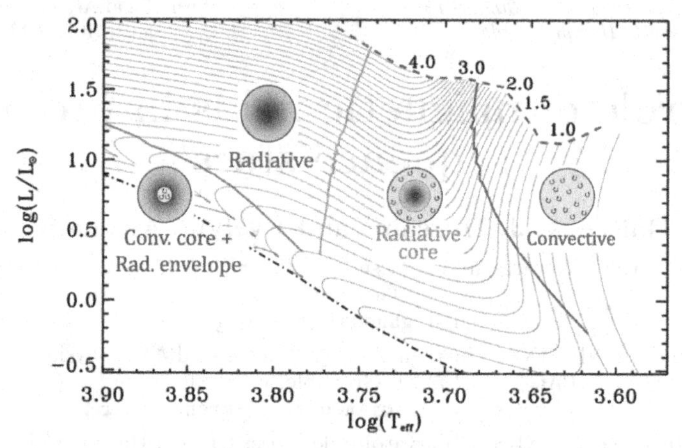

Figure 1. H-R diagram showing Behrend & Maeder (2001) pre-main sequence evolutionary tracks for stellar masses up to $4\,M_\odot$. The dashed blue line marks the position of the birthline. All stars with masses less than $3.5\,M_\odot$ will undergo a stage along their pre-main sequence evolution in which they have either partially or fully convective interiors. A star with a mass of $1.5\,M_\odot$ or more will be subject to several fundamental changes in their internal structure, having a fully convective interior near the birthline, to developing a radiative core, to becoming fully radiative and finally developing a convective core just before reaching the Zero Age Main Sequence (black dot-dashed line).

between 7–10d); these are much longer than predicted from angular momentum conservation considering the stars are contracting and actively accreting material from their circumstellar disks. Observations indicate discrete regions at which accretion streams impact the stellar surface; shocks form near the photosphere that can be detected at X-ray and UV wavelengths. These XUV diagnostics indicate material impacting at near-free fall velocities at or near the photosphere. The model that can best explanation most of the observational properties of classical T Tauri stars is the magnetospheric accretion model. In this model, a strong dipole field from the star extends several stellar radii, truncating the inner edge of the stellar disk and channelling disk material along these magnetic field lines in streams that then impact the stellar surface. From a study of the classical T Tauri star, AA Tau, Bouvier *et al.* (2007) find that both the photometric variability and the modulation seen in its Balmer line can be explained in terms of accretion funnels that pass along the line of sight and a magnetically warped disk that periodically occults the star.

The magnetospheric accretion environment can regulate the angular momentum of these young systems through accretion powered stellar winds, ejections through reconnection events and similar interactions between the stellar magnetosphere and the disk (e.g., Hartmann & Macgregor 1982; Matt & Pudritz 2005, 2008, Zanni & Ferreira 2013). Once accretion stops or becomes less efficient then stars are free to spin up as they contract towards the Main Sequence.

2. PMS star magnetic fields: detection techniques

Stellar magnetic fields can be directly detected and characterised using two main techniques, *Zeeman broadening* and *Zeeman Doppler imaging*. Both of these are spectroscopic techniques that require high resolution spectra ($R > 30\,000$) and rely on the Zeeman effect. In the presence of a magnetic field atomic and molecular lines can show broadening or even full splitting, depending on the magnetic sensitivity of the line and the size of the magnetic field. Lines with no magnetic sensitivity are magnetic null lines and are useful diagnostics of the non-magnetic photosphere. A brief description of these techniques as applied to T Tauri stars is given here, along with a list of the main advantages and disadvantages associated with each (Table 1). The interested reader is referred to the review by Donati & Landstreet (2009) for a more detailed description of the Zeeman effect and these techniques.

2.1. *Zeeman broadening*

The Zeeman broadening technique measures the broadening in intensity line profiles with different magnetic sensitivities (g, Landé-factors). It has been used to measure magnetic field strength distributions and magnetic fluxes in a range of T Tauri stars (e.g., Johns-Krull 2007, 2008, Yang *et al.* 2011 & references therein). The size of the broadening scales with the square of the wavelength (Eqn. 2.1), so longer wavelengths yield more robust measurements and measurements of the mean magnetic field. The surface magnetic fields of T Tauri stars have been measured using a set of Ti I lines near 2.2μ that have different magnetic sensitivities. Magnetic null ($g_{\text{eff}} = 0$) CO lines near 2.3μ are used to characterise non-magnetic photospheric parameters, e.g., veiling ,Teff, vsini, microturbulence. This technique is particularly effective when applied to T Tauri stars, as many are relatively slow rotators and have spectral types of K or later. Ti I 2.2μ lines are best suited to K-M spectral types as they weaken in hotter stars.

$$\Delta\lambda_B = 4.67\lambda_o^2 g_{\text{eff}} B. \tag{2.1}$$

The wavelength broadening, $\Delta\lambda_B$, scales with the stellar magnetic field, B, the wavelength of the line λ_o, and its effective Landé factor, g_{eff}, a measure of the mean magnetic sensitivity of the line. Zeeman Broadening studies have shown that multiple magnetic field strength components are often required to fit the observed splitting in T Tauri stars if applied to spectra encompassing lines with different magnetic sensitivities. A mean magnetic field strength can be computed from these different components as follows: $Bf = \Sigma B_i f_i$, with a typical range of field strengths (B_i) between 2 kG to 6 kG, and associated filling factors (f_i) for each component ranging between 20-50%.

2.1.1. *Zeeman Broadening: strengths and challenges*

The main strengths and challenges associated with this technique are summarised in the left column of Table 1 and expanded on in this section. A key strength is that it uses the information in Stokes I intensity profiles, hence the magnetic field measurements are sensitive to the strongest surface magnetic fields observable on the projected stellar disk, even if these are concentrated in complex small scale active regions. This technique is not subject to flux cancellation unlike techniques which utilise circularly polarised signatures (Stokes V profiles).

The modelling in Zeeman Broadening measurements assumes a purely radial field (e.g., Johns-Krull 2007, Yang *et al.* 2011). While extreme orientations could potentially be discerned (e.g., fields entirely aligned parallel to – or perpendicular to – the line-of-sight), very little information on the field topology can be obtained from these spectra alone. Yang *et al.* (2011) confirm previous findings that the Ti I spectra of T Tauri stars show little evidence for extreme orientations such as these. If the surface fields are composed of a mixture of radial, azimuthal (East-West) and meridional (North-South) orientations, the magnetic field measurements will be affected as different field orientations alter the strengths of the π and σ components of each line, and therefore the shapes of the magnetically sensitive line profiles. In the absence of further information radial field orientations are, however, the simplest assumption.

Table 1. Magnetic field measurement techniques: pros & cons (Sec. 2.1.1 & 2.2.1)

	Zeeman Broadening	**Zeeman Doppler Imaging**
Pros	Measure strongest magnetic fields Insensitive to field geometry	Recovers large-scale field topology Photospheric & accretion diagnostics
Cons	Snapshot single-epoch ***Bf*** measurements Assume uniform temperature 2.2μ studies limited to K-M type stars Slow rotation ($v_e \sin i < 30$km/s) Non-unique solution of multi-component ***B*** fields	Flux cancellation (circularly polarised spectra) Phase coverage & S:N affect magnetic flux strength Limited dark spot information Missing field information from inclined hemisphere Non-unique solution (regularising functions)

Zeeman Broadening measurements are best applied to low $v_e \sin i$ stars, typically less than 20 km/s. Significant Doppler broadening makes it difficult to measure the Zeeman broadening, $\Delta\lambda_B$ effectively. It should be noted that it has however been applied successfully up to $v_e \sin i \sim 55$ km/s (the M1.5 T Tauri star, TWA 5a; Yang *et al.* 2008). Further assumptions used in Zeeman broadening measurements are that the magnetic field regions are uniformly distributed over the stellar surface; clearly if the magnetic field regions were concentrated at the poles the required filling factors would change. The temperature structure associated with the magnetic field regions is also assumed to be uniform, i.e., not preferentially concentrated in cool spotted regions (Johns-Krull 2007). As these studies are usually based on single-epoch spectra, no information on the latitudinal positions or inhomogeneity of the surface fields can be obtained. Finally, the multi-component solutions are non-unique: multi-component fields are not always required to achieve similar levels of agreement with the data (Yang *et al.* 2011). This degeneracy can introduce a 10–15% uncertainty in the mean magnetic field strength measurements.

2.2. *Zeeman Doppler Imaging*

Doppler imaging techniques have been used to invert time-series of intensity spectra to recover surface maps of inhomogeneities, including brightness distributions, temperatures, abundances in a range of stars from Ap stars to cool M dwarfs (e.g., Vogt, Penrod & Hatzes 1987; Piskunov, Tuominen & Vilhu 1990; Barnes & Collier Cameron 2001). Zeeman Doppler imaging techniques apply Doppler imaging principles to circularly polarised (Stokes V) signatures in order to reconstruct the large scale magnetic fields on the surfaces of magnetically active stars (Semel 1989, Donati & Cameron 1997). Zeeman Doppler imaging exploits a key characteristic of Stokes V profiles, that they are predominantly sensitive to the line-of-sight component (longitudinal component) of the magnetic field. In cool stars these signatures are very weak, typically at 0.1% of the continuum level. A single-epoch Stokes V spectra enables us to robustly detect the presence of the longitudinal magnetic field. As demonstrated by Donati *et al.* (1997), the longitudinal component of B, B_l at a particular epoch can be estimated from the first moment of the Stokes V profile.

A time-series of Stokes V profiles, covering a full stellar rotation period, enables us to track the modulation of the line-of-sight component of the stellar magnetic field caused by magnetic regions crossing the projected stellar disk. This rotational modulation enables us to pinpoint not only the location of these regions in latitude and longitude, but also their relative field orientations.

When applied to T Tauri stars that are still accreting this technique can be applied simultaneously to both the photospheric and accretion line profiles (e.g., He I D_3 at 5876Å or the near-infrared Ca II triplet at 8500Å). This is possible thanks to the development of large-format high resolution échelle spectrographs† that cover over a thousand photospheric lines and several useful accretion diagnostics in T Tauri stars. It is necessary to sum up the signature from hundreds of photospheric lines using the cross-correlation technique, Least Squares Deconvolution (LSD) in order to detect stellar magnetic field signatures robustly (Donati *et al.* 1997, Kochukhov *et al.* 2010, Chen & Johns-Krull 2013). This is because the sizes of the signatures are typically small, at 0.1% of the continuum level, even in magnetically active stars.

In classical T Tauri stars strong emission lines often show significant circular polarisation in individual lines. In the majority of systems observed the Ca II NIR triplet shows strong Stokes V signatures indicating kG fields associated with the accretion. In some systems where the Ca II NIR Stokes V signatures are found to be weak (e.g., V4046 Sag), they support the picture of a complex large scale stellar magnetic field; with flux cancellation in Stokes V being caused by accretion spots with multiple polarities.

2.2.1. *ZDI: Strengths & Challenges*

The main limitations and challenges associated with this technique are summarised in Table 1 and expanded on in this section. The spatial resolution of the magnetic field maps obtained with Zeeman Doppler Imaging depends on the $v_e \sin i$ of the star, the spectroscopic resolution and the phase coverage obtained. In low $v_e \sin i$ stars ($< 20\,\mathrm{km/s}$), while it is possible to discriminate between simple and complex topologies ,flux cancellation results in an inability to measure the strongest fields at the stellar surface. Hussain *et al.* (2009) investigate whether it is possible to discriminate between simple dipole-dominant fields and more complex fields in low $v_e \sin i$ stars and find that even though the spatial resolution is more limited at low $v_e \sin i$'s, more complex large scale fields will be detected using Zeeman Doppler imaging.

Large gaps in phase coverage can result in smearing of magnetic field regions as their exact positions become harder to pinpoint. This also weakens the size of the magnetic flux size recovered in the map. The magnetic field maps obtained of T Tauri stars have been obtained using Doppler imaging codes that assume a two-component brightness model, consisting of an immaculate photosphere and a cool, dark spotted component. As spots are dark the relative flux contribution is limited and so it is inevitable that magnetic fields concentrated in the darkest spotted regions cannot be detected in photospheric line profiles. What is notable is that, in common with active G and K-type stars on the Main Sequence (e.g., Donati & Cameron 1997), a significant fraction of what appears to be the unspotted "immaculate" photosphere is found to be strongly magnetic. This may imply that the whole stellar surface is spotted at small scales that cannot currently be recovered using Doppler imaging techniques.

Finally, as with all Doppler imaging techniques, the solutions obtained are non-unique. It is possible to fit an observed time-series of Stokes I and V spectra with many different magnetic field and brightness distributions within a specified level of χ^2 agreement. It is therefore necessary to employ a regularising function, which enables a unique robust solution to be obtained. The maps presented here use Maximum Entropy, which minimises the amount of information needed to fit the observed spectroscopic time-series. Hussain *et al.* (2000) present a comparison of images obtained with two independent Maximum Entropy Zeeman Doppler imaging codes and find the images to be very similar though the exact form of the entropy used determines the sharpness of the structure reconstructed in the surface maps.

† The primary facilities optimised for these studies are CFHT/ESPaDOnS, TBL/NARVAL and the ESO 3.6-m/HARPS.

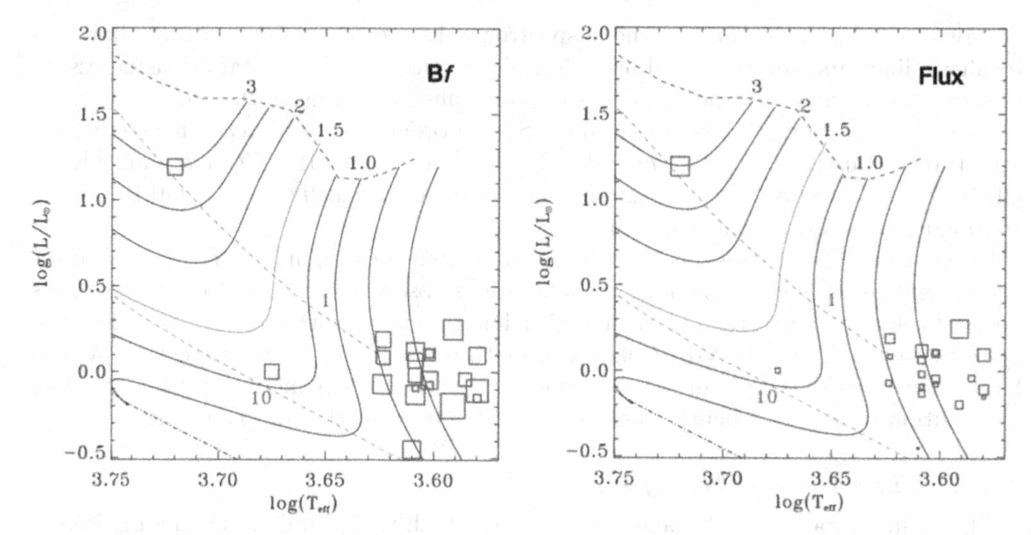

Figure 2. Surface magnetic fields in T Tauri stars measured using the Zeeman broadening technique: the symbol sizes scale with mean magnetic field strength (*left*) and mean magnetic flux (*right*); Bf-values are from Yang *et al.* (2011) and references therein. Behrend & Maeder (2001) evolutionary tracks are over-plotted for PMS stars with masses between 0.6 and 3 $M_\odot$.

3. Zeeman Broadening measurements of B and Bf

Over 30 T Tauri stars have been studied using the Zeeman Broadening technique, including several classical T Tauri stars. Figure 2 shows the resulting mean magnetic field strengths, Bf (left) and the mean magnetic fluxes plotted in an H-R diagram. These plots also show evolutionary tracks for PMS stars with masses ranging from 0.7 to 3 $M_\odot$, using the Behrend & Maeder (2001) tracks. All published measurements are shown here, excluding the lowest mass T Tauri stars with masses below 0.55 $M_\odot$.

In the left-hand plot of Fig.2, the symbol sizes reflect the mean magnetic field strengths, Bf, ranging from 1.1 to 3.5 kG, with 1 kG being the lower limit of the measurements. No clear trend with age or stellar mass is visible in the sizes of the mean Bf value. Indeed the highest and lowest average Bf are both found in stars with similar M0-1 spectral types: DE Tau and LO Ori, which have mean magnetic field strengths of 1.12 kG and 3.45kG respectively. The prototype of the class, T Tau, is the highest mass star in the sample shown here ($M_* \sim 3\,M_\odot$ according to the Behrend & Maeder tracks) and clearly has a similar average Bf to that found in its lower mass counterparts.

The large magnetic field strengths recovered are typically much larger than those required by pressure equipartition considerations, $B_{eq} = (8\pi P_g)^{1/2}$; where P_g is the gas pressure at the atmospheric height corresponding to the observed line formation (Johns-Krull 2007). As T Tauri stars have low surface gravities and relatively low gas pressures, flux tube equilibrium models predict lower magnetic fields than are found from Zeeman Broadening measurements. This implies that no equilibrium can exist between magnetic and "non-magnetic" regions (if they exist) in T Tauri stars. Caution must therefore be exercised in interpreting activity phenomena on these stars as simple scaled up versions of those observed on the Sun and solar-type stars.

The magnetic field strengths measured from these techniques are also generally significantly stronger than those predicted by simple analytic models of magnetospheric accretion (Johns-Krull 2007, 2008). These models estimate the sizes of the surface dipolar fields required given the radius, mass and accretion rate of a particular T Tauri star. Hence it is likely that even though T Tauri stars host adequately strong fields to channel

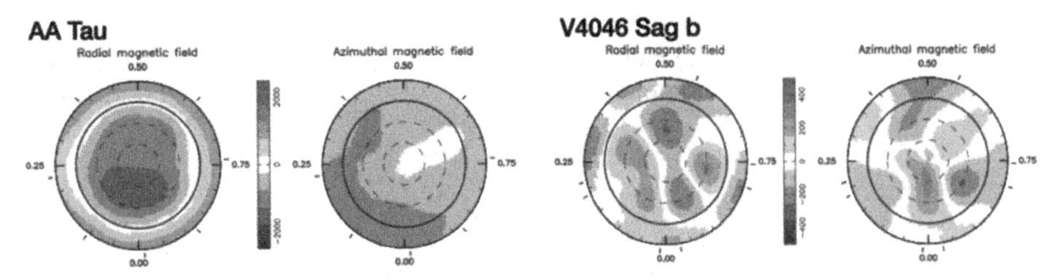

Figure 3. Radial and azimuthal field maps for two classical T Tauri stars; AA Tau (*left*) and V4046 Sag b (*right*). V4046 Sag b, has a significantly more complex field than AA Tau despite the similar spatial resolution of the maps. Magnetic flux maps obtained to date show that fully convective stars possess relatively simple strong AA Tau-type fields, while stars with large radiative cores display complex fields. These maps are polar projections of the stellar surface, with concentric rings marking rings of 30 degree latitudes down to -30 degrees (the equator is the thick solid circle). The colour scales represent magnetic fluxes in G - note the different scales in the two sets of maps. These maps are reproduced from Donati *et al.* (2010, 2011).

accretion from circumstellar disks, the fields are unlikely to be described in terms of a simple dipole field aligned with the rotation axis in the majority of the observed systems.

On the right-hand plot of Fig. 2 , the symbol sizes scale with the mean flux for each star, $F_B = 4\pi R^2 \bar{B}$; where $\bar{B}$ is the mean magnetic field strength and R is the stellar radius. As T Tauri stars age and contract towards the main sequence their changing stellar sizes drive the corresponding decrease in the mean magnetic flux. A possible interpretation is that these stars are moving from fossil-type simple unchanging fields to more complex fields with a dynamo origin as they approach the Zero Age Main sequence (Yang *et al.* 2011). On the other hand, temporal changes seen in magnetic field maps of accreting T Tauri stars obtained from Zeeman Doppler imaging techniques over a period of years imply a dynamo-type mechanism and appear to argue against a simple "frozen-in" fossil field origin in the youngest T Tauri stars (Sec. 4.1).

4. Zeeman Doppler Imaging: magnetic field maps

The surface magnetic field maps of twelve classical T Tauri stars have been reconstructed using Zeeman Doppler imaging techniques, mostly under the framework of an international campaign led by Jean-Francois Donati entitled, "Magnetic Protostars and Planets" (MaPP†). As these stars are actively accreting it is possible to study their surface magnetic fields and their accretion properties simultaneously using the same dataset, which contains thousands of photospheric lines as well as several accretion-sensitive diagnostics such as He I, Ca II H&K, several Balmer lines and the Ca II NIR triplet. The list of stars for which maps have been produced and published include: V2129 Oph, BP Tau, V2247 Oph, AA Tau, TW Hya, V4046 Sag a & b, GQ Lup, DN Tau, CR Cha, CV Cha, and MT Ori (Donati *et al.* 2007, 2008b, 2010a,b, 2011a,b,c, 2012, 2013; Hussain *et al.* 2009; Skelly *et al.* submitted).

While relatively few T Tauri star systems have been studied some clear trends are emerging. Their large scale fields can be characterised as lying between two extreme cases as shown in Fig. 3. This figure illustrates the radial field and azimuthal (east-west) oriented field maps for two classical T Tauri stars, AA Tau and V4046 Sag b. While they have different masses they have similar basic parameters (AA Tau: $v_e \sin i = 11.3\,\mathrm{km/s}$ & $T_{\mathrm{eff}} = 4000\,\mathrm{K}$; V4046 Sag b: $v_e \sin i = 13.5\,\mathrm{km/s}$ & $T_{\mathrm{eff}} = 4250\,\mathrm{K}$). As these stars have

† Further information at: http://lamwws.oamp.fr/magics/mapp/MappScience.

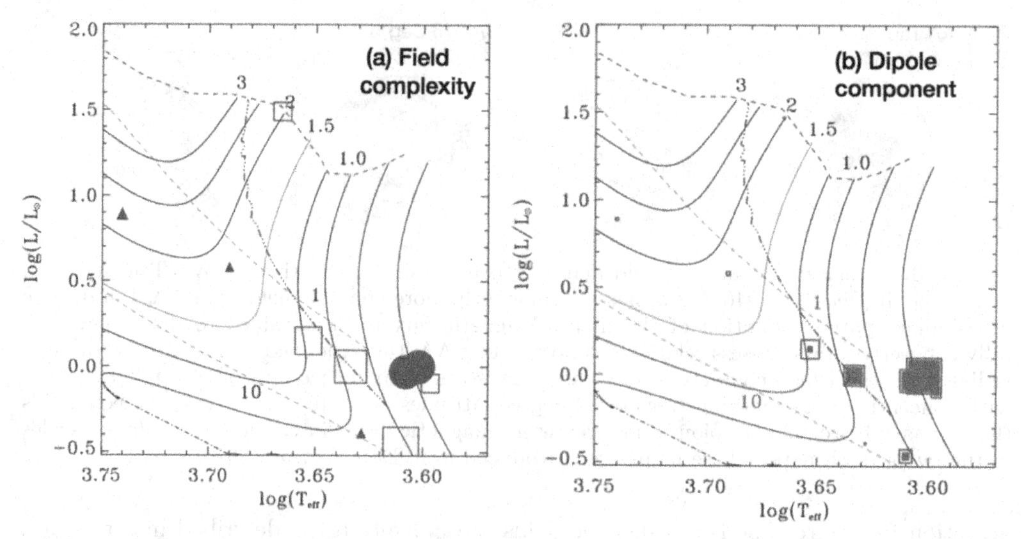

Figure 4. Surface magnetic fields in T Tauri stars from Zeeman Doppler imaging. *Left*: Symbol sizes scale with the magnetic intensity. Different symbols denote the dominant field orientation: circles – dipole-dominant fields; squares – octupole-dominant fields; triangles – complex multipolar fields. *Right*: Symbols scale with the strength of the dipole field component (Gregory *et al.* 2012); where multi-epoch observations were taken the stronger dipole field measurement is represented by an unfilled square for clarity. The largest change in the dipole field is found in maps of V2129 Oph taken 4 years apart. Behrend & Maeder (2001) evolutionary tracks are over-plotted as in Fig. 2. Stars develop a radiative core to the left of the dot-dashed line.

such similar $v_e \sin i$'s and the data used have the same spectroscopic resolution (from CFHT/ESPaDOnS) both sets of maps have comparable spatial resolution scales.

It is immediately apparent from Fig. 3 that V4046 Sag b has a significantly more complex field distribution than AA Tau, this is reflected in the multiple switches in polarity observed in both the radial and azimuthal surface field maps. AA Tau maps also show a stronger magnetic flux (±2kG) compared to V4046 Sag b (±0.4kG). While this may indicate that V4046 Sag b has a weaker surface magnetic field, V4046 Sag b maps are more likely to be affected by flux cancellation due to the star's more complex large scale field.

Fig. 4 shows a plot summarising key properties of the magnetic field maps obtained for all of the classical T Tauri stars maps that have been studied to date. The sizes of the symbols scale with the magnetic intensities in Fig. 4a, from 0.4-0.5 kG (in CR Cha and V4046 Sag a & b) to 4 kG (in GQ Lup).

The dominant mode of the large-scale field is represented by the symbol shape; circles show dipole-dominant fields, open squares are octupole-dominant fields and triangles are for maps with higher degrees of complexity. The dot-dashed line marks the division between fully convective stars and stars with radiative cores. It is clear that fully convective stars have the strongest simplest fields; the closer they get to the dividing line they become octupole-dominant; and as the radiative cores grow the fields become more complex.

Fully convective stars with masses lower than 0.7 $M_\odot$ may have more complex fields: DN Tau (0.65 $M_\odot$) is octupole-dominant even though fully convective (Donati *et al.* 2013). The fully convective star, V2247 Oph, has an even more complex field; it is not pictured in Fig. 4 as it has a very low mass (0.35 $M_\odot$). The reason for the increased complexity on the lowest mass T Tauri stars is discussed in Sec. 5.

Four stars show complex fields: CV Cha, CR Cha, V4046 Sag a and b (Hussain *et al.* 2009, Donati *et al.* 2011). These stars may cover a range of masses from 0.9 to 2.5 $M_\odot$ but they all have large radiative cores ($M_{\rm core} \gtrsim 0.4 M_*$). Their Stokes V signatures show complex structure and rotational modulation indicating the complex field maps recovered from Zeeman Doppler imaging; accretion diagnostics provide further support. The Ca II NIR Stokes V profiles of V4046 Sag a and b show weak polarisation despite having a similar mass accretion rate to AA Tau, TW Hya and GQ Lup. This may be caused by flux cancellation in multiple accretion streams with opposite polarities. No Stokes V signatures were detected in He I and Balmer line profiles of CV Cha and CR Cha. CV Cha has the highest mass accretion rate of all the stars studied to date ($\log \dot{M} \sim -7.5\,M_\odot\,{\rm yr}^{-1}$), a simple large scale field would easily have been detected in its strong emission lines.

Symbol sizes scale with the dipole field strengths in Fig. 4b (Gregory *et al.* 2012). As found in various studies, this property is particularly important in determining the disk truncation radius (e.g., Johnstone *et al.* 2013). It is clear that as complexity increases the dipole field strength drops, which should have a corresponding effect on the accretion state of the star.

4.1. *Temporal evolution*

Zeeman Doppler imaging maps have been acquired at multiple epochs for five T Tauri star systems, BP Tau, AA Tau, V2129 Oph, GQ Lup and DN Tau (Donati *et al.* 2008,2010b, 2012, 2013. All of the stars studied so far have shown changes in their large scale fields. Despite these changes the global properties of the fields of a particular star remain similar (i.e., a star with an octupole-dominant field does not become dipole-dominant).

DN Tau maps acquired two years apart show that there are changes in its large scale magnetic field as well as changes in the shape of the accretion spot (Fig. 5, Donati *et al.* 2013). The accretion spot appears more circular and polar in 2012; at the same time the magnetic field maps show that the relative strength of the octupole to dipole field component has increased. The changing shape of the accretion region from a crescent shape to a circular shape accompanied by a suggestion of an increasingly dominant octupole field is in agreement with predictions from 3-D MHD simulations of disc accretion (e.g., Romanova *et al.* 2011). However, an effort is needed to quantify how accurately the relative strengths of the different field components can be measured in ZDI maps.

Models can use these magnetic field maps as inputs to predict the locations of accretion spots; these can be compared directly with the observed accretion maps (e.g., the Ca II emission maps in Fig. 5). Observations of GQ Lup taken at three separate epochs over a period of two years suggest that the field can change significantly within one year (Johns-Krull *et al.* 2013, Donati *et al.* 2012). Further monitoring of the large scale fields of these stars is planned in the framework of a new large programme, MaTYSSE (see Sec. 5, this will enable us to monitor and characterise the variability in both the accretion and stellar magnetic fields in T Tauri stars over a 10-year period.

5. Summary & Conclusions

- Mean magnetic field strengths measured on 33 T Tauri stars range between 1-3.5 kG. While there are no clear trends in the mean Bf values with stellar parameter; the mean magnetic fluxes clearly decrease with age (by ~30% in the first Myr); this is predominantly due to the shrinking radii of the evolving PMS stars.
- The magnetic field strength measurements indicate the presence of very strong fields, with strengths up to 6 kG, at the surfaces of several T Tauri stars. Comparisons with analytic models suggest that T Tauri star magnetic fields are likely not organised in simple dipoles in most cases.

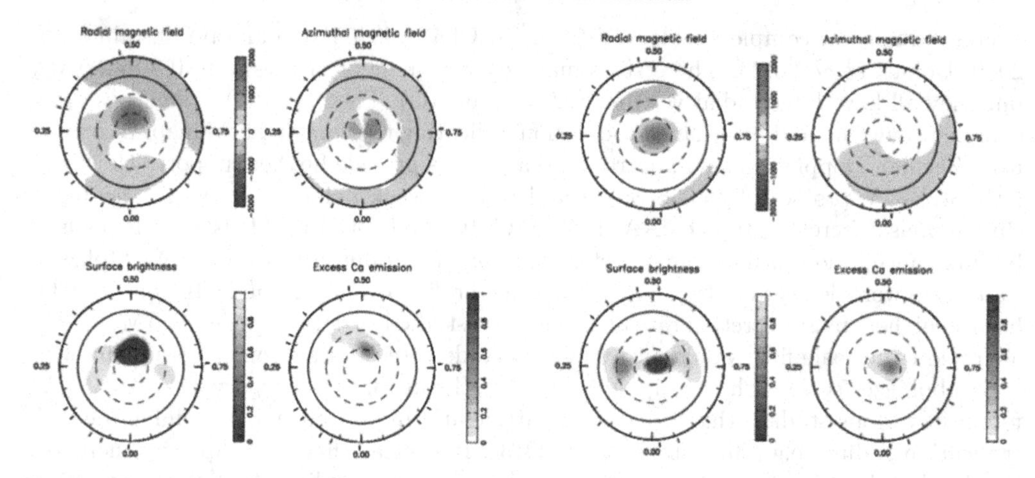

Figure 5. Temporal evolution in the large scale field of DN Tau. Radial and azimuthal field maps, brightness maps and excess Ca II emission maps of DN Tau acquired two years apart (Left: Dec 2010 & Right: Dec 2012). As in Fig. 3 these are polar projections of the stellar surface and the observed phases are denoted by tick marks around each map. While the field strength recovered is similar at both epochs the octupole:dipole field strength appears to increase in 2012 and may cause the changing shape and position of the accretion spot, as traced by excess Ca II emission. Figures from Donati *et al.* (2013).

- The large scale fields have been mapped on twelve accreting T Tauri stars as part of a larger campaign, MaPP, which aims to study the large scale magnetic fields in T Tauri stars and their influence on their accretion states.
- MaPP results indicate that stars with fully convective envelopes have large scale axisymmetric dipole dominant fields, these change to octupole-dominated fields once a radiative core develops, with even more complex large scale fields found in stars with large radiative cores ($M_{\rm core} \gtrsim 0.4M_*$). This dependence of the field on the internal structure has clear analogies with the magnetic field studies of main sequence M dwarfs, where fully convective M dwarfs tend to possess simpler large scale fields than their higher mass counterparts, which have a radiative core (Morin *et al.* 2011).
- The picture may change again in lower mass T Tauri stars ($M_* \lesssim 0.6\,{\rm M}_\odot$) which fall into a bistable dynamo regime and may have more complex fields(Gregory *et al.* 2012). Of the 4 fully convective stars studied, AA Tau and BP Tau ($M_* \sim 0.7 - 0.75\,{\rm M}_\odot$) are nearer the fully convective limit and show dipole-dominant fields. In contrast, the lower mass stars, DN Tau ($0.65\,{\rm M}_\odot$) and V2247 Oph ($0.35\,{\rm M}_\odot$), are deeper in the fully convective phase and show increasingly more complex fields that are similar to the lowest mass M dwarfs (Donati *et al.* 2013).
- The large scale fields of classical T Tauri stars show temporal changes over a period of years. MHD accretion models predict that changing field topologies have an impact on the shape and distribution on the accretion spots. More observations will enable us to characterise and quantify the changes in the large scale field and accretion over a period of ten years. At the same time 3D MHD models can use these maps to evaluate the impact of these changing fields on the accretion properties of the stars in detail.

The coming years will bring many more advances in our understanding of magnetism in PMS stars. In particular the recently started large programmes on the CFHT, MaTySSE†

† *MaTYSSE:* Magnetic Topologies of Young Stars & the Survival of close-in massive Exoplanets (PI: Donati).

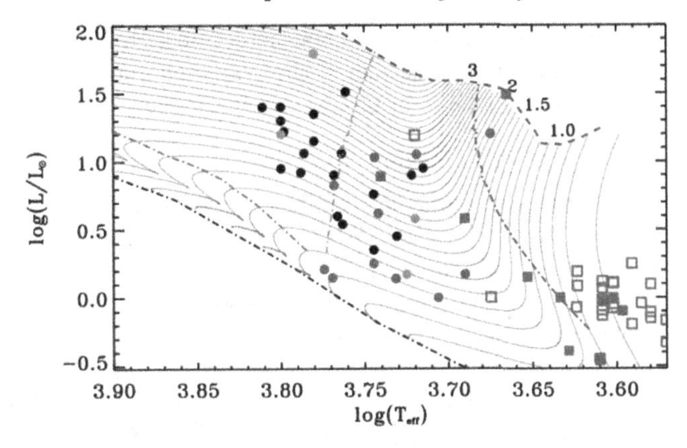

Figure 6. Going to higher mass: The incidence of magnetic fields in T Tauri stars across a range of masses. Evolutionary tracks are solid lines and the dot-dashed lines show where stars develop radiative cores (red line), become fully radiative (orange) and develop convective cores (green; also see Fig. 1). T Tauri stars for which magnetic fields have been detected are marked in red (open squares – Zeeman broadening measurements; filled squares – Zeeman Doppler imaging; circles – spectro-polarimetric observations described below). Marginal detections are denoted by orange circles.

and BinaMIcS‡ will contribute to this. The former will reconstruct the large scale magnetic fields in T Tauri stars that have stopped accreting to investigate how stellar magnetic fields and dynamos are affected by this change in accretion state. BinaMIcS will investigate magnetospheric interaction in close T Tauri binary star systems.

5.1. *Intermediate mass T Tauri stars*

From a spectro-polarimetric study analysing homogeneous data acquired of 70 Herbig Ae/Be stars Alecian *et al.* (2013) confirm the detection of five magnetic stars. They find a low incidence (10%) of stellar magnetic fields in Herbig AeBe systems. This is comparable to that seen in main sequence A and B-type stars: with between 5–10% of A and B-type stars host strong magnetic fields (Donati & Landstreet 2009). The properties of the magnetic Herbig Ae/Be stars are also found to be similar to those seen on their main sequence counterparts (Alecian *et al.* 2013, also see Grunhut *et al.*, *this volume*). We have recently started a study to characterise the magnetic field properties of the precursors to Herbig Ae stars, the intermediate mass T Tauri stars ($1.5 < M_* < 3$ M$_\odot$).

We are probing the magnetic field incidence for over 40 intermediate mass T Tauri stars, which have been identified from their effective temperatures and luminosities in the literature. Magnetic fields are detected using snapshot Stokes V spectra obtained from the CFHT and ESO 3.6-m telescopes. For stars for which we have acquired multi-epoch Stokes V spectra, we can begin to analyse the large scale field topology. The ultimate goal is to characterise the magnetic field in intermediate mass T Tauri stars as they evolve into fully radiative Herbig stars (Alecian, Hussain *et al.* in prep.).

Fig. 6 shows the first results from our programme; the intermediate mass T Tauri stars are filled circles. On this figure we over-plot the T Tauri stars for which magnetic fields have been measured and analysed using the Zeeman broadening and Zeeman Doppler imaging techniques. Dot-dashed lines denote where PMS stars interiors undergo key changes. It is clear that the incidence of stellar magnetic fields drops off with increasing $T_{\rm eff}$; i.e., as the stars lose their convective envelopes. A detailed spectroscopic analysis is underway to obtain better constraints on the fundamental properties of each of the target

‡ *BinaMIcS:* Binarity and Magnetic Interactions in various classes of Stars (PI: Alecian).

stars and to place strong upper limits on any non-detections. However, these first results support the idea that the internal structure of the star, not mass, is a key determinant of stellar dynamo behaviour – at least in G–M-type PMS stars.

5.2. *Multi-wavelength studies*

Multi-wavelength campaigns are particularly effective ways to test details of magnetospheric accretion models. Argiroffi *et al.* (2010) study the X-ray emission in the close binary T Tauri system, V4046 Sag, as part of a coordinated XMM-*Newton*/CFHT campaign. They find that the soft high density X-ray emission in the system shows significant orbital modulation, while the hotter plasma associated with the stellar corona shows no evidence of orbital modulation. This modulation is best explained in terms of high density plasma concentrated near the stellar photosphere in compact regions that are eclipsed as the system rotates. Fonseca *et al.* show the power of combining multi-wavelength photometry with Balmer line spectroscopy to investigate the relationship between the structure of the inner accretion disk and stellar magnetosphere.

Further detailed tests of magnetospheric accretion models will be possible by using large scale field maps of T Tauri stars as inputs to 3-D MHD models. These models are being used to investigate how different large scale field geometries affect the accretion geometries in systems and can be tested against observations through detailed modelling of profiles that form in the accretion funnels, e.g., Hβ and He I 10830Å, (e.g., Kurosawa *et al.* 2012, Adams & Gregory 2012).

References

Adams, F. C. & Gregory, S. G. 2012, 744, 55

Alecian, E., Wade, G. A., Catala, C., Grunhut, J. H., Landstreet, J. D., Bagnulo, S., Böhm, T., Folsom, C. P., Marsden, S., & Waite, I. 2013, *MNRAS*, 429, 1001

Argiroffi, C., Flaccomio, E., Bouvier, J., Donati, J.-F., Getman, K. V., Gregory, S. G., Hussain, G. A. J.., Jardine, M. M., Skelly, M. B., & Walter, F. M. 2011, *A&A*, 530, 1

Barnes J. R. & Collier Cameron, A. 2001, *MNRAS*, 326, 950

Behrend, R. & Maeder, A. 2001, *A&A*, 373, 190

Bouvier, J., Alencar, S. H. P.., Boutelier, T., Dougados, C., Balog, Z., Grankin, K., Hodgkin, S. T., Ibrahimov, M. A., Kun, M., Magakian, T. Yu, & Pinte, C. 2007, *A&A*, 463, 1017

Chen, W. & Johns-Krull, C. M., 2013, *ApJ*, 776, 113

Donati, J.-F. & Collier Cameron, A. 1997, *MNRAS*, 291, 1

Donati, J.-F., Semel, M., Carter, B. D., Rees, D. E., & Collier Cameron, A. 1997, *MNRAS*, 291, 658

Donati, J.-F. & Landstreet, J. D. 2009, *ARA&A*, 47, 333

Donati, J.-F., Jardine, M. M., Gregory, S. G., *et al.* 2007, *MNRAS*, 380, 1297

Donati, J.-F., Jardine, M. M., Gregory, S. G., *et al.* 2008, *MNRAS*, 386, 1234

Donati, J.-F., Skelly, M. B., Bouvier, J., *et al.* 2010a, *MNRAS* 402, 1426

Donati, J.-F., Skelly, M. B., Bouvier, J., *et al.* 2010b, *MNRAS* 409, 1347

Donati, J.-F., Bouvier, J., Walter, F. M. *et al.* 2011a, *MNRAS*, 412, 2454

Donati, J.-F., Gregory, S. G., Alencar, S. H. P., *et al.* 2011b, *MNRAS*, 417, 472

Donati, J.-F., Gregory, S. G., Montmerle, T., *et al.* 2011c, *MNRAS*, 417, 1747

Donati, J.-F., Gregory, S. G., Alencar, S. H. P., Hussain, G., Bouvier, J., Dougados, C., Jardine, M. M., Ménard, F. Romanova, M. M. *et al.* 2012, *MNRAS*, 425, 2948

Donati, J.-F., Gregory, S. G., Alencar, S. H. P., Hussain, G., Bouvier, J., Jardine, M. M., Ménard, F., Dougados, C., Romanova, M. M., *et al.* 2013, *MNRAS*, 436, 881

Gregory, S. G., Donati, J.-F. , Morin, J., Hussain, G. A. J.., Mayne, N. J., Hillenbrand, L. A., & Jardine, M. 2012, *MNRAS*, 755, 97

Hartmann, L., MacGregor, K. B.1982, *ApJ*, 259, 180

Hussain, G. A. J.., Donati, J.-F., Collier Cameron, A., & Barnes, J. R. 2000, *MNRAS*, 318, 961

Hussain, G. A. J., Collier Cameron, A., Jardine, M. M., Dunstone, N., Ramirez Velez, J., Stempels, H. C., Donati, J.-F., Semel, M., Aulanier, G., Harries, T., Bouvier, J., Dougados, C., Ferreira, J., Carter, B. D., & Lawson, W. A. 2009, *MNRAS*, 398, 189

Johns-Krull, C. M. 2007, *ApJ*, 664, 975

Johns-Krull, C. M., 2008, IAUS, 259, 345, Edited by K. G. Strassmeier, A. G. Kosovichev, J. E. Beckman

Johns-Krull, C. M., Chen, W., Valenti, J. A., Jeffers, S. V., Piskunov, N. E., Kochukhov, O., Makaganiuk, V., Stempels, H. C., Snik, F., Keller, C., & Rodenhuis, M. 2013, *ApJ*, 765, 11

Johnstone, C. P., Jardine, M., Gregory, S. G., Donati, J.-F., Hussain, G. 2013 MNRAS, *in press* (arXiv:1310.8194)

Kochukhov, O., Makaganiuk, V., & Piskunov, N. 2010, *A&A*, 524, 5

Kurosawa, R. & Romanova, M. M. 2013, *MNRAS*, 431, 2673

Kurosawa, R. & Romanova, M. M. 2012, *MNRAS*, 426, 2901

Matt, S. & Pudritz, R. E. 2005, *ApJ*, 632, 135

Matt, S. & Pudritz, R. E. 2008, *ApJ*, 678, 1109

Morin, J. Dormy, E., Schrinner, M. , & Donati, J.-F. 2011, *MNRAS*, 418, 133

Piskunov, N. E., Tuominen, I., & Vilhu, O. 1990, *A&A*, 230, 363

Romanova, M. M., Long, M., Lamb, F. K., Kulkarni, A. K., & Donati, J.-F. 2011, *MNRAS*, 411, 915

Semel M. 1989, *A&A* 225, 456

Siess, L., Dufour, E., & Forestini, M. 2000, *A&A*, 358, 593

Strassmeier, K. G. 2009, *A&AR*, 17, 251

Tognelli, E., Prada Moroni, P. G., & DeglInnocenti, S. 2011, *A&A*, 533, 109

Vogt, S. S., Penrod, G. D., & Hatzes, A. P. 1987, 321, 496

Yang H., Johns-Krull, C. M. & Valenti, J. A. 2008, *AJ*, 136, 2286

Yang, H. & Johns-Krull, C. M. 2011, *ApJ*, 729, 83

Zanni, C. & Ferreira, J. 2013, *A&A* 550, 99

Magnetic Fields throughout Stellar Evolution
Proceedings IAU Symposium No. 302, 2013
P. Petit, M. Jardine & H. Spruit, eds.

doi:10.1017/S1743921314001665

Magnetic Fields in 7 Young Stellar Objects Observed with Nançay Radio Telescope

Olga Bayandina, Alexei Alakoz and Irina Val'tts

Astro Space Center of the Lebedev Physical Institute, Moscow, Russia
email: bayandix@yandex.ru

Abstract. Magnetic fields (MF) can play an essential role in the evolution of the interstellar medium - especially at the early evolutionary stages. Small scale research related to the interaction of MF and pre-stellar condensations are unresolved issues. In quantitative terms, submissions about forming a full picture of gas-dust fragments evolution are far from complete, considering delay of their collapse caused by MF and the reverse effect of self-gravitating objects on the transformation of force lines and changing the values of local strength. The role of these interrelated processes is very important in the estimation of time of evolution of protostellar structures. In contrast to OH, in methanol molecule (most investigating at the moment) there is no unpaired electron, and the Zeeman splitting of the energy levels in CH_3OH regards only the levels caused by the nuclear spin. Therefore, Zeeman spectrum in methanol is certainly not going to be as effective as in OH. However, since many methanol masers - Class I (MMI - formed at the earliest stage of the evolution of gas and dust condensations) and Class II (MMII - the area around very young stars and protoplanetary disks) - are associated with OH masers, then from spectra of OH masers the parameters of MF can be estimated, at least, near different methanol masers classes, i.e. in condensations which are at different evolutionary stages. This report presents the results of polarization observations 7 OH maser sources at the NRT (France). The main goal is comparing similarities and differences in MF strength and orientation in these masers, which essentially different according to the type of methanol masers associated with them, i.e. the evolutionary type.

Keywords. Masers, magnetic fields, ISM: evolution

We carried out observations 7 OH maser sources in October 2003 at the Nançay Radio Telescope (NRT), France. The telescope and receiving equipment technical specifications and calibration standard methods you can find in Szymczak *et al.* (1997), Slysh *et al.* (2010).

Observations were made in the two main lines of OH - at 1665 and 1667 MHz. These observations were done in the circular polarization using a phase shifter. All four Stokes parameters in each autocorrelator channel can be obtained by combining polarization modes.

Gaussian parameters of OH spectra details are obtained and estimates of polarization parameters (the degree of circular polarization m_C, the linearly polarized flux density p, the degree of linear polarization m_L) were made. The calculated values obtained by the well-known formula, see Szymczak (2009).

Estimates of the magnetic field strength B were obtained according to the Zeeman splitting of the OH lines and the approximation the Stokes parameter V by the derivative from Stokes parameter I. We used standard formula, that can be found in Elitzur (1998), Crutcher (1999). It was shown that the value B fluctuates for different sources in the range of $\leqslant$0.5 mG to 1.4 mG.

The analysis of the association the OH masers with methanol emission was conducted: it was shown, that the magnetic field in OH masers associated with MMI can be determined more reliable, than in OH masers associated with MMII and has higher values. It was shown, that investigated areas may be linked structures as small clouds IRDC or the typical protoplanetary disks. Probably, one can consider, that the magnetic field obtained from the spectrum of the OH maser characterizes the magnetic field in the condensation forming a methanol maser - within the limits of errors in determining the distances between OH masers and MMII in these sources.

In the investigated sources areas have been allocated with the centers on the coordinates of observed OH maser condensation clusters. The question of a possible association of OH maser clusters and the closest MMI and MMII clusters was worked out in details. It was shown, that these associations are real, i.e. magnetic field acting within the OH clusters can be extended to the methanol maser groups.

Values of volume density of molecular hydrogen in OH condensations, its column density and the column OH density, the ratio of the mass to the magnetic flux in size of the investigated area between OH masers and methanol masers, the ratio of thermal to magnetic pressures and the virial ratios of the gravitational, kinetic, and magnetic energy were obtained.

It was shown, that in the majority of sources the ratio of the mass to the magnetic flux exceeds the critical value. On the other hand, it was shown, that the ratio of thermal to magnetic pressures in all cases <1 in hotter areas and $\ll 1$ in cold, i.e. clouds can be found in dominant magnetic mode. This conflict is associated with possible large errors as in determining the values of the magnetic field strength (in particular, the angle θ, characterizes its orientation) as well as in possible overvaluation of the distance to the source increases the size of the studied areas.

We obtained estimates of the gravitational collapse time in the considered sources 0.06 to 0.13×10^6 years which is less or comparable to typical lifetime of the star forming region 10^5 years.

This analysis shows that the question of the spatial overlap or not overlap of OH, I or MMII masers groups is open for today and it is fundamental for the assessment of the possibility of using the magnetic fields data obtained from the OH masers spectra for the solution problem of the magnetic fields influence on the process of stability or instability of methanol maser condensations, which is particularly important for the progress in the understanding of methanol maser models, especially Class I.

Online figures for all sources are available: http://www.asc.rssi.ru/OB/NP.pdf

Support for this work was provided by the Basic Research Program of the Division of Physical Sciences of the RAS-17, the Russian Foundation of Basic Research (project 13-02-00460) and the Education and Science Ministry (project 8405).

References

Szymczak, M. & Gérard, E. 1997, *Astron. and Astrophys*, 423, 209

Slysh, V. I., Pashchenko, M. I., & Rudnitskii, G. M. 2010, *Astron. Rep.*, 87, 655

Elitzur, M. 1998, *Astrophys. J.*, 504, 390

Crutcher, R. M. 1999, *Astrophys. J.*, 520, 706

Szymczak, M. & Gérard, E. 2009, *Astron. and Astrophys.*, 494, 118

Kurtz, S., Hofner, P., & Álvarez, C. V. 2004, *Astrophys. J. Suppl. Ser.*, 155, 149

Caswell, J. L., Green, J. A., & Phillips, C. J. 2013, *Monthly Not. Roy. Astron. Soc.*, 431, 1180

Caswell, J. L., Fuller, G. A., Green, J. A., *et al.*, 2010, *Monthly Not. Roy. Astron. Soc.*, 404, 1029

Green, J. A., Caswell, J. L., Fuller, G. A., *et al.*, 2010, *Monthly Not. Roy. Astron. Soc.*, 409, 913

Slysh, V. I., Kalenskii, S. V., Val'tts, I. E., & Otrupcek, R. 1994 *Monthly Not. Roy. Astron. Soc.*, 268, 464

Magnetic Fields throughout Stellar Evolution
Proceedings IAU Symposium No. 302, 2013
P. Petit, M. Jardine & H. Spruit, eds.

doi:10.1017/S1743921314001677

Can we predict the magnetic properties of PMS stars from their H-R diagram location?

S. G. Gregory[1], J.-F. Donati[2], J. Morin[3], G. A. J. Hussain[4], N. J. Mayne[5], L. A. Hillenbrand[6] and M. Jardine[1]

[1]School of Physics & Astronomy, University of St Andrews, St Andrews, KY16 9SS, U.K.
email: sg64@st-andrews.ac.uk

[2]UPS-Toulouse/CNRS-INSU, IRAP UMR 5277, Toulouse, F31400 France

[3]Inst. für Astrophysik, Univ. Göttingen, Friedrich-Hund-Platz 1, D-37077 Göttingen, Germany

[4]ESO, Karl-Schwarzschild-Str. 2, D-85748 Garching, Germany

[5]School of Physics, University of Exeter, Exeter EX4 4QL, U.K.

[6]California Institute of Technology, MC 249-17, Pasadena, CA 91125, U.S.A.

Abstract. Spectropolarimetric observations combined with tomographic imaging techniques have revealed that all pre-main sequence (PMS) stars host multipolar magnetic fields, ranging from strong and globally axisymmetric with $\gtrsim$kilo-Gauss dipole components, to complex and non-axisymmetric with weak dipole components ($\lesssim$0.1 kG). Many host dominantly octupolar large-scale fields. We argue that the large-scale magnetic properties of a PMS star are related to its location in the Hertzsprung-Russell diagram. This conference paper is a synopsis of Gregory *et al.* (2012), updated to include the latest results from magnetic mapping studies of PMS stars.

Keywords. stars: evolution, stars: interiors, stars: magnetic field, stars: pre-main sequence

---

1. Introduction

Since the 1990s it has been known that pre-main sequence (PMS) stars are capable of generating magnetic fields of a few kilo-Gauss in strength (e.g. Basri *et al.* 1992; Johns-Krull *et al.* 1999; Johns-Krull 2007; Yang & Johns-Krull 2011). Fields of this magnitude, provided that they are sufficiently globally ordered, are easily strong enough to truncate circumstellar disks during the classical T Tauri star phase (e.g. Königl 1991).

Zeeman-Doppler imaging, combined with the techniques of least squares deconvolution (LSD; Donati *et al.* 1997) and tomographic imaging, allows magnetic maps to be derived from circularly polarised spectra. For accreting PMS stars, the maps are constructed by simultaneously considering the rotationally modulated polarisation signature in both the LSD-averaged photospheric line and in the accretion-related emission lines (Donati *et al.* 2010b). The maps themselves can then be decomposed into the various ℓ and m-number spherical harmonic modes (where $\ell = 1, 2, 3\ldots$ are the dipole, quadrupole, octupole... field components).

The first magnetic maps of an accreting PMS star were published by Donati *et al.* (2007), and have since been obtained for a small sample of stars - see the tables in Gregory *et al.* (2012) for a list which has recently been expanded to include DN Tau (Donati *et al.* 2013). Most PMS star magnetic maps have been derived during the MaPP (Magnetic Protostars & Planets) project (PI: J.-F. Donati). The data acquisition phase of MaPP (2008-12) with ESPaDOnS at the Canada-France-Hawai'i Telescope (CFHT) and NARVAL at the Télescope Bernard Lyot, is now complete. A highlight of MaPP is the discovery of a clear link between the internal structure of a star and its external, large-scale, magnetic field topology.

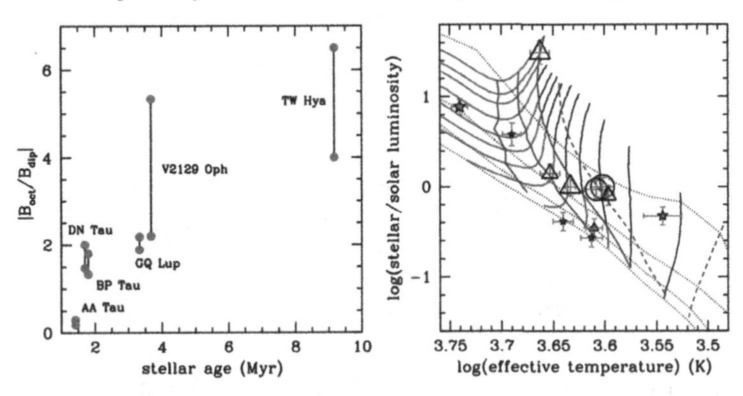

Figure 1. [left] Almost all fully convective plus partially convective PMS stars with small radiative cores with published magnetic maps have dipole plus octupole fields. The magnitude of the polar strength of the dipole to the octupole increases with age. All stars have been observed at least twice, at epochs one-to-a-few years apart. [right] A H-R diagram showing stars with published magnetic maps (updated from Gregory *et al.* 2012). Mass tracks (solid black/red during the fully/partially convective phase of evolution for $M_*/M_\odot = 0.3 - 1.9$ in steps of 0.2, then 2.2, 2.5, 2.7 & 3.0 from right to left) and isochrones (dotted lines; age $= 1, 5, 10$ & 15 Myr from upper right to lower left) from Siess *et al.* (2000) are shown. Solid blue lines are the fully convective limit (right), and the loci of stars with radiative core masses of $M_{\rm core}/M_* = 0.4$ & 0.8 (middle & left). Circles are fully convective stars with axisymmetric large-scale fields with strong (~kG) dipole components. Triangles are stars with small radiative cores ($M_{\rm core}/M_* < 0.4$; with the exception of DN Tau, see section 2) and large-scale fields that are mostly axisymmetric and (typically) dominantly octupolar. Asterisks are stars with non-axisymmetric large-scale magnetic fields with weak dipole components ($\lesssim$0.1 kG). Dashed blue lines are discussed in section 3.

2. PMS star magnetic topology & the link with stellar structure

About half of the accreting PMS stars with published magnetic maps have large-scale fields that are well described by slightly tilted dipole and octupole components (other field modes are present too but in almost all cases are less significant; Gregory & Donati 2011). AA Tau, BP Tau, DN Tau, GQ Lup, TW Hya & V2129 Oph have this sort of magnetic topology (Donati *et al.* 2007, 2008b, 2010b, 2011a,c, 2012, 2013), with the first (last) three listed having fully convective interiors (small radiative cores, $M_{\rm core}/M_* \lesssim 0.4$; see Gregory *et al.* 2012). Their large-scale fields are dominantly axisymmetric and it appears that the magnitude of the ratio of the polar strength of the octupole to the dipole component $|B_{\rm oct}/B_{\rm dip}|$ increases with age (Fig. 1). Fully convective stars are capable of generating strong kG dipole components, while those with small radiative cores have dominantly octupolar magnetic fields with dipole components that vary from a few times 0.1 kG to of order ~kG. The exception to this is the fully convective DN Tau (Donati *et al.* 2013), although its $|B_{\rm oct}/B_{\rm dip}|$ ratio, and field polarity distribution across the stellar surface, is similar to that of the other fully convective stars (see below regarding dynamo bistability). Intriguingly, although the large-scale fields of PMS stars are evolving between observing epochs (e.g. Donati *et al.* 2011a), their general magnetic topology features remain unchanged across year-long timescales e.g. if a star has a dominantly octupolar magnetic field at one epoch, the field still has this configuration at the next epoch.

The other half of the sample have complex non-axisymmetric large-scale magnetic fields with weak dipole components ($\lesssim$0.1 kG). CR Cha, CV Cha, V2247 Oph, V4046 Sgr A & V4046 Sgr B have such large-scale magnetic fields (Hussain *et al.* 2009; Donati *et al.* 2010a, 2011b). All apart from V2247 Oph have large radiative cores ($M_{\rm core}/M_* \gtrsim 0.4$).

A H-R diagram showing stars with published magnetic maps is shown in Fig. 1. The symbol type is related to the large-scale field topology of the star, as described in the figure caption. Donati *et al.* (2011c) and Gregory *et al.* (2012) argued that there appears

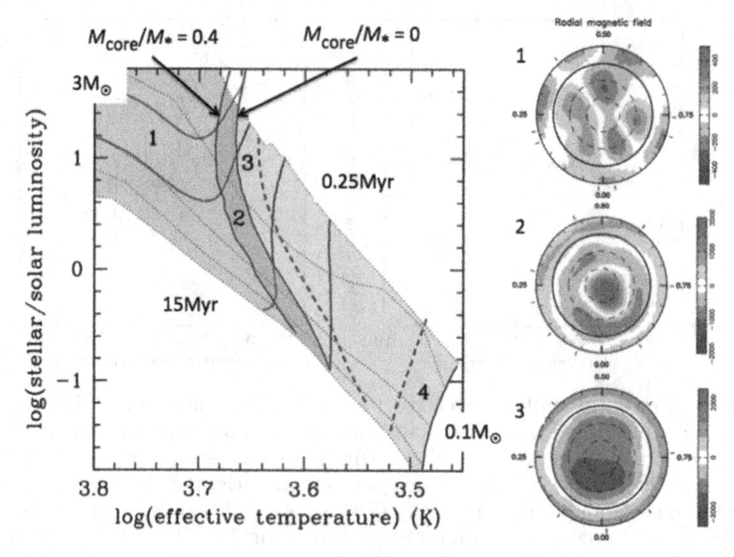

Figure 2. A magnetic H-R diagram (Gregory *et al.* 2012). The PMS is shown with mass tracks coloured as in Fig. 1 (0.1, 0.5, 1, 2 & 3 M⊙) and isochrones (0.25, 1, 5, 10 & 15 Myr) from Siess *et al.* (2000). Stars in the different numbered regions have differing internal structures and are observed to have different large-scale magnetic field topologies. Solid blue lines connect the loci of effective temperatures and luminosities where stars have the same internal structure, defined as the same ratio of radiative core mass to stellar mass. Stars in region 1 are largely radiative and host complex non-axisymmetric large-scale magnetic fields. Stars in region 2 have small radiative cores ($0 \leqslant M_{\rm core}/M_* \leqslant 0.4$) and axisymmetric magnetic fields that are (typically) dominantly octupolar. Stars in region 3 are fully convective and host axisymmetric fields with kilo-Gauss dipole components. A fourth magnetic topology region exists at the lowest masses (see text) with the dashed blue lines indicating upper/lower limits to the boundary between regions 3 & 4. The magnetic maps (right column) from top-to-bottom are V4046 Sgr B, V2129 Oph & AA Tau (Donati *et al.* 2011b,a, 2010b). They are shown in flattened polar projection. Blue/red denotes negative/positive field with fluxes labelled in Gauss. Dashed lines are lines of constant latitude, separated by 30°, with the bold circle the stellar equator. From top-to-bottom the stars are largely radiative with a complex multipolar magnetic field; partially convective with a small core and a dominantly octupolar field; and fully convective with a simple dipole magnetic field.

to be evidence for a magnetic evolutionary scenario. *Fully convective stars, which lie close to the fully convective limit, host simple axisymmetric large-scale fields and can have strong (∼kG) dipole components. The octupole component becomes more and more dominant once a radiative core develops, with the large-scale magnetic field eventually becoming complex and non-axisymmetric with a weak dipole (∼0.1 kG) once the core has grown to occupy a sufficient proportion of the stellar interior (empirically once $M_{\rm core}/M_* \gtrsim 0.4$).*

3. The magnetic Hertzsprung-Russell diagram & discussion

There is a limited sample of PMS stars with published magnetic maps. However, main sequence (MS) M-dwarfs follow similar magnetic topology trends (Donati *et al.* 2008a; Morin *et al.* 2008), with a weakening dipole component as we consider stars that have more and more radiative interiors. Their large-scale magnetic fields remain dominantly axisymmetric unless the radiative core mass exceeds ∼40% of the total stellar mass (Gregory *et al.* 2012); similar to what is found for the PMS star sample.

The similarities between the MS and PMS star samples prompted Gregory *et al.* (2012) to ask if we can predict the large-scale magnetic properties of a PMS star solely from its H-R diagram location. For example, will its large-scale field be dominantly axisymmetric or non-axisymmetric? Will it be dominantly octupolar, or more complex? Will the dipole component be of order ~kG or ~0.1 kG? There appears to be at least three (defined in the caption of Fig. 2) magnetic topology regimes across the PMS of the H-R diagram.

A fourth, bistable dynamo, regime may exist amongst the lowest mass fully convective PMS stars, similar to that discovered for late M-dwarfs (Morin *et al.* 2011). Amongst the fully convective PMS stars, DN Tau hosts a large-scale field that is dominantly octupolar (see section 2) while the large-scale field of V2247 Oph is more akin to those of substantially radiative PMS stars (Donati *et al.* 2010a). Thus, Gregory *et al.* (2012) speculated that a bistable dynamo regime exists for the lowest mass fully convective PMS stars, where stars with a variety of large-scale magnetic field topologies will be discovered. This is labelled as region 4 in Fig. 2 - the dashed blue lines show possible upper/lower limits below which bistable behaviour may be found. For MS stars dynamo bistability is found at $\lesssim 0.2\,M_\odot$, which sets the lower limit. The upper limit is set by noting that $0.2\,M_\odot$ is ~60% of the MS fully convective limit of $0.35\,M_\odot$. As the fully convective limit is a function of age (see Gregory *et al.* 2012), then the upper limit corresponds to a mass of 60% of the fully convective limit at a given age. DN Tau falls between the limits in the H-R diagram, and suggests that there may be a smooth transition from simple magnetic fields amongst the more massive fully convective stars, to the bistable dynamo regime of lower mass stars.

A more complete magnetic mapping survey (of fully convective PMS stars in particular) is required to confirm or refute the arguments herein, a task ideally suited for SPIRou, the under construction nIR spectropolarimeter for CFHT. The clearest magnetic topology trend is that PMS stars with shallow convective zone depths have far more complex large-scale magnetic fields with dipole components up to an order of magnitude below those found for those with deep convective zones.

Acknowledgements: SGG acknowledges support from the Science & Technology Facilities Council (STFC) via an Ernest Rutherford Fellowship [ST/J003255/1].

References

Basri, G., Marcy, G. W., & Valenti, J. A. 1992, *ApJ*, 390, 622
Donati, J.-F., Gregory, S. G., Alencar, S. H. P., *et al.* 2013, *MNRAS*, in press [astro-ph/1308.5143]
Donati, J.-F., Gregory, S. G., Alencar, S. H. P., *et al.* 2012, *MNRAS*, 425, 2948
Donati, J.-F., Gregory, S. G., Alencar, S. H. P., *et al.* 2011c, *MNRAS*, 417, 472
Donati, J.-F., Gregory, S. G., Montmerle, T., *et al.* 2011b, *MNRAS*, 417, 1747
Donati, J.-F., Bouvier, J., Walter, F. M., *et al.* 2011a, *MNRAS*, 412, 2454
Donati, J.-F., Skelly, M. B., Bouvier, J., *et al.* 2010b, *MNRAS*, 409, 1347
Donati, J.-F., Skelly, M. B., Bouvier, J., *et al.* 2010a, *MNRAS*, 402, 1426
Donati, J.-F., Jardine, M. M., Gregory, S. G., *et al.* 2008b, *MNRAS*, 386, 1234
Donati, J.-F., Morin, J., Petit, P., *et al.* 2008a, *MNRAS*, 390, 545
Donati, J.-F., Jardine, M. M., Gregory, S. G., *et al.* 2007, *MNRAS*, 380, 1297
Donati, J.-F., Semel, M., Carter, B. D., Rees, D. E., & Cameron, A. C. 1997, *MNRAS*, 291, 658
Gregory, S. G., Donati, J.-F., Morin, J., *et al.* 2012, *ApJ*, 755, 97
Gregory, S. G. & Donati, J.-F. 2011, *AN*, 332, 1027
Hussain, G. A. J., Collier Cameron, A., Jardine, M. M., *et al.* 2009, *MNRAS*, 398, 189
Johns-Krull, C. M. 2007, *ApJ*, 664, 975
Johns-Krull, C. M., Valenti, J. A., & Koresko, C. 1999, *ApJ*, 516, 900
Königl, A. 1991, *ApJL*, 370, L39
Morin, J., Dormy, E., Schrinner, M., & Donati, J.-F. 2011, *MNRAS*, 418, L133
Morin, J., Donati, J.-F., Petit, P., *et al.* 2008, *MNRAS*, 390, 567
Siess, L., Dufour, E., & Forestini, M. 2000, *A&A*, 358, 593
Yang, H. & Johns-Krull, C. M. 2011, *ApJ*, 729, 83

Magnetic Fields throughout Stellar Evolution
Proceedings IAU Symposium No. 302, 2013
P. Petit, M. Jardine & H. Spruit, eds.

doi:10.1017/S1743921314001689

The magnetosphere of the close accreting PMS binary V4046 Sgr AB

S. G. Gregory[1], V. R. Holzwarth[2], J.-F. Donati[3], G. A. J. Hussain[4], T. Montmerle[5], E. Alecian[6], S. H. P. Alencar[7], C. Argiroffi[8], M. Audard[9], J. Bouvier[10], F. Damiani[8], M. Güdel[11], D. P. Huenemoerder[12], J. H. Kastner[13], A. Maggio[8], G. G. Sacco[14] and G. A. Wade[15]

[1]School of Physics & Astronomy, University of St Andrews, St Andrews, KY16 9SS, U.K.
email: sg64@st-andrews.ac.uk
[2]Freytagstr. 7, D-79114 Freiburg i.Br., Germany
[3]Inst. de Recherche en Astrophysique et Planétologie UMR 5277, Toulouse, F31400 France
[4]ESO, Karl-Schwarzschild-Str. 2, D-85748 Garching, Germany
[5]Institut d'Astrophysique de Paris, 98bis bd Arago, FR-75014 Paris, France
[6]Obs. de Paris, LESIA, 5, place Jules Janssen, F-92195 Meudon Principal Cedex, France
[7]Dept. de Fìsica - UFMG, Av. Antônio Carlos, 6627, 30270-901 Belo Horizonte, MG, Brazil
[8]INAF-Osservatorio Astronomico di Palermo, Piazza del Parlamento 1, I-90134 Palermo, Italy
[9]ISDC Data Center for Astrophysics, Univ. of Geneva, CH-1290 Versoix, Switzerland
[10]UJF-Grenoble 1/CNRS-INSU, IPAG, UMR 5274, F-38041, Grenoble, France
[11]Dept. of Astrophysics, University of Vienna, Türkenschanzstrasse 17, A-1180 Vienna, Austria
[12]MIT, Kavli Inst. for Astrophysics & Space Research, Cambridge, MA 02139, U.S.A.
[13]CIS, Rochester Inst. of Technology, 54 Lomb Memorial Drive, Rochester, NY 14623, U.S.A.
[14]INAF-Arcetri Astrophysical Observatory, Largo Enrico Fermi 5, I - 50125 Florence, Italy
[15]Dept. of Physics, Royal Military College of Canada, Kingston, K7K 7B4, Canada

Abstract. We present a preliminary 3D potential field extrapolation model of the joint magnetosphere of the close accreting PMS binary V4046 Sgr. The model is derived from magnetic maps obtained as part of a coordinated optical and X-ray observing program.

Keywords. stars: formation, stars: interiors, stars: magnetic field, stars: pre-main sequence

1. Introduction - large multi-wavelength observing campaign

V4046 Sgr is a close (separation $\sim 9\,R_\odot$; Donati *et al.* 2011) circularised and synchronised PMS binary, accreting gas from a large circumbinary disk (Rosenfeld *et al.* 2012). It was observed as part of a coordinated X-ray and spectropolarimetric observing program with *XMM-Newton* and ESPaDOnS@CFHT during 2009. The observational highlights include: (i). the derivation of the first magnetic maps of a close accreting PMS binary system, see Fig. 1 (Donati *et al.* 2011). (ii). The detection of rotationally modulated soft X-ray emission associated with accretion shocks where accreting gas impacts the surface of the stars (Argiroffi *et al.* 2012). The modulation period is half of the binary orbital period. (iii). The realisation that V4046 Sgr may be a quadruple system, with GSC 07396-00759 a distant (projected separation ~12,350 au) companion to V4046 Sgr AB (Kastner *et al.* 2011). The companion itself is likely a non-accreting PMS binary.

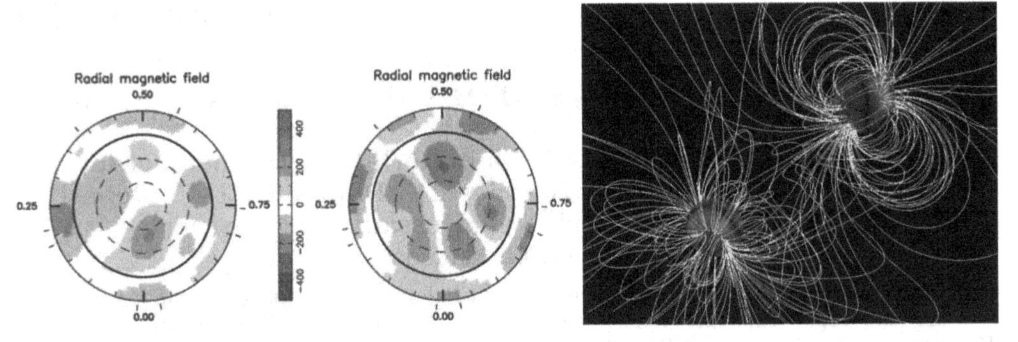

Figure 1. Magnetic maps of the primary/secondary star of V4046 Sgr (left/middle respectively). Blue/red is negative/postive field, with fluxes labelled in Gauss. Tick marks/numbers denote the phases of observation/rotation phase. The maps are shown in flattened polar projection with the bold circle/dashed lines the stellar equator/lines of constant latitude. Maps of the other field components & brightness/excess accretion-related emission maps can be found in Donati *et al.* (2011). The right panel shows a binary magnetic field extrapolation from the magnetic maps of V4046 Sgr. Only the large-scale field lines are shown. The magnetic fields are highly tilted with respect to the stellar rotation axes. Field lines connect through the interior of the binary from the nightside of one star to the dayside of the other.

2. The magnetic field of V4046 Sgr & field extrapolation

Magnetic maps of V4046 Sgr, derived from Zeeman-Doppler imaging, are shown in Fig. 1 (Donati *et al.* 2011). Only the radial field components are shown. Both stars are found to host complex large-scale magnetic fields with weak dipole components, consistent with their partially convective internal structure (Gregory *et al.* 2012).

A binary magnetic extrapolation is shown in Fig. 1. This has been constructed using a newly developed binary magnetic field extrapolation code that will be described in a forthcoming paper (Holzwarth in prep.). The code assumes that the large-scale field is potential and subject to three boundary conditions: the magnetic field is as measured from the maps at the surface of the each star, and a source surface boundary condition designed to mimic the pulling open of the large-scale magnetic loops by the stellar wind. The magnetic fields of both stars are linked, with loops connecting the dayside of one star to the nightside of the other. The field geometry, and the distribution of accretion columns and hot spots, will be detailed in a future paper (Gregory *et al.* in prep.).

Acknowledgements: SGG acknowledges support from the Science & Technology Facilities Council (STFC) via an Ernest Rutherford Fellowship [ST/J003255/1]. GAW is supported by a Discovery Grant from the Natural Science & Engineering Research Council of Canada (NSERC).

References

Argiroffi, C., Maggio, A., Montmerle, T., Huenemoerder, D. P., Alecian, E., Audard, M., Bouvier, J., Damiani, F., Donati, J.-F., Gregory, S. G., Güdel, M., Hussain, G. A. J., Kastner, J. H., & Sacco, G. G. 2012, *ApJ*, 752, 100

Donati, J.-F., Gregory, S. G., Montmerle, T., Maggio, A., Argiroffi, C., Sacco, G., Hussain, G., Kastner, J., Alencar, S. H. P., Audard, M., Bouvier, J., Damiani, F., Güdel, M., Huenemoerder, D., & Wade, G. A. 2011, *MNRAS*, 417, 1747

Gregory, S. G., Donati, J.-F., Morin, J., Hussain, G. A. J., Mayne, N. J., Hillenbrand, L. A., & Jardine, M. 2012, *ApJ*, 755, 97

Kastner, J. H., Sacco, G. G., Montez, R., Huenemoerder, D. P., Shi, H., Alecian, E., Argiroffi, C., Audard, M., Bouvier, J., Damiani, F., Donati, J.-F., Gregory, S. G., Güdel, M., Hussain, G. A. J., Maggio, A., & Montmerle, T. 2011, *ApJL*, 740, L17

Rosenfeld, K. A., Andrews, S. M., Wilner, D. J., & Stempels, H. C. 2012, *ApJ* 759, 119

Magnetic Fields throughout Stellar Evolution
Proceedings IAU Symposium No. 302, 2013
P. Petit, M. Jardine & H. Spruit, eds.

doi:10.1017/S1743921314001690

V4046 Sgr: X-rays from accretion shock

C. Argiroffi[1,2], A. Maggio[2], T. Montmerle[3], D. Huenemoerder[4], E. Alecian[5], M. Audard[6], J. Bouvier[7], F. Damiani[2], J.-F. Donati[8], S. Gregory[9], M. Güdel[10], G. Hussain[11], J. Kastner[12] and G. G. Sacco[13]

[1]Dip. di Fisica e Chimica, Università di Palermo, Palermo, Italy, email: argi@astropa.unipa.it; [2]INAF - Osservatorio Astronomico di Palermo, Piazza del Parlamento 1, 90134 Palermo, Italy; [3]Institut d'Astrophysique de Paris, France; [4]MIT, Kavli Institute for Astrophysics and Space Research, Cambridge, MA, USA; [5]Observatoire de Paris, LESIA, Meudon Principal Cedex, France; [6]ISDC Data Center for Astrophysics, University of Geneva, Switzerland; [7]UJF-Grenoble 1/CNRS-INSU, Institut de Planetologie et d'Astrophysique de Grenoble, France; [8]IRAP-UMR 5277, CNRS & Univ. de Toulouse, France; [9]School of Physics & Astronomy, University of St Andrews, U.K; [10]Department of Astronomy, University of Vienna, Austria; [11]ESO, Garching, Germany; [12]Center for Imaging Science, Rochester Institute of Technology, NY, USA; [13]INAF Osservatorio Astrofisico di Arcetri, Italy.

Abstract. We present results of the X-ray monitoring of V4046 Sgr, a close classical T Tauri star binary, with both components accreting material. The 360 ks long XMM observation allowed us to measure the plasma densities at different temperatures, and to check whether and how the density varies with time. We find that plasma at temperatures of $1 - 4$ MK has high densities, and we observe correlated and simultaneous density variations of plasma, probed by O VII and Ne IX triplets. These results strongly indicate that all the inspected He-like triplets are produced by high-density plasma heated in accretion shocks, and located at the base of accretion flows.

Keywords. Accretion, Stars: pre-main sequence, X-rays: stars

1. Introduction

Classical T Tauri Stars (CTTS) are young low mass stars still accreting material from their circumstellar disks. Accreted material, impacting onto the stellar surface, forms strong shocks, where it is heated up to temperatures of a few MK, and then cools down radiating in the X-ray band. Inspecting the X-ray emission from CTTS allows us to infer important properties on post-shock plasma, and, as a consequence, on the accretion process (e.g. Kastner *et al.* 2002, Brickhouse *et al.* 2010).

This work is part of a large campaign of quasi-simultaneous X-ray/optical observations targeting V4046 Sgr, aimed at simultaneously constrain the properties of the shock-heated X-ray emitting plasma, the large scale magnetic field, and the accretion geometry (Donati *et al.* 2011, Argiroffi *et al.* 2012). V4046 Sgr is a close binary in which both components are accreting material from a circumbinary disk (Stempels & Gahm 2004 and references therein). Here we present the results, obtained from high resolution X-ray spectra gathered with XMM, specifically aimed at investigating the properties of the dense plasma heated in the accretion shock.

2. Plasma Density

Plasma density can be measured by the f/i ratio of He-like triplets. In the XMM/RGS spectrum of V4046 Sgr we measured the fluxes of N VI, O VII, and Ne IX, from which we derived the density n_e of the plasma at ~ 1, ~ 2, and ~ 4 MK, respectively (see fig. 1).

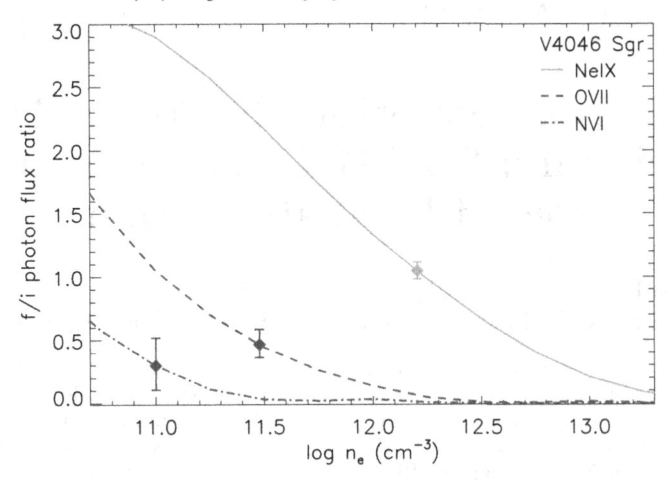

Figure 1. Predicted and observed f/i line ratios of He-like triplets vs plasma density.

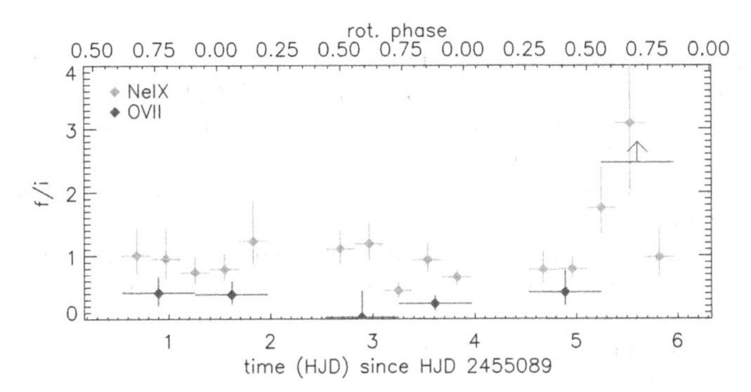

Figure 2. Time resolved f/i line ratios of O VII and Ne IX triplets.

Thanks to the very long XMM observation (360 ks) we searched for density variability by inspecting the f/i ratios of Ne IX and O VI collected on short time intervals (fig. 2). We found significant density variations in both O VII and Ne IX ratios. In particular, during the last part of the observation, both the f/i ratios increased, indicating that a simultaneous density decrease by a factor ~ 10 occurred in the plasma at 2 and 4 MK.

The high density observed indicates that the plasma at $1 - 4$ MK is plasma heated in the accretion shock. The correlated and simultaneous density variations of plasma probed by O VII and Ne IX strongly suggests that the two He-like triplets are produced by the same plasma component, likely located in the post-shock region at the base of the accretion stream.

References

Argiroffi, C., Maggio, A., Montmerle, T., *et al.* 2012, *ApJ*, 752, 100

Brickhouse, N. S., Cranmer, S. R., Dupree, A. K., Luna, G. J. M., & Wolk, S. 2010, *ApJ*, 710, 1835

Donati, J.-F., Gregory, S. G., Montmerle, T., *et al.* 2011, *MNRAS*, 417, 1747

Kastner, J. H., Huenemoerder, D. P., Schulz, N. S., Canizares, C. R., & Weintraub, D. A. 2002, *ApJ*, 567, 434

Stempels, H. C. & Gahm, G. F. 2004, *A&A*, 421, 1159

Magnetic Fields throughout Stellar Evolution
Proceedings IAU Symposium No. 302, 2013
P. Petit, M. Jardine & H. Spruit, eds.

doi:10.1017/S1743921314001707

X-rays from accretion shocks in classical T Tauri stars: 2D MHD modeling and the role of local absorption

C. Argiroffi[1,2], R. Bonito[1], S. Orlando[2], M. Miceli[2], F. Reale[1,2], G. Peres[1,2], T. Matsakos[3,4], C. Stehlé[5] and L. Ibgui[5]

[1]Dip. di Fisica e Chimica, Università di Palermo, Palermo, Italy, email: argi@astropa.unipa.it; [2]INAF - Osservatorio Astronomico di Palermo, Piazza del Parlamento 1, 90134 Palermo, Italy; [3]CEA, IRAMIS, Service Photons, Atomes et Molécules, Gif-sur-Yvette, France; [4]Laboratoire AIM, CEA/DSM - CNRS - Université Paris Diderot, IRFU/SAp, Gif-sur-Yvette, France; [5]LERMA, Observatoire de Paris, Universitè Pierre et Marie Curie, CNRS, 5 place J. Janssen, Meudon, France.

Abstract. In classical T Tauri stars (CTTS) strong shocks are formed where the accretion funnel impacts with the denser stellar chromosphere. Although current models of accretion provide a plausible global picture of this process, some fundamental aspects are still unclear: the observed X-ray luminosity in accretion shocks is order of magnitudes lower than predicted; the observed density and temperature structures of the hot post-shock region are puzzling and still unexplained by models.

To address these issues we performed 2D MHD simulations describing an accretion stream impacting onto the chromosphere of a CTTS, exploring different configurations and strengths of the magnetic field. From the model results we then synthesized the X-ray emission emerging from the hot post-shock, taking into account the local absorption due to the pre-shock stream and surrounding atmosphere.

We find that the different configurations and strengths of the magnetic field profoundly affect the hot post-shock properties. Moreover the emerging X-ray emission strongly depends also on the viewing angle under which accretion is observed. Some of the explored configuration are able to reproduce the observed features of X-ray spectra of CTTS.

Keywords. Accretion, MHD, Stars: pre-main sequence, X-rays: stars

---

1. MHD simulations

CTTS are young low mass stars still accreting material from their circumstellar disks. Accreting material forms strong shock impacting on the stellar surface, being therefore heated by these shocks up to $\sim 1-5$ MK, and emitting significantly in the X-rays. High resolution X-ray spectra of CTTS provide us several observational constraints on the accretion process. However some fundamental aspects, like the over-predicted L_X from accretion shocks, and the observed density vs temperature structures of the hot post-shock region, are puzzling and still unexplained by models.

To address these issues we performed 2D MHD simulations of the accretion shock region (fig. 1), using the PLUTO code (Mignone *et al.* 2007), exploring different configurations and strengths of the stellar magnetic field (Orlando *et al.* 2013). For each simulation we computed the total emerging X-ray spectrum produced by the hot post-shock plasma, taking into account the different absorption suffered by X-rays emitted by different post-shock portions, and the different possible viewing angles.

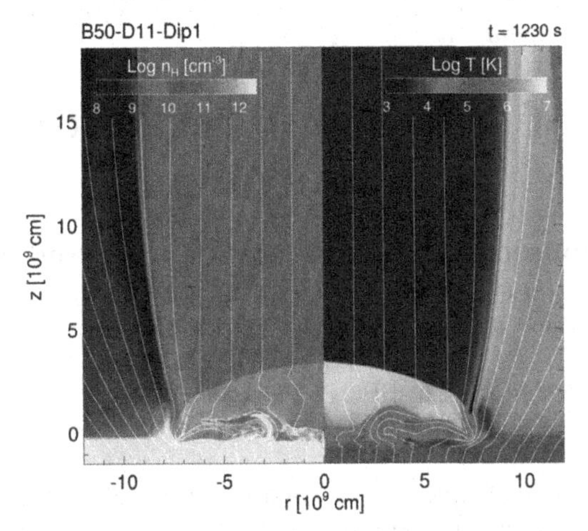

Figure 1. Density (left panels) and temperature (right panels) distributions of one of the simulations; the white lines mark magnetic field lines.

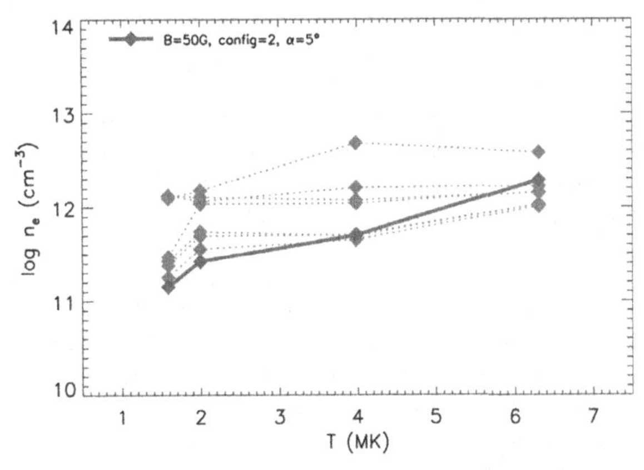

Figure 2. Densities vs temperature obtained from the He-like triplets diagnostics of the emerging X-ray spectra corresponding to different simulations and different viewing angles.

2. Results

We find that both magnetic field strength and configuration deeply affect the location and characteristics of post-shock plasma, and, as a consequence, affect the absorption suffered by the emitted X-rays. The emerging X-ray spectra for different cases and for different viewing angles strongly differ in shape and luminosity. In particular we find that some configurations produce X-ray emerging spectra that reproduce the density vs temperature pattern (fig. 2) observed in the X-ray spectra of CTTS, and up to date not explained by models.

On the one hand our results indicate that accretion shock models are able to reproduce the features observed in the X-ray band of CTTS. On the other hand our results evidence how magnetic field configuration, local absorption, and viewing angle are fundamental parameters to interpret the observed X-ray emission from accretion shock in CTTS.

References

Mignone, A., Bodo, G., Massaglia, S., Matsakos, T., Tesileanu, O., Zanni, C., & Ferrari, A. 2007, *ApJS*, 170, 228

Orlando, S., Bonito, R., Argiroffi, C., *et al.* 2012, *A&A*, in press

Magnetic Fields throughout Stellar Evolution
Proceedings IAU Symposium No. 302, 2013
P. Petit, M. Jardine & H. Spruit, eds.

doi:10.1017/S1743921314001719

Analysis of star-disk interaction in young stellar systems

Nathalia N. J. Fonseca[1,2], **Silvia H. P. Alencar**[1] **and Jérôme Bouvier**[2]

[1]Departamento de Física - ICEx - UFMG,
Av. Antônio Carlos, 6627, 30270-901, Belo Horizonte, MG, Brazil
email: nath@fisica.ufmg.br

[2]UJF-Grenoble 1/CNRS-INSU, Institut de Planétologie et d'Astrophysique de Grenoble (IPAG), UMR 5274, Grenoble, F-38041, France

Abstract. We present preliminary results of the study of star-disk interaction in the classical T Tauri star V354 Mon, a member of the young stellar cluster NGC 2264. As part of an international campaign of observations of NGC 2264 organized from December 2011 to February 2012, high resolution photometric and spectroscopic data of this object were obtained simultaneously with the Chandra, CoRoT and Spitzer satellites, and ground-based telescopes, such as CFHT and ESO/VLT. The optical and infrared light curves of V354 Mon show periodic brightness minima that vary in depth and width every 5.21 days rotational cycle. We found evidence that the Hα emission line profile changes according to the period of photometric variations, indicating that the same phenomenon causes both modulations. Such correlation was also identified in a previous observational campaign on the same object, where we concluded that material non-uniformly distributed in the inner part of the disk is the main cause of the photometric modulation. This assumption is supported by the fact that the system is seen at high inclination. It is believed that this distortion of the inner part of the disk results from the dynamical interaction between the stellar magnetosphere, inclined with respect to the rotation axis, and the circumstellar disk, as also observed in the classical T Tauri star AA Tau, and predicted by magnetohydrodynamic numerical simulations. A model of occultation by circumstellar material was applied to the photometric data in order to determine the parameters of the obscuring material during both observational campaigns, thus providing an investigation of its stability on a timescale of a few years. We also studied V422 Mon, a classical T Tauri star with photometric variations similar to those of V354 Mon at optical wavelengths, but with a distinct behavior in the infrared. The mechanism that produces such a difference is investigated, testing the predictions of magnetospheric accretion models.

Keywords. stars: pre-main sequence, techniques: photometric, techniques: spectroscopic, accretion, accretion disks

1. Introduction

During the pre-main sequence phase, young (~1 Myr), low-mass ($M \leqslant 2\ M_{\odot}$) stars that exhibit signs of active accretion of material from its surrounding disk are classified as classical T Tauri stars (CTTSs). They present photometric and spectroscopic irregular variability, emission excess with respect to the stellar photosphere at wavelengths from X-rays to the radio, broad emission lines with redshifted and blueshifted absorptions, and forbidden emission lines. These features are successfully explained by magnetospheric accretion models (Shu *et al.* 1994; Hartmann *et al.* 1994; Muzerolle *et al.* 2001; Kurosawa *et al.* 2006; Lima *et al.* 2010). In this scenario, the strong stellar magnetic field (~few kG) disrupts the disk at a distance of few stellar radii from the star. The inner disk material is then channeled at free-fall velocity along the dipole field lines, forming accretion funnels, where broad emission lines and redshifted absorption are generated. Hot spots are produced as material hits the photosphere, emitting an excess of optical and UV flux. An ionized disk wind is responsible for ejecting material from the system, producing blueshifted absorption and forbidden emission lines. The circumstellar disk reprocesses the radiation generated in the system, producing the observed infrared excess.

AA Tau is a well studied CTTS, observed for a month during three different campaigns (Bouvier *et al.* 1999, 2003, 2007). Its photometric modulation is characterized by an almost constant brightness level interrupted by quasi-cyclical and irregular episodes of attenuation. This peculiar behaviour is produced by a warp in the inner part of the circumstellar disk, which is caused by the interaction between the inner disk and the stellar magnetic field, tilted with respect to the rotation axis. The structure of the warp seems to evolve with time, as the photometric variation changes its shape on a timescale of a few weeks in each campaign to a few years from one campaign to the next. From an additional program of the CoRoT satellite that observed the young cluster NGC 2264 for 23 days uninterruptedly in March 2008, Alencar *et al.* (2010) identified that 28% of the observed CTTSs exhibited the same type of variability as AA Tau. This result indicates that occultation by circumstellar material as the main cause of photometric variability is common among young stars.

A few years ago, an international group of researchers organized a multiwavelength observational campaign of NGC 2264 with several satellites: CoRoT and MOST for 39 days in the optical, Spitzer for 29 days in the infrared, and Chandra for 3.5 days in the X-rays. The observations were simultaneously performed from December 2011 to January 2012. Data from ground-based telescopes were also obtained: spectroscopy with VLT/Flames for 20 nights, *ur* photometry with CFHT/Megacam for 15 nights in February 2012, I band with USNO for $\sim$70 nights from November 2011 to March 2012, and others. Covering simultaneously a wide range of wavelengths on a long timescale, this campaign will enable a comprehensive analysis of the physical processes that occur in young stars.

2. V354 Mon

V354 Mon, a K4V CTTS member of NGC 2264, was observed by CoRoT in both 2008 and 2011 campaigns and classified as AA Tau type. This star shows a periodic, but irregular photometric variability, as the depth and width of the minima change at each rotational cycle (Fig. 1, left). This morphological feature is present in the light

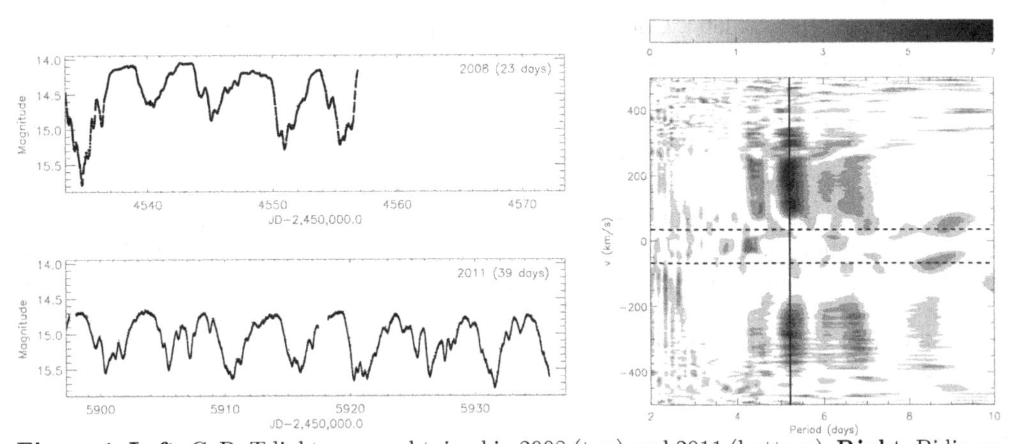

Figure 1. Left. CoRoT light curves obtained in 2008 (top) and 2011 (bottom). **Right.** Bidimensional periodogram of the Hα line flux. The horizontal dashed lines delimit the region dominated by nebular emission. The vertical solid line marks the period of 5.21 days.

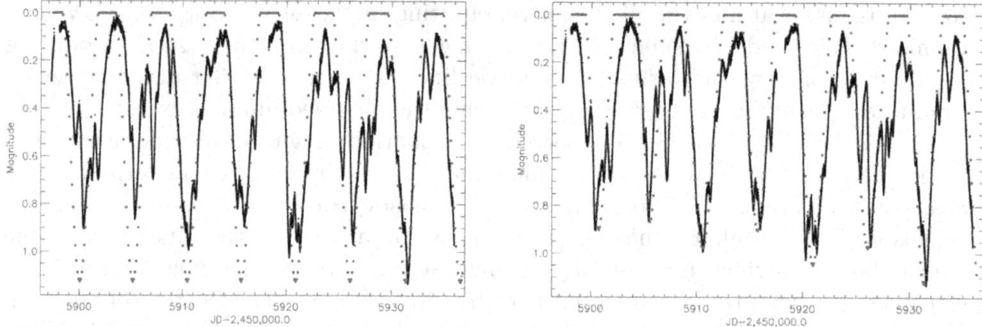

Figure 2. Best fit of the occultation model with fixed parameters (left) and individual fit of each CoRoT light curve minima (right) obtained in 2011.

curve of both campaigns. The photometric coverage for more than 100 days provided by combining CoRoT, CFHT, and USNO data led to a good determination of the period, 5.21 ± 0.04 days. This value is in agreement with the periods obtained from the 2008 CoRoT light curve, 5.26 ± 0.50 days, and by Lamm *et al.* (2004), 5.22 ± 0.87 days, indicating that the main cause of the photometric variability is stable over a few years.

During the 2008 CoRoT observations, simultaneous échelle spectroscopy was obtained for V354 Mon with the SOPHIE spectrograph at the Observatoire de Haute Provence (CNRS, France). We noticed that the emission lines vary in a cyclic manner according to the photometric modulation (Fonseca *et al.* in preparation). This correlation is also observed in the new campaign, as shown through the periodogram analysis of the Hα normalized flux (Fig. 1, right). Therefore, the same phenomenon produces the photometric and spectroscopic variability in V354 Mon.

Magnetohydrodynamic (MHD) simulations (Romanova *et al.* 2004) predict the formation of a warp in the inner disk when the magnetic axis is misaligned with respect to the rotation axis. As the stellar system rotates, the warp could occult periodically part of the stellar photosphere. This is a plausible scenario for V354 Mon, as this system is viewed at high inclination (~75°) and shows the same type of variability as AA Tau. We applied a model of occultation by circumstellar material, originally developed for AA Tau (Bouvier *et al.* 1999), to the light curve of V354 Mon obtained in the 2011 campaign in order to determine the general parameters of the obscuring material. The warp, located at the corotation radius (r_c), presented a maximum scale height h/r_c of 0.33 and an azimuthal extension of 320°(Fig. 2, left). Comparing with the parameters obtained in the model for the light curve of the 2008 campaign, 0.30 and 360°(Fonseca et al. in preparation), we notice that the warp remained stable on a timescale of a few years. It is also interesting to observe that these characteristics are very similar to the ones obtained in the model of the variability of AA Tau (Bouvier *et al.* 1999).

The depth and width of V354 Mon minima vary at each rotational cycle, indicating that the warp changes its shape with time. This reveals a dynamical interaction between the stellar magnetosphere and the inner part of the disk (Goodson & Winglee 1999). However, the parameters obtained from the individual fit of the model to the light curve minima are not very different from cycle to cycle (Fig. 2, right, and Table 1), indicating that the warp is a permanent structure.

3. V422 Mon

V422 Mon is a CTTS that presents light curve minima that vary in depth and width every rotational cycle, with a period of 8.93 ± 0.92 days. This similarity with V354

Table 1. Occultation model parameters from individual fit of 2011 light curve minima
$$h(\phi) = h_{max} \left| \cos \frac{\pi(\phi - \phi_0)}{2\phi_c} \right|$$

Minimum	1st	2nd	3rd	4th	5th	6th	7th
h_{max} (r_c)	0.31	0.30	0.31	0.31	0.32	0.31	0.33
$2\phi_c$ (°)	260	240	360	320	320	360	310

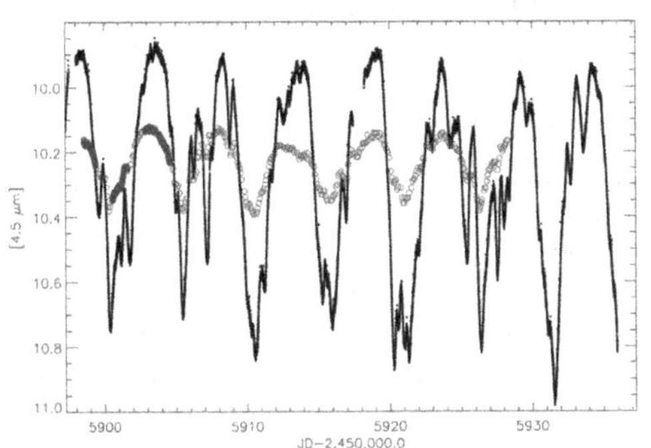

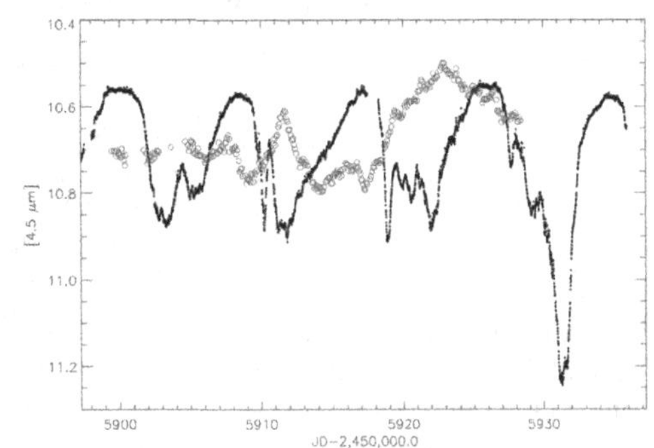

Figure 3. V354 Mon (left) and V422 Mon (right) CoRoT light curves (dots) obtained in 2011, compared with 4.5 μm Spitzer data (open circle). CoRoT data was shifted to the mean of the Spitzer data.

Mon would indicate that both present a similar geometry. However, while the V354 Mon photometric modulation in the infrared is very similar to that seen in the optical, there is no agreement between the optical and infrared photometric variations of V422 Mon (Fig. 3). The optical modulation may be produced by a warp in the inner disk, as in V354 Mon, but the infrared variation may be caused by changes in the inner disk emission.

Acknowledgement

N. N. J. F. acknowledges support from CAPES (fellowship process n_o. 18697-12-7), CNPq, CNES, and IAU. S. H. P. A. acknowledges support from CAPES, CNPq, and Fapemig. J. B. acknowledges support by ANR Toupies grant 2011 Blanc SIMI5-6 020 01.

References

Alencar, S. H. P.., Teixeira, P. S., Guimarães, M. M., McGinnis, P. T., Gameiro, J. F., Bouvier, J., Aigrain, S., Flaccomio, E., & Favata, F. 2010, *A&A*, 519, 88

Bouvier, J., Alencar, S. H. P.., Boutelier, T., Dougados, C., Balog, Z., Grankin, K., Hodgkin, S. T., Ibrahimov, M. A., Kun, M., Magakian, T. Y., & Pinte, C. 2007, *A&A*, 463, 1017

Bouvier, J., Chelli, A., Allain, S., Carrasco, L., Costero, R., Cruz-Gonzalez, I., Dougados, C., Fernández, M., Martín, E. L., Ménard, F., Mennessier, C., Mujica, R., Recillas, E., Salas, L., Schmidt, G., & Wichmann, R. 1999, *A&A*, 349, 619

Bouvier, J., Grankin, K. N., Alencar, S. H. P.., Dougados, C., Fernández, M., Basri, G., Batalha, C., Guenther, E., Ibrahimov, M. A., Magakian, T. Y., Melnikov, S. Y., Petrov, P. P., Rud, M. V., & Zapatero Osorio, M. R. 2003, *A&A*, 409, 169

Goodson, A. P. & Winglee, R. M. 1999, *ApJ*, 524, 159

Hartmann, L., Hewett, R., & Calvet, N. 1994, *ApJ*, 426, 669

Kurosawa, R., Harries, T. J., & Symington, N. H. 2006, *MNRAS*, 370, 580

Lamm, M. H., Bailer-Jones, C. A. L.., Mundt, R., Herbst, W., & Scholz, A. 2004, *A&A*, 417, 557

Lima, G. H. R.. A., Alencar, S. H. P.., Calvet, N., Hartmann, L., & Muzerolle, J. 2010, *A&A*, 522, 104

Muzerolle, J., Calvet, N., & Hartmann, L. 2001, *ApJ*, 550, 944

Romanova, M. M., Ustyugova, G. V., Koldoba, A. V., & Lovelace, R. V. E.. 2004, *ApJ*, 610, 920

Shu, F., Najita, J., Ostriker, E., Wilkin, F., Ruden, S., & Lizano, S. 1994, *ApJ*, 429, 781

Magnetic Fields throughout Stellar Evolution
Proceedings IAU Symposium No. 302, 2013
P. Petit, M. Jardine & H. Spruit, eds.

doi:10.1017/S1743921314001720

Magnetospheric Accretions and the Inner Winds of Classical T Tauri Stars

Ryuichi Kurosawa[1] and M. M. Romanova[2]

[1]Max-Planck-Institut für Radioastronomie,
Auf dem Hügel, 69, 53212 Bonn, Germany
email: kurosawa@mpifr-bonn.mpg.de

[2]Department of Astronomy, Cornell University,
Ithaca, NY 14853-6801, USA
email: romanova@astro.cornell.edu

Abstract. Recent spectropolarimetric observations suggest that young low-mass stars such as classical T Tauri stars (CTTSs) possess relatively strong (~kG) magnetic field. This supports a scenario in which the final accretion onto the stellar surface proceeds through a magnetosphere, and the winds are formed in magnetohydrodynamics (MHD) processes. We examine recent numerical simulations of magnetospheric accretions via an inclined dipole and a complex magnetic fields. The difference between a stable accretion regime, in which accretion occurs in ordered funnel streams, and an unstable regime, in which gas penetrates through the magnetosphere in several unstable streams due to the magnetic Rayleigh-Taylor instability, will be discussed. We describe how MHD simulation results can be used in separate radiative transfer (RT) models to predict observable quantiles such as line profiles and light curves. The plausibility of the accretion flows and outflows predicted by MHD simulations (via RT models) can be tested against observations. We also address the issue of outflows/winds that arise from the innermost part of CTTSs. First, we discuss the line formations in a simple disk wind and a stellar wind models. We then discuss the formation of the conically shaped magnetically driven outflow that arises from the disk-magnetosphere boundary when the magnetosphere is compressed into an X-type configuration.

Keywords. stars: low-mass, stars: magnetic fields, stars: winds, outflows, stars: pre–main-sequence, stars: variables: other, MHD, radiative transfer, line: formation, line: profiles

---

1. Introduction

In the standard picture of accretions in classical T Tauri stars (CTTSs), the stellar magnetic fields are assumed to be strong enough to truncate the accretion disks near the star-disk corotation radii (e.g., Camenzind 1990; Koenigl 1991). The matter originating in the vicinity of the disk truncation radius flows along the magnetic field lines to the stellar surface. The kinetic energy of the infalling matter is converted into the thermal energy of the plasma in the shock regions above the photosphere. This general picture of magnetospheric accretions successfully explains some of the important observational aspects of CTTSs, e.g., the UV-optical continuum excess (e.g. Calvet & Gullbring 1998) and the inverse P-Cygni profiles (e.g. Hartmann *et al.* 1994; Muzerolle *et al.* 2001). In the earlier magnetospheric accretion models (e.g. Koenigl 1991; Hartmann *et al.* 1994), the flows are often assumed to be steady and axisymmetric, as in the original models for accreting neutron stars (e.g. Ghosh *et al.* 1977) . While these axisymmetric models agree with some of the observational phenomena as mentioned above, they are not applicable to many other observational aspects. For example, (1) some CTTSs are known to exhibit complex line and continuum variability (e.g. Johns & Basri 1995; Petrov *et al.* 1996), and (2) recent time-series spectropolarimetric (Zeeman-Doppler Imaging) observations (Donati *et al.* 2007, 2011; Jardine *et al.* 2008; Gregory *et al.* 2008; Hussain *et al.* 2009) have revealed that some of CTTSs have complex stellar magnetic fields. Clearly, these observational aspects (complex variability and accretion geometry) cannot be explained by simple axisymmetric models. To address the issues of the complex variablity and non-axisymmetry of CTTSs, we will use three-dimensional (3-D) radiative transfer models of time-series line profiles using the results of 3-D magnetohydrodynamics (MHD) simulations of the accretion onto CTTSs through inclined magnetic multipoles. The results are presented in Sec. 2

Not only the accretion flows, but also the outflows/winds in CTTSs are most likely shaped by magnetic fields. The formation of the outflow itself likely occurs in MHD processes in which open magnetic fields are anchored to a star, an accretion disk or both (e.g., Ferreira *et al.* 2006). Observationally, there are ample of spectroscopic evidences that rather high velocity winds/outflows originate from the innermost part of the star-disk systems, e.g., the blueshifted absorption component in strong optical and near-UV lines such as Hα, Na I D, Ca II H&K (e.g., Reipurth *et al.* 1996; Alencar & Basri 2000; Ardila *et al.* 2002). In addition, the near-infrared He I λ10830 line has been recognized as a robust wind diagnostic line by e.g. Takami *et al.* (2002), Edwards *et al.* (2003), Edwards *et al.* (2006). In particular, Edwards *et al.* (2006) showed that the line is very sensitive to the presence of the wind, and about 70 per cent of CTTSs exhibit a blueshifted absorption (below continuum) in He I λ10830 while only about 10 per cent of CTTSs show a similar type of absorption component in Hα (Reipurth *et al.* 1996). Edwards *et al.* (2006) and Kwan *et al.* (2007) suggested that the blueshifted absorption component in He I λ10830 profiles is cased by a stellar wind in about 40 per cent, and by a disc wind in about 30 per cent of the samples in Edwards *et al.* (2006). In Sec. 3, we explore how the radiative transfer models of He I λ10830 can be used to probe the origin of the inner winds of CTTSs. We will also test the plausibility of the conically shaped magnetically driven wind ("the conical wind") solution found in the MHD simulations of Romanova *et al.* (2009).

2. Non-Axisymmetric Accretion Models and Complex Variablity

In this section, we examine the types of variability expected from (1) *steady* non-axisymmetric accretion flows (Sec. 2.1), and (2) *unsteady* non-axisymmetric accretion flows (Sec. 2.2).

2.1. *Rotationally Induced Modulations Caused by Steady Accretion Flows*

A set of radiative transfer models which simulate the rotationally induced line variability arising from a complex circumstellar environment of CTTSs was presented by Kurosawa *et al.* (2008) (see also Symington *et al.* 2005; Kurosawa *et al.* 2005). The results of the 3-D MHD simulations from Romanova *et al.* (2003) and Romanova *et al.* (2004), who considered accretion onto a CTTS with a misaligned dipole magnetic axis with respect to the rotational axis, were used in a separate 3-D radiative transfer model (e.g., Harries 2000; Kurosawa *et al.* 2006) to predict observed line profiles (e.g., Hβ, Paβ and Brγ) as a function of rotational phases. The accretion in those MHD models are in a steady state, and the accretion mainly proceeds in two funnel flows, i.e., one lands on the upper and the other on the lower hemispheres. In this study, the general dependencies of line variability on inclination angles (i) and magnetic axis misalignment angles (Θ) were examined.

Using the same method, we now examine the accretion flows with more complex magnetic field configurations (with a combination of an inclined dipole and an octupole

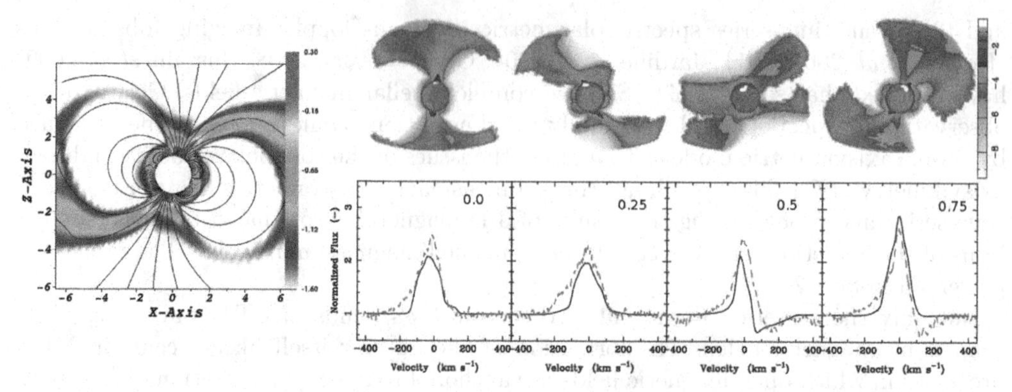

Figure 1. *Left panel*: A cross section of the density from the 3-D MHD simulations of accretion onto the CTTS V2129 Oph. The density is shown in logarithmic scale and in arbitrary units. The solid lines represent sample magnetic field (a combination of a dipole and an octupole). *Right panels*: The radiative transfer models computed for the accretion flows from the MHD simulation. *Upper panels* show the Hδ emission maps at four different rotation phases of the star (0.0, 0.25, 0.5 and 0.7 from left to right). *Lower panels* show the corresponding line profiles models of Hδ (solid) along with the data obtained by ESPaDOnS (dashed). The models and data are from Alencar *et al.* (2012).

components), similar to those in Romanova *et al.* (2011) and Long *et al.* (2011). In particular, the accretion flows around the CTTS V2129 Oph (K5) were modeled by the MHD simulations in which the information from the surface magnetic field map (by the Zeeman-Doppler imaging technique) obtained by Donati *et al.* (2011) was incorporated (see Alencar *et al.* 2012). For example, the observation by Donati *et al.* (2011) suggests that the inclinations of the magnetic dipole and the octupole are about 15° and 25° with respective to the rotation axis, and the octupole axis is about 0.1 ahead of the dipole in the rotational phase. A single time-slice of the MHD simulations was used to predict the dependency of the observed line profiles on the rotational phase by using the 3-D radiative transfer model mentioned earlier. The results are shown in Fig. 1. As one can see from this figure, near the stellar surface, the octupole component dominates and it redirect the accretion funnels to a higher latitude. On the other hand, the dipole component dominates in larger scales, and it interact with the innermost part of the accretion disk. The line profile models for Hβ are compared with the observations obtained by ESPaDOnS at CFHT (Alencar *et al.* 2012) at four different rotational phases (0.0, 0.25, 0.5 and 0.75). The figure shows that our model agrees with the time-series observed line profiles very well. More detailed comparisons of the model with observations can be found in Alencar *et al.* (2012).

2.2. *Irregular Variability Caused by Accretion in an Unstable Regime*

In the previous section (Sec. 2.1), we considered the cases in which the accretion flows are steady even though the flows themselves are non-axisymmetric. Here, we examine the unsteady accretion flows caused by the magnetic Rayleigh-Taylor (R-T) instability, which occurs at the interface between an accretion disk and a stellar magnetosphere. The instability tends to appear in a system with a relatively slowly rotating magnetosphere, a high mass-accretion rate and a small misalignment angle for the magnetic poles. The detail theoretical studies on the conditions for instability have been presented by e.g., Arons & Lea (1976); Spruit & Taam (1993); Li & Narayan (2004). More recently, the global three-dimensional (3D) magnetohydrodynamic (MHD) simulations by e.g., Romanova *et al.* (2008); Kulkarni & Romanova (2008, 2009) have shown that the R-T instability

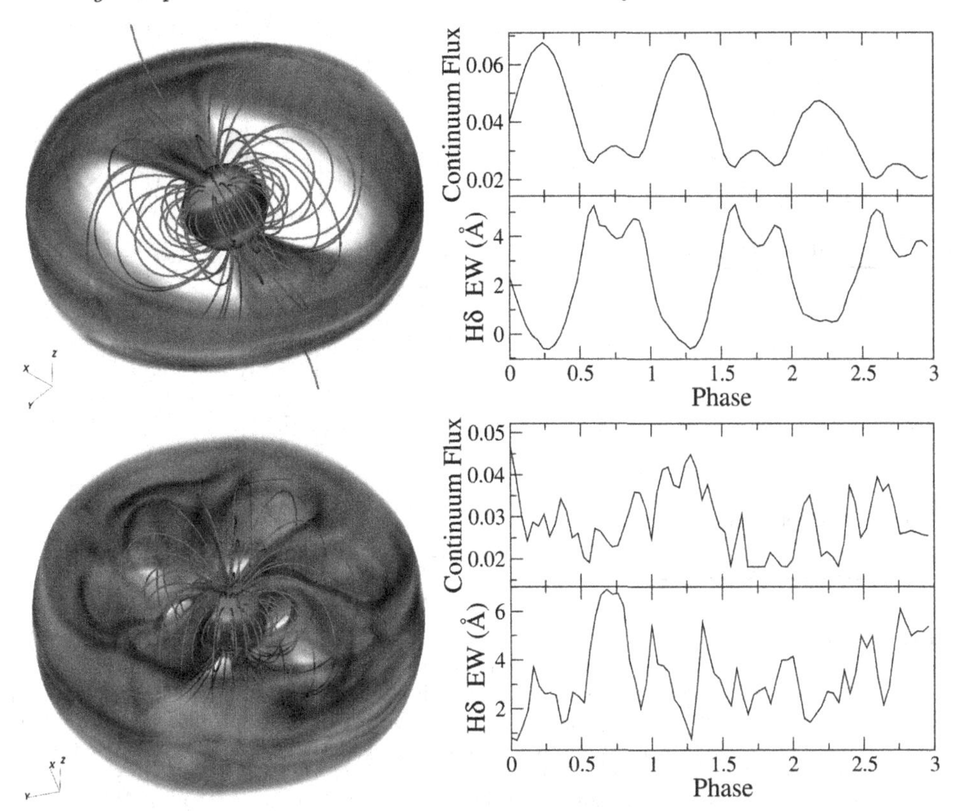

Figure 2. A comparison of accretions onto a CTTS in a stable (*upper panels*) and an unstable (*lower panes*) regimes. While the density distributions (volume renderings; in logarithmic scales; in arbitrary units) from the 3D MHD simulations are shown in the *left panels*, the radiative transfer models for the continuum flux and Hδ line equivalent widths plotted as a function of rotation phases are shown on the *right panels*. The continuum flux are evaluated near Hδ. The figures are from Kurosawa & Romanova (2013).

induces accretions in multiple (a few to several) unstable vertically elongated streams or 'tongues' which penetrate the magnetosphere (the lower left panel in Fig. 2). They found that the corresponding time-scale of the variability induced by the instability is typically a few times smaller than the rotation period of a star. Here, we examine the fundamental differences in the observational properties of the accretions in the stable and unstable regimes.

The results of global 3-D MHD simulations of matter flows in both stable and unstable accretion regimes are used in Kurosawa & Romanova (2013) to calculate time-dependent hydrogen line profiles and study their variability behaviors (see Fig. 2). In the stable regime, some hydrogen lines (e.g., Hβ, Hγ, Hδ, Paβ and Brγ) show a redshifted absorption component only during a fraction of a stellar rotation period, and its occurrence is periodic. However, in the unstable regime, the redshifted absorption component is present rather persistently during a whole stellar rotation cycle, and its strength varies non-periodically. In the stable regime, an ordered accretion funnel stream passes across the line of sight to an observer only once per stellar rotation period while in the unstable regime, several accreting streams/tongues, which are formed randomly, pass across the line of sight to an observer. The latter results in the quasi-stationarity appearance of the redshifted absorption despite the strongly unstable nature of the accretion. In the unstable regime, multiple hot spots form on the surface of the star, producing the

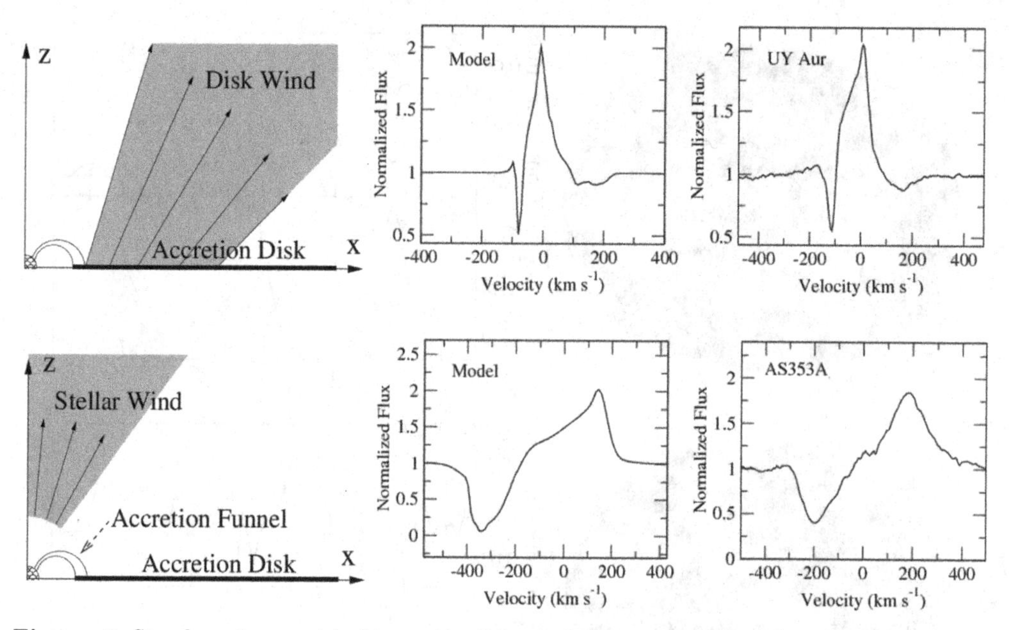

Figure 3. Simple axisymmetric kinematic disk wind (*upper panels*) and stellar wind (*lower panels*) models that are combined a dipolar magnetospheric accretion. The *left panels* show the flow configurations. The corresponding line profile models of He I λ10830 are shown in the *middle panels*. The inclination angles (i) used for computing the line profiles are $i = 48°$ and $10°$ for the models with a disk wind and a stellar wind, respectively. The observed He I λ10830 line profiles for the CTTSs UY Aur and AS 353 A (from Edwards *et al.* 2006) are shown in the *right panels*, for qualitative comparisons. The figures are from Kurosawa *et al.* (2011).

stochastic light curve with several peaks per rotation period. Interestingly, such irregular light-curves are frequently observed in CTTSs (e.g. Herbst *et al.* 1994; Rucinski *et al.* 2008; Alencar *et al.* 2010). Note that irregular light curves are found in about 39 per cent of the CTTS samples in Alencar *et al.* (2010). No clear periodicity in the line variability is found in many CTTSs. This study suggests a CTTS that exhibits a stochastic light curve and a stochastic line variability, with a rather persistent redshifted absorption component, may be accreting in the unstable accretion regime.

3. The Inner Winds of Classical T Tauri Stars

Understanding the origin of an outflow, whether it is a stellar wind, the X-wind (Shu *et al.* 1994), the conical wind (Romanova *et al.* 2009) or disk wind, is important as it is closely related to the angular momentum evolution of young stellar objects. Here, we demonstrate how the line profile models can be used to probe the origin of the inner winds.

3.1. *Simple Kinematics Wind Models*

As briefly mentioned in Sec. 1, Edwards *et al.* (2006) have demonstrated a robustness of the near-infrared He I λ10830 line for probing the inner winds of CTTSs. In particular, they have classified He I λ10830 line profiles into two types based on the shapes of the blueshifted wind absorption component. The first type has a relatively narrow and low wind absorption component (the upper right panel in Fig. 3). The second type has a wide and deep wind (P-Cygni like) absorption component which reaches the maximum velocity of 300–400 km s^{-1} (the lower right panel in Fig. 3). Edwards *et al.* (2006) suggested that the former is caused by a disk wind, and the star is viewed at a relatively high inclination angle at which a line of sight can intersect with the disk wind. On the other hand, the latter is cased by a stellar wind, and the star is viewed nearly pole-on.

To test the scenarios of Edwards *et al.* (2006) for the two distinctive wind absorption features, we have constructed the simple kinematic flow models around CTTSs. See Kurosawa *et al.* (2011) for the model descriptions. The first model consists of a disk wind and magnetospheric accretion funnels (the upper left panel in Fig. 3). The second model consists of a stellar wind and magnetospheric accretion funnels (the lower left panel in Fig. 3). The corresponding line profile models for He I λ10830 are also shown in the same figure (the middle panels). Our disk wind + magnetosphere model reproduces not only the narrow blueshifted wind absorption component, but also the redshifted absorption component as seen in the observed He I λ10830 line profile of the CTTS UY Aur. Note that the redshifted absorption component is caused by the infalling gas in the magnetospheric accretion funnel. Similarly, our stellar wind + magnetosphere model reproduces a rather wide and deep blueshifted absorption component as seen in the observed He I λ10830 in the CTTS AS 353 A. Interestingly, the narrow wind absorption caused by the disk wind is present in the line profile only when the inclination angle of the system is an intermediate to a high value because of the geometry of the disk wind. Similarly, the wide and deep P-Cygni like wind absorption is present only at a very low inclination angle (near pol-on). Hence, our simple kinematic wind models confirms the earlier finding of Edwards *et al.* (2006) who suggested two different types of wind for the narrow and wide blueshifted absorption components seen in He I λ10830. A similar analysis was also performed by Kwan *et al.* (2007) with simplified line emissivity and opacity in the winds.

In reality (as also suggested by Edwards *et al.* 2006), both types of the winds (a stellar and a disk winds) could coexist; however, it is likely that only one type of wind absorption component appears in a line profile because the line of sight to the stellar surface could intersect only one type of wind for a given viewing angle of the system.

3.2. *The Conical Wind Model*

To advance our understanding of the formation of the inner winds, we now apply our radiative transfer models to more realistic winds. For this purpose, we move our focus back to MHD simulation results. Recent MHD simulations by Romanova *et al.* (2009) have shown a new type of outflow configuration, so called "the conical wind" (the upper panel in Fig. 4), which is formed when a large scale stellar dipole magnetic field is compressed by the accretion disk into the X-wind (Shu *et al.* 1994) configuration. The outflows occur in a rather narrow conical shell, and the outflow speed of the matter is an order of the Keplerian rotation speed near the wind launching region. Interestingly, the conical wind model does not require the magnetospheric radius to be very close the corotation radius, unlike the X-wind model. Similar simulations with extended computational domain by Lii *et al.* (2012) have shown that the conical wind become collimated at large distance. The model was applied to the high mass-accretion rate FU Ori systems in which the wind can be strongly collimated (Königl *et al.* 2011).

We extended the original conical wind model of Romanova *et al.* (2009) to include a well-defined magnetospheric accretion funnel flow which is essential for modeling the strong optical and near-infrared emission lines in CTTSs. We achieved this by using a slightly stronger magnetic filed and a lower mass-accretion rate to reduced the compression of magnetosphere by the disk, hence to form a larger magnetospheric funnel flow. The resulting flow geometry is shown in Fig. 4 (the upper panel). We then used the density and velocity field from the MHD simulation, after it reached a semi-steady

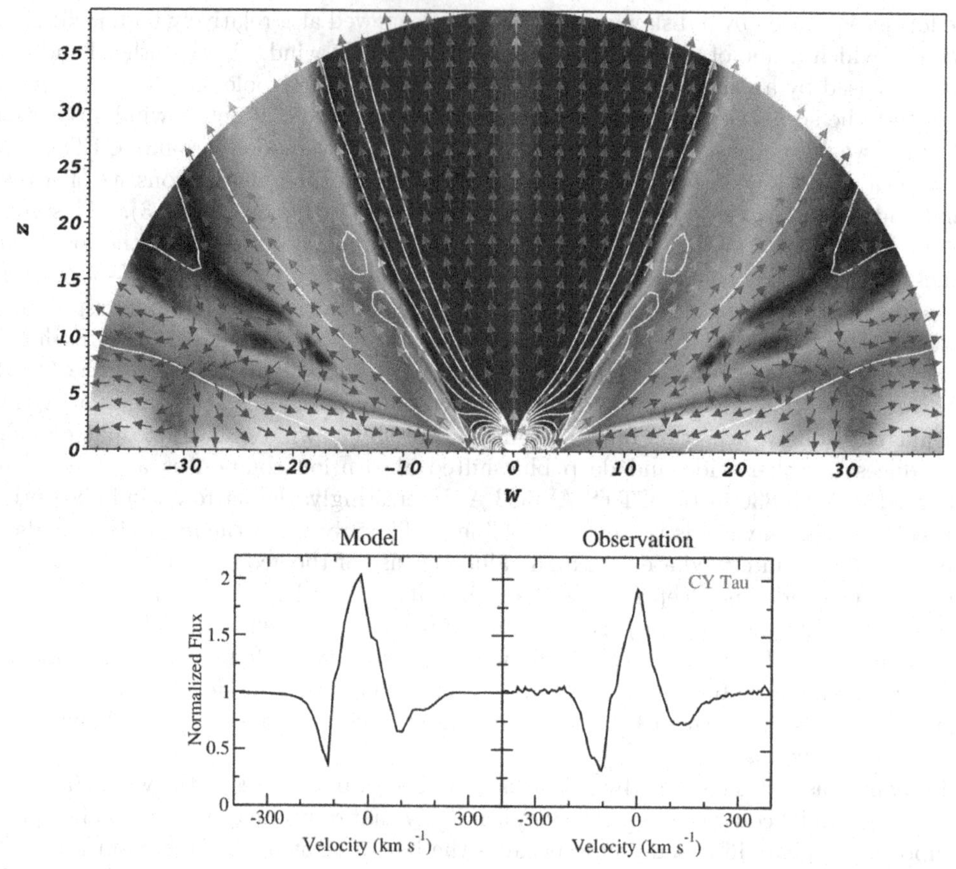

Figure 4. Summary of the conical wind model (Romanova *et al.* 2009) and a qualitative comparison of the model and observed He I λ10830 profiles. *Upper panel*: the poloidal matter flux map is over-plotted with the poloidal component of the velocity (*arrows*). The *solid contours* represent the magnetic field lines. *Lower panels*: a qualitative comparison of the model profile of He I λ10830 with the observation of the CTTS CY Tau (Edwards *et al.* 2006). The figures are from Kurosawa & Romanova (2012)

state, in our radiative transfer calculations to predict hydrogen and helium line profiles. The line profiles were computed with various combinations of X-ray fluxes (important for photoionizing He I), gas temperatures and inclination angles. A large variety of line profile morphology is found, and many of the model profiles are very similar to those found in observations. An example of He I λ10830 line profile calculation along with the observation of the CTTS CY Tau (from Edwards *et al.* 2006) is shown in Fig. 4 (the lower panels). More examples can be found in Kurosawa & Romanova (2012). As one can see in Fig. 4, the conical wind model can well reproduce the relatively narrow and low-velocity blueshifted wind absorption component in the observed line profile of He I λ10830. It also reproduces the redshifted absorption component, which is caused by the infalling material in the accretion funnel flow. In summary, based on our line profile models, we found that the conical wind model is a very plausible outflow model for a system which shows a narrow and low-velocity blueshifted absorption component in He I λ10830, and it is an important alternative model to a disk wind model which can also form the narrow and low-velocity blueshifted absorption component, as we have shown in Sec. 3.1.

4. Summary

Recent observations suggest that the geometry of the magnetospheric accretion may significantly deviate from an axisymmetry for some of CTTSs. We have briefly reviewed the 3-D MHD simulations which considered the accretion onto CTTSs with inclined magnetic multipoles (e.g., a dipole and an octupole). Non-axisymmetric accretion funnel flows are naturally found in those simulations. Using the output from the simulations in our radiative transfer models, observable quantities such as line profiles and light curves are computed. The predictions can be readily compared with observations. While the periodic/regular variability behaviors seen in observations of CTTSs can be well reproduced by non-axisymmetric accretion models that are in a steady state (Sec. 2.1), the irregular variability behaviors seen in some of CTTSs could be explained by the magnetospheric accretion that is in unstable regime due the magnetic Rayleigh-Taylor (R-T) instability (Sec. 2.2). In addition to the irregular light curves and line variability, the unstable accretion due to the R-T instability would produce a variable but rather persistent redshifted absorption component in higher Balmer lines (e.g., Hγ and Hδ) and in some near-infrared hydrogen lines such as Paβ and Brγ, due to the presence of many accretion streams caused by the instability.

We have also investigated the possible origins of the inner winds in CTTSs using the line profile models. Using the simple kinematic wind models in our radiative transfer models (Sec. 3.1), we confirmed the earlier finding by Edwards *et al.* (2006) who suggested that there are two types of inner winds of CTTSs. We confirmed that the relatively narrow and low-velocity blueshifted absorption component in He I λ10830 can be reproduced by a disk wind model, and the wide and deep blueshifted absorption component in the same line can be reproduced by a stellar wind model. For the stellar wind component to be visible in the absorption component, the inclination angle of the system must be rather small. We have also tested the plausibility of the conical wind solution found in the recent MHD simulations by Romanova *et al.* (2009) by using the simulation results in our radiative transfer models (Sec. 3.2). We found the conical wind model can also reproduce the relatively narrow and low-velocity blueshifted absorption component in He I λ10830 seen in some CTTSs. Finally, high resolution spectroscopic data and radiative transfer models are invaluable for testing different outflow and inflow scenarios since a direct imaging of the inner wind launching regions is still unattainable for most of the CTTSs.

Acknowledgment

We thank the conference organizers for the excellent meeting. RK thank Silvia Alencar, Suzan Edwards and Tim Harries for their support and discussions. Resources supporting this work were provided by the NASA High-End Computing (HEC) Program through the NASA Advanced Supercomputing (NAS) Division at Ames Research Center and the NASA Center for Computational Sciences (NCCS) at Goddard Space Flight Center. The research was supported by NASA grants NNX11AF33G and NSF grant AST-1211318.

References

Alencar, S. H. P. & Basri, G. 2000, *AJ*, 119, 1881

Alencar, S. H. P., Bouvier, J., Walter, F. M., Dougados, C., Donati, J.-F., Kurosawa, R., Romanova, M., Bonfils, X., Lima, G. H. R. A., Massaro, S., Ibrahimov, M., & Poretti, E. 2012, *A&A*, 541, A116

Alencar, S. H. P., Teixeira, P. S., Guimarães, M. M., McGinnis, P. T., Gameiro, J. F., Bouvier, J., Aigrain, S., Flaccomio, E., & Favata, F. 2010, *A&A*, 519, A88

Ardila, D. R., Basri, G., Walter, F. M., Valenti, J. A., & Johns-Krull, C. M. 2002, *ApJ*, 567, 1013

Arons, J. & Lea, S. M. 1976, *ApJ*, 207, 914

Calvet, N. & Gullbring, E. 1998, *ApJ*, 509, 802

Camenzind, M. 1990, in Reviews in Modern Astronomy, Vol. 3, Reviews in Modern Astronomy, ed. G. Klare, 234

Donati, J.-F., Bouvier, J., Walter, F. M., Gregory, S. G., Skelly, M. B., Hussain, G. A. J., Flaccomio, E., Argiroffi, C., Grankin, K. N., Jardine, M. M., Ménard, F., Dougados, C., & Romanova, M. M. 2011, *MNRAS*, 417, 472

Donati, J.-F., Jardine, M. M., Gregory, S. G., Petit, P., Bouvier, J., Dougados, C., Ménard, F., Collier Cameron, A., Harries, T. J., Jeffers, S. V., & Paletou, F. 2007, *MNRAS*, 380, 1297

Edwards, S., Fischer, W., Hillenbrand, L., & Kwan, J. 2006, *ApJ*, 646, 319

Edwards, S., Fischer, W., Kwan, J., Hillenbrand, L., & Dupree, A. K. 2003, *ApJ*, 599, L41

Ferreira, J., Dougados, C., & Cabrit, S. 2006, *A&A*, 453, 785

Ghosh, P., Pethick, C. J., & Lamb, F. K. 1977, *ApJ*, 217, 578

Gregory, S. G., Matt, S. P., Donati, J.-F., & Jardine, M. 2008, *MNRAS*, 389, 1839

Harries, T. J. 2000, *MNRAS*, 315, 722

Hartmann, L., Hewett, R., & Calvet, N. 1994, *ApJ*, 426, 669

Herbst, W., Herbst, D. K., Grossman, E. J., & Weinstein, D. 1994, *AJ*, 108, 1906

Hussain, G. A. J., Collier Cameron, A., Jardine, M. M., Dunstone, N., Ramirez Velez, J., Stempels, H. C., Donati, J.-F., Semel, M., Aulanier, G., Harries, T., Bouvier, J., Dougados, C., Ferreira, J., Carter, B. D., & Lawson, W. A. 2009, *MNRAS*, 398, 189

Jardine, M. M., Gregory, S. G., & Donati, J.-F. 2008, *MNRAS*, 386, 688

Johns, C. M. & Basri, G. 1995, *ApJ*, 449, 341

Koenigl, A. 1991, *ApJ*, 370, L39

Königl, A., Romanova, M. M., & Lovelace, R. V. E. 2011, *MNRAS*, 416, 757

Kulkarni, A. K. & Romanova, M. M. 2008, *MNRAS*, 386, 673

Kulkarni, A. K. & Romanova, M. M. 2009, *MNRAS*, 398, 701

Kurosawa, R., Harries, T. J., & Symington, N. H. 2005, *MNRAS*, 358, 671

Kurosawa, R., Harries, T. J., & Symington, N. H. 2006, *MNRAS*, 370, 580

Kurosawa, R. & Romanova, M. M. 2012, *MNRAS*, 426, 2901

Kurosawa, R. & Romanova, M. M. 2013, *MNRAS*, 431, 2673

Kurosawa, R., Romanova, M. M., & Harries, T. J. 2008, *MNRAS*, 385, 1931

Kurosawa, R., Romanova, M. M., & Harries, T. J. 2011, *MNRAS*, 416, 2623

Kwan, J., Edwards, S., & Fischer, W. 2007, *ApJ*, 657, 897

Li, L.-X. & Narayan, R. 2004, *ApJ*, 601, 414

Lii, P., Romanova, M., & Lovelace, R. 2012, *MNRAS*, 420, 2020

Long, M., Romanova, M. M., Kulkarni, A. K., & Donati, J.-F. 2011, *MNRAS*, 413, 1061

Muzerolle, J., Calvet, N., & Hartmann, L. 2001, *ApJ*, 550, 944

Petrov, P. P., Gullbring, E., Ilyin, I., Gahm, G. F., Tuominen, I., Hackman, T., & Loden, K. 1996, *A&A*, 314, 821

Reipurth, B., Pedrosa, A., & Lago, M. T. V. T. 1996, *A&AS*, 120, 229

Romanova, M. M., Kulkarni, A. K., & Lovelace, R. V. E. 2008, *ApJ*, 673, L171

Romanova, M. M., Long, M., Lamb, F. K., Kulkarni, A. K., & Donati, J.-F. 2011, *MNRAS*, 411, 915

Romanova, M. M., Ustyugova, G. V., Koldoba, A. V., & Lovelace, R. V. E. 2004, *ApJ*, 610, 920

Romanova, M. M., Ustyugova, G. V., Koldoba, A. V., & Lovelace, R. V. E. 2009, *MNRAS*, 399, 1802

Romanova, M. M., Ustyugova, G. V., Koldoba, A. V., Wick, J. V., & Lovelace, R. V. E. 2003, *ApJ*, 595, 1009

Rucinski, S. M., Matthews, J. M., Kuschnig, R., Pojmański, G., Rowe, J., Guenther, D. B., Moffat, A. F. J., Sasselov, D., Walker, G. A. H., & Weiss, W. W. 2008, *MNRAS*, 391, 1913

Shu, F., Najita, J., Ostriker, E., Wilkin, F., Ruden, S., & Lizano, S. 1994, *ApJ*, 429, 781
Spruit, H. C. & Taam, R. E. 1993, *ApJ*, 402, 593
Symington, N. H., Harries, T. J., & Kurosawa, R. 2005, *MNRAS*, 356, 1489
Takami, M., Chrysostomou, A., Bailey, J., Gledhill, T. M., Tamura, M., & Terada, H. 2002, *ApJ*, 568, L53

Magnetic Fields throughout Stellar Evolution
Proceedings IAU Symposium No. 302, 2013
P. Petit, M. Jardine & H. Spruit, eds.

doi:10.1017/S1743921314001732

Building a numerical relativistic non-ideal magnetohydrodynamics code for astrophysical applications

S. Miranda Aranguren, M. A. Aloy and Carmen. Aloy

Departament d'Astronomia i Astrofisica Universitat de Valencia,
E-46100 Burjassot (Valencia) Spains
email: sergio.miranda@uv.es

Abstract. Including resistive effects in relativistic magnetized plasmas is a challenging task, that a number of authors have recently tackled employing different methods. From the numerical point of view, the difficulty in including non-ideal terms arises from the fact that, in the limit of very high plasma conductivity (i.e., close to the ideal MHD limit), the system of governing equations becomes stiff, and the standard explicit integrating methods produce instabilities that destroy the numerical solution. To deal with such a difficulty, we have extended the relativistic MHD code MR-GENESIS, to include a number of Implicit Explicit Runge-Kutta (IMEX-RK) numerical methods. To validate the implementation of the IMEX-RK schemes, two standard tests are presented in one and two spatial dimensions, covering different conductivity regimes.

Keywords. MHD, Relativity, numerical methods.

1. Introduction

The equations of Relativistic Magnetohydrodynamics form a system of balance laws where stiffness can arise because of the unbound increase of the electric conductivity, σ, in the ideal limit, which makes numerically ill-defined the electric density current, $\mathbb{J} = \sigma W[\mathbb{E} + \mathbb{V} \times \mathbb{B} - (\mathbb{E} \cdot \mathbb{V})\mathbb{V}] + q\mathbb{V}$ (where W, $\mathbb{V}$, q $\mathbb{E}$, and $\mathbb{B}$ are the Lorentz factor, the velocity, the electric charge, the electric and the magnetic field, respectively). Inspired by the work of Dedner *et al.* (2002), we consider an augmented system of equations (see below), where two scalar potentials ψ and ϕ enforce the conservation of q and of the solenoidal constraint, $\nabla \cdot \mathbb{B} = 0$, respectively, to the extent determined by the order of the numerical method employed. The inclusion of these extra potentials adds extra flux terms in the equations of $\mathbb{E}$ and $\mathbb{B}$, as well as additional source terms in the energy density (τ) equation, though the energy flux $\mathbb{F}_\tau$ remains unchanged. The rest of the equations for the rest-mass density D, and for the momentum density $\mathbb{S}$, and their respective fluxes $\mathbb{F}_D$, and $\mathsf{F}_\mathbb{S}$ are the same as in the standard equations.

To deal with the stiffness of the system, a number of alternatives have been proposed (Komissarov 2007, Palenzuela *et al.* 2009). We have included several Implicit-Explicit Runge Kutta methods (IMEX-RK; Higueras *et al.* 2012, Pareschi *et al.* 2005) in the MRGENESIS code (Aloy *et al.* 1999) and present several test problems.

$$
\begin{aligned}
\partial_t \psi &= -\nabla \cdot \mathbb{E} + q - \kappa\psi & \partial_t q &= -\nabla \cdot \mathbb{J} \\
\partial_t \phi &= -\nabla \cdot \mathbb{B} - \kappa\phi & \partial_t D &= -\nabla \cdot \mathbb{F}_D \\
\partial_t \mathbb{E} &= \nabla \times \mathbb{B} - \nabla\psi - \mathbb{J} & \partial_t \tau &= -\nabla \cdot \mathbb{F}_\tau - \mathbb{B} \cdot \nabla\phi - \mathbb{E} \cdot \nabla\psi \\
\partial_t \mathbb{B} &= -\nabla \times \mathbb{E} - \nabla\phi & \partial_t \mathbb{S} &= -\nabla \cdot \mathsf{F}_\mathbb{S} - (\nabla \cdot \mathbb{B})\mathbb{B}
\end{aligned}
$$

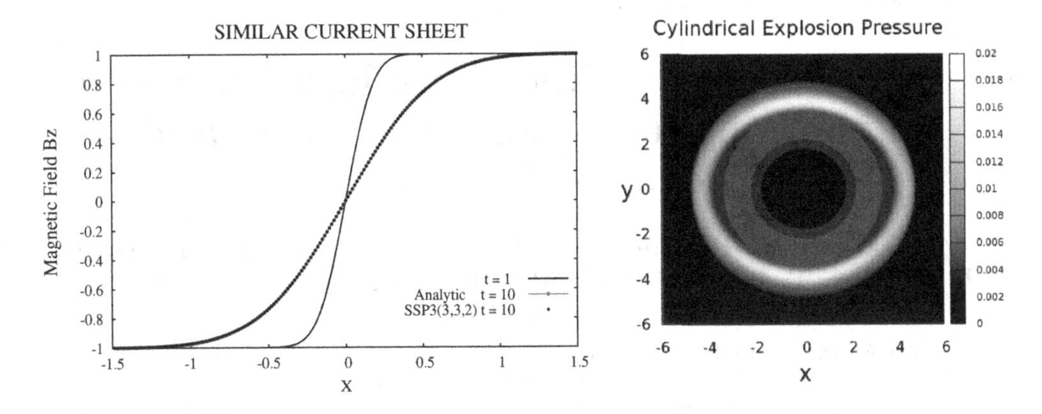

Figure 1. Left: Self-similar current sheet at $t = 1$ (initial state; solid line), analytic solution at $t = 10$ (open circles) and numerical solution computed with 800 points at $t = 10$ (filled circles). Right: Thermal pressure for the cylindrical explosion test computed with a uniform grid of 400×400 cells, and a CFL of 0.2, and using the DP2A(242) scheme.

2. Test Problems

Due to space restrictions, we only show a couple of standard tests also considered by other authors (Komissarov 2007, Palenzuela *et al.* 2009, Bucciantini & Del Zanna 2013). These tests encompass from one-dimensional, smooth flows driven by resistive effects (Sect. 2.1) to two-dimensional tests in the ideal limit (Sect. 2.2).

2.1. *Resistive Self-similar Current Sheet*

The initial conditions of this test are set at $t = 1$, when we take $\sigma = 100$, $\rho = 1$, $p = 50$, $\mathbb{E} = \mathbb{V} = 0$ and $\mathbb{B} = (0, 0, B^z(x, 1))$, such that $B^z(x, t) = B_0 \,\mathtt{erf}(\frac{1}{2}\sqrt{\frac{\sigma x^2}{t}})$, where $\mathtt{erf}$ is the error function. In Fig. 1 (left) we show that at $t = 10$, the numerical solution found with the SSP3(3,3,2) IMEX-RK method matches satisfactory the analytical one.

2.2. *Cylindrical Explosion*

The set up consists of a plane with dimensions $(x, y) \in [-6, 6]$ having a central circle with radius $r = 0.8$, where the pressure ($p = 1$) and the density $\rho = 0.01$ are higher than elsewhere ($p = \rho = 0.001$; $r > 1$). The central region is continuously connected with the surroundings using an exponentially decreasing pressure and density in the region $0.8 \leqslant r \leqslant 1$. The magnetic field, $\mathbb{B} = (0.05, 0, 0)$, is uniform, and $\mathbb{V} = \vec{0}$. This test is used to validate the new resistive code in the ideal limit (a uniform conductivity $\sigma = 10^6$ is set everywhere). Our results (Fig. 1 right) are comparable with the those obtained with our relativistic ideal MHD code (Leismann *et al.* 2005, Anton *et al.* 2010).

References

Aloy, M. A., *et al.* 1999, *ApJS*, 122,151
Anton, L., *et al.* 2010, *ApJS*, 188, 1
Bucciantini, N., Del Zanna L. 2013, *MNRAS*, 428, 71
Dedner, A., *et al.* 2002, *J. Comput. Phys.*, 175, 645
Higueras, I., *et al.* 2012, *ASC Report No. 14/2012*
Komissarov, S. S. 2007, *MNRAS*, 382, 995
Leismann, T. *et al.* 2005, *Astron. Astroph.*, 436 503
Palenzuela, C., *et al.* 2009, *MNRAS*, 394, 1727
Pareschi, L., *et al.* 2005, *J. Sci. Comput.*, 25, 112

Magnetic Fields throughout Stellar Evolution
Proceedings IAU Symposium No. 302, 2013
P. Petit, M. Jardine & H. Spruit, eds.

doi:10.1017/S1743921314001744

3D YSO accretion shock simulations: a study of the magnetic, chromospheric and stochastic flow effects

T. Matsakos[1,2,3], J.-P. Chièze[2], C. Stehlé[3], M. González[4], L. Ibgui[3], L. de Sá[2,3], T. Lanz[5], S. Orlando[6], R. Bonito[7,6], C. Argiroffi[7,6], F. Reale[7,6] and G. Peres[7,6]

[1]CEA, IRAMIS, Service Photons, Atomes et Molécules, 91191 Gif-sur-Yvette, France
email: titos.matsakos@cea.fr

[2]Laboratoire AIM, CEA/DSM - CNRS - Université Paris Diderot, IRFU/Service d'Astrophysique, CEA Saclay, Orme des Merisiers, 91191 Gif-sur-Yvette, France

[3]LERMA, Observatoire de Paris, Université Pierre et Marie Curie and CNRS, 5 Place J. Janssen, 92195 Meudon, France

[4]Université Paris Diderot, Sorbonne Paris Cité, AIM, UMR 7158, CEA, CNRS, 91191 Gif-sur-Yvette, France

[5]Laboratoire Lagrange, Université de Nice-Sophia Antipolis, CNRS, Observatoire de la Côte d'Azur, 06304 Nice cedex 4, France

[6]INAF - Osservatorio Astronomico di Palermo, Piazza del Parlamento 1, 90134 Palermo, Italy

[7]Dipartimento di Fisica e Chimica, Università degli Studi di Palermo, Piazza del Parlamento 1, 90134 Palermo, Italy

Abstract. The structure and dynamics of young stellar object (YSO) accretion shocks depend strongly on the local magnetic field strength and configuration, as well as on the radiative transfer effects responsible for the energy losses. We present the first 3D YSO shock simulations of the interior of the stream, assuming a uniform background magnetic field, a clumpy infalling gas, and an acoustic energy flux flowing at the base of the chromosphere. We study the dynamical evolution and the post-shock structure as a function of the plasma-beta (thermal pressure over magnetic pressure). We find that a strong magnetic field (~hundreds of Gauss) leads to the formation of fibrils in the shocked gas due to the plasma confinement within flux tubes. The corresponding emission is smooth and fully distinguishable from the case of a weak magnetic field (~tenths of Gauss) where the hot slab demonstrates chaotic motion and oscillates periodically.

Keywords. accretion, magnetohydrodynamics (MHD), radiative transfer, shock waves, instabilities

1. Introduction

The accretion process in young stars occurs by plasma streams that originate from the inner part of the surrounding disk and flow along the field lines of the magnetosphere before hitting the stellar surface. The plasma impacts onto the chromosphere with free-fall velocities resulting in the formation of strong shocks. The temperature that develops in the post-shock region is on the order of a few million Kelvin, which leads to the emission of soft X-ray photons. Recent observations of several protostars seem to have detected such radiation (e.g. Argiroffi *et al.* 2007).

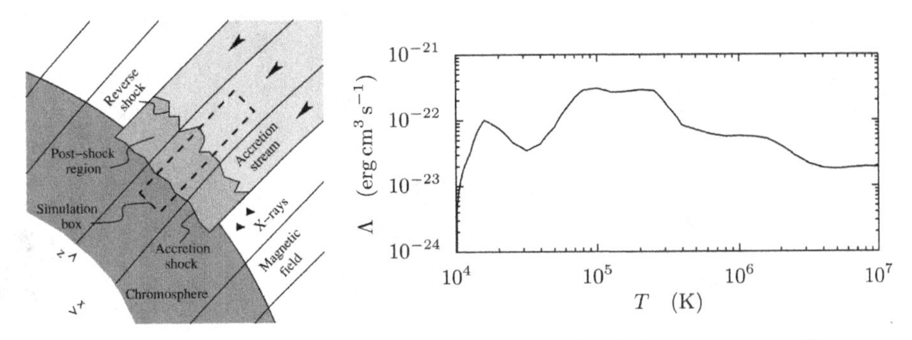

Figure 1. Left panel: a simplified sketch of an accretion shock for a strong, uniform magnetic field. The computational domain is indicated with dashed lines. Right panel: the prescribed cooling function Orlando *et al.* 2010, Matsakos *et al.* 2013.

Numerical studies have explored the dynamics of accretion shocks in 1D and 2D (e.g. Sacco *et al.* 2008, 2010; Koldoba *et al.* 2008; Orlando *et al.* 2010, 2013; Matsakos *et al.* 2013). To highlight the evolution of the system, consider the simplified configuration depicted in the left panel of Fig. 1 in 1D, namely focus on the interior of the stream and assume a uniform magnetic field. After the impact, a reverse shock travels along the stream as material accumulates building a hot slab. The optically thin radiation losses that are often adopted to describe the post-shock region, see right panel of Fig. 1, trigger cooling instabilities. Specifically, as the plasma cools, the infalling gas can no longer be supported due to the large pressure drop. As a result, the reverse shock collapses and a new hot slab starts to build up again (Sacco *et al.* 2008; Koldoba *et al.* 2008). This formation-collapse cycle of the post-shock region suggests a quasi-periodic behavior for the observable X-ray emission, with a frequency of 10^{-3}-10^{-1} Hz. However, observations do not yet find evidence of periodicity (Drake *et al.* 2009; Günther *et al.* 2010).

The absence of periodic patterns could be due to a variety of physical mechanisms. For instance, radiative transfer effects could complicate the dynamics (de Sá et al. in preparation) or non-uniform magnetic fields could break the 1D symmetry and lead to a complex evolution (Orlando *et al.* 2010, 2013). Here, we show that even in the simple case of a uniform field and optically thin cooling, the global post-shock oscillations can effectively be suppressed.

We perform 3D magneto-hydrodynamical (MHD) simulations of a vertical element of the interior of the accretion stream (left panel of Fig. 1) using the PLUTO code (Mignone *et al.* 2007). We introduce realistic perturbations in the system, i.e. clumps in the stream and a chromospheric acoustic flux, and we explore three different models by changing the value of the magnetic field. The computational domain in the horizontal directions spans x, $y = [0.0, 0.5 \times 10^{-3}]\, R_\odot$ and in the vertical $z = [0.0, 7.7 \times 10^{-2}]\, R_\odot$. We use a static grid and a resolution of $32 \times 32 \times 256$ cells.

On the top boundary, as well as the upper part of the domain initially, we prescribe a uniform flow of number density $n_{\rm acc} = 2 \cdot 10^{11}$ cm^{-3} and velocity $v_{\rm acc} = 500\,{\rm km\,s^{-1}}$. Moreover, we add clumps of random size and location on top of the accretion stream. Their density follows the gaussian distribution with the peak value set at $n_{\rm clm} = 10^{12}$ cm^{-3} and standard deviation equal to 10%-30% of the box width ($\sim 10^2$-10^3 km). On the bottom boundary as well as the lower part of the initial box, we specify an isothermal chromosphere with $T_{\rm chr} = 10\,000$ K that is in equilibrium with the imposed constant gravity. The protostar is assumed of solar mass and radius. For more information on the setup, see its 2D version in Matsakos *et al.* (2013).

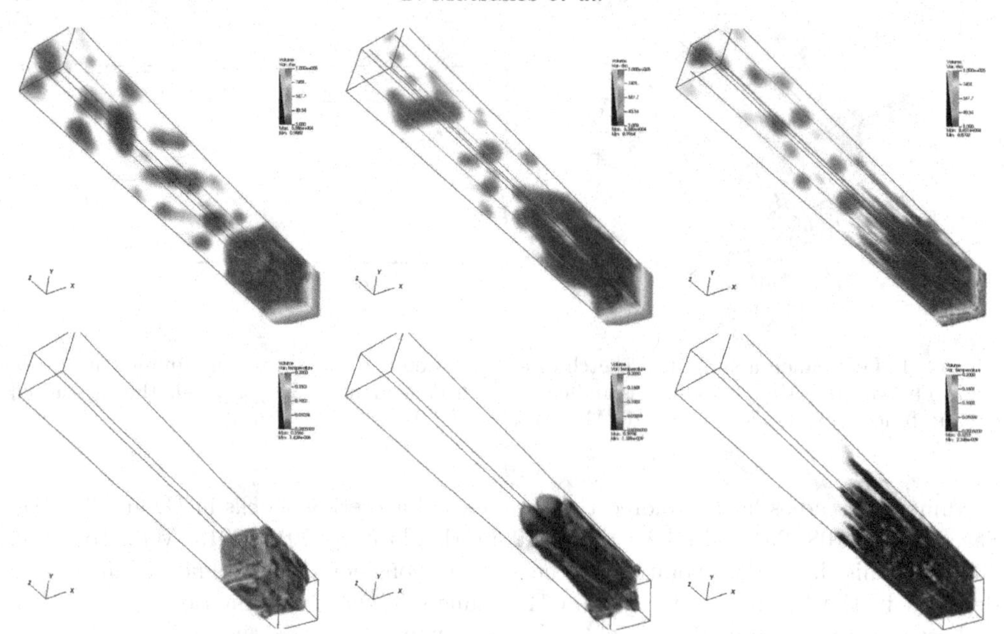

Figure 2. Volume plots of the density (top panels) and the temperature (bottom panels, only the million Kelvin regions are shown). From left to right, the magnetic field is assumed 30 G, 100 G, and 300 G. In each plot we can identify three regions: the clumpy stream, the hot post-shock region, and the chromosphere.

2. Results

In the case of a weak magnetic field (left panels of Fig. 2), the thermal pressure in the post-shock region dominates the magnetic (post-shock plasma-β larger than unity). Therefore, the inhomogeneities that appear in the hot slab, due to the clumps and the chromospheric variability, are quickly smoothed out. This leads to plasma mixing and a post-shock that shows chaotic motion with local velocities less but close to the infalling gas speed. Nevertheless, despite this turbulent structure, the system is found to retain globally the periodic formation-collapse pattern of the reverse shock.

When the post-shock has a plasma-β close to unity, the vertical magnetic flux tubes cannot be easily deformed and elongated structures form in the post-shock (middle panels of Fig. 2). This effect can be seen even more clearly in the case of a strong magnetic field (right panels of Fig. 2). Here, the plasma-β of the hot slab is smaller than unity and hence the plasma is well confined within the flux tubes. In turn, the post-shock acquires a rich structure that consists of a collection of fibrils, each one of which follows the 1D quasi-periodic behavior. In the absence of perturbations, the fibrils oscillate in phase and the emission of the accretion shock would show a global periodic signature. However, the perturbations introduced in the system bring quickly the almost independent oscillators out of phase. Consequently, the overall emission is expected to be smooth, hiding the individual oscillations.

In order to visualize the structure of the whole post-shock region, Fig. 3 displays a reconstruction of a 2D vertical cut of an accretion stream of radius $5 \cdot 10^4$ km. This is achieved by putting different simulation moments of the computational box side by side, since the fibrils are independent from each other. We have chosen to show only the model with a strong magnetic field because young stars have surface values on the order of ~kG (e.g. Johns-Krull 2007). Evidently, even in the case of a uniform magnetic field,

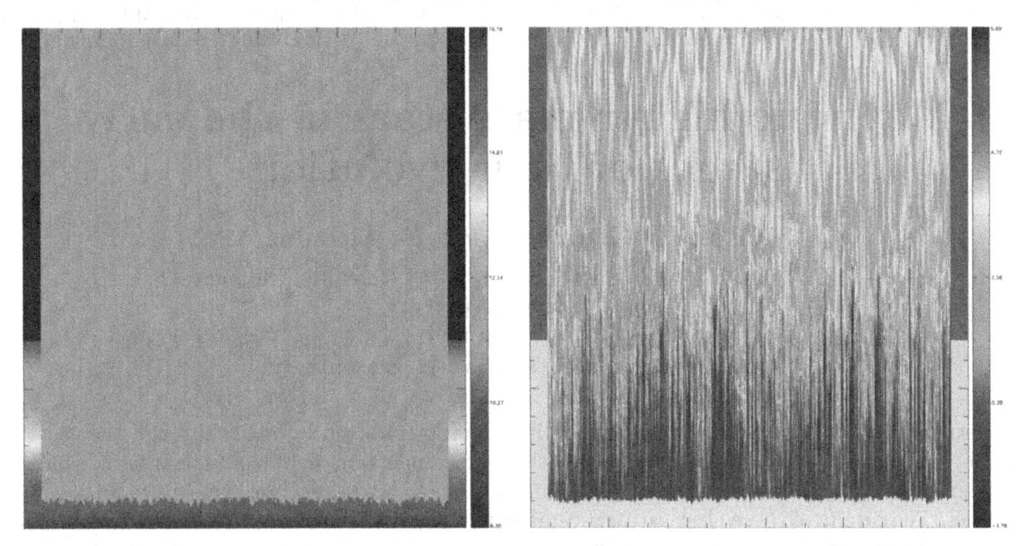

Figure 3. Density (left) and temperature (right) of the whole reconstructed accretion stream.

the presence of perturbations can create very rich structures in the post-shock region. Future work will focus on generating synthetic emission from such simulations in order to compare with observations.

Acknowledgements

This work was supported by the ANR STARSHOCK project (ANR-08-BLAN-0263-072009/2013) and was granted access to the HPC resources of CINES under the allocation 2012-c2012046943 and 2013-c2013046943 made by GENCI (Grand Equipement National de Calcul Intensif).

References

Argiroffi, C., Maggio, A., & Peres, G. 2005, *A&A*, , 465, L5

Drake, J. J., Ratzlaff, P. W., Laming, J. M., & Raymond, J., *ApJ*, , 703, 1224

Günther, H. M., Lewandowska, N., Hundertmark, M. P. G., *et al.* 2010, *A&A*, , 518, 54

Koldoba, A. V., Ustyugova, G. V., Romanova, M. M., & Lovelace, R. V. E. 2008, *MNRAS*, , 388, 357

Johns-Krull, C. M. 2007, *ApJ*, , 664, 975

Matsakos, T., Chièze, J. P., Stehlé, C., *et al.* 2013, *A&A*, , 557, 69

Orlando, S., Bonito, R., Argiroffi, C., *et al.* 2013, *A&A*, , in press

Mignone, A., Bodo, G., Massaglia, S., *et al.* 2007, *ApJS*, , 170

Orlando, S., Sacco, G. G., Argiroffi, C., *et al.* 2010, *A&A*, , 510, 71

de Sá, L. in preparation

Sacco, G. G., Argiroffi, C., Orlando, S., *et al.* 2008, *A&A*, , 491, L17

Sacco, G. G., Orlando, S., Argiroffi, C., *et al.* 2010, *A&A*, , 522, 55

Magnetic Fields throughout Stellar Evolution
Proceedings IAU Symposium No. 302, 2013
P. Petit, M. Jardine & H. Spruit, eds.

doi:10.1017/S1743921314001756

Magnetic higher-mass stars in the early stages of their evolution

Jason H. Grunhut[1] **and E. Alecian**[2]

[1]ESO, Karl-Schwarzschild-Str. 2, D-85748 Garching, Germany
email: jgrunhut@eso.org

[2]UJF-Grenoble 1 CNRS-INSU, IPAG, UMR 5274, Grenoble, F-38041, France
email: evelyne.alecian@obs.ujf-grenoble.fr

Abstract. Over the past decade, significant investigations have been made through the use of high-resolution spectropolarimetry to probe the surface magnetic field characteristics of young higher-mass ($M \gtrsim 1.5\,M_\odot$) stars from pre-main sequence to zero-age main sequence evolutionary phases. The results of these observational campaigns suggest that these young higher-mass stars host similar magnetic properties to their main sequence descendants - strong, stable, globally-ordered fields that are detected in approximately 10 percent of all stars. This strongly contrasts with lower-mass stars, where it is generally accepted that a solar-like dynamo is in operation that generates more complex, globally-weak fields that are ubiquitous. The consensus is magnetic fields in higher-mass stars are fossil remnants of a magnetic field present in the molecular cloud, or generated very early during stellar formation. This review discusses the spectropolarimetric observations of higher-mass stars and how these observations have guided our current understanding of the magnetic characteristics of young higher-mass stars.

Keywords. stars: magnetic fields, stars: ealry-type, stars: pre–main-sequence, stars: statistics, techniques: polarimetric

1. Introduction

1.1. *Magnetism in low-mass, main sequence stars*

Many articles can be found in these proceedings alone that discuss the magnetic properties of cool, low-mass stars (e.g. Gregory *et al.*), but a brief summary is provided here nonetheless. Magnetism in cool, low-mass stars is ubiquitous. Besides indirect proxies of magnetism (e.g. emission in Balmer lines, Ca H&K lines, or UV, X-ray or radio emission; e.g. Donati & Landstreet 2009), magnetometry of the Sun and other solar analogues reveal these stars to host topologically complex fields. It is also well-established that these same observations indicated the fields to show intrinsic variability on timescales of days, weeks and even years, some of which may be cyclical (analogous to the solar cycle; e.g. Berdyugina 2005).

It is generally well-accepted now that the observed fields in these stars are a result of contemporaneous dynamo processes that convert mechanical energy driven by convection into magnetic energy (Parker (1955)). While the details of dynamo mechanisms are not fully understood, its basic principles are well established over a large range of masses from planets to stars (e.g. Charbonneau 2005; Christensen *et al.* 2009). As a consequence of dynamo processes, a clear correlation between the rotation period of the star and the strength of the magnetic activity is observed (e.g. Donati & Landstreet 2009). Furthermore, as the size of the convective envelope increases, we see significant changes in the characteristics of magnetic field properties that vary from being topologically complex with globally weak magnetic fields (as observed in solar analogues) to being more organized with globally stronger fields (as observed in the fully convective M-dwarfs (e.g. Morin *et al.* 2010)). Therefore, in addition to a clear correlation between rotation and magnetic activity, there is also a clear correlation between stellar mass/internal structure and the magnetic properties (e.g. Donati & Landstreet 2009).

1.2. *Magnetism in intermediate-mass main sequence stars*

In contrast to cool, low-mass stars, more massive stars ($M \gtrsim 1.5\,M_\odot$) have a significantly different internal structure. Instead of an outer convective envelope (as found in stars with $M \lesssim 1.5\,M_\odot$), higher-mass stars host an outer radiative envelope. Without the necessary convective motion in their outer envelopes, higher-mass stars are therefore not expected to drive a dynamo. Indeed, classical observational tracers of dynamo activity fade and disappear among stars of spectral type F, at roughly the conditions predicting the disappearance of energetically-important envelope convection. Despite this fact, magnetic fields have been detected in a small population of (over 400) A- and B-type stars (e.g. Bychkov *et al.* 2009) dating back over 60 years to the first discovery by Babcock (1947). These stars are easily identifiable as they present distinctive photospheric chemical peculiarities compared to non-magnetic A- and B-type stars, presumably a result of the magnetic field (e.g. Folsom *et al.* 2007).

Compared to the magnetic fields found in low-mass stars, the incidence and characteristics of magnetic fields found in intermediate-mass stars ($1.5 \lesssim M \lesssim 8\,M_\odot$) are significantly different. While magnetism in low-mass stars is essentially ubiquitous, magnetic fields are only detected in 5-10% of the population of intermediate-mass stars (e.g. Bagnulo *et al.* 2006). Observations suggest that the majority of intermediate-mass stars present globally-ordered, topologically simple magnetic fields, often characterized by a dipole or a low-order multipole (e.g. Aurière *et al.* 2007), although a few stars show evidence for additional small-scale structure (e.g. Kochukhov *et al.* 2004; Kochukhov & Wade 2010). Furthermore, the magnetic axis is typically found to be oblique to the rotation axis.

In strong contrast to the topologically variable fields found in low-mass stars, observations of intermediate-mass stars suggest the fields to be stable over timescales of at least decades (e.g. Silvester *et al.* 2012). The interpretation is that the magnetic field is *frozen* into the star and any observed variations result from rotational modulation. Additionally, unlike low-mass stars, there appears to be no strong correlation between the fundamental physical and magnetic properties (e.g. Donati & Landstreet 2009). Curiously, there appears to be a puzzling magnetic dichotomy among intermediate-mass stars (Aurière *et al.* 2007) - these stars are either found to be highly magnetic (polar field strengths $> 300\,G$) or non-magnetic. However, this picture has changed slightly over the last few years with the discovery of very weak magnetic fields in a few bright stars (longitudinal field strengths $< 1\,G$; e.g. Lignières *et al.* 2009; Petit *et al.* 2011; see also Lignières *et al.* in these proceedings).

The surface polar field strengths of the highly-magnetic stars are generally strong, ranging from several hundred to tens of thousands of gauss (e.g. Donati & Landstreet 2009). The strongest fields among the intermediate-mass stars detected to date are in the range of $\sim 30 - 35\,kG$ (e.g. HD 215411 (Babcock 1960); HD 75049 (Freyhammer *et al.* 2008; Elkin *et al.* 2010)). While the overwhelming majority of magnetic intermediate-mass stars host dipolar magnetic fields, there are a few exceptions that host more complex fields (e.g. HD 32633 (Leone *et al.* 2000), HD 133880 (Landstreet 1990; Bailey *et al.* 2012), HD 137509 (Kochukhov 2006)).

The magnetic properties of intermediate-mass stars suggest a fundamentally different origin for the field - the observed fields cannot be generated by a contemporaneous dynamo. Despite our greatly improved understanding of the characteristics and statistical properties of magnetism among intermediate-mass main sequence stars, the fundamental question regarding the origin of the magnetic field is still debated.

2. The origin of magnetism in higher-mass stars

The reader is also encouraged to read the discussion of the origin of magnetism in higher-mass stars from Braithwaite *et al.* (these proceedings).

2.1. *Dynamos*

2.1.1. *Core-dynamo*

As already discussed, it is well established that dynamo processes are responsible for contemporaneously driving the magnetic fields observed in low-mass stars, so it is not surprising to imagine similar processes being responsible for the fields observed in higher-mass stars. Since higher-mass stars lack a significant convective envelope, it has been proposed that dynamo fields could instead be generated deep in the stellar cores (Charbonneau & MacGregor 2001; MacGregor & Cassinelli 2003; Brun *et al.* 2005). While it has been shown that a dynamo generated field is capable of being generated in the core, it is unlikely that those fields can be transported to the stellar surface where they can be observed and detected e.g. Walder *et al.* 2012. The two suggested mechanisms for transporting magnetic flux from the core to the surface are through buoyancy/diffusion and/or meridional circulations. However, MacGregor & Cassinelli (2003) point out that the timescales for buoyancy/diffusion are longer than the typical main sequence lifetime of intermediate-mass stars and therefore should not be observable at the surface. Furthermore, strong meridional currents are necessary to transport magnetic flux to the surface (Charbonneau & MacGregor 2001, but MacGregor & Cassinelli (2003) show that these required strong currents inhibit the dynamo process. Therefore, core-dynamo generated fields are unexpected to account for the observed magnetic fields.

2.1.2. *Sub-surface convection*

In recent years another theory has emerged suggesting that a thin, sub-surface iron convection zone may be responsible for generating magnetic fields (Cantiello *et al.* 2009; Cantiello & Braithwaite 2011). The sub-surface convection theory shares many parallels with the dynamos acting in solar-like stars. Because of this, one expectation is that there should be a strong correlation between stellar mass (which correlates with the size of the iron convection zone) and the field strength. Unfortunately, the known intermediate-mass magnetic stars don't span a large enough range of mass to test this hypothesis. However, as with other dynamo generated fields, the produced fields should be intrinsically variable, which would highly suggest that sub-surface convective dynamos cannot be responsible for the stable, large-scale fields observed in intermediate-mass stars.

2.2. *Fossil origin*

The leading hypothesis suggests that the observed strong and stable magnetic fields are *fossil* fields - the frozen remnants of the Galactic field accumulated and possibly enhanced by a dynamo field generated in an earlier phase of evolution (e.g. Cowling 1945; Mestel 2001; Moss 2001). Numerical and analytical studies show that the surviving fields can relax into configurations exhibiting long-term stability (Braithwaite 2009; Duez & Mathis 2010) and can approximately explain the field topology and other general characteristics of magnetism in intermediate-mass stars. The presence of instabilities (e.g. Tayler 1973; Spruit 1999 during the relaxation process can potentially account for the observed magnetic dichotomy, since only strong fields are expected to survive (Aurière *et al.* 2007). This may potentially explain why all higher-mass stars are not magnetic, but it does not naturally explain why only about 10% host strong magnetic fields.

2.3. *Stellar mergers*

A competing hypothesis that has been gaining significant traction in recent years suggests that the strongly magnetic higher-mass stars are the result of either a stellar merger or mass-transfer event (e.g. Tutukov & Fedorova 2010. Largely motivated by the dearth of magnetic higher-mass stars observed in close binary systems (Abt & Snowden 1973; Carrier *et al.* 2002), the merger hypothesis (e.g. Ferrario *et al.* 2009) suggests that the strong differential rotation or other velocity flows induced during the merger/mass-transfer episode drive a strong but short-lived dynamo field (e.g. Lacaze *et al.* 2006; Donati *et al.* 2008), which ultimately relaxes into a stable configuration. The dynamo action eventually ceases and the resulting field establishes a stable configuration as predicted by Braithwaite (2009) and Duez & Mathis (2010), similar to the fossil field hypothesis. A similar hypothesis is also proposed to account for the highly-magnetic white dwarfs (Tout *et al.* 2008). The attractiveness of this hypothesis is that it more naturally explains why only 10% of higher-mass stars (the mergers) host strong magnetic fields.

2.4. *Understanding magnetism in intermediate-mass main sequence stars*

While the currently favoured hypothesis is that the observed magnetic fields in intermediate-mass main sequence stars are of fossil origin, there are still some outstanding questions that need to be addressed. One particular question is at what stage of the PMS evolution these fields first appear. Furthermore, even though arguments against dynamo fields were already presented, further observational constraints are ideal. To this end, if dynamos are responsible for the observed fields then one would expect a correlation between mass and field properties, similar to what is observed in low-mass stars (or as predicted by Cantiello & Braithwaite (2011)). To address these and other issues we discuss the results of ongoing investigations to study the magnetic properties of a pre-main sequence stars and massive stars.

3. Magnetism in pre-main seqeunce intermediate-mass stars

3.1. *Pre-main sequence evolution*

The pre-main sequence phase of evolution is defined as the period of stellar formation after the proto-stellar phase, but before the main sequence phase where a star is undergoing core-hydrogen burning. At the beginning of the PMS phase the star is on the birthline - the locus of points in the HR diagram where the star is, for the first time, observable at optical wavelengths after shedding the majority of its proto-stellar envelope and is no longer undergoing significant accretion (Stahler 1983). At this point, nuclear-burning has not yet started in the star's core and slow gravitational collapse is the main source of energy. Just before the end of the PMS phase, nuclear reactions begin, contributing more and more energy to the luminosity of the star. As the star evolves closer to the Zero-Age-Main-Sequence (ZAMS), gravitational contraction slows, eventually coming to a stop (see Fig. 1).

Typically, the theoretical birthline computed by Palla (1990) for PMS intermediate-mass stars (the so-called Herbig Ae/Be (HAeBe) stars), using a single proto-stellar mass-accretion rate of $10^{-5}\,\mathrm{M}_\odot\,\mathrm{yr}^{-1}$, is adopted as it well-reproduces the upper envelope of observed distribution of intermediate-mass stars in an HR diagram. However, above $8\,\mathrm{M}_\odot$ this birthline intersects the ZAMS, suggesting that no pre-main sequence stars more

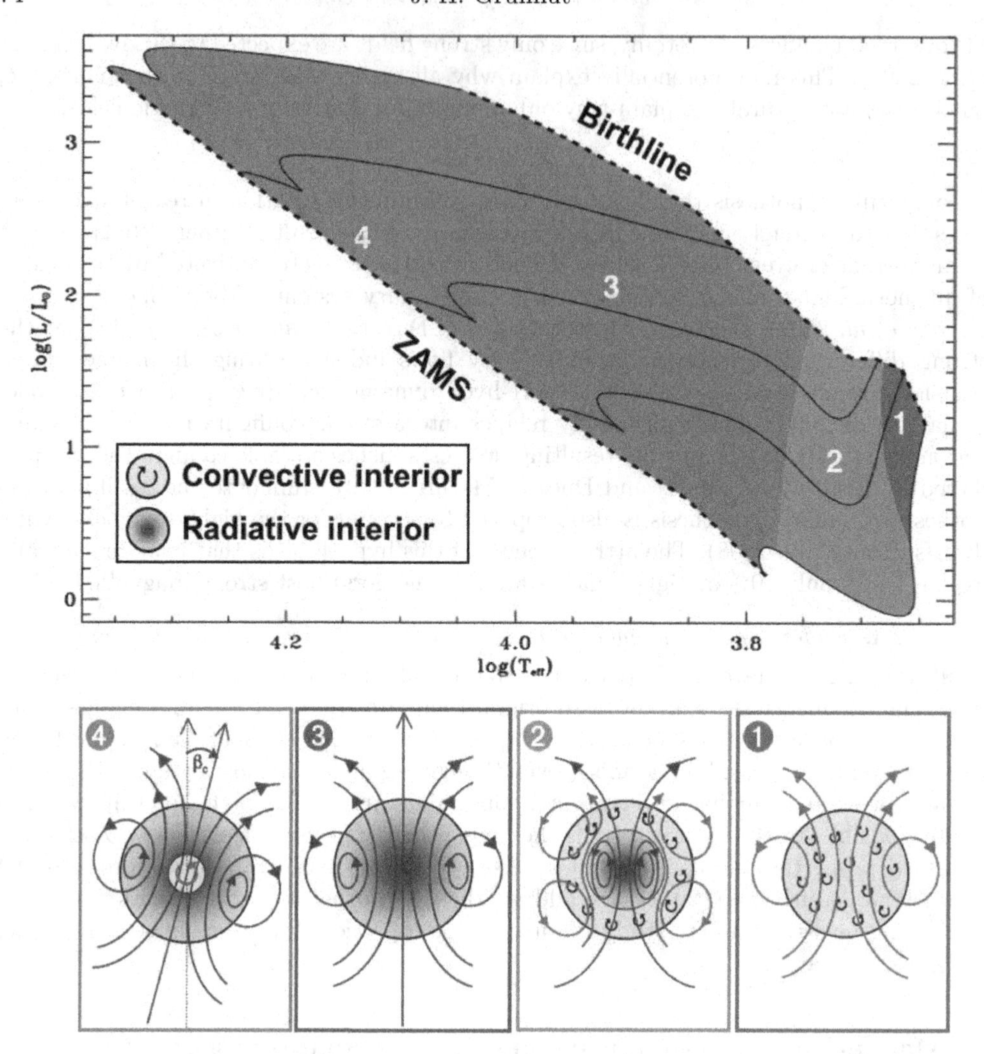

Figure 1. Top: Pre-main sequence evolutionary tracks (full lines) for 1.2, 2, 3, 5, and 8 $M_\odot$, computed with CESAM (Morel 1997). The birthline is from Behrend & Maeder (2001). The shaded area separate regions with different internal stellar structures: phase 1 (red) - fully convective interior, phase 2 (green) - radiative core + convective envelope, phase 3 (blue) - fully radiative interior, phase 4 (purple) - convective core + radiative envelope. **Bottom:** Schematic view of the possible structure of the magnetic field for the four different phases. Regions drawn in yellow with circular arrows represent convection where dynamo fields are generated (represented with red field lines). Blue regions represent radiative interiors containing large-scale fossil fields (represented with purple field lines).

massive than 8 $M_\odot$ should be visible in the optical domain. Therefore, a single mass-accretion rate for all masses might not be fully correct, as argued by Alecian *et al.* (2013), and instead an alternate birthline, such as that computed by Behrend & Maeder (2001) could be used. In computing their birthline, Behrend & Maeder (2001) assume that all stars, independent of their final mass, evolve along the birthline, increasing in mass (from the lower right to the upper left), during the proto-stellar phase during which accretion is the dominant source of energy. Since the star is still embedded in its parental cloud it accretes a large amount of material with an accretion rate that depends on the mass (luminosity) of the embedded star. Once the star has accreted most of its close, surrounding material, it leaves the birthline and evolves along the usual PMS track. In this simple view, all stars, independent of their mass, have been formed from a unique core of a certain mass, and the final mass of the star is determined from the mass of the original cloud from which the star forms.

Within this view of star formation, the energy transport within all higher-mass stars undergo significant variations. The usual PMS evolution of an intermediate-mass star begins with the star containing a fully convective interior and therefore the star initially evolves along the Hayashi track, almost vertically in the HR diagram. As the star continues to contract its core reaches a critical temperature at which point energy is more efficiently transported by radiation. From this point on the radiative core continues to grow until it reaches the surface. Eventually, nuclear reactions are ignited in the core and convection becomes the most efficient means of energy transport, right before the star reaches the ZAMS (Fig. 1). For higher mass stars, the evolution is very similar, but most of the internal structural changes occur as the star evolves along the brithline, and therefore begin their PMS life with either a partially or totally radiative interior.

It should be emphasized here that, according to this scenario of star formation, all stars, independent of their final mass, go through a fully convective phase. Therefore, the passage of all higher-mass stars through a fully convective phase needs to be seriously considered in any theory for the origin of magnetic fields in higher-mass stars, as convective motion can have a significant impact on the generation or enhancement of magnetic fields. To this end, the following section will review an ongoing project that has been undertaken over the last decade to determine the magnetic properties of PMS intermediate-mass stars during the four major phases of the PMS evolution (the fully convective phase, the core-radiative phase, the fully radiative phase, and the core-convective phase, respectively red, green, blue and purple in Fig. 1).

3.2. *Magnetism in intermediate-mass pre-main sequence stars*

A high-resolution ($R \sim 68000$) circular polarization (Stokes V) spectropolarimetric survey of 128 HAeBe stars was carried out using ESPaDOnS at the Canada-France-Hawaii Telescope and Narval and the Télescope Bernad Lyot. A sample of field HAeBe stars were selected from the catalogue of Thé *et al.* (1994) and Vieira *et al.* (2003), while an additional sample were selected from members of the following young clusters: NGC 2244 (Park & Sung 2002), NGC 2264 (Sung *et al.* 1997), and NGC 6611 (de Winter *et al.* 1997). It should be noted that the global magnetic properties of HAeBe stars were unknown before this survey. In order to increase the S/N and enhance our sensitivity to weak magnetic Zeeman signatures (used to diagnose and characterize the presence of magnetic fields), we employed the Least-Squares Deconvolution (LSD) multi-line technique of Donati *et al.* (1997) in our analysis. This survey obtained magnetic detections in 8 stars (e.g. Wade *et al.* 2005; Alecian *et al.* 2008b), some of which are fully radiative, and finds a magnetic incidence fraction of ~6%, similar to the incidence that is found for intermediate-mass main sequence stars. Spectropolarimetric monitoring of many of the detected stars was also carried out in order to characterize the strength and structure of their surface fields. Our results indicate that the magnetic properties of HAeBe star are similar to what is found on the main sequence - the fields are mainly dipolar, are strong (surface field strengths of 300 G to 4 kG), and stable over many years (e.g. Alecian *et al.* 2008a, 2009). This survey has therefore established a direct link between the magnetic properties of stars found on the PMS and on the main sequence. We can therefore conclude that the magnetic properties of A/B stars must have been shaped before the HAeBe phase of the stellar evolution (Alecian *et al.* 2013).

3.3. *Magnetism in young massive stars*

While the magnetic properties of main sequence intermediate-mass stars are well studied, there is a distinct lack of information about magnetism in main sequence massive stars ($M > 8\,M_{\odot}$). This largely reflects the fact that unlike intermediate-mass magnetic stars that show strong and distinctive photospheric chemical peculiarities compared to non-magnetic A- and B-type stars (and therefore can be used as a proxy for magnetism), the strong, radiatively-driven outflows of more massive stars generally inhibit these chemical peculiarities, making them difficult to identify. Furthermore, the relatively weak fields of these stars coupled with relatively few spectral lines from which to directly diagnose the presence of magnetism in the optical spectra of massive stars meant that these fields remained undetected by previous generations of instrumentation. Magnetic fields were detected in a small number of massive stars, the majority of these being He-peculiar B-type stars that are high-mass extensions of the chemically-peculiar intermediate-mass stars (e.g. Bohlender *et al.* 1987). Among the massive O-type stars, only 3 were known to be magnetic prior to 2009: θ^1 Ori C, HD 191612 and ζ Ori A† (Donati *et al.* 2002, 2006; Bouret *et al.* 2008).

Thanks to the new generation of spectropolarimeters and large international initiatives like the Magnetism in Massive Stars (MiMeS) project, the magnetic properties of young massive stars are now being thoroughly investigated (see Wade *et al.* (these proceedings) for more information about the MiMeS project and results). In particular, a large survey of ~550 Galactic B9-O4 stars were observed as part of the MiMeS project. A magnetic incidence fraction of $7 \pm 1\%$ is inferred from this sample of stars, which is fully consistent with the incidence fraction among intermediate-mass stars. Furthermore, an incidence fraction of $8 \pm 2\%$ is derived from sample of B-type stars, which is fully consistent with the $6 \pm 3\%$ that is found from the O-type stars.

Of particular interest to this article are the results for O-type stars, as these stars are much more massive than the intermediate-mass stars and are also very young. The MiMeS project has more than tripled the number of known magnetic O-type stars. Magnetic fields have now been firmly detected in 9 O-type stars: θ^1 Ori C, HD 191612, HD 108 (Martins *et al.* 2010), HD 57682 (Grunhut *et al.* 2009, 2012), HD 148937 (Wade *et al.* 2012a), Tr16-22 (Nazé *et al.* 2012), CPD -28 2561 (Wade *et al.* 2012c), NGC 1624-2 (Wade *et al.* 2012b), and Plaskett's star (Grunhut *et al.* 2013). Detailed investigations of these stars (e.g. Grunhut *et al.* 2012) have revealed that these stars host mainly dipolar, strong (surface polar field strengths from 300 G to ~22 kG), and stable fields, similar to the properties found for main sequence and PMS intermediate-mass stars.

4. Discussion

The results reported here establish that the statistical and magnetic properties of stars with outer radiative envelopes remain unchanged across 1.5 decades of mass and over a large range of ages from the pre-main sequence onwards. Systematic surveys and detailed observations reveal that between 5-10% of these stars host strong ($\gtrsim 300$ G), stable, globally-ordered (mainly dipolar) magnetic fields.

With this information we can now seriously address the question of the origin of magnetism in these stars. Based on the presence of magnetic fields in fully radiative stars, and the lack of correlations between the field properties (such as strength or complexity) with stellar properties, we can all but rule out dynamos as the origin of the large-scale

† New observations do not confirm the original claim by Bouret *et al.* (Bouret *et al.* (2008)) (Neiner *et al.* in prep)

fields that are observed. Unfortunately, at this point in time there are no real constraints regarding stellar mergers as the origin of the observed fields; however, further interest into this hypothesis is gathering with the discovery of a magnetic field in the secondary companion of the massive, close binary system known as Plaskett's star (Grunhut *et al.* 2013) and from the suggestion that the bipolar nebula around HD 148937 is remnant ejecta from a merging event (Langer 2012). Since the incidence fraction of magnetism amongst PMS stars is fully consistent with that of main sequence stars, strong constraints are placed on the time-frame for these events - the majority of these interactions must occur prior to the PMS phase, and are possibly the result of orbital decay (Korntreff *et al.* 2012). If the majority of these stellar interactions occur very early in the star's evolution then the observed fields (at a later stage of evolution) are essentially indistinguishable from fossil fields.

The fossil field hypothesis naively implies that all high-mass stars should display magnetic fields at their surface. This clearly disagrees with observations that find a magnetic incidence fraction of 5-10%. A natural explanation would be the existence of fundamental differences in the initial conditions of star forming regions (e.g. local density, local magnetic field strength, etc.). An efficient way to test this hypothesis is to study the magnetic properties of a large number ($\sim$150) of close binary systems, containing two stars formed at the same time and from the same environment. This is one of the aims of the recently initiated BinaMIcS project (Binarity and Magnetic Interaction in various classes of Stars; see Mathis *et al.* (these proceedings) for further details). Among other objectives, this project will acquire about 300 high-resolution spectropolarimetric spectra of 150 close binary systems, thanks to large programs obtained at CFHT (PI: Alecian/Wade) and TBL (PI: Neiner). These data will allow us to determine the incidence and to characterize the magnetic fields of both components of these binaries. Since close binaries are expected to be coeval, BinaMIcS will therefore help us to disentangle the effects of initial conditions from other effects (such as early evolution or rotation). This project provides one further step towards understanding the origin of magnetic fields and the magnetic properties of close binaries (Alecian, Wade, Mathis, Neiner *et al.*, in prep.).

In light of of this discussion our current view of the evolution of magnetic fields in PMS intermediate-mass stars is schematically described in Fig. 1. During phase 1 the star is expected to be fully convective, and just like main sequence fully-convective stars (M-dwarfs) that present globally-ordered magnetic fields at their surfaces, we propose that the PMS stars found at this stage should also generate similarly simple and strong large-scale fields. In phase 2, when the radiative core appears, we expect a solar-type dynamo to be driven by the convective envelope and therefore produce a complex surface field. At the same time, the field originally created in phase 1 relaxes in the radiative core. Once the star reaches phase 3 and becomes fully radiative, a dynamo no longer operates and the relaxed fossil field is now observable at the stellar surface. In the final stage when the convective core appears (phase 4), an interaction of the dynamo generated in the core and the relaxed fossil field in the radiative envelope could occur. To this end, Featherstone *et al.* (2009) performed simulations to study such an interaction. They find that an interaction could occur that could result in the change of the obliquity of the fossil field. This likely accounts for the various observed magnetic obliquities.

Acknowledgements

Some of the work described in this contribution was undertaken in collaboration with many other scientists. The authors would like to thank these collaborations and especially C. Catala, C. P. Folsom, G. Hussain, J. Landstreet, N. Langer, S. Mathis, J. Morin, C. Neiner, and G. A. Wade.

References

Abt, H. A. & Snowden, M. S. 1973, *ApJS*, 25, 137
Alecian, E., *et al.* 2008a, *MNRAS*, 385, 391
Alecian, E., *et al.* 2008b, *A&A*, 481, 99
Alecian, E., *et al.* 2009, *MNRAS*, 400, 354
Alecien, E., *et al.* 2013, *MNRAS*, 429, 1001
Aurière. *et al.* 2007, *A&A*, 475, 1053
Bailey, J. D., *et al.* 2012, *MNRAS*, 423, 328
Babcock, H. W. 1947, *ApJ*, 105, 105
Babcock, H. W. 1960, *ApJ*, 132, 521
Bagnulo, S., Landstreet, J. D., Mason, E., Andretta, V., Silaj, J., & Wade, G. A. 2006, *A&A*, 450, 777
Behrend, R. & Maeder, A. 2001, *A&A*, 373, 190
Berdyugina, S. V. 2005, *Living Reviews in Solar Physics*, 2, 8
Bohlender, D., Landstreet, J., Brown, D., & Thompson, I. 1987, *ApJ*, 323, 325
Bouret, J.-C., Donati, J.-F., Martins, F., Escolano, C., Marcolino, W., Lanz, T., & Howarth, I. 2008, *MNRAS*, 389, 75
Braithwaite, J. 2009, *MNRAS*, 397, 763
Brott, I., *et al.* 2011, *A&A*, 530, 115
Brun, A. S., Browning, M. K., & Toomre, J. 2005, *ApJ*, 629, 461
Bychkov, V. D., Bychkova, L. V., & Madej, J. 2009, *MNRAS*, 394, 1338
Cantiello, M., Langer, N., Brott, I., de Koter, A., Shore, S. N., Vink, J. S., Voegler, A., Lennon, D. J., & Yoon, S.-C. 2009, *A&A*, 499, 279
Cantiello, M. & Braithwaite, J. 2011, *A&A*, 534, 140
Carrier, F., North, P., Udry, S., & Babel, J. 2002, *A&A*, 394, 151
Charbonneau, P. & MacGregor, K. B. 2001, *ApJ*, 559, 1094
Charbonneau, P. 2005, *Living Reviews in Solar Physics*, 2, 2
Christensen, U. R., Holzwarth, V., & Reiners, A. 2009, *Nature*, 457, 167
Cowling, T. G. 1945, *MNRAS*, 105, 166
de Winter, D., Koulis, C., Theé, P. S., van den Ancker, M. E., Pérez, M. R., & Bibo, E. A. 1997, *A&AS*, 121, 223
Donati, J.-F., Semel, M., Carter, B., Rees, D., & Collier Cameron, A. 1997, *MNRAS*, 291, 658
Donati, J.-F., Babel, J., Harries, T., Howarth, I., Petit, P., & Semel, M. 2002, *MNRAS*, 333, 55
Donati, J.-F., Howarth, I. D., Bouret, J.-C., Petit, P., Catala, C., & Landstreet, J. 2006, *MNRAS*, 365, 6
Donati. *et al.* 2008, *MNRAS*, 390, 545
Donati, J.-F. & Landstreet, J. 2009, AR *A&A*, 47, 333
Duez, V. & Mathis, S. 2010, *A&A*, 517, 58
Elkin, V. G., Mathys, G., Kurtz, D. W., Hubrig, S., & Freyhammer, L. M. 2010, *MNRAS*, 402, 1883
Elkin, V. G., Kurtz, D. W., Mathys, G., & Freyhammer, L. M. 2010, *MNRAS*, 404, 1883
Featherstone, N. A., Browning, M. K., Brun, A. S., & Toomre, J. 2009, *ApJ*, 705, 1000
Ferrario, L., Pringle, J. E., Tout, C. A., & Wickramasinghe, D. T. 2009, *MNRAS*, 400, 71
Freyhammer, L. M., Elkin, V. G., Kurtz, D. W., Mathys, G., & Martinez, P. 2008, *MNRAS*, 389, 441
Grunhut, J. H., *et al.* 2012b, *MNRAS*, 426, 2208
Grunhut, J. H., *et al.* 2013, *MNRAS*, 428, 1686
Kochukhov, O. *et al.* 2004, *A&A*, 414, 613
Kochukhov, O. 2006, *A&A*, 454, 321
Kochukhov, O. & Wade, G. A. 2010, *A&A*, 513, 13
Korntreff, C., Kaczmarek, T., & Pfalzner, S. 2012, *A&A*, 543, 126
Landstreet, J. D. 1990, *ApJ*, 352, 5
Langer, N. 2012, AR *A&A*, 50, 107
Leone, F., Catanzaro, G., & Catalano, S. 2000, *A&A*, 355, 315

Lignières, F., Petit, P., Bohm, T., & Aurière, M. 2009, *A&A*, 500, 41
MacGregor, K. B. & Cassinelli, J. P. 2003, *ApJ*, 586, 480
Mestel, L. 2001, *ASP-CS*, 248, 3
Meynet, G., Eggenberger, P., & Maeder, A. 2011, *A&A*, 525, 11
Morel, P. 1997, *A&AS*, 597, 614.
Moss, D. 2001, *ASP-CS*, 248, 305
Nazé, Y., Bagnulo, S., Petit, V., Rivinius, Th., Wade, G., Rauw, G., & Gangé, M. 2012, *MNRAS*, 423, 3413
Palla, F. & Stahler, S. W. 1990, *ApJ*, 360, 47
Park, B.-G. & Sun, H. 2002, *AJ*, 123, 892
Parker, E. N. 1955, *ApJ*, 122, 293
Petit, P. *et al.* 2011, *A&A*, 532, 13
Silvester, J., Wade, G. A., Kochukhov, O., Bagnulo, S., Folsom, C. P., & Hanes, D. 2012, *MNRAS*, 426, 1003
Spruit, H. 1999, *A&A*, 349, 189
Stahler, S. W. 1983, *ApJ*, 274, 822
Sung, H., Bessell, M. S., & Lee, S.-W. 1997, *AJ*, 114, 2644
Tayler, R. J. 1973, *MNRAS*, 161, 365
Thé, P. S., de Winter, D., & Perez, M. R. 1994, *A&AS*, 104, 315
Tout, C. A., Wickramasinghe, D. T., LIebert, J., Ferrario, L., & Pringle, J. E. 2008, *MNRAS*, 387, 897
Tutukov, A. V. & Fedorova, A. V. 2010, *Astron. Rep.*, 51, 156
Vieira, S. L., *et al.* 2003, *AJ*, 126, 2971
Wade, G. A., *et al.* 2012a, *MNRAS*, 419, 2459
Wade, G. A., *et al.* 2012b, *MNRAS*, 425, 1278
Wade, G. A., Grunhut, J. H., & the MiMeS Collaboration, 2012c, *ASP-CS*, 464, 405
Walder, R., Folini, D., & Meynet, G. 2012, *Space Sci. Revs.*, 166, 145
Weber, E. & Davis, Jr., L., *ApJ*, 148, 217

Discussion

Magnetic Fields throughout Stellar Evolution
Proceedings IAU Symposium No. 302, 2013
P. Petit, M. Jardine & H. Spruit, eds.

doi:10.1017/S1743921314001768

What do weak magnetic fields mean for magnetospheric accretion in Herbig AeBe star+disk systems?

A. N. Aarnio[1], J. D. Monnier[1], T. J. Harries[2] and D. M. Acreman[2]

[1]Dept. of Astronomy, University of Michigan, 830 Dennison Building, 500 Church Street, Ann Arbor, MI, 48109, USA
email: aarnio@umich.edu

[2]School of Physics, University of Exeter, Stocker Road, Exeter, EX4 4QL, UK

Abstract. In the presently favored picture of star formation, mass is transferred from disk to star via magnetospheric accretion and out of the system via magnetically driven outflows. This magnetically mediated mass flux is a fundamental process upon which the evolution of the star, disk, and forming planetary system depends. Our current understanding of these processes is heavily rooted in young solar analogs, T Tauri Stars (TTS). We have come to understand recently, however, that the higher mass pre-main sequence (PMS) Herbig AeBe (HAeBe) stars have dramatically weaker dipolar fields than their lower mass counterparts. We present our current observational and theoretical efforts to characterize magnetospherically mediated mass transfer within HAeBe star+disk systems. We have gathered a rich spectroscopic and interferometric data set for several dozen HAeBe stars in order to measure accretion and mass loss rates, assess wind and magnetospheric accretion properties, and determine how spectral lines and interferometric visibilities are diagnostic of these processes. For some targets, we have observed spectral line variability and will discuss ongoing time-series spectroscopic efforts.

Keywords. stars: emission line, Be, stars: evolution, stars: winds, outflows, stars: planetary systems: protoplanetary disks

1. Introduction

HAeBe stars are identified as the higher mass pre-main sequence counterparts to TTS (Herbig, 1960). HAeBe stars are observed to have circumstellar disks which, as with the evolution of higher mass stars in generally, dissipate faster than the TTS case (Hernández, 2005). Excess UV luminosity from accretion shocks Calvet & Gullbring (1998) is observed, as well as P-Cygni absorption features from powerful winds. While radiative transfer modeling has been able to broadly determine physical parameters necessary for a magnetospheric accretion scenario to produce the observed spectral line profiles (Kurosawa, Harries, & Symington, 2006), the relative contributions of wind and magnetosphere to emission lines remain poorly constrained.

Fundamentally, as developed for TTS, the magnetospheric accretion model depends on a strong, highly ordered dipolar field capable of meeting circumstellar disk material a few stellar radii from the star, channeling it to the stellar surface via accretion streams. Indeed, sufficiently strong fields are observed on TTS (e.g., Basri *et al.* 1992, Johns-Krull *et al.* 1999, Johns-Krull, 2007, to name but a few studies), ~1 kG as estimated to be sufficient (Königl, 1991). Recent spectropolarimetric studies of magnetic fields for a range of pre-main sequence stars have shown that the polar dipole strength decreases dramatically with stellar mass (cf. Gregory *et al.* 2012); this is likely due to the shift from a convective dynamo to a shear-based dynamo (Tout & Pringle, 1995). As the dipolar

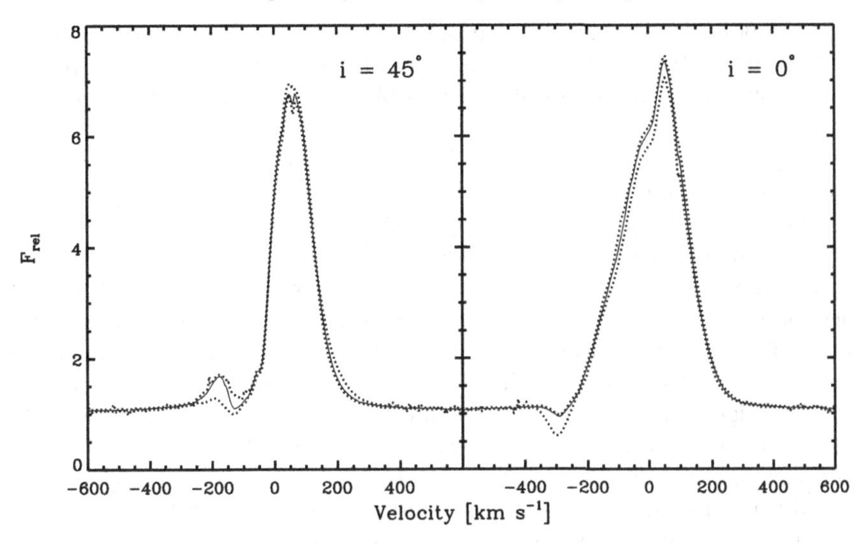

Figure 1. Our time-series optical spectra were taken for a Herbig Be star (spectral type B9, 45° inclination; left panel) and a Herbig Ae star (A2 spectral type, face-on inclination; right panel). Here we show the dramatic variation observed in the Hα line profiles over the course of 5 days' observation. Solid lines denote the median line profile, dashed lines show the minimum/maximum. Within the nightly observations, we observed strong variability: this underscores the need for simultaneity when comparing spectral lines.

component weakens with increasing stellar mass, higher order moments begin to dominate; in some cases, weak fields are detected on a handful of HAeBe stars (e.g., Hubrig *et al.* 2009), but a startling number of HAeBe stars have *no detected field* at all: Alecian *et al.* (2013) report ~90% non-detections of magnetic signatures in a spectropolarimetric survey of 70 HAeBe stars. In light of these observations, we ask how, then, are outflows and accretion, magnetically-driven phenomena, occurring on HAeBe stars?

2. Our HAeBe Observational/Theoretical Campaign

To address how accretion and outflow happen in HAeBe systems in the limit of weak (or absent) magnetic fields, we are conducting a combined observational and theoretical effort to characterize mass transfer within HAeBe star+disk systems.

2.1. *Spectroscopy and Interferometry*

We have obtained high resolution optical spectra of ~60 HAeBe stars. These data were taken using the MIKE (Magellan Inamori Kyocera Echelle) spectrograph; MIKE is a double echelle spectrograph that covers a wavelength range of 3200-9400Å at resolving powers R~65,000−83,000. For almost 90% of these stars, we also have interferometric data, and for another ~70% we have contemporaneous near-infrared spectra. Additionally, there exists a wealth of photometric data in the literature, which we have used to model spectral energy distributions (SEDs). For some of these targets, we have multi-epoch MIKE spectra, and for two objects, we followed up with time-series optical spectroscopy (Sec. 2.2). The optical data will be presented in a forthcoming publication (Aarnio *et al.* 2013, *in prep*), and here we present early results from that sample as well as the time series spectroscopic campaign.

Categorizing the HAeBe Hα profiles using the Reipurth, Pedrosa, & Lago (1996) morphologies, we find that the vast majority of the sample (~80%) shows simple, emission-only profiles centered at the stellar rest velocity or type-B morphologies with blueshifted absorption components in addition to emission components. The remaining line profiles were either II- or III-R; interestingly, in no cases did we see inverse P-Cygni profiles. The type-R line profiles occurred in objects with published high inclination angles.

2.2. *Time Series Spectroscopy*

For two targets that were observed to vary over multiple epochs, we conducted a pilot study, obtaining high cadence time series optical spectroscopy with MIKE to monitor the timescales of emission line variability. In order to assess short-term variability (potentially due to inhomogeneous mass loss or accretion), we observed each object for an hour at the highest possible cadence, ~6 min and ~4.5 min with the blue and red sides of MIKE, respectively. To search for variability on the order of the stellar rotation period (e.g., probing asymmetric geometries in the field/accretion streams; Long *et al.* 2012), we repeated these high-cadence "bursts" for 5 nights within the span of a week. Our final data set consists of ~100 spectra per object. In Fig. 1, we show the Hα profiles (minus the stellar photospheric component) of these two objects; in order to compare inclination and spectral type effects, we chose one Ae and one Be star, and the stars are face-on and inclined at 45°, respectively. In addition to seeing strong variability on very short timescales in blueshifted absorption features generally attributed to the wind, we also see variability in forbidden lines.

Perhaps surprisingly, we see little variability in the emission components. If the emission originates in a magnetospheric accretion scenario, does a lack of variability imply axisymmetric, steady accretion? Given the unlikeliness of symmetry and steady-state behavior, is the emission then not magnetospheric in origin? Is there some emission from the magnetosphere, but it contributes less than another source? Or, given what we're beginning to understand about magnetism across the H-R diagram (Gregory *et al.* 2012), is emission from multiple accretion stream components averaging out, making shorter timescale variability undetectable?

2.3. *Line Profile Modeling*

We are in the process of modeling spectral line profiles using TORUS (Transfer of Radiation Under Sobolev; Harries, 2000). For a set of parameters describing the star, disk, disk wind, and magnetosphere, TORUS can produce dust continuum images, SEDs, and atomic line transfer calculations (Fig. 2). Each of these three outputs can then be compared to the interferometric data, literature photometry, and spectroscopy. TORUS is a flexible, modular code which will allow us to modify separate system components individually, updating the disk wind model with information from recent observations, and basing the magnetospheric accretion configuration on extrapolated magnetic field maps from recent observational efforts. We have begun to test some novel magnetic field configurations, and will base these on field extrapolations from recent spectrointerferometric results. Finally, we will take these models and synthesize line profile evolution over a stellar rotation period to compare to time series spectra.

3. Discussion

Observationally and theoretically, magnetospheric accretion as it is presently envisaged for TTS has been well established. While it has been generally believed that this paradigm could be extended to HAeBe stars, recent advances in our understanding of stellar magnetic fields as a function of stellar mass have challenged this. Empirical calibrations of spectroscopic indicators and accretion rate, long used for TTS, have been found to break down at higher masses (Mendigutía *et al.* 2012).

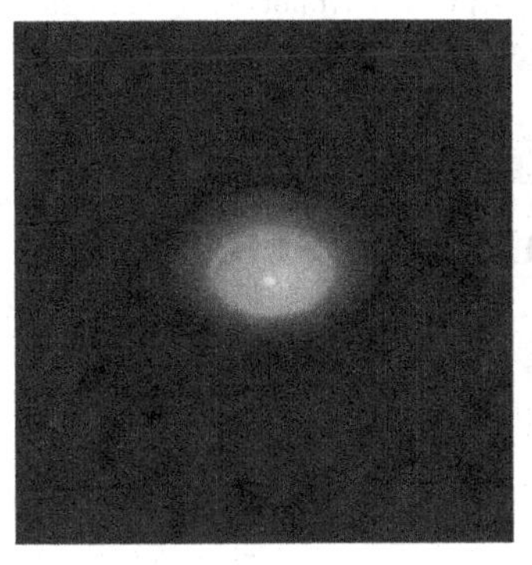  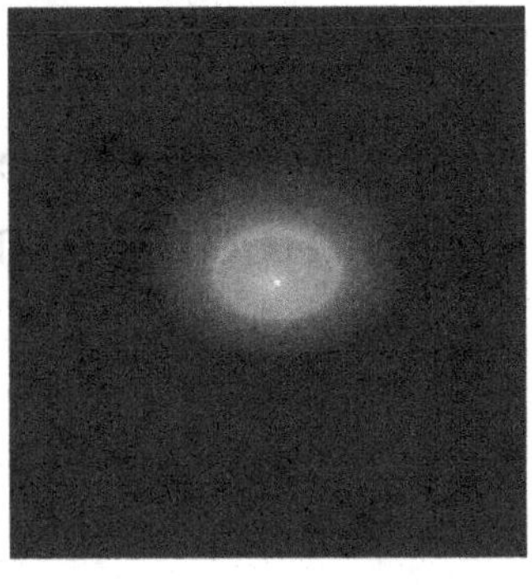

Figure 2. Example TORUS line transfer calculations: Hα, for a few magnetospheric configurations. For a fiducial HAeBe star, we show a TTS-like, large-scale magnetosphere that meets the inner gas disk (left), a scaled-down magnetosphere, the dipolar component only extending to $\sim$1.5 stellar radii (center), and no magnetosphere at all (right). All three models produce strong, centrally peaked Hα emission.

Our combined spectroscopic and interferometric data set allows us to approach the issues of accretion and outflow in HAeBe systems from a unique perspective: simultaneously, we will model line profiles across multiple species, using interferometry to break inclinations in system inclination and disk size which have plagued earlier efforts. We have also seen that looking into the time domain could prove to be a powerful probe of system dynamics, revealing processes occurring on many dynamical timescales and their relationships to one another.

References

Alecian, E., Wade, G. A., Catala, C., Grunhut, J. H., Landstreet, J. D., Bagnulo, S., Böhm, T., Folsom, C. P., Marsden, S., & Waite, I. 2013, *MNRAS*, 429, 1001.

Basri, G., Marcy, G. W., & Valenti, J. A., 1992, *ApJ*, 390, 622.

Calvet, N. & Gullbring, E., 1998, *ApJ*, 509, 802.

Gregory, S. G., Donati, J.-F., Morin, J., Hussain, G. A. J., Mayne, N. J., Hillenbrand, L. A., & Jardine, M., 2012, *ApJ*, 755, 97.

Harries, T. J., 2000, *MNRAS*, 315, 722.

Herbig, G. H., 1960, *ApJS*, 4, 337.

Hernández, J., Calvet, N., Hartmann, L., Briceño, C., Sicilia-Aguilar, A., & Berlind, P., 2005, *AJ*, 129, 856.

Hubrig, S., Stelzer, B., Schöller, M., Grady, C., Schütz, O., Pogodin, M. A., Curé, M., Hamaguchi, K., & Yudin, R. V., 2009, *A&A*, 502, 283.

Johns-Krull, C. M., Valenti, J. A., & Koresko, C., 1999, *ApJ*, 516, 900.

Johns-Krull, C. M., 2007, *ApJ*, 664, 975.

Königl, A., 1991, *ApJL*, 370, L39.

Kurosawa, R., Harries, T. J., & Symington, N. H., 2006, *MNRAS*, 370, 580.

Long, M., Romanova, M. M., & Lamb, F. K., 2012, *New Astronomy*, 17, 232.

Mendigutía, I., Mora, A., Montesinos, B., Eiroa, C., Meeus, G., Merín, B., & Oudmaijer, R. D. 2012, *A&A* 543, A59.

Reipurth, B., Pedrosa, A., & Lago, M. T. V. T. 1996, *A&AS*, 120, 229.

Tout, C. A. & Pringle, J. E., 1995, *MNRAS*, 272, 528.

Magnetic Fields throughout Stellar Evolution
Proceedings IAU Symposium No. 302, 2013
P. Petit, M. Jardine & H. Spruit, eds.

doi:10.1017/S174392131400177X

Searches for the new magnetic intermediate-mass stars on various stages of MS evolution

Evgeny A. Semenko

Special Astrophysical Observatory of the Russian Academy of Sciences,
Nizhny Arkhyz, Russia
email: sea@sao.ru

Abstract. A limited list of new results of the searches for the new magnetic stars among late B and early A stars is in this work. Continual observations with spectroscopic devices of the 6m Russian telescope BTA led to successful detection of about 10 new magnetic stars that occupy different parts of evolutional tracks for the stars of 2–3 solar masses. Measurements of the longitudinal magnetic field show weak and medium strength magnetic field in all program stars.

Keywords. star, magnetic field, spectropolarimetry

1. Introduction

The study of stellar magnetism goes on during more than 60 last years, and now more than 450 magnetic Ap/Bp stars are known. Only few dozens of them ever became the objects of detailed study. Better understanding of an evolution of stellar magnetic fields require the extension of the list of known mCP.

The current work is pointed to detection of new magnetic stars among poorly studied B8–A3 stars. By their physical properties these stars are considered as an intermediate class of magnetic stars that occur on different stages of MS evolution.

Below the results of longitudinal field measurements are presented. Most of stars were discovered as magnetic for the first time within the current work during the last 3 years.

2. Observations. Data reduction

Spectral material for our work was collected with the Main Stellar Spectrograph (MSS) installed in the Nasmyth-2 focus of the Russian 6m telescope BTA. This spectrograph equipped with the differential circular polarization analyzer and allows to observe the stars up to 11-12 magnitude in spectropolarimetric mode. Long-slit spectra have a mean resolution of 0.12 Å pix^{-1}.

The raw data were reduced with a set of programs mentioned by Kudryavtsev *et al.* (2006). The same set is capable to make the measurements of longitudinal magnetic field by fitting a gaussian function to spectral lines profile. Additionally, using a custom implementation of method, described by Bagnulo *et al.* (2002) we control the positional magnetic measurements. In order to check the right polarity of magnetic field and take probable instrumental polarization into account we observed every night one or two standard stars with well-measured magnetic field along with cool zero-field star.

3. The results

3.1. *HD 50341*

This star was selected for observation due to its photometric IR excess typical for the young stars. Despite of the low accuracy of measurements, caused by the fast rotation, six individual measurements allow us to make the conservative conclusion about magnetic nature of HD 50341. Further study of this star will consider also the detailed analysis of hi-res spectra.

3.2. *HD 63347*

HD 63347 was selected for observation due to its IR excess. Additional indication of its young age was found in Tetzlaff *et al.* (2011). On the figure our measurements of B_z phased with period 1.74984 days (Koen, Eyer 2002). Detailed study of echelle spectra, obtained with BTA, confirm SrCrEu abundance anomalies that together with young age make this star unique.

3.3. *HD 96003*

Bright star ($V = 6\overset{m}{.}06$) was characterised as SrCr by Renson & Manfroid (2009). Seven measurements of polarized, rich on the lines, spectra demonstrate approximately constant (about -150–-180 G) value of B_z. HD 96003 is a binary system with angular separation of about $2''$ between components. An individual measurement of each component should be attractive and may have a high importance for the understanding of stellar magnetic field evolution.

3.4. *HD 201174*

In the current list HD 201174 is the only one early known magnetic star. The first measurements of its longitudinal field were made in SAO RAS by D. Kudryavtsev. Strong cross-over effect and sharp spectral lines allow us to make a detailed study of physical properties of the star. Location of HD 201174 on the HR diagram is typical for young stars. Rotational period is still unknown but 16 individual measurements of B_z imply relatively short value of P_{rot}.

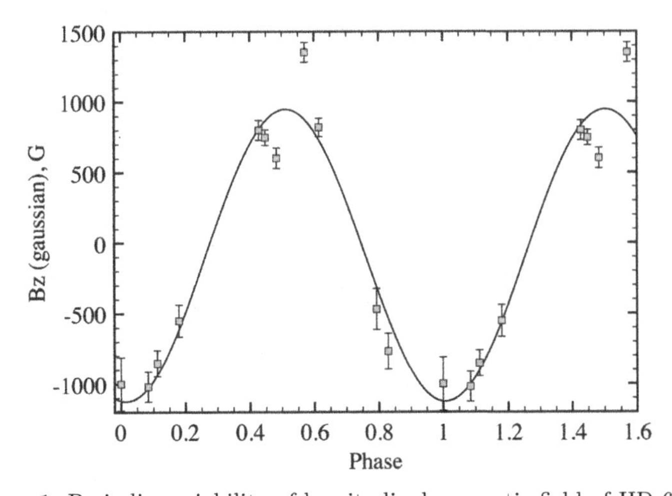

Figure 1. Periodic variability of longitudinal magnetic field of HD 63347.

Acknowledgments

The author is grateful to the staff of the Laboratory for stellar magnetism study of the Special Astrophysical Observatory for assistance in collecting observational data and any useful discussions. This work is supported partly by the Russian Fund for Basic Researches (grants RFBR No 12-02-00009-a and 12-02-31246_mol-a) and by the programs of the Ministry of Education and Science of the Russian Federation.

References

`http://www.sao.ru/hq/lizm/mss/en/`
Kudryavtsev, D. O., Romanyuk, I. I., Elkin, V. G., & Paunzen, E. 2006, *MNRAS*, 372, 1804
Bagnulo, S., Szeifert, T., Wade, G. A., Landstreet, J. D., & Mathys, G. 2002, *A&A*, 389, 191
Renson, P. & Manfroid, J. 2009, *A&A*, 498, 961
Tetzlaff, N., Neuhäuser, R., & Hohle, M. M. 2011, *MNRAS*, 410, 190
Koen, C. & Eyer, L 2002, *MNRAS*, 331, 45

Magnetic Fields throughout Stellar Evolution
Proceedings IAU Symposium No. 302, 2013
P. Petit, M. Jardine & H. Spruit, eds.

doi:10.1017/S1743921314001781

Chemical peculiarities in magnetic and non-magnetic pre-main sequence A and B stars

C. P. Folsom[1,2], S. Bagnulo[2], G. A. Wade[3], J. D. Landstreet[2,4] and E. Alecian[5]

[1]Institut de Recherche en Astrophysique et Planétologie, Toulouse, France
email: colin.folsom@irap.omp.eu
[2]Armagh Observatory, Armagh, Northern Ireland
[3]Department of Phyics, Royal Military College of Canada, Kingston, Canada
[4]Physics & Astronomy Department, University of Western Ontario, London, Canada
[5]Observatoire de Paris, Meudon, France

Abstract. In A- and late B-type stars, strong magnetic fields are always associated with Ap and Bp chemical peculiarities. However, it is not clear at what point in a star's evolution those peculiarities develop. Strong magnetic fields have been observed in pre-main sequence A and B stars (Herbig Ae and Be stars), and these objects have been proposed to be the progenitors of Ap and Bp stars. However, the photospheric chemical abundances of these magnetic Herbig stars have not been studied carefully, and furthermore the chemical abundances of 'normal' non-magnetic Herbig stars remain poorly characterized. To investigate this issue, we have studied the photospheric compositions of 23 Herbig stars, four of which have confirmed magnetic fields. Surprisingly, we found that half the non-magnetic stars in our sample show λ Bootis chemical peculiarities to varying degrees. For the stars with detected magnetic fields, we find one chemically normal star, one star with λ Boo peculiarities, one star displaying weak Ap/Bp peculiarities, and one somewhat more evolved star with somewhat stronger Ap/Bp peculiarities. These results suggests that Ap/Bp peculiarities are preceded by magnetic fields, and that these peculiarities develop over the pre-main sequence lives of A and B stars. The incidence of λ Boo stars we find is much higher than that seen on the main sequence. We argue that a selective accretion model for the formation of λ Boo peculiarities is a natural explanation for this remarkably large incidence.

1. Introduction

Recently, strong magnetic fields have been found in some Herbig Ae and Be (HAeBe) stars. These are pre-main sequence A- and B-type stars ($\sim$2 to $\sim$10 $M_\odot$). Magnetic fields have been found in 5-10% of HAeBe stars: 3/50 stars by Wade *et al.* (2007), and 5/70 stars by Alecian *et al.* (2013) in currently the most comprehensive and accurate study. The magnetic fields found in these stars are geometrically simple, predominately dipolar with dipole strengths on the order of 1 kG (e.g. Alecian *et al.* 2008, Folsom *et al.* 2008).

The incidence and morphology of magnetic fields in HAeBe stars has led to the conclusion that these objects are the pre-main sequence progenitors of main sequence Ap and Bp stars. Ap and Bp stars are main sequence chemically peculiar stars with strong magnetic fields. Their magnetic strengths range from 300 G to $\sim$10 kG (Aurière *et al.* 2007), and they have an incidence of 5-10%. The strong chemical peculiarities in these stars are understood to be a consequence of atomic diffusion operating in the stable radiative stellar envelope. However, the timescale for the development of these peculiarities remains unknown. Measurements of chemical abundances in the progenitors of Ap and

Bp stars, magnetic HAeBe stars, would provide important constraints on this timescale. However, the chemical abundances of HAeBe stars in general are poorly studied (e.g. Acke & Waelkens 2004). Therefore, we performed a precise investigation of chemical abundances in both magnetic and non-magnetic HAeBe stars.

The λ Bootis stars are chemically peculiar A-type stars, but unlike Ap stars they are characterized by underabundances of iron-peak elements and approximately normal abundances for C, N, O, and often S. These stars are much rarer than Ap stars, appearing with roughly a 2% incidence on the main sequence (Paunzen 2001). The origin of these abundances is unknown, but they are unlikely to be a result of atomic diffusion (e.g. Charbonneau 1993). A leading hypothesis is that λ Boo peculiarities are the result of selective accretion, in which lighter elements are accreted more readily than heavier elements, building up a layer of apparent underabundances at the surface of the star (Venn & Lambert 1990). If the formation of λ Boo peculiarities depends on an accretion process, then naively one would expect to find these peculiarities more often in HAeBe stars, objects which were recently accreting, and may still be. Thus a second goal of this study is to search for the presence of λ Boo peculiarities among HAeBe stars.

Here we present results from Folsom *et al.* (2012), and from Folsom *et al.* (2008), providing a sample of 23 HAeBe stars, 4 of which have confirmed magnetic fields.

2. Observations

To investigate these questions we used observations obtained with the ESPaDOnS instrument at the Canada-France-Hawaii Telescope. This is a high resolution (R = 65000) spectropolarimeter with a wavelength range of 3700-10500 Å. Observations for one southern target (HD 101412) were obtained with the SEMLPOL polarimeter attached to the University College London Echelle Spectrograph at the Anglo-Australian Telescope. For the analysis of Balmer lines, archival observations from the FORS1 spectrograph at the Very Large Telescope were used. This lower resolution instrument obtains spectra in a single order, which removes some ambiguities in the normalization of Balmer lines.

The magnetic properties of the observations used here were analyzed by Alecian *et al.* (2013), thus we have precise self-consistent diagnostics of presence or absence of magnetic fields in these stars. The sample was chosen to include 23 stars, covering a range of 7500 to 15000 K in $T_{\rm eff}$, and spanning a range of $v \sin i$ up to 200 km s^{-1}. The stars were also chosen to have only modest amounts of emission, or shell absorption, in their spectra. This bias was necessary, as we required a large number of uncontaminated photospheric lines in order to determine photospheric chemical abundances. The sample includes all well established magnetic HAeBe stars cooler than 15000 K.

3. Abundance analysis

The abundance analysis proceeded by directly fitting synthetic spectra, produced with the ZEEMAN spectrum synthesis code (Landstreet 1988, Wade *et al.* 2001), to the observed spectra. Initial estimates of $T_{\rm eff}$ and $\log g$ were made by fitting the wings of Balmer lines, far from contamination by emission. However, there is a substantial degeneracy between $T_{\rm eff}$ and $\log g$ in these estimates, thus they were refined using ionization and excitation balances, by simultaneously fitting lines of an element with different ionization states and a wide range of excitation potentials. Chemical abundances, $v \sin i$, microturbulence, $T_{\rm eff}$, and $\log g$ were fit simultaneously by χ^2 minimization. Six spectral regions, ~500 Å long, were independently fit for each star. The average and standard deviation of these results were taken as the final best value and its uncertainty, respectively.

Great care was taken to avoid fitting lines contaminated with circumstellar emission or absorption. With multiple observations of the stars, we could often identify lines contaminated by small amounts of emission using unexpected line variability. By comparing line shapes to the synthetic spectra, we could identify lines that departed from simple rotation broadening, and exclude those. Finally, by examining lines with very low excitation potentials for inconsistencies, we could identify potential emission infilling and exclude those lines from the fit.

For further details on the analysis methodology see Folsom *et al.* (2012).

4. Results

From the abundance analysis we find 10 (out of 23) stars are chemically normal, with approximately solar abundances. However, we find another 11 stars that show underabundances of iron-peak elements and roughly solar abundances of C, N, and O, thus they display λ Boo peculiarities with varying strengths. This represents roughly a 50% incidence of λ Boo peculiarities, while these peculiarities only appear in about 2% of main sequence A stars (Paunzen 2001).

Among the four magnetic stars in our sample, we find one that is chemically normal (HD 190073), one that displays λ Boo peculiarities (HD 101412), one with weak marginal Bp peculiarities (V380 Ori A), and one with strong Bp peculiarities (HD 72106 A). The presence of λ Boo peculiarities in HD 101412 appears to be simply a consequence of the very high incidence of these peculiarities among HAeBe stars, thus we consider both HD 190073 and HD 101412 to be chemically indistinguishable from the non-magnetic stars. Examining the binary system HD 72106 in detail, we find all the Herbig star characteristics of the system are associated with the secondary. Placing the system on the H-R diagram, the primary is consistent with the zero age main sequence (ZAMS). Thus we conclude that HD 72106 A has probably reached the main sequence, but it is likely still a very young object since the secondary still appears to be a pre-main sequence star. For the SB2 system V380 Ori, both components are clearly still on the pre-main sequence. In both binary systems with a magnetic primary, we find abundances for the secondary consistent with solar.

The full set of final atmospheric parameters and chemical abundances derived for all stars are presented by Folsom *et al.* (2012), and by Folsom *et al.* (2008) for HD 72106.

5. Discussion and Conclusions

We find roughly 50% of the stars in our sample display λ Boo chemical peculiarities, an incidence rate dramatically larger than the $\sim$2% seen on the main sequence. We interpret this as evidence in favor of a selective accretion hypothesis for the formation of λ Boo peculiarities. If the mechanism for forming λ Boo peculiarities depends on accretion, then one would expect to find such peculiarities frequently in HAeBe stars, objects which have recently been accreting, and may still be. Over time these surface chemical peculiarities would become mixed into the stars, and thus on the main sequence the stars would likely display normal abundances.

Among the magnetic stars, we find two stars with Bp chemical peculiarities and two stars with abundances matching the non-magnetic HAeBe stars (i.e. chemically 'normal'). Thus Bp peculiarities can appear on the pre-main sequence, but magnetic fields precede the presence of Ap/Bp peculiarities. This is in contrast to the main sequence where all strongly magnetic A and late B stars are Ap and Bp stars. Placing the magnetic stars on the H-R diagram, there appears to be a rough progression, with the two chemically

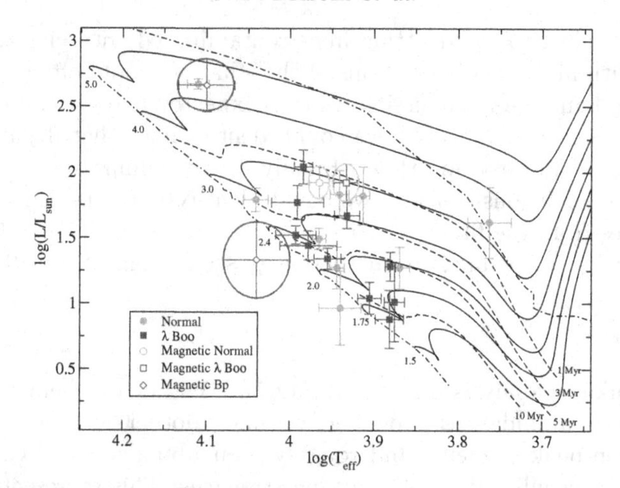

Figure 1. H-R diagram for the stars in this study. Solid lines are evolutionary tracks (labeled by mass in $M_{\odot}$), dashed lines are isochrones (labeled by age), and the birthline and ZAMS are dashed-dotted lines. The stars are classified by their chemical abundances and magnetic properties, and the four magnetic stars are highlighted with ovals.

'normal' stars being further from the ZAMS, the weak Bp star being closer to the ZAMS, and the strong Bp star being on the ZAMS. This may represent the development of chemical peculiarities over time during the pre-main sequence, however a larger sample size is needed to draw firm conclusions.

Finally, we find no evidence for other types of chemical peculiarities among the HAeBe stars in our sample. Specifically, we see no Am or HgMn chemical peculiarities. These chemical peculiarities occur in 10 to 20% of main sequence A- and late B-type stars, and thus naively we would expect to have found between 2 and 4 such stars. A much larger sample is needed to draw any firm conclusions, but it may be that these peculiarities do not have time to form on the pre-main sequence.

References

Acke, B. & Waelkens, C. 2004, *A&A* 427, 1009

Alecian, E., Catala, C., Wade, G. A., Donati, J.-F., Petit, P., Landstreet, J. D., Böhm, T., Bouret, J.-C., Bagnulo, S., Folsom, C., Grunhut, J., & Silvester, J. 2008, *MNRAS* 385, 391

Alecian, E., Wade, G. A., Catala, C., Grunhut, J. H., Landstreet, J. D., Bagnulo, S., Böhm, T., Folsom, C. P., Marsden, S., & Waite, I. 2013, *MNRAS* 429, 1001

Aurière, M., Wade, G. A., Silvester, J., Lignières, F., Bagnulo, S., Bale, K., Dintrans, B., Donati, J. F., Folsom, C. P., Gruberbauer, M., *et al.* 2007, *A&A* 475, 1053

Charbonneau, P. 1993, *ApJ* 405, 720

Folsom, C. P., Wade, G. A., Kochukhov, O., Alecian, E., Catala, C., Bagnulo, S., Böhm, T., Bouret, J.-C., Donati, J.-F., Grunhut, J., Hanes, D. A., & Landstreet, J. D. 2008, *MNRAS* 391, 901

Folsom, C. P., Bagnulo, S., Wade, G. A., Alecian, E., Landstreet, J. D., Marsden, S. C., & Waite, I. A. 2012, *MNRAS* 422, 2072

Landstreet, J. D. 1988, *ApJ* 326, 967

Paunzen, E. 2001, *A&A* 373, 633

Venn, K. A. & Lambert, D. L. 1990, *ApJ* 363, 234

Wade, G. A., Bagnulo, S., Kochukhov, O., Landstreet, J. D., Piskunov, N., & Stift, M. J. 2001, *A&A* 374, 265

Wade, G. A., Bagnulo, S., Drouin, D., Landstreet, J. D., & Monin, D. 2007, *MNRAS* 376, 1145

Magnetic Fields throughout Stellar Evolution
Proceedings IAU Symposium No. 302, 2013
P. Petit, M. Jardine & H. Spruit, eds.

doi:10.1017/S1743921314001793

Angular momentum evolution of young stars

S. P. Littlefair

Dept. of Physics & Astronomy, University of Sheffield,
Sheffield, S6 2NE, UK
email: s.littlefair@shef.ac.uk

Abstract. In recent years, rotation periods for large numbers of pre-main-sequence stars have become available, covering a wide range of ages and star forming environments. Simultaneously, theoretical developments in the physics of the star-disc interaction have been carried out, and observational measurements of the magnetic field geometry of both fully convective, and pre-main-sequence stars have become available. This review discusses these recent developments, and the extent to which the observational data fits within the existing theoretical frameworks.

Keywords.

1. Introduction

If the protostellar collapse process was dominated by gravity, one would expect to see pre-main-sequence (PMS) stars rotating at close to their breakup velocities. However, initial studies of rotational velocities in PMS stars revealed typical rotation rates below 40 $\mathrm{km\,s^{-1}}$, around one-fifth of the breakup velocity (Vogel & Kuhi 1981; Bouvier, Bertout, Benz *et al.* 1986; Hartmann, Hewett, Stahler *et al.* 1986). This finding is even more puzzling when one considers that these stars are accreting material from their circumstellar disc which carries with it large amounts of specific angular momentum. If accretion is sustained at moderate rates for a few Myr, even a slowly rotating star will be spun up to more than half it's breakup velocity (Bouvier 2013). The slow rotation rates of pre-main-sequence stars therefore requires a mechanism which efficiently removes angular momentum from the central star.

Using a model first developed for accreting neutron stars (Ghosh & Lamb 1979), Koenigl (1991) was first to suggest that the magnetic interactions between the inner disc and the magnetic field of the star could efficiently extract angular momentum. Very shortly thereafter, evidence for a correlation between rotation rate and accretion was revealed (Edwards, Strom, Hartigan *et al.* 1993; Bouvier, Cabrit, Fernandez *et al.* 1993). The correlation revealed that accreting young stars rotated more slowly, on average, than non-accreting stars, providing strong observational support for Koenigl's framework. This review looks at how theoretical and observational breakthroughs have affected this picture in the intervening twenty years. We start by examining the support for the disc locking framework, as set out by (Koenigl 1991).

1.1. *The disc locking framework*

The calculation of torques on the star, exerted by magnetic field lines anchored to the star and connected to the disc has been carried out by many authors (e.g. Ghosh & Lamb 1979; Lovelace, Romanova & Bisnovatyi-Kogan 1995; Wang 1995; Yi 1995; Armitage & Clarke 1996; Rappaport, Fregeau & Spruit 2004). These models are quite similar in the whole. Here we use the prescription developed by Matt & Pudritz (2005), which itself follows Armitage & Clarke (1996) - see figure 1. The star's magnetic field connects to the disc between radii R_t and R_{out}. The rotation rate of the disc differs from that of

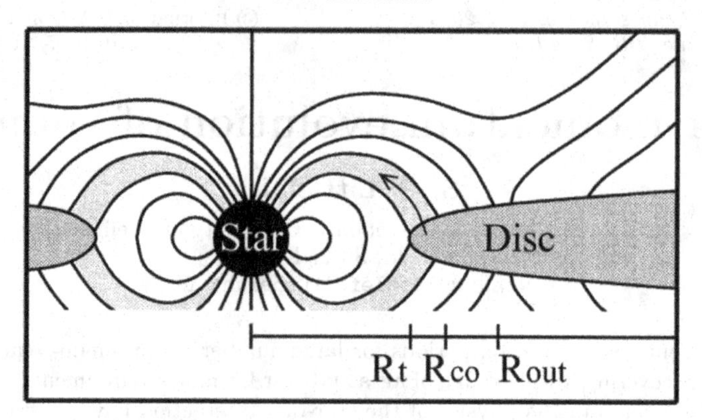

Figure 1. Magnetic star-disc interaction under the disc-locking framework. The stellar field connects to the accretion disc between radii R_t and R_{out}. The stellar field dominates the accretion flow onto the star (arrow). Taken from Matt & Pudritz 2005.

the star at all radii, except at R_{co}, the co-rotation radius. For $r < R_{co}$ the disc rotates faster than the star, whilst for $r > R_{co}$ the star rotates faster than the disc. At all radii in the disc apart from R_{co}, differential rotation twists the field azimuthally. Inside R_{co}, the field lines are twisted so that they lead the star. Torques from this region spin the star up. Outside R_{co}, the field lines are twisted so that they trail the star. Torques from this region spin the star down. In certain circumstances, this can lead to an equilibrium situation, any increase in stellar spin would move R_{co} inwards, increasing the spin-down torques, and vice-versa. This is the disc-locking framework as put forward by Koenigl (1991). How well does it hold up against modern observational results?

2. Rotation rates of pre-main-sequence stars

The pioneering works on early stellar rotation focused on measurement of projected rotational velocities. However, the development of large-format CCDs in the mid 90's allowed large scale photometric monitoring campaigns that provided rotational period distributions for thousands of low-mass stars in the PMS stage, across a wide range of masses, ages and environments. Reference lists for most of the available studies are available in the excellent reviews of Irwin & Bouvier (2009) and Bouvier (2013). Figure 2 (from Irwin & Bouvier 2009) shows a sample of some of these results.

From the extensive rotation rates in the literature, some clear trends have emerged. At very young ages (~ 1 Myr) the initial distribution is very broad, with rotation periods typically spanning the range 1–10 days at all masses. For stars more massive than $M > 0.3 M_{\odot}$ the period distribution is bi-modal, with peaks at 2 and 8 days (Herbst, Bailer-Jones & Mundt 2001). For lower masses the distribution is still broad, but is unimodal with a peak around 2 days. During early PMS evolution (1–5 Myr) there is little evolution in period for the higher-mass objects, despite the contraction of these stars towards the main sequence, whilst the lowest mass objects spin up significantly. Furthermore, this spin-up is mass dependent - indeed, Henderson & Stassun (2012) have suggested the period-mass slope for lower mass stars can be a useful age indicator for very young clusters.

At later stages of evolution the period distribution is flat over the range 0.5–1 $M_{\odot}$. This is shown most clearly in the period distribution for the 13 Myr old h-Per association (Moraux, Artemenko, Bouvier *et al.* 2013). In spite of the fact that stellar contraction towards the main sequence is continuing, the slow rotators still show periods in the

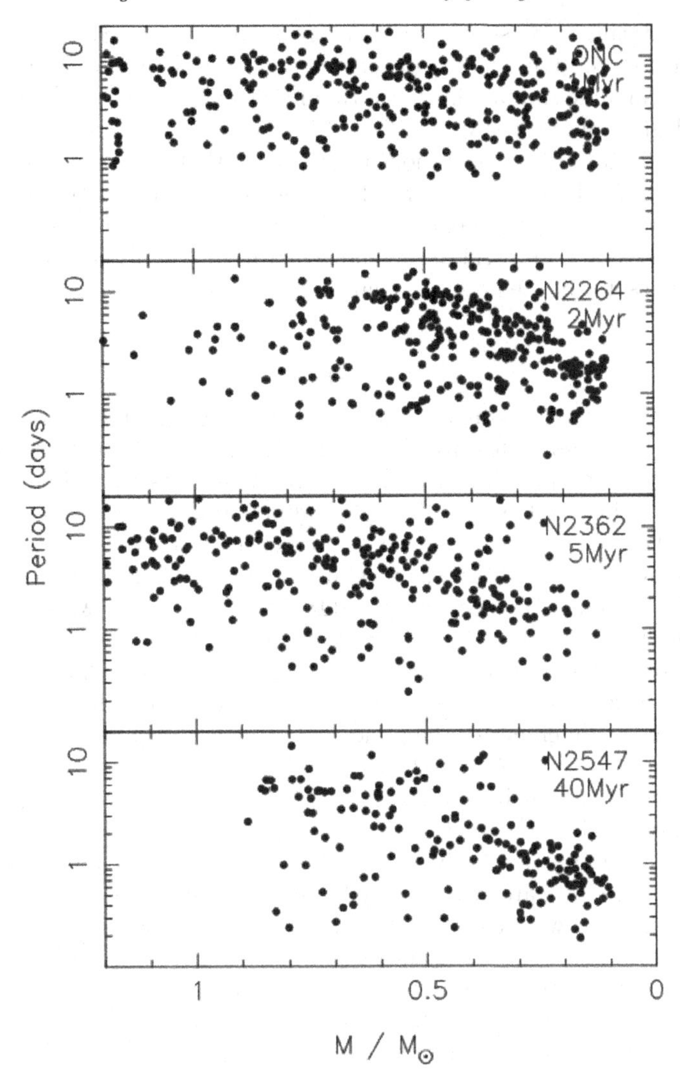

Figure 2. Compilation of rotation periods for young stars with masses $M \leqslant 1.2M_\odot$. References for the original studies can be found in Irwin & Bouvier 2009.

range 8–10 days, whilst the most rapid rotators have spun up to 0.2–0.3 days. This later evolution is reasonably well described by the combined effects of magnetic stellar wind braking and core-envelope decoupling for the slow rotators, which explains the lack of spin-up of the slow rotators by hiding angular momentum in their cores. For a more detailed review of later evolution, we refer the reader to the review of Bouvier (2013). Instead, we focus on explaining the very early evolution, and the origin of the period distribution at ~ 1 Myr.

2.1. *The link between discs and rotation rate*

If the star-disc interaction is responsible for the slow rotation of PMS stars, a correlation between discs and rotation is expected. Despite very early confirmation of this correlation (Edwards *et al.* 1993; Bouvier *et al.* 1993), this area remained controversial for nearly ten years. This was largely due to some studies (e.g. Littlefair, Naylor, Burningham *et al.* 2005) finding no correlation between rotation and the presence of a disc. The situation improved dramatically with surveys of disc presence from the Spitzer space telescope, which gave much more robust indicators of disc presence. Repeated studies using reliable disc indicators and large sample sizes (e.g. Rebull, Stauffer, Megeath *et al.* 2005; Cieza & Baliber 2007; Dahm, Slesnick & White 2012; Affer, Micela, Favata *et al.* 2013) found clear indication of a disc-rotation correlation, in the sense that stars with discs rotate more slowly than those which do not. This has led to general consensus that some mechanism related to the presence of an accretion disc is responsible for the slow rotation of PMS stars. In the next section, I present arguments why this mechanism cannot be disc-locking, but first I outline some of the observational challenges and uncertainties which affect the picture presented above.

2.2. *Caveats*

The rotational evolution of young stars has been understood largely through the period-mass diagrams for clusters with a range of ages. The ages of young stars in particular are subject to large, and possibly systematic, uncertainties (see e.g. Bell, Naylor, Mayne *et al.* 2013). In addition, it is not clear that applying a single age to a young cluster is appropriate. Whilst the interpretation of luminosity spreads in young clusters as age spreads is controversial (e.g. Jeffries, Littlefair, Naylor *et al.* 2011; Baraffe, Vorobyov & Chabrier 2012; Hartmann, Zhu & Calvet 2011; Hosokawa, Offner & Krumholz 2011), it is certainly true that some clusters with rotational studies could be composed of multiple stellar populations. A good example of this is Cepheus OB3b, which has two separate sub-clusters which may have different ages (Allen, Gutermuth, Kryukova *et al.* 2012). This may explain the difference in the period-mass diagram between Cepheus OB3b and the similarly aged cluster NGC 2362, which has also been ascribed to environmental differences between the two clusters (Littlefair, Naylor, Mayne *et al.* 2010).

One might expect the photometric period to be the most accurately determined of these quantities, and this is largely true. However, even here caution is advised. The majority of studies to date are ground based and suffer from gaps in coverage due to the day-night cycle and bad weather. One cluster where this is not true is NGC 2264, where a 28 day continuous monitoring campaign was carried out by the COROT satellite. Affer *et al.* (2013) present rotational periods from this campaign, and show that around 20% of periods from the previous ground based studies were unreliable. Moreover, even when a photometric period is reliably detected, it may not be due to the rotation of the young star (e.g. Artemenko, Grankin & Petrov 2012). Other sites of emission in the star-disc system (e.g. the inner rim of the accretion disc) can potentially give rise to periodic variability. By using their unbroken lightcurves to make subjective decisions about the origin of variability, Affer *et al.* (2013) estimate that 10% of all PMS stars in NGC 2264 show periodic variability from obscuration by the inner disc. Amongst stars with discs, this fraction rises to 35%, raising serious concerns about biasses in photometric samples. However, Affer *et al.* (2013) also found that the link between rotation and discs is still present, even after this contamination is removed.

3. Challenges to disc locking

In the disc locking picture, torques from the twisted field lines transfer angular momentum from the star to the disc. Theoretical challenges to this picture means that is unlikely to be the cause of slow rotation in young stars.

If we define the degree of twisting $\gamma = \frac{B_\phi}{B_z}$ then γ itself is a function of the rate at which the magnetic field slips through the disc. Matt & Pudritz (2005) parameterise the field slippage with a diffusion parameter $\beta \equiv \frac{\eta_t}{h\nu_k}$, where η_t is the effective magnetic diffusivity, h is the disc thickness and ν_k is the Keplerian velocity. For high values of

β, the field slips through the disc and γ is small, leading to small torques on the star. As β falls, the degree of twisting, and the torque, increases. However, a critical value of the field twist, $\gamma_c \approx 1$, exists, beyond which the field lines inflate and open (e.g. Aly 1985; Lovelace *et al.* 1995). At this point they can no longer transfer angular momentum between the star and the disc. Therefore, for small values of $\beta \sim 0.01$, the field is well coupled to the disc which leads to significant opening of the field lines and a *reduction* in the star-disc torques. Matt & Pudritz (2005) show that the maximal torque actually occurs for $\beta \approx 1$, representing significant slippage of the magnetic field through the disc.

For the disc locking model, the equilibrium spin rate can be written as

$$\Omega_{eq} = C(\beta, \gamma_c) G^{5/7} M_*^{5/7} \dot{M}_a^{3/7} \mu^{-6/7}, \tag{3.1}$$

where $\mu = BR^3$ is the stellar magnetic moment, M is the stellar mass, $\dot{M}_a$ is the accretion rate and C is a dimensionless constant. For the disc locking model to explain the rotation rates seen in PMS stars requires C to be of order unity, and hence $\beta \sim 1$. This is problematic, since in realistic accretion discs we expect *beta* ~ 0.01, which reduces the spin-down torque by two orders of magnitude (Matt & Pudritz 2005). These simple theoretical arguments are supported by detailed 2D simulations (e.g. Zanni & Ferreira 2009).

Of course, a star may not necessarily reach spin equilibrium. To calculate the rotation period of a star at 1 Myr, it is necessary to follow the full evolution, taking account of the spin-down torques from the star disc interaction, the angular momentum accreted from disc material, and the contraction of the star towards the main sequence. This calculation has been performed (Matt, Pinzón, de la Reza *et al.* 2010). The end result is that for $\beta \sim 1$, the star-disc torques are strong, and the stellar spin is close to the equilibrium values, which range from 1–10 days at 1 Myr. For realistic values of $\beta \sim 0.01$, the star-disc torques are weak; instantaneous spin rates differ greatly from the equilibrium values, and rotation rates at 1 Myr range from 0.6–2 days. In short, disc locking is not expected to operate in physically realistic accretion discs.

Disc locking faces observational challenges as well. For all stars in a disc-locked state, the truncation radius of the accretion disc is close to the co-rotation radius (e.g. Matt & Pudritz 2005). Le Blanc, Covey & Stassun (2011) found that this does not hold for the young stars of IC 348, with many slow rotators having truncation radii larger than the co-rotation radius, and no obvious difference in truncation radii between slow and fast rotators. It is worth mentioning, however, that measuring the truncation radius of the accretion disc is not easy (e.g. Carr 2007; Pinte, Ménard, Berger *et al.* 2008). Perhaps more seriously for disc-locking, Littlefair, Naylor, Mayne *et al.* (2011) found a link between stellar radius and rotation in four young associations, in the sense that the slow rotators were, on average, smaller than the fast rotators. The stellar radius enters equation 3.1 for the equilibrium period through the magnetic moment, such that $P_{eq} \propto R^{2.5}$. Therefore, the observed correlation is in the opposite sense to that expected under disc locking. It is also in the opposite sense to that expected if the young stars shrink with age, and spin up as they are released from disc locking. In conclusion, it is both theoretically and observationally unlikely that the slow rotation rates seen in PMS stars are explained by disc locking, i.e. the transfer of angular momentum between the star and the disc along closed field lines.

4. Alternatives to disc locking

The challenges faced by disc locking have led different groups to examine alternatives in which outflows along open field lines remove angular momentum from the star-disc system as a whole. These outflows fall into three main categories: winds from the accretion disc (the X-wind), accretion-powered stellar winds, and magnetospheric ejections. For a fuller review of the different mechanisms, see Ferreira (2013).

4.1. *Disc Winds*

In considering the impact of disc winds on the spin evolution of young stars, we can neglect extended, magnetised, disc winds (e.g. Blandford & Payne 1982; Ferreira 1997). This is because such winds are not causally connected to the star and thus cannot extract angular momentum from it (Ferreira 2013). The same is not true of the X-wind model (Ostriker & Shu 1995), originally proposed with the twin aims of explaining both the presence of jets around PMS stars and the slow stellar rotation within single theory. This model shares some similarities with the disc-locking model; again, the accretion disc is truncated just inside the co-rotation radius. However, it is assumed that the magnetic pressure forces a narrow region straddling the co-rotation radius, the X-region, to rotate as a solid body at the stellar rotation rate. Material just inside the x-region sub-rotates and threads easily onto inward-leaning field lines and is accreted onto the star. Material just outside super-rotates and threads easily onto outward-leaning field lines, escaping in a wind. Torques from the accretion funnels transfer angular momentum to the disc material, forcing it outwards, whilst torques from wind extract angular momentum from the disc material, forcing it inwards. Thus the material is 'pinched' into the X-region, causing truncation of the disc. As a result, much or all of the angular momentum of the accretion flow is transferred to the wind, allowing the star to maintain slow rotation.

Recent work on the X-wind model has extended it to non-dipolar field configurations (Mohanty & Shu 2008) and these models can re-create the observed stellar spin rates in the few cases where the magnetic and accretion properties are measured well enough to allow a comparison (V2129 Oph and BP Tau; Donati, Jardine, Gregory *et al.* 2007, 2008). However, the X-wind model does face theoretical and observational challenges. For example, detailed MHD simulations of the star-disc interaction have never produced X-winds, although it is not clear that they would have been expected, given the way the simulations were set up (Ferreira 2013). Observationally the X-wind model also predicts disc truncation near the co-rotation radius, which is not observed (Le Blanc *et al.* 2011), and jet kinematics of young stars are not consistent with predictions from the X-wind model (Ferreira, Dougados & Cabrit 2006; Cabrit 2007).

4.2. *Accretion Powered Stellar Winds*

Stellar winds, as opposed to winds from the accretion disc, may be a significant contributor to the outflows from PMS stars (e.g. Fendt, Camenzind & Appl 1995; Hirose, Uchida, Shibata *et al.* 1997; Romanova, Ustyugova, Koldoba *et al.* 2009). Provided the outflow rate is high enough, the stellar winds can also extract enough angular momentum to explain the slow rotation rates of PMS stars (e.g. Hartmann & Stauffer 1989; Matt & Pudritz 2005). By following the full evolution of an accreting PMS star, including accreted angular momentum and contraction towards the main sequence, Matt, Pinzón, Greene *et al.* (2012) showed that the rotation rates observed at $\sim$1 Myr can be explained by a stellar wind, provided that the outflow rate is approximately 10% of the accretion rate. To obtain these high outflow rates, Matt & Pudritz (2005) suggested that a fraction of the potential energy from the accreted matter was used to drive an accretion powered stellar wind. There is some evidence for this; some emission lines in PMS stars are best explained by stellar wind kinematics, and they generally correlate with accretion rates (e.g. Johns-Krull 2007; Kurosawa, Romanova & Harries 2011). However, questions remain as to whether PMS stars have enough accretion power to drive such strong outflows (Zanni & Ferreira 2011), or as to how the accretion power is used to drive the wind. Matt & Pudritz (2008) suggested that waves generated in the photosphere by the impact at the base of the accretion funnels could transport energy to the open field lines, where it would drive enhanced MHD activity which in turn drives a stellar wind. Such a mechanism has been studied in a simplified, 1D environment by Cranmer (2008), who found that outflow rates only reached around one percent of the accretion rate, not enough to significantly affect the rotation of the PMS star.

4.3. *Magnetospheric Ejections*

As we saw in section 3, differential rotation between the star and disc leads to the opening of field lines, limiting the effectiveness of disc locking. However the inflation, opening and reconnection of these field lines means that we expect to see magnetospheric ejections in PMS stars. These magnetospheric ejections may regulate stellar angular momentum in two ways; not only do they exert a braking torque directly on the star, but they also carry angular momentum from the disc, thus reducing the spin-up torque from accretion. In fact, the combination of magnetospheric ejections and a stellar wind can exert a net spin-down torque on the star (Zanni & Ferreira 2013). The extreme limit of magnetospheric ejections are propeller phases, when the magnetic field disrupts accretion entirely and drives outflow from the star-disc system (e.g. Romanova, Ustyugova, Koldoba *et al.* 2005; D'Angelo & Spruit 2011). Propeller phases can exert strong spin-down torques on the central star. Whilst simulations of magnetospheric ejections and propeller phases are promising, it is unknown if, in practise, they occur with sufficient frequency to have a significant effect on the stellar rotation . The simulations of Zanni & Ferreira (2013) require strong kilo-Gauss dipolar fields and it is not clear if these are common amongst PMS stars (e.g. Donati *et al.* 2008; Donati, Skelly, Bouvier *et al.* 2010). Meanwhile, no observational evidence exists for a PMS star in a propeller phase.

5. Summary

It is clear that disc locking does not explain the slow rotation of PMS stars; opening of field lines in realistic star-disc systems means that the torques on the star are too small. Whilst several alternative descriptions of the star-disc interaction can in principle provide significant spin-down torques, it is not clear if any of them can successfully spin down PMS stars to rotation periods of 10 days at $\sim$ 1 Myr, or maintain slow rotation for the next few Myr. It is of course quite possible that some or all of these mechanisms act in concert to spin down the star (Ferreira 2013). Possibly, modifications to the theories above, such as and X-wind where the stellar and disc magnetic fields interact (Ferreira, Pelletier & Appl 2000) could be more efficient.

Littlefair *et al.* (2011) put forward a more radical, and speculative, hypothesis. The need for a spin-down torque during the PMS phase is driven by the absence of spin up for slowly rotating PMS stars between 1–5 Myr. Since these stars are believed to be contracting towards the main sequence, and are fully convective, this implies angular momentum loss from the stars. Littlefair *et al.* (2011) show, however, that the slowly rotating stars are also the smallest in any given cluster. The cause of radius spreads in young clusters is highly controversial. It may reflect a spread in ages, but it may also be caused by the effects of accretion driving stars from thermal equilibrium (e.g. Baraffe *et al.* 2012; Hartmann *et al.* 2011; Hosokawa *et al.* 2011). If the latter is true, then the small, slowly rotating PMS stars in young clusters will not contract significantly between 1–5 Myr, and the need for an efficient spin-down torque during this phase is dramatically reduced. Of course, this picture does not explain the slow rotation of stars at $\sim$ 1 Myr and so one, or all, of the processes above will still be needed at earlier phases, when the young star was still embedded. The importance of the star-disc interaction during the embedded phase would explain the present day link between discs and rotation as a 'fossil' of earlier processes.

One question which remains unanswered is the origin of the large spread in rotation rates at 1 Myr, with stars in the $\sim$ 1 Myr Orion Nebula Cluster showing an order of magnitude spread in rotation rate, corresponding to four orders of magnitude in angular momenta. This wide spread in initial rotation rates probably reflects processes occurring during the embedded phase. For example, Gallet & Bouvier (2013) suggest that variations in the protostellar disc mass could lead to a wide dispersion in angular momenta. Another question, perhaps linked to the wide spread in rotation rates, is the origin of the bi-modality seen in the Orion Nebula Cluster. One possibility, which has been only briefly explored, is the effects of varying geometries in the stellar magnetic field (e.g. Morin, Donati, Petit *et al.* 2010; Gregory, Donati, Morin *et al.* 2013; Gastine, Morin, Duarte *et al.* 2013). There is no doubt that many of the processes described above will behave differently for fields which are dominantly dipolar, compared to fields dominated by e.g. octopolar components. As an example, Batygin & Adams (2013) find that the accretion torque for a star with a predominantly octopolar field is an order of magnitude lower than for a dipolar field.

The last ten years has has seen tremendous development in our understanding of the processes governing the rotation of young stars. We now recognise that disc locking cannot be effective, and have a number of plausible alternative theories which may present a solution. In parallel we have also reached a firm conclusion regarding the link between rotation and the presence of accretion discs, and have a new observational constraint, in the observed link between stellar rotation and radius. Coupled with advances in our knowledge of the magnetic field geometries of young stars it is hoped that the next ten years will bring new understanding of the processes responsible for the slow rotation of pre-main-sequence stars.

References

Affer, L., Micela, G., Favata, F., *et al.* 2013, *MNRAS* 430, 1433

Allen, T. S., Gutermuth, R. A., Kryukova, E., *et al.* 2012, *ApJ* 750, 125

Aly, J. J., 1985, in: M. R. Kundu, G. D. Holman (eds.), *Unstable Current Systems and Plasma Instabilities in Astrophysics*, volume 107 of *IAU Symposium*, pp. 217–219

Armitage, P. J. & Clarke, C. J., 1996, *MNRAS* 280, 458

Artemenko, S. A., Grankin, K. N., & Petrov, P. P., 2012, *Astronomy Letters* 38, 783

Baraffe, I., Vorobyov, E., & Chabrier, G., 2012, *ApJ* 756, 118

Batygin, K. & Adams, F. C., 2013, *ArXiv e-prints*

Bell, C. P. M., Naylor, T., Mayne, N. J., *et al.* 2013, *MNRAS* 434, 806

Blandford, R. D. & Payne, D. G., 1982, *MNRAS* 199, 883

Bouvier, J., 2013, in: *EAS Publications Series*, volume 62 of *EAS Publications Series*, pp. 143–168

Bouvier, J., Bertout, C., Benz, W., *et al.* 1986, *A&A* 165, 110

Bouvier, J., Cabrit, S., Fernandez, M., *et al.* 1993, *A&A* 272, 176

Cabrit, S., 2007, in: J. Bouvier, I. Appenzeller (eds.), *IAU Symposium*, volume 243 of *IAU Symposium*, pp. 203–214

Carr, J. S., 2007, in: J. Bouvier, I. Appenzeller (eds.), *IAU Symposium*, volume 243 of *IAU Symposium*, pp. 135–146

Cieza, L. & Baliber, N., 2007, *ApJ* 671, 605

Cranmer, S. R., 2008, *ApJ* 689, 316

Dahm, S. E., Slesnick, C. L., & White, R. J., 2012, *ApJ* 745, 56

D'Angelo, C. R. & Spruit, H. C., 2011, *MNRAS* 416, 893
Donati, J., Jardine, M. M., Gregory, S. G., *et al.* 2008, *MNRAS* 386, 1234
Donati, J.-F., Jardine, M. M., Gregory, S. G., *et al.* 2007, *MNRAS* 380, 1297
Donati, J.-F., Skelly, M. B., Bouvier, J., *et al.* 2010, *MNRAS* 409, 1347
Edwards, S., Strom, S. E., Hartigan, P., *et al.* 1993, *AJ* 106, 372
Fendt, C., Camenzind, M., & Appl, S., 1995, *A&A* 300, 791
Ferreira, J., 1997, *A&A* 319, 340
Ferreira, J., 2013, in: *EAS Publications Series*, volume 62 of *EAS Publications Series*, pp. 169–225
Ferreira, J., Dougados, C., & Cabrit, S., 2006, *A&A* 453, 785
Ferreira, J., Pelletier, G., & Appl, S., 2000, *MNRAS* 312, 387
Gallet, F. & Bouvier, J., 2013, *A&A* 556, A36
Gastine, T., Morin, J., Duarte, L., *et al.* 2013, *A&A* 549, L5
Ghosh, P. & Lamb, F. K., 1979, *ApJ* 234, 296
Gregory, S. G., Donati, J.-F., Morin, J., *et al.* 2013, *ArXiv e-prints*
Hartmann, L., Hewett, R., Stahler, S., *et al.* 1986, *ApJ* 309, 275
Hartmann, L. & Stauffer, J. R., 1989, *AJ* 97, 873
Hartmann, L., Zhu, Z., & Calvet, N., 2011, *ArXiv e-prints*
Henderson, C. B. & Stassun, K. G., 2012, *ApJ* 747, 51
Herbst, W., Bailer-Jones, C. A. L., & Mundt, R., 2001, *ApJ* (Letters) 554, 197
Hirose, S., Uchida, Y., Shibata, K., *et al.* 1997, *PASJ* 49, 193
Hosokawa, T., Offner, S. S. R., & Krumholz, M. R., 2011, *ApJ* 738, 140
Irwin, J. & Bouvier, J., 2009, in: E. E. Mamajek, D. R. Soderblom, R. F. G. Wyse (eds.), *IAU Symposium*, volume 258 of *IAU Symposium*, pp. 363–374
Jeffries, R. D., Littlefair, S. P., Naylor, T., *et al.* 2011, *MNRAS* 418, 1948
Johns-Krull, C. M., 2007, *ApJ* 664, 975
Koenigl, A., 1991, *ApJ* (Letters) 370, 39
Kurosawa, R., Romanova, M. M., & Harries, T. J., 2011, *MNRAS* 416, 2623
Le Blanc, T. S., Covey, K. R., & Stassun, K. G., 2011, *AJ* 142, 55
Littlefair, S. P., Naylor, T., Burningham, B., *et al.* 2005, *MNRAS* 358, 341
Littlefair, S. P., Naylor, T., Mayne, N. J., *et al.* 2010, *MNRAS* 403, 545
Littlefair, S. P., Naylor, T., Mayne, N. J., *et al.* 2011, *MNRAS* (Letters) 413, 56
Lovelace, R. V. E., Romanova, M. M., & Bisnovatyi-Kogan, G. S., 1995, *MNRAS* 275, 244
Matt, S. & Pudritz, R. E., 2005, *ApJ* (Letters) 632, 135
Matt, S. & Pudritz, R. E., 2008, *ApJ* 681, 391
Matt, S. P., Pinzón, G., de la Reza, R., *et al.* 2010, *ApJ* 714, 989
Matt, S. P., Pinzón, G., Greene, T. P., *et al.* 2012, *ApJ* 745, 101
Mohanty, S. & Shu, F. H., 2008, *ApJ* 687, 1323
Moraux, E., Artemenko, S., Bouvier, J., *et al.* 2013, *ArXiv e-prints*
Morin, J., Donati, J.-F., Petit, P., *et al.* 2010, *MNRAS* 407, 2269
Ostriker, E. C. & Shu, F. H., 1995, *ApJ* 447, 813
Pinte, C., Ménard, F., Berger, J. P., *et al.* 2008, *ApJ* (Letters) 673, 63
Rappaport, S. A., Fregeau, J. M., & Spruit, H., 2004, *ApJ* 606, 436
Rebull, L. M., Stauffer, J. R., Megeath, T., *et al.* 2005, *American Astronomical Society Meeting Abstracts* 207,
Romanova, M. M., Ustyugova, G. V., Koldoba, A. V., *et al.* 2005, *ApJ* (Letters) 635, 165
Romanova, M. M., Ustyugova, G. V., Koldoba, A. V., *et al.* 2009, *MNRAS* 399, 1802
Vogel, S. N. & Kuhi, L. V., 1981, *ApJ* 245, 960
Wang, Y.-M., 1995, *ApJ* (Letters) 449, 153
Yi, I., 1995, *ApJ* 442, 768
Zanni, C. & Ferreira, J., 2009, *A&A* 508, 1117
Zanni, C. & Ferreira, J., 2011, *ApJ* (Letters) 727, 22
Zanni, C. & Ferreira, J., 2013, *A&A* 550, A99

Magnetic Fields throughout Stellar Evolution
Proceedings IAU Symposium No. 302, 2013
P. Petit, M. Jardine & H. Spruit, eds.

doi:10.1017/S174392131400180X

The Effects of Magnetic Activity on Lithium-Inferred Ages of Stars

Aaron J. Juarez[1,2], Phillip A. Cargile[2], David J. James[3] and Keivan G. Stassun[1,2]

[1]Dept. of Physics, Fisk University,
Nashville, TN 37208, USA

[2]Dept. of Physics and Astronomy, Vanderbilt University,
Nashville, TN 37235, USA
email: a.juarez@vanderbilt.edu

[3]Cerro Tololo Inter-American Observatory,
Casilla 603, La Serena, Chile

Abstract. In this project, we investigate the effects of magnetic activity on the Lithium Depletion Boundary (LDB) to recalibrate the measured ages for star clusters, using the open cluster Blanco 1 as a pilot study. We apply the LDB technique on low-mass Pre-Main-Sequence (PMS) stars to derive an accurate age for Blanco 1, and we consider the effect of magnetic activity on this inferred age. Although observations have shown that magnetic activity directly affects stellar radius and temperature, most PMS models do not include the effects of magnetic activity on stellar properties. Since the lithium abundance of a star depends on its radius and temperature, we expect that LDB ages are affected by magnetic activity. After empirically accounting for the effects of magnetic activity, we find the age of Blanco 1 to be ~100 Myr, which is ~30 Myr younger than the standard LDB age of ~130 Myr.

Keywords. Magnetic activity, fundamental parameters, pre-main sequence stars, open clusters

---

1. Introduction

The Lithium Depletion Boundary (LDB) technique is currently the most accurate method in estimating the age of a stellar cluster, to an uncertainty of ~10%. The LDB age is robust due to its insensitivity to input physics and thus gives consistent, precise ages among different PMS models (Jeffries & Naylor 2001). Observations have shown that magnetic activity directly affects stellar temperature and radius; however, it is not accounted for in most PMS lithium depletion models (Morales *et al.* 2008). We use empirical corrections from Stassun *et al.* (2012) to account for the effect of magnetic activity on the LDB to recalibrate the measured ages for star clusters. Correcting these benchmark LDB ages for magnetic activity will improve our overall ability to measure stellar ages by comparing them to more commonly used chronometers.

2. Analysis

Having doubled our sample set to nearly 50 Gemini Multi-Object Spectrograph (GMOS) spectra from the Gemini-North telescope, we sought to completely reanalyze the data in a consistent manner by developing software that automates the analysis. Our data reduction pipeline reduces GMOS spectra using standard reduction routines from the IRAF Gemini-GMOS package, such as bias removal, aperture extraction, and wavelength calibration. Our recently developed spectral analysis software determines the Hα Equivalent Width (EW), Lithium I (6707.8 Å) EW, radial velocity, and spectral type of the Blanco 1 candidates.

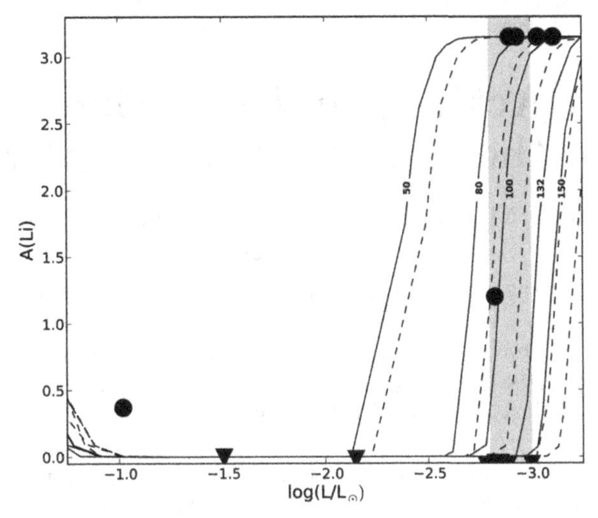

Figure 1. BCAH98 Li isochrones for various ages (lines) and our new Li abundances for Blanco 1 (points). Solid lines are the standard (non-activity) Li tracks and dashed lines are the shifted tracks which account for magnetic activity.

3. Updated LDB Age, Empirical Correction, and Implications

Our membership criteria was based on Hα emission, radial velocity, and proper motion of the cluster (Platais *et al.* 2011). Fourteen stars were considered high-confidence low-mass Blanco 1 members. The majority of these stars exhibit Hα emission as expected from chromospheric activity in young stars. To account for the magnetic activity, the BT-Settl Models of Allard *et al.* (2012) were used along with our measurements of Hα EW to determine changes in $T_{\rm eff}$ and radius via the empirical relationships.

In Fig. 1, we plot new Li abundances from our sample, where circles are $>2\sigma$ detections and triangles are 3σ lower limits, determined from Zapatero Osorio *et al.* (2002). We plot BCAH98 Li isochrones for luminosity, spanning 50–150 Myr. Solid lines are standard (non-activity) Li tracks and dashed lines are shifted tracks for average changes in stellar parameters, which correspond to roughly 10% $T_{\rm eff}$ suppression and 10% radius inflation.

We define the LDB in Blanco 1 to be bounded by the shaded region, where we observe the Li abundance to go to zero. Using the BCAH98 standard predictions (solid line), we infer an updated LDB age to be near 132 Myr, which is the previously determined LDB age (Cargile *et al.* 2010). Including activity changes the LDB age of Blanco 1 to $\sim$100 Myr – a difference of $\sim$30 Myr younger – indicated by the dashed line in Fig. 1. It is, therefore, worthwhile to consider the effects of magnetic activity on the inferred LDB age for other clusters with known LDBs since PMS stars often exhibit magnetic activity, which can change the timescale over which Li is depleted.

References

Allard, F., Homeier, D., & Freytag, B. 2012, *Phil. Trans. R. Soc. A*, 370, 2765

Baraffe, I., Chabrier, G., Allard, F., & Hauschildt, P. H. 1998, *A&A*, 337, 403

Cargile, P. A., James, D. J., & Jeffries, R. D. 2010, *ApJ* (Letters), 725, 111

Jeffries, R. D. & Naylor, T. 2011, *ASPC*, 243, 633

Morales, J. C., Ribas, I., & Jordi, C. 2008, *A&A*, 478, 507

Platais, I. Girard, T. M. and Vieira, K., Lopez, C. E., Loomis, C., McLean, B. J., Pourbaix, D., Moraux, E., Mermilliod, J.-C., James, D. J., Cargile, P. A., Barnes, S. A., & Castillo, D. J. 2011, *MNRAS*, 413, 1024

Stassun, K. G., Kratter, K. M., Scholz, A., & Dupuy, T. J. 2012, *ApJ*, 756, 47

Zapatero Osorio, M. R., Bejar, V. J. S.., Pavlenko, Ya., Rebolo, R., Allende Prieto, C., Martin, E. L., & Garcia Lopez, R. J. 2002, *A&A*, 384, 937

Magnetic Fields throughout Stellar Evolution
Proceedings IAU Symposium No. 302, 2013
P. Petit, M. Jardine & H. Spruit, eds.

doi:10.1017/S1743921314001811

Activity and Rotation in the Young Cluster h Per

C. Argiroffi[1,2], M. Caramazza[2], G. Micela[2], E. Moraux[3] and J. Bouvier[3]

[1]Dip. di Fisica e Chimica, Università di Palermo, Piazza del Parlamento 1, 90134 Palermo, Italy, email: argi@astropa.unipa.it

[2]INAF - Osservatorio Astronomico di Palermo, Piazza del Parlamento 1, 90134 Palermo, Italy

[3]UJF-Grenoble 1/CNRS-INSU, Institut de Plantologie et d'Astrophysique de Grenoble (IPAG) UMR 5274, Grenoble, F-38041, France

Abstract. We study the rotation-activity relationship for low-mass members of the young cluster h Persei, a ~ 13 Myr old cluster. h Per, thanks to its age, allows us to link the rotation-activity relation observed for main-sequence stars to the still unexplained activity levels of very young clusters.

We constrained the activity levels of h Per members by analyzing a deep Chandra/ACIS-I observation pointed to the central field of h Per. We combined this X-ray catalog with the catalog of h Per members with measured rotational period, presented by Moraux *et al.* (2013). We obtained a final catalog of 202 h Per members with measured X-ray luminosity and rotational period. We investigate the rotation-activity relation of h Per members considering different mass ranges. We find that stars with $1.3\,M_\odot < M < 1.4\,M_\odot$ show significant evidence of supersaturation for short periods. This phenomenon is instead not observed for lower mass stars.

Keywords. Dynamo, Stars: activity, Stars: pre-main sequence, X-rays: stars

1. Introduction

It is known that late type stars can be magnetically active, becoming therefore bright in the X-rays because of the hot coronal plasmas heated and confined by the stellar magnetic field. Magnetic fields in late type stars are produced by dynamo processes, whose efficiency is related to plasma motions in the stellar interior.

Pallavicini *et al.* (1981) initially evidenced how stellar activity correlates with the stellar rotational velocity. To take into account also the role of the convective envelope in the magnetic field production, Noyes *et al.* (1984) proved that magnetic activity is better determined by the Rossby number, *Ro*, defined as the ratio between the rotational period $P_{\rm rot}$ and the convective turnover time τ.

Pizzolato *et al.* (2003) and Wright *et al.* (2011), studying large samples of late-type main-sequence (MS) stars, definitely showed that the stellar dynamo is characterized by different regimes. In the non-saturated regime, i.e. for $Ro > 0.1$, the stellar X-ray luminosity, and hence the dynamo efficiency, anticorrelates with *Ro*. For $Ro < 0.1$, in the so-called saturated regime, MS stars show a constant X-ray emission level, with on average $L_{\rm X}/L_{\rm bol} \approx 10^{-3}$. Both the non-saturated and saturated regimes were well constrained by large sample of MS stars. Randich *et al.* (1996) found that a third regime probably occurs at very low *Ro* values ($Ro < 0.01$), the supersaturation regime, in fact very fast rotators show $L_{\rm X}/L_{\rm bol}$ lower than the saturated level. This behavior was observed only for very few stars belonging to young cluster ($\sim$30$-$50 Myr), likely because in older cluster the longer rotational periods make such small *Ro* values not accessible.

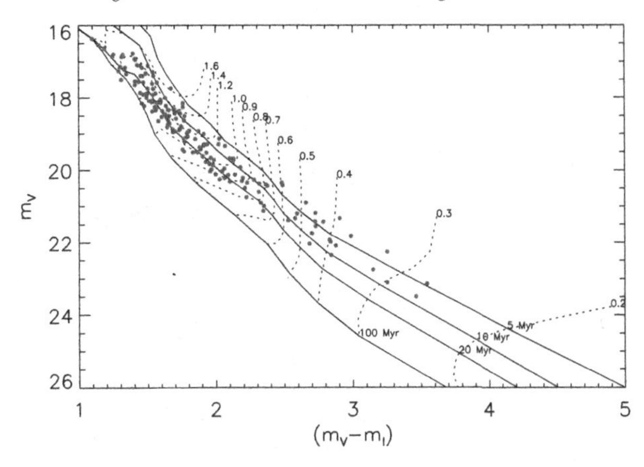

Figure 1. Color magnitude diagrams of the 170 sources that compose our final catalog. Isochrones (solid lines) and evolutionary tracks (dotted lines, with the corresponding mass in solar units) are from Siess *et al.* (2000) and are scaled to the distance of 2300 pc.

It is known that pre main-sequence (PMS) stars, like MS stars, are magnetically active, having strong magnetic field and showing intense coronal emission. The relation between *Ro* and fractional X-ray emission in PMS, studied in very young cluster (age < 5 Myr), shows much larger scatter than in MS stars (e.g. Preibisch *et al.* 2005). It still unclear which stellar parameters, other than $P_{\rm rot}$ and τ, determine the magnetic activity levels in PMS stars. A substantial difference between MS and PMS stars is that, at very young ages, PMS stars do not yet have a radiative core, being therefore fully convective. This different internal structures likely indicates that different dynamo mechanisms determine magnetic activity in PMS and MS stars.

To bridge the gap between the well constrained case of MS stars and the puzzling case of very young PMS stars, we studied the activity-rotation relation in young cluster h Persei. h Per is a rich cluster, ~ 13 Myr old, located at 2300 pc, and characterized by a $E(B-V) \sim 0.55$. Because of its age the h Per cluster offers us several advantages: it contains both fast and slow rotators, allowing us therefore to test the different regimes of stellar dynamo; accretion processes already ended; all the stars with $0.5\,{\rm M}_\odot < M < 1.5\,{\rm M}_\odot$ already developed a radiative core, having therefore an inner stellar structures similar to that of MS stars.

2. Analysis

The h Per cluster was observed with the Chandra satellite for 200 ks. This observation, in which we detected 1010 X-ray sources, allowed us to constrain the magnetic activity level of h Per members.

We compared the X-ray source catalog with the catalog of h Per members with measured period presented by Moraux *et al.* (2013). We obtained a catalog of 202 h Per members with detected X-ray flux and measured rotational period. 170 of these 202 members have also measured optical photometry from Currie *et al.* (2010).

We derived stellar X-ray luminosities $L_{\rm X}$ from the observed X-ray flux. Stellar masses and rotational periods are from Moraux *et al.* (2013). Bolometric luminosities, $L_{\rm bol}$, needed to compute the fractional X-ray luminosity, were evaluated starting from the V magnitude and $V-I$ color, and correcting for interstellar absorption. We estimated the empirical turnover times τ inferring the $B-V$ colors from the observed $V-I$, and then applying the empirical relation derived by Wright *et al.* (2011).

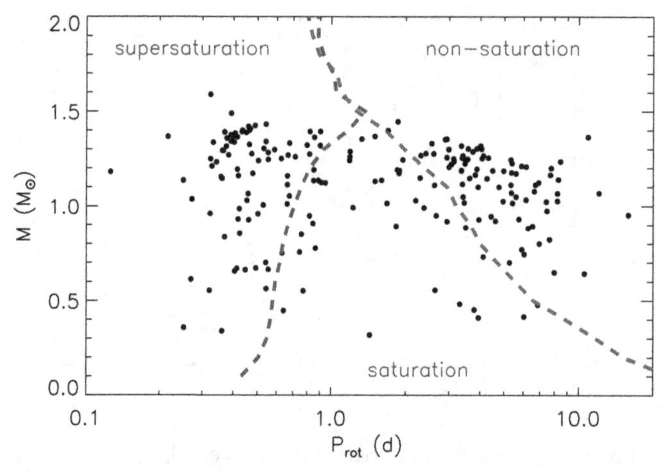

Figure 2. Mass vs period of the h Per members with measured $P_{\rm rot}$ and $L_{\rm X}$. The regions corresponding to the different dynamo regimes are separated dotted lines. The red dotted line marks the transition from saturated to supersaturated regime. This threshold is obtained assuming that supersaturation is caused by coronal stripping occurring at $R_{\rm cor}/R_{*} = 3$, note that a similar threshold is obtained considering the coronal updrafts mechanism instead of coronal stripping, as indicated by Wright *et al.* (2011). The blue dotted line marks the transition from saturated to non-saturated regime expected to occur at $Ro = 0.13$.

Our final sample is composed of stars with: X-ray luminosity ranging 2.4×10^{29} to 5.8×10^{30} erg s^{-1}; mass ranging from ~ 0.3 up to $\sim 1.6\,{\rm M}_{\odot}$; rotational period ranging from ~ 0.13 and ~ 15.9 d; convective turnover time τ ranging from 9 and 96 d.

In fig. 1 we show the color magnitude diagram of our catalog. In in fig. 2 we show the mass vs period plot, with superimposed the line marking the transitions between the different regimes. From this plot it is clear how our stellar sample allows us to investigate the behavior of stars expected to be in the supersaturated, saturated, and non-saturated regimes.

3. Results

We investigated the relation between $\log L_{\rm X}/L_{\rm bol}$ vs Rossby number Ro. We separated our stellar sample into different mass bins because the predicted Ro values separating the different regimes of stellar dynamo are expected to vary with stellar mass. In fig. 3 we show the fractional X-ray luminosity vs Rossby number for different mass bins.

Supersaturated regime: We find that stars with $1.3\,{\rm M}_{\odot} < M < 1.4\,{\rm M}_{\odot}$ and with $Ro \leqslant 0.07$ display a significant correlation between their fractional X-ray luminosity and Rossby number. In fig. 3 the dotted gray line indicate the observed correlation. The limiting Ro value of 0.07 perfectly fits with the value predicted by coronal stripping assuming $R_{\rm cor} = 3R_{\star}$. Conversely lower mass stars do not show any clear evidence of supersaturation, even at the lowest Ro values, as instead expected considering the threshold shown in fig. 3. This difference indicates that for different stellar masses something different happens in the threshold between saturation and supersaturation.

Saturated regime: We observe that the fractional X-ray luminosity of saturation regime varies for different stellar masses, with lower masses displaying the larger fractional X-ray luminosity. In fact for $Ro < 0.13$ stars in the range $0.3-0.9\,{\rm M}_{\odot}$ show $\log(L_{\rm X}/L_{\rm bol})_{\rm mean} = -2.8$, while stars in the range $0.9 - 1.3\,{\rm M}_{\odot}$ show $\log(L_{\rm X}/L_{\rm bol})_{\rm mean} = -3.3$.

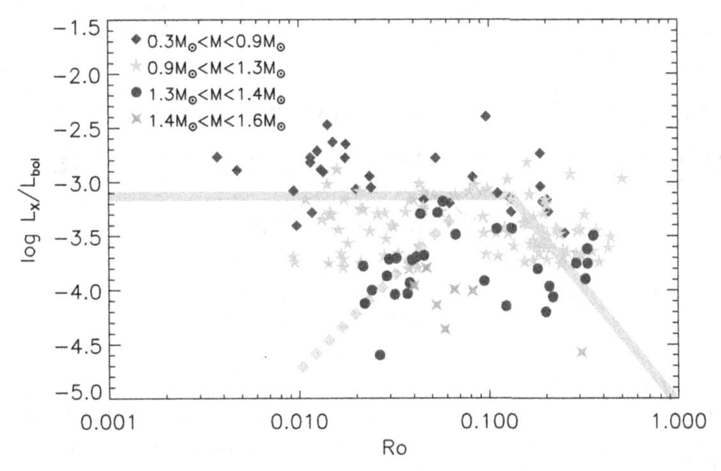

Figure 3. $\log L_X/L_{\rm bol}$ vs Rossby number *Ro*. Solid gray line marks the relation derived by Wright *et al.* (2011) for saturated and non-saturated region. Dotted gray line indicated the correlation found for stars with $1.3\,{\rm M}_\odot < M < 1.4\,{\rm M}_\odot$ in the supersaturation region.

Non-Saturated regime: Even if we have a significant fraction of stars having *Ro* corresponding to the non-saturated regime we do not find any clear trend indicating non-saturation, probably because of the large intrinsic scatter of the fractional X-ray luminosity.

4. Conclusion

In this work we present the study of the activity-rotation relation for PMS stars at 13 Myr. We find that the activity-rotation relation at this age varies significantly for different stellar masses. We observe that at 13 Myr the intrinsic scatter in the $\log L_X/L_{\rm bol}$ vs *Ro* plot is lower than that observed at younger ages. The very low *Ro* values accessible at this age allowed us to observe the first clear detection of supersaturation for PMS stars with mass ranging between 1.3 and $1.4\,{\rm M}_\odot$.

References

Currie, T., Hernandez, J., Irwin, J., *et al.* 2010, *ApJS*, 186, 191
Moraux, E., Artemenko, S., Bouvier, J., *et al.* 2013, *A&A*, in press
Noyes, R. W., Hartmann, L. W., Baliunas, S. L., Duncan, D. K., & Vaughan, A. H. 1984, *ApJ*, 279, 763
Pallavicini, R., Golub, L., Rosner, R., *et al.* 1981, *ApJ*, 248, 279
Pizzolato, N., Maggio, A., Micela, G., Sciortino, S., & Ventura, P. 2003, *A&A*, 397, 147
Preibisch, T., Kim, Y.-C., Favata, F., *et al.* 2005, *ApJS*, 160, 401
Randich, S., Schmitt, J. H. M. M., Prosser, C. F., & Stauffer, J. R. 1996, *A&A*, 305, 785
Siess, L., Dufour, E., & Forestini, M. 2000, *A&A*, 358, 593
Wright, N. J., Drake, J. J., Mamajek, E. E., & Henry, G. W. 2011, *ApJ*, 743, 48

Magnetic Fields throughout Stellar Evolution
Proceedings IAU Symposium No. 302, 2013
P. Petit, M. Jardine & H. Spruit, eds.

doi:10.1017/S1743921314001823

X-ray emission regimes and rotation sequences in the M34 open cluster

Philippe Gondoin

European Space Agency, ESTEC,
Postbus 299, 2200 AG, the Netherlands
email: pgondoin@rssd.esa.int

Abstract. I report on a correlation between the saturated and non-saturated regimes of X-ray emission and the rotation sequences that have been observed in the M34 open cluster. An interpretation of this correlation in term of magnetic activity evolution in the early stage of evolution on the main sequence is presented.

1. X-ray observations of the M34 open cluster

The M34 open cluster was observed with the *XMM – Newton* space observatory on 12 February 2003. Detection was made of 189 X-ray sources that are listed in the *XMM – Newton* Serendipitous Source Catalog (Watson *et al.* 2009). This list of X-ray sources was correlated with lists of M34 cluster members with known rotation periods established by Meibom *et al.* (2011), Irwin *et al.* (2006), and James *et al.* (2010). In total, 41 single stars in the M34 open cluster have been found that have known rotational periods and detected X-ray emissions (Gondoin 2012).

X-ray fluxes were derived from the source count rates using energy conversion factors (ECF) calculated in the 0.5-4.5 keV range (Gondoin 2006). The X-ray fluxes were converted into stellar X-ray luminosities assuming a distance of 470 pc (Jones & Prosser 1996). The X-ray luminosity distribution of the sample stars rolls off at luminosities lower than $L_X \approx 10^{29}$ erg s^{-1}, which provides a sensitivity limit estimate of the *XMM – Newton* observation. The mass of the sample stars ranges from 0.4 $M_\odot$ to 1.3 $M_\odot$ and reaches a maximum around 0.8 $M_\odot$. Their rotation periods are between 0.49 days and 11 days.

Recent studies have shown that stars tend to group into two main sub-populations that lie on narrow sequences in diagrams where the measured rotation periods of the members of a young stellar cluster are plotted against their B – V colors. Figure 1 (left) shows the rotational periods P of the sample stars as a function of their reddening corrected $(B - V)_0$ indices. The color-period diagram also displays the I and C rotational sequences of M34 along the form established by Barnes (2007) and Meibom *et al.* (2011). The proximity of the M34 data points to these curves was used to determine their membership to the I sequence, to the C sequence or to the gap. Figure 1 (right) displays the X-ray to bolometric luminosity ratio L_X/L_{bol} of the sample stars as a function of their Rossby number ($Ro = P/\tau_c$) distinguishing members of the I sequence, of the C sequence and of the gap.

Figure 1 shows a correlation between the X-ray activity regimes and the rotation sequences. Indeed, members of the C sequence have small Rossby numbers ($Ro < 0.1$), and an X-ray to bolometric luminosity level close to the 10^{-3} saturation level. Members of the I sequence, in contrast, have larger Rossby numbers ($Ro \geqslant 0.17$), and an X-ray to bolometric luminosity ratio significantly smaller than the saturation limit. Remarkably,

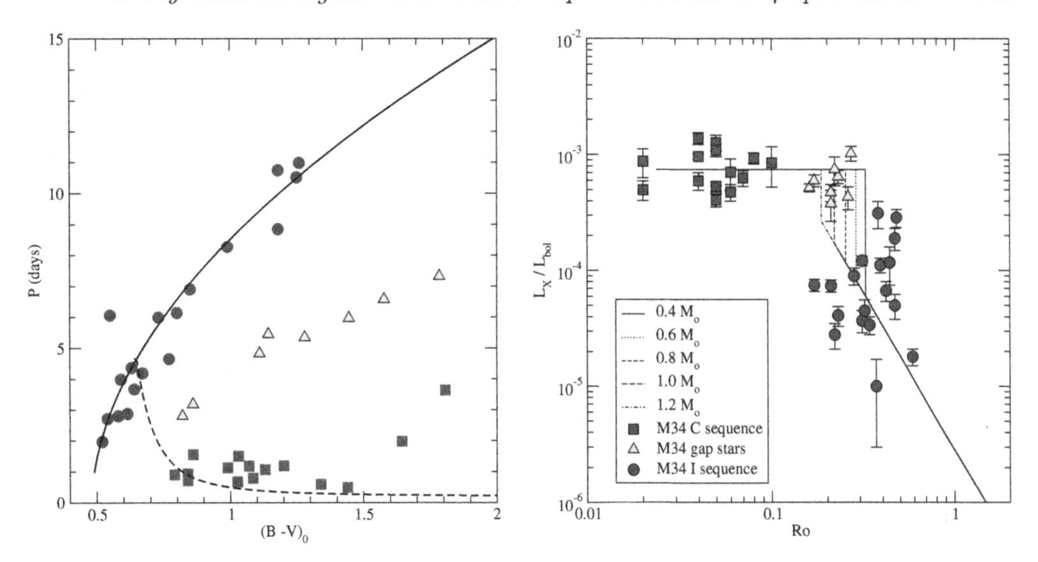

Figure 1. Left: rotation periods vs. (B−V) indices of the M34 sample stars. The solid line and the dashed lines represent the I and the C sequence, respectively. The proximity of the M34 data points to these curves was used to determine their membership to the I sequence (blue circles), to the C sequence (red square) or to the gap (grey triangles). Right: X-ray to bolometric luminosity ratio vs Rossby number of the sample stars compared models of X-ray activity evolution for stars with masses between 0.4 and 1.2 $M_\odot$ having an initial period of rotation of 1.1 days on the ZAMS.

gap stars occupy an intermediary position with Rossby numbers in the same range as those of some I sequence stars but with X-ray to bolometric luminosity ratio similar to those of C sequence stars, i.e., close to the saturation level.

2. A model of X-ray activity evolution

Based on this correlation, I derived a model of X-ray activity evolution (Gondoin 2013) assuming that stars on the C sequence and in the gap emits X-rays at the saturation level while I-sequence stars exhibit a power-law dependence of the X-ray to bolometric luminosity ratio as a function of the Rossby number. This model can be expressed as

$$\frac{L_{\rm X}}{L_{\rm bol}} = \begin{cases} R_{\rm X,sat} & \text{if star} \in \text{C sequence or gap,} \\ C^{st} \times Ro^{\beta} & \text{if star} \in \text{I sequence.} \end{cases} \tag{2.1}$$

Barnes (2010) proposed one particularly simple formulation of the rotation period evolution on the main-sequence star as a function of the convective turnover time $\tau_{\rm c}$ and the initial period of rotation on the ZAMS P_0,

$$t = \frac{\tau_{\rm c,B}}{k_{\rm C}} \times \ln\left(\frac{P(t)}{P_0}\right) + \frac{k_{\rm I}}{2\tau_{\rm c,B}} \times (P(t)^2 - P_0^2). \tag{2.2}$$

The combination of Eq. 2.1 and 2.2 provides a time evolution model of the stellar X-ray emission on the main sequence. The dependence of the X-ray luminosity on stellar mass is implicitly contained in the bolometric luminosity and in the convective turnover time. A parameterisation of the convective turnover time as a function of stellar mass has been provided by Wright *et al.* (2011). The bolometric luminosity is related to stellar mass on the main sequence by $M/{\rm M}_\odot = (L_{\rm bol}/{\rm L}_\odot)^{1/4}$ (e.g. Duric 2003).

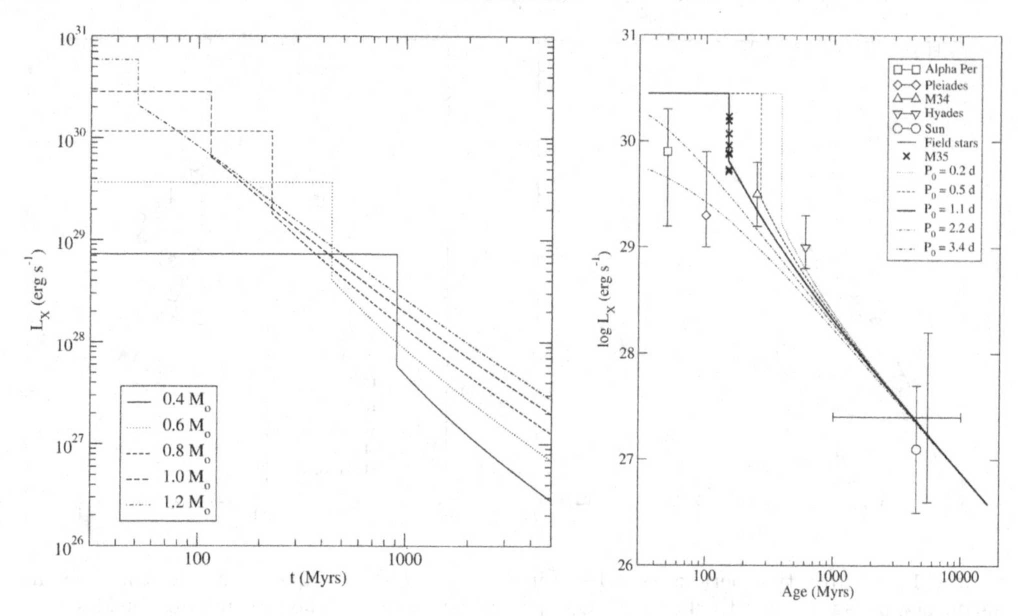

Figure 2. Left: simulated evolution of the X-ray luminosity of main-sequence stars with masses between 0.4 and 1.2 $M_\odot$ that have an initial period of rotation of 1.1 days on the ZAMS. Right: X-ray luminosity evolution models of solar mass stars with initial rotation periods on the ZAMS ranging from 0.2 to 3.4 days compared with median luminosities of stars with similar masses in the α Per, Pleiades, M34, and the Hyades. The vertical error bars indicate the 25 % and 75 % quartiles of the distributions (Micela 2002).

One graphical representation of the model is shown in Fig. 1 (right). It displays the X-ray to bolometric luminosity ratio as a function of Rossby number for stars with $0.4 \leqslant M/M_\odot \leqslant 1.2$ and $P_0 = 1.1$ days on the ZAMS. One interesting output of the model is that the hypothesis of a correlation between rotation sequences and X-ray emission regimes leads to a transition from the saturated to the non-saturated regime of X-ray emission in a range of Rossby number between 0.32 and 0.19 for stars with $0.4 \leqslant M/M_\odot \leqslant 1.2$ and $P_0 = 1.1$ day. Figure 1 (right) shows that gap stars in M34 are located in this domain.

Using the above model, I calculated the time evolution of the X-ray luminosities of Sun-like stars with a range of initial rotation periods observed in young stellar clusters. These stars are assumed to reach the ZAMS at an age of 35 Myrs, approximately equal to the Kelvin-Helmholtz timescale of a one solar mass star. Figure 2 (right) compares the simulated evolutionary track of their X-ray emissions with the measured X-ray luminosities of Sun-like stars in various open clusters. The model shows, in agreement with the measurements, a large dispersion of X-ray luminosities among young Sun-like stars. This large dispersion in the early phase of evolution on the main sequence occurs because fast rotating Sun-like stars emit X-ray in the saturated regime while slower rotators with similar masses already operate in the non-saturated regime of X-ray emission.

3. Physical interpretation

Numerical simulations (e.g. Käpylä *et al.* 2009) indicate that large-scale dynamos are excited in rapidly rotating convection, i.e. in the absence of shear. Such turbulent dynamos probably operate in young stars providing that their rotation is rapid enough. The transition from the C to the I rotation sequence appears to be associated with a sharp decrease of the X-ray to bolometric luminosity ratio. Since this ratio is a lower limit of the ratio between the surface magnetic flux and the outer convective flux, its steep decrease around $Ro \approx 0.14$ - 0.4 is indicative of a significant drop in dynamo efficiency possibly associated with a quenching of the turbulent dynamo in rapidly rotating stars as their rotation rate decays due to rotational braking by stellar winds.

The evolution model of X-ray activity (see Fig. 2 left) provides an estimate of the age $t_{\rm sat}$ at which a star changes from the saturated to the non-saturated regime of X-ray emission. This age can be compared with the age $t_{\rm gap}$ taken by a star to evolve through its C rotation phase, and to reach the nominal rotational gap g marking the onset of the I rotation phase. The comparison shows that $t_{\rm sat} \gg t_{\rm gap}$. The transition from saturated to non-saturated X-ray emission thus occurs well after the stellar evolution through the rotational gap between the C and I sequences, which corresponds to a maximum of the rotation deceleration. If the associated redistribution of angular momentum is the result of a nascent interface dynamo due to a developing gradient in angular velocity at the base of the convection zone, the model suggests that during the time interval between the rotation sequence transition and the X-ray regime transition two different dynamo regimes operate simultaneously within the interior of Sun-like stars.

According to the above scenario, the angular momentum redistribution mechanism responsible for the transition from the C to the I rotation sequence results in a changing mixture of two dynamo processes occurring side by side, i.e. a boundary-layer interface dynamo and a convective envelope turbulent dynamo. This last process dominates in rapidly rotating young stars. As the shear between the fast spinning radiative interior and the convective envelope increases, another process strengthens in which dynamo action occurs in the boundary region between the radiative core and the convective envelope. This dynamo process relies on differential rotation, but also induces important redistributions of angular momentum. As the rotation of the convective envelope decays, the turbulent dynamo is quenched and the interface dynamo becomes dominant, decreasing progressively at later stages of evolution when rotation dies away.

References

Barnes, S. A. 2007, *ApJ*, 669, 1167
Barnes, S. A. 2010, *ApJ*, 722, 222
Durney, B. R. & Latour, J. 1978, *Geophys. and Astrophys. Fluid Dyn.*, 9, 241
Flower, P. J. 1996, *ApJ*, 469, 335
Gondoin, P. 2006, *AA*, 454, 595
Gondoin, P. 2012, *AA*, 546, A117
Gondoin, P. 2013, *AA*, 556, A14
Irwin, J., Aigrain, S., Hodgkin, S., *et al.* 2006, *MNRAS*, 370, 954
James, D. J., Barnes, S. A., Meibom, S., *et al.* 2010, *AA*, 515, A100
Jones, B. F. & Prosser, C. F. 1996, *AJ*, 111,1193
Käpylä, P. J., Korpi, M. J., & Brandenburg, A. 2009, *ApJ*, 697, 1153
Micela, G. 2002, Stellar Coronae in the Chandra and XMM-Newton Era, ASP Conference Proceedings, Vol. 277, p.263
Meibom, S., Matthieu, R. D., Stassun, K. G., *et al.* 2011, *ApJ*, 733, 115
Watson, M. G., Schröder, A. C., Fyfe, D., *et al.* 2009, *AA*, 493, 339
Wright, N. J., Drake, J. J., Mamajek, E. E., & Henry, G. W. 2011, *ApJ*, 743, 48

Magnetic Fields throughout Stellar Evolution
Proceedings IAU Symposium No. 302, 2013
P. Petit, M. Jardine & H. Spruit, eds.

doi:10.1017/S1743921314001835

The evolution of surface magnetic fields in young solar-type stars

C. P. Folsom[1], P. Petit[1], J. Bouvier[2], J.-F. Donati[1] and J. Morin[3]

[1]Institut de Recherche en Astrophysique et Planétologie, Toulouse, France
email: colin.folsom@irap.omp.eu
[2]Institut de Planétologie et d'Astrophysique de Grenoble, France
[3]Institute of Astrophysics, Georg-August-University, Göttingen, Germany

Abstract. The surface rotation rates of young solar-type stars decrease rapidly with age from the end of the pre-main sequence though the early main sequence. This suggests that there is also an important change in the dynamos operating in these stars, which should be observable in their surface magnetic fields. Here we present early results in a study aimed at observing the evolution of these magnetic fields through this critical time period. We are observing stars in open clusters and stellar associations to provide precise ages, and using Zeeman Doppler Imaging to characterize the complex magnetic fields. Presented here are results for six stars, three in the in the β Pic association (~10 Myr old) and three in the AB Dor association (~100 Myr old).

1. Introduction

Solar-type stars undergo a dramatic evolution in their rotation rates as they leave the pre-main sequence and settle into their main sequence lives. Since these stars have dynamo driven magnetic fields, there is likely an important evolution in their magnetic properties over the same time period. In turn, stellar magnetic fields play a key role in angular momentum loss. Thus to fully understand the angular momentum evolution of these stars, it is critical to characterize their magnetic evolution. No study has yet traced the evolution of these magnetic properties in solar-type stars from the pre-main sequence through the early main sequence with any precision.

In order to investigate this, we are obtaining a series spectropolarimetric observations of solar type stars in open clusters using the ESPaDOnS instrument at the Canada France Hawaii Telescope. Focusing on stars in open clusters allows us to place precise ages on the targets. Using a time series of observations and the Zeeman Doppler Imaging (ZDI; Donati & Brown 1997; Donati *et al.* 2006) technique we can characterize the complex magnetic field geometries of the targets. Here we present early results from this project, focusing on on 3 stars in the β Pic association (~10 Myr old; Torres *et al.* 2008) and 3 stars in the AB Dor association (~100 Myr old; Torres *et al.* 2008; Barenfeld *et al.* 2013).

2. Analysis

Least Squares Deconvolution (LSD) was applied to our observations, and the resulting 'mean' line profiles were used to measure longitudinal magnetic fields and radial velocities. The longitudinal magnetic fields were used to determine a rotational period for each star, which was checked against the phasing of the LSD profiles, and the radial velocity variability. Stellar parameters: $T_{\rm eff}$, $\log g$, microturbulence, and $v \sin i$, were derived by fitting synthetic spectra computed with the ZEEMAN code to the observed spectra. These parameters, together with the rotation period, provide us with accurate self consistent values upon which to base ZDI maps.

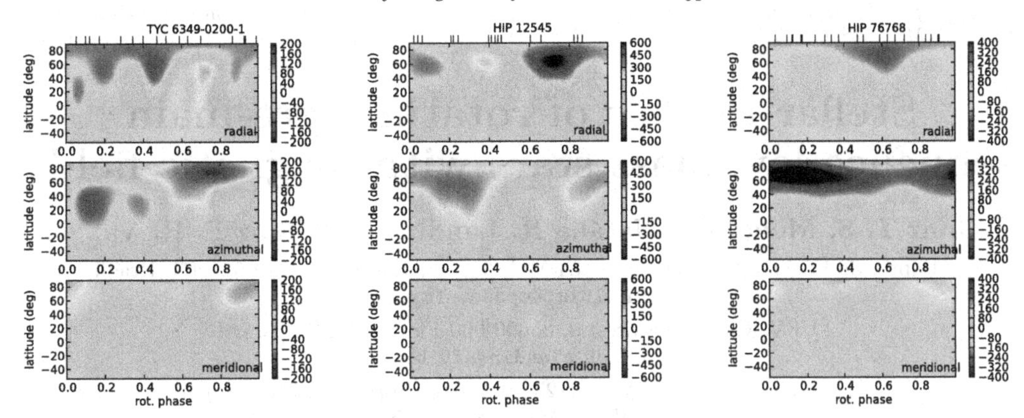

Figure 1. Sample magnetic field maps, from ZDI, for the stars TYC 6349-0200-1, HIP 12545, and HIP 76768. Plotted are the radial, azimuthal, and meridional components of the magnetic field. Tick marks at the top of the figure indicate phases at which observations were obtained.

ZDI was performed for the six stars studied so far, using a maximum entropy image reconstruction procedure, and representing the magnetic field as a combination of spherical harmonics (Donati *et al.* 2006). The maps were based off of the computed LSD profiles, which was necessary to provide sufficient S/N.

3. Early results

Of the six stars analyzed so far, three are in the β Pic association (~10 Myr): HIP 12545, TYC 6349-0200-1, and TYC 6878-0195-1, and three are in the AB Dor association (~100 Myr): HIP 76768, TYC 0486-4943-1, and TYC 5164-0567-1. We find similar rotation periods for all six stars, from 3.4 days (TYC 6349-0200-1) to 5.7 days (TYC 6878-0195-1), with no clear distinction between the two associations. We also find similar effective temperatures for the stars, from 4400 K (TYC 6349-0200-1) to 5100 K (TYC 5164-0567-1). However, we find significant differences in the mean magnetic field modulus from the ZDI maps, from 40 G (TYC 0486-4943-1) to 180 G (HIP 12545). We also find significant differences in the geometries of the reconstructed magnetic fields. Three sample ZDI maps are presented in Fig. 1.

Trends in the evolution of the magnetic field with rotation period or age are weak at best for the small range of parameter space probed by these stars. However, as this project progresses, it will be extended to associations with older ages, and to stars with a wider range of rotation periods. This will enable us to draw much stronger conclusions than we can from the limited dataset currently available.

References

Barenfeld, S. A., Bubar, E. J., Mamajek, E. E., & Young, P. A. 2013, *ApJ* 766, 6

Donati, J.-F. & Brown, S. F. 1997, *A&A* 326, 1135

Donati, J.-F., Howarth, I. D., Jardine, M. M., Petit, P., Catala, C., Landstreet, J. D., Bouret, J.-C., Alecian, E., Barnes, J., Forveille, T., Paletou, F., & Manset, N. 2006, *MNRAS* 370, 629

Torres, C. A. O., Quast, G. R., Melo, C. H. F., & Sterzik, M. F. 2008, in: Reipurth, B. (ed.), *Handbook of Star Forming Regions, Volume II The Southern Sky*, (ASP Monograph Publications), 5, 757

Magnetic Fields throughout Stellar Evolution
Proceedings IAU Symposium No. 302, 2013
P. Petit, M. Jardine & H. Spruit, eds.

doi:10.1017/S1743921314001847

Stellar models of rotating, pre-main sequence low-mass stars with magnetic fields

Luiz T. S. Mendes[1,3], **Natália R. Landin**[2,3] **and Luiz P. R. Vaz**[3]

[1]Depto. de Engenharia Eletrônica, UFMG, 31270-901 Belo Horizonte, MG, Brazil
email: luizt@cpdee.ufmg.br

[2]UFV, Campus Florestal, 35690-000 Florestal, MG, Brazil
email: nlandin@ufv.br

[3]Depto. de Física, UFMG, 31270-901 Belo Horizonte, MG, Brazil
email: vaz@fisica.ufmg.br

Abstract. We report our present efforts for introducing magnetic fields in the ATON stellar evolution code code, which now evolved to truly modifying the stellar structure equations so that they can incorporate the effects of an imposed, large-scale magnetic field. Preliminary results of such an approach, as applied to low-mass stellar models, are presented and discussed.

Keywords. Stars : Magnetic fields – Stars : Evolution – Stars : Rotation.

---

1. Introduction and Method

The mean-field $\alpha - \Omega$ dynamo model has been the most explored theoretical tool for a better understanding of magnetic field generation in the Sun and probably also in sun-like stars. However, dynamo models are inherently 3-D, making them difficult to be integrated with 1-D stellar evolution codes. Fortunately, a method first proposed by Lydon & Sofia (1995, hereafter LS95) allows a self-consistent modification of the stellar structure equations in order to incorporate the effects of a large-scale magnetic field to 1-D stellar evolution codes. Here we report the application of that method to the ATON 2.3 evolution code and some preliminary results obtained with it.

The LS95 method treats the magnetic field as a perturbation on the stellar structure equations, by means of a new state variable $\chi = B^2/(8\pi\rho)$ representing the magnetic energy density. The true 3-D nature of the corresponding magnetic pressure P_χ is crudely simplified to represent only the radial dependence of the magnetic pressure. The transition from the intrinsic 3-D magnetic field geometry to this 1-D approximation is described by a numerical factor γ, so that $P_\chi = (\gamma - 1)\,\chi\,\rho$. The reader is referred to the LS95 work for full details regarding this technique.

2. The Models

The main features of the ATON 2.3 evolution code such as opacities, diffusive mixing, overshooting, convection treatment and structural effects of rotation are described elsewhere (Ventura *et al.* 1998; Mendes *et al.* 1999). We computed rotating stellar models of 1 and 0.6 $M_\odot$ with solar chemical composition, mixing-length convection treatment with $\alpha = \Lambda/H_P = 1.5$, and initial rotation rate obtained from Kawaler's (1987) mass-radius and mass-moment of inertia relations for low-mass stars. A surface magnetic field strength < 100 G was adopted, that can be viewed as an "average" between the mean solar dipole field of 1 G and the kG values observed in T Tauri stars. The field scales throughout the stellar interior preserving the surface ratio between the magnetic and gas energy densities (D'Antona *et al.* 2000).

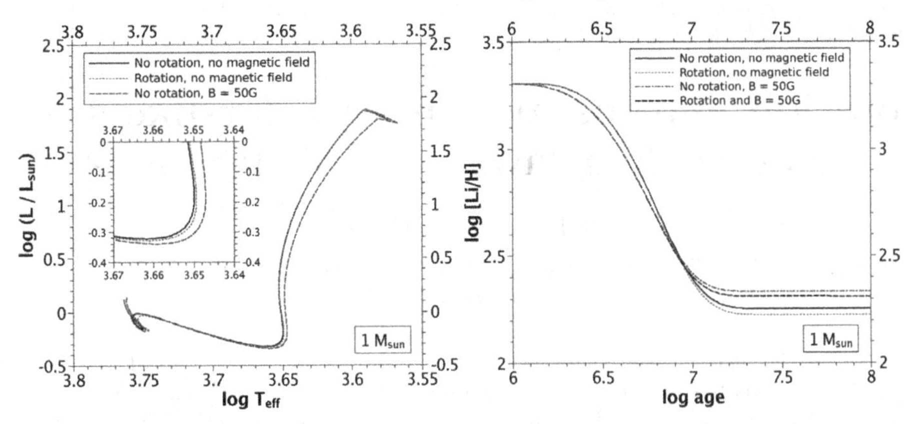

Figure 1. *Left panel:* Evolutionary tracks for 1.0 $M_\odot$ models corresponding to a standard model with no rotation and no magnetic field, a rotating model and a "magnetic" model. *Right panel:* Lithium depletion as a function of age for the 1 $M_\odot$ model. The long-dashed line represents the combined effects of rotation and magnetic fields.

3. Preliminary results

The effects of a magnetic field on the stellar structure are shown in the left panel of Fig. 1. We confirm previous findings that the magnetic field perturbation reduces the surface convection zone of the models (e.g. Mullan & MacDonald 2001; Feiden & Chaboyer 2012). At first, one could think that such a reduction could mimic a slightly more massive star resulting in a hotter track, but the role played by the excess overadiabacity induced by the magnetic perturbation actually makes the tracks slightly cooler, as explained in D'Antona *et al.* (2000). We also see that the magnetic perturbation has a larger effect than rotation on the stellar models, especially during the Hayashi tracks.

Another long-dicussed physical effect of a magnetic field on the structure of low-mass stars concerns the lithium depletion: as the magnetic field suppresses thermal convection, the surface lithium abundance should be higher as confirmed by Fig. 1 (right panel). However, our abundance levels under the presence of a magnetic field differ significantly from those of other researchers (e.g. Li & Bi 2012) and this deserves further investigation.

We plan to extend the model computations to a larger range of stellar masses and surface magnetic field strengths in the immediate future, in order to use the available observations of stellar magnetic fields as constraints to our models.

Acknowledgments. The authors thank Drs. I. Mazzitelli and F. D'Antona (INAF, Italy) for allowing them to implement physical improvements in their `ATON 2.0` evolution code. L. T. S. Mendes acknowledges the partial support of IAU for attending this meeting. N. R. Landin and L. P. R. Vaz acknowledge the support of CAPES, FAPEMIG and CNPq Brazilian agencies.

References

D'Antona, F., Ventura, P., & Mazzitelli, I. 2000, *ApJL*, 543, L77
Feiden, G. A. & Chaboyer, B. 2012, *ApJ*, 761, 30
Kawaler, S. 1987, *PASP*, 99, 1322
Li, T. & Bi, S. 2012, in Solar and Astrophysical Dynamos and Magnetic Activity, ed. A. G. Kosovichev, E. M. de Gouveia dal Pino, & Y. Yan (Cambridge University Press), 201
Lydon, T. J. & Sofia, S. 1995, *ApJS*, 101, 357
Mendes, L., D'Antona, F., & Mazzitelli, I. 1999, *A&A*, 341, 174
Mullan, D. J., & MacDonald, J. 2001, *ApJ*, 559, 353
Ventura, P., Zeppieri, A., Mazzitelli, I., & D'Antona, F. 1998, *A&A*, 334, 953

Magnetic Fields throughout Stellar Evolution
Proceedings IAU Symposium No. 302, 2013
P. Petit, M. Jardine & H. Spruit, eds.

doi:10.1017/S1743921314001859

Rotation and magnetism of solar-like stars: from scaling laws to spot-dynamos

Allan Sacha Brun

AIM-IRFU/SAp, CEA/CNRS/University of Paris 7, CEA-Saclay, 91191 Gif-sur-Yvette, France
email: sacha.brun@cea.fr

Abstract. The Sun is the archetype of magnetic star and its proximity coupled with very high accuracy observations has helped us understanding how solar-like stars (e.g with a convective envelope) redistribute angular momentum and generate a cyclic magnetic field. However most solar models have been so fine tuned that when they are applied to other solar-like stars the agreement with observations is not good enough. I will thus discuss, based on theoretical considerations and multi-D MHD stellar models, what can be considered as robust properties of solar-like star dynamics and magnetism and what is still speculative. I will derive scaling laws for differential rotation and magnetic energy as a function of stellar parameters, discuss recent results of stellar dynamo models and define the new concept of spot-dynamo, e.g. global dynamo that develops self-consistent magnetic buoyant structures that emerge at the surface.

Keywords. Sun: convection, rotation, magnetism; dynamo; stellar magnetism and rotation

1. Introduction

The Sun exhibits fascinating magnetic phenomena, with sunspot emergence, flares, prominences and CME's most of which varies in number and intensity during the 11-yr activity cycle. Being able to understand the origin of such a large variety of magnetic manifestations and their link to the underlying solar dynamo has been challenging. Many observations show for instance that the symmetric (quadrupolar-like) and antisymmetric (dipolar-like) dynamo families come into play to modulate the 11-yr cycle, make one hemisphere lag the other during reversals (DeRosa *et al.* (2012)) and sometimes even lead to grand minima of activity (Tobias(1997)). It is thus important in order to progress in our current understanding of the solar dynamo to characterize how dynamo actions varies as a function of stellar parameters. Thanks to improved instrumentations, observations of the magnetism of solar-type stars, i.e. stars possessing a deep convective envelope and a radiative interior (late F, G, K and early M spectral type) are becoming more and more available Giampapa (2005). One difficulty of such observational programs is that they require long term observations since stellar cycle periods are likely to be commensurate to the solar 11-yr sunspots cycle period. Thanks to the data collected at Mount Wilson Observatory since the late 60's, such data is available (Wilson (1978), Baliunas *et al.* (1995)). Among the sample of 111 stars (including the Sun as a star) originally observed between the F2 to M2 spectral types, it is found that about 50% of the stars possess a cyclic activity, with cycle (starspot) periods varying roughly between 5 to 25 yrs, i.e. between half to twice the sunspot cycle period. They further indicate that among the inactive stars of the sample some are likely to be in a quiet phase (as was the Sun during the Maunder minimum). Overall activity cycles seem to be more frequent for less massive K stars than for F stars. More recent observational programs have been pursued that now even provide information on the field topology as a function of the rotation rate, such as the one using the Espadons and Narval instruments and the Zeeman Doppler

Imaging technique (Donati *et al.* (1997)). Applying this observational technique over a sample of four solar analogues with rotation rate Ω_0 varying from one to three times solar $\Omega_\odot$, Petit *et al.* (2008) have shown that the field amplitude increases as a function of the star's rotation rate and, more importantly, becomes more and more dominated by its toroidal component (modulo possible bias in the observational technique used). If such a trend is confirmed, i.e. that the field topology is becoming more toroidal with increasing rotation rate, it is a very important and instructive result and puts strong constraints on the dynamo models. In a more recent study Morgenthaler *et al.* (2011) have continued to monitor these stars over several years and have observed that some of them underwent a reversal of their global magnetic field, confirming the tendancy of solar-like stars to have time varying (cyclic?) global field polarity.

The systematic analysis of stellar magnetism data revealed that for solar type stars there is a good correlation between the cycle and rotation periods of the stars and that correlation is even stronger when using the Rossby number ($Ro = P_{rot}/\tau$) that takes into account the convection turnover time τ at the base of the stellar convective envelope (Noyes *et al.* (1984), Baliunas *et al.* (1996)). As the star rotates faster, its cycle period is found to be shorter. Typically, Noyes *et al.* (1984) found that $P_{cyc} \propto P^n_{rot}$, with $n = 1.25 \pm 0.5$. Based on an extended stellar sample Saar & Brandenburg (1999) and Saar (2002) have argued that there is actually two branches when plotting the cycle period vs the rotation period of the stars (results later confirmed by Böhm-Vitense(2007)). They make the distinction between the primary (starspot) cycle and Gleissberg or grand minima type modulation of the stellar activity. For the active branch they found an exponent $n \sim 0.8$ and for the inactive stars $n \sim 1.15$. It is also found that this correlation breaks at high rotation rate with the possible appearance of a super active branch. Recent progress based on asteroseismic data from Corot and *Kepler* have also started to put new constraints on stellar magnetism by increasing the number of observed stars (Mathur *et al.* (2013)) thanks to the change of frequency of oscillations (mostly acoustic modes) induced by magnetic field. There too evidence for cyclic activity are found (García *et al.* (2010)). It was also noticed that at very high rotation rate, the chromospheric (soft-X ray) activity level usually used as a good proxy for stellar magnetism, is saturating. The saturation of the X-ray luminosity seems to limit the validity of the scaling found at more moderate rotation rates (Pizzolato *et al.* (2003), Wright *et al.* (2011)). For G type stars this saturation is found for rotation rate above 35 kms^{-1}, for K type stars at about 10 kms^{-1} and for M dwarfs around 3-4 kms^{-1}, so about a Rossby number of 0.1. Note that the observed quantity Bf, where B is the field amplitude and f the surface filling factor, does not allow to distinguish if B actually saturates with Ω_0 or if only the filling factor f does (Reiners (2012)). It is thus also important to understand through dynamo simulations how stellar magnetic flux scales with rotation rate since it is telling us how the magnetic field generated by dynamo action inside the stars emerges and imprints the stellar surface and how it varies.

2. Stellar Dynamo: theoretical concepts

Stars possess a priori all the ingredients necessary to the development of a dynamo instability (Weiss (1994)), such as a large-scale shear (or differential rotation), turbulent motions, helicity (thanks to rotation and its associated Coriolis force) and low diffusivity. All these properties are favourable to the emergence by dynamo action of a magnetic field. Observations in the Sun and in most solar-like stars (which much less details of course), of phenomena such as starspots or flares, clearly hint to the presence of magnetic fields. Their temporal dependence and properties has led naturally to consider that the origin of this magnetism is indeed dynamo action. However, we also know that stellar magnetic activity manifests itself in a multitude of facets (irregular, cyclic, modulated), certainly indicating the presence of several types of dynamo or magnetism. Until the recent advent of massively parallel super computers, astrophysicists were especially interested with the cyclic and large-scale dynamo and developed simplified models based on mean field dynamo theory and have put forward the fundamental concept of α and ω effects (Moffatt(1978), Charbonneau (2010)).

Theoretical considerations to interpret stellar magnetism based on classical mean field α-ω dynamo models (Durney & Latour (1978), Baliunas *et al.* (1996), Montesinos *et al.*(2001)) naturally yield correlation between rotation rate and stellar activity. In particular it is found that both magnetic field generation and the dynamo number D (i.e. a Reynolds number characterizing the mean field α and ω dynamo effects used in the models) vary with the rotation period of the star $D \propto 1/Ro^2$. This is due to the fact that in these models both effects are sensitive to the rotation rate of the star. The ω-effect is a direct measure of the differential rotation $\Delta\Omega$ established in the star. It is well known both theoretically and observationally that the differential rotation in the convective envelope of solar-type stars is directly connected to the star's rotation rate Ω_0 (Donahue *et al.*(1996), Barnes *et al.*(2005), Ballot *et al.* (2007), Brown *et al.* (2008), Küker *et al.*(2011), Matt *et al.* (2011), Augustson *et al.* (2012), Gastine *et al.* (2013)). However, the exact scaling exponent n_r (i.e. $\Delta\Omega \propto \Omega_0^{n_r}$) is still a matter of debate among both the observers and the theoreticians, being sensitive to both the observational techniques used and to the modelling approach. Likewise since the α-effect is a parameterization of the mean electromotive force (*emf*), it was actually shown to be directly related to helical turbulence (Moffatt(1978), Pouquet *et al.*(1976)), thus naturally connected to the rotation rate of the star and the amount of kinetic helicity present in its convective envelope. So this explains why in α-ω dynamo model it is straigthforward to related rotation and dynamo action.

However the currently preferred solar dynamo model, e.g. the so called flux transport Babcock-Leigthon dynamo model (Dikpati *et al.* (2004)), relies not on the α-effect to regenerate the poloidal component of the magnetic field but on the so-called Babcock-Leighton effect (Babcock(1961), Leighton(1969)), e.g. the tendancy for active regions or sunspot bipoles to be tilted with respect to the east-west direction (Joy's law). So as the active regions decay away over several weeks, the poloidal component of the diffuse field plays the actual role of a source term. This tilt is thought to be due to the action of the Coriolis force during the rise and emergence of the toroidal structures as active regions (D'Silva & Choudhuri(1993)). So here too a simple link to rotation can be obtained. Note however that recent 3-D simulations in spherical shells with developed convection motions (Jouve *et al.*(2013), and references therein) indicate that this is not the only effect responsible for the observed tilt and that the twist and arching of the toroidal structures as well as the continuous action of the surface convection during the emergence have some influence on the resulting tilt. So it may not be as simple as anticipated to relate a Babcock-Leigthon like source term to rotation and one may anticipate a different scaling between D and Ro than is standard α-ω dynamos. Further another important ingredient in flux transport models is the large scale meridional circulation (MC) used to connect the surface source term generating the poloidal field to the region of strong shear at the base of the convection zone (i.e. the tachocline) where it will be subsequently sheared by the ω-effect in order to close the global dynamo loop (i.e. $B_{pol} \rightarrow B_{tor} \rightarrow B_{pol}$). The meridional flow (or "conveyor belt") thus plays an important role in setting the cycle period of the global dynamo in this class of models. As a direct consequence it is natural to ask how the meridional circulation amplitude and profile change with the rotation rate and how these may influence the magnetic cycle period. Several authors have thus looked at the influence of the meridional circulation on the butterfly diagram and activity cycle period (Dikpati *et al.*(2001), Charbonneau & Saar(2001), Nandy(2004), Jouve & Brun (2007), Nandy & Martens(2007)). They all reached the same conclusion: only a positive scaling of the amplitude of meridional flows with the rotation rate can reconcile the models with observations of magnetism of solar-like stars. Unfortunately as we will discuss in the next sections, 3-D simulations actually find the opposite, the meridional circulation actually weakens with faster rotation rates, and this has important consequences for current stellar dynamo model as demonstrated by Jouve *et al.* (2010). So while it is easy to find a link between magnetic field amplitude, cycle period and rotation rate, observations of stellar magnetism actually impose that these relationship follow very specific trends not necessarily fitting our current solar dynamo paradigm.

3. 2-D mean field models of stellar magnetism

As we have seen, the observational correlation between rotation and activity obtained by Noyes *et al.* (1984), Baliunas *et al.* (1996), Saar & Brandenburg (1999), Böhm-Vitense(2007) could be due to the influence of rotation on dynamo action in stellar convective envelopes. However, flux transport dynamo model are in difficulty because their cycle period P_{cyc} depends strongly on the meridional circulation amplitude (as well as its profile, Jouve & Brun (2007)):

$$P_{cyc} \propto \Omega_0^{0.05} s_0^{0.07} v_0^{-0.83} \tag{3.1}$$

As will be seen in §4 the meridional circulation is found to decrease with the rotation rate as $v_0 \propto \Omega_0^{-0.45}$. This is not intuitive as one could expect that the meridional circulation increases with the rotation rate. A careful study of the vorticity equation shows that it actually weakens with rotation rate as more and more kinetic energy is being transferred to longitudinal motions at the expense of meridional kinetic energy. The fact that in recent 3-D simulations the meridional circulation is found to weaken as the models is rotated faster directly implies that standard advection dominated flux transport dynamo models yield the opposite dependency with rotation than the one observed, e.g. activity cycles are found to be longer for faster rotating stars (Jouve *et al.* (2010)). This fact alone impose to revise our current dynamo paradigm for solar-like stars. One way is to shortcircuit the advection path by for instance adding more cells in latitude or increase the radial diffusion as was done in Jouve *et al.* (2010), Hazra *et al.*(2013). Considering several cells either in latitude and/or radius for the meridional circulation is actually in better agreement with numerical simulations of rotating convection zone at low Rossby number and seems to also be observed in the Sun (Zhao *et al.*(2013)). An alternative is to consider another transport process such as magnetic turbulent pumping. We will now discuss this new class of dynamo models in more details (see also Guerrero & de Gouveia Dal Pino (2008), DoCao & Brun (2011)).

Magnetic pumping refers to transport of magnetic fields in convective layers that does not result from bulk motion. One particular case is turbulent pumping. In inhomogeneous convection due to density stratification, convection cells take the form of broad hot upflows surrounded by a network of downflow lanes Miesch *et al.* (2008). In such radially asymmetric convection, numerical simulations show that the magnetic field is preferentially dragged downward (Tobias *et al.*(2001)). This effect has been demonstrated to operate in the bulk of the solar convection zone. A significant equatorward latitudinal component also arise when rotation becomes important, i.e. when the Rossby number is less than unity. Turbulent pumping speeds of a few ms^{-1} can be reached according to the numerical simulations of Käpylä *et al.* (2006). Therefore, its effects are expected to be comparable to those of meridional circulation.

In the mean field dynamo framework in order to model the global dynamo operating in solar-like stars, we start from the induction equation, which after the usual scales separation between mean and fluctuating fields, e.g. $\mathbf{B} = \langle \mathbf{B} \rangle + \mathbf{b}$, becomes (Moffatt(1978)):

$$\frac{\partial \langle \mathbf{B} \rangle}{\partial t} = \nabla \times (\langle \mathbf{V} \rangle \times \langle \mathbf{B} \rangle) + \nabla \times \langle \mathbf{v} \times \mathbf{b} \rangle - \nabla \times (\eta_m \nabla \times \langle \mathbf{B} \rangle). \tag{3.2}$$

A closure relation must then be used to express the mean electromotive force (*emf*) $\epsilon = \langle \mathbf{v} \times \mathbf{b} \rangle$ in terms of mean magnetic field, leading to a simplified mean-field equation. If the mean magnetic field varies slowly in time and space, the *emf* can be represented in terms of $\langle \mathbf{B} \rangle$ and its gradients

$$\epsilon_i = a_{ij} B_j + b_{ijk} \frac{\partial B_j}{\partial x_k} + ... \tag{3.3}$$

where a_{ij} and b_{ijk} are in the general case tensors containing the transport coefficients (for simplicity we dropped $\langle \rangle$). The tensors a_{ij} and b_{ijk} cannot, in general, be expressed from first principles due to the lack of a comprehensive theory of convective turbulence. In the kinematic regime where the magnetic energy is negligible in comparison to the kinetic energy, the most simple approximation is to neglect all correlations higher than second order in the fluctuations. This is the so-called first order smoothing approximation (FOSA); Charbonneau (2010). In most studies isotropic turbulence is assumed and the pseudo tensor a_{ij} reduces into a single scalar giving rise to the α-effect. However in the full tensor non-isotropic case for a, the *emf* can be expressed as:

$$\epsilon = (\alpha \mathbf{B} + \gamma \times \mathbf{B}) - \beta \nabla \times \mathbf{B} \tag{3.4}$$

where α is a scalar referring to the standard α-effect. The term γ is the turbulent pumping and β is defined such that $b_{ijk} = \beta \epsilon_{ijk}$ (with ϵ_{ijk} the Levi-Civita tensor) and represents the turbulent enhancement of magnetic diffusion. Note that as we work in the framework of Babcock-Leighton flux transport models, we will replace the $\alpha \mathbf{B}$ term by a non local source term S representative of flux emergence.

Stellar mean field dynamo models have been studied in the case of a shallow MC by Guerrero & de Gouveia Dal Pino (2008). We have expanded in DoCao & Brun (2011) their results by considering both a deeper MC and various rotation rates, applying this pumping dominated dynamo models to other stars. Under certain conditions we showed that turbulent pumping can shorten the advection path driven by MC. We refer to DoCao & Brun (2011) for detailed analytical expression of the turbulent pumping. On Figure 1 we show buttefly diagrams obtained with this new dynamo model. We clearly see that for faster rotation the cycle period is shorter and the butterfly diagram remains solar-like with the correct phase relationship between the poloidal and toroidal field components.

By studying a large range of paremeters, DoCao & Brun (2011) were able to derive the following dependancy for the cycle period P_{cyc}:

$$P_{cyc} \propto v_0^{-0.40} \gamma_{r0}^{-0.30} \gamma_{\theta 0}^{-0.15} \tag{3.5}$$

We found that the turbulent pumping becomes a major player in setting the magnetic period, but its influence is not as large as in Guerrero & de Gouveia Dal Pino (2008). First, the MC is still the dominant effect and the radial pumping component is not as important. Second, the effect of γ_θ is not negligible. This supports the idea that the latitudinal advection process, and especially at the BCZ, is an important ingredient in advection dominated BL models, capable of transporting the toroidal magnetic field from the pole toward the equator. This difference may come from their choice of a shallow MC with almost zero velocity at the BCZ.

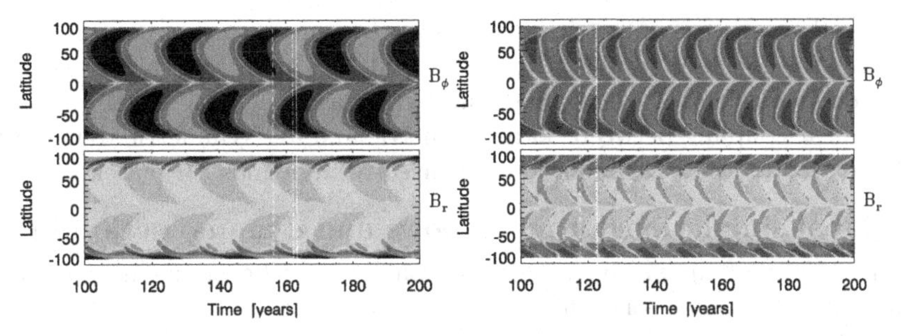

Figure 1. Butterfly diagram for 2 representative cases : $0.7\Omega_\odot$ and $3.0\Omega_\odot$. Both figures share the same color scale : between -510^3G and 510^3G for B_r and between -910^5G and 910^5G for B_ϕ. We also show the phase relations between B_{pol} and B_{tor}, (DoCao & Brun (2011))

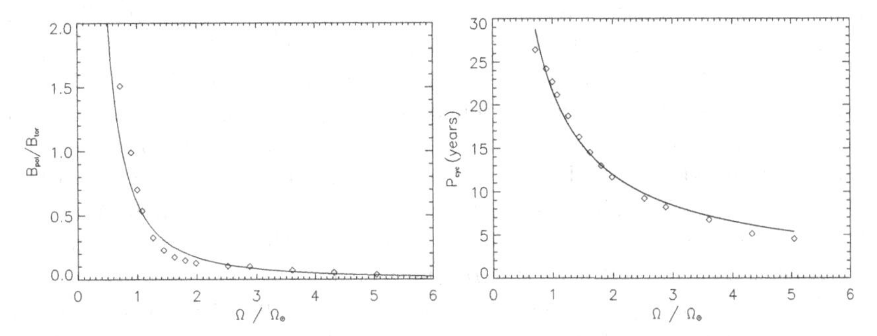

Figure 2. Left: $B_{\rm pol}/B_{\rm tor}$ ratio as function of the rotation rate. Solid line is a least square fit of the data. Right: Magnetic cycle period as function of the rotation rate in models including turbulent pumping. Solid line is a least square fit of the simulated data (DoCao & Brun (2011))

On Figure 2 we show the ratio of the poloidal to toroidal field and the cycle period as a function of the star's rotation rate. We clearly find the observed tendancy of a more and more dominant toroidal component and of shorter cycle. A simple look at the scaling law 3.5 gives that if we want to recover these observational trends, in particular the shorter cycle period (again, we assume that $v_0 \propto \Omega^{-0.45}$), and assuming that γ_r/γ_θ remains constant, the pumping effect should roughly scales as Ω_0^2. Such a scaling may be too extreme and only systematic 3-D numerical simulations will tell us if this is the case or not. Nevertheless, pumping dominated stellar dynamos are a plausible solution to explain observations.

Note that recent observations of the Sun and of solar analogues also point to the important role played by the symmetric family dynamo modes (such as the axisymmetric quadrupole) and that succesful dynamo models must also possess symmetric and not just antisymmetric equatorial symmetry. Indeed DeRosa *et al.* (2012) have shown that in the Sun the symmetric modes contribute to about 25% of the overall magnetic energy and that during reversals the quadrupolar mode actually dominates. This is certainly at the origin of the time lag of 1 to 2 years between the north and southern hemisphere in the Sun, since a dipole plus a quadrupole of equal amplitude would lead to an hemispherical dynamo. During grand minima activity phase, such as during the Maunder minimum, an hemispherical state with mostly sunspots in the southern hemisphere has been observed, there too, pointing for a strong contribution of the symmetric dynamo family Tobias(1997). It is certainly also the case that solar-like stars do not possess a purely antisymmetric state of their magnetic field and that both dipolar and quadrupolar-like symmetries are found Petit *et al.* (2008). Simple kinematic dynamos usually do not couple both dynamo familiies due to simple symmetry considerations of their main ingredients.

Only the introduction of asymmetric flows or source terms at the level of 0.1% can couple the families to the adequate level (DeRosa *et al.* (2012)). An alternative is to use dynamo coefficients deduced from 3-D numerical simulations as in Dubé & Charbonneau(2013). Indeed nonlinear coupling between both symmetric and antisymmetric dynamo families can easily be achieved in 3-D simulation of convective dynamos Strugarek *et al.* (2013).

4. 3-D global simulations of mean flows and dynamo action in stars

With the advent of massively parallel computers it is becoming more tractable to attack the difficult problem of stellar convection and dynamo with full 3-D MHD non linear simulations.

4.1. *Differential rotation and Meridional Circulation in Stars*

Systematic studies of rotating convection in spherical shells to model solar-like stars have been undertaken over the last 10 years by several groups and codes (Brun & Toomre (2002), Ballot *et al.* (2007), Brown *et al.* (2008), Käpylä *et al.* (2011), Augustson *et al.* (2012), Gastine *et al.* (2013), and references therein). The general trend is that the Coriolis force modifies convection such as to establish a large scale differential rotation $\Omega(r,\theta)$ (Brun & Rempel (2009)). Depending on the influence of the Coriolis force, usually mesured by the turbulent Rossby number $Ro = \omega_{conv}/2\Omega_0 \sim v_{conv}/2\Omega_0 d$, or a variant, with ω_{conv}, v_{conv} characteristic vorticity and velocity in the convection zone and d the convection zone depth, the resulting differential rotation can be anti-solar (high $Ro > 1$, with fast poles-slow equator), solar-like ($0.2 < Ro < 0.9$, with fast equator, slow poles and some constancy at mid latitude of the isocontours of Ω) or Jupiter-like ($Ro < 0.1$, with cylindrical profile with alternance of prograde and retrograde jets) Matt *et al.* (2011). On Figure 3 we represent these different profiles in models for various masses (0.5, 0.7 and 1.1 Msol) that also include the coupling to a stably stratified radiative interior, thus possessing a tachocline Matt & Brun (2013). For each stellar mass these various states can be achieved but for a different effective rotation rate (or $v \sin i$). Indeed, we find that the convective velocity v_{conv} roughly scales as $(L_*/(\bar{\rho}_{cz} R_*^2))^{1/3}$. Hence, more massive is the stars, higher is its luminosity and lower is its average density as the base of the convective envelope moves outward in relative mass (the stellar radius variation for a masse range between 0.5 to 1.2 solar mass is a factor of 2 at most). The direct consequence is that the convective velocity increases significantly with stellar mass and so does the Rossby number for a fixed rotation rate. So we anticipate that the transition between prograde and retrograde differential rotation does not occur at the same rotation rate for a given spectral type. Searching for this limit observationaly would be most useful to theoreticians. We also find that the differential rotation amplitude from the equator to 60 deg increases with rotation rate but not as fast as the rotation rate such that the relative differential rotation $\Delta\Omega/\Omega_0$ reduces. 3-D numerical simulations also predict that $\Delta\Omega$ should be larger for more massive stars as shown in Figure 4 left panel. This is in qualitative agreement with observations of Donahue *et al.*(1996) and Barnes *et al.*(2005).

Likewise the meridional circulation is influenced by rotation. We generally found that for simulations with $Ro < 1$, meridional circulations possess many cells in radius and/or latitude per hemisphere. Only for anti-solar differential rotation cases, is the meridional circulation uni-cellular Matt *et al.* (2011). We also find that the amplitude decreases as the rotational influence is increased as shown on Figure 4 right panel. This comes about by having more energy diverted toward differential rotation kinetic energy reservoir as motions tends to become more horizontal than to the meridional circulation kinetic energy reservoir that requires motions to go across surfaces aligned with the rotation axis which becomes harder as Ω_0 increases. As we have seen in §3, both multi-cellular

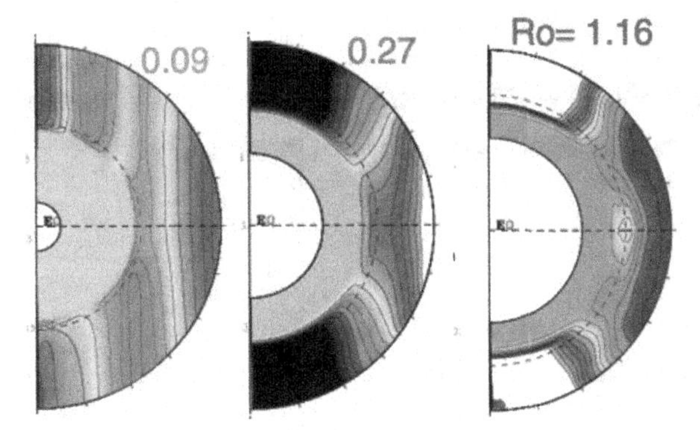

Figure 3. Differential rotation profiles realized in 3-D ASH simulations of solar-like G and K stars, chosen such as to emphasize the 3 rotation regimes as the Rossby number changes from < 0.1 to > 1: banded-cylindrical, solar-like-conical, anti-solar (slow equator-fast poles).

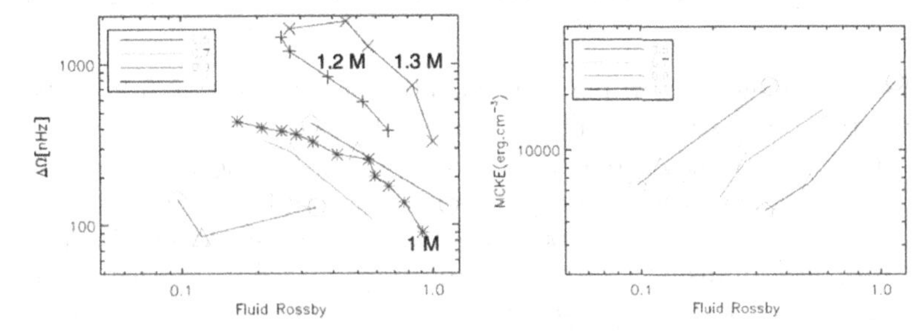

Figure 4. Left: Latitudinal variation of angular velocity contrast with Rossby number. We note that the contrast increases with faster rotation and that more massive is the star higher for a given Rossby number is its differential rotation. Right: Variation of kinetic energy in meridional circulation flow with Rossby number. We note that as the star rotates faster less energy is channeled to the meridional circualtion (Matt *et al.* (2011), Matt & Brun (2013))

meridional circulations and weaker flow amplitudes imply that we must reconsider the standard dynamo model for solar-like stars if we want to reproduce observations.

Up to now we have discussed results that did not take into account the retroaction of the magnetic field. This is of course correct as long as the feedback of the Lorentz force on motions is negligible. We now discuss in which conditions this is or not the case.

4.2. *Stellar nonlinear dynamo action, scaling laws of magnetic energy and spot-dynamo*

3-D numerical simulations of dynamo action in solar-like stars have revealed a large range of behavior, from steady dynamo, to irregular and cyclic ones (Brun *et al.* (2004), Brown *et al.* (2010), Brown *et al.* (2011), Racine *et al.*(2011), Gastine *et al.*(2012), Augustson *et al.* (2013), Käpylä *et al.*(2013), Nelson *et al.* (2013), and references therein). In particular in model with a dominant influence of the rotation, large scale magnetic wreaths (see Figure 6 left panel) have been obtained without requiring the presence of a tachocline Brown *et al.* (2010), Brown *et al.* (2011). We show on Figure 5, 4 butterfly diagrams (time-latitude plots of the azimuthally averaged toroidal magnetic field near the base of the CZ) realized in such simulations for of a solar-like star rotating at 3 times the solar rate Nelson *et al.* (2013). We remark that as the model is made more turbulent, the steady magnetic wreaths become more time dependent and can lead to cyclic activity (bottom right panel).

In such stars, along with the degree of turbulence, rotation plays an important role in determining the global properties of their magnetism. This is due to a shift in the

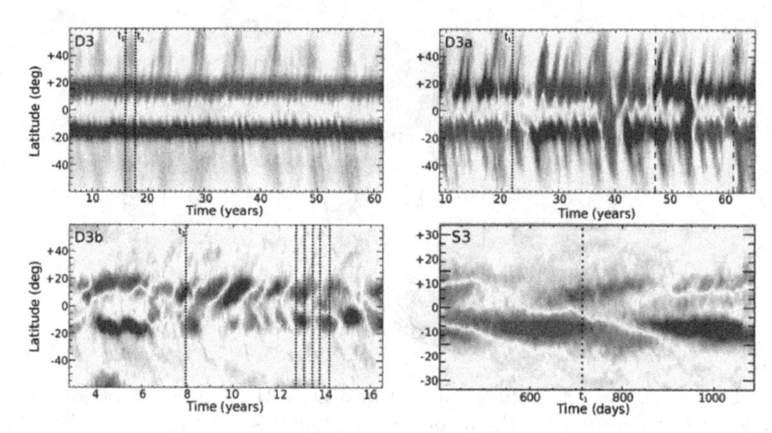

Figure 5. Magnetic wreaths yielding in turn steady (D3), irregular (D3a) and quasi cyclic (D3b & S3) magnetic butterfly diagrams (Brown *et al.* (2010), Nelson *et al.* (2013))

balance of forces driving the flow between the advection, Coriolis and Lorentz terms. As the rotation rate increases the Lorentz force tends to balance the Coriolis force yielding larger magnetic energy in superequipartion with the kinetic energy of the flow (a direct consequence of a magnetostrophic state; c.f. strong scaling below) as in the Earth's iron core. The Elsasser number $\Lambda = B^2/4\pi\bar{\rho}_{cz}\eta\Omega_0$, with $\bar{\rho}_{cz}$ mean density in the convective envelope, η magnetic diffusivity, Ω_0 stellar rotation rate, B a characteristic magnetic field of the CZ, is useful to discuss this balance of terms in the Navier-Stokes (N. V.) equation. Depending on the amplitude of this number and on the balance assumed in the Navier-Stokes equation, various scaling of the magnetic field amplitude can be expected (Fauve & Pétrélis(2007), Christensen(2010)):

• First, let's recall that an order of magnitude of an equilibrium magnetic field (assuming ideal gas law) can easily be obtained: $B_{eq} \sim \sqrt{8\pi P_{gas}} \sim \sqrt{\bar{\rho}_{cz}}$, since T_{eff} varies by a factor 2 to 3 between early F and late K stars, whereas $\bar{\rho}_{cz}$ varies by more than a factor 100 (Matt *et al.* (2011)). Now let's assume that the magnetic Reynolds number ~1 such that a characteristic velocity is given by $v \sim \eta/d$, and let's study the balance of terms in N. V. eq.:

• Laminar (weak) scaling: Lorentz ~ viscous diffusion
  $\Rightarrow B^2_{weak} \sim \bar{\rho}_{cz}\nu v/d \sim \bar{\rho}_{cz}\nu\eta/d^2$

• Turbulent (equipartition) scaling: Lorentz ~ advection
  $\Rightarrow B^2_{turb} \sim \bar{\rho}_{cz}v^2 \sim \bar{\rho}_{cz}\eta^2/d^2 \Leftrightarrow |B_{weak}| \sim |B_{turb}|P_m^{1/2}$

• Magnetostrophic (strong) scaling (e.g. Elsasser nb $\Lambda \sim 1$): Lorentz ~ Coriolis
  $\Rightarrow B^2_{strong} \sim \bar{\rho}_{cz}\Omega_0\eta$

with v, d characteristic velocity and length scales, $P_m = \nu/\eta$ the magnetic Prandtl nb. Of course there is an upper limit to the magnitude of the magnetic energy ultimately set by the amount of energy (likely the star's outward energy flux) than can be made available to the dynamo process. We recall here that dynamo action does not exist for any class of motions due to its intrinsic 3-D character (Moffatt(1978)).

So a possible scenario is the following: Stars rotating at moderate rate (such that their Elsasser number is small), have a level of magnetic energy (or averaged global field strenght) that are less than or of the order of the equipartition field given by either the weak or turbulent scalings. As stars rotate faster and get closer to be in a magnetostrophic state with an Elssasser number of order 1 or larger, the formation of large and intense magnetic wreaths starts. The magnetic field is more an more dominated by

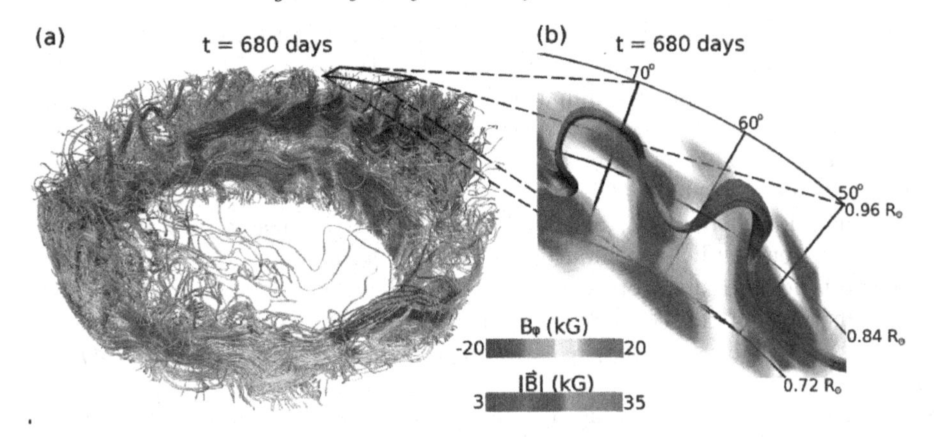

Figure 6. Turbulent magnetic wreaths (panel a) leading to the generation of buoyant loops (panel b) in case S3 (Nelson *et al.* (2013))

its toroidal component and the magnetic energy becomes larger and larger going above the equipartition value and follows the strong scaling. The consequence is the following. As the magnetic energy (or field amplitude) becomes large, the associated Lorentz force starts back reacting strongly on the mean flow. The first consequence is what can be called an "omega-quenching", e.g. the differential rotation reduces in strength and an almost solid body rotation state in the convective zone (envelope or core) is established (Brun (2004), Brun *et al.*(2005)). In mean-field classification this means that the stellar dynamo transits from being an $\alpha - \omega$ or $\alpha^2 - \omega$ to being an α^2 dynamo, i.e. helical turbulence is solely responsible for field generation and maintenance, the large scale shear now plays a marginal role. At that stage what remains of the magnetic wreaths is still unclear, more work must be done. As the rotation is made even faster, quenching of the α effect, due to the large scale magnetic field being more and more intense, occurs. The link to the L_x saturation (Pizzolato *et al.* (2003), Wright *et al.* (2011)) is not as straightforward to deduce as one must also assess how the filling factor of the magnetic field on the star's surface evolves with stellar parameter not just the field strenght. As we have seen the field amplitude does not easily saturate. So one can supposely think that the first saturation is due to ω-quenching and limitation of overall spot coverage and the second "over-saturation" may be due to "α-quenching", so of the actual field strength (see Gondoin(2012) for an alternative explanation). In order to be able to properly set the transition and the saturation of B and f independently, *spot-dynamo*, e.g. dynamo generating self consitently rising omega-loop must be developped and the parameter space explored systematically. We show a first step toward that goal in Figure 6 where we see a magnetic wreath-like structure becoming turbulent and intermittent enough, that intense bundles of fields reach 50 kG and start becoming buoyant, forming omega-loop like structures (Nelson *et al.* (2011), Nelson *et al.* (2013)). We believe that such simulations are the progenitor of future more realistic *spot-dynamos*.

5. conclusion

By extending the concept of dynamo to other stellar spectral type ($F \rightarrow M$), it would seem that a transition occurs from $\alpha - \omega$ + flux transport dynamos, to α^2 dynamos and a turbulent dynamo. We have seen that theory and numerical simulations can explain qualitatively the general trends. We now summarize the most important results:

- Convective velocities v_{conv} roughly scales with cubic root of $L_*/(\bar{\rho}_{cz} R_*^2)$ (stars luminosity devided by mean density in CZ and stellar radius squared). So it implies that prograde vs retrograde state changes at different Ω_0 as spectral type is changed (since $Ro = v_{conv}/(2\Omega_0 d)$ and v_{conv} changes with spectral type),

• Cylindrical vs conical vs shellular differential profiles depends on Reynolds stresses and thermal (baroclinic) effects (see Miesch *et al.* (2006) and Ballot *et al.* (2007) for more details). Larger absolute differential rotation for both more massive stars and higher rotation rates are recovered,

• The meridional circulation is found to be weaker for faster rotation rate, due to relatively more energy being channeled to longitudinal motions,

• Magnetic field B reduces or can even supress differential rotation $\Omega(r,\theta)$ (ω-quenching),

• at high rotation rate we get magnetic wreaths that generate omega-loops as we lower diffusivity, cyclic dynamos are easier to get, and a new concept of *spot-dynamo* has emerged,

• Strength of field (weak/strong) depends on balance of forces in N.V eq. and Multipolar or Dipolar magnetic bi-stability can exist but multipolar fields seem to dominate at high stratification,

• Observed stellar cycle period becomes shorter for faster rotation, implies to modify the standard flux transport mean field dynamo model to include either multi-cellular flow or turbulent pumping,

• Stratification and/or a tachocline and/or a low P_m may help getting equatorward butterfly diagram (results in a shift of location of $\Omega(r,\theta)$ and α-like effects and hence of their phase relationship).

This work was partly supported by ERC STARS2 207430 grant and ANR Toupies grant. I am thankful to University of Kyoto, RIMS and Prof. M. Yamada & S. Takehiro for their hospitality during Fall 2013. I thank K. Augustson, N. Featherstone, B. Brown, J. Toomre, M. Miesch, N. Nelson, A. Strugarek and S. Matt for useful discussions and for providing some of the figures.

References

Augustson, K. C., Brown, B. P., Brun, A. S., Miesch, M. S., & Toomre, J. 2012, *ApJ*, 756, 169
Augustson, K. C., Brun, A. S., & Toomre, J. 2013, *ApJ*, 777, 153
Babcock, H. W. 1961, *ApJ*, 133, 572
Baliunas, S. L., Donahue, R. A., Soon, W. H., *et al.* 1995, *ApJ*, 438, 269
Baliunas, S. L., Nesme-Ribes, E., Sokoloff, D., & Soon, W. H. 1996, *ApJ*, 460, 848
Ballot, J., Brun, A. S., & Turck-Chièze, S. 2007, *ApJ*, 669, 1190
Barnes, J. R., Collier Cameron, A., Donati, J.-F., *et al.* 2005, *MNRAS*, 357, L1
Böhm-Vitense, E. 2007, *ApJ*, 657, 486
Brown, B. P., Browning, M. K., Brun, A. S., Miesch, M. S., & Toomre, J. 2008, *ApJ*, 689, 1354
Brown, B. P., Browning, M. K., Brun, A. S., Miesch, M. S., & Toomre, J. 2010, *ApJ*, 711, 424
Brown, B. P., Miesch, M. S., Browning, M. K., Brun, A. S., & Toomre, J. 2011, *ApJ*, 731, 69
Browning, M. K., Miesch, M. S., Brun, A. S., & Toomre, J. 2006, *ApJL*, 648, L157
Brun, A. S. 2004, *Solar Phys.*, 220, 333
Brun, A. S. & Toomre, J. 2002, *ApJ*, 570, 865
Brun, A. S., Miesch, M. S., & Toomre, J. 2004, *ApJ*, 614, 1073
Brun, A. S., Browning, M. K., & Toomre, J. 2005, *ApJ*, 629, 461
Brun, A. S. & Rempel, M. 2009, *Space Sci. Rev.*, 144, 151
Brun, A. S., Miesch, M. S., & Toomre, J. 2011, *ApJ*, 742, 79
Charbonneau, P. & Saar, S. H. 2001, Magnetic Fields Across the H-R Diagram, 248, 189
Charbonneau, P. 2010, Living Reviews in Solar Physics, 7, 3
Christensen, U. R. 2010, *Space Sci. Rev.*, 152, 565
D'Silva, S. & Choudhuri, A. R. 1993, *A&A*, 272, 621
DeRosa, M. L., Brun, A. S., & Hoeksema, J. T. 2012, *ApJ*, 757, 96

Dikpati, M., Saar, S. H., Brummell, N., & Charbonneau, P. 2001, Magnetic Fields Across the H-R Diagram, 248, 235

Dikpati, M., de Toma, G., Gilman, P. A., Arge, C. N., & White, O. R. 2004, *ApJ*, 601, 1136

Do Cao, O. & Brun, A. S. 2011, Astron. Nach., 332, 907

Donahue, R. A., Saar, S. H., & Baliunas, S. L. 1996, *ApJ*, 466, 384

Donati, J.-F., Semel, M., Carter, B., Rees, D., & Collier Cameron, A. 1997, *MNRAS*, 291, 658

Dubé, C. & Charbonneau, P. 2013, *ApJ*, 775, 69

Durney, B. R. & Latour, J. 1978, Geophysical and Astrophysical Fluid Dynamics, 9, 241

Fauve, S. & Pétrélis, F. 2007, Comptes Rendus Physique, 8, 87

García, R. A., Mathur, S., Salabert, D., *et al.* 2010, Science, 329, 1032

Gastine, T., Duarte, L., & Wicht, J. 2012, *A&A*, 546, A19

Gastine, T., Yadav, R., Morin, J *et al.* 2013, *MNRAS*, in press

Giampapa, M. S. 2005, Saas-Fee Adv. Course 34: The Sun, Solar Analogs and the Climate, 307

Giampapa, M. S., Hall, J. C., Radick, R. R., & Baliunas, S. L. 2006, *ApJ*, 651, 444

Gondoin, P. 2012, *A&A*, 546, A117

Guerrero, G. & de Gouveia Dal Pino, E. M. 2008, *A&A*, 485, 267

Hazra, G., Karak, B. B., & Choudhuri, A. R. 2013, arXiv:1309.2838

Jouve, L. & Brun, A. S. 2007, *A&A*, 474, 239

Jouve, L., Brown, B. P., & Brun, A. S. 2010, *A&A*, 509, A32

Jouve, L., Brun, A. S., & Aulanier, G. 2013, *ApJ*, 762, 4

Käpylä, P. J., Korpi, M. J., Ossendrijver, M., & Stix, M. 2006, *A&A*, 455, 401

Käpylä, P. J., Mantere, M. J., Guerrero, G., Brandenburg, A., & Chatterjee, P. 2011, *A&A*, 531, A162

Käpylä, P. J., Mantere, M. J., Cole, E., Warnecke, J., & Brandenburg, A. 2013, *ApJ*, 778, 41

Küker, M., Rüdiger, G., & Kitchatinov, L. L. 2011, *A&A*, 530, A48

Leighton, R. B. 1969, *ApJ*, 156, 1

Mathur, S., García, R. A., Morgenthaler, A., *et al.* 2013, *A&A*, 550, A32

Matt, S. P., Do Cao, O., Brown, B. P., & Brun, A. S. 2011, Astron. Nach., 332, 897

Matt, S. P. & Brun, A. S. 2013, ApJ submitted

Miesch, M. S., Brun, A. S., & Toomre, J. 2006, *ApJ*, 641, 618

Miesch, M. S., Brun, A. S., De Rosa, M. L., & Toomre, J. 2008, *ApJ*, 673, 557

Moffatt, H. K. 1978, Cambridge, England, Cambridge University Press, 1978. 353 p.

Montesinos, B., Thomas, J. H., Ventura, P., & Mazzitelli, I. 2001, *MNRAS*, 326, 877

Morgenthaler, A., Petit, P., Morin, J., *et al.* 2011, Astronomische Nachrichten, 332, 866

Nandy, D. 2004, *Solar Phys.*, 224, 161

Nandy, D. & Martens, P. C. H. 2007, Advances in Space Research, 40, 891

Nelson, N. J., Brown, B. P., Brun, A. S., Miesch, M. S., & Toomre, J. 2011, *ApJL*, 739, L38

Nelson, N. J., Brown, B. P., Brun, A. S., Miesch, M. S., & Toomre, J. 2013, *ApJ*, 762, 73

Noyes, R. W., Weiss, N. O., & Vaughan, A. H. 1984, *ApJ*, 287, 769

Petit, P., Dintrans, B., Solanki, S. K., *et al.* 2008, *MNRAS*, 388, 80

Pizzolato, N., Maggio, A., Micela, G., Sciortino, S., & Ventura, P. 2003, *A&A*, 397, 147

Pouquet, A., Frisch, U., & Leorat, J. 1976, Journal of Fluid Mechanics, 77, 321

Racine, É., Charbonneau, P., Ghizaru, M., Bouchat, A., & Smolarkiewicz, P. 2011, *ApJ*, 735, 46

Reiners, A. 2012, Living Reviews in Solar Physics, 9, 1

Saar, S. 2002, Stellar Coronae in the Chandra and XMM-NEWTON Era, 277, 311

Saar, S. H. & Brandenburg, A. 1999, *ApJ*, 524, 295

Strugarek, A., Brun, A. S., Mathis, S., & Sarazin, Y. 2013, *ApJ*, 764, 189

Tobias, S. M. 1997, *A&A*, 322, 1007

Tobias, S. M., Brummell, N. H., Clune, T. L., & Toomre, J. 2001, *ApJ*, 549, 1183

Weiss, N. O. 1994, Lectures on Solar and Planetary Dynamos, 59

Wilson, O. C. 1978, *ApJ*, 226, 379

Wright, N. J., Drake, J. J., Mamajek, E. E., & Henry, G. W. 2011, *ApJ*, 743, 48

Zhao, J., Bogart, R., Kosovichev, A., Duvall, T., Jr., & Hartlep, T. 2013, *ApJL*, 774, L29

Magnetic Fields throughout Stellar Evolution
Proceedings IAU Symposium No. 302, 2013
P. Petit, M. Jardine & H. Spruit, eds.

doi:10.1017/S1743921314001860

Probing the structure of local magnetic field of solar features with helioseismology

Khalil Daiffallah

Observatory of Algiers, CRAAG,
Route de l'Observatoire, BP 63, Bouzaréah 16340, Algiers, Algeria
email: k.daiffallah@craag.dz

Abstract. Motivated by the problem of local solar subsurface magnetic structure, we have used numerical simulations to investigate the propagation of waves through monolithic magnetic flux tubes of different sizes. A cluster model can be a good approximation to simulate sunspots as well as solar plage regions which are composed of an ensemble of compactly packed thin flux tubes. Simulations of this type are powerful tools to probe the structure and the dynamics of various solar features which are directly related to solar magnetic field activity.

Keywords. Magnetohydrodynamics, waves, magnetic field, sunspots, plages.

1. Introduction

Understanding the origin of the Sun's magnetic field is the most important topic in solar physics today. Sunspots are a manifestation of strong magnetic fields at the surface. Other magnetic features in the form of magnetic flux tubes can be distinguished, like plages which are concentrations of small-scale magnetic flux tubes (bright area in solar surface) and pores which are isolated vertical magnetic flux tubes. Constraining the subsurface structure, dynamics and evolution of these magnetic features is essential to establish a relationship between internal solar properties and magnetic activity in the photosphere. Helioseismology is a powerful tool to probe the structure and dynamics of the Sun through the observation of solar oscillations. However, the method of local helioseismology is still limited since the magnetic field is not included in the theory. We only interpret the observations in the quiet Sun in terms of temperature variation or velocity flow, but not in sunspots where there is a strong magnetic field. Therefore, numerical simulations are needed to infer the structure of the magnetic field by modeling the interaction of waves with magnetic features.

In the first part of this study (Section 2), we investigate the propagation of a linear surface gravity wave packet (f-mode) through monolithic structures of magnetic flux tube of different sizes. In the second part (Section 3), we explore the helioseismic response of a cluster model which consideres the subsurface magnetic field of sunspots and plages as a bundle of small-scale magnetic flux tubes in a spaghetti-like configuration.

2. Helioseismic signature of monolithic magnetic flux tubes

We use the SLiM code (Cameron *et al.* 2007) to propagate a linear and Gaussian f-mode wave packet ($\nu = 3$ mHz) through a three dimensional enhanced polytropic atmosphere (Cally & Bogdan 1997).

In this section, we explore the helioseismic response of magnetic flux tubes with radii from 200 km (e.g., internetwork magnetic field) to 3 Mm (e.g., pores or small sunspots). We considere a 4820 G, purely vertical flux tube along the z-direction. The flux tube is

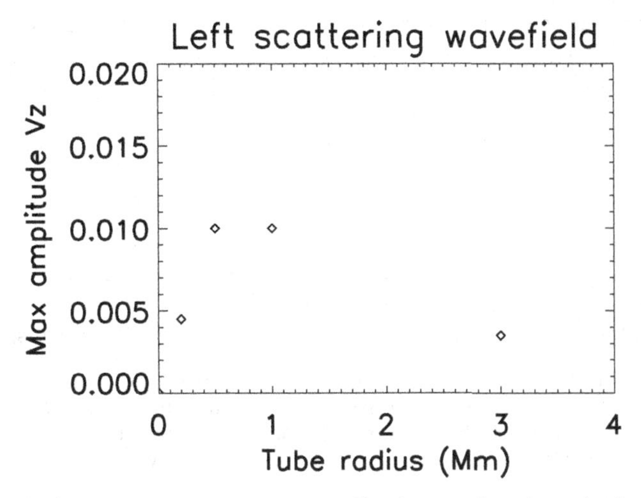

Figure 1. A plot of the maximum scattering amplitude as a function of tube radius measured in the far field.

almost evacuated and it is superposed on the background atmosphere. The scattered wave field is constructed as the difference between the simulation with and without the flux tube. Different scattered wave field patterns were observed for the different flux tubes. When the flux tube is small compared to the value of $\lambda/2\pi$ where λ is the wavelength of the incident wave, mainly the $m = 1$ kink modes are excited (m is the azimutal number of the wave). For mid-ranged tubes, the oscillations are a mixture of $m = 1$ kink modes, and $m = 0$ sausage modes. For larger tubes, numerous modes with various m are excited (Daiffallah *et al.* 2011).

If we plot the maximum scattering amplitude versus tube radius measured in the far field (Fig. 1), we find that tubes with radii around $\lambda/2\pi$ are the largest scatterers. We explain the decrease of scattering after this value of radius by the excitation of m-modes and the absorption of waves by these large tubes. For example, in the case of tubes with $R = 3$ Mm, part of f-mode is converted to a slow magneto-acoustic-gravity mode which propagates along magnetic field lines in the z-direction (e.g., Cally 2005). This process can explain the observations of Braun *et al.*(1987) and Braun *et al.*(1988) which reveal a deficiency in the power of the outgoing wave compared to the incoming wave from a typical sunspot on the solar surface (Fourier-Hankel analysis).

3. Helioseismic response of a cluster of small magnetic flux tubes

In this section, we want to know what is the structure of the magnetic field beneath sunspots: is it like a monolithic model as in section 2 or like a cluster model? how to distinguish from the observed oscillations between these two models? The cluster model can be a good approximation to simulate solar plage regions which are composed of an ensemble of compactly packed thin flux tubes (Hanasoge & Cally 2009).

Motivated by these problems, we investigate the propagation of waves (f-mode) through a cluster of small identical magnetic flux tubes of 200 km radius (e.g., Hindman & Jain 2012; Daiffallah 2013; Felipe *et al.* 2013).

We have studied two cases, one is a compact cluster which consists of seven identical magnetic flux tubes in a hexagonal close-packed configuration, and the other case is a loose cluster of nine tubes.

The Fig. 2 shows the horizontal displacement of the central tube axis as a function of depth z (dot-dashed line) for the compact cluster in the left panel, and the loose

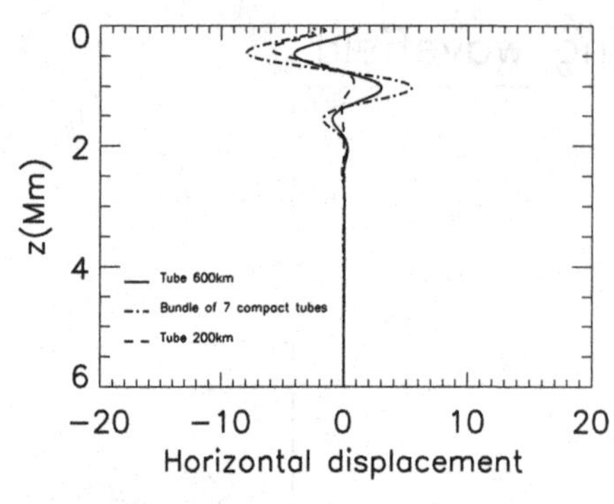

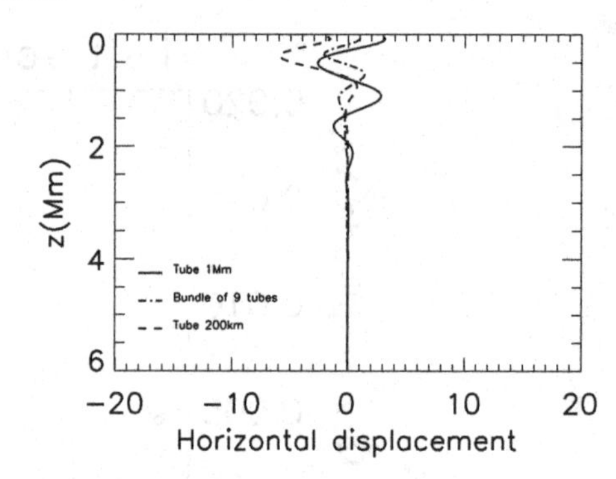

Figure 2. Horizontal displacement of the central tube axis of the compact cluster in the left panel, and the loose cluster in the right panel (dot-dashed line), as a function of depth z. The dashed and the solid lines show the oscillations of a single tube of 200 km radius and the monolithic equivalent tube, respectively. The separation between the tubes inside the loose cluster is about $\lambda/2\pi$.

cluster in the right panel. We can observe that the upper part of the compact cluster oscillates more like a single tube of 200 km radius (dashed line) than like the monolithic equivalent tube of 600 km (solid line). Furthermore, the large amplitude of the compact cluster compared to the amplitude of the 200 km single tube confirms that the oscillation concerns the totality of the compact cluster and not individual tubes. This can be seen in the scattered wave field of the compact cluster in the Fig. 3 where the compact cluster seems to oscillate like a single object and where there is no evidence of multiple scattering in the near field.

The amplitude of the loose cluster in the right panel of Fig. 2 is smaller than the amplitude of the 200 km single tube, which means that the incident wave energy is converted to tube oscillation and the oscillation of the loose cluster in Fig. 2 corresponds to that of the individual tubes. Therefore, the loose cluster will display multiple scattering from the individual tubes in the near field. In this case, the absorption of the incident wave by the cluster will be enhanced.

4. Discussion

Sunspots are manifestations of strong magnetic field at the solar surface. They represent a major relation between internal magnetic field and solar activity in the photosphere. Local helioseismology is a powerful tool to investigate the substructure of the Sun. However, interpretation of data have been somewhat ambiguous in solar active regions where the magnetic field is strong. Numerical simulations provide an efficient and direct way to understand the helioseismic signature of solar magnetic features, which have recently begun to be observed. We simulate in this study the propagation of a linear f-mode wave packet through different magnetic features. The aim is to get informations about the characteristics and the structure of these features by studying the scattered waves observed at the solar surface. The principal results of the simulations can be summarized in the following way:

- Magnetic flux tubes are strong absorbers and scatterers of f-mode waves.
- Different scattered wave field patterns were observed for different monolithic magnetic flux tube radii. This will allow us to infer the typical size of magnetic structures, but also magnetic field strength, vector orientation, profile, ...
- A cluster model of magnetic flux tubes (compact or loose) scatters waves in a different way from a monolithic model. This will allow us to infer the magnetic structure

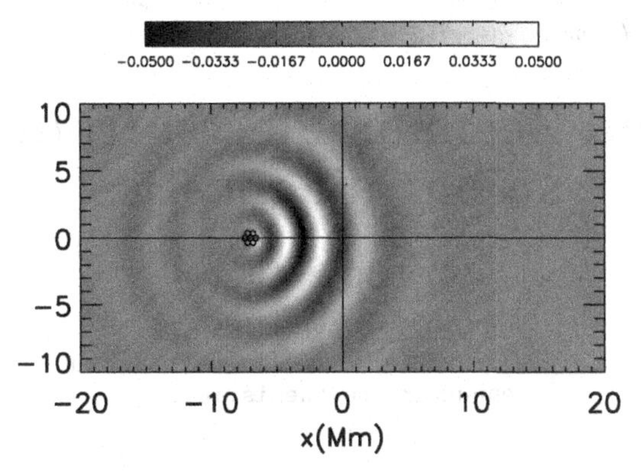

Figure 3. A snapshot at $t = 3300$ s of the scattered wave field (Vz) of a compact cluster of seven identical tubes of 200 km radius.

of more complex solar features (sunspots, plages, ...) and distinguish between monolithic or cluster models, compact or loose clusters.

• Determining the parameters of these magnetic features, that is, structure, typical size, field strength, ... will help to reveal details of the processes at work in the solar dynamo and how the magnetic field is transported up through the convection zone.

• The subsurface structure of sunspots is still poorly understood, highlighting the need for (i) high-resolution space observations, (ii) improved numerical simulations.

References

Braun, D. C., Duvall, T. L., Jr., & LaBonte, B. J. 1987, *ApJL*, 319, L27
Braun, D. C., Duvall, T. L., Jr., & LaBonte, B. J. 1988, *ApJ*, 335, 1015
Cally, P. S. & Bogdan, T. J. 1997, *ApJ*, 486, L67
Cally, P. S. 2005, *MNRAS*, 358, 353
Cameron, R., Gizon, L., & Daiffallah, K. 2007, *AN*, 328, 313
Daiffallah, K., Abdelatif, T., Bendib, A., Cameron, R., & Gizon, L. 2011, *Solar Phys.*, 268, 309
Daiffallah, K. 2013, *Solar Phys.*, 212
Felipe, T., Crouch, A., & Birch, A. 2013, *ApJ*, 775, 74
Giles, P. M., Duvall, T. L., Scherrer, P. H., & Bogart, R. S. 1997, *Nature*, 390, 52
Hanasoge, S. M. & Cally, P. S. 2009, *ApJ*, 697, 651
Hindman, B. W. & Jain, R. 2012, *ApJ*, 746, 66

Magnetic Fields throughout Stellar Evolution
Proceedings IAU Symposium No. 302, 2013
P. Petit, M. Jardine & H. Spruit, eds.

doi:10.1017/S1743921314001872

Hanle and Zeeman effects: from solar to stellar diagnostics

A. López Ariste

THEMIS - CNRS UPS 853
C/ Vía Láctea s/n 38205 - La Laguna, Spain
email: arturo@themis.iac.es

Résumé. We suggest the use of the area asymmetries of the Stokes V profile of a line sensitive to the Zeeman effect to diagnose variatios of the magnetic field along the line of sight in stellar atmospheres. This tool could allow to disentangle the magnetic topology of the observed stellar features in analogy to the solar case : a fibril topology as in plage and netwrok magnetic fields vs. a homogeneous and strong field as in sunspots. We also suggest the use of the Hanle effect as a means to observe weak global dipoles.

Keywords. Polarimetry. Magnetic Fields.

Spectropolarimetry of the solar atmosphere enjoys two advantages over that of other stars. First, the greater photon flux allows higher signal-to-noise ratios at higher spectral resolution and in consequence encourages the study and analysis of subtler polarization effects in spectral lines. Second the spatial resolution of the solar atmosphere allows the observer to distinguish localized features and structures and to propose more detailed models for, among others, the magnetic fields in those regions. Because of those advantages solar spectropolarimetry has explored and taken advantage of subtle polarization signatures that, we suggest, can now be exported to stellar spectropolarimetry. In this contribution I present and propose two of those subtle signatures as tools : the asymmetries of the Zeeman-due Stokes V profile and the Hanle effect.

Under the Zeeman effect, the spectral profile in Stokes V of a magnetically sensitive line shows a characteristic antisymmetric profile with two identical lobes of opposite sign. In the usual case (usual in the observation of solar and stellar photospheres) of a Zeeman splitting smaller than or comparable to the width of the spectral line (a width due to thermal, rotational or other broadening effects) the lobes are fixed in position and its amplitude is proportional to the longitudinal magnetic flux, where by longitudinal we mean the projection of the magnetic vector over the line of sight. This is what is often called the weak-field regime of the Zeeman effect and it is commonly used in Zeeman Magnetic Imaging of stellar photospheres or in many solar magnetographs. The previous description is true as long as the magnetic field is constant over the formation region of the line observed. This may be a good starting approximation, but it is clear that eventually we should consider that the magnetic field can change along the path of the photon through the stellar atmosphere (Westendorp Plaza *et al.*, 2001a, Westendorp Plaza *et al.* 2001b). The effect of this variability on the Stokes V profile is to make the two lobes asymmetric : they will no longer have the same amplitude and the areas enclosed by them will be different with the result of a net circular polarization when integrated in wavelength (López Ariste, 2002). The observation of such asymmetries in Stokes V profiles is therefore a potential diagnostic of the variability of the magnetic field in the stellar atmosphere.

Two kinds of asymmetry appear : the amplitudes of the lobes may be different, and the areas enclosed by the lobes may be different. Of these two kinds it is the last one that carries the diagnostic potential. It is possible to make the amplitudes of the lobes differ by, for example, simple addition of different Stokes profiles with different Doppler shifts arising from different points in the photosphere. On the other hand there is only one way to produce an area asymmetry δA : to have simultaneously non-zero gradients of the magnetic B and velocity v fields with opacity τ along the path of the photon .

$$\delta A = \frac{\int V d\lambda}{\int \|V\| d\lambda} \propto \frac{dB}{d\tau}\frac{dv}{d\tau}$$

The addition of symmetric Stokes V profiles, of different amplitudes and at different Doppler shifts will not change the integral over wavelength, it will not introduce any area asymmetry. Thus, the observation of area asymmetry unambiguously point towards gradients of magnetic and velocity fields along the path of the photon. There is a second remarkable aspect of area asymmetries of Stokes V profiles. One can define an area sign by always subtracting, let's say, the red lobe from the blue lobe. Is there a preferred sign for the area asymmetry? In the solar photosphere the answer is yes (Grossmann-Doerth *et al.*, 2000). Due to the thermal structure of the solar atmosphere, the expected increase of magnetic fields with depth often results in the red lobe having a larger area than the blue one. This has two immediate implications when translating this diagnostic tool to stellar observations : first it is difficult to erase the area asymmetry from the disk-integrated Stokes V profile if one lobe is larger than the other over large spans of the atmosphere; second it suggests a direct comparison with the solar case. Is the sign the same as in the solar photosphere? Are amplitudes comparable? Is this what the Sun would look like?

Observations of stellar Stokes V profiles have shown the presence of area asymmetry (Petit *et al.*, 2013, Tsvetkova *et al.*, 2013). The interpretation of those will pass through the introduction of gradients of magnetic field. The solar case has produced models of magnetic field topology that can be told apart in terms of the area asymmetry of Stokes V profiles. As a suggested path towards the interpretation of area asymmetries in stellar Stokes V profiles we can suggest the following dictionary :

Solar plages, magnetic network : These solar features are characterized by relatively weak fields covering large areas of the photosphere with the same polarity. In stellar observations, one is tempted to associate them with the so-called small-scale magnetic structures. The magnetic topology of those features is best described as fibril : collections of vertical flux tubes where magnetic field concentrates and filling just a fraction of the atmosphere with no or very weak field outside them. In almost every ray path that one can trace through such a topology, the photon will cross several times those high-field

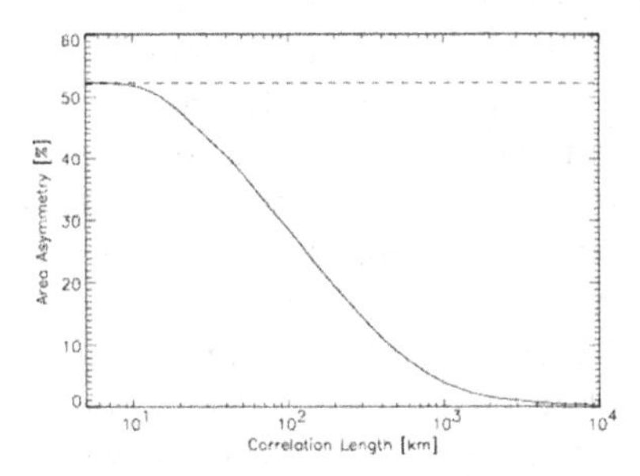

Fig. 1. Relationship between the area asymmetry of the Stokes V profile and the characteristic scales of change of the magnetic field (from Carroll & Staude, 2005)

flux tubes. The Stokes V profile will therefore show area asymmetries. If the comparison is correct, we should expect small-scale magnetic fields in stars to show clear area asymmetries.

Sunspots : The magnetic field in sunspots is homogeneous, but their size is relatively small when compared to plages or network fields. Because of this it is tempting to say that they are not observed in most of the stellar observations. The homogeneity of the magnetic field results in a zero or near-zero area asymmetry

Global field : This is the magnetic field of the star as a whole. Although in times of solar maximum its structure may be complex, the global field can be thought of as made of a low-order polar composition with scales of change comparable to the radius of the star. This is referred to as the large scale field in the stellar context. Because of the large scale of these fields, the photon has few opportunities to see gradients and one expects low or zero area asymmetries in the respective Stokes V profiles

The observation of area asymmetries in stars then points to the presence of plage and network fields, with their fibril topologies, as the ones more often seen in stars.

Zeeman effect is sensitive to magnetic fields in the range of hundreds to thousands of G. Although fields of the order of few G are often mentioned as the result of the measurement of Zeeman effect, it must be stressed that those measurements refer to flux densities which should have been measured in Mx/cm^2. In a homogeneous field, flux density is equal to intensity and Mx/cm^2 are G. But in a non homogeneous field the equivalence breaks and it is an abuse of language to use G as the measurement unit. However distributed, whatever is final flux density, the fields to which our Zeeman effect tools are sensitive are in the range of hundreds to thousands of G. In order to measure weaker field strengths another physical effect is used in solar physics : the Hanle effect.

Hanle effect is a modification of the atomic polarization by the magnetic field (Landi Degl'Innocenti and Landolfi, 2004). By atomic polarization we refer to the presence of quantum coherences among the different atomic levels and/or terms. Since we modify it, it has to be nonzero to start with. The most common source of atomic polarization is illumination by an anisotropic radiation field. Hanle effect is therefore related to lines formed by scattering. The best example is that of solar prominencesCasini *et al.*, 2003 : clouds of cold and dense plasma supported by magnetic fields in the midst of the rarefied and hot corona. Prominence plasma is illuminated by a cone of radiation from the photosphere which induces atomic polarization. Hanle effect modifies it in the presence of magnetic fields of strength such that induces a Zeeman splitting comparable to the natural width of the atomic level. Upon re-emission, the photon is linearly polarized, the amount and plane of polarization depend on the anisotropy of the radiation field and on the magnetic field. But it also depends enormously on the actual atomic structure, what makes Hanle effect always difficult to describe.

Since the polarization is attached to a scattering process, when integrated over a stellar disk the net result is zero. In order to use the Hanle effect in stellar observations there must be some break in that spherical symmetry. The presence of accretion disks or huge prominences may be one reason for that symmetry breaking. Non-spherical stars may also break that symmetry. Here we present another possibility introduced by López Ariste *et al.* (2011) in which it is the Hanle effect itself that breaks that symmetry when it is due to a dipole global field. Hanle effect diminishes the polarization rate at certain positions of the disk, depending on the field strength but mostly on the geometry of the field respect to the radiation field from the photosphere and its position on the disk. These local modifications of the scattering polarization break the spherical symmetry and result in a net polarization in those lines sensitive to the Hanle effect. The mechanism

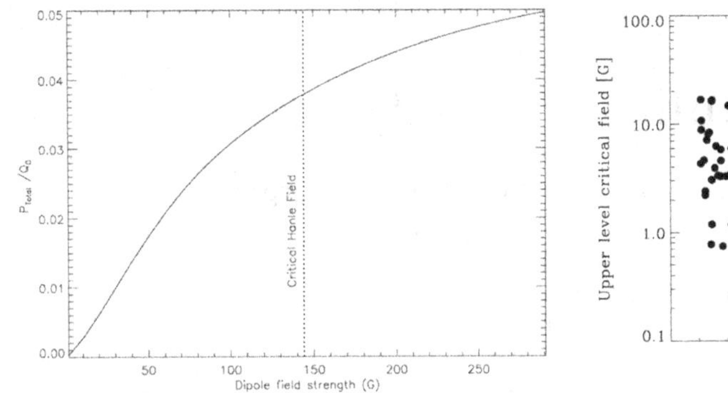

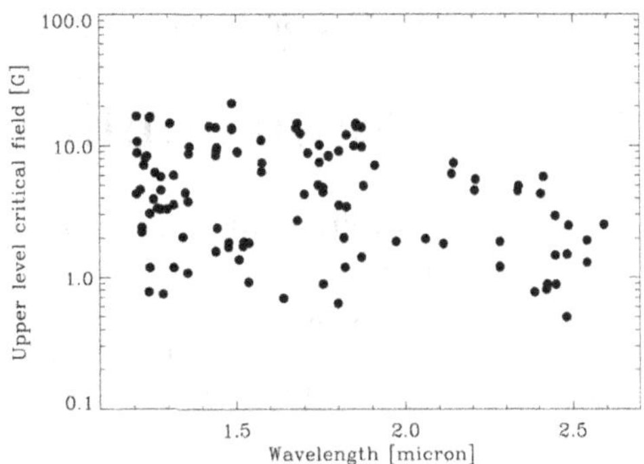

Fig. 2. Left : Variation of expected integrated linear polarization due to the Hanle effect of a dipole as a function of its magnetic strength, for a spectral line with critical Hanle field of around 150G. Right : Compendium of critical Hanle fields for lines in the solar spectrum (from López Ariste *et al.*, 2011)

works for dipoles with strengths well below 100G and though interesting it carries the disadvantage that the actual polarization signals are very weak, perhaps in the order of 10^{-4} times the intensity of the continuum.

We have presented two diagnostic tools in use in solar magnetism that can be exported to the stellar case. The first one is the area asymmetry of the Stokes V profile. This asymmetry depends on the simultaneous presence of gradients in the magnetic and velocity fields along the path of each photon. The observation of these asymmetries reveals therefore the presence of non-constant magnetic fields. Sunspots are not expected to show much asymmetries, but the more extended plage and network fields with their fibril field topology do. If the small scale fields observed in stars correspond to these solar magnetic topologies, they should be identifiable through area asymmetries in their Stokes V profiles, even after integration over the stellar disk. The second diagnostic tool is Hanle effect, which is sensitive to weak fields. Hanle effect is associated to scattering processes and therefore cancels out when integrated over the stellar disk unless something breaks the spherical symmetry. One such case is illustrated, as a global dipole field resulting in a net linear polarization which carries the signature of the dipole field.

References

Carroll, T. A. & Staude, J. 2005. *Astronomische Nachrichten* 326, 296–300.

Casini, R., López Ariste, A., Tomczyk, S., & Lites, B. W. 2003. *The Astrophysical Journal Letters* 598, L67–L70.

Grossmann-Doerth, U., Schüssler, M., Sigwarth, M., & Steiner, O. 2000. *Astronomy and Astrophysics* 357, 351–358.

Landi Degl'Innocenti, E. & Landolfi, M. 2004. *Polarization in Spectral Lines.* volume 307. Kluwer Academic Publishers, Dordrecht.

López Ariste, A. 2002. *The Astrophysical Journal* 564, 379–384.

López Ariste, A., Asensio Ramos, A., & González Fernández, C. 2011. *Astronomy and Astrophysics* 527, 120.

Petit, P., Aurière, M., Konstantinova-Antova, R., Morgenthaler, A., Perrin, G., Roudier, T., & Donati, J. F. 2013, p. 231.

Tsvetkova, S., Petit, P., Aurière, M., Konstantinova-Antova, R., Wade, G. A., Charbonnel, C., Decressin, T., & Bogdanovski, R. 2013. *Astronomy and Astrophysics* 556, 43.

Westendorp Plaza, C., del Toro Iniesta, J. C., Ruiz Cobo, B., & Martínez Pillet, V. 2001a. *The Astrophysical Journal* 547, 1148–1158.

Westendorp Plaza, C., del Toro Iniesta, J. C., Ruiz Cobo, B., Martínez Pillet, V., Lites, B. W., & Skumanich, A. 2001b. *The Astrophysical Journal* 547, 1130–1147.

Magnetic Fields throughout Stellar Evolution
Proceedings IAU Symposium No. 302, 2013
P. Petit, M. Jardine & H. Spruit, eds.

doi:10.1017/S1743921314001884

Coronal influence on dynamos

Jörn Warnecke[1,2]† and Axel Brandenburg[1,2]

[1]NORDITA, KTH Royal Institute of Technology and Stockholm University, Roslagstullsbacken 23, SE-10691 Stockholm, Sweden,
email: joern@nordita.org
[2]Department of Astronomy, Stockholm University, SE-10691 Stockholm, Sweden

Abstract. We report on turbulent dynamo simulations in a spherical wedge with an outer coronal layer. We apply a two-layer model where the lower layer represents the convection zone and the upper layer the solar corona. This setup is used to study the coronal influence on the dynamo action beneath the surface. Increasing the radial coronal extent gradually to three times the solar radius and changing the magnetic Reynolds number, we find that dynamo action benefits from the additional coronal extent in terms of higher magnetic energy in the saturated stage. The flux of magnetic helicity can play an important role in this context.

Keywords. MHD, Sun: magnetic fields, Sun: activity, Sun: rotation, turbulence, Sun: corona

1. Introduction

The solar magnetic field is produced by a dynamo operating beneath the solar surface. In the convection zone, the turbulent motions driven by convection and shear from the differential rotation are able to amplify and organize the magnetic field. These fields manifest themselves at the solar surface in form of sunspots, in which the field is so strong that the heat transported by convection is suppressed, leading to dark spots on the solar disk. One important feature of these sunspots is their latitudinally dependent occurrence. Averaging over longitude, one finds the typical behavior of equatorward migration of the underlying mean magnetic field. This behavior gives clear evidence for the existence of a dynamo mechanism in the Sun. In dynamo theory the α-effect plays an important role, because this effect describes the amplification of large-scale magnetic field in the absent of shear. In the Sun, it is believed that the α-effect produces new poloidal field from the toroidal field. How strong its contribution for the production of toroidal field is, is currently under debate (see e.g. Käpylä *et al.* 2013).

Numerical simulations of turbulent dynamos have shown that the α-effect can be catastrophically quenched at high magnetic Reynolds numbers (see Brandenburg & Subramanian 2005, for a detailed discussion). One possible loophole to alleviate the quenching is to allow for magnetic helicity fluxes (Blackman & Field 2000; Subramanian & Brandenburg 2006; Brandenburg *et al.* 2009). In this context, it is very important to choose a realistic boundary condition for the dynamo, which allows for magnetic helicity fluxes. Preventing a transport of helicity out of the simulation domain may influence the dynamo solution and the strength of the amplified magnetic field. Besides the magnetic helicity fluxes, commonly used boundary conditions for the magnetic field such as vertical field or perfect conductor restrict the dynamo and the magnetic field to certain solutions. This led to the development of the so-called "two-layer model", where we combine the lower layer, in which the magnetic field is generated by dynamo action, representing the solar convection zone with a upper, force-free layer, representing the solar corona.

† supported by the ERC AstroDyn Research Project No. 227952.

Our first application of the two-layer model led to the formation of structures reminiscent of plasmoid- and CME-like ejections, driven by a forced turbulent dynamo (Warnecke & Brandenburg 2010; Warnecke *et al.* 2011, 2012a) and, subsequently, by a self-consistently driven convective dynamo (Warnecke *et al.* 2012b) in the lower layer. This indicates that the dynamo can be directly responsible for producing coronal ejections and form structures in the solar corona. But in the recent work of Warnecke *et al.* (2013a), the authors find that differential rotation and the migration of the mean magnetic field can be also influenced by the presence of a coronal layer. In this paper, we investigate how the corona influences the dynamo action.

2. Model

We use spherical polar coordinates, (r, θ, ϕ). The setup is the same as that of Warnecke *et al.* (2011), where we use a spherical wedge with $0.7R_\odot \leqslant r \leqslant R_{\rm C}$, $\pi/3 \leqslant \theta \leqslant 2\pi/3$, corresponding to $\pm 30°$ latitude, and $0 < \phi < 0.3$, corresponding to a longitudinal extent of $17°$. $R_\odot$ is the radius of the Sun and $R_{\rm C}$ is the outer radius of the coronal layer. At $r = R$ the domain is divided at into two parts. The lower layer mimics the convection zone, where a magnetic field gets generated by turbulent dynamo action. The upper layer is a nearly force-free part, which mimics the solar corona. We solve the following equations of compressible magnetohydrodynamics,

$$\frac{\partial \boldsymbol{A}}{\partial t} = \boldsymbol{U} \times \boldsymbol{B} + \eta \boldsymbol{\nabla}^2 \boldsymbol{A}, \tag{2.1}$$

$$\frac{{\rm D}\boldsymbol{U}}{{\rm D}t} = -\boldsymbol{\nabla} h + \boldsymbol{g} + \frac{1}{\rho}\left(\boldsymbol{J} \times \boldsymbol{B} + \boldsymbol{\nabla} \cdot 2\nu\rho\mathsf{\mathbf{S}}\right) + \boldsymbol{F}_{\rm for}, \tag{2.2}$$

$$\frac{{\rm D}h}{{\rm D}t} = -c_s^2 \boldsymbol{\nabla} \cdot \boldsymbol{U}, \tag{2.3}$$

where $\boldsymbol{A}$ is the magnetic vector potential and the magnetic field is defined by $\boldsymbol{B} = \boldsymbol{\nabla} \times \boldsymbol{A}$, which makes Equation (2.2) obey $\boldsymbol{\nabla} \cdot \boldsymbol{B} = 0$ at all times. η and ν are the magnetic diffusivity and the kinematic viscosity, respectively. ${\rm D}/{\rm D}t = \partial/\partial t + \boldsymbol{U} \cdot \boldsymbol{\nabla}$ is the advective derivative, $\boldsymbol{g} = GM\boldsymbol{r}/r^3$ is the gravitational acceleration, G is Newton's gravitational constant, and M is the mass of the Sun. We choose $GM/R_\odot c_{\rm s}^2 = 3$. $\boldsymbol{J} \times \boldsymbol{B}$ is the Lorentz force and $\boldsymbol{J} = \boldsymbol{\nabla} \times \boldsymbol{B}/\mu_0$ is the current density, where μ_0 is the vacuum permeability. The traceless rate-of-strain tensor is defined as $\mathsf{S}_{ij} = \frac{1}{2}(U_{i;j} + U_{j;i}) - \frac{1}{3}\delta_{ij}\boldsymbol{\nabla} \cdot \boldsymbol{U}$, where the semi-colons denote covariant differentiation, $h = c_{\rm s}^2 \ln \rho$ is the specific pseudo-enthalpy, with $c_{\rm s} = {\rm const}$ is the isothermal sound speed. As in the work of Warnecke *et al.* (2012a), the forcing function is only present in the lower layer of the domain. This means that the forcing function goes smoothly to zero in the upper layer of the domain ($r \gg R_\odot$). The function $\boldsymbol{f}$ consists of random plane helical transverse waves with relative helicity $\sigma = (\boldsymbol{f} \cdot \boldsymbol{\nabla} \times \boldsymbol{f})/k_{\rm f}\boldsymbol{f}^2$ and wavenumbers that lie in a band around an average forcing wavenumber of $k_{\rm f} R_\odot \approx 63$. The forcing function also has a dependence on the helicity which is here chosen to be $\sigma = -\cos\theta$ such that the kinetic helicity of the turbulence is negative in the northern hemisphere and positive in the southern. More detailed descriptions can be found in Warnecke *et al.* (2011) and Haugen, Brandenburg, and Dobler (2003). The magnetic field is expressed in units of the equipartition value, $B_{\rm eq} = \mu_0 u_{\rm rms}\overline{\rho}$, where $\overline{\rho} = \langle\rho\rangle_{r \leqslant R_\odot, \theta, \phi}$, $u_{\rm rms} = \langle u_r^2 + u_\theta^2 + u_\phi^2\rangle^{1/2}_{r \leqslant R_\odot, \theta, \phi}$, and $\langle\cdot\rangle_{r \leqslant R_\odot, \theta, \phi}$ denotes an average over θ, ϕ and $r \leqslant R_\odot$, i.e., over the whole dynamo in region. The fluid and the magnetic Reynolds numbers are defined as,

$$\mathrm{Re} = u_{\rm rms}/\nu k_{\rm f}, \quad \mathrm{Re}_M = u_{\rm rms}/\eta k_{\rm f}. \tag{2.4}$$

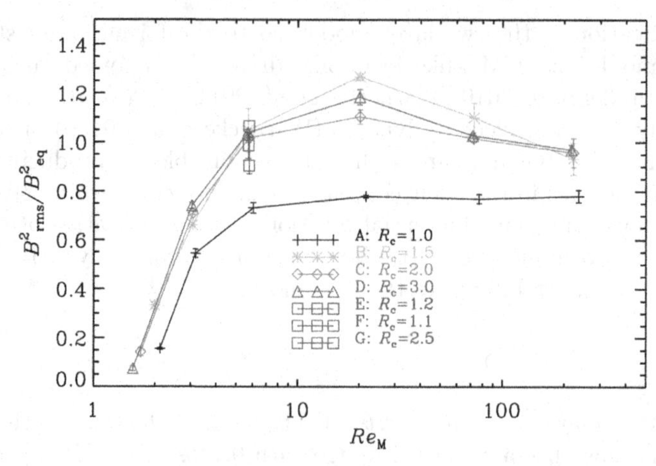

Figure 1. Dependence of magnetic field energy normalized by the equipartition value $B^2_{\rm rms}/B^2_{\rm eq}$ with coronal radial extent $R_{\rm C}$ and magnetic Reynolds number Re_M. The solid black line indicates the dynamo region without corona.

Their ratio is expressed by the magnetic Prandtl number $\mathrm{Pr}_M = \mathrm{Re}_M/\mathrm{Re}$.

As an initial condition we use Gaussian noise as seed magnetic field in the dynamo region. Our domain is periodic in the azimuthal direction. For the velocity field we use a stress-free boundary condition on all other boundaries. For the magnetic field we apply a perfect conductor conditions in both θ boundaries and the lower radial boundary ($r = 0.7\,R_\odot$). On the outer radial boundary ($r = R_{\rm C}$), we employ vertical field conditions. We use the PENCIL CODE† with sixth-order centered finite differences in space and a third-order accurate Runge-Kutta scheme in time; see Mitra *et al.* (2009) for the extension of the PENCIL CODE to spherical coordinates.

3. Dynamo action

We perform 27 runs where we change $R_{\rm C}$ and Re_M, but keep Pr_M constant. The letters for different sets indicate the coronal extents: $R_{\rm C}/R_\odot = 1$, 1.5, 2, 3, 1.2, 1.1, and 2.5 for Sets A–F. In the first four sets, we vary Re_M from 1.5 to 220, for the last three sets we use $\mathrm{Re}_M = 6$.

For all runs the turbulent motion in the lower layer of the domain drives dynamo action, which amplifies the magnetic field. After exponential growth, the field saturates and shows cycles. The field shows an equatorward migration of the all three magnetic field components, as described in Warnecke *et al.* (2011). This is caused by an α^2 dynamo, where α changes sign over the equator (Mitra *et al.* 2010a). In Figure 1, we show for all the 27 runs the normalized magnetic field energy $B^2_{\rm rms}/B^2_{\rm eq}$ as function of magnetic Reynolds number Re_M. The value for $B^2_{\rm rms}/B^2_{\rm eq}$ is obtained by averaging in space over the lower layer of the domain $r \leqslant R_\odot$ and in time over many hundred turnover times in the saturated stage. The error bars in Figure 1 reflect the quality of the temporal averaging. From Figure 1, we can deduct two important results. First, for runs with a corona the magnetic energy peaks at $\mathrm{Re}_M \approx 20$. This seems to be not the case for runs without a corona. On the other hand, the magnetic energy declines for larger Re_M, as was also found by Käpylä *et al.* (2010), which could be related to a change in the onset conditions for the different cases. Second, the magnetic energies for all runs with a coronal extent are larger by a factor of ≈ 1.5. It seems that the actual radial size of the coronal

† `http://pencil-code.googlecode.com`

extension is not that important as long as there exists a coronal layer. The run of Set F has just a coronal extent of $R_C = 1.1\,R_\odot$, but the magnetic energy is closer to runs with larger coronal extent than to the one without corona.

Magnetic helicity fluxes might be a key to solving this riddle. However, the outer radial boundary condition in the runs without a corona also allow for magnetic helicity fluxes. We recall that the simulation with a corona generates large ejections of magnetic helicity (Warnecke *et al.* 2011, 2012a). Without possessing a coronal extent the dynamo might be not able to produce ejection of magnetic helicity and therefore has a much lower magnetic helicity flux through the boundary. Studies on the nature of helicity fluxes in these runs are already on the way (Warnecke *et al.* 2013c, in preparation). However, magnetic helicity fluxes might be important only much larger magnetic Reynolds numbers (Del Sordo, *et al.* 2013).

4. Conclusions

We have shown that a coronal layer on the top of a dynamo region can support dynamo action. This is visible through an increase in magnetic energy by adding a corona at the top of the domain and leaving all other parameters the same. However, it will be necessary to study magnetic helicity fluxes through the surface of the lower layer for these cases to derive any further conclusions. The two-layer model has been used before to show the impact of a dynamo on coronal properties and generating CME-like ejections (Warnecke & Brandenburg 2010; Warnecke *et al.* 2011, 2012a). With this model is was also possible to generate spoke-like differential rotation and equatorward migration in global convective dynamo simulations (Warnecke *et al.* 2013a), whereas models without a corona have not been able to reproduce these features in the same parameter regime (Käpylä *et al.* 2013). Besides dynamo models, this two-layer approach is successful in combination with stratified turbulence in producing a bipolar magnetic region (Warnecke *et al.* 2013b) as a possible mechanism of sunspot formation.

References

Blackman, E. G. & Field, G. B. 2003, *ApJ,* 534, 984
Brandenburg, A., Candelaresi, S., & Chatterjee, P. 2009, *MNRAS,* 398, 1414
Brandenburg, A., & Subramanian, K. 2005, *Astron. Nachr.,* 326, 400
Del Sordo, F., Guerrero, G., & Brandenburg, A. 2013, *MNRAS,* 429, 1686
Haugen, N. E. L., Brandenburg, A., & Dobler, W. 2004, *Phys. Rev. E,* 70, 016308
Käpylä, P. J., Korpi, M. J., & Brandenburg, A. 2010, *A&A,* 518, A22
Käpylä, P. J., Mantere, M. J., Cole, E., Warnecke, J., & Brandenburg, A. 2013, *ApJ,* to be published arXiv:1301.2595
Mitra, D., Tavakol, R., Brandenburg, A., & Moss, D. 2009, *ApJ,* 697, 923
Mitra, D., Tavakol, R., Käpylä, P. J., & Brandenburg, A. 2010a, *ApJL,* 719, L1
Subramanian, K. & Brandenburg, A. 2006, *ApJ,* 648, L71
Warnecke, J. & Brandenburg, A. 2010, *A&A,* 523, A19
Warnecke, J., Brandenburg, A., & Mitra, D. 2011, *A&A,* 534, A11
Warnecke, J., Brandenburg, A., & Mitra, D. 2012a, *JSWSC,* 2, A11
Warnecke, J., Käpylä, P. J., Mantere, M. J., & Brandenburg, A. 2012b, *Solar Phys.,* 280, 299
Warnecke, J., Käpylä, P. J., Mantere, M. J., & Brandenburg, A. 2013a, *ApJ,* to be published arXiv:1301.2248
Warnecke, J., Losada, I. R., Brandenburg, A., Kleeorin, N. & Rogachevskii, I. 2013b, *ApJ,* submitted arXiv:1308.1080

Magnetic Fields throughout Stellar Evolution
Proceedings IAU Symposium No. 302, 2013
P. Petit, M. Jardine & H. Spruit, eds.

doi:10.1017/S1743921314001896

A Bcool spectropolarimetric survey of over 150 solar-type stars

Stephen Marsden,[1] Pascal Petit,[2,3] Sandra Jeffers,[4]
Jose-Dias do Nascimento,[5] Bradley Carter[1] and Carolyn Brown[1]
on behalf of the Bcool project team

[1]Computational Engineering and Science Research Centre, University of Southern Queensland, Toowoomba, 4350, Australia
email: `Stephen.Marsden@usq.edu.au`

[2]Université de Toulouse, UPS-OMP, Institut de Recherche en Astrophysique et Planétologie, Toulouse, France

[3]CNRS, Institut de Recherche en Astrophysique et Planétologie, 14 Avenue Edouard Belin, F-31400 Toulouse, France

[4]Institut für Astrophysik, Georg-August-Universität Göttingen, Friedrich-Hund-Platz 1, 37077 Göttingen, Germany

[5]Departamento de Fisica Teórica e Experimental, Universidade Federal do Rio Grande do Norte, CEP: 59072-970 Natal, RN, Brazil

Abstract. As part of the Bcool project, over 150 solar-type stars chosen mainly from planet search databases have been observed between 2006 and 2013 using the NARVAL and ESPaDOnS spectropolarimeters on the Telescope Bernard Lyot (Pic du Midi, France) and the Canada France Hawaii Telescope (Mauna Kea, USA), respectively. These single "snapshot" observations have been used to detect the presence of magnetic fields on 40% of our sample, with the highest detection rates occurring for the youngest stars. From our observations we have determined the mean surface longitudinal field (or an upper limit for stars without detections) and the chromospheric surface fluxes, and find that the upper envelope of the absolute value of the mean surface longitudinal field is directly correlated to the chromospheric emission from the star and increases with rotation rate and decreases with age.

Keywords. line : profiles - stars : activity - stars : magnetic fields

1. Introduction

The Bcool† project is an international collaboration looking at the magnetic activity of low-mass stars, predominately through the use of spectropolarimetry. One of the main research areas of Bcool is the study of solar-type stars to help understand how the magnetic dynamo operates in such stars. To this end the Bcool project has obtained high-resolution spectropolarimetric observations of 167 solar-type stars using the twin spectropolarimeters NARVAL and ESPaDOnS on the Telescope Bernard Lyot and the Canada France Hawaii Telescope, respectively. The observations were taken over 25 observing runs from late 2006 through to mid 2013. The targets are predominately dwarf stars (~90%) although a number (17) were identified to be subgiants, see Figure 1 (left-hand side). The sample covers stellar masses from 0.6 up to 2.5 $M_\odot$ with a range of ages. A significant fraction of the sample (44%) have stellar effective temperatures around the solar value (i.e. 5750 ± 125 K). The main aims of this survey are to determine which stars have detectable magnetic fields for follow-up study through the use of Zeeman Doppler Imaging (ZDI, Donati *et al.* 1997) to map their surface magnetic topologies and also to

† http://bcool.ast.obs-mip.fr

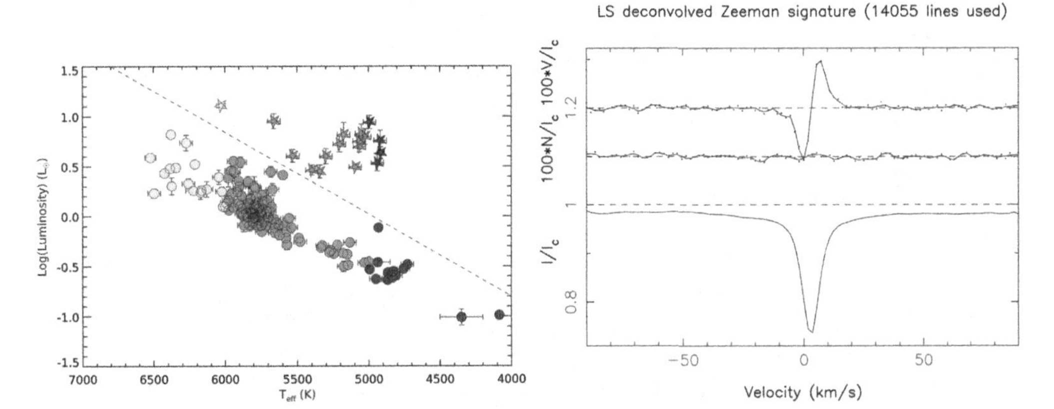

Figure 1. The left-hand plot shows the HR diagram for our stars, with stars in red (dark grey) having $T_{eff} < 5000$ K, those in orange (mid grey) having $5000\text{ K} \leqslant T_{eff} \leqslant 6000$ K and those in yellow (light grey) having $T_{eff} > 6000$ K. Filled circles represent dwarf stars while five-pointed stars represent subgiants, with the dashed line showing the dividing line between the two. The right-hand plot shows the Stokes V (upper), Null (middle) and Stokes I (lower) LSD profiles of a sample star with the Stokes V profile showing a magnetic detection.

determine if the large-scale magnetic field properties of solar-type stars vary with basic stellar parameters. This paper describes the preliminary results from this spectropolarimetric solar-type star "snapshot" survey with the main results to be presented in a later paper (Marsden *et al.* in prep.).

2. Targets and Observations

The targets in the sample have been primarily taken from the Spectroscopic Properties of Cool Stars (SPOCS) sample (Valenti & Fischer 2005, Takeda *et al.* 2007) with additional stars taken from Wright *et al.* (2004) and Baliunas *et al.* (1995).

Each star was observed at least once using the NARVAL and ESPaDOnS spectropolarimeters and from each observation both an intensity (Stokes I) and a circularly polarsied (Stokes V) spectrum was recovered using LIBRE-ESPRIT, an automatic reduction software package based on ESPRIT (Donati *et al.* 1997). As the signal-to-noise ratio of the observed spectra was not high enough to detect Zeeman signatures in individual lines it was necessary to apply the technique of Least-Squares Deconvolution (LSD, Donati *et al.* 1997) to the polarimetric signals from the data. A magnetic field can then be detected through the variations in the Stokes V LSD profile, see Figure 1 (right-hand side).

3. Results

For our sample of 167 solar-type stars, we have been able to detect a magnetic field on the surface of 40% of our targets (67 stars). This includes the detection of magnetic fields on 3 of our 17 subgiant stars. There appears to be a slight increase in the detection rate for K-stars (12/21 = 57%) over G-stars (48/128 = 38%) and F-stars (8/18 = 44%), but the numbers of K- and F-stars in our sample are small compared to the G-stars. Those stars with magnetic detections have a wide range of rotation rates ($v\sin i = 0.0$ to 54.7 km s^{-1}) and cover all ages, with 21 of our stars with magnetic field detections being classified as mature solar-type stars with ages greater than 2 Gyr.

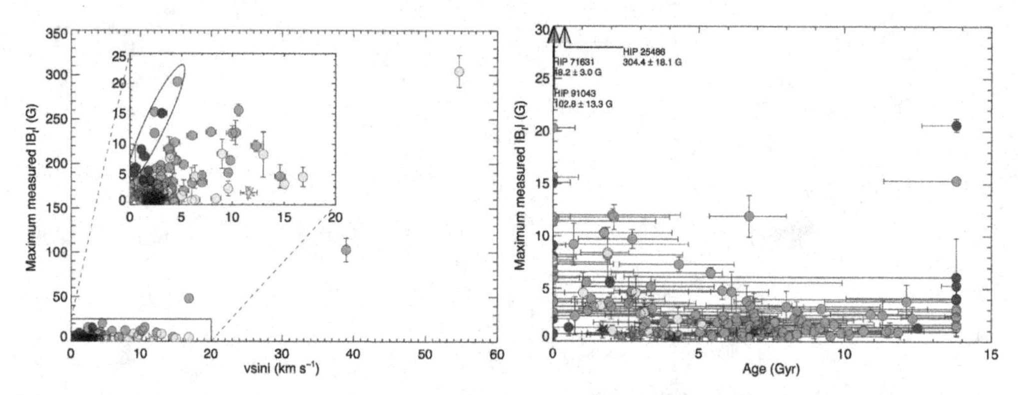

Figure 2. Plot of the absolute value of the mean longitudinal magnetic field ($|B_l|$) against $v\sin i$ (left-hand side) and stellar age (right-hand side). The symbols are the same as in Figure 1 (left-hand side).

3.1. *The Longitudinal Magnetic Field*

A measure of the strength of the surface magnetic field on a star can be derived from the star's Stokes V and I LSD profiles using (Donati *et al.* 1997):

$$B_l = -2.14 \times 10^{11} \frac{\int vV(v)dv}{\lambda gc \int [I_c - I(v)]dv}, \tag{3.1}$$

where B_l is the line-of-sight-component of the stellar magnetic field integrated over the visible stellar disc, c is the speed of light and λ and g are the mean wavelength and Landé factor of the LSD profile. I_c is the continuum level of the intensity profile and both the Stokes V and I LSD profiles are integrated over velocity space (v).

The upper envelope of the absolute value of the longitudinal magnetic field ($|B_l|$) is found to increase with rotation rate and decrease with age (see Figure 2), as would be expected of an indicator of the surface magnetic activity. However, unlike traditional activity indicators which are sensitive to the magnetic field strength, $|B_l|$ is also sensitive to the polarity of the magnetic field and it thus very dependent upon the distribution and polarity mix of magnetic regions across the stellar surface. Therefore, due to cancellation effects, the measure of $|B_l|$ can be lower than expected. This explains why it is the upper envelope of $|B_l|$ that correlates with stellar parameters.

3.2. *Calcium HK emission*

In addition to the longitudinal magnetic field we have also analysed more traditional activity measures, including the Calcium HK emission from the star. The Ca HK emission was standardised to the Mt. Wilson survey's S-index (cf., Baliunas *et al.* 1995) using the method described in Wright *et al.* (2004), by calibrating our results to those of Wright *et al.* (2004) using common stars between the two samples.

Since B_l is based on an average measure of the amount of magnetic flux on the visible stellar surface it is expected that B_l should correlate with other activity indicators, such as the Ca HK emission. $|B_l|$ is plotted against the simultaneously obtained Ca HK emission for our stars in Figure 3 (left-hand side) with the plot showing that there is a clear correlation between the upper envelope of $|B_l|$ and the Ca HK emission.

The right-hand side plot in Figure 3 shows a histogram of the magnetic field detections / non-detections against the Ca HK-index. As can be seen the detection rate increases with the Ca HK-index and for stars with a Ca HK-index greater than 0.3 the detection

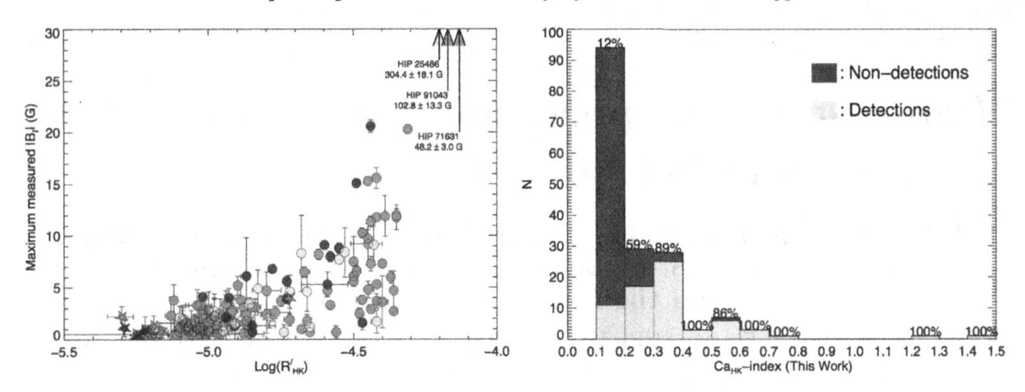

Figure 3. Plot of $|B_l|$ against Ca HK emission (left-hand side) and a Histogram of magnetic field detections/non-detections against Ca HK emission (right-hand side). The symbols in the left-hand side plot are the same as in Figure 1 (left-hand side), while the percentages above each column in the histogram gives the detection rate for each bin.

rate is almost 90%, while for those stars with Ca HK-index values less than 0.2 the detection rate is down near 10%.

4. Conclusions

One of the most active research areas in the Bcool project is the study of solar-type stars. We have observed a large sample (167) of such stars using high-resolution spectropolarimetric observations in order to detect and characterise their magnetic fields. Prior to this survey, the majority of the magnetic field detections on solar-type stars have been for young stars (i.e. Donati *et al.* 1997, Donati *et al.* 2003, Marsden *et al.* 2006, Marsden *et al.* 2011). This project has detected magnetic fields on 40% of our sample (67 stars), with 21 of our detected stars being classified as mature age solar-type stars with an age greater than 2 Gyr. This is a four-fold increase in the number of mature-age stars with magnetic field detections discovered so far. In addition, we have detected magnetic fields on 3 of our 17 subgiant stars.

We have shown that the upper envelope of the absolute value of the longitudinal magnetic field ($|B_l|$) increases with rotation rate and decreases with age and is strongly correlated to other more traditional activity indicators. We have also shown that the detection rate for magnetic fields is strongly linked to the Ca HK emission of the star.

This survey represents a unique dataset on the magnetic fields of solar-type stars that will provide the basis for further detailed study into the magnetic dynamos of these stars.

References

Baliunas S. L., Donahue R. A., Soon W. H., *et al.* 1995, *ApJ*, 438, 269
Donati J.-F., Semel M., Carter B. D., Rees D. E., & Cameron A. C. 1997, *MNRAS*, 291, 658
Donati J.-F., Collier Cameron A., Semel M., *et al.* 2003, *MNRAS*, 345, 1145
Marsden S. C., Donati J.-F., Semel M., Petit P., & Carter B. D. 2006, *MNRAS*, 370, 468
Marsden S. C., Jardine M. M., Ramírez Vélez J. C., *et al.* 2011, *MNRAS*, 413, 1922
Takeda G., Ford E. B., Sills A., Rasio F. A., Fischer D. A., & Valenti J. A. 2007, *ApJS*, 168, 297
Valenti J. A. & Fischer D. A. 2005, *ApJS*, 159, 141
Wright J. T., Marcy G. W., Butler R. P., & Vogt S. S. 2004, *ApJS*, 152, 261

Magnetic Fields throughout Stellar Evolution
Proceedings IAU Symposium No. 302, 2013
P. Petit, M. Jardine & H. Spruit, eds.

doi:10.1017/S1743921314001902

High-resolution spectropolarimetry of κ Cet: A proxy for the young Sun

J. D. do Nascimento[1], P. Petit[2], M. Castro[1], G. F. Porto de Mello[3], S. V. Jeffers[4], S. C. Marsden[5], I. Ribas[6], E. Guinan[7] and the Bcool Collaboration

[1]UFRN, Brazil, [2]IRAP, France, [3]UFRJ, Brazil, [4]IAG, Germany, [5] CESRC, Australia, [6]CSIC, Spain, [7]Villanova Univ., USA

Abstract. κ^1 Cet (HD 20630, HIP 15457, d = 9.16 pc, V = 4.84) is a dwarf star approximately 30 light-years away in the equatorial constellation of Cetus. Among the solar proxies studied in the Sun in Time, κ^1 Cet stands out as potentially having a mass very close to solar and a young age. On this study, we monitored the magnetic field and the chromospheric activity from the Ca II H & K lines of κ^1 Cet . We used the technique of Least-Square-Deconvolution (LSD, Donati *et al.* 1997) by simultaneously extracting the information contained in all 8,000 photospheric lines of the echelogram (for a linelist matching an atmospheric model of spectral type K1). To reconstruct a reliable magnetic map and characterize the surface differential rotation of κ^1 Cet we used 14 exposures spread over 2 months, in order to cover at least two rotational cycles (Prot $\sim$ 9.2 days). The Least Square deconvolution (LSD) technique was applied to detect the Zeeman signature of the magnetic field in each of our 14 observations and to measure its longitudinal component. In order to reconstruct the magnetic field geometry of κ^1 Cet , we applied the Zeeman Doppler Imaging (ZDI) inversion method. ZDI revealed a structure in the radial magnetic field consisting of a polar magnetic spot. On this study, we present the fisrt look results of a high-resolution spectropolarimetric campaign to characterize the activity and the magnetic fields of this young solar proxy.

Keywords. κ^1 Cet , HD 20630, HIP 15457, solar analogs, magnetic field

---

1. Introduction

The observational programme *Sun in Time* (Ribas *et al.* 2005) is focused on a small sample of carefully-selected and well-studied stellar proxies that well represent some key stages in the evolution of the Sun. These authors used X-ray, EUV, and FUV domains to characterize some solar proxies. Among the solar proxies studied in the Sun in Time, κ^1 Cet stands out as potentially having a mass and metallicity very close to solar with an estimated age of ~0.7 Gyr (Ribas *et al.* 2005). This could be a very good analog of the Sun at the critical time when life is thought to have originated on Earth 3.8 Gyr ago. The star was discovered to have a rapid rotation, roughly once every nine days. κ^1 Cet is also considered a good candidate to contain terrestrial planets. In spite of our in-depth knowledge of κ^1 Cet , including its radiative properties, abundances, atmospheric parameters, and evolutionary state, we know little or nothing about the magnetic field properties of this important solar proxy. On this study we present the first look results from a regularly observational campaign using the NARVAL spectropolarimeter at the Télescope Bernard Lyot (Pic du Midi, France) to observe κ^1 Cet , and infer the intensity and nature of its magnetic field. The Zeeman-Doppler Imaging technique is employed to reconstruct the large-scale photospheric magnetic field structure of κ^1 Cet and to investigate its short-term temporal evolution.

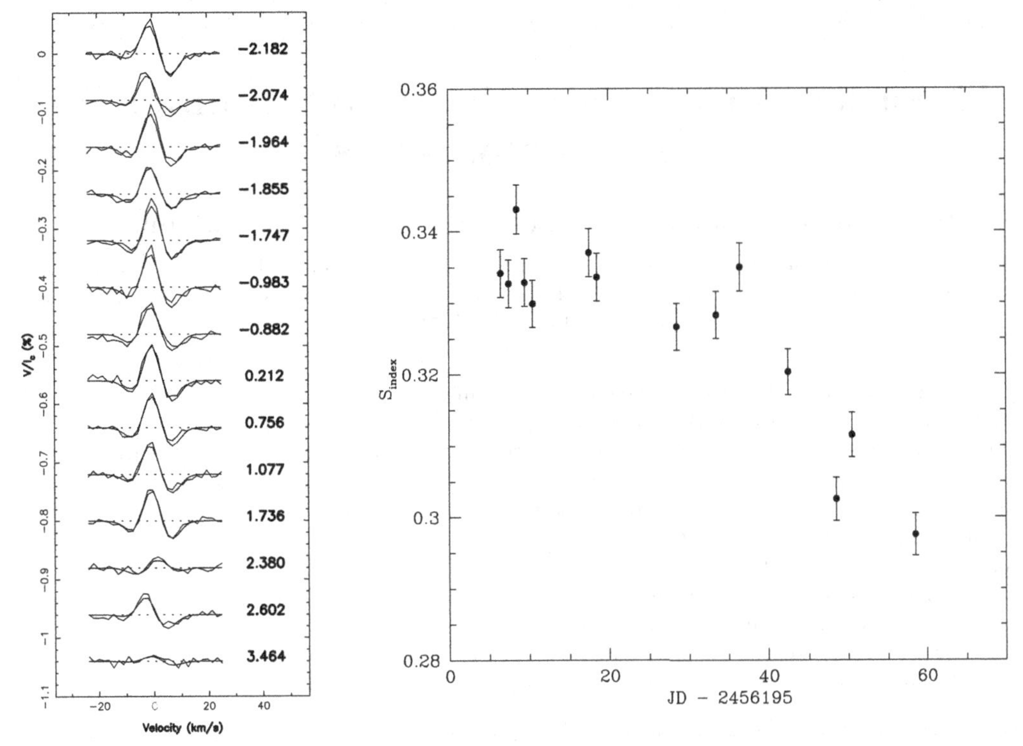

Figure 1. *Left*: The normalized Stokes V profiles of κ^1 Cet . Continuum line represent the data and dashed line correspond to synthetic profiles of our magnetic model. Successive profiles are shifted vertically for display clarity. Rotational phases of observations are indicated in the right part of the plot. *Right*: From the Stokes I spectra, we determined the S_{index} (calibrated from the Mount Wilson S_{index}) to quantify the chromospheric emission changes in the Ca II H line. The complete pipeline of the computation is described in Morgenthaler *et al.* (2012) and Wright *et al.* (2004).

2. First look results

From a careful spectral analysis by Ribas *et al.* (2010) and the comparison of different methods these authors give for κ^1 Cet the following atmospheric parameters: Teff = 5665 $\pm$ 30 K (Hα profile and energy distribution), log g = 4.49 $\pm$ 0.05 dex (evolutionary model and spectroscopy) and [Fe/H] = +0.10 0.05 (Fe II lines). In the left panel of Fig. 1, we show the normalized Stokes V profiles of κ^1 Cet . The Least Square deconvolution (LSD) technique was applied to detect the Zeeman signature of the magnetic field in each of our 14 observations to measure its longitudinal component, leading to B_l =+7.0 $\pm$0.9 Gauss. The main structure observed in the ZDI model is a structure in the radial magnetic field consisting of a polar magnetic spot. In this study, we present the first look results of a high-resolution spectropolarimetric campaign to characterize the activity and the magnetic fields of this young solar proxy.

References

Donati, J.-F., Semel, M., Carter, B. D., *et al.* 1997, *MNRAS*, 291, 658
Morgenthaler, A., Petit, P., Saar, S., Solanki, S. K., *et al.* 2012, *ApJ*, 540, 138
Ribas, I., Guinan, E., & Güdel, Audard, M. 2005, *ApJ*, 722, 680
Ribas, I., Porto de Mello, G. F., Ferreira, L. D., Hebrard, E., *et al.* 2010, *ApJ*, 714, 384
Wright, J. T., Marcy, G. W., Butler, R. P., Vogt, S. S., *et al.* 2004, *ApJS*, 152, 261

Magnetic Fields throughout Stellar Evolution
Proceedings IAU Symposium No. 302, 2013
P. Petit, M. Jardine & H. Spruit, eds.

doi:10.1017/S1743921314001914

Theoretical evolution of Rossby number for solar analog stars

Matthieu Castro, Tharcísyo Duarte and José Dias do Nascimento Jr.

Departamento de Física Teórica e Experimental (DFTE), Universidade Federal do Rio Grande do Norte (UFRN), Campus Universitário Lagoa Nova, CEP: 59078-970, Natal (RN), Brazil
email: mcastro@dfte.ufrn.br

Abstract. Magnetic fields of late-type stars are presumably generated by a dynamo mechanism at the interface layer between the radiative interior and the outer convective zone. The Rossby number, which is related to the dynamo process, shows an observational correlation with activity. It represents the ratio between the rotation period of the star and the local convective turnover time. The former is well determined from observations but the latter is estimated by an empirical iterated function depending on the color index $(B - V)$ and the mixing-length parameter. We computed the theoretical Rossby number of stellar models with the TGEC code, and analyze its evolution with time during the main sequence. We estimated a function for the local convective turnover time corresponding to a mixing-length parameter inferred from a solar model, and compare our results to the estimated Rossby number of 33 solar analogs and twins, observed with the spectropolarimeters ESPaDOnS@CFHT and Narval@LBT.

Keywords. solar analogs, magnetic field, Rossby number

1. Evolutionary models

Stellar evolution calculations were computed with the Toulouse-Geneva stellar evolution code TGEC. Details of the input physics used in our models can be found in do Nascimento *et al.* (2013). The evolution of the angular momentum is calculed with the Kawaler (1988) law, and for the initial rotation rates, we adopted the relation (3) in Landin *et al.* (2010). The calibration method of the models is based on the Richard *et al.* (2004) prescription: a solar model is calibrated to match the observed solar radius and luminosity at the solar age. The parameter K of the angular momentum evolution law is adjusted to give the solar rotation period ($P_{\rm rot} = 27.1$ d) at the solar age. Evolutionary models of masses 0.81, 0.85, 0.90, 0.95, 1.00, 1.05, and 1.10 $M_\odot$ were calculated from the ZAMS to the top of the RGB. The input parameters for the other masses are the same as for the 1.00 $M_\odot$ model.

2. Stars sample

To compare with our models, we use a sample of 33 sun-like stars observed with the spectropolarimeters ESPaDOnS at the Canada-France-Hawaii Telescope (Mauna Kea, USA) and NARVAL at the Télescope Bernard Lyot (Pic du Midi, France). Longitudinal magnetic field of the stars were determined with the technique of Least-Square-Deconvolution (LSD, Donati *et al.* 1997). Rotation periods of the stars, determined from chromospheric activity, were found in the literature (Noyes *et al.* 1984, Wright *et al.* 2004, Lovis *et al.* 2011).

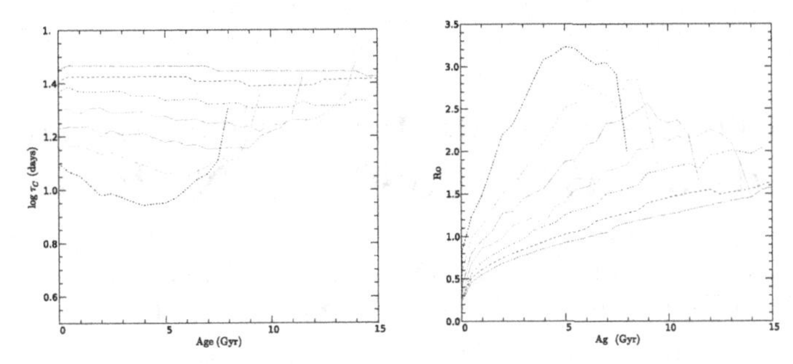

Figure 1. $\log \tau_{\rm C}$ (*left*) and Rossby number *Ro* (*right*) as a function of age for our models of 0.81 (dotted), 0.85 (dashed), 0.90 (dot-dashed), 0.95 (continuous), 1.00 (thick dotted), 1.05 (thick dashed), and 1.10 (thick dot-dashed) $M_\odot$.

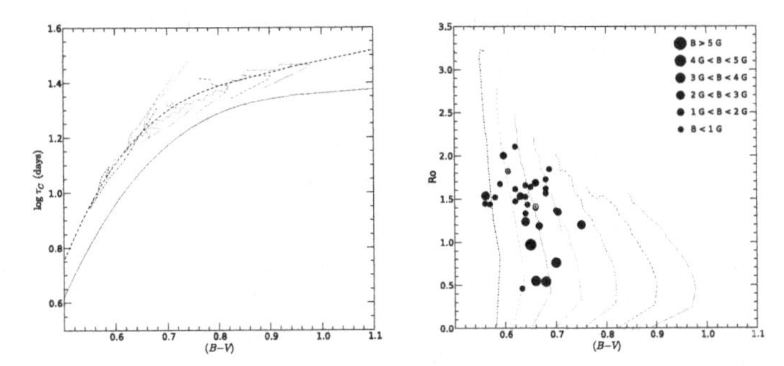

Figure 2. *Left*: $\log \tau_{\rm c}$ as a function of $(B-V)$. The dashed line is the polynomial regression of the models tracks. The dotted line is the polynomial regression found by Noyes *et al.* (1984). *Right*: Rossby number *Ro* as a function of $(B-V)$ for our models limited to the main sequence. Filled black circles are observed stars, the size depends on the strength of the magnetic field.

3. Rossby number

For all the models, we calculated the Rossby number $Ro = P_{\rm rot}/\tau_{\rm c}$ where $P_{\rm rot}$ is the rotation period, and $\tau_{\rm c}$ the local convective turnover time, calculated at a distance of one pressure height scale $H_{\rm P}$ above the base of the convective zone. Its value is computed through the equation $\tau_{\rm c} = \alpha H_{\rm P}/v$, where v is the convective velocity. In Fig. 1, we plot the evolution of $\log \tau_{\rm c}$ (left panel) and *Ro* (right panel) as a function of the stellar age.

In the left panel of Fig. 2, we show the $\log \tau_{\rm c}$ of our models as a function of $(B-V)$. A cubic fit through these curves (dashed line) defines an empirical function $\log \tau_{\rm C}(B-V)$, given by: $\log \tau_{\rm c} = -5.468 + 22.810(B-V) - 25.637(B-V)^2 + 9.796(B-V)^3$. From this equation, we determine the $\tau_{\rm c}$ of the stars of our sample, and then a Rossby number *Ro*. In the right panel of Fig. 2 we plot the Rossby number as a function of $(B-V)$ for our models and for the stars of our sample, determined from the observed rotation period and the convective turnover time inferred from the above equation.

References

do Nascimento, J.-D., da Costa, J. S., & Castro, M. 2013, *A&A*, 548, L1
Donati, J.-F., Semel, M., Carter, B. D. *et al.* 1997, *MNRAS*, 291, 658
Kawaler, S. D. 1988, *ApJ*, 333, 236
Landin, N. R., Mendes, L. T. S., & Vaz, L. P. R. 2010, *A&A*, 510, A46
Lovis, C., Dumusque, X., Santos, N. C. *et al.* 2011, eprint arXiv:1107.5325
Noyes, R. W., Hartmann, S. W., Baliunas, S. *et al.* 1984, *ApJ*, 279, 763
Richard, O., Thado, S., & Vauclair, S. 2004, *SoPh*, 220, 243
Wright, J. T., Marcy, G. W., Paul Butler, R., & Vogt, S. S. 2004, *ApJS*, 152, 261

Magnetic Fields throughout Stellar Evolution
Proceedings IAU Symposium No. 302, 2013
P. Petit, M. Jardine & H. Spruit, eds.

doi:10.1017/S1743921314001926

The large scale magnetic field of the G0 dwarf HD 206860 (HN Peg)

Sudeshna Boro Saikia[1], Sandra V. Jeffers[1], Pascal Petit[2], Stephen Marsden[3], Julien Morin[1], Ansgar Reiners[1] and the Bcool project

[1]Institut für Astrophysik,Universität Göttingen
Friedrich Hund Platz 1, 37077, GöttingenGermany
email: sudeshna@astro.physik.uni-goettingen.de,
jeffers@astro.physik.uni-goettingen.de, Ansgar.Reiners@phys.uni-goettingen.de

[2]CNRS, Institut de Recherche en Astrophysique et Planétologie
14 Avenue Edouard Belin, F-31400 Toulouse, France
email: ppetit@irap.omp.eu

[3]Faculty of Sciences, University of Southern Queensland
Toowoomba 4350, Australia
email:Stephen.Marsden@usq.edu.au

Abstract. HD 206860 is a young planet (HN Peg b) hosting star of spectral type G0V and it has a potential debris disk around it. In this work we measure the longitudinal magnetic field of HD 206860 using spectropolarimetric data and we measure the chromospheric activity using Ca II H&K, H-alpha and Ca II infrared triplet lines.

1. Introduction

It is widely accepted that stellar activity is directly related to the magnetic field, which in turn is related to the underlying dynamo. Our understanding of the dynamo can greatly benefit from the surface magnetic activity measurements. HD 206860 is a young solar type star of spectral type G0V which makes it an ideal candidate for spectropolarimetric observations.

2. Observations and Data Analysis

We use data collected as part of the BCOOL project by the NARVAL spectropolarimeter at the TBL at Observatoire Pic du Midi,France. We have six epochs of observations from 2007 to 2012. HD 206860 is a young (0.2 Gyr) solar analouge and we carry out direct magnetic field measurements and chromospheric activity measurement

2.1. *Activity indicators*

To measure the S-index (Baliunas *et al.* (1985)) we use two triangular bandpasses at H and K lines and two bandpasses R and V at 400.107 and 390.107 nm respectively. We use a rectangular bandpass at the Hα line and two bandpasses at 655.885 and 656.730 nm respectively(Morgenthaler *et al.* (2012)) to measure the Hα-index variability. To calculate the CaIRT-index we use rectangular bandpasses at the line cores of the Ca IRT lines and we take two bandpasses at 870.49 nm and 847.58 nm. The variability of the three indices is shown in Fig. 1(*Top*).

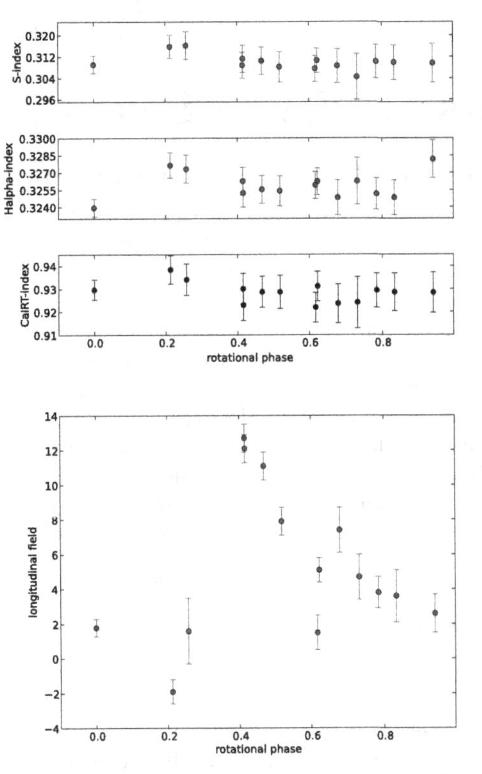

Figure 1. *Top:* The variation of different activity indicators as a function of rotational phase for 2012.*Bottom:* The variation of the longitudinal magnetic field as a function of the rotational phase for 2012.

2.2. *Longitudinal magnetic field*

The longitudinal magnetic field of HN Peg is measured from Stokes V data. The (signal-to-noise) SN in the raw spectra is not high enoguh to obtain the Zeeman signature for individual spectral lines. Hence we apply LSD Donati *et al.* (1997) technique to the lines to boost the SN. The variation of the magnetic field is seen in Fig. 1(*Bottom*).

3. Summary

We carry out magnetic field measurements for the G0 dwarf HN Peg. The chromospheric activity measurements indicate the presence of a variable magnetic field which is supported by the longitudinal magnetic field measurements. The initial results show HN Peg is an active young star with a variable magnetic field.

References

Baliunas, S. L., Horne, J. H., Porter, A., Duncan, D. K., Frazer, J., Lanning, H., Misch, A., Mueller, J., Noyes, R. W., Soyumer, D., Vaughan, A. H., & Woodard, L. 1985, *ApJ*, 294, 310

Morgenthaler, A., Petit, P., Saar, S., Solanki, S. K., Morin, J., Marsden, S. C., Aurière, M., Dintrans, B., Fares, R., Gastine, T., Lanoux, J., Lignières, F., Paletou, F., Ramírez Vélez, J. C., Théado, S., & Van Grootel, V. 2012, *A&A*, 540, 138

Donati, J. F., Semel, M., Carter, B. D., Rees, D. E., & Collier Cameron, A. 1997, *MNRAS*, 291, 658

Magnetic Fields throughout Stellar Evolution
Proceedings IAU Symposium No. 302, 2013
P. Petit, M. Jardine & H. Spruit, eds.

doi:10.1017/S1743921314001938

Starspots on Young Solar-Type Stars

Carolyn Brown, Brad Carter, Stephen Marsden and Ian Waite

Computational Engineering and Science Research Centre, University of Southern Queensland, Toowoomba, Australia
email: carolyn.brown@usq.edu.au

Abstract. Doppler Imaging of starspots on young solar analogues is a way to investigate the early history of solar magnetic activity by proxy. Doppler images of young G-dwarfs have yielded the presence of large polar spots, extending to moderate latitudes, along with measurements of the surface differential rotation. The differential rotation measurement for one star (RX J0850.1-7554) suggests it is possibly the first example of a young G-type dwarf whose surface rotates as almost a solid body, in marked contrast to the differential rotation of other rapidly rotating young G-dwarfs and the present-day Sun. Overall, our Doppler imaging results show that the young Sun possessed a fundamentally different dynamo to today.

Keywords. stars : activity - stars : imaging - stars: spots

---

1. Introduction

Active young solar-type stars provide proxies for studying the early evolution of the Sun's dynamo and activity. Here we present preliminary Doppler Imaging (DI) results and differential rotation measurements for three rapidly rotating ($v\sin i > 20$ km s^{-1}) active young G-type stars with similar radii (1.09 - 1.22 $R_\odot$) and masses (1.15 - 1.20 $M_\odot$) to the Sun. These stars were chosen as they also represent different stages of early solar evolution, with 17 Myr old RX J0850.1-7554 being a Pre Main Sequence star, 25 Myr old LQ Lup a post-T Tauri star, and 35 Myr old R58 being on the Zero Age Main Sequence.

2. Methodology

The observations presented here were all taken with the same instrumentation setup at the Anglo-Australian Telescope utilising the high resolution échelle spectrograph, UCLES, to output the spectroscopic data on the CCD. Least-Squares Deconvolution (LSD, Donati *et al.* 1997) was utilised to increase the signal-to-noise to a level where DI could be carried out for the selected targets using the maximum entropy image reconstruction code described in Brown *et al.* (1991). The surface differential rotation of the stars was determined using the the χ^2-minimisation technique of Petit *et al.* (2002).

3. Results and Discussion

DI of the target stars produced reconstructed images that showed evidence of a large polar spot as well as lower latitude features on all target stars (Figure 1), a result consistent with observations of other young solar type stars (i.e. Marsden *et al.* 2006).

The differential rotation of these spot features can assist in the study of the stellar dynamo, and thus differential rotation measurements were determined for our targets (see Table 1). The results of these measurements are generally consistent with the results of Marsden *et al.* (2011) which show an increase in differential rotation ($\delta\Omega$) with a

Table 1. Preliminary differential rotation ($\delta\Omega$) measurements determined for our targets. $\delta\Omega$ ranges from moderate (for R58 in 2005) to almost solid-body rotation (for RX J0850.1-7554).

Star (year)	$\delta\Omega$ (rad d^{-1})
R58 (2003)	0.109 ± 0.014
R58 (2005)	0.195 ± 0.024
LQ Lup (2002)	0.097 ± 0.004
RX J0850.1-7554 (2006)	0.004 ± 0.046

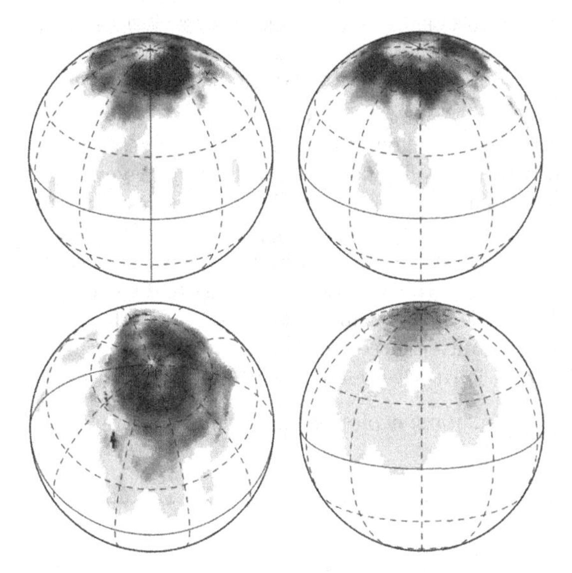

Figure 1. Doppler Imaging maps for our target stars in spherical projection. Clockwise from Top Left: R58 (2003), R58 (2005), RX J0850.1-7554 (2006) and LQ Lup (2002).

decrease in convective zone depth for solar-type stars. In addition, both R58 and LQ Lup show temporal evolution of their differential rotation compared to previous studies (Marsden *et al.* 2005 and Donati *et al.* 2000), a result also discovered for the active young K-dwarf AB Dor (Collier Cameron & Donati 2002). These temporal changes thus suggest G-dwarfs may be winding up and down over long periods similar to lower-mass K-dwarfs, although more observations are needed to confirm such a suggestion.

In marked contrast to the moderate differential rotation of R58 and LQ Lup, the preliminary measurement for the star RX J0850.1-7554 shows a differential rotation near zero, and hence a near-solid body rotation that does not fit the trend presented in Marsden *et al.* (2011). However, due to the large errors associated with our measurement, RX J0850.1-7554 requires more observations to confirm such a low differential rotation rate.

References

Brown, S. F., Donati, J.-F., Rees, D. E., & Semel, M. 1991, *A&A*, 250, 463

Collier-Cameron, A. & Donati, J.-F. 2002, *MNRAS*, 329, L23

Donati, J.-F., Semel, M., Carter, B. D., Rees, D. E. & Cameron, A. C. 1997, *MNRAS*, 291, 658

Donati, J.-F., Mengel M., Carter, B. D., Marsden, S., Collier Cameron, A., & Wichmann, R. 2000, *MNRAS*, 316, 699

Marsden, S. C., Waite, I. A., Carter, B. D., & Donati, J.-F. 2005, *MNRAS*, 359, 711

Marsden, S. C., Donati, J.-F., Semel, M., Petit, P., & Carter, B. D. 2006, *MNRAS*, 370, 468

Marsden, S. C., Jardine, M. M., Ramírez, Vélez J. C., *et al.* 2011, *MNRAS*, 413, 1939

Petit, P., Donati, J.-F., & Collier Cameron, A. 2002, *MNRAS*, 334, 374

Magnetic Fields throughout Stellar Evolution
Proceedings IAU Symposium No. 302, 2013
P. Petit, M. Jardine & H. Spruit, eds.

doi:10.1017/S174392131400194X

Do Magnetic Fields Actually Inflate Low-Mass Stars?

Gregory A. Feiden[1,2] and Brian Chaboyer[2]

[1]Dept. of Physics & Astronomy, Uppsala University, Box 516, Uppsala 751 20, Sweden.

[2]Dept. of Physics & Astronomy, Dartmouth College, 6127 Wilder Laboratory, Hanover, NH 03755, USA.

email: gregory.a.feiden.gr@dartmouth.edu

Abstract. Magnetic fields have been hypothesized to inflate the radii of low-mass stars—defined as less than $0.8\,M_\odot$—in detached eclipsing binaries (DEBs). We evaluate this hypothesis using the magnetic Dartmouth stellar evolution code. Results suggest that magnetic suppression of thermal convection can inflate low-mass stars that possess a radiative core and convective outer envelope. A scaling relation between X-ray luminosity and surface magnetic flux indicates that model surface magnetic field strength predictions are consistent with observations. This supports the notion that magnetic fields may be inflating these stars. However, magnetic models are unable to reproduce radii of fully convective stars in DEBs. Instead, we propose that model discrepancies below the fully convective boundary are related to metallicity.

Keywords. binaries: eclipsing, stars: evolution, stars: interiors, stars: low-mass, stars: magnetic field

---

1. Introduction

It has been well-documented over the past decade that stellar evolution models are unable to accurately predict radii and effective temperatures—so called "fundamental properties"—of low-mass stars ($M < 0.8\,M_\odot$) in detached eclipsing binaries (DEBs; see, e.g., Ribas 2006, Feiden & Chaboyer 2012a). Model radii have been shown to be too small and effective temperatures too hot, particularly in the most well-studied systems. Since stellar evolution models are heavily used to aid with the interpretation of observational data, it is crucial that these modeling errors be addressed. Magnetic fields, maintained by spin-orbit synchronization of DEB components, have been hypothesized to be the culprit. Magnetic activity indicators appear to correlate with radius discrepancies, supporting this hypothesis (e.g., López-Morales 2007). We aim to test this hypothesis using the recently developed magnetic Dartmouth stellar evolution code.

2. Method

We use models generated as a part of the Dartmouth Magnetic Evolutionary Stellar Tracks and Relations program (DMESTAR; Feiden & Chaboyer 2012b, 2013a) to test whether model radii may be inflated by interactions between a magnetic field and thermal convection. Two techniques are used to incorporate magneto-convection in the models: (1) stabilization of convection, and (2) inhibition of convective efficiency (Feiden & Chaboyer 2013a). Method one alters the Schwarzschild convective stability criterion by assuming the magnetic field is in equilibrium with the surrounding gas. An upper limit to the magnetic field strength occurs when the magnetic field is in thermal equipartition with the gas, approximately when the magnetic pressure is equal to the gas pressure. Method

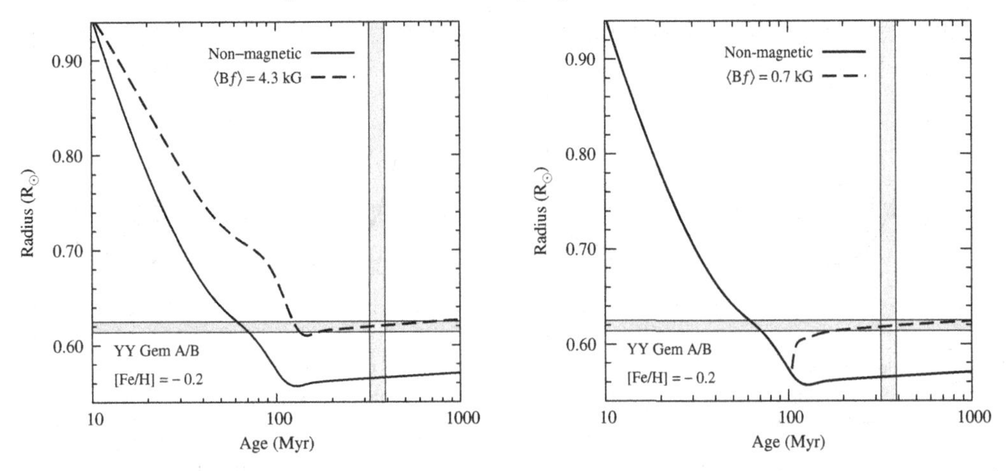

Figure 1. Radius evolution of standard (black, solid line) and magnetic (maroon, dashed line) models of the equal-mass DEB YY Gem. (*Left*) Magnetic model using stabilization of convection introduced at 10 Myr. (*Right*) Magnetic model invoking inhibition of convective efficiency initialized at 100 Myr. The perturbation age has no effect on results along the main sequence. The blue horizontal shaded region indicates the observed radius with associated 1σ uncertainties and the vertical grey region highlights the estimated age of the system.

two instead assumes that the energy required to create the magnetic field is drawn from the kinetic energy of convective flows. Therefore the magnetic field strength has an upper limit determined by equipartition with convective flows, $B_{\rm eq} = (4\pi\rho u_{\rm conv}^2)^{1/2}$.

Standard stellar evolution models (i.e., non-magnetic models) were run for both components of DEB systems to assess the level of radius inflation required of our magnetic models. A series of magnetic models were then computed with varying magnetic field strengths, using the magneto-convection techniques outlined above, until the model fundamental properties agreed with observationally determined values. We performed this procedure for several well-studied DEB systems to ascertain whether results were robust.

3. Results

Effects of magneto-convection on fundamental stellar properties predicted by stellar evolution models may be summarized for two different low-mass stellar populations: stars that have a radiative core with a convective outer envelope and fully convective stars. We will focus on model radius predictions, as stellar radii are more reliably measured from observations than effective temperatures.

The influence of magneto-convection on partially convective stars is demonstrated in Figure 1. The only significant difference between the two panels is the adopted magneto-convection technique with stabilization of convection used in the left panel and inhibition of convective efficiency in the right panel. It is clear that accounting for magneto-convection can inflate model radius predictions at a level required to reconcile models with observations. In general, we observe that both magneto-convection techniques provide a qualitatively correct solution for partially convective stars. However, the two magneto-convection techniques predict significantly different surface magnetic field strengths. Inhibition of convective efficiency typically requires weaker surface magnetic field strengths—by roughly a factor of 5—than stabilization of convection. For example, Figure 1 shows that stabilization of convection (left panel) requires a 4.3 kG surface magnetic field while inhibition of convective efficiency (right panel) requires a 0.7 kG surface magnetic field

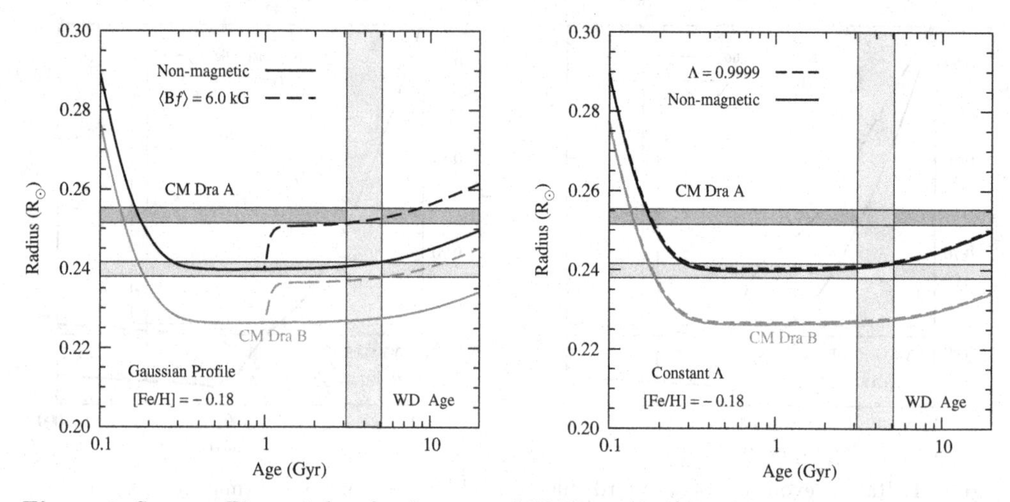

Figure 2. Same as Figure 1, but for the stars of CM Dra. (*Left*) Models invoking the stabilization of convection initialized at 1 Gyr. (*Right*) Models that have inhibited convective efficiency initialized at 100 Myr.

to correct models of YY Gem to observations. In both cases the peak interior magnetic field strengths are roughly on the order of 1 – 10 kG.

Results are quite different for fully convective stars, as illustrated with a representative system (CM Draconis) in Figure 2. The left panel shows that stabilization of convection is sufficient to reconcile model radii with observations. Models presented in the left panel of Figure 2 have a surface magnetic field strength of 6.0 kG with a peak interior magnetic field strength of roughly 50 MG. Inhibiting convective efficiency, however, does not radically alter model predictions. Models in the right panel of Figure 2 have magnetic field strengths equal to 99.99% of the equipartition value—roughly 3 kG at the surface and 50 kG deep within the star. What is not apparent from Figure 2 is that we adjusted the interior magnetic field strength within the models invoking stabilization of convection to provide greater radius inflation. Fully convective models are sensitive to the deep interior magnetic field strength, which can be set arbitrarily to achieve the desired inflation.

4. Discussion

Introducing magneto-convection into stellar models appears to provide at least a qualitative solution to the problem of inflated low-mass stars in DEBs. What must be addressed is whether the magnetic field properties (surface and interior field strengths) are physically realistic. Using a scaling relation between stellar coronal X-ray luminosity and surface magnetic flux (Feiden & Chaboyer 2013a), we find that models invoking the stabilization of convection require surface magnetic field strengths that are likely too strong. This is particularly evident for partially convective stars where estimated surface magnetic field strengths are too strong by about a factor of 5. However, models using inhibition of convective efficiency predict surface magnetic field strengths consistent with X-ray luminosity estimates. Assuming a "turbulent dynamo" is primarily responsible for the generation of magnetic fields in low-mass partially convective stars, it appears plausible that magnetic fields are inflating stellar radii.

Surface magnetic field strengths for fully convective stars invoking stabilization of convection appear plausible based on X-ray luminosity estimates. Models using inhibition of convective efficiency, while providing the best agreement with observed surface magnetic

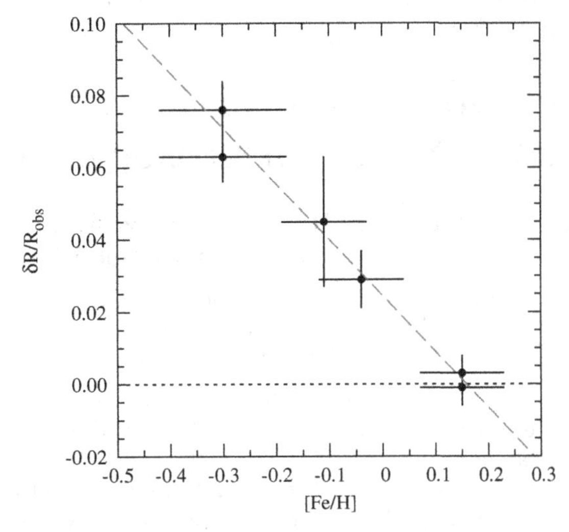

Figure 3. Relative radius error between stellar evolution models and observations of fully convective stars plotted against observationally determined metallicities.

field strengths (Reiners 2012), fail to reproduce observed stellar properties. Recall, we mentioned that the deep interior field strength in fully convective stars was important. Magnetic field strengths must be nearly 50 MG to influence the structure of low-mass fully convective stars. This appears too strong to be physically plausible. A turbulent dynamo mechanism, such as that operating in fully convective stars, cannot produce a magnetic field strength of such magnitude. Instead, a field of that strength must be of primordial origin, which appears inconsistent with estimated diffusion timescales and additional observational evidence (Feiden & Chaboyer 2013b). Therefore, it seems unlikely, given current evidence, that magnetic fields are responsible for inflating fully convective stars.

Ultimately, we must then view the result for partially convective stars with some skepticism. Further theoretical modeling of magneto-convection and additional observational constraints on properties like metallicity, surface magnetic flux, and star spot coverage are required. Yet, it is not too early to begin searching for other explanations. Among fully convective stars in DEBs, for instance, we find a strong anti-correlation between the level of radius inflation of real stars and estimated stellar metallicities (see Figure 3). At the moment, we have no convincing explanation for this anti-correlation, just that we see it in the DEB data and hints of it in interferometric data (Boyajian *et al.* 2012).

This work was supported by NSF grant AST-0908345 and the William H. Neukom 1964 Institute for Computational Science at Dartmouth College.

References

Boyajian, T. S., *et al.* 2012, *ApJ*, 757, 112
Feiden, G. A. & Chaboyer, B. 2012a, *ApJ*, 757, 42
Feiden, G. A. & Chaboyer, B. 2012b, *ApJ*, 761, 30
Feiden, G. A. & Chaboyer, B. 2013a, arXiv:1309.0033
Feiden, G. A. & Chaboyer, B. 2013b, in preparation
López-Morales, M. 2007, *ApJ*, 660, 732
Reiners, A. 2012, *Living Reviews in Solar Physics*, 8, 1
Ribas, I. 2006, *Ap&SS*, 304, 89

Magnetic Fields throughout Stellar Evolution
Proceedings IAU Symposium No. 302, 2013
P. Petit, M. Jardine & H. Spruit, eds.

doi:10.1017/S1743921314001951

A new spectropolarimeter for the San Pedro Martir National Observatory

E. Iñiguez-Garín, D. Hiriart, J. Ramirez-Velez, J. M. Núñez-Alfonso, J. Herrera and J. Castro-Chacón

Instituto de Astronomía, Universidad Nacional Autónoma de México,
Ensenada, B. C., México
email: einiguez@astro.unam.mx

Abstract. We present the design of a stellar spectropolarimeter to measure the magnetic field of point sources. The polarization module interfaces the Boller and Chivens (B&Ch) intermediate-low resolution (R $\sim$ 500 – 4000) spectrograph to the 2.1-m telescope of the San Pedro Martir National Astronomical Observatory in Mexico. The module uses a Savart plate to split the beam into two orthogonal states of polarization and a quarter (half) waveplate to measure circular (linear) polarization. The module is mounted to the telescope and it feeds the spectrograph through a set of four fibers, two for the polarized star images and two for the spectrograph calibration lamp. The instrument will be capable of measuring polarization in spectral lines to determine the longitudinal and transversal fields in magnetic stars.

Keywords. Astronomical instrumentation: polarimetry – magnetic fields – stars: magnetic stars

---

1. Introduction

We present a polarization module operating at optical wavelengths that interfaces the B&Ch spectrograph and the 2.1m telescope of the San Pedro Martir National Astronomical Observatory(SPM-NAO). The polarization module feeds the spectrograph through a set of four optical fibers that allows the instrument to be mechanically decoupled from the telescope in arrange used in similar instruments (Kim *et al.* (2007)). This new polarization module at SPM-NAO will replace the actual one that it is directly coupled to the spectrograph (Hiriart *et al.* (2011)). An advantage of using fiber optics in this new version is that the images of the star remain fixed at the spectrograph slit, reducing problems due to the error in telescope guiding and seeing effects.

2. Design

Figure 1 presents an schematic of the new proposed spectropolarimeter. The polarization module consists of either a rotatable quarter-wave plate (QWP) or a rotatable half wave plate (HWP) followed by a Savart plate (SVP). The beam emerging from the waveplate is split by the SVP in two beams of orthogonal polarized states. Each of the beams is injected into a fiber optic that brings the light to the slit entrance of the spectrograph.

The optical system is capable of feeding the comparison lamp thru the same optical train by reflecting it at the back of the slit entrance of the module. The retarder plate plus the SVP system generates two orthogonal polarized states of the star simultaneously with the orthogonal polarized states of the calibration lamp. The polarization elements could be removed from the optical axis of the telescope when it is required the regular use of the spectrograph.

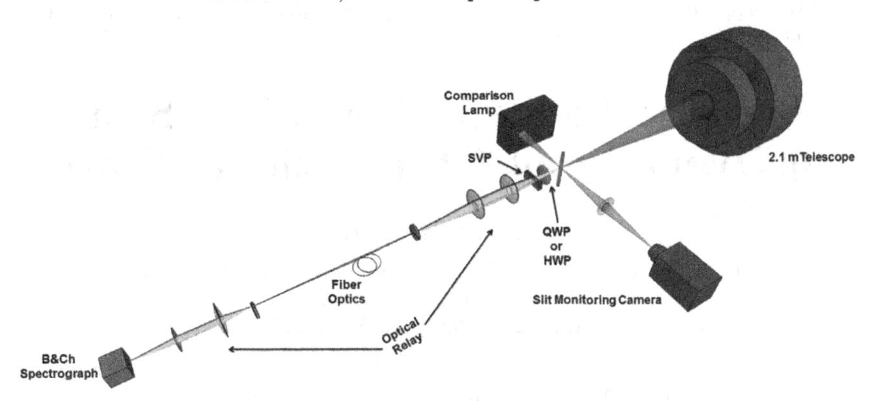

Figure 1. Optical layout of the polarimeter module for the B&Ch spectrograph at the 2.1-m telescope of the San Pedro Martir Observatory.

The retarder plates are made of birefringent polymers by Astropribor Company. These QWP and HWP are true zero–order retarders with a super–achromatic response that retards the wave at a rate of $0.25 \pm 0.001\lambda$in 400–860 nm and $0.50 \pm 0.001\lambda$ in 380–810 nm, respectively. The SVP was manufactured by United Crystals Company. The emergent beams from the SVP are parallel with a displacement of 600 μm at 633 nm. The SVP has a broadband anti–reflective coating from 300 to 900 nm on both surfaces. The optical fibers are manufactured by Polymicro Technologies. Each fiber has a core diameter of 200 μm, a cladding thickness of 20 μm, and an external polymer coating of 20 μm. The fibers have a total length of 13 m. These fibers, manufactured specially for astronomical applications, have a very small focal ratio degradation. The coupling lens system to the fibers will be made with a 1:1 optical relay to preserve the focal ratio of the telescope. Optical relays are composed by two achromatic doublets separated by 100 mm.

3. Conclusions

We presented the design of a new spectropolarimeter module for the 2.1-m telescope at San Pedro Martir National Astronomical Observatory. The instrument is intended to measure the magnetic fields of point sources thru their spectra. This instrument will be capable of measure polarization in spectral lines to determine the longitudinal (circular polarization) and transversal (linear polarization) magnetic fields in magnetic stars.

Acknowledgements

The authors acknowledge financial support from CONACyT, Mexico, thru grant 180817.

References

Hiriart, D., *et al.* 2011, *Magnetic Stars. Proceedings of the International Conference, held in the Special Astrophysical Observatory of the Russian AS, August 27- September 1 , 2010, Eds: I. I. Romanyuk and D. O. Kudryavtsev, p. 220-223*

Kim, K.-M., *et al.* 2007, *PASP*, 119, 1052

Magnetic Fields throughout Stellar Evolution
Proceedings IAU Symposium No. 302, 2013
P. Petit, M. Jardine & H. Spruit, eds.

doi:10.1017/S1743921314001963

Magnetic Fields in Low-Mass Stars: An Overview of Observational Biases

Ansgar Reiners

Institut für Astrophysik, Georg-August Universität Göttingen,
37077 Göttingen, Germany
email: Ansgar.Reiners@phys.uni-goettingen.de

Abstract. Stellar magnetic dynamos are driven by rotation, rapidly rotating stars produce stronger magnetic fields than slowly rotating stars do. The Zeeman effect is the most important indicator of magnetic fields, but Zeeman broadening must be disentangled from other broadening mechanisms, mainly rotation. The relations between rotation and magnetic field generation, between Doppler and Zeeman line broadening, and between rotation, stellar radius, and angular momentum evolution introduce several observational biases that affect our picture of stellar magnetism. In this overview, a few of these relations are explicitly shown, and the currently known distribution of field measurements is presented.

1. Introduction

An important difference between massive and low-mass stars is the presence of an outer convection zone. In this article, we distinguish between high- and low-mass stars on the basis of the presence of an outer convective zone; high-mass stars have no outer convective envelopes. While they may have convective cores, radiative energy transport dominates in the outer zones of these stars. Because dissipation timescales are long, strong magnetic fields may survive there, but fields are not generated, and fields that may be generated in the core find no easy way to the surface.

Low-mass stars, on the other hand, have outer convective envelopes in which magnetic fields decay within only a few decades or centuries (Chabrier & Küker, 2006), and where motion of ionized particles apparently manage to generate strong magnetic fields as for example in the Sun. The efficiency of magnetic field generation through a dynamo process depends on several conditions, but the details of these are not well known (e.g., Charbonneau, 2010). The Sun is one anchor for our models of stellar dynamos. It is probably a fairly common representative of its type (Basri *et al.*, 2013), but we know many stars that are a lot younger and more active. These stars produce orders of magnitude more non-thermal radiation (activity). The reason for this is probably their faster rotation leading to enhanced dynamo action powering non-thermal heating of the chromosphere and corona.

Observations of magnetic fields in stars other than the Sun require relatively high data quality. More important, the signatures of magnetic fields must be disentangled from other effects, which is often difficult because the characteristic properties of low-mass stars evolve in time (and differently for different stellar masses). In this article, I introduce the main characters important for spectroscopic measurements of magnetic fields and their interpretation, and I present the currently known distribution of field measurement using different techniques. A more comprehensive review about observations of low-mass star magnetic fields can be found in Reiners (2012) where the data used here are also presented.

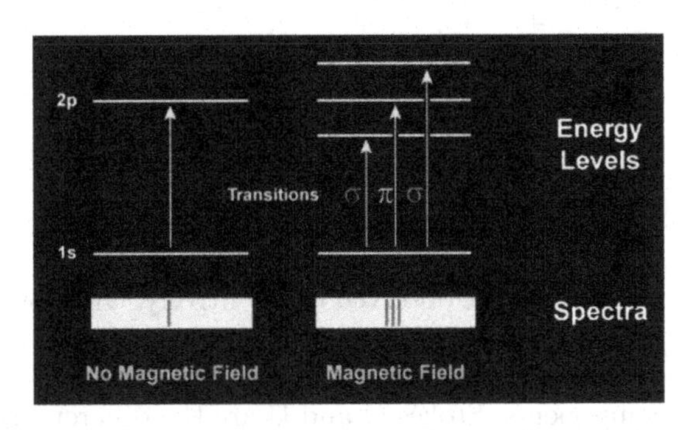

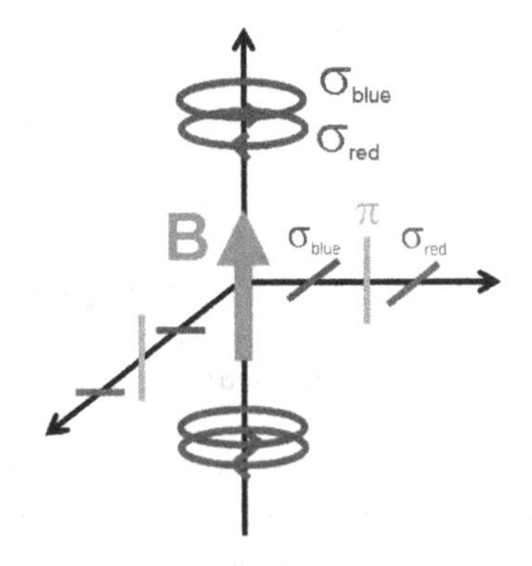

Figure 1. Simplified Zeeman splitting mechanism. The upper energy level is split into three levels in the presence of a magnetic field (*right*). The three different components have different polarizations and produce very different signatures depending on the direction of observation (*right*).

2. Cast of Characters

Four main characters are conspiring in our picture of stellar dynamos and their spectroscopic observations. They are the following:

2.1. *Zeeman*

The Zeeman effect is the most obvious and most direct consequence of the presence of a magnetic field in a star. As we know from the Sun, magnetic fields can lead to enhanced non-thermal radiation that we call activity, but only the direct detection through the Zeeman effect can show that stars other than the Sun really follow similar rules, and that other stars do indeed produce average magnetic fields orders of magnitude stronger than the Sun does.

The principle of the Zeeman effect is shown in Fig. 1. The energetic degeneracy between energy levels can be lifted by the presence of a magnetic field, which typically leads to three different groups of transitions, two σ-groups and one π-group. The groups have different polarizations and are selectively emitted into certain directions depending on the orientation of the magnetic field. The displacement of the σ-groups with respect to the non-displaced π-group is

$$\frac{\Delta\lambda}{\text{mÅ}} = 46.67\, g \left(\frac{\lambda_0}{\mu\text{m}}\right)^2 \frac{B}{\text{kG}}; \tag{2.1}$$

written in units of wavelength, the Zeeman effect is a function of λ^2. In units of velocity, the Zeeman effect can be written as

$$\frac{\Delta v}{\text{ms}^{-1}} = 1.4\, g\, \frac{\lambda_0}{\mu\text{m}} \frac{B}{\text{G}}, \tag{2.2}$$

which still depends on wavelength. At a wavelength of $\lambda = 1\,\mu$m, the typical Zeeman displacement is $\Delta v = 1\,\text{m}\,\text{s}^{-1}$ for a field strength of $B = 1\,$G.

2.2. *Stokes*

The polarization states of the π- and σ-components are different. This provides great potential for the detection and measurement of magnetic fields because different polarizations can be compared with each other differentially. Individual polarization vectors, however, cannot simply be observed but need to be filtered out, e.g., by the use of polarizing beamsplitters (see, e.g., Tinbergen, 1996). Together with retarding waveplates,

$$\begin{array}{lcccc} \mathrm{I} & = & \updownarrow & + & \leftrightarrow \\ \mathrm{Q} & = & \updownarrow & - & \leftrightarrow \\ \mathrm{U} & = & \searrow & - & \swarrow \\ \mathrm{V} & = & \circlearrowleft & - & \circlearrowright \end{array}$$

Figure 2. The four Stokes parameters.

combinations of linear and circular polarizations can be observed consecutively. One possible choice for observable combinations of polarization states is defined by the so-called Stokes vectors, shown in Fig. 2. Stokes I is simply the integrated light, i.e. the sum of of the two perpendicular linear or circular components. Stokes Q and U are the differences of the two perpendicular linear polarization components, the reference frames of Q and U are rotated by 45° with respect to each other. Stokes V is the difference between left- and right-handed circular polarization.

Depending on the direction of observation, the linear and circular polarization vectors carry different parts of the magnetic field information. What is worse, regions of opposite polarity produce circular polarization that can entirely cancel out each other. This is because the blue-shifted circularly polarized component of a "positive" magnetic field has exactly the same shift and amplitude as the analog component caused by a "negative" magnetic field, but the sign of that component is opposite. The sum of the two Stokes V components is therefore exactly zero. Note that this cancellation does not occur in the linearly polarized components because the direction of polarization is identical for the two σ-components. It is important to realize that the information in integrated light, Stokes I, depends on the direction of observation, too, mainly because the linearly polarized π-components are invisible if the magnetic field direction is parallel to the direction of observation.

2.3. *Doppler*

In a rotating star, light emitted from the side of the star that is approaching the observer is blueshifted, and light from the other side of the star is redshifted. This leads to net broadening of spectral lines and can be used to determine the projected rotational velocity, $v \sin i$, of the star (e.g., Gray, 2008). Low-mass stars as defined here (all stars with outer convective envelopes) include all stars with masses and radii between approximately 1.2 and 0.1 times the solar values, i.e., their characteristic properties vary over more than one order of magnitude. Young stars can also possess outer convective envelopes and are a lot larger than main sequence (MS) stars adding to the great variety of targets. In Fig. 3, the equatorial velocities of four different MS stars are shown as a function of their rotation periods. For example, a G2 star with a rotation period of $P = 10\,\mathrm{d}$ will have an equatorial velocity of $v_{\mathrm{eq}} = 5\,\mathrm{km\,s^{-1}}$, but an M4 star of the same period will only show a maximum line broadening corresponding to the equatorial velocity of $v_{\mathrm{eq}} = 1\,\mathrm{km\,s^{-1}}$. This difference has severe consequences for the observability of spectroscopic line diagnostics, as for example Zeeman broadening, because Zeeman broadening must be disentangled from Doppler broadening.

2.4. *Rossby*

A prediction from Dynamo theory is that the efficiency of a convective stellar dynamo may depend on the ratio between Coriolis force and field dissipation (e.g., Ossendrijver, 2003). This ratio can be expressed in terms of typical convective and rotation timescales. For the rotation timescale, an obvious choice is the rotational period. For the convective timescale, a timescale used quite often is the convective overturn time, which is defined

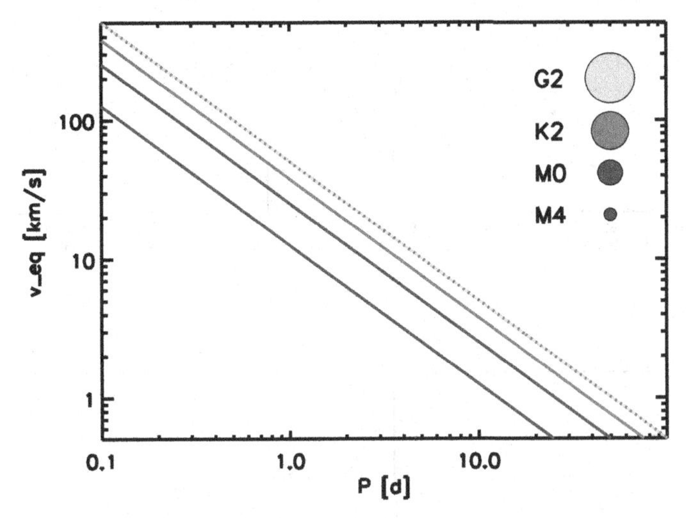

Figure 3. Equatorial Doppler velocities for four MS stars as a function of their rotation periods. The spectral types and relative sizes are shown in the upper right legend. The vertical position of the relations follows the ordering in the legend (and colors of the lines match legend colors).

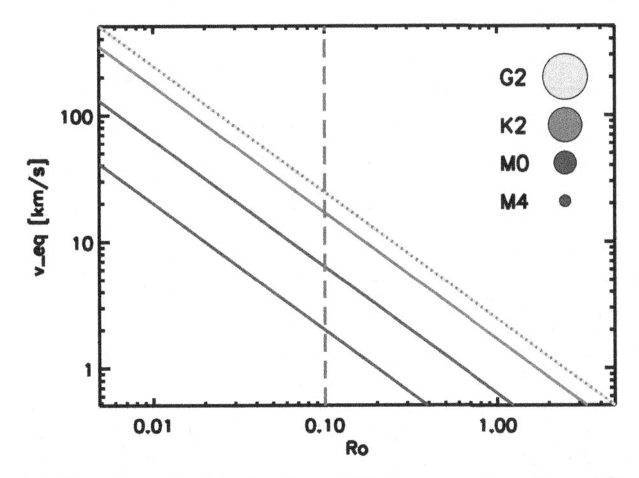

Figure 4. Equatorial Doppler velocities for four MS stars as a function of Rossby number. The spectral types and relative sizes are shown in the upper right legend. The vertical position of the relations follows the ordering in the legend (and colors of the lines match legend colors).

as the typical convective velocity divided by the size of the convection zone (Durney & Latour, 1978). The ratio of the two is called the Rossby number, $Ro = P/\tau_{\rm conv}$.

The convective overturn time is a slowly varying function of stellar mass, and therefore the Rossby number is mostly determined by the value of the rotation period. Nevertheless, if activity in different stars should be compared, it is often useful to use the Rossby number instead of comparing rotation periods. Figure 4 shows the equatorial velocity on the surface of a star as a function of Rossby number. It is similar as Fig. 3 but the differences in $v_{\rm eq}$ are even larger for stars of different mass because not only the radius but also the Rossby number is different.

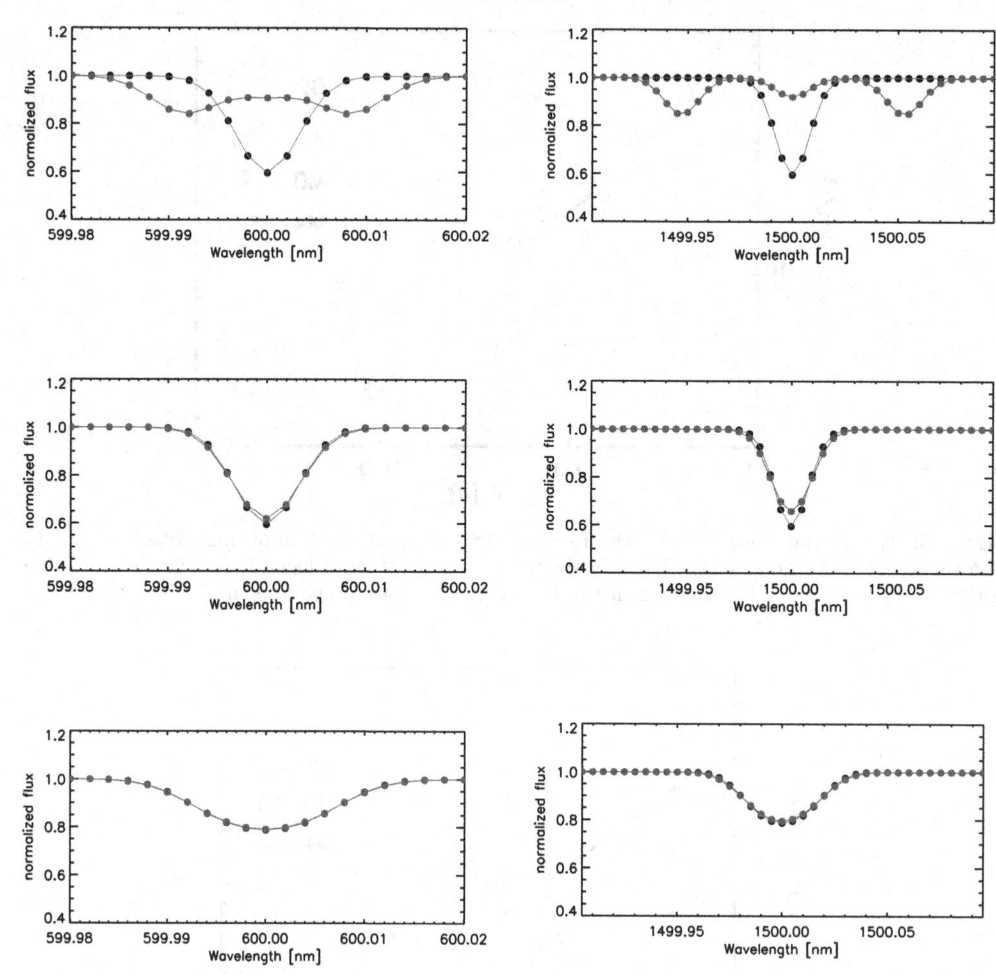

Figure 5. The effect of Zeeman broadening on a single spectral line. *Left:* A typical line at $\lambda = 600\,\mathrm{nm}$; *right:* a typical line at $\lambda = 1500\,\mathrm{nm}$; $g = 2$ in all cases. *Upper panel:* Zeeman broadening in a star with an average magnetic field of $B = 2000\,\mathrm{G}$ (red) compared to a line with no $B = 0\,\mathrm{G}$. *Centre panel:* Effect of a field of $B = 200\,\mathrm{G}$ (red) compared to zero field strength. *Bottom panel:* Effect of $B = 200\,\mathrm{G}$ observed in a star rotating at $v\,\sin i = 5\,\mathrm{km\,s^{-1}}$.

3. Zeeman or Doppler?

The relation between rotation and activity is a well-established observational fact; slow rotators produce little activity, faster rotators produce more (Pizzolato *et al.*, 2003; Wright *et al.*, 2011). The ratio between activity seen in non-thermal emission and the star's bolometric luminosity is a function of the rotation period, but it differs between different stars. For equal Rossby numbers, however, it is expected that this ratio is similar for all stars. Therefore, convective overturn times are sometimes motivated empirically by searching for the function of τ_{conv} that minimizes the scatter in the activity-rotation relation (Noyes *et al.*, 1984; Kiraga & Stepien, 2007; Wright *et al.*, 2011). A problem for the theoretical calculations of τ_{conv} is that it is not obvious what definition of τ_{conv} one should use – is it the convective overturn time at the bottom of the convective envelope, or the weighted mean throughout the convection zone, or something different?

Measuring a magnetic field in a rotating star requires Zeeman broadening to be a significant fraction of the total line broadening that is dominated by rotation. In slow

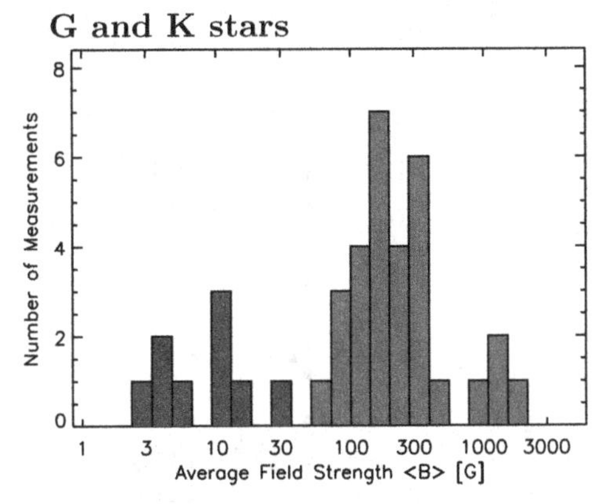

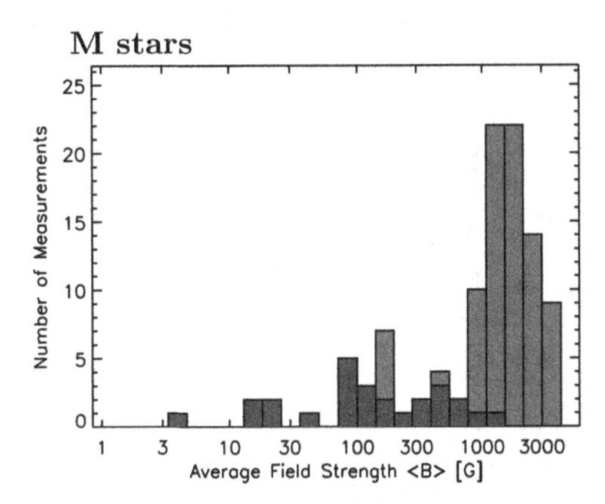

Figure 6. Magnetic field measurements from Stokes I (grey histograms) and Stokes V (blue histograms) in G- and K-type stars (*left* panel) and M stars (*right* panel).

rotators ($Ro > 0.1$), the magnetic field grows with rotation, in fast rotators, the field seems to be saturated and the field does not grow further with rotation (Reiners *et al.*, 2009). The ratio between Zeeman and Doppler broadening is therefore smaller in the regime of saturated dynamos.

The effect of Zeeman broadening in the presence of rotation is displayed in Fig. 5; the bottom panel of this figure shows a typical situation for a sun-like star at $Ro \approx 0.5$.

In the non-saturated part of the rotation-activity or rotation-magnetic field relation, Zeeman broadening ($\Delta v_{\mathrm{Zeeman}}$) is approximately proportional to Doppler broadening (v_{eq}). Based on the heterogeneous sample of Zeeman measurements collected in Reiners (2012), an estimate of the ratio for G dwarfs is

$$\frac{\Delta v_{\mathrm{Zeeman}}}{v_{\mathrm{eq}}} \approx 0.07 \left(\frac{\lambda_0}{1\,\mu\mathrm{m}} \right) g. \tag{3.1}$$

This ratio is approximately valid for all stars with non-saturated activity. In more rapidly rotating stars, the ratio is smaller because rotational broadening is larger but Zeeman broadening is saturated.

4. Stokes' Choices

Most direct measurements of magnetic fields were carried out either in Stokes I or Stokes V (but see Kochukhov *et al.*, 2011). The observational systematics of the methods lead to significant biases that need to be understood if we want to interpret the results. For example, Stokes I measurements have a hard time detecting weak magnetic fields in rapid rotators, but they capture almost all field components. Stokes V measurements, on the other hand, can detect very small fields but cancellation of opposite field directions can make significant field components invisible.

A collection of Stokes I and Stokes V average magnetic field measurements is shown in Fig. 6 (data collection from Reiners, 2012). Stokes I measurements find magnetic fields of several 100 G and more, smaller fields cannot be detected because of limited sensitivity. Stokes V measurements in G- and K-type stars are limited to field strengths of a few 10 G, which is probably because of cancellation effects. In M-stars, the detected fields are significantly larger.

In Fig. 7, Stokes I measurements for stars of spectral types G0–M9 are shown as a function of spectral type. Field stars are shown together with pre-main sequence stars.

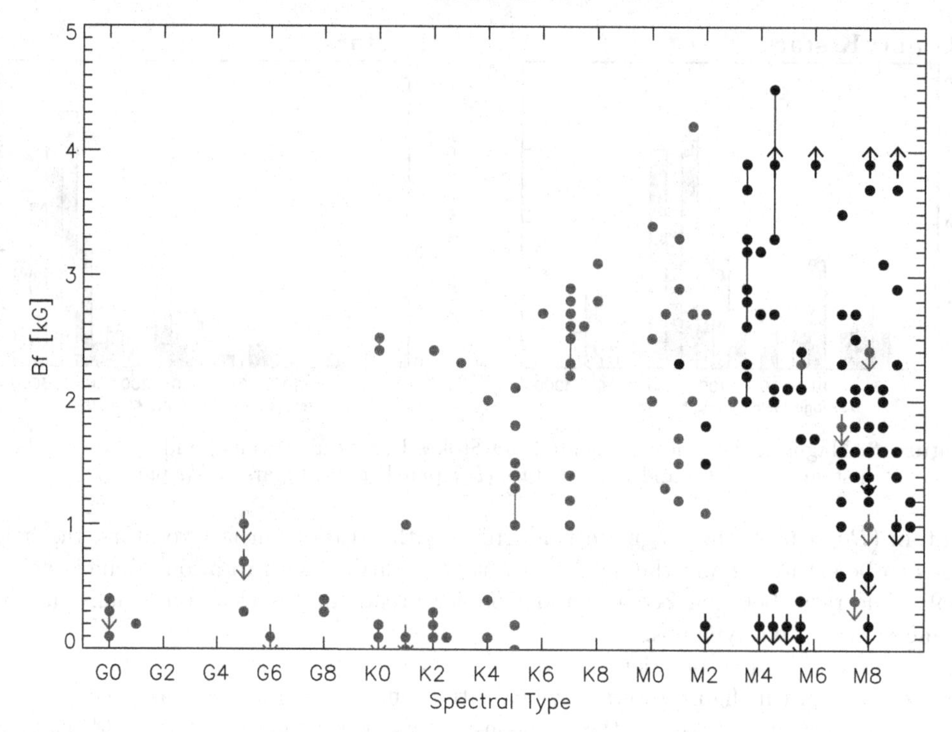

Figure 7. Observations of Stokes I average magnetic fields in different types of stars. *Blue:* G- and K-dwarfs; *black:* M-dwarfs; *red:* pre-MS stars.

The magnetic fields found in late-type stars are higher than in hotter stars, but not all late-type stars have strong fields. The driver of magnetic field generation is rotation; the distribution in Fig. 7 reflects the distribution of rotation velocities in stars of different spectral type and age. Furthermore, it reflects the fact that strong fields are much more difficult to observe in G stars than in M stars. The reason is the following: A field strength of ca. 1000 G can be expected in stars with Rossby numbers $Ro \approx 0.2$. According to Fig. 4, the equatorial rotation velocity of a G2-star at this rotation rate is $v_{\rm eq} \approx 10\,{\rm km\,s^{-1}}$. In an M0 star with the same Rossby number, the equatorial rotation velocity is only $v_{\rm eq} \approx 3\,{\rm km\,s^{-1}}$, and in an M4 star we find $v_{\rm eq} \approx 1\,{\rm km\,s^{-1}}$. The signature of magnetic fields at low Rossby numbers is therefore much more obvious in low-mass stars.

References

Basri, G., Walkowicz, L. M., & Reiners, A. 2013, *ApJ*, 769, 37

Chabrier, G. & Küker, M. 2006, *A&A*, 446, 1027

Charbonneau, P. 2010, *Living Rev. Solar Phys.*, 7

Durney, B. R. & Latour, J. 1978, *Geophys. Astrophys. Fluid Dyn.*, 9, 241

Gray, D. F. 2008, *The Observation and Analysis of Stellar Photospheres*, Cambridge University Press

Kiraga, M. & Stepien, K. 2007, *Acta Astron.*, 57, 149

Kochukhov, O., Makaganiuk, V., Piskunov, N., Snik, F., Jeffers, S. V., Johns-Krull, C. M., Keller, C. U., Rodenhuis, M., & Valenti, J. A., *ApJ*, 732, 19

Noyes, R. W., Hartmann, L. W., Baliunas, S. L., Duncan, D. K., & Vaughan, A. H. 1984, *ApJ*, 279, 763

Ossendrijver, M. A. J. H. 2003, *AAR*, 11, 287

Pizzolato, N., Maggio, A., Micela, G., Sciortino, S., & Ventura, P. 2003, *A&A*, 397, 147

Reiners, A., Basri, G., & Browning, M. 2009, *ApJ*, 692, 538
Reiners, A. 2012, *Living Rev. Solar Phys.*, 8, 1
Tinbergen, J. 1996, *Astronomical Polarimetry*, Cambridge University Press
Wright, N. J., Drake, J. J., Mamajek, E. E., & Henry, G. W. 2011, *ApJ*, 743, 48

Magnetic Fields throughout Stellar Evolution
Proceedings IAU Symposium No. 302, 2013
P. Petit, M. Jardine & H. Spruit, eds.

doi:10.1017/S1743921314001975

On the spectropolarimetric signature of FeH in the laboratory and in sunspots

Patrick Crozet[1], Amanda J. Ross[1], Nathalie Alleq[1], Arturo López Ariste[2], Claude Le Men[2] and Bernard Gelly[2]

[1]Institut Lumière Matière, UMR5306 Université Lyon 1-CNRS, Université de Lyon, 69622 Villeurbanne Cedex, France.

[2]THEMIS Télescope Héliographique pour l'Etude du Magnétisme et des Instabilités Solaire, UPS853 (CNRS/Observatorio del Teide, C/Via Lactea s/n 38200 La Laguna, Tenerife, Spain.
email: `patrick.crozet@univ-lyon1.fr`

Résumé. Laboratory spectra showing Zeeman patterns in some FeH lines susceptible to be used as magnetic probes in cool stellar atmostpheres have been recorded in the laboratory, and molecular Landé factors obtained from analysis. These Landé factors have been used to model some lines recorded in Stokes V polarisation in sunspot spectra at the solar telescope THEMIS.

Keywords. electronic spectrum of FeH, molecular Landé factor, sunspot spectrum, Stokes V profile

---

1. Laboratory determination of Landé factors

Zeeman-broadened lines in the near-IR bands of FeH have been suggested as a magnetic probe of cool stellar atmospheres (Afram *et al.*, (2008), but the Landé factors for these transitions are notoriously difficult to predict because of electronic configuration complexity in the FeH radical. Laser magnetic resonance spectroscopy (Brown *et al.* (2006))has established g_J values for the v = 0 in the X $^4\Delta_\Omega$ ground state. This work aimed to supply reliable g_J Landé factors for some electronically excited levels in the F $^4\Delta_\Omega$ state of FeH, refining estimates made by Harrison & Brown (2008) from Zeeman profiles of FeH transitions seen in the sunspot atlas of Wallace *et al* (1998). FeH is formed at $\sim$ 500 K in a hollow-cathode sputtering source described elsewhere, Vallon *et al.* (2009). Excitation spectra were recorded as total fluorescence signals, first in zero field, to secure the energy origins and Λ-doubling separations for each line of interest, and then at magnetic fields between 0.2 and 0.5 Tesla (calibrated against Ar I lines) provided by permanent magnets. Laser polarisation was selected to record either $\Delta M_J = 0$ or $\pm$ 1 transitions. In most cases the Zeeman profiles of the lines are only partially resolved (see below). Effective Landé factors g'_J have been determined from peak positions (when deconvolution is possible) or by profile fits, relying on literature values from Brown *et al.* (2006) for parameters in the ground electronic state, fitting measured wavenumbers to the expression

$$\nu = \nu_0 + \mu_B\ B\ (g'_J\ M'_J - g''_J\ M''_J),$$

where μ_B is the Bohr magneton (46.68645 $m^{-1}T^{-1}$) and B is the longitudinal magnetic field intensity. Fig. 1 illustrates the results, showing behaviour close to Hund's coupling case (b) for the ground state, but a very different trend (between limiting cases (a) and (b)) for rotational levels of F $^4\Delta_{7/2,\ 5/2}$, although both are nominally $^4\Delta$ electronic states, and could be expected to demonstrate similar angular momentum coupling.

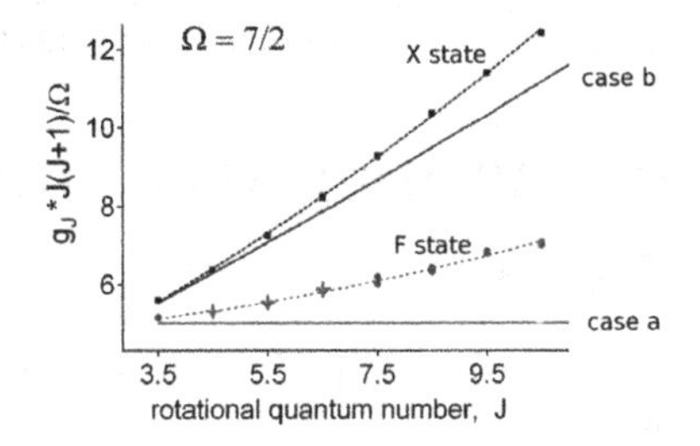

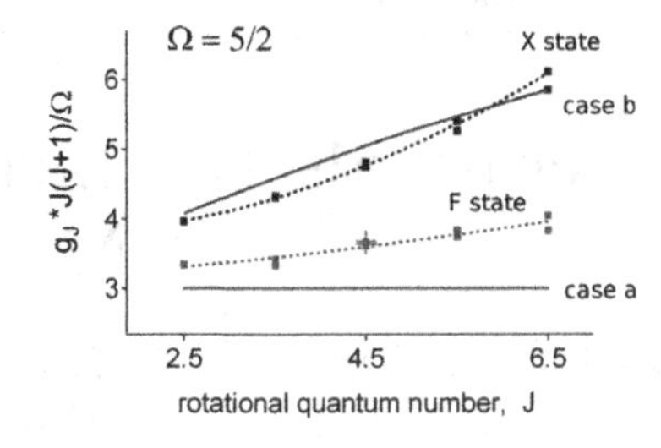

Fig. 1. Some electronic Landé factors in FeH. F $^4\Delta_\Omega$ v = 0, 1 data indicated by ∘ and +.

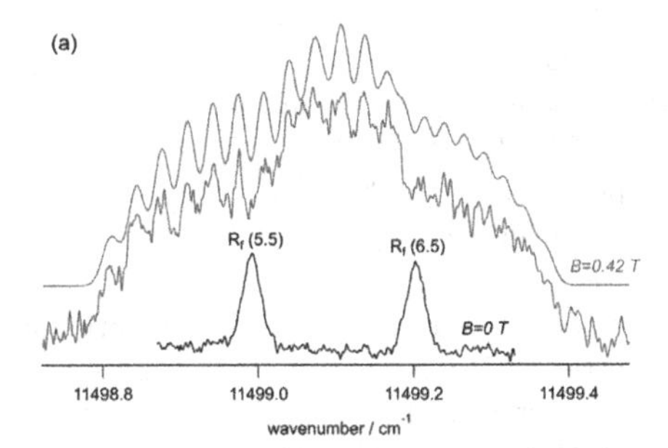

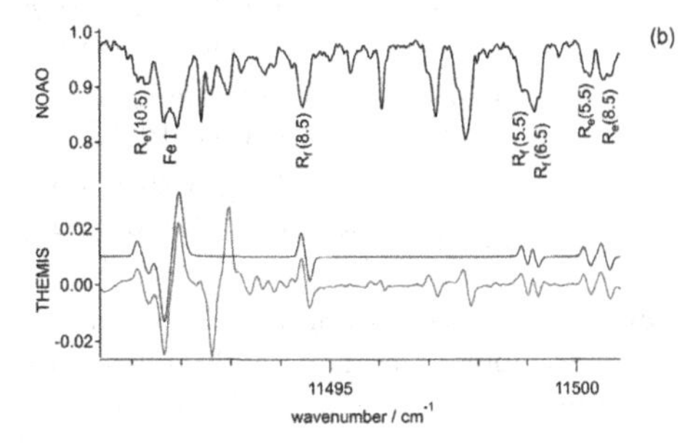

Fig. 2. a) Laboratory spectrum of overlapped R_1 (5.5) and R_1 (6.5) lines showing partially-resolved Zeeman structures. b) Sunspot (NOA0 11582) Stokes V spectrum from Themis with some calculated profiles drawn with vertical offset. Solar atlas absorption from Wallace *et al* (1998) is shown above for comparison, identifying the molecular transitions.

2. Determination of magnetic fields in sunspots

Polarimetric spectra have been recorded at the solar telescope THEMIS in Tenerife in the regions corresponding to the lines with best-resolved Zeeman structures in the laboratory. With a 0.5" slit and high order Echelle spectrograph, lines around 8730 Å in sunspot NOA0 11582 were recorded at a resolution of 90 mÅ. They are not the strongest contributions to the FeH bands in sunspot spectra, but the excellent S/N ratio ($\sim 10^4$) of the Stokes-V signals from THEMIS allows weak features to be examined and modeled. Our spectral simulations optimized linewidths to match effective instrumental resolution for six molecular transitions and one atomic line. Landé factors were constrained in these profile fits. Fig. 2 shows some lines in the F_2 F-X 1-0 band of ^{56}FeH. The laboratory spectrum shows the structure of two close-lying lines in this region corresponding to a blended feature in the solar absorption spectrum. Analysis of molecular line profiles leads to an average value for the longitudinal magentic field of -0.26(1)T. The profile of the Fe I line, however, leads to a larger value, -0.286 T. Atoms are assumed to be preferentially probed at deeper regions of the photosphere, where field lines are tighter. Molecular radicals become more abundant in cooler and higher-altitude regions of the sunspot.

References

Afram, N., Berdyugina, S. V., Fluri, D. M., Solanki , S. K., & Lagg, A. 2008, *A.&A.* 482(2), 387

Harrison, J. J. & Brown, J. M., *ApJ.* 686 (2008) 1426

Vallon, R., Ashworth, S. H., Crozet, P., Field, R. W., Forthomme, D., Harker, H., Richard, C., & Ross, A. J. 2009, *J. Phys. Chem. A*, 113, 13159

Brown, J. M., Körsgen, H., Beaton, S. P., & Evenson, K. M. 2006, *J. Chem. Phys.* 124(23) 234309

Wallace, L., Livingstone, W., Bernath, P., & Ram, R. S. 1998, *N. S. O. Technical Report 1998-002*, Available online ftp ://nsokp.nso.edu/pub/atlas/spot3alt

Magnetic Fields throughout Stellar Evolution
Proceedings IAU Symposium No. 302, 2013
P. Petit, M. Jardine & H. Spruit, eds.

doi:10.1017/S1743921314001987

What controls the large-scale magnetic fields of M dwarfs?

T. Gastine[1], J. Morin[2,3], L. Duarte[1], A. Reiners[2], U. Christensen[1] and J. Wicht[1]

[1]Max Planck Institut für Sonnensystemforschung,
Max Planck Straße 2, 37191 Katlenburg-Lindau, Germany
email: gastine@mps.mpg.de

[2]Institut für Astrophysik, Georg-August-Universität Göttingen,
Friedrich-Hund Platz, 37077 Göttingen, Germany

[3]LUPM, Université de Montpellier and CNRS,
Place E. Bataillon, 34090 Montpellier, France

Abstract. Observations of active M dwarfs show a broad variety of large-scale magnetic fields encompassing dipole-dominated and multipolar geometries. We detail the analogy between some anelastic dynamo simulations and spectropolarimetric observations of 23 M stars. In numerical models, the relative contribution of inertia and Coriolis force –estimated by the so-called local Rossby number– is known to have a strong impact on the magnetic field geometry. We discuss the relevance of this parameter in setting the large-scale magnetic field of M dwarfs.

Keywords. MHD, stars: magnetic field, stars: low-mass, turbulence

1. Introduction

The magnetic fields of planets and rapidly-rotating stars are maintained by convection-driven dynamos operating in their interiors. Scaling laws recently derived from geodynamo-like models successfully predict the magnetic field strength of a wide range of astrophysical objects from Earth and Jupiter to some rapidly-rotating stars (e.g. Christensen & Aubert 2006; Christensen *et al.* 2009; Yadav *et al.* 2013a,b). This emphasises the similarities between the dynamo mechanisms at work in planets and active M dwarfs.

Spectropolarimetric observations of rapidly-rotating M stars show a broad variety of large-scale magnetic fields encompassing dipole-dominated and multipolar geometries (Donati *et al.* 2008; Morin *et al.* 2008a,b,2010). Combining global-scale numerical dynamo models and observational results, we want to better understand the similarities of dynamos in planets and low-mass stars. To study the physical mechanisms that control the magnetic field morphology in these objects, we have explored the influence of rotation rate, convective vigor and density stratification on the magnetic field properties in anelastic dynamo models (Gastine *et al.* 2012,2013).

In such models, the relative importance of inertia and Coriolis force in the force balance –quantified by the local Rossby number Ro_l– is thought to have a strong impact on the magnetic field geometry (Christensen & Aubert 2006). A sharp transition between dipole-dominated and multipolar dynamos is indeed observed at $Ro_l \simeq 0.1$. However, Simitev & Busse (2009) find that both dipolar and multipolar magnetic fields are two possible solutions at the same parameter regime, depending on the initial condition of the system. As shown by Schrinner *et al.* (2012), this dynamo bistability challenges the Ro_l criterion as the multipolar dynamo branch can extend well below the threshold value $Ro_l \simeq 0.1$.

Here we discuss the analogy between the anelastic dynamo models by Gastine *et al.*

(2012) and the spectropolarimetric observations of 23 M stars. The reader is referred to (Gastine *et al.* 2013) for a more comprehensive description of the results.

2. Dynamo models and spectropolarimetric observations

We consider MHD simulations of a conducting anelastic fluid in spherical shells rotating at a constant rotation rate Ω. A fixed entropy contrast Δs between the inner and the outer boundary drives the convective motions. Our numerical models are computed using the anelastic spectral code MagIC (Wicht 2002, Gastine & Wicht 2012) that has been validated against several hydrodynamical and dynamo benchmarks (Jones *et al.* 2011). The governing MHD equations are non-dimensionalised using the shell thickness $d = r_o - r_i$ as the reference lengthscale and Ω^{-1} as the time unit.

The solution of a numerical model is then characterised by several diagnostic parameters. The rms flow velocity is given by the Rossby number $Ro = u_{\rm rms}/\Omega d$, while the magnetic field strength is measured by the Elsasser number $\Lambda = B_{\rm rms}^2/\rho\mu\lambda\Omega$, where ρ is the density, and μ and λ are the magnetic permeability and diffusivity. The typical flow lengthscale l is defined as $l = \pi d/\bar{\ell}_u$, where $\bar{\ell}_u$ is the mean spherical harmonic degree obtained from the kinetic energy spectrum (Christensen & Aubert 2006; Schrinner *et al.* 2012). Following Christensen & Aubert (2006), a *local Rossby number* $Ro_l = u_{\rm rms}/\Omega l$, can then be used to evaluate the impact of inertia on the magnetic field geometry. Finally, the geometry of the surface magnetic field is quantified by its dipolarity $f_{\rm dip} = \boldsymbol{B}^2_{\ell=1,m=0}(r = r_o)/\sum_{\ell,m}^{\ell_{\rm max}} \boldsymbol{B}^2_{\ell,m}(r = r_o)$, the ratio of the magnetic energy of the dipole to the magnetic energy contained in spherical harmonic degrees up to $\ell_{\rm max} = 11$.

We compare these dynamo models with spectropolarimetric observations of 23 active M dwarfs with rotation period ranging from 0.4 to 19 days. The data reduction and analysis is detailed by Donati *et al.* (2006) and Morin et al. (2008a,b,2010). We derive observation-based quantities aimed to reflect the diagnostic parameters employed in the numerical models. The *empirical Rossby number* $Ro_{\rm emp} = P_{\rm rot}/\tau_c$ is our best available proxy for Ro_l, where τ_c is the turnover timescale of convection based on the rotation-activity relation (Kiraga & Stepien 2007). We define an Elsasser number based on the averaged unsigned large-scale magnetic field $\langle {\rm B_V} \rangle$ that roughly characterises the ratio between Lorentz and Coriolis forces. We also consider the fraction of the magnetic energy that is recovered in the axial dipole mode in Zeeman-Doppler imaging maps (ZDI, Semel 1989). The spatial resolution of such maps mostly depends on the projected rotational velocity $v \sin i$. The actual degree and order $\ell_{\rm max}$ up to which the reconstruction can be performed ranges from 4 to 10. We directly compare this quantity to the dipolarity employed in numerical models and term them both $f_{\rm dip}$ in Figs. 1-2.

3. Results and discussion

Figure 1 shows $f_{\rm dip}$ versus Ro_l in the numerical models, while Fig. 2 displays the relative dipole strength of M stars against $Ro_{\rm emp}$ derived from spectropolarimetric observations. The numerical models cluster in two distinct dynamo branches: the upper branch corresponds to the dipole-dominated regime ($f_{\rm dip} > 0.6$), while the lower branch contains the multipolar dynamos ($f_{\rm dip} < 0.2$). Fig. 3 shows two selected cases of these two kinds of dynamo action. The dipolar branch is limited by a maximum $Ro_l \simeq 0.1$, beyond which all the models become multipolar. In contrast to earlier Boussinesq studies (e.g. Christensen & Aubert 2006), the multipolar branch also extends well below $Ro_l \simeq 0.1$, where both dipolar and multipolar solutions are stable (see Schrinner *et al.* 2012). Bistability of the magnetic field is in fact quite common in the parameter range explored here, meaning that both dipole-dominated and multipolar fields are two possible stable configurations at the same set of parameters (Simitev & Busse 2009). The multipolar branch

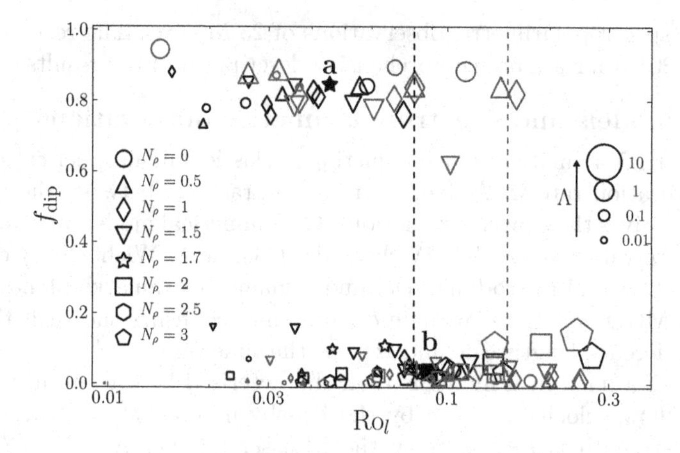

Figure 1. $f_{\rm dip}$ plotted against Ro_l in the anelastic dynamo models computed by Gastine *et al.* (2012). Red (grey) symbols correspond to numerical simulations in thick (thin) shells ($r_i/r_o = 0.2$ and $r_i/r_o = 0.6$). The symbol sizes scale with the amplitude of the surface field, given in units of the square-root of the Elsasser number. The two vertical lines mark the possible upper-limits of the dipole-dominated dynamos. The two filled symbols are further discussed in Fig. 3.

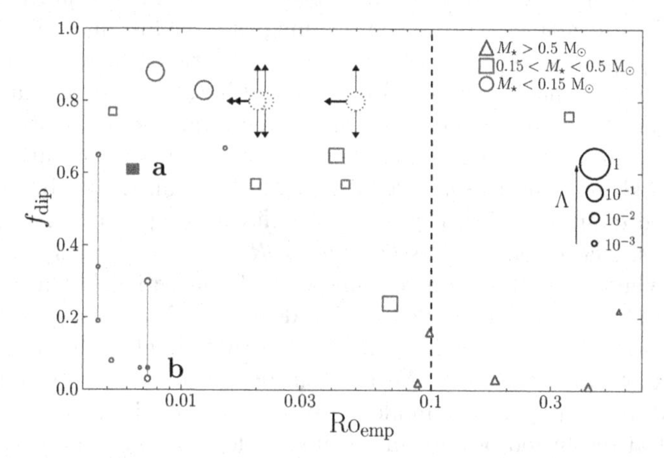

Figure 2. $f_{\rm dip}$ plotted against $Ro_{\rm emp}$. The symbol sizes scale with the square root of the Elsasser number based on the large-scale magnetic field derived from spectropolarimetric observations. The vertical dashed line marks the possible upper bound of the dipolar regime. For the two stars with the largest temporal variation, individual epochs are connected by a vertical red line. Dotted red circles with errorbars correspond to some stars from Morin *et al.* (2010) for which a definite ZDI reconstruction was not possible.

at low Ro_l is partly composed by the anelastic models with $\rho_{\rm bot}/\rho_{\rm top} > 7$ (Gastine *et al.* 2012) and partly by the multipolar attractors of these bistable cases. Note that different assumptions in the numerical models (for instance variable transport properties) help to extend the dipolar regime towards higher density contrasts (Duarte *et al.* 2013).

Although it is difficult to directly relate the diagnostic parameters employed in numerical models to their observational counterparts, the separation into two dynamo branches seems to be relevant to the sample of active M dwarfs displayed in Fig. 2. In particular, the late M dwarfs (with $M_\star < 0.15\,{\rm M}_\odot$) seem to operate in two different dynamo regimes: the first ones show a strong dipolar field, while others present a weaker multipolar magnetic field with a pronounced time-variability.

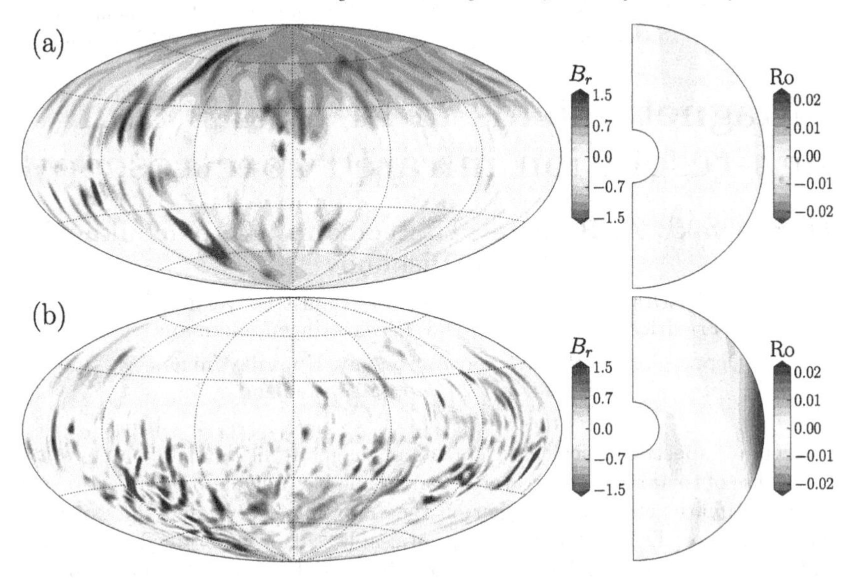

Figure 3. Snapshot of the radial component of the surface magnetic field and the axisymmetric zonal flow $\bar{u}_\phi$ for a dipolar dynamo model (**a**), and a multipolar case (**b**). Magnetic field are given in units of the square root of the Elsasser number and velocities in units of the Rossby number.

This analogy between numerical models and observations of active M dwarfs could be further assessed by additional observations. Indeed, if the analogy holds, stars with a multipolar field are expected over a continuous range of Rossby number where dipole-dominated large-scale fields are also observed (i.e. $0.01 < Ro_{\mathrm{emp}} < 0.1$).

Acknowledgements

TG and LD are supported by the Special Priority Program 1488 "PlanetMag" of the German Science Foundation.

References

Christensen, U. R. & Aubert, J. 2006, *Geophys. J. Int.*, 166, 97
Christensen, U. R. 2010, *Space Sci. Rev.*, 152, 565
Donati, J.-F., Forveille, T., Cameron, A. C., *et al.* 2006, *Science*, 311, 633
Donati, J.-F., Morin, J., Petit, P., *et al.* 2008, *MNRAS*, 390, 545
Duarte, L., Gastine, T., & Wicht, J., 2013, *Physics of the Earth and Planetary Interiors*, 222, 22
Gastine, T. & Wicht, J. 2012, *Icarus*, 219, 428
Gastine, T., Duarte, L., & Wicht, J. 2012, *A&A*, 546, A19
Gastine, T., Morin, J., Duarte, L., Reiners, A., Christensen, U. R., & Wicht, J., 2013, *A&A*, 549, L5
Jones, C. A., Boronski, P., Brun, A. S., *et al.* 2011, *Icarus*, 216, 120
Kiraga, M. & Stepien, K. 2007, *Acta Astronomica*, 57, 149
Morin, J., Donati, J.-F., Forveille, T., *et al.* 2008a, *MNRAS*, 384, 77
Morin, J., Donati, J., Petit, P., *et al.* 2008b, *MNRAS*, 390, 567
Morin, J., Donati, J.-F., Petit, P., *et al.* 2010, *MNRAS*, 407, 2269
Schrinner, M., Petitdemange, L., & Dormy, E. 2012, *ApJ*, 752, 121
Semel, M. 1989, *A&A*, 225, 456
Simitev, R. D. & Busse, F. H. 2009, *Europhysics Letters*, 85, 19001
Wicht, J. 2002, *Physics of the Earth and Planetary Interiors*, 132, 281
Yadav, R. K., Gastine, T., Christensen, U. R., 2013a, *Icarus*, 225, 185
Yadav, R. K., Gastine, T., Christensen, U. R., & Duarte, L. D. V., 2013b, *ApJ*, 774, 6

Magnetic Fields throughout Stellar Evolution
Proceedings IAU Symposium No. 302, 2013
P. Petit, M. Jardine & H. Spruit, eds.

doi:10.1017/S1743921314001999

Magnetic fields in M-dwarfs from high-resolution infrared spectroscopy

D. Shulyak[1], A. Reiners[1], U. Seemann[1], O. Kochukhov[2] and N. Piskunov[2]

[1]Institute of Astrophysics, Georg-August University, Friedrich-Hund-Platz 1, D-37077 Göttingen, Germany

[2]Department of Physics and Astronomy, Uppsala University, Box 515, 751 20, Uppsala, Sweden

Abstract. Accurate spectroscopic measurements of magnetic fields in low mass stars remain challenging because of their cool temperatures, strong line blending, and often fast rotation. This is why previous estimates were based either on the analysis of only a few lines or made use of some indirect techniques. This frequently led to noticeable scatter in obtained results. In this talk I will present and discuss new results on the determination of the intensity and geometry of the magnetic fields in M-dwarfs using IR observations obtained with CRIRES@VLT. The instrument provides unprecedented data of high resolution ($R = 100\,000$) which is crucial for resolving individual magnetically broadened molecular and atomic lines. Such an in-depth analysis based on direct magnetic spectral synthesis opens a possibility to deduce both field intensity and geometry avoiding most of the limitation and assumptions made in previous studies.

Keywords. stars: atmospheres – stars: low-mass – stars: magnetic field

1. Introduction

Low-mass stars of spectral type M often show high level of activity accompanied by strong X-ray fluxes, appearance of emission lines, and global magnetic fields of the order of a few kilogauss detected in many stars, and the scales of these fields are comparable to the size of the stars themselves (see, e.g., Morin *et al.* 2010).

Stellar evolution predicts that stars of spectral types later than M3.5 become fully convective and do not host an interface layer of strong differential rotation. Both partially and fully convective stars can host magnetic fields of similar intensities but likely with different dynamo mechanisms operating in their interiors, and this asks for observational confirmations or otherwise.

There have been a number of attempts to measure magnetic fields in M-dwarfs and more information can be found in the review by Reiners (2012). Because in stars cooler than mid - M atomic line intensity decay rapidly, the molecular lines of FeH Wing-Ford $F^4\,\Delta - X^4\,\Delta$ transitions around 0.99 μm were proposed as alternative magnetic field indicators (Valenti *et al.* 2001; Reiners & Basri 2006).

Unfortunately, theoretical attempts to compute Zeeman patterns of FeH lines have not achieved much success. This is because the Born-Oppenheimer approximation, which is usually used in theoretical descriptions of level splitting, fails for the FeH molecule (for more details see Asensio Ramos & Trujillo Bueno (2006), Berdyugina & Solanki (2002), and poster #28 by P. Crozet at al., this meeting). As a consequence, empirical (Afram *et al.* 2008) and experimental (Harrison & Brown 2008; Crozet *et al.* 2012) attempts were carried out to estimate the Landé g-factor of FeH lines.

In this work we make use of very high resolution infrared spectra of M-dwarfs obtained with CRIRES@VLT to measure the complexity of their surface magnetic fields by studying individual line profiles. We attempt to derive distributions of filling factors of local magnetic field components that provide a best agreement between observed and theoretical spectra and to address the question whether there are any differences between fully and partially convective stars solely from spectroscopic analysis.

2. Fitting methods

The approach to compute g-factors was described in Shulyak *et al.* (2010), and is based on numerical libraries from the MZL (Molecular Zeeman Library) package originally written by B. Leroy (Leroy 2004), and adopted by us for the particular case of FeH.

Using Stokes I spectra we measure the intensity and complexity of surface magnetic fields, i.e. the minimum number of magnetic field components required to fit the observed line profiles.

For each spectrum we apply a chi-square Levenberg-Marquardt minimization algorithm with filling factors f_i as fit parameters. We consider 21 filling factors which correspond to the magnetic fields ranging from 0 kG to 10 kG in steps of 0.5 kG. The whole procedure is applied for different sets of atmospheric parameters: $T_{\rm eff}$, α(Fe), and $\upsilon \sin i$.

3. Results

Figure 1 illustrates examples of theoretical fits to a few selected FeH features in four M dwarfs GJ 388, GJ 729, GJ 285, and GJ 406, as well as recovered distribution of magnetic fields. The computations presented on the figure were carried out assuming magnetic field model with the dominating radial component of the magnetic field. Results for alternative configuration, as well as more detailed explanations will be given in Shulyak *et al.* (2013).

The values of the mean sufrace magnetic field $\langle B_{\rm s} \rangle$ derived in our study are, in general, consistent with the measurements taken at different times and considering the substantial uncertainties reported by all authors and the different spectral indicators used (see Saar & Linsky 1985; Johns-Krull & Valenti 1996, 2000; Kochukhov *et al.* 2009; Reiners & Basri 2007).

All four stars show three distinct groups of different magnetic field strength. We did not find solutions with homogeneous field distributions ($f = 1$ for the component that equals the average field), and we did not find solutions in which different field strengths are equally represented on the stellar surfaces ($f = const$ for all values of B up to $B_{\rm max}$). This is an interesting result because the *average* fields on the four stars are on the order of the *maximum* field strength we observe in sunspots, while the local magnetic flux densities occur to be much larger than those found in sunspots, and they co-exist with groups of much lower flux densities.

It is remarkable that very similar magnetic field distributions are found in all analized stars. Spectropolarimetric observations of stars in this spectral range indicate the existence of two distinct magnetic field geometries (with a few exceptions, see, e.g., Morin *et al.* (2010)): partially convective objects seem to harbor non-axisymmetric, toroidal fields, while fully convective objects prefer axisymmetric, poloidal fields. However, if such a dichotomy of geometries exists, we would expect it affecting the distribution of magnetic field components among our sample stars. The pattern of the magnetic field distributions we find show no evidence for such a transition at the convection boundary (where GJ 388 and GJ 729 are partially convective and GJ 285 and GJ 406 are fully convective, see Fig. 1). We therefore conclude that our Stokes I measurements of the entire magnetic field cannot confirm a difference in magnetic field geometries between partially and fully convective stars.

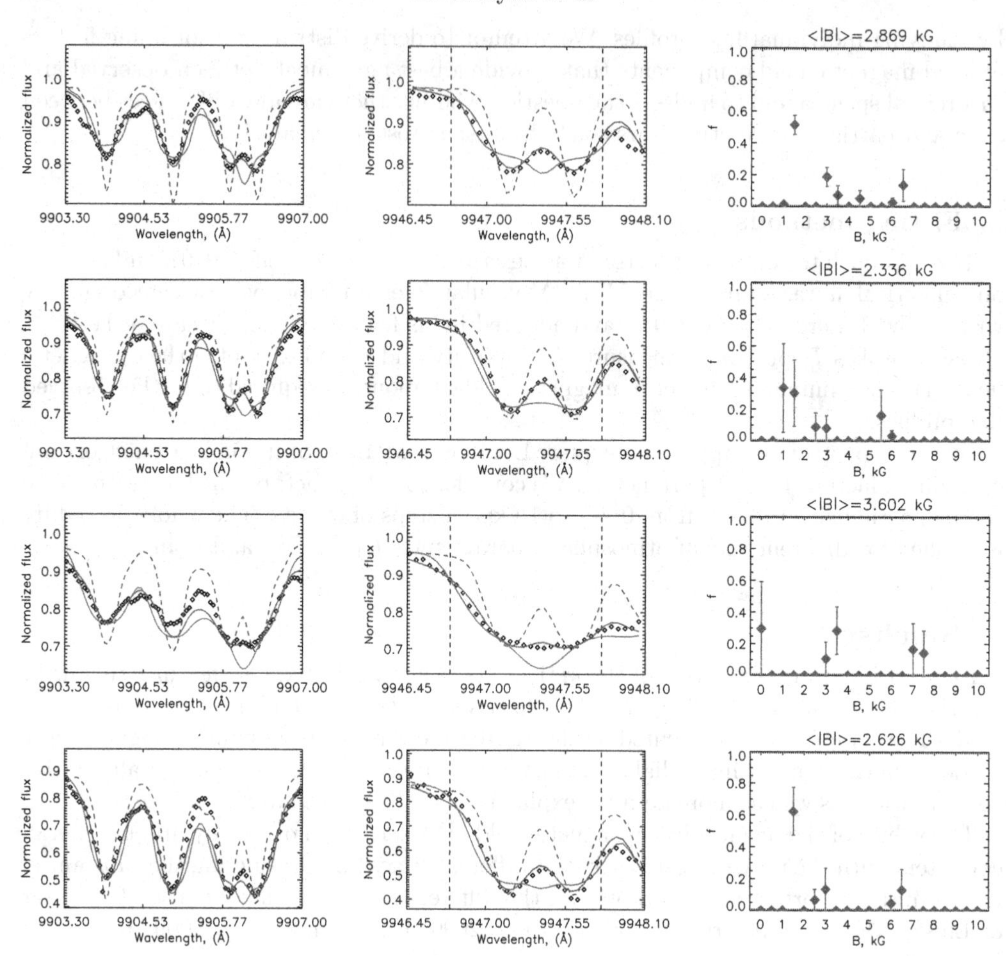

Figure 1. Examples of theoretical fit to selected FeH lines and resulting distributions of filling factors for (from top row to bottom) GJ 388 ($T_{\rm eff}$ = 3400 K, $v \sin i$ = 3 km/s), GJ 729 ($T_{\rm eff}$ = 3400 K, $v \sin i$ = 4 km/s), GJ 285 ($T_{\rm eff}$ = 3300 K, $v \sin i$ = 6 km/s), and GJ 406 ($T_{\rm eff}$ = 3100 K, $v \sin i$ = 3 km/s). Violet dashed line – computation with zero magnetic field, red line – computation with multi-component magnetic field shown on the right bottom plot, blue line – computations with homogeneous magnetic field (i.e. $f = 1$) of the same intensity as multi-component magnetic field. The strength of the resulting surface magnetic field is given on top of each plot of field distributions.

In our sample, the maximum field strength correlates with (projected) rotation rate. While the average field strength saturates at Rossby numbers $Ro \sim 0.1$ and $B \approx 4\,$kG (Reiners *et al.* 2009), the local field strength found in the present work does not saturate.

The detection of localized field components with the strength of up to 7.5 kG (GJ 285) suggests that if these structures are stable over long time intervals they must be in equipartition with the surrounding plasma. The equipartition field at the level of photosphere of a mid-M dwarf is of the order of only 4 – 5 kG. That means that, similar to sunspots, the high localized field strength can be in equipartition with the surrounding plasma if the regions of strong magnetic fields have cooler temperatures compared to the rest of the atmosphere. Unfortunately, no accurate estimate of the temperature contrast is available for M-dwarfs because their surfaces remain unresolved, and high quality time series of spectroscopic observations in Stokes I are not yet available. This is a very interesting problem to be addressed in future research.

Acknowledgements

We wish to thank Prof. Manfred Schuessler for his useful comments on the paper. We acknowledge financial support from CRC 963 – Astrophysical Flow Instabilities and Turbulence to DS (project A16-A17) and funding through a Heisenberg Professorship, RE 1664/9-1 to AR.

We also acknowledge the use of electronic databases (VALD, SIMBAD, NASA's ADS) and cluster facilities at the computing centre of Georg August University Göttingen (GWDG). This research has made use of the Molecular Zeeman Library (Leroy, 2004).

References

Afram, N., Berdyugina, S. V., Fluri, D. M., Solanki, S. K., & Lagg, A. 2008, *A&A*, 482, 387
Asensio Ramos, A. & Trujillo Bueno, J. 2006, *ApJ* , 636, 548
Berdyugina, S. V. & Solanki, S. K. 2002, *A&A*, 385, 701
Crozet, P., Tourasse, G., Ross, A., Paletou, F., & López Ariste, A. 2012, EAS Publications Series, 58, 63
Johns-Krull, C. M. & Valenti, J. A. 2000, *Stellar Clusters and Associations: Convection, Rotation, and Dynamos*, 198, 371
Johns-Krull, C. M. & Valenti, J. A. 1996, *ApJ*, 459, L95
Harrison, J. J. & Brown, J. M. 2008, *ApJ* , 686, 1426
Kochukhov, O., Heiter, U., Piskunov, N., *et al.* 2009, 15 th Cambridge Workshop on Cool Stars, Stellar Systems, and the Sun, 1094, 124
Leroy, B. 2004, Molecular Zeeman Library Reference Manual (avalaible on-line at http://bass2000.obspm.fr/mzl/download/mzl-ref.pdf)
Morin, J., Donati, J.-F., Petit, P., *et al.* 2010, *MNRAS* , 407, 2269
Reiners, A. 2012, Living Reviews in Solar Physics, 9, 1
Reiners, A., Basri, G., & Browning, M. 2009, *ApJ* , 692, 538
Reiners, A. & Basri, G. 2007, *ApJ*, 656, 1121
Reiners, A. & Basri, G. 2006, *ApJ* , 644, 497
Shulyak, D., Reiners, A., Seemann, U., Kochukhov, O., & Piskunov, N. 2013, *A&A*, submitted
Saar, S. H. & Linsky, J. L. 1985, *ApJ*, 299, L47
Shulyak, D., Reiners, A., Wende, S., *et al.* 2010, *A&A*, 523, A37
Valenti, J. A., Johns-Krull, C. M., & Piskunov, N. E. 2001, 11th Cambridge Workshop on Cool Stars, Stellar Systems and the Sun, 223, 1579

Magnetic Fields throughout Stellar Evolution
Proceedings IAU Symposium No. 302, 2013
P. Petit, M. Jardine & H. Spruit, eds.

doi:10.1017/S1743921314002002

Bridging planets and stars using scaling laws in anelastic spherical shell dynamos

R. K. Yadav[1,2], T. Gastine[1], U. R. Christensen[1] and L. Duarte[1,3]

[1]Max-Planck-Institut für Sonnensystemforschung, 37191 Katlenburg-Lindau, Germany
[2]Institut für Astrophysik, Georg-August-Universität, 37077 Göttingen, Germany
[3]Technische Universität Braunschweig, Germany

Abstract. Dynamos operating in the interiors of rapidly rotating planets and low-mass stars might belong to a similar category where rotation plays a vital role. We quantify this similarity using scaling laws. We analyse direct numerical simulations of Boussinesq and anelastic spherical shell dynamos. These dynamos represent simplified models which span from Earth-like planets to rapidly rotating low-mass stars. We find that magnetic field and velocity in these dynamos are related to the available buoyancy power via a simple power law which holds over wide variety of control parameters.

Keywords. stars: low-mass, brown dwarfs; stars: magnetic field; convection; methods: numerical

---

Introduction: In the last decade or so some qualitative agreement has been found in geodynamo simulations and observations (see e.g. Jones 2011). However, a direct and quantitative comparison of simulations and observations is not possible because of the large diffusivities used in numerical simulations as compared to the astrophysical values. To better connect numerical simulations with observations it is thus of great importance to find out generic scaling laws which are valid for both.

Christensen & Aubert (2006) found consistent scaling laws for magnetic field and velocity as a function of the available buoyancy power in Boussinesq spherical-shell dynamo simulations. Christensen *et al.* (2009) extended the magnetic field scaling law to physically relevant parameter regime and found good agreement with magnetic field observed on Earth, Jupiter, and some rapidly rotating low-mass stars.

Numerical scaling studies mentioned above ignored compressibility as they were geared to model dynamos operating in liquid metal interiors of Earth like planets. Giant planets and low-mass stars on the other hand might have highly compressible interiors with radially varying diffusivities (French *et al.* 2012). Assessing the effect of compressibility on various scaling laws in dynamos is very important to better understand the dynamo mechanism in giant planets and rapidly rotating low-mass stars.

Results: In recent years few extensive parameter-studies have been performed to study various aspects of compressible dynamos using the anelastic approximation (Gastine *et al.* 2012; Gastine *et al.* 2013; Duarte *et al.* 2013). We use this dataset along with Boussinesq dynamos (Yadav *et al.* 2013a) to explore the scaling of different quantities. Lorentz number $Lo = (\int (\mathbf{B}\cdot\mathbf{B})\, dV / \int \tilde{\rho}\, dV)^{1/2}$ represents the mean magnetic field and convective Rossby number $Ro_{conv} = (\frac{1}{V}\int (\mathbf{u}_{na}\cdot\mathbf{u}_{na})\, dV)^{1/2}$ represents the mean convective velocity, where $\mathbf{B}$, $\mathbf{u}_{na}$ is non-dimensional magnetic field and non-axisymmetric velocity, respectively, and $\tilde{\rho}$ is radially varying density. Note that such averaging can be described as magnetic and kinetic energy per-unit-mass, which is more appropriate for density varying interiors.

Despite radially-varying properties, Lo and Ro_{conv} still scale consistently as a function of the available buoyancy power per-unit-mass P as shown in Fig. 1. Empirical

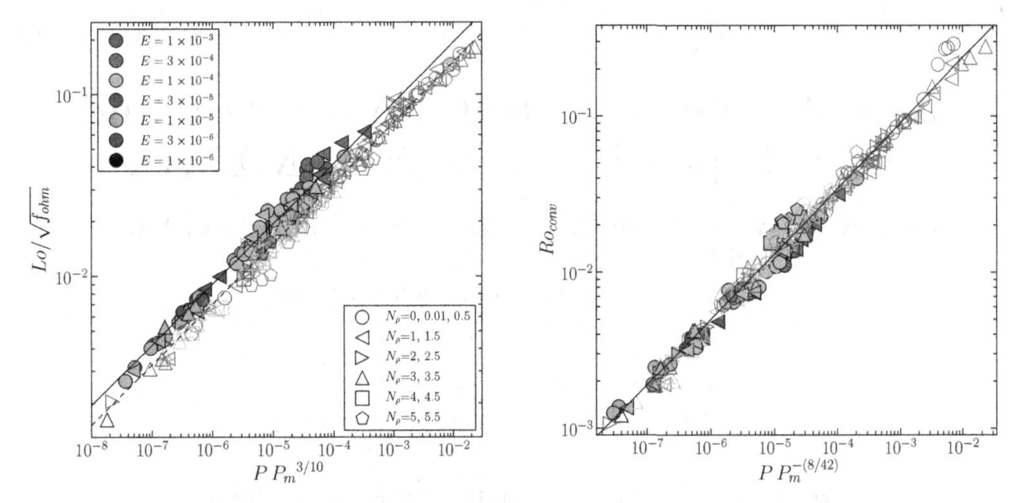

Figure 1. Scaling of non-dimensional magnetic field Lo (left panel) and convective velocity Ro_{conv} (right) as a function of the buoyancy power per-unit-mass P. Filled (empty) symbols are dipolar (multipolar) dynamos. Symbol color represents Ekman number E (right panel; top left legend) and symbol shape represents number of density scale heights N_ρ in the convecting fluid shell (right panel; bottom right legend).

power-law describing magnetic field scaling is $Lo/\sqrt{f_{ohm}} = c\,P^{0.33}\,{P_m}^{0.1}$, with $c = 0.9$ for dipolar dynamos and $c = 0.7$ for multipolar dynamos, and convective velocity is $Ro_{conv} = 1.6\,P^{0.42}\,P_m^{-0.08}$ (Yadav *et al.* 2013b). The Lo scaling requires f_{ohm} (fraction of total energy lost as ohmic heating) for a consistent scaling behaviour. Both scalings also require inclusion of P_m (magnetic Prandtl number) for optimum fit quality. However, such P_m dependence might be only due to the rather large diffusivities employed in present numerical models and may not be important for natural objects where $P_m \ll 1$ (Christensen *et al.* 2010).

In summary, we generalize the scaling laws found in earlier studies to compressible dynamos and support the hypothesis that magnetic field and velocity are related to the available buoyancy power by power-laws in dynamos. Decent observational evidence exists for the magnetic field scaling (Christensen *et al.* 2009) but comparison of velocity scaling with observations has not been possible so far (except for Earth's core).

Acknowledgements: We acknowledge funding from the DFG through Project SFB 963/A17 and through the special priority program 1488 (PlanetMag). Simulations were run on GWDG and HLRN computing facilities.

References

Jones, C. A. 2011, *Annual Rev. of Fluid Mech.*, 43, 583

Christensen. 2010, *Spa. Sci. Rev.*, 152, 565

Christensen, U. R. & Aubert, J. 2006, *Geophys. J. Int.*, 166, 97

Christensen, U. R., Holzwarth, V., & Reiners, A. 2009, *Nature*, 457, 167

Duarte, L. D., Gastine, T., & Wicht, J. 2013, *Phys. Earth and Planet Int.*, 222, 22

French, M., Becker, A., Lorenzen, W., *et al.* 2012, *ApJS*, 202, 5

Gastine, T., Duarte, L., & Wicht, J. 2012, *A&A*, 546, A19

Gastine, T., Morin, J., Duarte, L., *et al.* 2013, *A&A*, 549, L5

Yadav, R. K., Gastine, T., & Christensen, U. R. 2013a, *Icarus*, 225, 185

Yadav, R. K., Gastine, T., Christensen, U. R., & Duarte, L. D. 2013b, *ApJ*, 774, 6

Magnetic Fields throughout Stellar Evolution
Proceedings IAU Symposium No. 302, 2013
P. Petit, M. Jardine & H. Spruit, eds.

doi:10.1017/S1743921314002014

Age, Activity and Rotation in Mid and Late-Type M Dwarfs from MEarth

Andrew A. West[1], Kolby L. Weisenburger[1], Jonathan Irwin[2], David Charbonneau[2], Jason Dittmann[2] and Zachory K. Berta-Thompson[3]

[1]Department of Astronomy, Boston University,
725 Commonwealth Ave, Boston, MA 02215, USA
email: aawest@bu.edu, kolbylyn@bu.edu

[2]Harvard-Smithsonian Center for Astrophysics,
60 Garden St., Cambridge, MA 02138, USA
email: jirwin@cfa.harvard.edu,
dcharbonneau@cfa.harvard.edu, jdittmann@cfa.harvard.edu

[3]MIT, Kavli Institute for Astrophysics and Space Research,
77 Massachusetts Ave., Bldg. 37-673, Cambridge, MA 02139, USA
email: zkbt@mit.edu

Abstract. Using spectroscopic observations and photometric light curves of 280 nearby M dwarfs from the MEarth exoplanet transit survey, we examine the relationships between magnetic activity (quantified by Hα emission), rotation period, and stellar age (derived from three-dimensional space velocities). Although we have known for decades that a large fraction of mid-late-type M dwarfs are magnetically active, it was not clear what role rotation played in the magnetic field generation (and subsequent chromospheric heating). Previous attempts to investigate the relationship between magnetic activity and rotation in mid-late-type M dwarfs were hampered by the limited number of M dwarfs with measured rotation periods (and the fact that $v\sin i$ measurements only probe rapid rotation). However, the photometric data from the MEarth survey allows us to probe a wide range of rotation periods for hundreds of M dwarf stars (from less than one to over 100 days). Over all M spectral types we find that magnetic activity decreases with longer rotation periods, including late-type, fully convective M dwarfs. We find that the most magnetically active (and hence, most rapidly rotating) stars are consistent with a kinematically young population, while slow-rotators are less active or inactive and appear to belong to an older, dynamically heated stellar population.

Keywords. stars: activity, stars: low-mass, brown dwarfs, stars: late-type, stars: magnetic fields, stars: rotation, stars: kinematics, Galaxy: kinematics and dynamics

1. Introduction

Many M dwarfs have strong magnetic fields that can heat their stellar chromospheres and coronae, creating "magnetic activity" that is observed from the radio to the X-ray (e.g. Hawley *et al.* 1996; West *et al.* 2004; Reiners & Basri 2008; Berger *et al.* 2011). Although magnetic activity has been observed in M dwarfs for decades, the exact mechanism that gives rise to the chromospheric and coronal heating is still not well-understood.

In solar-type stars, magnetic field generation and subsequent heating is closely tied to stellar rotation and age; the faster a star rotates, the stronger its surface activity and younger its age (e.g. Skumanich 1972; Soderblom *et al.* 1991; Mamajek & Hillenbrand 2008). All indications suggest that this connection between age, rotation and activity extends to early-type M dwarfs (< M4), where rotation and activity are strongly correlated (e.g. Kiraga & Stepien 2007). The finite active lifetimes of early-type M dwarfs observed

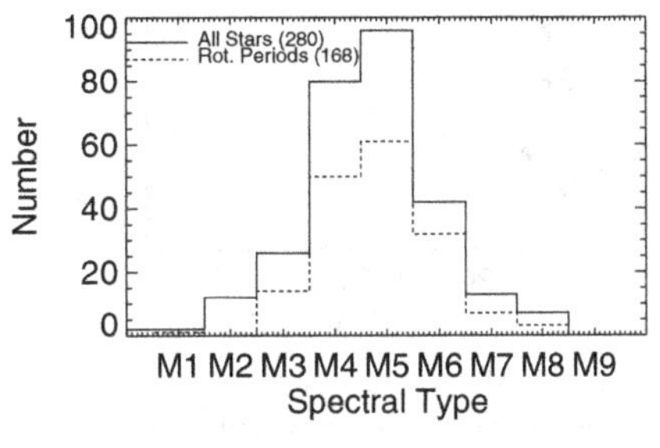

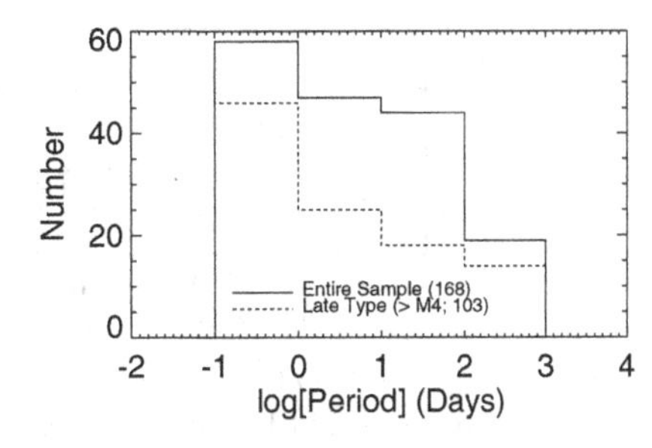

Figure 1. Distribution of MEarth stars with FAST spectroscopic observations (left), including stars with measured rotation periods (dashed). Distribution of rotation periods for all MEarth M dwarfs (solid) and late-type M dwarfs (dashed) with spectroscopic observations (right).

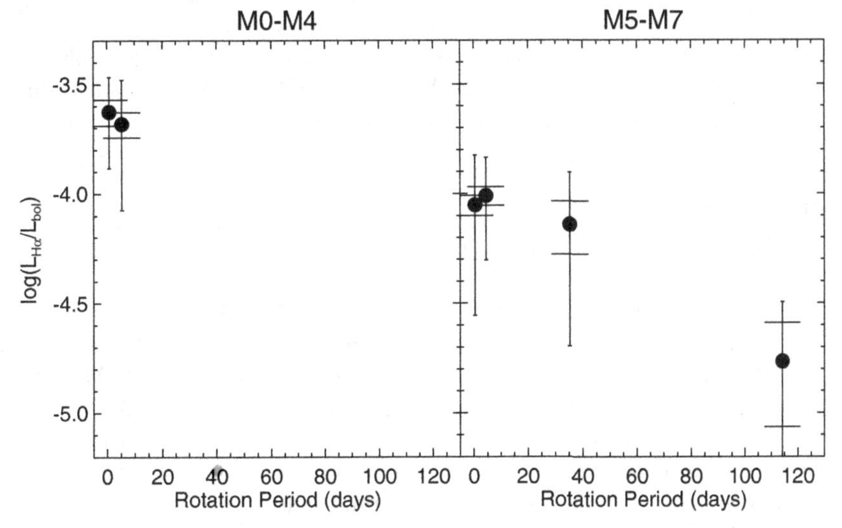

Figure 2. $(L_{\mathrm{H}\alpha}/L_{\mathrm{bol}})$ as a function of rotation period for early-type (left) and late-type (right) M dwarfs. Each data point corresponds to the mean value (of activity and rotation period) for the stars in the bin. The vertical error bars represent the spread of the data (short bars) and the uncertainty in the mean values (long bars).

in nearby clusters suggest that age continues to play an important role in the rotation and magnetic activity evolution of low-mass stars (Stauffer *et al.* 1994).

At a spectral type of ~M3 (0.35 $M_{\odot}$; Chabrier & Baraffe 1997), stars become fully convective, a property that may affect how magnetic field (and the resulting heating) is generated. Despite this change, magnetic activity persists in late-type M dwarfs; the fraction of active M dwarfs peaks around a spectral type of M7 before decreasing into the brown dwarf regime (Hawley *et al.* 1996; West et al. 2004).

One method for studying stellar rotation is to use photometrically derived rotation periods (e.g. Kiraga & Stepien 2007). Periodic signals result from brightness variations caused by long-lived spots on the stellar surface rotating in and out of view. Recent programs to search for transiting planets around late-type M dwarfs have produced large catalogs of time-domain photometry from which can be gleaned several important stellar properties, including rotation periods.

One of these transit programs, MEarth (Nutzman & Charbonnaeu 2008; Berta *et al.* 2012) is surveying ~2000 nearby, late-type M dwarfs at the Fred Lawrence Whipple Observatory (FLWO) at Mount Hopkins, Arizona using eight 0.4 m telescopes. Irwin *et al.* (2011) measured the rotation periods for 41 of the MEarth M dwarfs and found

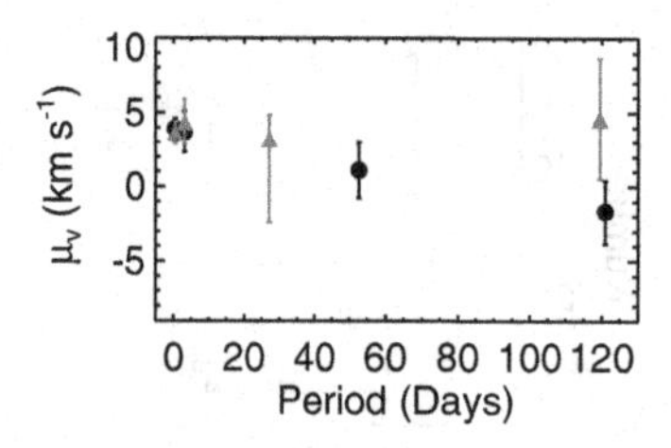

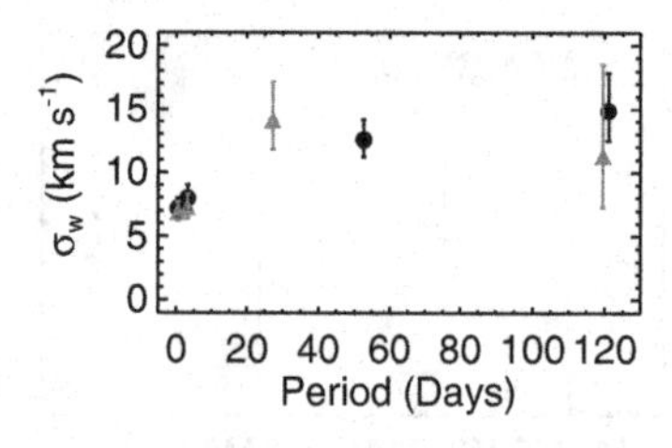

Figure 3. The mean V velocity (left) and the W velocity dispersion (right) as a function of rotation period for active (gray) and the entire FAST sample (black).

periods ranging from 0.28 to 154 days. The MEarth sample is an ideal sample of nearby (and therefore relatively bright) late-type M dwarfs from which to extract additional rotation periods and magnetic activity (from follow-up observations).

In this contribution, we use an expanded sample of MEarth rotation periods, along with low-resolution spectroscopy (taken from FLWO) and full 3D space motions to investigate the relation between rotation, activity and age in mid and late-type M dwarfs.

2. Observations and Methods

The photometric data used for measuring rotation periods come from the MEarth transit survey, which is searching for transiting exoplanets around ~2000 nearby M dwarfs (see Nutzman & Charbonnaeu 2008; Berta *et al.* 2012). Rotation periods were extracted using a least-squares periodogram fitting algorithm that is described in detail in Irwin *et al.* (2011). We also obtained R ~2000-3000 optical spectra for 298 MEarth targets using the FAST spectrograph on the 1.5m Tillinghast Telescope at FLWO.

Parallaxes from MEarth astrometry were derived for 198 of the stars in our spectroscopic sample (see Dittmann *et al.* 2013). For stars without MEarth parallaxes, we adopted distances from previous parallax determinations or photometric parallax techniques (Lèpine 2005). Proper motions for all of the stars were measured as part of the LSPM-North catalog (Lèpine & Shara 2005).

The FAST spectra were processed with the Hammer spectral-typing facility (Covey *et al.* (2007) - we identified and classified 280 M dwarfs (see Figure 1). Of the 280 M dwarfs for which we obtained FAST spectroscopy, 168 have measured rotation periods.

From the spectra we measured Hα equivalent widths and radial velocities as defined by West *et al.* (2004). We identified 145 magnetically active stars in our sample. For those (active) stars with Hα emission, we calculated the magnetic activity strength, $L_{\mathrm{H}\alpha}/L_{\mathrm{bol}}$ using the χ factor from Walkowicz, Hawley & West (2004).

Using stellar positions, proper motions, distances and radial velocities we measured (U, V, W) space velocities for 231 M dwarfs in our sample. We calculated the mean motions and velocity distributions (in U, V and W) for the M dwarfs within our sample as a function of rotation period and magnetic activity state. We used a Bayesian framework and calculated the posterior probabilities that sub-groups of stars were selected from a grid of Gaussian velocity distributions.

3. Results

For the 114 active stars with measured rotation periods, we examined the relation between magnetic activity and rotation period in both early-type (M0-M4) and late-type (M5-M7) M dwarfs. Figure 2 shows the magnetic activity strength (quantified by $L_{\mathrm{H}\alpha}/L_{\mathrm{bol}}$) as a function of rotation period. Each data point corresponds to the mean value (of activity and rotation period) for the stars in the bin. The vertical error bars represent the spread of the data (short bars) and the uncertainty in the mean values (long bars). We find that late-type M dwarfs are less active than their early-type counterparts and show a clear trend of decreasing activity level with increasing rotation period.

Our kinematic analysis suggests clear ties between both age and activity and age and rotation. Figure 3 shows the mean V velocity (left) and the W velocity dispersion (right) as a function of rotation period for active (gray) and the entire FAST sample (black). The W velocity dispersion increases with increasing period, indicating that stars with longer rotation periods have undergone more dynamical heating and are therefore older. For the entire FAST sample, the decrease in V motions as a function of rotation period is consistent the model that as stars age, their orbits become more elliptical (through dynamical encounters) and they exhibit asymmetric drift. However, the active stars do not appear to show asymmetric drift and are consistent with a younger population.

4. Discussion

Our results suggest that there is a correlation between activity and rotation in fully convective late-type M dwarfs and that in general, fast rotators are younger than slow rotators. Our kinematic analyses confirm previous findings that active stars appear to be younger than their inactive counterparts. However, we find surprisingly a small number of apparently young, active, slow rotators, which will be the topic of future investigations.

Acknowledgements

MEarth is supported by the Packard Foundation Fellowship for Science and Engineering and by the NSF under grant numbers AST-0807690 and AST-1109468. AAW acknowledges the support of NSF grants AST-1109273 and AST-1255568 and the Research Corporation for Science Advancement's Cottrell Scholarship.

References

Berger, E., *et al.* 2008, *ApJ*, 673, 1080
Berta, Z. K., Irwin, J., Charbonneau, D., Burke, C. J., & Falco, E. E., 2012, *AJ*, 144, 145
Chabrier, G. & Baraffe, I. 1997, *A&A*, 327, 1039
Covey, K. R., *et al.* 2007, *AJ*, 134, 2398
Dittmann, J., *et al.* 2013, *ApJ*, submitted
Hawley, S. L., Gizis, J. E., & Reid, I. N. 1996, *AJ*, 112, 2799
Irwin, J. and Berta, Z. K., Burke, C. J., Charbonneau, D., Nutzman, P., West, A. A., & Falco, E. E. 2011, *ApJ*, 727, 56
Kiraga, M. & Stepien, K. 2007, *AcA*, 57, 149
Lépine, S. 2005, *AJ*, 130, 1680
Lépine, S. & Shara, M. M. 2005, *AJ*, 129, 1483
Mamajek, E. E. & Hillenbrand, L. A. 2008, *ApJ*, 687, 1264
Nutzman, P. & Charbonneau, D. 2008, *PASP*, 120, 317
Pizzolato, N., Maggio, A., Micela, G., Sciortino, S., & Ventura, P. 2003, *A&A*, 397, 147
Reiners, A. & Basri, G. 2008, *ApJ*, 684, 1390
Skumanich, A. 1972, *ApJ*, 171, 565
Soderblom, D. R., Duncan, D. K., & Johnson, D. R. H. 1991, *ApJ*, 375, 722
Stauffer, J. R., Liebert, J., Giampapa, M., Macintosh, B., Reid, N., & Hamilton, D. 1994, *AJ*, 108, 160
Walkowicz, L. M., Hawley, S. L., & West, A. A. 2004, *PASP*, 116, 1105
West, A. A., *et al.* 2004, *AJ*, 128, 426

Magnetic Fields throughout Stellar Evolution
Proceedings IAU Symposium No. 302, 2013
P. Petit, M. Jardine & H. Spruit, eds.

doi:10.1017/S1743921314002026

Magnetic fields of Sun-like stars

Rim Fares
School of Physics and Astronomy, University of St Andrews, KY16 9SS, Scotland, UK
email: rim.fares@st-andrews.ac.uk

Abstract. Magnetic fields play an important role at all stages of stellar evolution. In Sun-like stars, they are generated in the outer convective layers. Studying the large-scale magnetic fields of these stars enlightens our understanding of the field properties and gives us observational constraints for the field generation models. In this review, I summarise the current observational picture of the large-scale magnetic fields of Sun-like stars, in particular solar-twins and planet-host stars. I will discuss the observations of large-scale magnetic cycles, and compare these to the solar cycle.

Keywords. stars: magnetic fields – stars: activity – stars: individual – techniques: spectropolarimetry

1. Introduction

Magnetic fields are present at different scales in the universe, from planets to stars, galaxies and galaxy clusters. In the case of stars, they play an important role at all stages of stellar evolution, from the collapse of the molecular cloud, through the pre-main sequence and main sequence phases to more evolved stages including supernovae, white dwarfs and neutron stars. They influence and control a number of physical processes, such as accretion, diffusion, mass-loss, angular momentum loss, and turbulence. Thus, studying the characteristics and generation of stellar magnetic fields is a necessary step to increase our understanding of stellar evolution (and also planetary formation and evolution).

The Sun is the closest - and thus the best studied - star. The discovery of its magnetic field goes back to the beginning of the 20th century when Hale, using the newly discovered Zeeman effect (Zeeman 1897), found that sunspots are magnetic features (Hale 1908). We now know that many observed features are due to magnetic fields, such as spots, faculae, and coronal mass ejection.

The solar magnetic field evolves in time. Sunspots emerge in mid-latitude activity belts, and the latitudes of these activity belts migrate towards the equator on a timescale of 11 years. This gives the well know butterfly diagram. I will refer to this cycle as the activity cycle. However, the large-scale magnetic field of the Sun varies on a different timescale. The polarity of the field flips every 11 years, meaning that the large-scale cycle is actually 22 years.

Activity cycles are observed on a number of other stars by studying activity proxies for magnetic fields such as CaII H&K and X-rays (for reviews, see Baliunas *et al.* 1995; Baliunas *et al.* 1997; Metcalfe *et al.* 2010; Sanz-Forcada *et al.* 2013; Berdyugina 2005). Sun-like stars with outer convective layers like the Sun generate their magnetic field by dynamo mechanisms active in these outer layers (e.g. Brown *et al.* (2011), Charbonneau(2010) for a review on solar dynamo models). The study of the magnetic field of Sun-like stars therefore brings new insights and constraints to the current dynamo theories. This enhances our knowledge of the large-scale magnetic fields of stars, which allows us to test how 'normal' the Sun is in a sample of Sun-like stars. The results I review here are for Sun-like stars of spectral types F, G and K, having masses between ~ 0.7 and 1.5 $M_\odot$. They have different depths of the outer convective envelope. Comparing stars with different properties can lead to a better understanding of the dynamo generation of the field.

2. Magnetic Mapping

In order to study large-scale stellar magnetic fields, one can examine the polarisation in the spectral lines. If a magnetic field is present where those lines are formed, due to the Zeeman effect, spectral lines will be polarised (the polarisation level depends on the magnetic sensitivity of the particular line). The polarisation properties depend on the position of the observer relative to the orientation of the magnetic field. For example, circular polarisation is sensitive to the line-of-sight component of the magnetic field (see Landi Degl'Innocenti & Landolfi 2004).

The aim of magnetic mapping is to reconstruct stellar large-scale magnetic field orientation, geometry and strength. When the star rotates, the observer sees different parts of the stellar disc. If those parts have different magnetic field distributions, the polarisation in the spectral lines will not be the same in spectra taken at different rotational phases. Thus, the technique used to reconstruct the large-scale magnetic field is a tomographic technique, like the one used in Magnetic Resonence Imaging. It is called Zeeman-Doppler Imaging (ZDI), and consists of inverting series' of circular polarised spectra into a magnetic topology, i.e. the distribution of magnetic fluxes and field orientations (Semel 1989). Since the inversion problem is ill-posed, regularisation techniques are used to get a unique map. These techniches include maximum entropy (Brown *et al.* 1991; Hussain *et al.* 2000) and Tikhonov regularisation (Piskunov & Kochukhov 2002). The results presented here are mostly obtained using the maximum entropy method. The magnetic field is described by its radial poloidal, non-radial poloidal and toroidal components, all described using spherical harmonics expansions (Donati *et al.* 2006a). ZDI is a powerful technique in recovering the large-scale magnetic field of the star, as well as its differential rotation. However, it has its limitations, because the small-scale fields are not resolved up to a certain limit, their signatures cancel out in some field geometries, and the field in dark spots is suppressed (see, e.g. Johnstone *et al.* 2010).

Collecting polarised spectra is possible using spectropolarimeters, such ESPaDOnS on CFHT, its twin instrument NARVAL on TBL, and HARPSpol on the 3.6-m in La Silla (Donati *et al.* 2006b; Piskunov *et al.* 2011; Snik *et al.* 2011). The polarisation signature is extremely small ($\sim 10^{-4}$). In order to increase the S/N ratio of the data, a multi-line technique called Least-Square Deconvolution (LSD) is used. It produces a mean profile with a higher S/N ratio than in single lines, depending on the number of lines used to calculate the mean profile (Donati *et al.* 1997; Kochukhov *et al.* 2010).

3. The Magnetic Topologies of Sun-like Stars

In this section, I will present the results of different studies targeting Sun-like stars. Some of these stars were observed by the Bcool project†, a project aiming at studying the magnetic fields of Sun-like stars and solar twins. Another campaign targeted hot-Jupiter hosting stars. It aimed to investigate interactions between the planet and the star, as well as studying how the stellar field influences the environment in which the planet evolves.

HD 179949 is a F star with $T_{\rm eff} = 6120$ K (Nordström *et al.* 2004), $M \sim 1.18\ M_\odot$,

† `http://bcool.ast.obs-mip.fr`

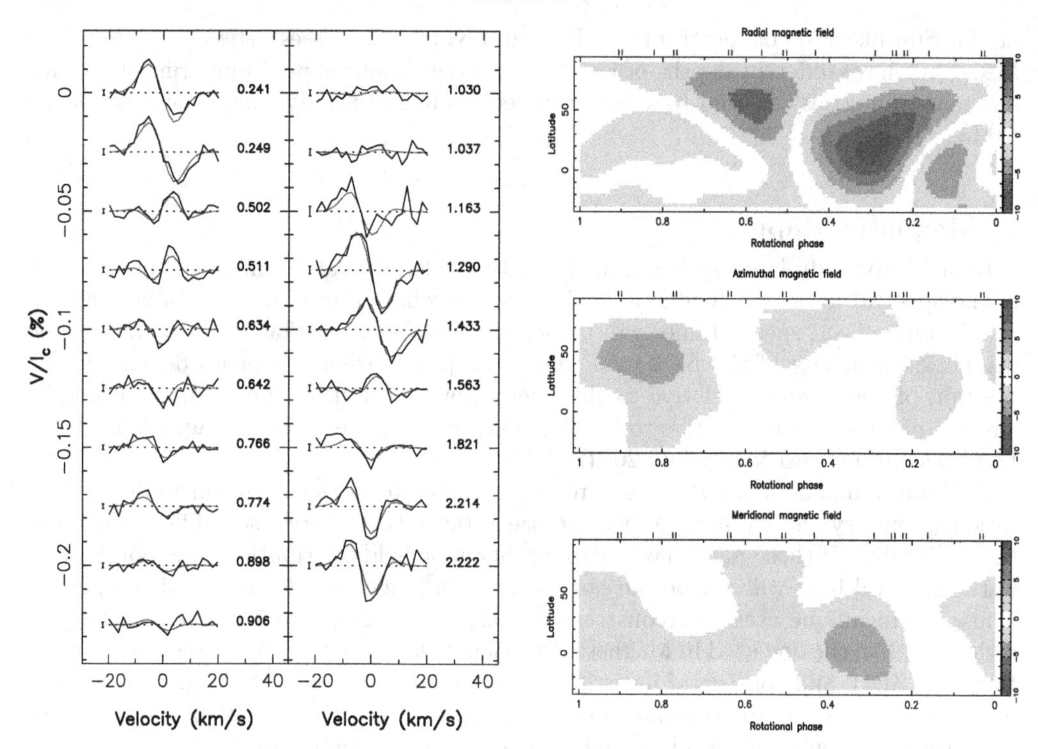

Figure 1. Left panel: Circular polarization profiles of HD 179949 for 2009 September. The observed and synthetic profiles are shown in black and red respectively. On the left of each profile we show a $\pm 1\ \sigma$ error bar, while on the right, the rotational cycles are indicated. Right panel: The three components of the field in spherical coordinates are presented. Adapted from Fares *et al.* (2012)

vsini $= 7.0 \pm 0.5$ km s^{-1} (Valenti & Fischer 2005), P$_{\rm rot}$ $= 7.6 - 10.3$ days, $d\Omega = 0.22$ rad day^{-1} (Fares *et al.* 2012), and hosts a hot-Jupiter. The stellar activity was reported to be modulated by the planetary orbital period instead of the stellar rotation period during some epochs (Shkolnik *et al.* 2003; Shkolnik *et al.* 2005; Shkolnik *et al.* 2008). This was interpreted as a possible stellar activity enhancement by the planet due star-planet interactions. Fares *et al.*(2012) observed this star during two epochs. Figure 1 shows the circular polarisation LSD profiles they obtained in September 2009 (left panel) and the reconstructed map (right panel). The three components of the magnetic field in spherical coordinates are shown. The mean magnetic field is 4 G, with 90% of the energy in the poloidal component (mainly radial).

Sun-like stars do not all exhibit the same magnetic field characteristics. For example, the G dwarf ξ Bootis A - T$_{\rm eff}$ $= 5570$ K, M ~ 0.86M$_\odot$, vsini $= 3.0 \pm 0.5$ km s^{-1} (Valenti & Fischer 2005), P$_{\rm rot}$ $= 6.4$ days (Toner & Gray 1988) - observed in July 2007 (Morgenthaler *et al.* 2011, 2012), shows a stronger magnetic field (80 G) with 80% of the energy in the toroidal component of the field. The circular polarisation profiles and the reconstructed magnetic map are shown in Fig. 2.

Those two previous examples show that there is a variety of magnetic field topologies and strengths among Sun-like stars. One can examine if there is a trend with stellar properties (e.g. rotation, temperature, mass). To answer this question, we have adapted the graph of Fig. 3 in Donati & Landstreet (2009) to include all the data for Sun-like stars. Fig. 3 represents the general characteristics of the reconstructed magnetic fields: the reconstructed magnetic energy density (i.e. the integral of $\langle B^2 \rangle$ over the surface),

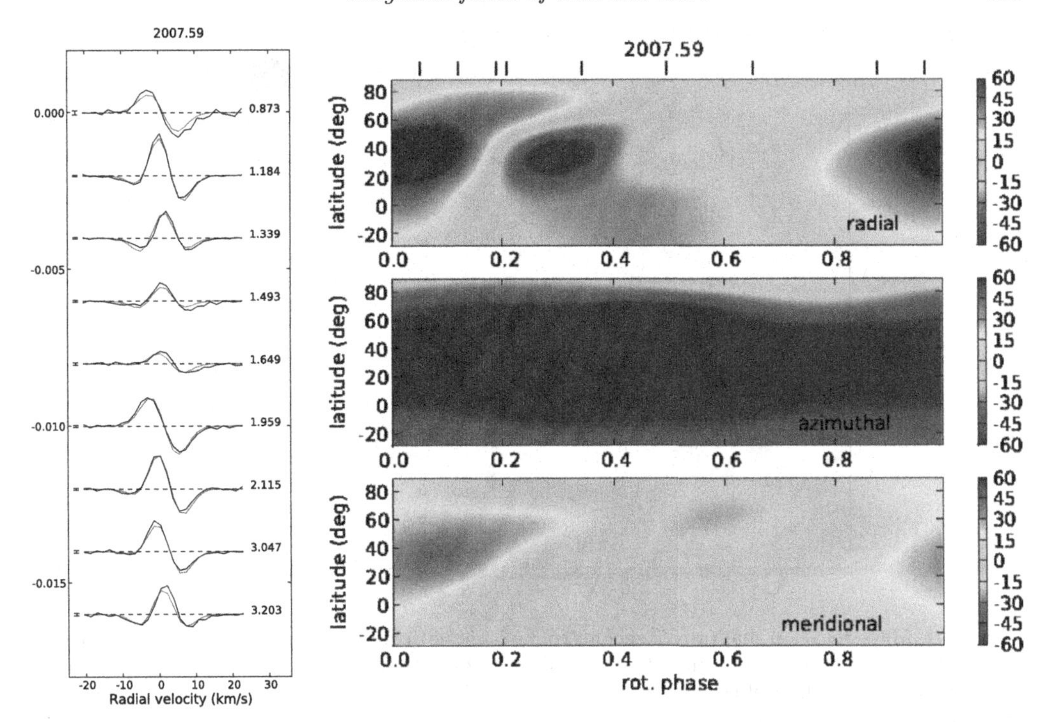

Figure 2. Circular polarisation profiles (left panel) and magnetic map (right panel) of ξ Boo A in July 2007 from Morgenthaler *et al.* (2012).

the percentage of poloidal field, and finally the fractional energy density in axisymmetric modes (i.e., with $m<l/2$, m and l being the order and degree of the spherical harmonic modes describing the reconstructed field). Each observed star for which a magnetic map has been reconstructed is indicated by a symbol at its rotation period and mass, with the symbol size reflecting the magnetic energy density, the symbol color reflecting if the field is mainly poloidal or mainly toroidal, and the symbol shape indicating how axisymmetric the poloidal component is. The data are from Catala *et al.*(2007); Moutou *et al.* (2007); Donati *et al.* (2003, 2008a); Petit *et al.* (2008, 2009); Jeffers & Donati (2008); Donati & Landstreet (2009); Fares *et al.* (2009, 2010, 2012, 2013); Morgenthaler *et al.* (2011,2012) and Marsden *et al.* (2011) .

Fig. 3 shows a variety of observed topologies. The fields of these stars can be either poloidal or toroidal, axisymmetric or non-axisymmetric, and of different strengths. The field strengths are in general smaller than those of M dwarfs (see Reiners & Basri 2006, 2010; Morin *et al.* 2008, 2010 and Donati *et al.* 2008b)

However, if we overplot the Rossby number equal to one, we see the main trend (the Rossby number is the ratio of the rotation period of the star to the convective turnover time):

stars having a Rossby number greater than one seem to have a weak, mainly poloidal and axisymmetric magnetic fields; while stars with Rossby number smaller than one (in the range of masses considered in this work) seem to have a stronger, mainly toroidal and non-axisymmetric magnetic fields.

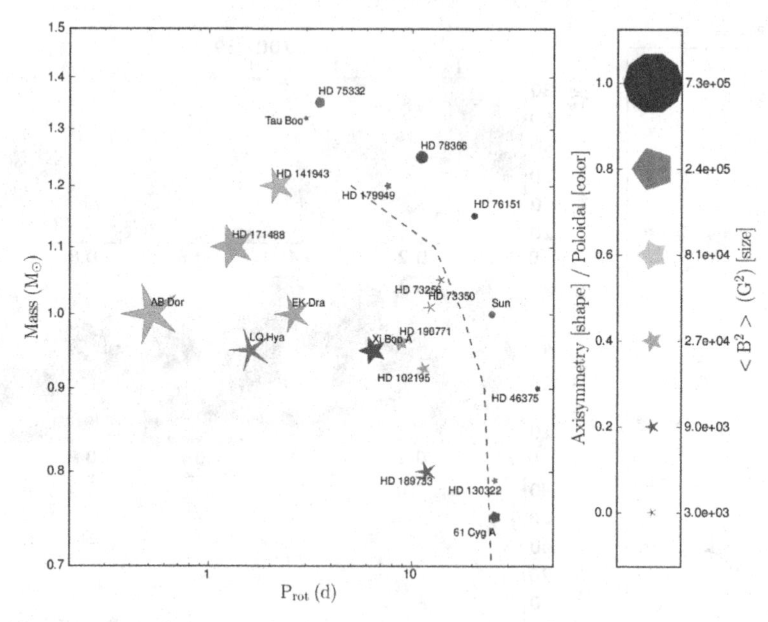

Figure 3. Mass-rotation diagram of reconstructed magnetic fields of Sun-like stars. The dashed line shows Rossby number of unity. The size of the symbol represents the field strength, its color the contribution of the poloidal component to the field, and its shape the degree axisymmetry of the poloidal component.

4. Magnetic Cycles

The Sun goes through activity and magnetic cycles, as discussed in Section 1. Monitoring the activity and searching for cycles of Sun-like stars has been the subject of many studies (e.g. Henry *et al.*1996; Baliunas *et al.* 1995, 1997). Some stars exhibit activity cycles, which vary in duration between almost a year to 25 years (e.g. Metcalfe *et al.* 2010). In addition, asteroseismology can also reveal activity cycles, by measuring the variation of the amplitude of the modes and the frequency shifts (e.g. García *et al.* 2010).

Knowing that activity cycles exist for Sun-like stars, one can ask if large-scale magnetic cycles exist as well. In order to investigate that, some stars were monitored over many epochs of observation. For example, τ Boo, an F star with $T_{\rm eff} = 6387$ K, $M \sim 1.33 M_\odot$, vsini $= 15.0 \pm 0.5$ km s^{-1} (Valenti & Fischer 2005), $P_{\rm rot} = 3.0 - 3.9$ days, $d\Omega = 0.4$ rad day^{-1} (Fares *et al.* 2009), and host to a hot-Jupiter, has shown a reversal of the polarity of the polar field every year between 2006 and 2009 (see Fig. 4). The star exhibits strong differential rotation, is orbited by a very massive planet, and has a rotation period similar to the planetary orbital period (Fares *et al.* 2009). The magnetic field is mainly poloidal and axisymmetric for these epochs, with a mean field strength of about 3-5 G.

The star was also observed in January 2008 half way between the reversal. The field is mainly azimuthal for this epoch (toroidal field). The field thus changes from a mainly poloidal to a mainly toroidal configuration, and than goes back to the poloidal configuration but with a different polarity. The magnetic cycle for this star is of 2 years, however a shorter cycle of 8 months cannot be ruled out. The effect of tidal interactions between the massive planet and the shallow outer convective envelope of the star was suggested as a possible cause for the short magnetic cycle. However, more recent studies have found a fast magnetic cycle of 3 years on another F star HD78366 (Morgenthaler *et al.* 2011).

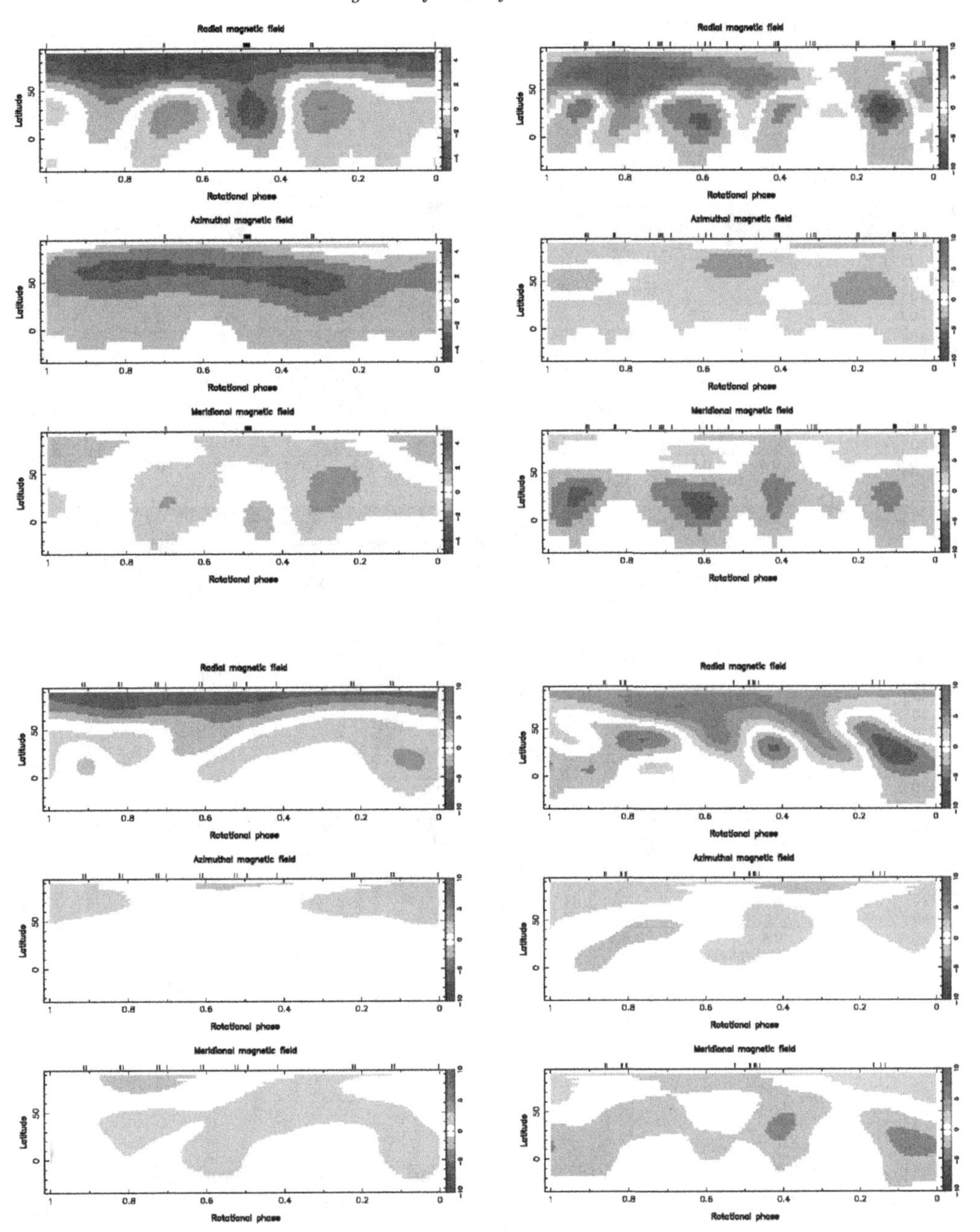

Figure 4. Magnetic maps of τ Boo. Top left: June 2006; top right: June 2007; bottom left: May 2008; bottom right July 2009. The maps show polarity flips every year, observed in all components of the field. Data adapted from Catala *et al.* (2007); Donati *et al.* (2008a); Fares *et al.* (2009,2013).

Complex Cycle

The magnetic field evolution of stars does not always show a simple change in polarity. For instance, HD190771 shows a complex cycle (T_{eff} = 5834 K, M $\sim 0.96M_{\odot}$, vsini = 4.3 ± 0.5 km/s (Valenti & Fischer 2005), P_{rot} = 8.8 days (Toner & Gray 1988), dΩ = 0.12 rad/d (Petit *et al.* 2009). This star was observed every year from 2007 until 2011

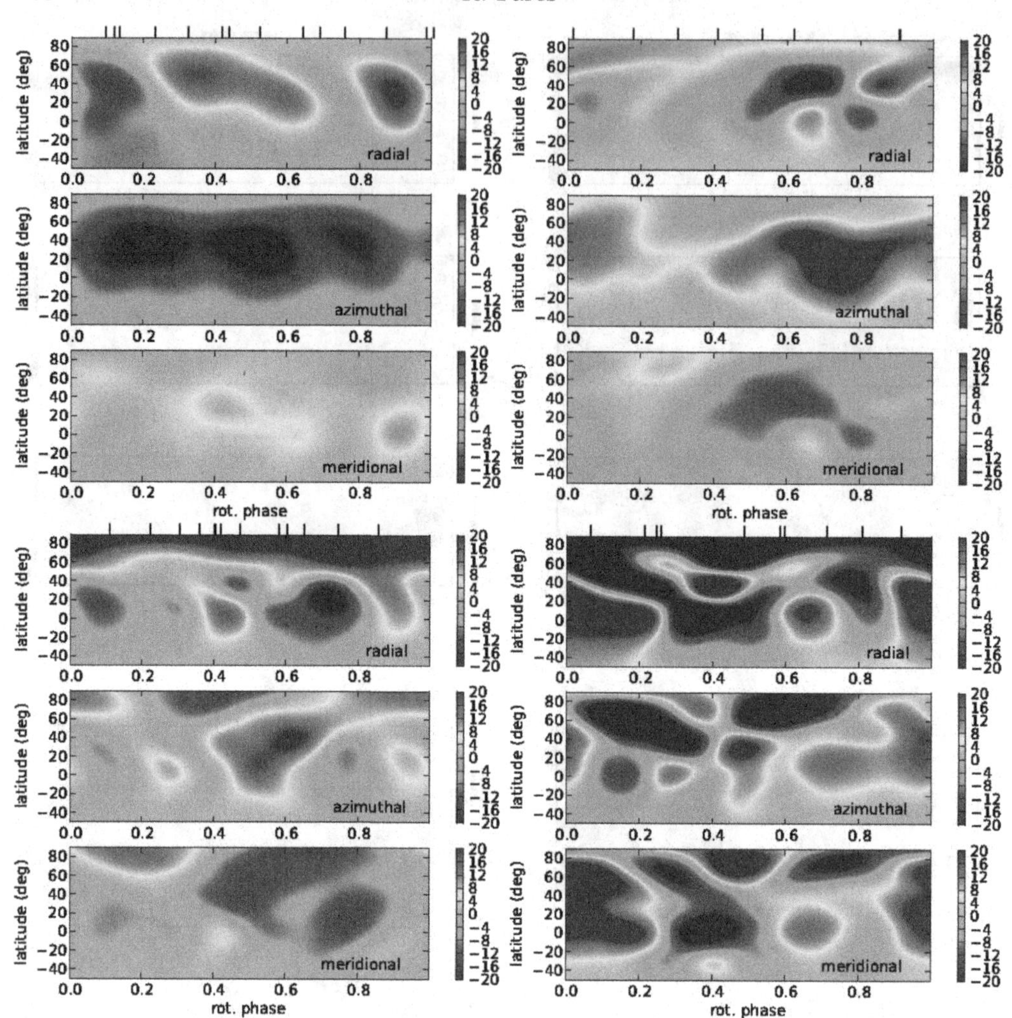

Figure 5. Magnetic maps of HD 190771 for 2007.59, 2008.67, 2009.47, and 2010.50 data sets from left to right and top to bottom (Morgenthaler *et al.* 2011).

(Petit *et al.* (2009), Morgenthaler *et al.* (2011) and Petit private com). The azimuthal field changes polarity between 2007 and 2008 (see Fig. 5), the field has a simple topology for both epochs. However, the topology of the field becomes more complex after 2008. A polarity switch is observed in 2010 when compared to 2008, but the field is much more complex in 2010. In 2011, the field goes back to a simple topology, having the same polarity as it had in 2008. The magnetic cycle of this star is complex when compared to the solar cycle.

The magnetic fields of Sun-like stars evolve with time. Stars that do not exhibit magnetic cycles still show an evolution of the poloidal and toroidal components of the field. Yet they do not show a polarity switch, e.g. ξ Boo A (Morgenthaler *et al.* 2012). These observations, in case of failure in detecting magnetic polarity flips, indicate that 'failed cycles' can also happen, i.e. variability without polarity flips.

Stellar Cycles versus the Solar Cycle

For the Sun, the length of the activity cycle is half that of the magnetic cycle. One can question if this relation holds for stars, and what we can learn about the generation of their magnetic fields.

Let us consider a stellar magnetic cycle with polarity flips, like the one of τ Boo. Although it is a fast cycle when compared to the solar cycle, it is considered normal since it shows regular polarity flips. This star was observed by the HK project of Mount Wilson. It has a long activity cycle of 11.6 years (Baliunas *et al.* 1995). If we compare this to the solar case, the activity cycle is much longer than the magnetic cycle. This system is an interesting one. Henry *et al.* (2000) found a persistent 116 day period over 30 years of observations in the CaII fluxes. However, this period does not appear in radial velocity nor photometric data. They say that it cannot be explained by the familiar phenomena of rotation, growth and decay of surface features, or an activity cycle. However, the spectropolarimetric data cannot rule out a magnetic cycle of 240 days (8 months; Fares *et al.* 2013). If this is the real period of the cycle, not only is it much shorter than the solar one, but it is also almost twice the 116 day period found in the CaII data.

Poppenhaeger *et al.* (2012) attempted to detect an X-ray cycle for τ Boo by observing the star over 6 epochs. Although the star exhibits variability in X-ray, it does not show a cyclic behavior. However, the lack of an X-ray cycle is not inconsistent with the existence of a mangetic cycle. Theoretical work shows that a magnetic cycle does not necessarily imply an X-ray one (e.g. McIvor *et al.* 2006). Also, Vidotto *et al.*(2012) simulated its stellar wind through the magnetic cycle, using the reconstructed maps as boundary conditions for the stellar magnetic field. They calculated the X-ray emission measure and find that this does not vary during the cycle, which agrees with the findings of Poppenhaeger *et al.* (2012).

The relations between activity cycles, magnetic cycles, and X-ray cycles should thus be investigated for stars. The picture we have currently is not similar to the solar one. Understanding the difference will improve our understanding on magnetic field generation.

5. Conclusions

The study of the magnetic fields of Sun-like stars gives new insights into stellar magnetism and provides constraints for dynamo theories. The magnetic fields in these stars are generated in the outer convective layers. However, a full understanding of the dynamo mechanisms acting in these layers has not yet been reached.

Sun-like stars have a wide range of magnetic properties. Magnetic field strengths and geometries vary between mainly poloidal and mainly toroidal fields. Field strengths vary between a few Gauss to a few hundred Gauss. Despite this variety, there are trends with stellar properties, especially the Rossby number. Sun-like stars with Rossby numbers smaller than unity have mainly toroidal magnetic fields, while stars with Rossby numbers greater than unity seem to have weaker magnetic fields, dominated by the poloidal components.

However, one should consider the existence of magnetic cycle. Some Sun-like stars do indeed have cycles; we observe solar-like cycles with polarity flips, more complex cycles, and some cyclic variations without polarity flips. The length of the current discovered magnetic cycles was surprising, as these cycles are very short when compared to the solar cycle.

This leads us to a set of important questions: is the solar cycle unusual? How do magnetic and activity cycles correlate in Sun-like stars? What are the stellar characteristics that drive the cycles (differential rotation, rotation,...)? Observations and theoretical works are still needed to answer these open questions.

Acknowledgments

RF acknowledges support from the STFC.

References

Aurière, M. 2003, *EAS Publications Series*, 9, 105
Baliunas, S. L., Donahue, R. A., Soon, W. H., *et al.* 1995, ApJ, 438, 269
Baliunas, S. L., Henry, G. W., Donahue, R. A., Fekel, F. C., & Soon, W. H. 1997, *ApJ*(Letters), 474, L119
Berdyugina, S. V. 2005, *Living Reviews in Solar Physics*, 2, 8
Brown, B. P., Miesch, M. S., Browning, M. K., Brun, A. S., & Toomre, J. 2011, *ApJ*, 731, 69
Brown, S. F., Donati, J.-F., Rees, D. E., & Semel, M. 1991, *A&A* , 250, 463
Catala, C., Donati, J.-F., Shkolnik, E., Bohlender, D., & Alecian, E. 2007, *MNRAS*, 374, L42
Charbonneau, P. 2010, *Living Reviews in Solar Physics*, 7, 3
Donati, J.-F., Semel, M., Carter, B. D., Rees, D. E., & Collier Cameron, A. 1997,*MNRAS* , 291, 658
Donati, J.-F., Collier Cameron, A., Semel, M., *et al.* 2003, *MNRAS*, 345, 1145
Donati, J.-F., Howarth, I. D., Jardine, M. M., *et al.* 2006a, *MNRAS*, 370, 629
Donati, J.-F., Catala, C., Landstreet, J. D., & Petit, P. 2006b, *Astronomical Society of the Pacific Conference Series*, 358, 362
Donati, J.-F., Moutou, C., Farès, R., *et al.* 2008a, *MNRAS*, 385, 1179
Donati, J.-F., Morin,J., Petit, P., *et al.* 2008b, *MNRAS*, 390, 545
Donati, J.-F. & Landstreet, J. D. 2009, *ARAA*, 47, 333
Fares, R., Donati, J.-F., Moutou, C., *et al.* 2009, *MNRAS*, 398, 1383
Fares, R., Donati, J.-F., Moutou, C., *et al.* 2010, *MNRAS*, 406, 409
Fares, R., Donati, J.-F., Moutou, C., *et al.* 2012, *MNRAS*, 423, 1006
Fares, R., Moutou, C., Donati, J.-F., *et al.* 2013, *MNRAS*, 2010
García, R. A., Mathur, S., Salabert, D., *et al.* 2010, *Science*, 329, 1032
Hale, G. E. 1908, *ApJ*, 28, 315
Henry, G. W., Baliunas, S. L., Donahue, R. A., Fekel, F. C., & Soon, W. 2000, *ApJ*, 531, 415
Hussain, G. A. J., Donati, J.-F., Collier Cameron, A., & Barnes, J. R. 2000, *MNRAS*, 318, 961
Jeffers, S. V. & Donati, J.-F. 2008, *MNRAS*, 390, 635
Johnstone, C., Jardine, M., & Mackay, D. H. 2010, *MNRAS*, 404, 101
Kochukhov, O., Makaganiuk, V., & Piskunov, N. 2010, *A&A*, 524, A5
Landi Degl'Innocenti, E. & Landolfi, M. 2004, *Astrophysics and Space Science Library*, 307
Marsden, S. C., Jardine, M. M., Ramírez Vélez, J. C., *et al.* 2011, *MNRAS*, 413, 1922
Metcalfe, T. S., Basu, S., Henry, T. J., *et al.* 2010, *ApJ*(Letters), 723, L213
McIvor, T., Jardine, M., Mackay, D., & Holzwarth, V. 2006, *MNRAS*, 367, 592
Morgenthaler, A., Petit, P., Morin, J., *et al.* 2011, *AN*, 332, 866
Morgenthaler, A., Petit, P., Saar, S., *et al.* 2012, *ApJ*, 540, A138
Morin, J., Donati, J.-F.,Petit, P., *et al.* 2008, *MNRAS*, 390, 567
Morin, J., Donati, J.-F.,Petit, P., *et al.* 2010, *MNRAS*, 407, 2269
Moutou, C., Donati, J.-F., Savalle, R., *et al.* 2007, *A&A*, 473, 651
Nordström, B., Mayor, M., Andersen, J., *et al.* 2004, *A&A*, 418, 989
Petit, P., Dintrans, B., Solanki, S. K., *et al.* 2008, *MNRAS*, 388, 80
Petit, P., Dintrans, B., Morgenthaler, A., *et al.* 2009, *ApJ*, 508, L9
Piskunov, N. & Kochukhov, O. 2002, *A&A*, 381, 736

Piskunov, N., Snik, F., Dolgopolov, A., *et al.* 2011, *The Messenger*, 143, 7
Poppenhaeger, K., Günther, H. M., & Schmitt, J. H. M. M. 2012, *Astronomische Nachrichten*, 333, 26
Reiners, A. & Basri, G. 2006, *ApJ*, 644, 497
Reiners, A. & Basri, G. 2010, *ApJ*, 710, 924
Sanz-Forcada, J., Stelzer, B., & Metcalfe, T. S. 2013, *A&A*, 553, L6
Semel, M. 1989, *A&A*, 225, 456
Shkolnik, E., Walker, G. A. H., & Bohlender, D. A. 2003, *ApJ*, 597, 1092
Shkolnik, E., Walker, G. A. H., Rucinski, S. M., Bohlender, D. A., & Davidge, T. J. 2005, *AJ*, 130, 799
Shkolnik, E., Bohlender, D. A., Walker, G. A. H., & Collier Cameron, A. 2008, *ApJ*, 676, 628
Snik, F., Kochukhov, O., Piskunov, N., *et al.* 2011, *Solar Polarization* 6, 437, 237
Valenti, J. A. & Fischer, D. A. 2005, *VizieR Online Data Catalog*, 215, 90141
Vidotto, A. A., Fares, R., Jardine, M., *et al.* 2012, *MNRAS*, 423, 3285
Toner, C. G. & Gray, D. F. 1988, *ApJ*, 334, 1008
Zeeman, P. 1897, *ApJ*, 5, 332

Magnetic Fields throughout Stellar Evolution
Proceedings IAU Symposium No. 302, 2013
P. Petit, M. Jardine & H. Spruit, eds.

doi:10.1017/S1743921314002038

Stellar Magnetic Dynamos and Activity Cycles

Nicholas J. Wright

Centre for Astrophysics Research, University of Hertfordshire, Hatfield, AL10 9AB

Abstract. Using a new uniform sample of 824 solar and late-type stars with measured X-ray luminosities and rotation periods we have studied the relationship between rotation and stellar activity that is believed to be a probe of the underlying stellar dynamo. Using an unbiased subset of the sample we calculate the power law slope of the unsaturated regime of the activity – rotation relationship as $L_X/L_{bol} \propto Ro^{\beta}$, where $\beta = -2.70 \pm 0.13$. This is inconsistent with the canonical $\beta = -2$ slope to a confidence of 5σ and argues for an interface-type dynamo. We map out three regimes of coronal emission as a function of stellar mass and age, using the empirical saturation threshold and theoretical super-saturation thresholds. We find that the empirical saturation timescale is well correlated with the time at which stars transition from the rapidly rotating convective sequence to the slowly rotating interface sequence in stellar spin-down models. This may be hinting at fundamental changes in the underlying stellar dynamo or internal structure. We also present the first discovery of an X-ray unsaturated, fully convective M star, which may be hinting at an underlying rotation - activity relationship in fully convective stars hitherto not observed. Finally we present early results from a blind search for stellar X-ray cycles that can place valuable constraints on the underlying ubiquity of solar-like activity cycles.

Keywords. stars: activity, X-rays: stars, stars: late-type, stars: coronae, stars: magnetic fields

1. Introduction

The stellar magnetic dynamo is thought to be driven by the interplay between convection and differential rotation (Parker 1955) in stars with radiative cores and convective envelopes. The observational manifestation of the dynamo is the relationship between rotation and X-ray activity observed in main sequence F, G, K and M stars. This was first quantified by Pallavicini *et al.* (1981), who found that X-ray luminosity scaled as $L_X \propto (v \sin i)^{1.9}$, providing the first evidence for the dynamo-induced nature of stellar coronal activity. For very fast rotators the relationship was found to break down with X-ray luminosity reaching a saturation level of $L_X/L_{bol} \sim 10^{-3}$ (Micela *et al.* 1985), independent of spectral type. This saturation level is reached at a rotation period that increases toward later spectral types (Pizzolato *et al.* 2003), but it is unclear what causes this. Despite much work there is yet to be a satisfactory dynamo theory that can explain both the solar dynamo and that of rapidly rotating stars and the continued lack of a sufficiently large and unbiased sample has no doubt contributed to this.

We have produced a new catalog of stars with stellar rotation periods and X-ray luminosities, as described in Wright *et al.* (2011). The catalog includes 824 solar- and late-type stars, 445 field stars and 379 stars in nearby open clusters (ages 40–700 Myrs). The sample was homogenised by recalculating all X-ray luminosities and converting them onto the ROSAT 0.1 – 2.4 keV band. To minimise biases we removed all sources known to be X-ray variable, those that exhibit signs of accretion, or those in close binary systems. The sample is approximately equally distributed across the colour range $V - K_s = 1.5 - 5.0$ (G2 to M4) with ~30 stars per subtype, dropping to ~10 stars per subtype from F7 to M6.

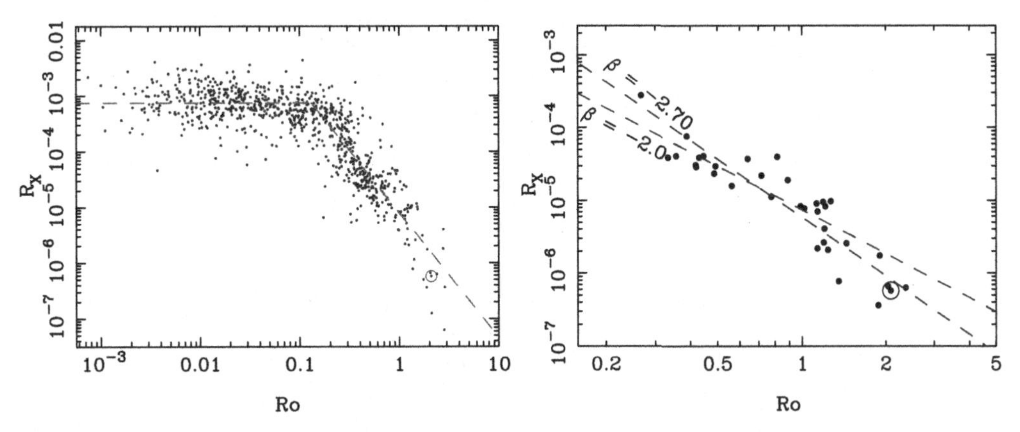

Figure 1. $R_X = L_X/L_{bol}$ versus $Ro = P_{rot}/\tau$ for all stars in our sample (left) and for the unbiased sample of 36 stars with unsaturated X-ray emission (right). The Sun is shown with a solar symbol. *Left:* The best-fitting saturated and non-saturated activity–rotation relations are shown as a dashed red line. *Right:* The log-log ordinary least squares bisector fit, $\beta = -2.70$, and a fit with the canonical slope of $\beta = -2.0$ are shown as dashed lines.

Here we highlight some of the results derived from this sample, and work that followed from it, with particular focus on results that probe the underlying stellar dynamo.

2. Dynamo efficiency in the unsaturated regime

The relationship between X-ray activity, represented by the ratio of X-ray to bolometric luminosity, $R_X = L_X/L_{bol}$ and the rotation period, parameterised by the spectral-type independent Rossby number, $Ro = P_{rot}/\tau$, the ratio of the rotation period to the convective turnover time (Noyes *et al.* 1984) is shown in Fig. 1 for all the stars in our sample. The diagram shows the two main regimes of coronal activity: the *unsaturated* regime where R_X increases with decreasing Ro, and a *saturated* regime where the X-ray luminosity ratio is constant with $\log R_X = -3.13 \pm 0.08$. The transition between these two regimes is found to occur at $Ro = 0.13 \pm 0.02$ from a two-part power-law fit (Fig. 1).

The sample used here suffers from a number of biases, most importantly an X-ray luminosity bias due to the selection only of stars detected in X-rays. To overcome this we used an X-ray unbiased subset of our sample, the 36 Mt. Wilson stars with measured rotation periods (Donahue *et al.* 1996), all of which are detected in X-rays. While some biases may still exist due to the ability to measure rotation periods, Donahue *et al.* (1996) conclude that any such biases are unlikely to affect the rotation period distribution. A single-part power-law fit in the linear regime (where $R_X \propto Ro^{\beta}$) to this sample is shown in Fig. 1 with a fit of $\beta = -2.70 \pm 0.13$. This is a steeper slope than the canonical value of $\beta \simeq -2$ from Pallavicini *et al.* 1981, though their use of projected rotation velocities instead of rotation periods represents a different relationship than that fitted here. Our slope is inconsistent with the canonical value to a confidence of 5σ, which argues against a distributed dynamo operating throughout the convection zone, the efficiency of which scales as Ro^{-2} (Noyes *et al.* 1984), and instead argues for an interface dynamo (e.g., Parker 1993), which has a more complex dependency where $\beta \neq -2$.

3. The evolution of stellar coronal activity

Since all solar and late-type stars are known to emit X-rays (e.g., Vaiana *et al.* 1981, Wright *et al.* 2010, Hynes *et al.* 2012), and X-ray emission is dependent on rotation,

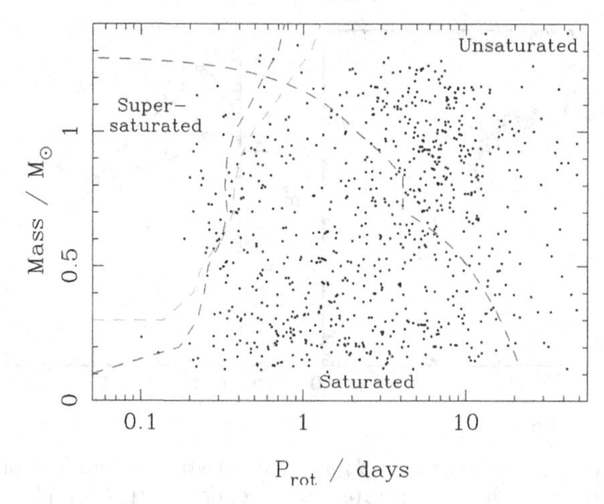

Figure 2. The three regimes of coronal X-ray emission shown in Mass-P_{rot} space, the latter of which is an age indicator. All the stars from Wright *et al.* (2011) are shown as black dots. The empirical saturation threshold of Ro < 0.13 (blue dashed line), and the theoretical supersaturation thresholds for coronal stripping (Jardine & Unruh 1999, red dashed line) and coronal updrafts (Stępień *et al.* 2001, green dashed line) are shown.

which is itself an age indicator (stars spin down as they age and lose angular momentum, Skumanich 1972), stellar coronal activity should evolve in a well-defined manner as stars age. This is shown in Figure 2, which can be used to follow the coronal evolution of a star from super-saturated, through the saturation regime, to the unsaturated regime. A number of interesting features are revealed by this diagram, notably that F-type stars do not pass through a saturated regime, going straight from super-saturated to unsaturated coronal emission. This explains why previous authors (e.g., Pizzolato *et al.* 2003) noted a lower R_X saturation level for F-type stars than for other solar- and late-type stars. It now appears that these stars were not saturated but super-saturated, and therefore their X-ray emission levels should not be considered indicative of saturated X-ray emission. It is now the case that *all* solar- and late-type stars exhibit the same (within the uncertainties) level of saturated X-ray emission of $\log R_X = -3.13$, suggesting a common mechanism for coronal saturation. Furthermore we note a correlation between the empirical timescale for coronal saturation and the timescale at which stars transition from the rapidly rotating convective sequence to the slowly rotating interface sequence in stellar spin-down models (e.g., Barnes 2003). That these two critical changes in stellar rotation and dynamo activity occur at similar times may be hinting at a common origin from a fundamental change in the underlying stellar dynamo or internal structure.

4. Slowly rotating fully convective stars

Late-type M dwarfs become fully convective at approximately a spectral type of M4-5 and as such should not have the interface region between the radiative core and convective envelope (known as the *tachocline*) that is believed to be necessary for the generation of an $\alpha\Omega$ dynamo (Parker 1993). As such it is expected that fully convective M dwarfs should not exhibit a rotation – activity relationship, though this is hard to verify since most late-type stars rotate very rapidly and have large convective turnover times, such that they are all found in the saturated regime of coronal X-ray emission. To fully test this idea we are observing a number of slowly rotating and fully convective M dwarfs with the *Chandra* X-ray Observatory to search for evidence of a rotation – activity relationship.

Only one target has been observed so far and was observed to have log $R_X \sim -4.5$, significantly below the saturation level of log $R_X \sim -3$. This is the first evidence of non-saturated X-ray emission in a fully convective M-type star and might provide the first evidence for a rotation – activity relationship in fully convective stars with considerable implications for the stellar dynamo at work in fully convective stars.

5. A blind search for stellar X-ray cycles

While chromospheric activity cycles are a well-studied phenomenon amongst solar- and late-type stars, only a handful of coronal activity cycles have been identified and studied (e.g., Favata *et al.* 2008). To rectify this situation we are performing blind searches for stellar X-ray cycles using deep and long-baseline X-ray observations. The initial dataset used for this project is the *XMM-Newton* archive, from which Hoffman *et al.* (2012) identified nine stars in six fields with data of sufficient depth to extract reliable photometry and long-enough baselines to identify cyclic behaviour. Using a Lomb-Scargle periodogram they searched for cycles, but none were found. From Monte Carlo simulations they simulated their detection capabilities and, assuming a uniform distribution of cycle periods and strengths over the domain searched, conclude with 95% confidence that <72% of moderately active stars have 5–13 year coronal cycles with an amplitude similar to that of the Sun. Further ongoing studies will provide better constraints on the ubiquity of coronal cycles in solar- and late-type stars.

6. Summary

A new catalog of stars with measured X-ray luminosities and rotation periods is used to study the rotation – activity relation. The power-law slope of the unsaturated regime is fit as $\beta = -2.7 \pm 0.13$, inconsistent with the canonical $\beta = -2$ value with a confidence of 5σ and arguing for an interface-type dynamo. We present the first discovery of an X-ray unsaturated fully convective M-type star, which may be hinting an an underlying rotation – activity relationship in fully convective stars, and present early results from a blind search for stellar X-ray cycles.

References

Barnes, S. A. 2003, *ApJ*, 586, 464
Donahue, R. A., Saar, S. H., & Baliunas, S. L. 1996, *ApJ*, 466, 384
Favata, F., Micela, G., Orlando, S., *et al.* 2008, *A&A*, 490, 1121
Hoffman, J., Günther, H. M., & Wright, N. J. 2012, *ApJ*, 759, 145
Hynes, R. I., *et al.*, 2012, *ApJ*, 761, 162
Jardine, M. & Unruh, Y. C. 1999, *A&A*, 346, 883
Micela, G., *et al.*, 1985, *ApJ*, 292, 172
Noyes, R. W., Hartmann, L. W., Baliunas, S. L., Duncan, D. K., & Vaughan, A. H. 1984, *ApJ*, 279, 763
Pallavicini, R., Golub, L., Rosner, R., Vaiana, G. S., Ayres, T., & Linsky, J. L. 1981, *ApJ*, 248, 279
Parker, E. N. 1955, *ApJ*, 122, 293
Parker, E. N. 1993, *ApJ*, 408, 707
Pizzolato, N., Maggio, A., Micela, G., Sciortino, S., & Ventura, P. 2003, *A&A*, 397, 147
Skumanich, A. 1972, *ApJ*, 171, 565
Stępień, K., Schmitt, J. H. M. M., & Voges, W. 2001, *A&A*, 370, 157
Vaiana, G. S., *et al.*, 1981, *ApJ*, 245, 163
Wright, N. J., Drake, J. J., & Civano, F. 2010, *ApJ*, 725, 480
Wright, N. J., Drake, J. J., Mamajek, E. E., & Henry, G. W., 2011, *ApJ*, 743, 48

Magnetic Fields throughout Stellar Evolution
Proceedings IAU Symposium No. 302, 2013
P. Petit, M. Jardine & H. Spruit, eds.

doi:10.1017/S174392131400204X

Differential rotation and meridional flows in stellar convection zones

Manfred Küker and Günther Rüdiger

Leibniz-Institut für Astrophysik Potsdam
An der Sternwarte 16, 14482 Potsdam, Germany
email: mkueker@aip.de, gruediger@aip.de

Abstract. Differential rotation and meridional flow are key ingredients in flux transport dynamo models of the solar activity cycle. As the subsurface flow pattern is not sufficiently constrained by observations, it is a major source of uncertainty in solar and stellar dynamo models. We discuss the current mean field theory of stellar differential rotation and meridional flows and its predicitons for the Sun and stars on the lower main sequence.

Keywords. Sun: activity, Stars: activity, Sun: rotation, Stars: rotation

1. Introduction

Differential rotation is the main field generator in both the α-Ω dynamo and the flux transport dynamo. In the flux transport dynamo, which is currently the most favored model, the meridional flow determines the cycle period. While the internal rotation of the solar convection zone is known from helioseismology, the meridional flow has so far been measured reliably at the surface, where it is directed towards the poles. Mass conservation requires a return flow towards the equator but the depth at which that occurs is not known. While Schad *et al.* 2012 did not find a return flow for radii down to $0.8R_\odot$, Zhao *et al.* report a reverse flow direction between 0.91 $R_\odot$ and 0.82 $R_\odot$.

For stars even the surface differential rotation has to be inferred from observations rather than measured directly and the meridional flow is unknown. The combined data from various methods methods suggests that the surface differential rotation of main sequence stars is only weakly dependent on the stellar rotation rate but strongly depends on the effective temperature. Barnes *et al.* (2005) presented a power law fit of the form $\delta\Omega \propto T^{8.92}$ for the temperature dependence of the surface differential rotation which was supported by the findings of Reiners (2006).

2. Mean field theory: predictions for low mass stars

Mean field theory uses averages to describe the large-scale behavior of the gas in the convection zones of the Sun and other low-mass main sequence stars. The model described in Küker *et al.* (2011) solves the equations of angular momentum transport, meridional flow, and convective heat transport using transport coefficients derived with the second order correlation approximation and model convection zones computed with the MESA stellar evolution code (Paxton *et al.* 2011.)

Our model predicts solar-type differential rotation in all cases and meridional flow with one flow cell per hemisphere. The surface flow is directed towards the poles and the return flow towards the equator is located at the bottom of the convection zone. The left panel in Figure 1 shows the surface differential rotation, $\delta\Omega = \Omega_{\rm max} - \Omega_{\rm min}$, predicted by the model for a sequence of ZAMS stars rotating with a period of 2.5d, the power law fit of

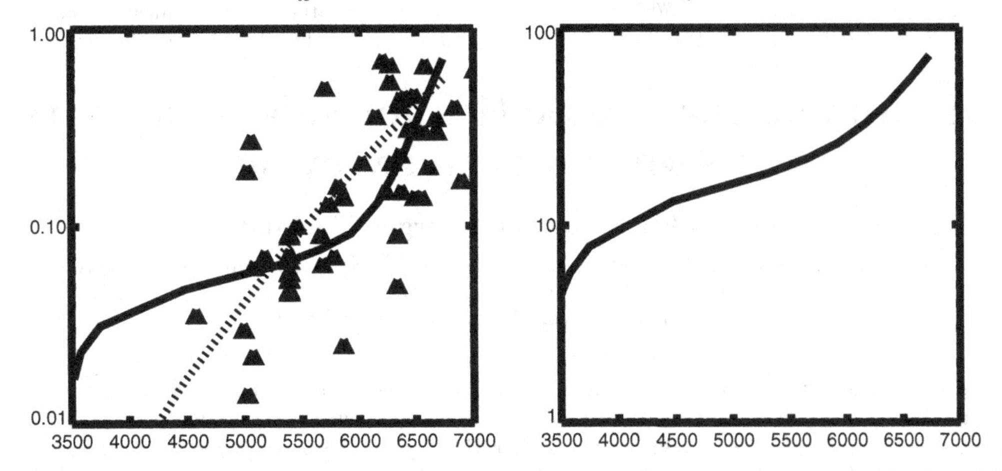

Figure 1. Left:Surface differential rotation in rad/day vs. effective temperature in K. The solid line indicates the prediction of the mean field model for zero age main sequence stars rotating with a period of 2.5d. The dotted line indicates the power law found by Barnes *et al.* (2005). The triangles represent surface differential rotation values of derived from observations. Right: Surface meridional flow in m/s vs. effective temperature in K.

Barnes *et al.* (2005), and the observed differential rotation values as compiled in Reiners (2006). The mean field model predicts a temperature dependence that is quite different from the power law. The curve is much flatter than the power law for temperatures below 6000 K and steeper for temperatures above that value. While Barnes *et al.* list several lower main sequence stars with differential rotation much weaker than predicted by the mean field model, the latter is strongly supported by data recorded by the Kepler spacecraft (Reinhold & Reiners 2013).

The right part of Figure 1 shows the surface meridional flow vs. the effective temperature corresponding to the differential rotation shown in the left part. Like the differential rotation, the meridional flow increases with temperature and the increase is steeper for temperatures above 6000 K than below. The steepening if the curve is less pronounced though. Assuming a flux transport dynamo for all main sequence stars in the temperature range shown we would therefore expect a the cycle period to decrease with increasing temperature, i.e. F stars should have cycle periods substantially shorter than the 11 year solar cycle.

References

Schad, A., Timmer, J., & Roth, M. 2012, *AN* 333, 991

Zhao, J., Bogart, R. S., Kosovichev, A. G., Duvall, T. L. Jr., & Hartlepp, Th. 2013, *ApJL*, 774, L29

Barnes, J. R., Collier Cameron, A., Donati, J.-F., James, D. J., Marsden, S. C., & Petit, P. 2005, *MNRAS*, 357, L1

Reiners, A. 2006, *A&A*, 446, 267

Küker, M., Rüdiger, G., & Kitchatinov, L. L. 2011, *A&A*, 530, A48

Paxton, B., Bildsten, L., Dotter, A., Herwig, F., Lesaffre, P., & Timmes, F. 2011, *ApJL*, 192, 3

Reinhold, T. & Reiners, A. 2013, *A & A* 557, A11

Magnetic Fields throughout Stellar Evolution
Proceedings IAU Symposium No. 302, 2013
P. Petit, M. Jardine & H. Spruit, eds.

doi:10.1017/S1743921314002051

Meridional flow velocities for solar-like stars with known activity cycles

Dilyara Baklanova and Sergei Plachinda

Crimean Astrophysical Observatory, Taras Shevchenko National University of Kyiv,
Nauchny 98409, Crimea, Ukraine
email: dilyara@crao.crimea.ua

Abstract. The direct measurements of the meridional flow velocities on stars are impossible today. We suppose that the matter on a surface of solar-like stars with stable activity period passes the way equal to $2\pi R_\star$ during the stellar Hale cycle. We present here the dependence of meridional flow velocity on Rossby number, which is an effective parameter of the stellar magnetic dynamo.

Keywords. stars: late-type, stars: activity

---

1. Introduction

The dependence of the mean meridional flow velocity of the Sun on number of the Hale cycle (Plachinda *et al.* 2011) was obtained under the assumption that during the Hale cycle the pass track of the poloidal magnetic dipole axis is equivalent to the circumference of the Sun. The velocity $\langle v \rangle = 6.29$ m s^{-1}, which gives $P_{Hale} = 22$ years for the Sun, for solar-like star 61 Cyg A gives the activity period 7.3 years that is in full agreement with observations. Therefore we supposed that the matter of meridional flows on a surface of solar-like stars with stable activity period also passes the way equivalent to $2\pi R_\star$, during their own Hale cycle. We use this approach to draw the dependence of the value of meridional flow velocity on Rossby number.

2. The dependence of meridional flow velocities versus Rossby number

We select from literature a sample of 28 stars with spectral types ranging from F9 to K7 using the following criteria:

- The period of the main activity cycle, P_{cyc}, should be available.
- The sample must not contain objects that have equiprobable periods of activity.
- Published by direct observations or theoretical relations stellar radii, color index $B - V$, and known periods of rotation, P_{rot}.

Periods of activity, P_{cyc}, were obtained from Ca II H and K emission and from photometric observations.

The Rossby number, $Ro = P_{rot}/\tau_c$, is the ratio of the stellar rotation period P_{rot} and the convective turnover time τ_c. To find the convective turnover time we used the empirical dependence of τ_c from color index $B - V$ (Noyes *et al.* 1984, eq. 4).

Meridional flow velocities have been calculated using the equation $\langle v \rangle = 2\pi R_\star / P_{Hale}$, where $R_\star$ is the radius of star in meters, P_{Hale} is the stellar magnetic activity period in seconds, $P_{Hale} = 2P_{cyc}$. The activity period of the Sun is a mean value, which was obtained by using sunspot numbers for all years of observations from 1755 to 2008 and it equals to $P_{cyc\odot} = 11$ years.

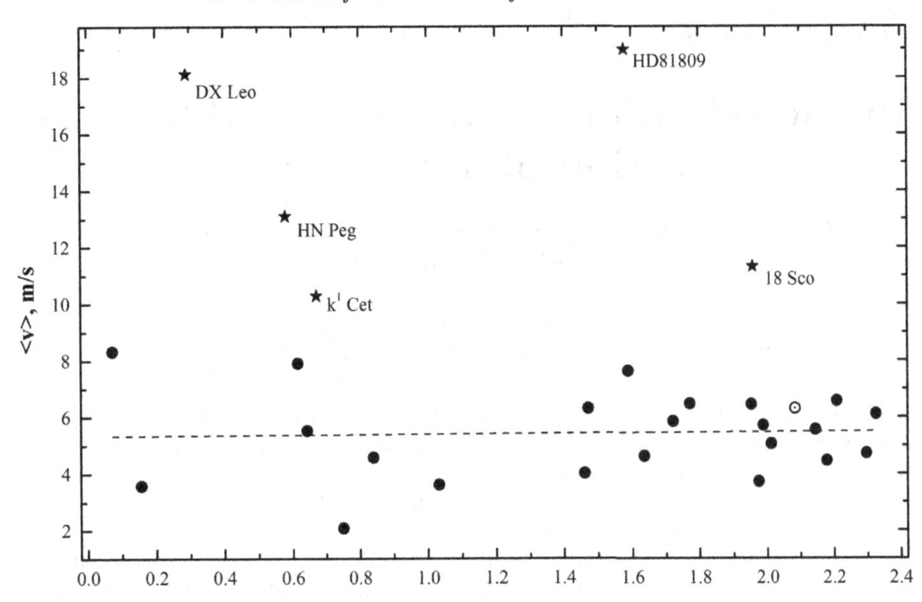

Figure 1. Meridional flow velocities vs. Rossby number. The dotted line is the approximation without 5 points (stars symbols) which lie out of 10 m s^{-1}.

The Figure 1 shows that meridional flow velocities $\langle v \rangle$ for solar-type stars lie near 5.4 ± 1.5 m s^{-1} that is in good agreement with the mean value of the meridional flow velocity of the Sun (6.29 m s^{-1}) obtained in the same manner. We suggest that the meridional flow velocity does not depends on the Rossby number. Only five stars out of 28 display greater values of the meridional flow velocity.

3. Conclusions

We have found that the mean value of the meridional flow velocity $\langle v \rangle$ of $\sim 80\%$ of the selected solar-like stars lies near 5.4 ± 1.5 m s^{-1} and does not depend on the Rossby number. Only five stars out of 28 display greater values of the meridional flow velocity. Id est, as in the case of the Sun, the meridional flow determines the duration of the Hale's cycle for stars with stable activity periods.

References

Noyes, R. W., Hartmann, L. W., Baliunas, S. L., Duncan, D. K., & Vaughan, A. H. 1984, *ApJ*, 279, 763

Plachinda, S., Pankov, N., & Baklanova, D. 2011, *Astron. Nachr.*, 332, 918

Magnetic Fields throughout Stellar Evolution
Proceedings IAU Symposium No. 302, 2013
P. Petit, M. Jardine & H. Spruit, eds.

doi:10.1017/S1743921314002063

On the reliability of measuring differential rotation of spotted stars

Zsolt Kővári, János Bartus, Levente Kriskovics, Krisztián Vida and Katalin Oláh

Konkoly Observatory,
Konkoly Thege út 15-17., H-1121, Budapest, Hungary
email: kovari, bartus, kriskovics, vida, olah@konkoly.hu

Abstract. Cross-correlation of consecutive Doppler images is one of the most common techniques used to detect surface differential rotation (hereafter DR) on spotted stars. The disadvantage of a single cross-correlation is, however, that the expected DR pattern can be overwhelmed by sudden changes in the apparent spot configuration. Another way to reconstruct the image shear using Doppler imaging is to include a predefined latitude-dependent rotation law in the inversion code ('sheared image method'). However, special but not unusual spot distributions, such like a large polar cap or an equatorial belt (e.g., small random spots evenly distributed along the equator), can distort the rotation profile similarly as the DR does, consequently, yielding incorrect measure of the DR from the sheared image method. To avoid these problems, the technique of measuring DR from averaged cross-correlations using time-series Doppler images ('ACCORD') is introduced and the reliability of this tool is demonstrated on artificial data.

Keywords. stars: activity, stars: imaging, stars: spots, stars: late-type

---

1. Profile distortion by spots vs. differential rotation

Compared to an unspotted star, the theoretical shape of a broadened spectral line profile from a spotted star is perturbed by starspots and surface differential rotation (DR). Indeed, in some cases these two effects cause very similar distortions, thus, their separation can be difficult, if possible at all. This difficulty may result in biased observation of surface DR, when applying the 'sheared image method' (hereafter SIM, Weber 2004), where the line profile fit is minimized for a given range of the image shear parameter α. In Fig. 1 we demonstrate how a polar cap/equatorial belt on a rigidly rotating star imitates solar-type/antisolar DR, respectively. That is, assuming realistic conditions, SIM yields better fits for non-rigid solutions for these two special examples (i.e., α of 0.08 and -0.02 instead of zero for the supposed rigid rotation). We note, that the line profile distortions due to polar cap/equatorial belt are similar in shape to those that are attributed to the respective solar/antisolar DR (cf. Reiners & Schmitt 2002). In Sect. 2 we give a tool to resolve this problem.

2. Measuring surface DR from average cross-correlations

Cross-correlation of two consecutive 'snapshots' of the stellar surface (i.e., Doppler images) can reveal the DR pattern (Donati & Collier Cameron 1997). The advantage of this method, as against the sheared image, is that no predefined rotation law is assumed. The disadvantage of a *single* cross-correlation is, however, that the DR pattern in the correlation map is often overwhelmed by sudden changes in the apparent spot configuration. Our method called ACCORD (acronym from 'Average Cross-CORrelation of consecutive Doppler images') is based on averaging as many correlation maps as possible to suppress the effect of such stochastic spot changes (see Kővári *et al.* 2004,

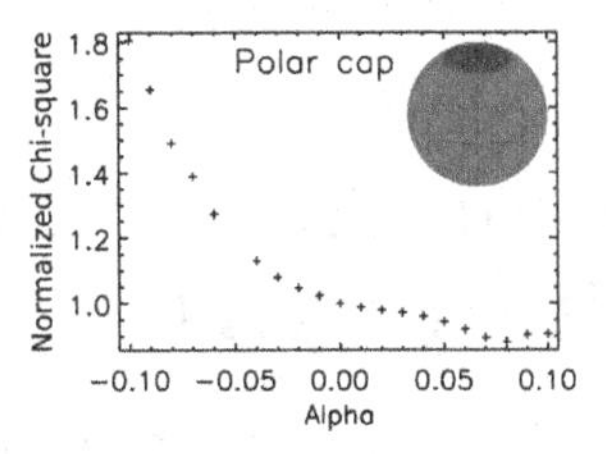

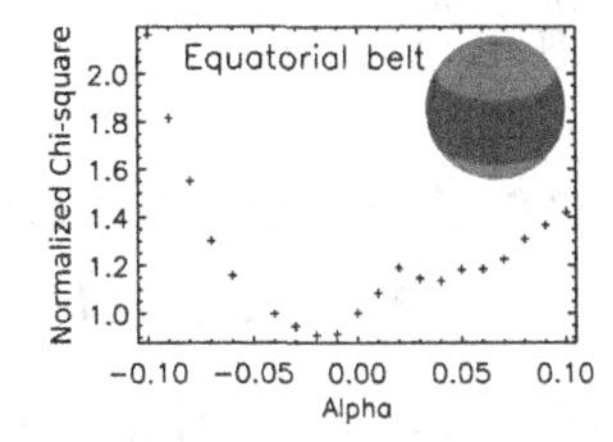

Figure 1. Biased measure of DR from SIM: a polar cap (left) and an equatorial belt (right) on a rigidly rotating star imitate solar-type and antisolar-type DR, respectively.

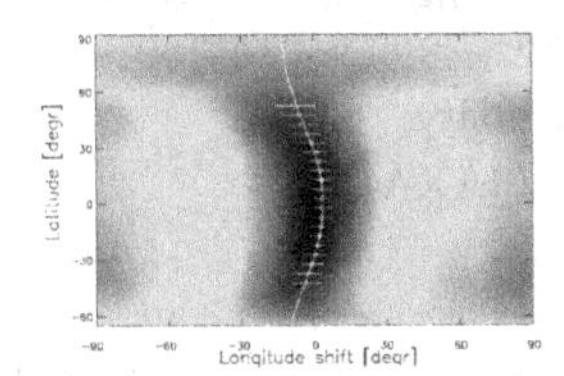

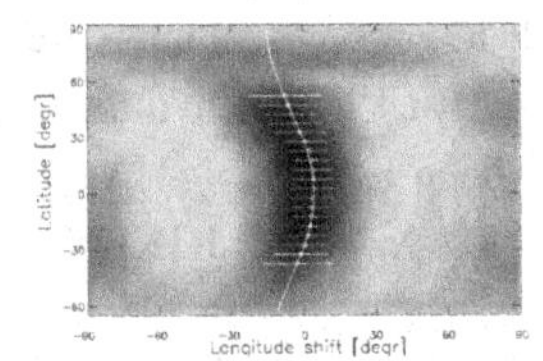

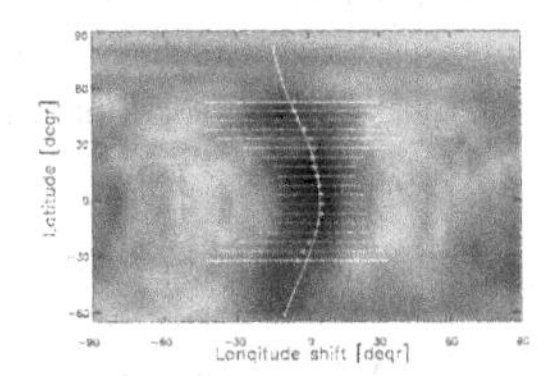

Figure 2. Averaged cross-correlations reveal DR laws with correct α parameters close to the original one of 0.006, either when assuming no noise (left), or assuming S/N = 200 (middle) and S/N = 100 (right), yielding $\alpha = 0.0058 \pm 0.0004$, $\alpha = 0.0067 \pm 0.0033$ and $\alpha = 0.0068 \pm 0.0048$, respectively. Note that uncertainty comes also from imperfect image reconstruction due to the necessarily incomplete phase coverage, thus errors are expected even when no noise is added.

2012 for details). To demonstrate the reliability of this method we generated artificial time-series observations from a test star with the stellar parameters of the fast rotating active dwarf LQ Hya. Even the data sampling (phase coverage) followed our actual 70-night long observing run in 1996/97 (Kővári *et al.* 2004). We placed a polar cap, as well as high- and low latitude spots on the differentially rotating surface with an assumed surface shear α of 0.006 (i.e., a weak solar-type DR, cf. Kővári *et al.* 2004). Visit `www.konkoly.hu/solstart/misc/testDR.gif` to see the surface evolution of the test star. After generating artificial spectra (for a given noise level) we followed the data handling as described in Kővári *et al.* (2004). Fig. 2 demonstrates how ACCORD reveals the correct DR law. Contrarily, *single* cross-correlations resulted in much scattered α values with relative errors of 23% (no noise), 48% (S/N = 200) and 150% (S/N = 100).

Finally, we performed a test assuming rigid rotation with either a polar cap or an equatorial belt, as described in Sect. 1, to see how specious are these configurations for ACCORD. Again, this method proved to be sufficiently robust, resulting in basically the expected zero shear for both cases: $\alpha = 0.000 \pm 0.001$ for the polar cap and $\alpha = 0.003 \pm 0.002$ for the equatorial belt, i.e., far better results than the respective values of 0.08 and -0.02 from SIM (cf. Fig. 1).

Acknowledgements

This work has been supported by the Hungarian Science Research Program OTKA K-81421, the Lendület-2009 and Lendület-2012 Young Researchers Programs of the Hungarian Academy of Sciences and by the HUMAN MB08C 81013 grant of the MAG Zrt.

References

Donati, J.-F. & Collier Cameron, A. 1997, *MNRAS*, 291, 1
Kővári, Zs., Strassmeier, K. G., Granzer, T., *et al.* 2004, *A&A*, 417, 1047
Kővári, Zs., Korhonen, H., Kriskovics, L., *et al.* 2012, *A&A*, A50
Reiners, A. & Schmitt, J. H. M.. M. 2002, *A&A*, 384, 155
Weber, M. 2004, *Ph.D. dissertation, Universität Potsdam*

Magnetic Fields throughout Stellar Evolution
Proceedings IAU Symposium No. 302, 2013
P. Petit, M. Jardine & H. Spruit, eds.

doi:10.1017/S1743921314002075

Hyper X-ray Flares on Active Stars Detected with MAXI

Masaya Higa[1], Yohko Tsuboi[1], Hitoshi Negoro[2], Satoshi Nakahira[3], Hiroshi Tomida[3], Masaru Matsuoka[4] and The MAXI team[2]

[1]Department of Physics, Faculty of Science and Engineering, Chuo University
email: higa@phys.chuo-u.ac.jp

[2]Department of Physics, Nihon University

[3]Japan Aerospace Exploration Agency (JAXA)

[4]MAXI team, RIKEN

Abstract. MAXI started its operation in 2009 August. Owing to its unprecedentedly high sensitivity as an all-sky X-ray monitor and to its capability of real-time data transfer, we have detected 56 strong flares from twenty-one active stars (eleven RS CVn systems, one Algol system, seven dMe stars, one dKe star and one Young Stellar Object). These flares have large X-ray luminosity of 6 $\times 10^{30}$–5 $\times 10^{33}$ ergs s^{-1} in the 2–20 keV band. The flares can be thought to be high ends among their own categories. During the flare from AT Mic on 2012 April 18th, one of the largest X-ray luminosities was recorded as a dMe star, 6 $\times 10^{32}$ ergs s^{-1} in the 2–20 keV band. It is larger than its bolometric luminosity by 4 times. The total energy emitted during the flare is 10^{36} ergs in the same band. Such total energy can be obtained on large flares from RS CVn system, but not on any other flares from dMe stars. In this proceeding, we report on the present situation in characteristics of hyper X-ray flares on each stellar categories.

Keywords. stars: flare - stars: RS-CVn type - stars: Algol - stars: dMe stars - stars: YSO

1. Introduction

Cool stars, which have spectral types of F, G, K, and M, are known to show X-ray flares (e.g., Favata & Micela 2003; Güdel 2004). Generally, the larger flare occurs less frequently. Therefore, the observed sample (or number) of large flares is fairly limited, and the physical parameters such as the peak temperature, the emission measure (EM), the e-folding time in the large flares have been poorly understood. This situation will be much improved if we can do un-biased all-sky survey with a large field of view and high sensitivity, which increases the chance to catch large flares.

The Monitor of All-Sky X-ray Image (MAXI; Matsuoka *et al.* 2009) is a mission of an all-sky X-ray monitor operated in the Japanese Experiment Module (JEM; Kibo) on the International Space Station (ISS) since 2009 August. It enables us to search for stellar flares effectively. In this proceeding, we report the results of survey for X-ray flares from stellar sources using the MAXI data taken by the first four-years operation from 2009 August to 2013 August.

2. Observation

The MAXI carries two scientific instruments: the Gas Slit Camera (GSC; Mihara *et al.* 2011; Sugizaki *et al.* 2011) covering the energy range of 2–30 keV and the Solid-state Slit Camera (SSC; Tsunemi *et al.* 2010; Tomida *et al.* 2011) covering that of 0.5–7 keV. This report is mainly based on the results obtained with the GSC, which has larger effective

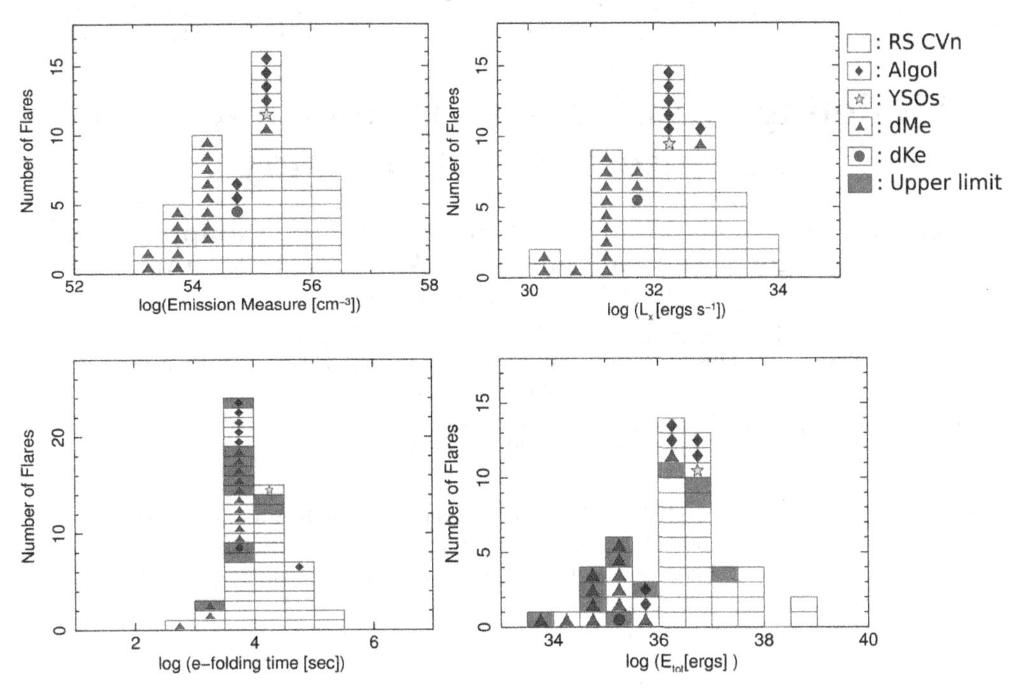

Figure 1. Distribution of the emission measure (top left), X-ray luminosity in the 2–20 keV band (top right), *e*-folding time at the flare decaying phase (bottom left), and total energy released during the flare (bottom right). The total energy is derived by multiplying the L_X by the *e*-folding time. The open squares, filled diamonds, filled star, filled triangles, and filled circle show RS-CVn type stars, Algol, YSO, dMe stars and a dKe star, respectively. The shaded data mean upper limit.

area than the SSC, although the results of the dMe star AT Mic introduced here was mainly obtained with the SSC.

3. Results

We detected 56 large flares from twenty-one active stars (eleven RS CVn systems, one Algol system, seven dMe stars, one dKe star and one Young Stellar Object). The X-ray luminosities are higher than 10^{30} ergs s^{-1} in the 2–20 keV band, and EMs are higher than 10^{53} cm^{-3}. These MAXI flares can be thought to be high ends among their own categories.

In figure 1, we show the distribution of the derived properties, EM, L_X, *e*-folding time and E_{tot}. In the all the parameters, dMe flares are localized at smaller values than the RS CVn flares. As for the EM, L_X, and E_{tot}, the high-ends in the flares from RS CVn stars are roughly two orders of magnitudes higher than those in the flares from dMe stars. The difference might originate from the area of starspots; RS CVn systems and dMe stars have similar coverage ratio of the starspots to the stellar surface, but the stellar radius of RS CVn stars is about one order of magnitude larger than that of dMe stars.

Among the flares of dMe stars, an outstanding flare was detected from AT Mic on 2012 April 18th. The EM, L_X, and E_{tot} are 3 $\times 10^{55}$ cm^{-3}, 6 $\times 10^{32}$ ergs s^{-1} in the 2–20 keV band, and 10^{36} ergs in the same band, respectively, although the *e*-folding time is typical value for dMe stars, 2.6 ksec. The L_X is four times larger than L_{bol}.

Magnetic Fields throughout Stellar Evolution
Proceedings IAU Symposium No. 302, 2013
P. Petit, M. Jardine & H. Spruit, eds.

doi:10.1017/S1743921314002087

Modeling transiting exoplanet and spots For interferometric study

Roxanne Ligi[1], Denis Mourard[1], Karine Perraut[2], Philippe Bério[1], Lionel Bigot[1], Andrea Chiavassa[1], Anne-Marie Lagrange[2] and Nicolas Nardetto[1]

[1]Laboratoire Lagrange, UMR 7293 UNS-CNRS-OCA
Boulevard de l'Observatoire, B. P. 4229 F, 06304 NICE Cedex 4, France.
email: roxanne.ligi@oca.eu

[2]UJF-Grenoble1/CNRS-INSU, Institut de Planétologie et d'Astrophysique de Grenoble, UMR 5274, Grenoble, F-38041, France

Abstract. Up to now, many techniques have been developed to detect and observe exoplanets, the radial velocity (RV) method being the most prolific one. However, stellar magnetic spots can mimic an exoplanet transit signal and lead to a false detection. A few models have already been developed to constrain the different signature of exoplanets and spots, but they only concern RV measurements or photometry. An interferometric approach, with high angular resolution capabilities, could resolve this problem.

Optical interferometry is a powerful method to measure accurate stellar diameters, and derive fundamental parameters of stars and exoplanets minimum masses. We have built an analytical code able to calculate visibility moduli and closure phases of stars with a transiting exoplanet, to be compared with a star with no exoplanet. From the difference of interferometric signal, we can derive the presence of the exoplanet, but this requires that the star is resolved enough. We have tested this code with current available facilities like VEGA/CHARA and determined which already discovered exoplanets systems can be resolved enough to test this method.

To make a more general study, we also tested different parameters (exoplanet and stellar diameters, exoplanet position) that can lead to a variation of the minimum baseline length required to see the exoplanet signal on the visibility modulus and the phase. Stellar spots act in the same way, but the difference of local intensity between an exoplanet transit and a spot can easily be studied thanks to the interferometric measurements.

Keywords. Exoplanets, magnetic spots, optical interferometry

---

1. Introduction

The detection and characterization of exoplanets has been one of the fastest developing fields in Astrophysics since the discovery of the first exoplanet by Mayor & Queloz (1995). However, this detection can be disturbed by stellar activity. Magnetic spots and bright plages can mimic an exoplanetary signals, leading to a false detection.

For example, Lagrange *et al.* (2010) have studied the impact of magnetic spots on RV measurements of a Solar-like star to verify if an Earth-like planet located on the habitable zone could be detected. They found that a good temporal sampling is necessary to remove the spots' signatures from RV measurements as they can be of higher ampltitude than the exoplanet's one. Meunier *et al.* (2010) made an equivalent study but on bright plages and found that it does not affect the exoplanets detection.

Understanding stellar activity is not only important for exoplanets search, but also because it gives information about stellar structure and evolution, without forgetting that the study of exoplanets, spots and stars are complementary. We could cite for example Sanchis-Ojeda *et al.* (2012)'s work, who used magnetic spots to derive the equatorial alignement of Kepler−30's three exoplanets. Contrary to them, Silva-Valio & Lanza (2012) used the planetary transit of Corot−2*a* to study spots at the surface of its host star.

Most of previous interferometric studies on exoplanets have been performed in the infrared domain. For instance, Matter *et al.* (2010) made an attempt to detect the signal of Gliese 86*b* using MIDI/VLTI and AMBER/VLTI, but a lack of precision prevented us to measure it. Zhao *et al.* (2008, 2011) have simulated the effects of exoplanets on closure phases using MIRC/CHARA. Again, instrumental issues prevented us to succeed. Some attempts with MIRC/CHARA have also been made to do direct imaging of spots (e.g., Monnier *et al.* 2012; Baron *et al.* 2012).

Most limitations in this field are then the angular resolution and instrumental limitations. Optial interferometry can be a solution. Direct imaging with big telescopes or infrared interferometry is resolving enough to separate planet and star but not to characterize the star. Optical interferometry has a much better angular resolution since the wavelength is smaller than in IR. As photometry and spectroscopy are operating in the optical domain, optical interferometry would also be a complementary method.

We propose to probe the capabilities of an optical interferometer, VEGA/CHARA, to measure the signal of an exoplanet and a magnetic spot. We also present a more general study using a fictive interferometer. For this, we have built a numerical code called `COMETS` (*COde for Modeling ExoplaneTs and Spots*) using analytical formulae allowing to model a dark transiting exoplanet (modelled as a dark disk) and/or a magnetic spot (umbra and penumbra modelled as a dark disk and a dark ring respectively). Thus we can measure the interferometric observables (visibility, phase and closure phase) for the different cases.

2. VEGA/CHARA capabilities

2.1. *VEGA/CHARA*

VEGA (Mourard *et al.* 2009) is a Visible spEctoGrAph and interferometer located at Mount Wilson, California. It uses the CHARA array's telecopes, which longest separation between one pair reaches 331 m. The six 1 meter-telescopes arranged in a Y-shape allows a good repartition of baselines and thus a good u,v coverage. VEGA operates at high (30000) or medium (6000) resolution in the wavelength range 450 – 850 nm. It has a maximum angular resolution of 0.3 milliarcseconds (mas), which gives the opportunity to resolve a large number of targets. Particularly, we have observed ten stars hosting planets, which are part of a catalog of 42 exoplanet hosts stars, measured their angular diameters with a precision of $\sim$ 2.4% and derived their parameters (mass, effective temperature) (Ligi *et al.* 2014, in prep.). Reaching the first zero of the visibility function is the first step to measure stellar activity in interferometry. Figure 1 shows the squared visibility curve of HD190360, which hosts two exoplanets and is part of our catalog. We can see that the measurements reach the first zero of visibility, thus we could constrain its limb-darkened diameters: 0.78 ± 0.01 mas.

2.2. *Exoplanet and spot signatures*

`COMETS` simulates a transiting exoplanet as a dark disk of intensity 0, whereas spots are considered hotter ($\sim$ 500 – 1000K colder than the star's photosphere) and are thus represented by an umbra (a dark disk) and a penumbra (a ring sourrounding the umbra) of intensity proportional to their temperature. Using CHARA baselines and VEGA specificities, the code measures the corresponding observables. Taking an ideal case, i.e.

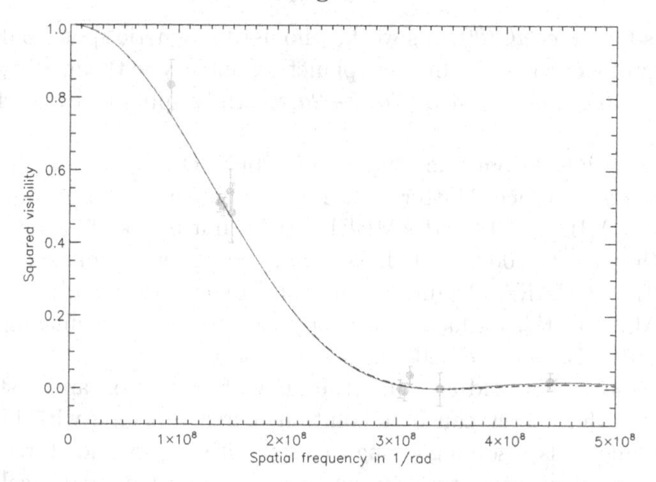

Figure 1. Squared visibility of HD190360 obtained with VEGA/CHARA. The points represent the data and the solid and dotted-dashed lines represent the uniform disk and the limb-darkened disk model respectively (Ligi *et al.* 2014, in prep.).

a star of 1 mas of angular diameter, so it is well resolved by VEGA, we measured the corresponding interferometric observables with one of the two features. Even with a very small exoplanet or spot (e.g., ~ 0.015 mas), a weak signal is seen on the closure phase but not on the visibility curve. However, with an exoplanet or spot ten times bigger, the signal can be observed on both observables. Thus, we calculated the minimum baseline length required to measure such signals. CHARA baselines are sufficient as long as the exoplanet is bigger than 0.09 and 0.13 mas, which provoques a signal of 1% or 2% on the visibilities respectively. Closure phases measurements allow to detect smaller exoplanets with a minimum signal of 2°. Resolving the star is of course the first condition to measure the signal of an extra feature, but the closure phase measurement is also an important condition. Unfortunately, VEGA cannot measure closure phases yet. We see that the limiting factor is also the accuracy of the measurements.

3. Disentangling between exoplanets and magnetic spots

Both exoplanet and spot have a signature on interferometric observables, which differs because of their different shapes. To disentangle between an exoplanet and a spot signature, we have plotted the Airy function in the u,v plane of a star with a transiting exoplanet and a spot (Figure 2). We can see that the Airy figure is disturbed by the presence of a spot and we can notice two important points. Firstly, the first lobe of visibility is a quite perfect disk, which means that the visibility is not perturbed by the presence of an exoplanet or a spot. Thus, one has to measure beyond the first lobe to measure the signature of an the exoplanet or spot. Secondly, the source is not well resolved beyong the first lobe (the visibility does not reach the zero value), and this phenomenon is emphasized by the presence of the spot.

With a fictive interferometer, we cannot measure closure phases but phases only (as triplets of telescopes cannot be defined anymore). However, phases react the same. When the system is not resolved, the signal on the phase is maximum. Spots and exoplanets also induce different variations of the phase, which also help desentangling between both.

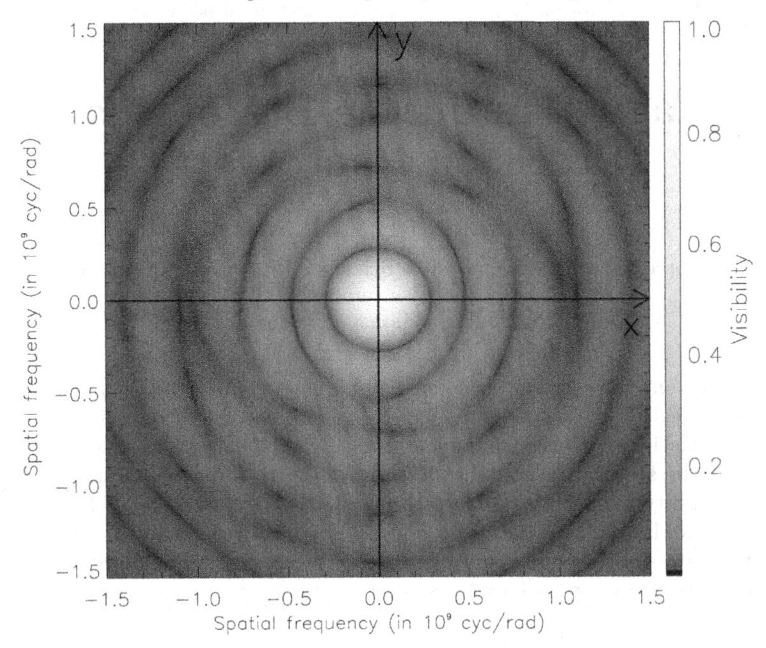

Figure 2. Airy figure of a 1 mas star with a 0.15 mas transiting exoplanet and a 0.15 mas magnetic spot (Ligi *et al.* 2013, in prep.).

4. Conclusion

Transiting exoplanets and magnetic spots have a signature in optical interferometry. Even if they are not measurable with VEGA/CHARA yet, it will be possible soon. VEGAS, *VEGA Second generation*, will measure closure phases and will be be equiped with adaptative optics. It will also be able to observe brighter objects with a better SNR. All this combined will allow to measure exoplanets and spots signatures.

References

Baron, F., Cotton, W. D., Lawson, P. R., *et al.* 2012, *SPIE*, Vol. 8445
Lagrange, A.-M., Desort, M., & Meunier, N. 2010, *A&A*, 512, A38
Matter, A., Vannier, M., Morel, S., *et al.* 2010, *A&A*, 515, A69
Mayor, M. & Queloz, D. 1995, *Nature*, 378, 355
Meunier, N., Desort, M., & Lagrange, A.-M. 2010, *A&A*, 512, A39
Monnier, J. D., Baron, F., Anderson, M., *et al.* 2012, *SPIE*, Vol. 8445
Mourard, D., Clausse, J. M., Marcotto, A., *et al.* 2009, *A&A*, 508, 1073
Sanchis-Ojeda, R., Fabrycky, D. C., Winn, J. N., *et al.* 2012, *Nature*, 487, 449
Silva-Valio, A. & Lanza, A. F. 2012, *A&A*, 529, A36
Zhao, M., Monnier, J. D., Che, X., *et al.* 2011, *PASP*, 123, 964
Zhao, M., Monnier, J. D., ten Brummelaar, T., Pedretti, E., & Thureau, N. 2008, *IAU*, Vol. 249

Magnetic Fields throughout Stellar Evolution
Proceedings IAU Symposium No. 302, 2013
P. Petit, M. Jardine & H. Spruit, eds.

doi:10.1017/S1743921314002099

Stellar Magnetism in the Era of Space-Based Precision Photometry

Lucianne M. Walkowicz

Department of Astrophysical Sciences, Princeton University,
4 Ivy Lane, Princeton NJ 08544
email: `lucianne@astro.princeton.edu`

Abstract. The advent of precision space-based photometric missions such as MOST, CoRoT and *Kepler* has revealed stellar magnetic activity in unprecedented detail. These observations enable new investigations into the fundamental nature of stellar magnetism by furthering our understanding of the stellar rotation and differential rotation that generate the field, and the photometric variability caused by the surface manifestations of the field. In the case of stars with planetary candidates, these data also offer synergy between studies of stars and planets. Here, I review the possibilities and challenges for deepening our understanding of magnetism in solar-like stars in the era of space-based precision photometry.

Keywords. stars: activity, stars: spots

---

1. Introduction

Magnetism in solar-like stars is thought to operate via an $\alpha\Omega$ dynamo, where shearing at the interface between the inner radiative and outer convective zones (or "tachocline") creates ropes of magnetic flux (see Reiners 2012 for a recent review of magnetism in solar-like stars). These flux tubes rise buoyantly through the convective zone and protrude from the stellar surface, creating a variety of observable phenomena, from dark starspots at the footpoints of these protruding flux tubes, to flares as magnetic field lines snap and reorganize in the outer stellar atmosphere (Strassmeier 2009). These phenomena create variability on a variety of timescales: from second to minutes to hours for flares, to days, weeks and months as stars rotate and magnetic surface features (such as spots) evolve. The appearance, evolution, temporal and spatial behavior of these features provides direct feedback to theory of the magnetic field generation: the stellar rotation creates the periodic variability that allows observers to measure the rotation and differential rotation that are responsible for the generation of the field itself, and the geometrical distribution of spots, their evolutionary timescales are also predictions of theory (Berdyugina 2005).

Stellar rotation periods are also interesting as a proxy for stellar age via gyrochronology relations, or the relationship between the stellar rotational spindown and age (Barnes 2007). As age is a difficult quantity to measure for individual field stars, the rotation period is often the best (and sometimes the only) existing method for constraining age. In addition to its intrinsic interest to stellar astrophysics, stellar age is also of particular interest for stars hosting exoplanetary systems. Knowing the age of an exoplanet host star provides a timestamp on the exoplanetary system, informing our understanding of the evolutionary history of the system and potential habitability for any planets that lie in the habitable zone.

On the Sun, surface magnetic phenomena are visible in striking detail, both in a spatial and temporal sense. The launch of Solar Dynamics Observatory in 2010 provides the most recent examples of exquisite fine structure evident as these events evolve. However, data for the Sun is truly unique, in that spatial resolution of these phenomena is difficult or often impossible to achieve for stars besides our Sun. Our knowledge of stellar magnetism is therefore faced with the fact that we study the Sun– the star upon which our understanding of stellar magnetism is based– in a very different way than we study magnetism in the rest of the stars in the Universe. For almost all other stars, only integrated quantities are accessible, and all spatial information must be inferred.

The era of time domain astronomy has brought opportunities to at least crack the temporal resolution of these highly variable events, and in some cases, to obtain surface maps as well. The Mount Wilson survey carved out great first steps in this area, monitoring activity for a number of solar like stars over a very long time baseline from the ground (Duncan *et al.* 1991). However, the advent of space-based time domain surveys in the past decade has breathed new life into studies of stellar magnetism. While photometric surveys themselves are not a new idea, the recent emphasis on such surveys has been partly driven by the search for transiting exoplanets, requiring high precision, high cadence, continuous and long-duration photometry. The potential synergy of such surveys was of course recognized early on: a very early seed of the eventual *Kepler* mission, the abstract of Borucki *et al.*(1985) states "The high precision, multiple-star photometric system required to detect planets in other stellar systems could be used to monitor flares, starspots, and global oscillations."

2. Overview

In the past decade, three space-based photometric surveys have allowed us to see the effects of stellar variability in unprecedented detail: MOST (Walker *et al.* 2003), launched in 2003 and still currently operating; CoRoT, launched in 2006 and completed in 2012 Auvergne *et al.* 2009; and NASA's *Kepler* mission Koch *et al.* 2010, launched in 2009 and in limbo since early 2013 due to a reaction wheel failure† In this proceeding, I focus primarily on recent results from Kepler, though the reader is advised to also seek out the ground-breaking results from the CoRoT and MOST missions.

Stellar rotation can be measured through a variety of methods. In the case of the Sun, rotation and differential rotation are readily apparent by monitoring the passage of sunspots as a function of solar latitude. This surface resolution is also what has revealed the now-familiar "butterfly diagram" where the preferred latitude for the appearance of sunspots changes over the course of the solar cycle, moving from high latitudes during the solar minimum to lower latitudes at the solar maximum. For stars, such detailed surface features are impossible, but the effects of stellar rotation are evident in both spectroscopy and photometry. Line broadening in spectra provides a measurement of $v\sin i$, or the rotational velocity v modified by the sine of the stellar inclination, i. If the star is viewed equator-on, the velocity can be turned into a true stellar period if the stellar radius is also known. However, stars are not always conveniently equator-on towards Earth, and typically the stellar inclination is unknown. In these cases, $v\sin i$ provides only a limit on the rotation period. In photometry, stellar rotation is revealed through its modulation of the integrated stellar brightness as surface features rotate into and out of view. In the case of the precise, nearly uninterrupted photometry available from MOST, CoRoT and Kepler, the stellar rotation in many cases is readily evident from even a short sequence of data, provided the starspots modulating the stellar brightness are

† The loss of *Kepler*'s third reaction wheel compromises its ability to perform the fine pointing that made its original photometric precision possible. As of this writing, plans for possible repurposing of the *Kepler* telescope are being considered.

long-lasting enough that they create a regular pattern of dimming and brightening over the course of several rotation periods. Precise rotation periods can then be determined by applying a Lomb-Scargle periodogram (Scargle 1982), autocorrelation (McQuillan *et al.* 2013), global fitting (Reinhold Reiners & Basri 2013), or other techniques. If spots exist at multiple latitudes, drift in phase due to differential rotation, and are similarly stable over multiple rotations, differential rotation will create clear beat patterns between the different periods in the photometry, and multiple peaks may also be evident in the periodogram. In these cases the primary point of confusion is aliasing in the periodogram, where the periodogram peak with highest power may actually be that of half the true rotation period.

Of course, these cases are ideal, and so are only true for a small number of stars. More typically, the stellar brightness is modulated by spots that evolve on timescales approaching or comparable to the rotation timescale, as well as other surface features such as plage. These more ambiguous cases often require confirmation of the period by eye, to confirm the period found by algorithm (e.g. periodogram or autocorrelation function). In addition, wIthout an absolute measurement of the unspotted stellar brightness, the observed variabiity will be relative, and it is therefore difficult to assess the relative contributions of bright and dark features. There is additional ambiguity created by the geometry of spot distribution over the stellar surface: isolated spot features that rotate into and out of view create a regular pattern of variability, but in the case of a more distributed spot geometry, where stellar spots mottle the surface rather than being in isolated groups, there may not be any time during the stellar rotation where the star is unspotted. In these cases, the variability may be relatively low amplitude, and so without an additional measure of activity (such as spectroscopic observations of the Ca II K line) there may be ambiguity between a star with relatively low levels of activity (and correspondingly few spots) and a star whose surface actually has many spots. One of the benefits of constant monitoring provided by space-based photometry missions is that, even if the spot pattern is not conducive to finding the stellar rotation period at one time, the evolution of the spot pattern may yield a more fortuitous arrangement at some point during the long baseline of the observations (see Fig. 1).

Magnetic activity is intimately related to the stellar rotation and differential rotation, and manifests not only in chromospheric emission but throughout the stellar atmosphere, as well as across the electromagnetic spectrum. Previous surveys of the stellar activity-rotation relation have shown a relationship between the level of chromospheric activity (measured by the Ca II K measure log R'_{HK}) and the Rossby number Ro, or the ratio of the rotation period to the convective overturn time (Ro $= P_{rot}/\tau_{conv}$). As shown in Figure 7 in Mamajek & Hillenbrand (2008), activity in solar-type stars decreases as stars spin down over the course of their main sequence lifetimes. Optical photometry traces magnetic effects in the stellar photosphere, where it creates the dark starspots that allow one to measure the stellar rotation in the first place. One might therefore ask whether the photospheric activity, as captured by the amplitude of optical variability, also shows a similar relationship to the stellar rotation as chromospheric activity tracers. Walkowicz & Basri (2013) recently used rotation periods determined for ~950 host stars of the *Kepler* exoplanet targets to showed that the correlation of photometric variabitliy amplitude and the Rossby number are loosely correlated, but with much greater scatter than the relationship between log R'_{HK} and Rossby number (see Figure 2 of Walkowicz & Basri (2013)). The authors attribute this scatter to differing spot geometries, such that rapidly rotating, active stars may appear to have low amplitude optical variability despite being quite magnetically active. The range of photometric amplitudes for the *Kepler* exoplanet candidate host stars is comparable to the Sun for similar Rossby numbers, as inferred

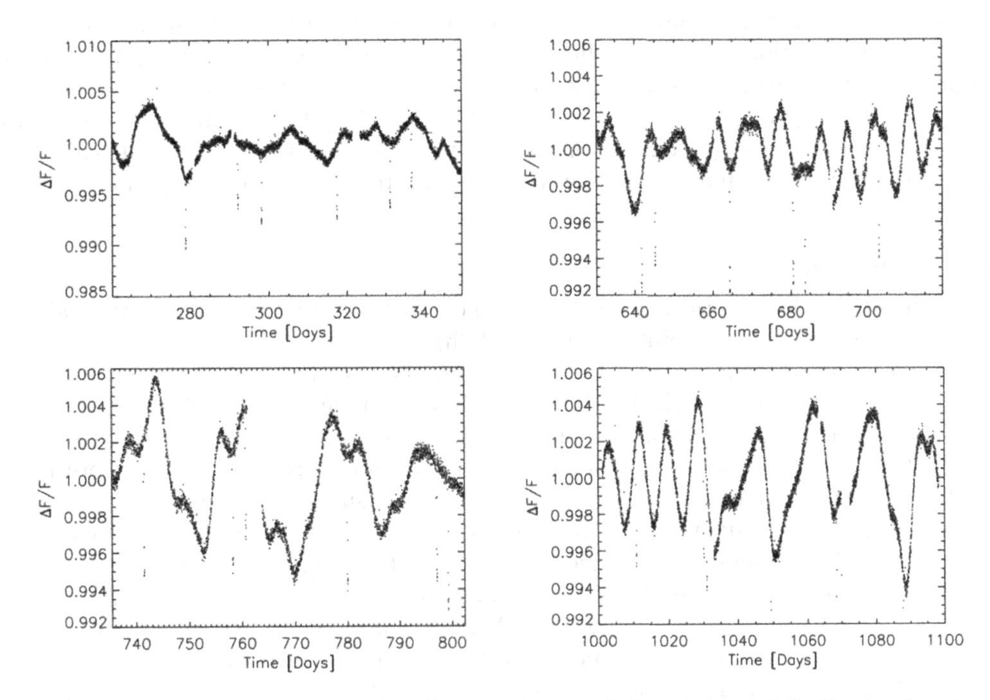

Figure 1. *Kepler* lightcurves for four different quarters of observations of the Kepler-9 system. The evolving spot distribution over the stellar surface causes dramatic changes in the morphology of the lightcurve, making some quarters more amenable to determination of the stellar period than others.

from the amplitude of white-light variability in SOHO Virgo g+r lightcurves (see Basri *et al.* 2010 for details).

Once solar-like stars have arrived on the main sequence, they converge to a well-defined age-rotation relationship, such that the stellar rotation period may be used to infer an age for the star. Indeed, Epstein & Pinsonneault (2012) compare a variety of common dating methods (such as asteroseismology or isochrone fitting), and find that rotation-based ages provide some of the best constraints on stellar age for older (age > 550 Myr) main sequence stars, even when accounting for uncertainties due to differential rotation. At present, gyrochronology relationships are best calibrated for stars younger than a Gyr, due to the fact that rotation periods for older stars with known ages are considerably fewer and further between (with the exception of a single star: the Sun). Clusters observed by *Kepler* (e.g. NGC 6811, Meibom *et al.* 2011) may provide additional constraints on existing gyrochronology relations, increasing confidence in ages derived from rotation in the future. At present, *Kepler* target stars for which rotation periods have been measured tend to be around rotation periods of 30 days or less. This is for a variety of reasons: first and foremost, active, rapidly rotating stars tend to have higher amplitude variability, and their short rotation periods are well-sampled in a single quarter (90 day interval) of *Kepler* data. In addition, the *Kepler* data are detrended to remove systematics caused by spacecraft motion; while the detrending pipeline preserves periodic signals of $\sim$20 days or less, longer periodic signals tend to be attenuated or removed entirely from the lightcurves (Smith *et al.* 2012, Stumpe *et al.* 2012). Therefore, most of the stars that are amenable to rotation period determination thus far are concentrated in stars whose rotation periods are shorter than that of the Sun. For rotation periods beyond $\sim$45 days, or half of the lightcurve, multiple quarters of data must be used to derive the period securely. Even so, it may not always be possible to derive a trustable period from the detrended data, as long drifts in stellar brightness that approach the length of a quarter begin to resemble instrumental effects, and are readily removed by the pipeline. For the many more slowly rotating stars amongst the *Kepler* targets, it may be necessary to re-extract the photometry from the pixels to mitigate systematics (e.g. Kinemuchi *et al.* 2012), and to join multiple quarters of data together to fully sample the period.

3. Implications

Ultimately, we must return to the heart of how we relate observable quantities about stellar magnetism, such as rotation and variability in integrated measures of activity, to a physical understanding of what these imply for the generation of the magnetic field. Here again we face the challenge that we lack the detailed surface observations that have so informed our understanding of the Sun, and so it remains challenging to relate observations of other stars to that of our magnetic Rosetta Stone. Space-based precision photometry has provided a new opportunity to map the surface features of stars through modeling of the spot distribution. In the case of stars hosting transiting exoplanets, deformations in the transit shape due to the planet transiting stellar spots can permit a detailed mapping of the stellar surface under the planetary transit chord (Nutzman *et al.* 2011, Sanchis-Ojeda *et al.* 2011, Silva-Valio & Lanza 2011), but only some transits show well-defined features due to spot transits. More often, distributed, evolving spots just create noise in the transit depth of the folded lightcurve, heightening uncertainty in the planet parameters without yielding useful information about the star.

Spot modeling has enjoyed a renaissance over the past decade due to the availability of new precise photometry, but these models can be highly degenerate (see Walkowicz, Basri & Valenti 2013 and references therein), making the problem "ill-posed even when the signal to noise approaches infinity" (K. Strassmeier, as stated during IAU SS13). If the starspot distribution is amenable, the morphology of the lightcurve may provide constraints on the stellar inclination (in that stars seen from lower, more pole-on inclinations will have more gradual spot ingresses and egresses). However, the presence of multiple spots, differential rotation and spot evolution can often confound uniqueness. Even if a resulting model is not unique, such models do provide robust estimates of the total spot coverage as a function of longitude and amount of differential rotation (e.g. Fröhlich *et al.* 2009). When available, complementary methods, such as modeling the complete lightcurve together with detailed mapping of the transit chord for stars with transiting planets, can reveal details of the stellar surface on a variety of scales (Silva-Valio & Lanza 2011). In some cases, it may be possible to know the inclination of the system from spectroscopy (yielding a measurement of vsini and the stellar radius) in combination with the photometric period. In these cases, the degeneracy between latitude and inclination may be broken, yielding information on the latitudinal distribution of starspots.

Rotation periods and differntial rotation have been measured for numerous stars observed by MOST, CoRoT and most recently Kepler. Currently, the most easily accessible rotation periods are for stars with relatively clear cut variability, where the effect of spots is obvious and the rotation period can often be guessed at just by eye. Unfortunately, the variability of our own Sun does not resemble these stars! Viewed in integrated optical light, the Sun's own variability appears far more erratic, with spots that evolve on timescales comparable to the stellar rotation period, as opposed to being stable over the course of several rotations (and thus yielding an uncomplicated determination of the solar rotation period). Faculae also play a large role in the solar variability, modulating the light as they pass across the stellar limb (where they are most visible, unlike spots whose projected area and thus greatest effect appears as they cross disc center). Numerous stars in the *Kepler* dataset bear a strong resemblance to our own Sun, but these are as yet largely unmined for rotation periods and differential rotation. However, bringing the solar and stellar views of magnetic activity closer together requires that we embrace this challenge.

References

Auvergne, M., Bodin, P., Boisnard, L., *et al.* 2009, *AAP*, 506, 411
Barnes, S. A. 2007, *ApJ*, 669, 1167
Basri, G., Walkowicz, L. M., Batalha, N., *et al.* 2010, *ApJL*, 713, L155
Berdyugina, S. *Living Rev. Solar Phys.* 2 (2005), 8
Borucki, W. J., Scargle, J. D., & Hudson, H. S. 1985, *ApJ*, 291, 852
Duncan, D. K., Vaughan, A. H., Wilson, O. C., *et al.* 1991, *ApJS*, 76, 383
Epstein, C. R. & Pinsonneault, M. H. 2012, arXiv:1203.1618
Fröhlich, H.-E., Küker, M., Hatzes, A. P., & Strassmeier, K. G. 2009, *AAP*, 506, 263
Garcia, R. A., Ballot, J., Mathur, S., Salabert, D., & Regulo, C. 2010, arXiv:1012.0494
Kinemuchi, K., Barclay, T., Fanelli, M., *et al.* 2012, *PASP*, 124, 963
Koch, D. G., Borucki, W. J., Basri, G., *et al.* 2010, *ApJL*, 713, L79
Mamajek, E. E. & Hillenbrand, L. A. 2008, *ApJ*, 687, 1264
McQuillan, A., Aigrain, S., & Mazeh, T. 2013, *MNRAS*, 432, 1203
Meibom, S., Barnes, S. A., Latham, D. W., *et al.* 2011, *ApJL*, 733, L9
Nutzman, P. A., Fabrycky, D. C., & Fortney, J. J. 2011, *ApJL*, 740, L10
Reiners, A. *Living Rev. Solar Phys.* 9 (2012), 1
Reinhold, T. & Reiners, A. 2013, *AAP*, 557, A11
Sanchis-Ojeda, R. & Winn, J. N. 2011, *apj*, 743, 61
Scargle, J. D. 1982, *ApJ*, 263, 835
Silva-Valio, A. & Lanza, A. F. 2011, *AAP*, 529, A36
Smith, J. C., Stumpe, M. C., Van Cleve, J. E., *et al.* 2012, *PASP*, 124, 1000
Strassmeier, K. G. 2009, *A&AR*, 17, 251
Stumpe, M. C., Smith, J. C., Van Cleve, J. E., *et al.* 2012, *PASP*, 124, 985
Walker, G., Matthews, J., Kuschnig, R., *et al.* 2003, *PASP*, 115, 1023
Walkowicz, L. M. & Basri, G. S. 2013, *MNRAS*, 436, 1883
Walkowicz, L. M., Basri, G., & Valenti, J. A. 2013, *ApJS*, 205, 17

Magnetic Fields throughout Stellar Evolution
Proceedings IAU Symposium No. 302, 2013
P. Petit, M. Jardine & H. Spruit, eds.

doi:10.1017/S1743921314002105

The new age of spotted star research using *Kepler* and CHARA

Rachael M. Roettenbacher[1], **John D. Monnier**[2], **Robert O. Harmon**[3] **and Heidi H. Korhonen**[4]

[1]Department of Astronomy, University of Michigan,
500 Church Street, Ann Arbor, MI, United States
email: rmroett@umich.edu

[2]Department of Astronomy, University of Michigan,
500 Church Street, Ann Arbor, MI, United States
email: monnier@umich.edu

[3]Department of Physics and Astronomy, Ohio Wesleyan University,
61 S. Sandusky Street, Delaware, OH, United States
email: roharmon@owu.edu

[4]Finnish Center for Astronomy with ESO, University of Turku,
Väisäläntí 20, FI-21500 Piikkiö, Finland
email: heidi.h.korhonen@utu.fi

Abstract. With the precise, nearly-continuous photometry from the Kepler satellite and the sub-milliarcsecond resolving capabilities of the CHARA Array, astronomy is entering a new age for the imaging and understanding of stellar magnetic activity. We present first results from our Guest Observer Program, where 180 single-epoch surface image reconstructions of KIC 5110407 have revealed differential rotation and hints of magnetic activity cycles based on both spot and flare variations. Analysis of our larger, full dataset will establish in unprecedented detail how surface magnetic activity correlates with stellar age and spectral type. In addition to Kepler work, we have harnessed the power of the world's largest infrared interferometer to "directly" image the spotted surfaces of a few of the closest RS CVn systems, allowing a comparison of contemporaneous Doppler and light-curve inversion imaging techniques.

Keywords. stars: spots, stars: activity, stars: imaging, stars: low-mass

---

1. Introduction

Starspots, analogous to sunspots, are areas of the stellar photosphere that are significantly cooler than their surroundings, appearing darker. In hot stars, these spots are regions of metals accumulating near the magnetic poles (e.g. Krtička *et al.* 2007). In cool stars, these spots are due to strong magnetic fields quenching convection in the outer layers of the atmosphere. Because these spots are the clearest manifestation of magnetic fields their structure and evolution is important to understanding stellar magnetic dynamos. Tracing spot motions over time allows for observations of spot evolution and measurements of differential rotation, important concepts in the understanding stellar activity and magnetic behavior (Strassmeier 2009).

To better understand starspots, we aim push the boundaries of current imaging techniques in order to study surface spot structures and evolution. We compare several state-of-the-art imaging techniques to validate methods and compare results. Additionally, we use one of these imaging techniques for analyzing stars with rapidly-evolving spot structures using the nearly-continuous precision photometry of the *Kepler* satellite.

2. Starspot Imaging

2.1. *Long-baseline Interferometry*

Starspots have been imaged with a variety of techniques involving observations obtained via interferometry, spectroscopy, and photometry. With interferometry, the starspots of bright, nearby stars have begun to be observed (e.g. Parks *et al.* 2011). With the recently upgraded Michigan Infrared Combiner (MIRC) at Georgia State University's Center for High Angular Resolution Astronomy (CHARA) Array, it is now possible to use all six CHARA telescopes to directly image stars with spotted surfaces via aperture synthesis modeling. Stars to be imaged interferometrically are required to be bright and have large angular diameters. Spectroscopically, Doppler imaging allows for reconstructions of surface features based upon variations of absorption line profiles as the star rotates (e.g. Vogt & Penrod 1983). A method that can only be applied to rapidly-rotating stars, Doppler imaging requires high-resolution spectroscopy to trace the motions of spots across a stellar surface through changing spectral features. Photometric light-curve inversion techniques allow for reconstructions based upon flux variations (e.g. Roettenbacher *et al.* 2011). We aim to image stars simultaneously with all three of these techniques to compare the results and better understand the surface features of spotted stars.

Because each technique has strict requirements for targets our sample is limited. Four observable spotted star candidates have been identified: ζ And, σ Gem, ϵ UMa, and o Dra. o Dra, ζ And, and σ Gem are all the primary stars of RS CVn systems and known to be active and periodically variable, suggesting spot activity (e.g. Wilson 1976, Gratton 1950, and Eberhard & Schwarzschild 1913, respectively). ϵ UMa is a chemically peculiar Ap star with periodic light curve variations (e.g. Struve & Hiltner 1943). In the first study of its type, we have successfully coordinated interferometric observations with high-resolution spectroscopy and photometry targeting these stars. We aim to compare the aperture synthesis imaging using interferometric data with the results of spectroscopic Doppler imaging and photometric light curve inversion. To date, spotted stellar surfaces have been mapped with each technique, but the three techniques have not been verified against one another with simultaneous data.

For our observations, we obtain data from a variety of sources. The interferometric data are all collected at the CHARA Array with MIRC. Doppler imaging reconstructions for this work are obtained at the European Southern Observatory's Very Large Telescope (VLT), the Nordic Optical Telescope (NOT), the STELLA robotic telescope, and the Belgian Mercator telescope. The photometric data are obtained by the Tennessee State University 0.4-m Automated Photometric Telescope (APT) at Fairborn Observatory, AZ and the SMARTS 1.3-m telescope at Cerro Tololo, Chile.

Preliminary analyses of our data sets show no evidence of spots in the interferometric data for ϵ UMa and o Dra, and our σ Gem data were plagued by bad weather at CHARA. A very preliminary study of ζ And from 2011 July indicates the consistency between the Doppler imaging and the interferometric aperture synthesis mapping. We hope to compare the 2011 data to the our observing campaign of ζ And occurring during 2013 September and October. While directly comparing the results of the three image reconstructions for the first time, we will also be able to identify significant changes in the stellar surface over the course of two years.

2.2. *High-cadence Photometry*

In addition to the simultaneous imaging techniques of select spotted stars, we analyze the light curve of a population of stars with rapidly-changing spot features to characterize spot evolution and differential rotation observed by the *Kepler* satellite. The advantage

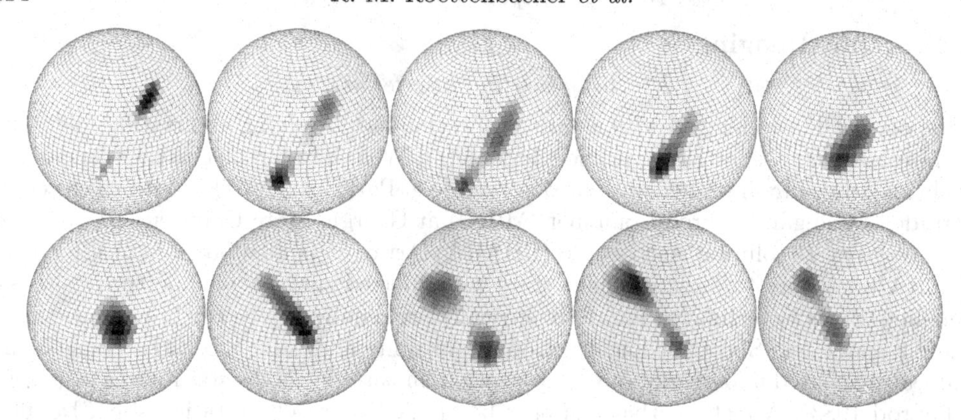

Figure 1. Panel of ten reconstructed surfaces of KIC 5110407 with an inclination of $i = 60°$. Each sequential surface (starting in the upper left with the first data point at BJD 2455124.43) is centered on the same latitude and longitude to illustrate spot evolution from one rotation period to the next (plot reprinted from Roettenbacher *et al.* 2013).

of photometry from the *Kepler* satellite is the the high-precision, nearly-continuous data of 30-minute cadence in comparison to ground-based observations.

We utilize a constrained, non-linear inversion algorithm, Light-curve Inversion (LI; see Harmon & Crews 2000), for mapping starspots from the *Kepler* photometric light curves. LI makes few *a priori* assumptions about the spotted stellar surface. No assumptions are made on spot shape, number, or size. Estimates for spot and photospheric temperatures, limb darkening coefficients, and rotation rate are made based on published values and the results of our analyses. LI has been tested extensively with simulations (Harmon & Crews 2000) and used for ground-based photometry (including comparison to Doppler images at some epochs; Roettenbacher *et al.* 2011) and for space-based photometry (Roettenbacher *et al.* 2013).

As a preliminary study and test of the method, we have applied LI to the light curve of a star in the *Kepler* field, KIC 5110407, a K-type pre-main sequence candidate that exhibits rapidly-changing spot structures. Our inverted surfaces show spot evolution and differential rotation (see Figure 1). We quantified the stellar differential rotation using the differential rotation law

$$\Omega(\theta) = \Omega_{\rm eq}\left(1 + k\sin^2\theta\right),$$

where $\Omega(\theta)$ is the angular velocity as a function of latitude, $\Omega_{\rm eq}$ is the angular velocity at the equator, θ is spot latitude, and k is the differential rotation parameter (as in Henry *et al.* 1995). Regardless of inclination, we found a value of k to be less than a scaled solar model (see Figure 2), consistent with stellar dynamo theories (e.g. Hall 1991).

With KIC 5110407 and other stars in our *Kepler* program, we aim to quantify differential rotation as well as analyze the presence of short-period activity cycles. For example, we saw evidence in the flares of KIC 5110407 to indicate the possible presence of an activity cycle. In the data (Quarters 2–5 and 7–9), we identified seventeen flares. We found an unusual concentration of flares between Q4-5, and a period without any flares in Q7-9 lasting over 200 days. By repeating our analysis of KIC 5110407 on other spotted stars in the *Kepler* field, we will be able to identify activity cycles, furthering our knowledge of the mechanisms of spot activity. With an understanding of activity cycles on a variety of stars, we hope to establish a relationship between stellar age and spectral type.

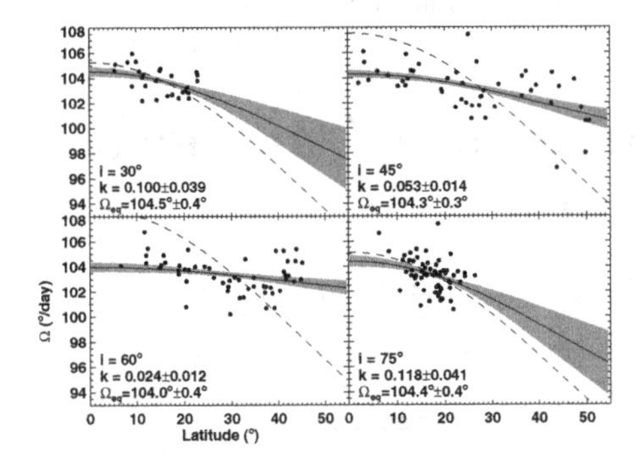

Figure 2. Derived differential rotation parameter, k, varies with the assumed angle of inclination, highlighting the degeneracy of the solutions of Light-curve Inversion (LI). KIC 5110407 has sub-solar differential rotation (solid line with $1 - \sigma$ errors in gray; fit with the solar k value, dashed line; plot reprinted from Roettenbacher *et al.* 2013).

3. Summary and Future Work

We are working on a campaign to cross-compare and validate several state-of-the-art imaging methods. By obtaining simultaneous interferometric, spectroscopic, and photometric data we will be able to directly compare reconstructed images of spotted stellar surfaces. In the first study of this type, we aim to obtain data on four spotted stars. The most promising data set will be obtained in 2013 September and October on the known spotted RS CVn ζ And, combining many resources to ensure the best temporal coverage for directly comparing images from different techniques and data sets.

While comparing three imaging methods, we are also focusing efforts on reconstructing spotted surfaces from the precision photometry of the *Kepler* satellite. Our first results on *Kepler* star KIC 5110407 show evidence of spot evolution, differential rotation, and potential evidence of an activity cycle (Roettenbacher *et al.* 2013). We will apply the techniques used in our test case to many other stars with rapidly-evolving spots with the ultimate goal of establishing a relationship between stellar age and spectral type based upon activity.

References

Eberhard, G. & Schwarzschild, K. 1913, *ApJ*, 38, 292

Gratton, L. 1950, *ApJ*, 111, 31

Hall, D. S.1991, in The Sun and Cool Stars: Activity, Magnetism, Dynamos, ed. I. Tuominen, D. Moss, & G Rüdiger (Lecture Notes in Physics, Vol. 380; Berlin: Springer), 353

Harmon, R. O. & Crews, L. J. 2000, *AJ*, 120, 3274

Henry, G. W., Eaton, J. A., Hamer, J., & Hall, D. S. 1995, *ApJS*, 97, 513

Krtička, J., Mikulášek, Z., Zverko, J., & Žižňovský, J. 2007, *A&A*, 470, 1089

Parks, J. R., White, R. J., Schaefer, G. H., Monnier, J. D., & Henry, G. W. 2011, in ASP Conf. Ser. 448, 16th Cambridge Workshop on Cool Stars, Stellar Systems, and the Sun, ed. C. M. Johns-Krull, M. K. Browning, & A. A. West (San Francisco, CA: ASP), 1217

Roettenbacher, R. M., Harmon, R. O., Vutisalchavakul, N., & Henry, G. W. 2011, *AJ*, 141, 138

Roettenbacher, R. M., Monnier, J. D., Harmon, R. O., Barclay, T., & Still, M. 2013, *ApJ*, 767, 60

Strassmeier, K. G. 2009, *A&ARv*, 17, 251

Struve, O. & Hiltner, W. A. 1943, *ApJ*, 98, 225

Vogt, S. S. & Penrod, G. D. 1983, *PASP*, 95, 565

Wilson, O. C. 1976, *ApJ*, 205, 823

Magnetic Fields throughout Stellar Evolution
Proceedings IAU Symposium No. 302, 2013
P. Petit, M. Jardine & H. Spruit, eds.

doi:10.1017/S1743921314002117

Rotation & differential rotation of the active Kepler stars

Timo Reinhold[1], Ansgar Reiners[2] and Gibor Basri[3]

[1,2]Institut für Astrophysik, Universität Göttingen, Friedrich-Hund-Platz 1,
37077 Göttingen, Germany
[1]email: reinhold@astro.physik.uni-goettingen.de
[2]email: areiners@astro.physik.uni-goettingen.de
[3]Astronomy Department, University of California, Berkeley, CA 94720, USA
email: basri@berkeley.edu

Abstract. Stellar rotation is a well-known quantity for tens of thousands of stars. In contrast, differential rotation (DR) is only known for a handful of stars because DR cannot be measured directly. We present rotation periods for more than 24,000 active stars in the Kepler field. Thereof, more than 18,000 stars show a second period, which we attribute to surface differential rotation. Our rotation periods are consistent with previous measurements and the theory of magnetic braking. Our results on DR paint a rather different picture: The temperature dependence of the absolute shear $\Delta\Omega$ is split into two groups separated around 6000 K. For the cooler stars $\Delta\Omega$ only slightly increases with temperature, whereas stars hotter than 6000 K show large scatter. This is the first time that DR has been measured for such a large number of stars.

Keywords. Rotation, differential rotation, Kepler, activity

1. Introduction

Stellar rotation strongly correlates with the age of the star and its level of activity. Active stars use to have a strong magnetic field responsible for various activity phenomena, e.g., spots, flares, etc.. Differential rotation (DR) triggers the magnetic field generation of the Sun ($\alpha\Omega$-dynamo) and is believed to be a major ingredient of the field generation in other active stars. Stellar variability due to star spots can be used as tracer of the stellar rotation period, esp. offering the possibility to detect DR. The Kepler space telescope delivers high quality data, with unprecedented duty cycle and number of targets, making it an ideal source for the search of periodic stellar variability.
In this study we summarize the most important results from Reinhold *et al.* (2013). We present rotation period measurements for more than 24,000 stars in the Kepler field. More than 75% of them showed a second period with we attribute to surface differential rotation. Furthermore, we show that our rotation periods and DR measurements are consistent with previous rotation measurements and theoretical predictions from mean-field theory, respectively.

2. Analysis

We analyzed Kepler Q3 data, which was reduced by the PDC-MAP pipeline to search for periodic stellar variability. Active stars were selected from the whole sample by filtering out stars whose *range of variability amplitude* $R_{var} > 0.3\%$ (see Basri *et al.* 2010; Basri *et al.* 2011). To extract the most significant periods from a light curve we used

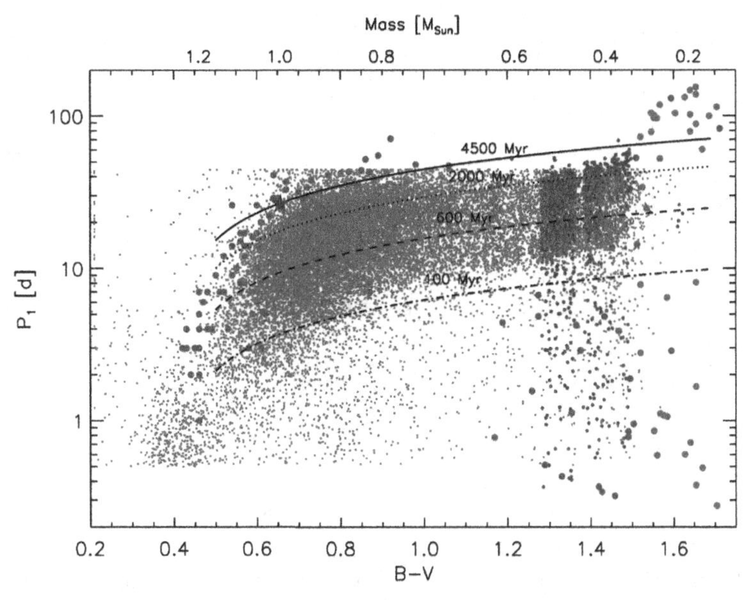

Figure 1. $B-V$ color vs. rotation period P_1 of the 24,124 stars with at least one detected period (gray dots). The filled circles represent data from Baliunas (1996) (olive), Kiraga, M. & Stepien, K. (2007) (purple), Irwin *et al.* (2011) (red), and McQuillan *et al.* (2013) (blue). Towards cooler stars we found an increase in rotation periods with a steep rise around $B-V \approx 0.6$ supplying evidence of rotational braking. The black lines represent a color-period relation found by Barnes (2007) for different isochrones. The blue star marks the position of the Sun for comparison.

the Lomb-Scargle periodogram in a prewhitening approach, yielding five periods for a global sine fit. From the global sine fit we selected the period with the highest power found in the prewhitening process, and call it P_1. Furthermore, we applied different filters, for instance, a lower and upper limit according $0.5\,d < P_1 < 45\,d$. This procedure yielded 24,124 rotation periods P_1. If a second period within $\pm 30\%$ of P_1 was found by our prewhitening analysis we called this period P_2. In 18,616 cases we found a second period P_2, which we attribute to surface DR. For details of the method and the selection process we refer the reader to Reinhold & Reiners (2013) and Reinhold *et al.* (2013).

3. Results

Rotation periods of the 24,124 stars are shown in Fig. 1. We see that our results are consistent with previous measurements. A strong increase of the rotation periods is evident around $B-V \approx 0.6$ supplying evidence of rotational braking. Furthermore, our results show good agreement with isochrones supplied by Barnes (2007) using empirical relations between color (mass) and rotation period.

For all stars with two detected periods, P_1 and P_2, we calculated the absolute surface shear $\Delta\Omega := 2\pi\,|1/P_1 - 1/P_2|$ as a measure of DR. We show the dependence of $\Delta\Omega$ on effective temperature in Fig. 2. In contrast to Barnes *et al.* (2005) we cannot reproduce the strong temperature dependence of $\Delta\Omega$ over the whole temperature range. For cool stars (3500–6000 K) we find that $\Delta\Omega$ shows weak dependence on temperature with a shallow increase towards hot stars. A different behavior of $\Delta\Omega$ is found for stars hotter than 6000 K. Our measurements show large scatter with a strong increase of $\Delta\Omega$ with temperature. These trends can be seen best by looking at the mean values $\langle\Delta\Omega\rangle$ of our measurements for different temperature bins, shown as open black circles.

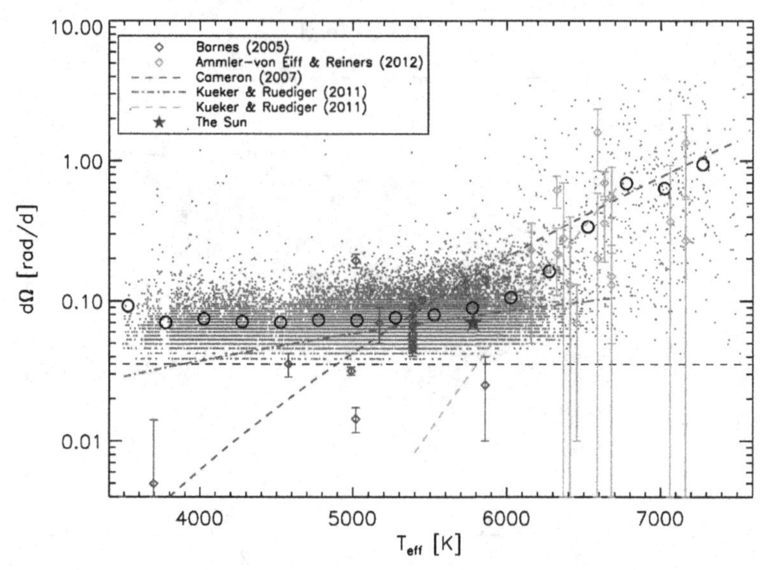

Figure 2. Effective temperature vs. horizontal shear $\Delta\Omega$ summarizing different measurements: The olive diamonds and error bars were taken from Barnes *et al.* (2005), the olive dashed curve from Collier Cameron (2007). Orange diamonds and error bars show measurements from Ammler-von Eiff & Reiners (2012). Our measurements are shown as gray dots. The red dash-dotted line and the light blue dashed line show theoretical predictions from Küker & Rüdiger (2011). The black dashed line marks our detection limit. The blue star marks the position of the Sun for comparison. The open black circles represent weighted means of our measurements for different temperature bins. From 3500–6000 K, $\Delta\Omega$ shows only weak dependence on temperature. Above 6000 K the shear strongly increases as reported by Barnes *et al.* (2005) and Collier Cameron (2007). The different behavior of $\Delta\Omega$ in these two temperature regions was supported by theoretical predictions from Küker & Rüdiger (2011) (red and light blue lines).

Furthermore, we see that these trends are in good agreement with theoretical prediction from mean-field theory by Küker & Rüdiger (2011). These authors found that the temperature dependence of $\Delta\Omega$ cannot be represented by a single power law over the whole temperature range but should rather be fitted separately for two groups separated around 6000 K (red dash-dotted and light blue dashed line).

For stars hotter than 7000 K the results should be treated with caution because there might probably be a contamination by long-period pulsators, e.g., γ Dor or extreme δ Scuti stars among our periods. It is challenging to distinguish between pulsations and rotation periods shorter than half a day only from the light curve.

This task will partly be covered in our future work when we tackle the full Kepler time coverage to see how stable the derived periods are over many years. Analyzing the remaining quarters of Kepler data will definitely shrink our sample, but will also provide a condensed set of reliable rotation periods of tens of thousands of stars.

References

Ammler-von Eiff, M. & Reiners, A. 2012, *A&A*, 542, A116

Baliunas, S., Sokoloff, D., & Soon, W. 1996, *ApJ*, 457, L99

Barnes, J. R., Collier Cameron, A., Donati, J.-F., James, D. J., Marsden, S. C., & Petit, P. 2005, *MNRAS*, 357 ,L1

Barnes, S. A. 2007, *ApJ*, 669, 1167

Basri, G., Walkowicz, L. M., Batalha, N., Gilliland, R. L., Jenkins, J., Borucki, W. J., Koch, D., Caldwell, D., Dupree, A. K., Latham, D. W., Meibom, S., Howell, S., & Brown, T. 2010, *ApJ*, 713, L155

Basri, G., Walkowicz, L. M., Batalha, N., Gilliland, R. L., Jenkins, J., Borucki, W. J., Koch, D., Caldwell, D., Dupree, A. K., Latham, D. W., Marcy, G. W., Meibom, S., & Brown, T. 2011, *AJ*, 141, 20

Collier Cameron, A. 2007, *AN*, 328, 1030

Irwin, J., Berta, Z. K., Burke, C. J., Charbonneau, D., Nutzman, P., West, A. A., & Falco, E. E. 2011, *ApJ*, 727, 56

Kiraga, M. & Stepien, K. 2007, *Acta Astron.*, 57, 149

Küker, M. & Rüdiger, G. 2011, *AN*, 332, 933

McQuillan, A., Aigrain, S., & Mazeh, T. 2013, *MNRAS*, 432, 1203

Reinhold, T. & Reiners, A. 2013, *A&A*, 557, A11

Reinhold, T., Reiners, A., & Basri, G. 2013, ArXiv e-prints

Magnetic Fields throughout Stellar Evolution
Proceedings IAU Symposium No. 302, 2013
P. Petit, M. Jardine & H. Spruit, eds.

doi:10.1017/S1743921314002129

Starspots Magnetic field by transit mapping

Adriana Válio[1] and Eduardo Spagiari[2]

[1]CRAAM, Mackenzie University, São Paulo, Brazil, email: avalio@craam.mackenzie.brl

[2]PGEE, Mackenzie University, São Paulo, Brazil, email: eduardo@spagiari.net

Abstract. Sunspots are important signatures of the global solar magnetic field cycle. It is believed that other stars also present these same phenomena. However, today it is not possible to observe directly star spots due to their very small sizes. The method applied here studies star spots by detecting small variations in the stellar light curve during a planetary transit. When the planet passes in front of its host star, there is a chance of it occulting, at least partially, a spot. This allows the determination of the spots physical characteristics, such as size, temperature, and location on the stellar surface. In the case of the Sun, there exists a relation between the magnetic field and the spot temperature. We estimate the magnetic field component along the line-of-sight and the intensity of sunspots using data from the MDI instrument on board of the SOHO satellite. Assuming that the same relation applies to other stars, we estimate spots magnetic fields of CoRoT-2 and Kepler-17 stars.

Keywords. Sunspots, star spots, stellar magnetic field

1. Sunspots

Dark spots observed on the surface of the Sun are important signatures of its global magnetic field. It is believed that other stars also behave the same way. However, currently it is not possible to directly observe spots on the surface of other stars due to their small sizes, which are a fraction of an arc sec. The method proposed by Silva (2003) to study spots on other stars is based on the small variations detected during planetary transits. During a planetary transit, as the planet passes in front of its parent star, there is the possibility it will occult, at least partially, a region darker or brighter than the stellar surface. A detailed analysis of the star light curve may show small variations caused by these occultations. Since the spot is darker than the rest of the stellar disk, an increase in the light intensity during the transit is seen. The fit to these light curve variations yields physical characteristics of the spots, such as the size, temperature and location on the surface of the star. Therefore, these planets can be used as probes to study the characteristics of stellar surface features such as spots. To estimate the starspots magnetic field, we assume that starspots follow the same relation between intensity and magnetic field known for sunspots.

The first step is to obtain an empirical relation between the intensity of sunspots and their magnetic field. The solar data used here were obtained from the MDI instrument on board of the SOHO (Solar and Heliospheric Observatory) satellite. A total of 226 spots were identified within the solar images for the whole year of 2003. Their area and average intensity with respect to the central disk intensity were determined as in Spagiari *et al.* (2012) from the white light images. The magnetic field was estimated from the corresponding magnetograms, both the maximum and minimum magnetic intensity were recorded within the same area used in the white light images. For better precision of the data, and to avoid projection effects, only spots located between longitudes $-40°$ and $40°$ were analyzed, totaling 137 sunspots.

Next, the spot relative intensity is converted to temperature by assuming that both the photosphere and the spots radiate like a blackbody. Thus, the spot temperature is

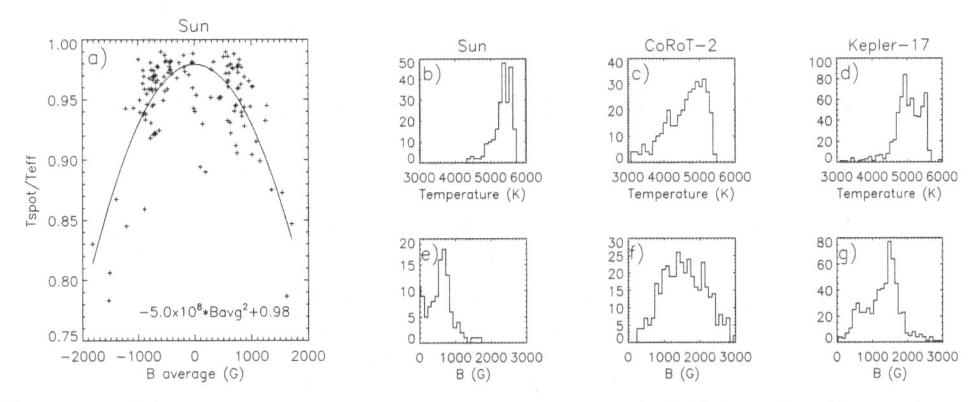

Figure 1. a) Sunspots temperature versus average magnetic field intensity. Temperature of spots on the Sun (b), CoRoT-2 (c), and Kepler-17(d). Maximum magnetic fields of sunspots (e), and spots on CoRoT-2 (f) and Kepler-17 (g) stars.

obtained from the equation $T_{spot} = \left[h\nu \ln\left(\frac{I_c}{I_{spot}}\left(\exp^{h\nu/KT_{eff}} - 1\right) + 1\right)/K\right]^{-1}$, where h and K are Planck and Boltzmann constants, ν is the observing frequency (equivalent to $\lambda = 600$ nm), T_{eff} is the solar effective temperature ($T_{eff} = 5780$ K), I_{spot} is the spot intensity and I_c the maximum intensity of the Sun at disk center.

2. Spots Results and Conclusions

Figure 1a shows the values of sunspots temperature (Figure 1b) as a function of their average magnetic field intensities (Figure 1e). From observation of sunspots, Dicke (1970) obtained the empirical relation: $T_{spot}/T_{eff} = -3.21 \times 10^{-8}B^2 + 0.95$, or $T_{spot}/T_{eff} = \alpha B^2 + \sigma$. Following Dike (1970), quadratic fit of the sunspot data are shown for spots located between longitudes $\pm 40°$ as a solid curve on Figure 1a. The estimated values from the fit on Figure 1a are $\alpha = (-5.0 \pm 0.4) \times 10^{-8} G^{-2}$ and $\sigma = 0.980 \pm 0.003$.

The next step is to use this empirical relation to estimate magnetic fields of stellar spots. Two stars with transiting planets were analyzed: CoRoT-2 ($T_{eff} = 5575$ K) and Kepler-17 ($T_{eff} = 5781$ K). Small variations in the transit light curves of these stars have been fitted and a total of 392 spots for CoRoT-2 (Silva-Valio & Lanza 2011) and 615 spots on Kepler-17 where characterized and their areas and intensities obtained. These spot intensities where then converted to temperatures (Figures 1c and d for the stars CoRoT-2 and Kepler-17, respectively).

Assuming that the decrease in intensity, or temperature, is caused by more intense magnetic fields, and that the spot intensity-magnetic field relation follows that of the Sun, the empirical relation defined above can be inverted to yield the magnetic field for a given relative temperature. Histograms of the derived maximum magnetic field for the stars CoRoT-2 and Kepler-17 are shown in Figures 1f and g, respectively.

The average fields obtained this way for sunspots are 700 ± 350 G, whereas the spots on the stars CoRoT-2 and Kepler-17 have average magnetic field intensities of 1700 ± 700 G and $1400 + 500$ G, respectively. As can be seen from the results, the youngest star CoRoT-2, that is also the most active, has the most intense magnetic fields.

References

Dicke, R. H. 1970, *ApJ*, 159, 25

Silva, A. V. R. 2003, *ApJ*, 585, L147

Silva-Valio, A. & Lanza, A. F. 2011, *A&A*, 529, 36

Spagiari, A. E., Santos, I. F., Costa, W. L. , Valio, A., & Marengoni, M. 2012, in: *Avancos em Visao Computacional* (Omnipax Editor), pp. 345-364.

Magnetic Fields throughout Stellar Evolution
Proceedings IAU Symposium No. 302, 2013
P. Petit, M. Jardine & H. Spruit, eds.

doi:10.1017/S1743921314002130

Investigating magnetic activity of F stars with the *Kepler* mission

S. Mathur[1], R. A. García[2], J. Ballot[3], T. Ceillier[1], D. Salabert[2], T. S. Metcalfe[1], C. Régulo[4], A. Jiménez[4] and S. Bloemen[5]

[1]Space Science Institute, Boulder, CO, USA, email: smathur@spacescience.org
[2]Laboratoire AIM, CEA Saclay, Gif-sur-Yvette, France [3]Institut de Recherche en Astrophysique et Planétologie, Toulouse, France [4]Instituto de Astrofísica de Canarias, Tenerife, Spain [5]Radboud University Nijmegen, The Netherlands

Abstract. The dynamo process is believed to drive the magnetic activity of stars like the Sun that have an outer convection zone. Large spectroscopic surveys showed that there is a relation between the rotation periods and the cycle periods: the longer the rotation period is, the longer the magnetic activity cycle period will be. We present the analysis of F stars observed by *Kepler* for which individual p modes have been measure and with surface rotation periods shorter than 12 days. We defined magnetic indicators and proxies based on photometric observations to help characterise the activity levels of the stars. With the *Kepler* data, we investigate the existence of stars with cycles (regular or not), stars with a modulation that could be related to magnetic activity, and stars that seem to show a flat behaviour.

Keywords. Stellar activity, Seismology, F stars

1. Introduction

Stellar magnetic activity is important to understand the solar magnetic cycle. For stars like the Sun, it results from the interaction between rotation, convection and magnetic field. An important aspect of measuring magnetic activity in other stars is to improve the rotation-age-activity relationship, which would allow us to measure the age of a star by measuring its magnetic activity level and its rotation period in a model-independent way. The last four years, the *Kepler* mission has been providing very good quality data allowing us to probe the structure (e.g. Beck *et al.* 2012) and the dynamics (e.g. Deheuvels *et al.* 2012) of stars using asteroseismology. Asteroseismology has also contributed in a more accurate determination of radius and mass of the exoplanets (Howell *et al.* 2012) and the detection of magnetic activity (García *et al.* 2010). With long and continuous datasets provided by the mission, we now have the opportunity to study the magnetic activity of the stars.

2. Analysis

We used the long-cadence data (sampling of 29.42 min) corrected as described in García *et al.* (2011). We analysed 22 stars of spectral type F that have been observed by the *Kepler* mission for almost 4 years. They were selected based on their effective temperature ($T_{\rm eff} \geqslant 6000$ K) and their rotation period measurements from García *et al.* (in prep.) ($P_{\rm rot} \leqslant 12$ days). The magnetic activity measurement is based on the presence of spots or active regions on the surface of the stars. As the stars rotate, the regular passage of these dark spots produce a modulation in the light curve related to the surface rotation period of the star. We measured the magnetic index, $\langle S_{\rm ph} \rangle$ based on our knowledge of

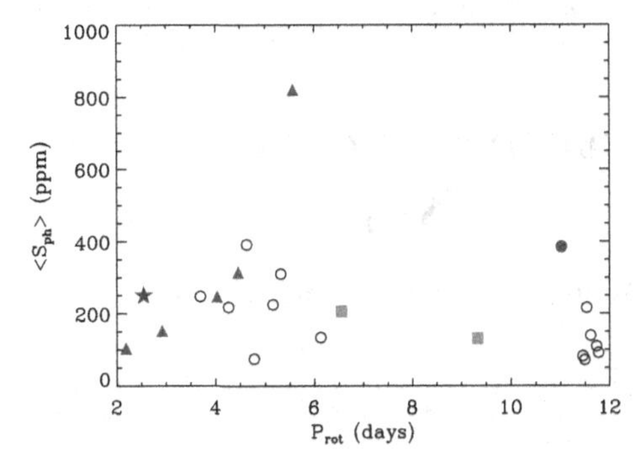

Figure 1. Magnetic index $< S_{\rm ph} >$ as a function of surface rotation period for the 22 stars: long-lived features (triangles), cycle-like (filled circle), trend (squares), and others (open circles). The star represents a star with long-lived features and a cycle-like behaviour.

rotation rates of the stars. Briefly, this is the mean value of standard deviations measured on subseries of length $5 \times P_{\rm rot}$ Mathur *et al.* (2013b). We also performed a time-frequency analysis with the wavelets (Mathur *et al.* 2010) to compute a magnetic proxy. This allowed us to look for signature of magnetic activity cycles.

3. Results

The time-frequency analysis led to different types of magnetic activity in our sample of stars: long-lived features on the surface suggesting the existence of active longitudes, cycle-like behaviours, some trends, and flat behaviours. We also looked for correlation between the magnetic index and other stellar parameters. We show in Fig. 1 $\langle S_{\rm ph} \rangle$ as a function of the rotation period. We do not see any correlation for the whole sample but a hint of correlation for the stars with long-lived features (triangles). But we remind that the index is taken at a given moment in the magnetic cycle of the star and might be biased depending on the observation during a maximum of minimum of magnetic activity. More details can be found in Mathur *et al.* (2013a). The next step will be to run 3D dynamo models of a few of these stars opening the path to our understanding of the dynamics of the stars.

Acknowledgements

SM acknowledges support from the NASA grant NNX12AE17G. T. C., RAG, and SM acknowledge the support of the European Community's Seventh Framework Programme (FP7/2007-2013) under grant agreement no. 269194 (IRSES/ASK).

References

Beck, P. G., Montalban, J., Kallinger, T., *et al.* 2012, *Nature*, 481, 55
Deheuvels, S., García, R. A., Chaplin, W. J., *et al.* 2012, *ApJ*, 756, 19
García, R. A., Hekker, S., Stello, D., *et al.* 2011, *MNRAS*, 414, L6
García, R. A., Mathur, S., Salabert, D., *et al.* 2010, *Science*, 329, 1032
Howell, S. B., Rowe, J. F., Bryson, S. T., *et al.* 2012, *ApJ*, 746, 123
Mathur, S., García, R. A., Ballot, J., *et al.* 2013a, *A&A*, submitted
Mathur, S., García, R. A., Régulo, C., *et al.* 2010, *A&A*, 511, A46
Mathur, S., Salabert, D., García, R. A., & Ceillier, T. 2013b, *J. of Space Weather & Climate*, submitted

Magnetic Fields throughout Stellar Evolution
Proceedings IAU Symposium No. 302, 2013
P. Petit, M. Jardine & H. Spruit, eds.

doi:10.1017/S1743921314002142

Detecting activity cycles of late-type dwarfs in Kepler data

Krisztián Vida and Katalin Oláh

Konkoly Observatory of the Hungarian Academy of Sciences,
H-1121 Budapest, Konkoly Thege M. str. 15–17. Hungary
email: vidakris@konkoly.hu, olah@konkoly.hu

Abstract. Using data of fast-rotating active dwarf stars in the Kepler database, we perform time-frequency analysis of the light curves in order to search for signs of activity cycles. We use the phenomenon that the active region latitudes vary with the cycle (like the solar butterfly diagram), which causes the observed rotation period to change as a consequence of differential rotation. We find cycles in 8 cases of the 39 promising targets with periods between of 300–900 days.

Keywords. stars:activity, stars: low-mass, stars: magnetic fields, stars: rotation, stars: spots

---

1. Introduction

Activity cycles on the Sun and other active stars are known for a long time. The difficulty of searching such cycles is that one needs long-term measurements from a few years to decades, as these are the typical time scales of the phenomenon. For that reason, long-term all-sky surveys seem a good option in the future, given they run for many years. Now, however, the *Kepler* satellite observes the most precise quasi-continuous light curves of about 150 000 targets, operating for a few years. Baliunas *et al.* (1996), and later Oláh & Strassmeier (2002) showed that the cycle length depends on rotation: cycles on fast-rotating stars are known to be shorter. Vida *et al.* (2013) found, that on ultrafast-rotating active stars it is possible to detect activity cycles with lengths of about a year, already within the reach of *Kepler*.

The light curves of *Kepler* are unfortunately not free from instrumental, long-term trends and shorter glitches, and although there are methods intended to correct for these, getting a homogeneous trustworthy dataset spanning for multiple quarters seems very hard, if even possible. Thus, when searching for activity cycles in spottedness simply using photometry, these trends interfere with the actual signal we are looking for.

Beside the change of spottedness, stellar activity cycles have another property – the change of latitude over the cycle, which effect is known on the Sun as the "butterfly diagram". Işık, Schmitt, & Schüssler (2011) showed, that the shape of the butterfly diagram changes fundamentally with the spectral type and rotation period, but on fast-rotating late-type stars the features corresponding to the "wings" of the solar butterfly diagram might be still distinguishable. This effect, together with the differential rotation of the stellar surface (i.e., that the equatorial region rotates faster than the poles), can help us to detect activity cycles. From the butterfly diagram we can learn, that the typical latitude of the active regions change with the cycle: in the case of the Sun, at maximum activity the spots appear up to 30° latitude, while at minimum they appear close to the equator. As a consequence, the differential rotation of the surface causes the typical observable rotation period (periodic light curve modulation as a result of spottedness) to change with the activity cycle. This effect – the observed rotational period – is not

Table 1. Basic parameters of the *Kepler*-targets with detected activity cycles, their determined rotation period, and the cycle period. The "L" sign denotes long-term changes with the possible length of the cycle shown in parentheses. Last line shows the comparison star without detected cyclic changes.

KeplerID	Kp. mag.	T_{eff}	$\log g$	P_{rot} (d)	P_{cyc} (d)
03541346	15.379	4194	4.503	0.9082	500
04953358	15.487	3843	4.608	0.6490	L (800)
05791720	14.067	3533	4.132	0.7651	320
07592990	15.788	4004	4.632	0.4421	480
08314902	15.745	4176	4.480	0.8135	580
10515986	15.592	3668	4.297	0.7462	L (780)
11087527	15.603	4303	4.556	0.4110	300
12365719	15.843	3735	4.473	0.8501	L (820)
10063343	13.164	3976	4.433	0.3326	-

influenced by the long-term trends of the light curves, thus we can use it to detect the cycles also in *Kepler* data.

2. Data and methods

To find objects with sufficiently short activity cycles, similar to the ones in Vida *et al.* (2013), we searched the Kepler Input Catalogue (KIC) for dwarfs ($\log g \approx 4.5$) cooler than 4500 K. We automatically analyzed the Q1 light curves of these 8826 objects to find the main period in the data using the discrete Fourier-transformation option of *TiFrAn* (Time-Frequency Analysis package, see Csubry & Kolláth 2004 and Kolláth & Oláh 2009). Supposing that the largest peak in the Fourier-spectrum corresponds to the rotation periods of the stars, we inspected visually all the short-period ($P_{\text{rot}} \lesssim 1d$) targets to select 39 objects, where the changes in the light curves indicated spottedness. For these targets, all the available data were downloaded (until Q13), and the long-term instrumental changes were removed from the light curves by fitting a third-order polynomial to each quarter.

The resulting light curves were cleaned from extremely outlying points (possibly flares), and were analyzed using the Short-Term Fourier-Transform (STFT) option in *TiFrAn* to find promising targets with possible activity cycles – i.e., where the STFT shows periodic changes.

3. Results and discussion

Of the 39 promising cool dwarf stars with fast rotation, we found signs of an activity cycle in 8 cases. The STFTs and light curves of these targets are shown in Fig. 1. The lengths of the cycles are between 300–900 days (the rotation and cycle periods, and basic stellar parameters are summarized in Table 1). The cycle lengths were determined both by visual inspection and using 1D cross-correlation and Fourier-analysis of the STFSs to give a quantitative result (Table 1 shows the latter values).

Five stars show definite cycles with lengths between 300–580 days (1–1.5 years). The remaining three, denoted with "L", show also systematic period change, but the time-base, about 4 years, is not long enough to find periods with great confidence over 2 years.

In Fig. 2 we plotted our results on the rotation period–cycle length diagram based on Oláh & Strassmeier (2002), showing only those 5 targets, where the length of the cycle is short enough compared to the length of the dataset to reasonably estimate the

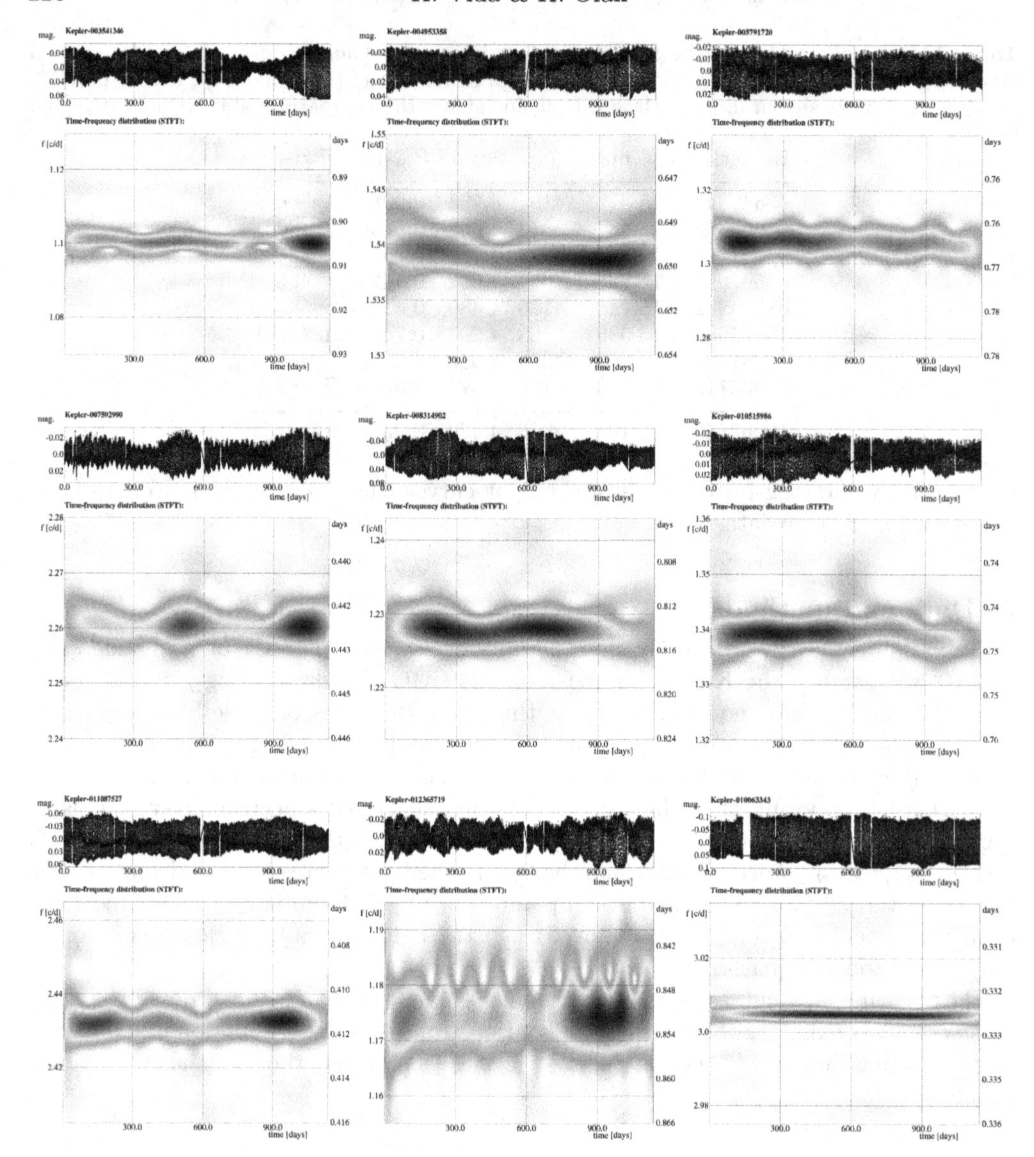

Figure 1. Light curves and their Short-Term Fourier-Transforms of the 8 *Kepler*-targets that possibly have an activity cycle in a range of 400–900 days. The last plot shows a comparison, where no sign of cyclic variation has been found.

cycle length. The stars with definite cycles of about 1–1.5 years are quite similar to the original sample from Vida *et al.* (2013), on which the target search was based on, and fill in the currently unmapped part of the rotation–cycle length diagram. In that paper the following activity cycles were reported: EY Dra ($P_{\mathrm{rot}} = 0.459d$, $P_{\mathrm{cyc}} = 348$d), V405 And ($P_{\mathrm{rot}} = 0.465d$, $P_{\mathrm{cyc}} = 305$d), and GSC 3377-0296 ($P_{\mathrm{rot}} = 0.445d$, $P_{\mathrm{cyc}} = 530$d). Using the activity cycles found in the present paper and those from Vida *et al.* (2013), we can refine the correlation (see Fig. 2). With the targets of the present paper, where the activity cycles were short enough to give a fair estimate for their lengths, for the shortest cycles (several objects, including the Sun show multiple cycles), we get the following

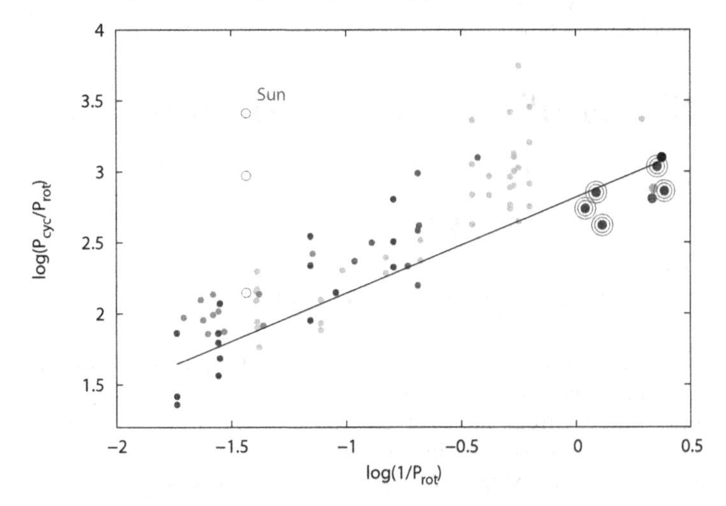

Figure 2. Relation between the rotation period and cycle lengths, bases on the plot of Oláh & Strassmeier (2002). Different colors denote different surveys, the line indicates the fit to the shortest cycle lengths. Encircled points show targets from this paper, where the cycle length was short enough to allow a reasonable estimation.

relation:

$$\log(P_{\rm cyc}/P_{\rm rot}) = 0.68 \log(1/P_{\rm rot}) + 2.82$$

for the sample of both single and binary stars from different surveys. Vida *et al.* (2013) found, that the lengths of the activity cycles for ultrafast-rotating dwarfs seems to be somewhat shorter than the previous samples would indicate. The objects from Vida *et al.* (2013) served as template to search for interesting *Kepler* targets, those results were the base of the present study.

At present we have altogether 8 stars with rotational periods between 0.4–1.0 days showing activity cycles of 1–1.5 years. As seen from Fig. 2, these objects continue the trend showing shorter cycle lengths when rotating faster.

Acknowledgement

The financial support of the OTKA grant K-81421, is acknowledged. This work was also supported by the "Lendület-2009", and "Lendület-2012" Young Researchers' Programs of the Hungarian Academy of Sciences, and by the HUMAN MB08C 81013 grant of the MAG Zrt. Funding for the *Kepler* mission is provided by NASA's Science Mission Directorate.

References

Baliunas, S. L., Nesme-Ribes, E., Sokoloff, D., & Soon, W. H., *Astrophysical Journal v.460*, 1996, 460, 848

Csubry, Z. & Kolláth, Z., 2006 *ASP Conference Series*, 349, 215C

Işık, E., Schmitt, D., & Schüssler, M. 2011, *Astronomy and Astrophysics*, 528, A135

Kolláth, Z. & Oláh, K., *Astronomy and Astrophysics*, 2009, 501, 695–702

Oláh, K. & Strassmeier, K. G. 2002, *Astronomische Nachrichten*, 323, 361

Vida, K., Kriskovics, L., & Oláh, K. *Astronomische Nachrichten*, accepted

Discussion

Magnetic Fields throughout Stellar Evolution
Proceedings IAU Symposium No. 302, 2013
P. Petit, M. Jardine & H. Spruit, eds.

doi:10.1017/S1743921314002154

The effects of stellar winds and magnetic fields on exoplanets

A. A. Vidotto

SUPA, University of St Andrews, North Haugh, KY16 9SS, UK
email: Aline.Vidotto@st-andrews.ac.uk

Abstract. The great majority of exoplanets discovered so far are orbiting cool, low-mass stars whose properties are relatively similar to the Sun. However, the stellar magnetism of these stars can be significantly different from the solar one, both in topology and intensity. In addition, due to the present-day technology used in exoplanetary searches, most of the currently known exoplanets are found orbiting at extremely close distances to their host stars (< 0.1 au). The dramatic differences in stellar magnetism and orbital radius can make the interplanetary medium of exoplanetary systems remarkably distinct from that of the Solar System. To constrain interactions between exoplanets and their host-star's magnetised winds and to characterise the interplanetary medium that surrounds exoplanets, more realistic stellar wind models, which account for factors such as stellar rotation and the complex stellar magnetic field configurations of cool stars, must be employed. Here, I briefly review the latest progress made in data-driven modelling of magnetised stellar winds. I also show that the interaction of the stellar winds with exoplanets can lead to several observable signatures, some of which that are absent in our own Solar System.

Keywords. MHD, stars: magnetic fields, stars: mass loss, stars: planetary systems, stars: rotation, stars: winds, outflows

1. Introduction

Stars experience mass loss in the form of winds during their lives and, as the star evolves, its wind changes characteristics. For some stars, at certain evolutionary phases, the amount of mass lost through stellar winds can amount to a significant portion of its own mass. This is the case, for example, of supergiant stars, which present massive winds that are believed to be mainly driven by radiation pressure on dust grains (for the red supergiants) or on spectral lines (for the blue supergiants). On the other hand, for cool stars at the main sequence phase, we believe their winds are significantly less massive, carrying only a small amount of mass compared to the total mass of the star. Although these winds are quite rarefied, they carry a significant amount of angular momentum, which affects the stellar rotational evolution (e.g., Bouvier *et al.* 1997).

P-Cygni profiles, the traditional mass-loss signatures observed in the denser winds of giant and supergiant stars, are not formed in the rarefied winds of cool dwarf stars. To detect these winds, other indirect methods have been proposed (e.g., Lim and White 1996; Wood *et al.* 2001; Wargelin and Drake 2002). So far, the method developed by Wood *et al.* (2001) has been the most successful one, enabling estimates of mass-loss rates for about a dozen cool dwarf stars.

Due to the lack of observational constraints on winds of cool low-mass stars, many works assume these winds to be identical to (or a scaled version of) the solar wind. In the next Section, I highlight a few observed properties of cool dwarf stars that suggest that their winds can actually be significantly different from the solar wind.

1.1. *What can we infer from winds of cool stars?*

1.1.1. *Temperatures*

We believe the solar coronal heating is ultimately caused by the Sun's magnetism, although it is still debatable which mechanism heats the solar corona. In the stellar case, coronal heating is much less understood. Stellar coronae can be much hotter than the solar corona. The temperature of the X-ray emitting coronae of solar analogs can exceed 10 MK (Guedel 2004), more than one order of magnitude larger than the solar coronal temperature of ~ 1 MK. If the temperature of the X-ray emitting (closed) corona is related to the temperature of the stellar wind (flowing along open field lines), as one would naively expect, then we may expect stellar winds of cool dwarf stars to have temperatures that could be much larger than the solar wind temperature. Because the terminal velocities of thermally-driven winds are very sensitive to the choice of temperature of the stellar wind (Parker 1958), the winds of cool dwarf stars might have a variety of terminal velocities.

1.1.2. *Magnetism*

Thanks to our privileged position immersed in the solar wind, we have access to a great quantity of data that allow a detailed understanding of the physics that is operating in the Sun. Measurements of mass flows and magnetic field of the solar wind have revealed an asymmetric solar wind (McComas *et al.* 1995; Suess and Smith 1996; Jones *et al.* 1998; Wilhelm 2006). In particular, Ulysses measurements showed that the solar wind structure depends on the characteristics of the solar magnetic field (McComas *et al.* 2008). The solar wind characteristics change along the solar cycle, presenting a simple bimodal structure of fast and slow flows during solar minimum, when the geometry of the solar magnetic field is closest to that of an aligned dipole. At solar maximum, when the the axis of the large-scale dipole becomes nearly perpendicular to the solar axis of rotation, the solar wind shows a more complex structure.

Although the richness of details of the magnetic field configuration is only known for our closest star, modern techniques have made it possible to reconstruct the large-scale surface magnetic fields of other stars. The Zeeman-Doppler Imaging (ZDI) technique is a tomographic imaging technique (Donati and Brown 1997) that allows us to reconstruct the large-scale magnetic field (intensity and orientation) at the surface of the star from a series of circular polarisation spectra. This method has now been used to investigate the magnetic topology of stars of different spectral types (Donati and Landstreet 2009) and has revealed fascinating differences between the magnetic fields of different stars. For example, solar-type stars that rotate about two times faster than our Sun show the presence of substantial toroidal component of magnetic field, a component that is almost non-existent in the solar magnetic field (Petit *et al.* 2008). The magnetic topology of low-mass ($< 0.5\ M_\odot$) very active stars seem to be dictated by interior structure changes: while partly convective stars possess a weak non-axisymmetric field with a significant toroidal component, fully convective ones exhibit strong poloidal axisymmetric dipole-like topologies (Morin *et al.* 2008; Donati *et al.* 2008a).

1.1.3. *Rotation*

In addition to magnetic field characteristics, the symmetry of a stellar wind also depends on the rotation rate of the star. In the presence of fast rotation, a magnetised wind can lose spherical symmetry, as centrifugal forces become more important with the increase of rotation rate (Washimi and Shibata 1993). The distribution of rotation periods in cool main-sequence stars is very broad, ranging from stars rotating faster than once per day to indefinitely long periods.

1.2. *Summary*

The variety of observed rotation rates, intensities and topologies of the magnetic fields of cool, dwarf stars indicate that their winds might come in different flavours and might be significantly different from the solar one. One might also bear in mind that, similarly to the Sun, low-mass stars are also believed to host magnetic and activity cycles (see Fares contribution, this volume), which imply that the characteristics of their winds can also vary in a time scale of the cycle periods.

2. Models and simulations of winds of cool stars

There are two most commonly used ways to model stellar winds, which I summarise below.

(*a*) One approach consists of computing the detailed energetics of the wind, starting from the photosphere, passing through the chromosphere, until it reaches the stellar corona (e.g., Hollweg 1973; Holzer *et al.* 1983; Hartmann and MacGregor 1980; Jatenco-Pereira and Opher 1989; Vidotto and Jatenco-Pereira 2006; Falceta-Gonçalves *et al.* 2006; Cranmer 2008; Cranmer and Saar 2011; Suzuki *et al.* 2012). Depending on the physics that is included in such models, this approach can become computationally expensive. As a result, it has been limited to analytical, one- and two-dimensional solutions. In addition, it is also usually focused in the inner most part of the corona and usually only adopts simplified magnetic field topologies.

(*b*) The second approach commonly used to model winds of cool stars consists of adopting a simplified energy equation, usually assuming the wind to be isothermal or described by a polytropic equation of state (in which thermal pressure p is related to density ρ as $p \propto \rho^{\gamma}$). In this case, one is allowed to perform multi-dimensional numerical simulations of stellar winds and can incorporate more complex magnetic field topologies (e.g., Mestel 1968; Pneuman and Kopp 1971; Tsinganos and Low 1989; Washimi and Shibata 1993; Keppens and Goedbloed 2000; Lima *et al.* 2001; Vidotto *et al.* 2009b,a, 2010b; Pinto *et al.* 2011; Jardine *et al.* 2013)

Because of the simplified energetics that are considered in the second approach, its domain can extend considerably farther out than the first one. As a result, polytropic winds can be useful in the characterisation of the interplanetary medium and also to characterise interactions between exoplanets and the winds of their host-stars. In the next Section, I present how to make more realistic stellar wind simulations by incorporating recent insights acquired on the magnetic topology of different stars into stellar wind models.

2.1. *Data-driven wind simulations*

To illustrate how observationally reconstructed surface maps (see Section 1.1.2) can be incorporated in the simulations of stellar winds, I will present the work done in Vidotto *et al.* (2012), where we performed numerical simulations of the stellar wind of the planet-hosting star τ Boo (spectral type F7V). τ Boo is a remarkable object, not only because it hosts a giant planet orbiting very close to the star (located at 0.046 au from the star), but also because it is the only star other than the Sun for which a full magnetic cycle has been reported in the literature (Donati *et al.* 2008b; Fares *et al.* 2009, 2013). These observations suggest that τ Boo undergoes magnetic cycles similar to the Sun, but with

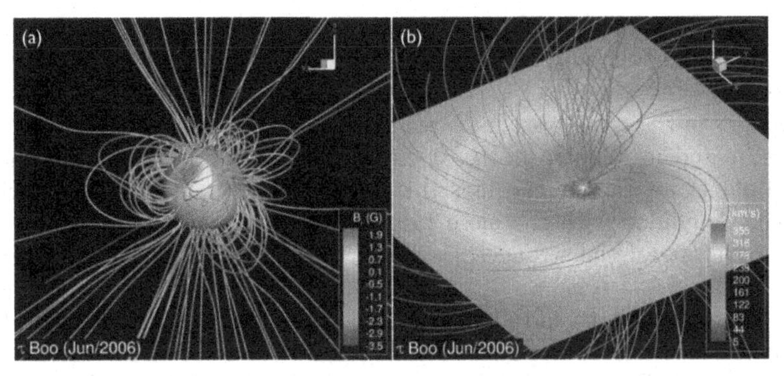

Figure 1. (a) Potential field extrapolation that is used as initial condition for the stellar wind simulations. The distribution of the observationally reconstructed surface magnetic field (used as boundary condition) is shown colour-coded. (b) Self-consistent solution of the magnetised stellar wind after the solution relaxed in the grid. The wind velocity is shown at the equatorial plane. Figure adapted from Vidotto (2013).

a cycle period that is about one order of magnitude smaller than the solar one (about 2 years as opposed to 22 years for the solar magnetic cycle).

The surface magnetic maps of τ Boo reconstructed by Catala *et al.* (2007), Donati *et al.* (2008b) and Fares *et al.* (2009) were used as boundary conditions for the stellar wind simulations. A potential field extrapolation is adopted at the initial state of the simulation (Fig. 1a). As the simulation evolves in time, stellar wind particles and magnetic field lines are allowed to interact with each other. Figure 1b shows the self-consistent solution found for the magnetic field lines, after the wind solution relaxed in the grid. Note that the magnetic field lines become stressed, wraping around the rotational axis of the star (pointing towards positive z). Colour-coded are the reconstructed large-scale surface field of τ Boo at Jun/2006 (Fig. 1a) and the wind velocity at the equatorial plane of the star (Fig. 1b).

Vidotto *et al.* (2012) found that variations of the stellar magnetic field during the cycle directly influence the outflowing wind. Therefore, the rapid variation of the large-scale magnetic field of τ Boo implies that the environment surrounding the close-in planet should be varying quite rapidly. In addition, Vidotto *et al.* (2012) estimated the mass-loss rate ($\dot{M}$) of τ Boo, showing that this star seems to have a denser wind than that of the Sun, with $\dot{M}$ that are 2 orders of magnitude larger than the solar value $\dot{M}_\odot$ ($\dot{M} \approx 135\ \dot{M}_\odot$).

3. Interaction between stellar winds and exoplanets

When the wind outflows from the star, it permeates the entire extrasolar system, interacting with any body that it encounters on its way. The interaction between stellar winds and exoplanets can lead to observable signatures, some of which are absent in our own solar system.

3.1. *Planetary radio emission*

In the solar system, the giant planets and the Earth emit at radio wavelengths. Such planetary radio emission is due to the interaction between the magnetic planets and the solar wind, and its power is proportional to the stellar wind energy dissipated in the wind-planet interaction. By analogy to what is observed in the solar system, it is expected that exoplanets also interact with the winds of their host stars and should, therefore, also generate radio emission. Because the energy dissipated by the stellar wind is larger at closer distances to the star (because the wind density and magnetic fields are significantly larger than further out from the star), radio emission of close-in planets, such as τ Boo b, is expected to be several orders of magnitude larger than the emission from the planets in the solar system (Zarka 2007). In addition, the relatively dense wind of τ Boo estimated in Vidotto *et al.* (2012) implies that the energy dissipated in the stellar wind-planet interaction can be significantly higher than the values derived in the solar system. Combined with the close proximity of the system ($\sim$ 16 pc), the τ Boo system has been one of the strongest candidates to verify exoplanetary radio emission predictions.

Using the detailed stellar wind model developed for its host-star, Vidotto *et al.* (2012) estimated radio emission from τ Boo b, exploring different values for the *assumed* planetary magnetic field. For example, they showed that, for a planet with a magnetic field similar to Jupiter's ($\simeq$ 14 G), the radio flux is estimated to be $\simeq 0.5-1$ mJy, occurring at an emission frequency of $\simeq$ 34 MHz. Although small, this emission frequency lies in the observable range of current instruments, such as LOFAR. To observe such a small flux, an instrument with a sensitivity lying at a mJy level is required. The same estimate was done considering the planet has a magnetic field similar to the Earth ($\simeq$ 1 G). Although the radio flux does not present a significant difference to what was found for the previous case, the emission frequency ($\simeq$ 2 MHz) falls at a range below the ionospheric cut-off, preventing its possible detection from the ground. In fact, due to the ionospheric cutoff at $\sim$ 10 MHz, radio detection with ground-based observations from planets with magnetic field intensities $\lesssim$ 4 G should not be possible (Vidotto *et al.* 2012).

3.2. *Bow shock signatures in transit observations*

Despite many attempts, radio emission from exoplanets has not been detected so far (e.g., Smith *et al.* 2009; Lazio *et al.* 2010; Lecavelier des Etangs *et al.* 2013). Such a detection would not only constrain local characteristics of the stellar wind, but would also demonstrate that exoplanets are magnetised. Fortunately, there may be other ways to probe exoplanetary magnetic fields, in particular for transiting systems, through signatures of bow shocks during transit observations.

Based on Hubble Space Telescope/Cosmic Origins Spectrograph (HST/COS) observations using the narrow-band near-UV spectroscopy, Fossati *et al.* (2010b) showed that the transit lightcurve of the close-in giant planet WASP-12b presents both an early ingress when compared to its optical transit, as well as excess absorption during the transit, indicating the presence of an asymmetric distribution of material surrounding the planet. Motivated by these transit observations, Vidotto *et al.* (2010a) suggested that the presence of bow shocks surrounding close-in planets might lead to transit asymmetries at certain wavelengths, such as the one observed in WASP-12b at near-UV wavelengths.

The main difference between bow shocks formed around exoplanets and the ones formed around planets in the solar system is the shock orientation, determined by the net velocity of the particles impacting on the planet's magnetosphere. In the case of the Earth, the solar wind has essentially only a radial component, which is much larger than the orbital velocity of the Earth. Because of that, the bow shock surrounding the Earth's magnetosphere forms facing the Sun. However, for close-in exoplanets that possess high orbital velocities and are frequently located at regions where the host star's wind velocity is comparatively much smaller, a shock may develop ahead of the planet. In general, we expect that shocks are formed at intermediate angles (see also Vidotto *et al.* 2011b; Llama *et al.* 2013).

Due to their high orbital velocities, close-in planets offer the best conditions for transit observations of bow shocks. If the compressed shocked material is able to absorb stellar

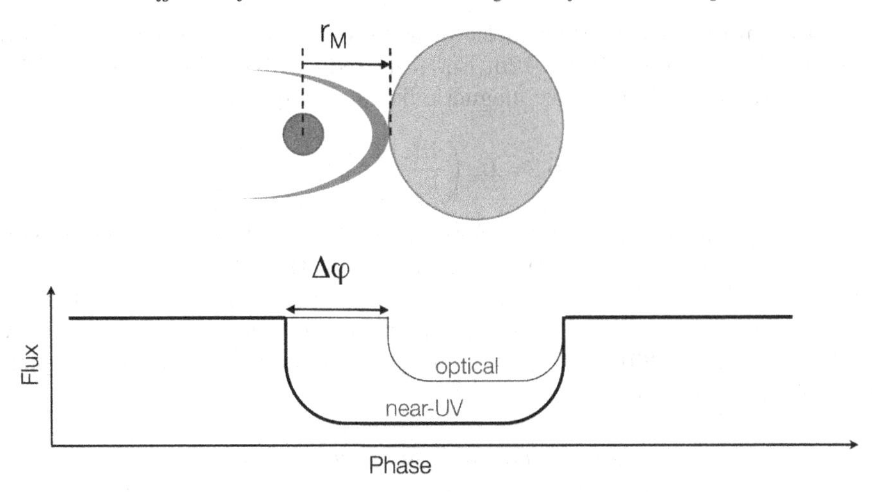

Figure 2. Sketch of the light curves obtained through observations in the optical and in certain near-UV lines, where the bow shock surrounding the planet's magnetosphere is also able to absorb stellar radiation. The stand-off distance from the shock to the centre of the planet is assumed to trace the extent of the planetary magnetosphere r_M, which can be estimated by measuring the difference between the phases ($\Delta\varphi$) at which the near-UV and the optical transits begin.

radiation, then the signature of bow shocks may be observed through both a deeper transit and an early-ingress in some spectral lines with respect to the broadband optical ingress (Vidotto *et al.* 2010a). The sketch shown in Figure 2 illustrates this idea.

In the case of WASP-12b, Vidotto *et al.* (2010a) suggested that the shocked material, which is able to absorb enough stellar radiation in the near-UV, can cause an asymmetry in the lightcurve observed (see lightcurve sketches in Figure 2), where the presence of compressed material ahead of the planetary orbit causes an early ingress, while the lack of compressed material behind the planetary orbit causes simultaneous egresses both in the near-UV transit as well as in the optical one. This suggestion was verified by Llama *et al.* (2011), who performed Monte Carlo radiation transfer simulations of the near-UV transit of WASP-12b. They confirmed that the presence of a bow shock indeed breaks the symmetry of the transit lightcurve

3.3. *Planetary magnetic fields: a new detection method?*

An interesting outcome of the observations of bow shocks around exoplanets is that it permits one to infer the magnetic field intensity of the transiting planet. By measuring the difference between the phases at which the near-UV and the optical transits begin ($\Delta\varphi$ illustrated in Figure 2), one can derive the stand-off distance from the shock to the centre of the planet, which is assumed to trace the extent of the planetary magnetosphere r_M. At the magnetopause, pressure balance between the coronal total pressure and the planet total pressure requires that

$$\rho_c \Delta u^2 + \frac{[B_c(a)]^2}{8\pi} + p_c = \frac{[B_p(r_M)]^2}{8\pi} + p_p, \tag{3.1}$$

where ρ_c, p_c and $B_c(a)$ are the local coronal mass density, thermal pressure, and magnetic field intensity at orbital radius a, and p_p and $B_p(r_M)$ are the planet thermal pressure and magnetic field intensity at r_M. The relative velocity between the material surrounding the planet and the planet itself is Δu. In the case of a magnetised planet with a magnetosphere of a few planetary radii, the planet total pressure is usually dominated by the contribution from the planetary magnetic pressure (i.e., $p_p \sim 0$). Vidotto *et al.* (2010a) showed that, in the case of WASP-12b, Eq. (3.1) reduces to $B_c(a) \simeq B_p(r_M)$. Further assuming that stellar and planetary magnetic fields are dipolar, we have

$$B_p = B_\star \left(\frac{R_\star / a}{R_p / r_M} \right)^3 , \tag{3.2}$$

where $B_\star$ and B_p are the magnetic field intensities at the stellar and planetary surfaces, respectively. Eq. (3.2) shows that the planetary magnetic field can be derived directly from *observed* quantities. For WASP-12, using the upper limit of $B_\star < 10$ G (Fossati *et al.* 2010a) and the stand-off distance obtained from the near-UV transit observation $r_M = 4.2\ R_p$ (Lai *et al.* 2010), we predicted an upper limit for WASP-12b's planetary magnetic field of $B_p < 24$ G.

3.4. *Searching for magnetic fields in other exoplanets*

In theory, the suggestion that through transit observations one can probe the planetary magnetic field is quite straightforward - all it requires is a measurement of the transit ingress phase in the near-UV. In practice, however, acquisition of near-UV transit data requires the use of space-borne facilities, making follow-ups and new target detections rather difficult.

In order to optimise target selection, Vidotto *et al.* (2011a) presented a classification of the known transiting systems according to their potential for producing shocks that could cause observable light curve asymmetries. The main considered assumption was that, once the conditions for shock formation are met, planetary shocks absorb in certain near-UV lines, in a similar way as WASP-12b. In addition, for it to be detected, the shock must compress the local plasma to a density sufficiently high to cause an observable level of optical depth. This last hypothesis requires the knowledge of the local ambient medium that surrounds the planet.

By adopting simplified hypotheses, namely that up to the planetary orbit the stellar corona can be treated as in hydrostatic equilibrium and isothermal, Vidotto *et al.* (2011a) predicted the characteristics of the ambient medium that surrounds the planet for a sample of 125 transiting systems, and discussed whether such characteristics present favourable conditions for the presence and detection of a bow shock. Excluding systems that are quite far ($\gtrsim 400$ pc), the planets that were top ranked are: WASP-19b, WASP-4b, WASP-18b, CoRoT-7b, HAT-P-7b, CoRoT-1b, TrES-3, and WASP-5b.

4. Conclusion

As the wind outflows from the star, it permeates the interplanetary medium, interacting with any planet encountered on its way. The proper characterisation of stellar winds is therefore crucial to constrain interactions between exoplanets and their surrounding environments and also essential for the study of space weather events on exoplanets. Stellar winds are affect by the stellar rotation, magnetism and coronal temperature, properties that can vary significantly from star to star. As a consequence, stellar winds of cool stars can actually be significantly different from the solar wind. In this talk, I illustrated how one can take an extra step towards more realistic models of stellar winds of low-mass stars by incorporating observationally reconstructed surface magnetic maps into simulations of stellar winds. I also showed that dramatic differences in stellar magnetism and orbital radius can make the interplanetary medium of exoplanetary systems remarkably distinct from the one present in the solar system. In addition, I showed that the interaction of the stellar winds with exoplanets can lead to observable signatures that are absent in our own solar system.

Acknowledgements

AAV acknowledges support from a Royal Astronomical Society Fellowship and thanks the IAU for a travel support grant to attend the Symposium.

References

Bouvier, J., Forestini, M., & Allain, S., 1997, *A&A* **326**, 1023

Catala, C., Donati, J.-F., Shkolnik, E., Bohlender, D., & Alecian, E., 2007, *MNRAS* **374**, L42

Cranmer, S. R., 2008, *ApJ* **689**, 316

Cranmer, S. R. & Saar, S. H., 2011, *ApJ* **741**, 54

Donati, J. & Landstreet, J. D., 2009, *ARA&A* **47**, 333

Donati, J., Morin, J., Petit, P., Delfosse, X., Forveille, T., Aurière, M., Cabanac, R., Dintrans, B., Fares, R., Gastine, T., Jardine, M. M., Lignières, F., Paletou, F., Velez, J. C. R., & Théado, S., 2008a, *MNRAS* **390**, 545

Donati, J.-F. & Brown, S. F., 1997, *A&A* **326**, 1135

Donati, J.-F., Moutou, C., Farès, R., Bohlender, D., Catala, C., Deleuil, M., Shkolnik, E., Collier Cameron, A., Jardine, M. M., & Walker, G. A. H., 2008b, *MNRAS* **385**, 1179

Falceta-Gonçalves, D., Vidotto, A. A., & Jatenco-Pereira, V., 2006, *MNRAS* **368**, 1145

Fares, R., Donati, J., Moutou, C., Bohlender, D., Catala, C., Deleuil, M., Shkolnik, E., Cameron, A. C., Jardine, M. M., & Walker, G. A. H., 2009, *MNRAS* **398**, 1383

Fares, R., Moutou, C., Donati, J.-F., Catala, C., Shkolnik, E., Jardine, M., Cameron, A., & Deleuil, M., 2013, *MNRAS* in press (arXiv:1307.6091)

Fossati, L., Bagnulo, S., Elmasli, A., Haswell, C. A., Holmes, S., Kochukhov, O., Shkolnik, E. L., Shulyak, D. V., Bohlender, D., Albayrak, B., Froning, C., & Hebb, L., 2010a, *ApJ* **720**, 872

Fossati, L., Haswell, C. A., Froning, C. S., Hebb, L., Holmes, S., Kolb, U., Helling, C., Carter, A., Wheatley, P., Cameron, A. C., Loeillet, B., Pollacco, D., Street, R., Stempels, H. C., Simpson, E., Udry, S., Joshi, Y. C., West, R. G., Skillen, I., & Wilson, D., 2010b, *ApJ* **714**, L222

Guedel, M., 2004, *A&A Rev.* **12**, 71

Hartmann, L. & MacGregor, K. B., 1980, *ApJ* **242**, 260

Hollweg, J. V., 1973, *ApJ* **181**, 547

Holzer, T. E., Fla, T., & Leer, E., 1983, *ApJ* **275**, 808

Jardine, M., Vidotto, A. A., van Ballegooijen, A., Donati, J.-F., Morin, J., Fares, R., & Gombosi, T. I., 2013, *MNRAS* **431**, 528

Jatenco-Pereira, V. & Opher, R., 1989, *A&A* **209**, 327

Jones, G. H., Balogh, A., & Forsyth, R. J., 1998, *Geophys. Res. Lett.* **25**, 3109

Keppens, R. & Goedbloed, J. P., 2000, *ApJ* **530**, 1036

Lai, D., Helling, C., & van den Heuvel, E. P. J., 2010, *ApJ* **721**, 923

Lazio, T. J. W., Carmichael, S., Clark, J., Elkins, E., Gudmundsen, P., Mott, Z., Szwajkowski, M., & Hennig, L. A., 2010, *AJ* **139**, 96

Lecavelier des Etangs, A., Sirothia, S. K., & Gopal-Krishna, and Zarka, P., 2013, *A&A* **552**, A65

Lim, J. & White, S. M., 1996, *ApJ* **462**, L91

Lima, J. J. G., Priest, E. R., & Tsinganos, K., 2001, *A&A* **371**, 240

Llama, J., Vidotto, A. A., Jardine, M., Wood, K., Fares, R., & Gombosi, T. I., 2013, *MNRAS*, in press (arXiv: 1309.2938)

Llama, J., Wood, K., Jardine, M., Vidotto, A. A., Helling, C., Fossati, L., & Haswell, C. A., 2011, *MNRAS* **416**, L41

McComas, D. J., Barraclough, B. L., Gosling, J. T., Hammond, C. M., Phillips, J. L., Neugebauer, M., Balogh, A., & Forsyth, R. J., 1995, *J. Geophys. Res.* **100**, 19893

McComas, D. J., Ebert, R. W., Elliott, H. A., Goldstein, B. E., Gosling, J. T., Schwadron, N. A., & Skoug, R. M., 2008, *Geophys. Res. Lett.* **35**, 18103

Mestel, L., 1968, *MNRAS* **138**, 359

Morin, J., Donati, J., Petit, P., Delfosse, X., Forveille, T., Albert, L., Aurière, M., Cabanac, R., Dintrans, B., Fares, R., Gastine, T., Jardine, M. M., Lignières, F., Paletou, F., Ramirez Velez, J. C., & Théado, S., 2008, *MNRAS* **390**, 567

Parker, E. N., 1958, *ApJ* **128**, 664

Petit, P., Dintrans, B., Solanki, S. K., Donati, J.-F., Aurière, M., Lignières, F., Morin, J., Paletou, F., Ramirez Velez, J., Catala, C., & Fares, R., 2008, *MNRAS* **388**, 80

Pinto, R. F., Brun, A. S., Jouve, L., & Grappin, R., 2011, *ApJ* **737**, 72

Pneuman, G. W. & Kopp, R. A., 1971, *Sol. Phys.* **18**, 258

Smith, A. M. S., Collier Cameron, A., Greaves, J., Jardine, M., Langston, G., & Backer, D., 2009, *MNRAS* **395**, 335

Suess, S. T. & Smith, E. J., 1996, *Geophys. Res. Lett.* **23**, 3267

Suzuki, T. K., Imada, S., Kataoka, R., Kato, Y., Matsumoto, T., Miyahara, H., & Tsuneta, S., 2012, *PASJ* in press (arXiv:1212.6713)

Tsinganos, K. & Low, B. C., 1989, *ApJ* **342**, 1028

Vidotto, A., 2013, *Astronomy and Geophysics* **54(1)**, 010001

Vidotto, A. A., Fares, R., Jardine, M., Donati, J.-F., Opher, M., Moutou, C., Catala, C., & Gombosi, T. I., 2012, *MNRAS* **423**, 3285

Vidotto, A. A., Jardine, M., & Helling, C., 2010a, *ApJ* **722**, L168

Vidotto, A. A., Jardine, M., & Helling, C., 2011a, *MNRAS* **411**, L46

Vidotto, A. A., Jardine, M., & Helling, C., 2011b, *MNRAS* **414**, 1573

Vidotto, A. A. & Jatenco-Pereira, V., 2006, *ApJ* **639**, 416

Vidotto, A. A., Opher, M., Jatenco-Pereira, V., & Gombosi, T. I., 2009a, *ApJ* **703**, 1734

Vidotto, A. A., Opher, M., Jatenco-Pereira, V., & Gombosi, T. I., 2009b, *ApJ* **699**, 441

Vidotto, A. A., Opher, M., Jatenco-Pereira, V., & Gombosi, T. I., 2010b, *ApJ* **720**, 1262

Wargelin, B. J. & Drake, J. J., 2002, *ApJ* **578**, 503

Washimi, H. & Shibata, S., 1993, *MNRAS* **262**, 936

Wilhelm, K., 2006, *A&A* **455**, 697

Wood, B. E., Linsky, J. L., Müller, H., & Zank, G. P., 2001, *ApJ* **547**, L49

Zarka, P., 2007, *Planet. Space Sci.* **55**, 598

Magnetic Fields throughout Stellar Evolution
Proceedings IAU Symposium No. 302, 2013
P. Petit, M. Jardine & H. Spruit, eds.

doi:10.1017/S1743921314002166

Planetary protection in the extreme environments of low-mass stars

A. A. Vidotto[1], M. Jardine[1], J. Morin[2], J.-F. Donati[3], P. Lang[1] and A. J. B. Russell[4]

[1]SUPA, University of St Andrews, North Haugh, KY16 9SS, UK
email: Aline.Vidotto@st-andrews.ac.uk

[2]Georg-August-Universität, Friedrich-Hund-Platz 1, D-37077, Goettingen, Germany

[3]Observatoire Midi-Pirénées, 14 Av. E. Belin, F-31400, Toulouse, France

[4]SUPA, University of Glasgow, University Avenue, G12 8QQ, Glasgow, UK

Abstract. Recent results showed that the magnetic field of M-dwarf (dM) stars, currently the main targets in searches for terrestrial planets, is very different from the solar one, both in topology as well as in intensity. In particular, the magnetised environment surrounding a planet orbiting in the habitable zone (HZ) of dM stars can differ substantially to the one encountered around the Earth. These extreme magnetic fields can compress planetary magnetospheres to such an extent that a significant fraction of the planet's atmosphere may be exposed to erosion by the stellar wind. Using observed surface magnetic maps for a sample of 15 dM stars, we investigate the minimum degree of planetary magnetospheric compression caused by the intense stellar magnetic fields. We show that hypothetical Earth-like planets with similar terrestrial magnetisation ($\sim$1 G) orbiting at the inner (outer) edge of the HZ of these stars would present magnetospheres that extend at most up to 6.1 (11.7) planetary radii. To be able to sustain an Earth-sized magnetosphere, the terrestrial planet would either need to orbit significantly farther out than the traditional limits of the HZ; or else, if it were orbiting within the life-bearing region, it would require a minimum magnetic field ranging from a few G to up to a few thousand G.

Keywords. stars: magnetic fields, stars: planetary systems, stars: rotation, astrobiology

---

Due to technologies currently adopted in exoplanet searches, dM stars have been the main targets in searches for terrestrial planets. For these stars, the orbital region where a planet should be able to retain liquid water at its surface (known as the habitable zone, HZ) is located significantly closer than the HZ of solar-type stars (Kasting *et al.* 1993). These factors make dM stars the prime targets for detecting terrestrial planets in the potentially life-bearing region around the star.

However, in addition to the retention of liquid water, other factors may be important in assessing the potential for a planet to harbour life. For example, the presence of a relatively strong planetary magnetic field is very likely to play a significant role in planetary habitability. A relatively extended planetary magnetosphere can deflect the stellar wind and other ejecta, protecting the planetary atmosphere against erosion.

In steady state, the extent of a planet's magnetosphere is determined by force balance at the boundary between the stellar coronal plasma and the planetary plasma. For the planets in the solar system, this is often reduced to a pressure balance at the dayside, the most significant contribution to the external (stellar) wind pressure being the solar wind ram pressure. However, for planets orbiting stars that are significantly more magnetised than the Sun or/and are located at close distances, the stellar magnetic pressure may play an important role in setting the magnetospheric limits (Ip *et al.* 2004; Lanza 2009; Vidotto *et al.* 2009, 2010, 2012, 2011).

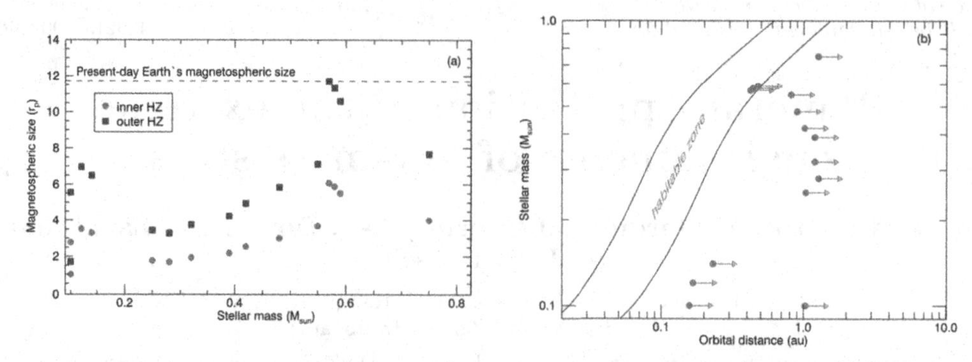

Figure 1. (a) The minimum degree of planetary magnetospheric compression caused by the intense stellar magnetic fields. (b) Closest orbital distance at which an Earth-like planet orbiting the stars in our sample would be able to sustain the present-day Earth's magnetospheric size, assuming it has the same magnetic field as the Earth. Adapted from Vidotto *et al.* (2013).

In the case of dM stars, their magnetic fields is very different from the solar one (in topology and intensity). Therefore, the magnetised environment surrounding a planet orbiting in the HZ of dM stars can differ substantially to the one encountered around the Earth. In the present work, we quantitatively evaluate the sizes of planetary magnetospheres resulting from the pressure exerted by the intense stellar magnetic fields found around dM stars. Our approach only invokes a stellar magnetic field, neglecting effects such as dynamic pressures. Figure 1a shows the minimum degree of planetary magnetospheric compression caused by the intense stellar magnetic fields for a sample of 15 dM stars whose magnetic fields have been observationally reconstructed Donati *et al.* (2008) and Morin *et al.* (2008, 2010). Hypothetical Earth-like planets with similar terrestrial magnetisation orbiting at the inner (outer) edge of the HZ of these stars would present magnetospheres that extend at most up to 6.1 (11.7) planetary radii (r_p). To be able to sustain an Earth-sized magnetospheres ($\sim 12\ r_p$), such planets would require a minimum magnetic field ranging from a few G to up to a few thousand G. Figure 1b shows the closest orbital distance at which an Earth-like planet orbiting the stars in our sample would be able to sustain the present-day Earth's magnetospheric size, assuming it has the same magnetic field as the Earth. Planets orbiting at a closer orbital radius would experience a stronger stellar magnetic pressure, which could reduce the size of the planet's magnetosphere significantly, exposing the planet's atmosphere to erosion by the stellar wind.

References

Donati, J., Morin, J., Petit, P., *et al.*, 2008, *MNRAS* **390**, 545
Ip, W.-H., Kopp, A., & Hu, J.-H., 2004, *ApJ* **602**, L53
Kasting, J. F., Whitmire, D. P., & Reynolds, R. T., 1993, *Icarus* **101**, 108
Lanza, A. F., 2009, *A&A* **505**, 339
Morin, J., Donati, J., Petit, P., *et al.*, 2008, *MNRAS* **390**, 567
Morin, J., Donati, J., Petit, P., Delfosse, X., *et al.*, 2010, *MNRAS* **407**, 2269
Vidotto, A. A., Fares, R., Jardine, M., *et al.*, 2012, *MNRAS* **423**, 3285
Vidotto, A. A., Jardine, M., Morin, J., *et al.*, 2013, *A&A* **557**, A67
Vidotto, A. A., Jardine, M., Opher, M., *et al.*, 2011, *MNRAS* **412**, 351
Vidotto, A. A., Opher, M., Jatenco-Pereira, V., & Gombosi, T. I., 2009, *ApJ* **703**, 1734
Vidotto, A. A., Opher, M., Jatenco-Pereira, V., & Gombosi, T. I., 2010, *ApJ* **720**, 1262

Magnetic Fields throughout Stellar Evolution
Proceedings IAU Symposium No. 302, 2013
P. Petit, M. Jardine & H. Spruit, eds.

doi:10.1017/S1743921314002178

Planets spinning up their host stars: a twist on the age-activity relationship

K. Poppenhaeger and S. J. Wolk

Harvard-Smithsonian Center for Astrophysics,
60 Garden Street,
02138 Cambridge, 02138 MA, USA
email: kpoppenhaeger@cfa.harvard.edu

Abstract. It is a long-standing question in exoplanet research if Hot Jupiters can influence the magnetic activity of their host stars. While cool stars usually spin down with age and become inactive, an input of angular momentum through tidal interaction, as seen for example in close binaries, can preserve high activity levels over time. This may also be the case for cool stars hosting a Hot Jupiter. However, selection effects from planet detection methods often dominate the activity levels seen in samples of exoplanet host stars, and planet-induced, systematically enhanced stellar activity has not been detected unambiguously so far. We have developed an approach to identify planet-induced stellar spin-up avoiding the selection biases from planet detection, by using visual proper motion binaries in which only one of the stars possesses a Hot Jupiter. This approach immediately rids one of the ambiguities of detection biases: with two co-eval stars, the second star acts as a negative control. We present results from our ongoing observational campaign at X-ray wavelengths and in the optical, and present several outstanding systems which display significant age/activity discrepancies presumably caused by their Hot Jupiters.

Keywords. planetary systems, stars: activity, binaries: visual, stars:evolution, stars:magnetic fields X-rays: stars

---

1. Introduction

Almost all planet-hosting stars known today are cool stars (spectral types F-M). All cool stars display stellar activity – a summarizing term for the occurrance of magnetic phenomena including flares, spots, and coronal high-energy emission. The magnetic activity of planet-hosting stars is an important factor to understand the evolution of exoplanets. High-energy irradiation of close-in planets can lead to atmospheric evaporation (Vidal-Madjar *et al.* 2003; Lecavelier Des Etangs *et al.* 2010), and coronal mass ejections and the stellar wind can strip away parts of the planetary atmosphere (Penz *et al.* 2008). The stability and chemistry of exoplanetary atmospheres are both influenced by these stellar activity phenomena. Recent transit observations in the UV (Bourrier *et al.* 2013) and in X-rays (Poppenhaeger *et al.* 2013) have demonstrated that atmospheres of Hot Jupiters are strongly extended. Modelling of such atmospheres suggests that time-variable phenomena like bow-shocks and variations in the stellar wind may play a role, too (Vidotto *et al.* 2012).

Stellar activity levels of stars with and without planets have received much scrutiny in recent years. Activity is well-known to be a function of stellar rotation and therefore, due to magnetic breaking, of stellar age. Stellar X-ray luminosity therefore declines with stellar age, as shown in Fig. 1, left panel.† Several studies have noted that exoplanets

† For a summary of age-activity relations using the fractional X-ray luminosity L_X/L_{bol} instead of L_X, see Jackson *et al.* (2012).

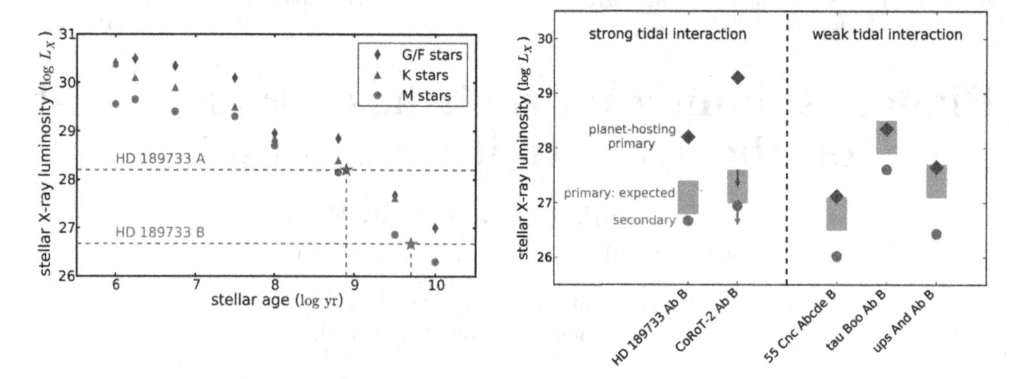

Figure 1. Left: Magnetic breaking causes cool stars to spin down over time, which can be measured as a decline in magnetic activity indicators such as X-ray emission.* **Right:** In wide stellar binaries in which one of the stars hosts an exoplanet, some of the planet-hosting stars display a much higher magnetic activity level than expected for their age. This is the case for systems with strong tidal interaction between planet and host star.

* Age-activity data from Preibisch & Feigelson (2005), Jeffries *et al.* (2006), Schmitt & Liefke (2004), Damiani *et al.* (2004), Engle & Guinan (2011), Hawley & Feigelson (1994).

may also have an influence on the stellar magnetic activity. The underlying idea is that massive close-in exoplanets should be able to interact with their host star magnetically and/or tidally (Cuntz *et al.* 2000; Shkolnik *et al.* 2005; Kashyap *et al.* 2008). However, because stellar activity effectively masks signals used to detect exoplanets, one has to be careful not to interpret intrinsic biases in samples of planet-hosting stars as a physical interaction effect (Poppenhaeger *et al.* 2010; Poppenhaeger & Schmitt 2011; Miller *et al.* 2012).

A useful way to test for activity enhancements caused by Hot Jupiters is to look at wide stellar binaries, in which only one star hosts a planet massive and close enough for strong tidal interaction. The star without a planet then acts as a negative control, because its activity evolution should be unperturbed by any planetary effects. One can thus test if both stars display magnetic activity levels appropriate for their common stellar age (taking into account slight differences due to spectral type). We are conducting an observational program using X-ray data and optical spectra to test for such activity differences.

2. Initial results

We have collected observational data for five exoplanet systems with wide stellar binaries so far, using *XMM-Newton* and *Chandra* for X-ray observations of the stellar corona, as well as FLWO's TRES spectrograph for optical high-resolution spectra. We list the observed X-ray luminosities for both stellar components of each system in Table 1. We estmate the age of the system from the X-ray brightness of the secondary (non-planet-hostimg) star. This yields an estimate for the X-ray luminosity of the primary (planet-hosting) star, assuming that both stars are co-eval. More detailed analyses of the spectral X-ray properties of some of the stars can be found in Pillitteri *et al.* (2010) and Poppenhaeger *et al.* (2013) for HD 189733 A and B, Schröter *et al.* (2011) for CoRoT-2 A, Poppenhaeger *et al.* (2012) for τ Boo and Poppenhaeger & Schmitt (2011) for υ And A.

For three of the systems we find that the expected and observed X-ray luminosities for the planet-hosting stars are in agreement; for two systems, however, the expected and

System	SpT A	SpT B	a_{sem} (AU)	M_P (M_{Jup})	$\log L_{X,B}$ (erg/s)	age_B (Gyr)	$\log L_X^{A\,(exp.)}$ (erg/s)	$\log L_X^{A\,(obs.)}$ (erg/s)
HD 189733 Ab B	K1V	M4V	0.03	1.138	26.67	5	27.1	28.2
CoRoT-2 Ab B	G9V	K9V	0.01	2.0	<26.95	3	27.3	29.3
55 Cnc Abcde B	K2V	M3V	0.03	>0.0001	26.02	10	26.8	27.12
τ Boo Ab B	F6V	M3V	0.03	>1.0	27.61	2	28.2	28.35
υ And Ab B	F7V	M4V	0.03	>1.0	26.43	7	27.2	27.65

Table 1. Observed X-ray luminosities of the planet-hosting primaries and the secondaries which are not known to possess planets. The X-ray luminosity of the secondary is used to estimate the age of the system. Assuming that both stelar components are co-eval, we can derive an *a priori* estimate how X-ray luminous the primary should be (column $\log L_{X,\,A\,(exp.)}$). The actual observed X-ray luminoities of the primaries are given in column $\log L_{X,\,A\,(obs.)}$.

observed X-ray luminosities disagree by an order of magnitude or more; see Fig. 1, right panel. These systems both host a Hot Jupiter in a very close orbit, and the host star is a late G or K dwarf. For the systems in which no discrepancy is found, the innermost planet is either small (55 Cnc) or the host star is an F star with a very thin outer convective envelope (τ Boo and υ And).

We interpret our initial findings as a manifestation of tidal interaction between the planet and the host star. Theoretical studies show that Hot Jupiters induce tidal bulges on the host star; cool stars can dissipate the energy contained in the bulges much more effectively than hot stars due to turbulent eddies in the convective envelopes, see Zahn (2008) and Torres *et al.* (2010). Observationally, stars with substantial convective envelopes have also been found to have exoplanets in orbits with low orbital obliquity, which is also believed to be a consequence of tidal interaction (Winn *et al.* 2010; Albrecht *et al.* 2012).

The Hot Jupiters in the HD 189733A and CoRoT-2A systems may therefore have inhibited the spin-down of their host stars, by transferring angular momentum from the planetary orbit into the stellar spin.

To test this hypothesis further, we are collecting high-resolution optical spectra for our sample to measure the projected rotational velocities $v \sin i$ of the stars as well as chromospheric activity indicators such as Ca II H and K and H α.

3. Conclusion

Our initial data suggests that there is indeed a large difference in activity levels for systems in which the planet exerts a strong tidal interaction on its host star, while systems with weak tidal interaction do not show elevated activity levels. Observations of a larger number of systems which can be tested for possible tidal effects from the planet are under way.

The consequences of planet-induced stellar high activity are substantial for understanding exoplanetary systems and, on a larger scale, stellar evolution. For systems with Hot Jupiters, age estimates can no longer be derived from stellar rotation and activity; also, the time-integrated high-energy irradiation of exoplanets will be much higher due to the prolonged high activity of the host stars, so that planetary evaporation may have had a much stronger influence on the planet zoo we observe today than current models suggest. In terms of stellar evolution, stellar population synthesis models are just beginning to include the star's spin as a parameter. First results show that the stellar rotation is a fundamental parameter for its evolution (Levesque *et al.* 2012), so that it is crucial to know if Hot Jupiters significantly change the spin of their host stars, which comprise ca. 1% of all cool stars (Wright *et al.* 2012).

References

Albrecht, S., Winn, J. N., Johnson, J. A., *et al.* 2012, *ApJ*, 757, 18

Bourrier, V., Lecavelier des Etangs, A., Dupuy, H., *et al.* 2013, *A&A*, 551, A63

Cuntz, M., Saar, S. H., & Musielak, Z. E. 2000, *ApJL*, 533, L151

Damiani, F., Flaccomio, E., Micela, G., *et al.* 2004, *ApJ*, 608, 781

Engle, S. G. & Guinan, E. F. 2011, in Astronomical Society of the Pacific Conference Series, Vol. 451, Astronomical Society of the Pacific Conference Series, ed. S. Qain, K. Leung, L. Zhu, & S. Kwok, 285

Hawley, S. L. & Feigelson, E. D. 1994, in Astronomical Society of the Pacific Conference Series, Vol. 64, Cool Stars, Stellar Systems, and the Sun, ed. J.-P. Caillault, 89

Jackson, A. P., Davis, T. A., & Wheatley, P. J. 2012, *MNRAS*, 422, 2024

Jeffries, R. D., Evans, P. A., Pye, J. P., & Briggs, K. R. 2006, *MNRAS*, 367, 781

Kashyap, V. L., Drake, J. J., & Saar, S. H. 2008, *ApJ*, 687, 1339

Lecavelier Des Etangs, A., Ehrenreich, D., Vidal-Madjar, A., *et al.* 2010, *A&A*, 514, A72

Levesque, E. M., Leitherer, C., Ekstrom, S., Meynet, G., & Schaerer, D. 2012, *ApJ*, 751, 67

Miller, B. P., Gallo, E., Wright, J. T., & Dupree, A. K. 2012, *ApJ*, 754, 137

Penz, T., Micela, G., & Lammer, H. 2008, *A&A*, 477, 309

Pillitteri, I., Wolk, S. J., Cohen, O., *et al.* 2010, *ApJ*, 722, 1216

Poppenhaeger, K., Günther, H. M., & Schmitt, J. H. M. M. 2012, Astronomische Nachrichten, 333, 26

Poppenhaeger, K., Robrade, J., & Schmitt, J. H. M. M. 2010, *A&A*, 515, A98+

Poppenhaeger, K. & Schmitt, J. H. M. M. 2011, *ApJ*, 735, 59

Poppenhaeger, K., Schmitt, J. H. M. M., & Wolk, S. J. 2013, *ApJ*, 773, 62

Preibisch, T. & Feigelson, E. D. 2005, *ApJS*, 160, 390

Schmitt, J. H. M. M. & Liefke, C. 2004, *A&A*, 417, 651

Schröter, S., Czesla, S., Wolter, U., *et al.* 2011, *A&A*, 532, A3+

Shkolnik, E., Walker, G. A. H., Bohlender, D. A., Gu, P., & Kürster, M. 2005, *ApJ*, 622, 1075

Torres, G., Andersen, J., & Giménez, A. 2010, *AARev*, 18, 67

Vidal-Madjar, A., Lecavelier des Etangs, A., Désert, J., *et al.* 2003, *Nature*, 422, 143

Vidotto, A. A., Fares, R., Jardine, M., *et al.* 2012, *MNRAS*, 423, 3285

Winn, J. N., Fabrycky, D., Albrecht, S., & Johnson, J. A. 2010, *ApJL*, 718, L145

Wright, J. T., Marcy, G. W., Howard, A. W., *et al.* 2012, *ApJ*, 753, 160

Zahn, J.-P. 2008, in EAS Publications Series, Vol. 29, EAS Publications Series, ed. M.-J. Goupil & J.-P. Zahn, 67–90

Magnetic Fields throughout Stellar Evolution
Proceedings IAU Symposium No. 302, 2013
P. Petit, M. Jardine & H. Spruit, eds.

doi:10.1017/S174392131400218X

Constraining Stellar Winds of Young Sun-like Stars

Colin P. Johnstone, Theresa Lüftinger, Manuel Güdel and Bibiana Fichtinger

University of Vienna, Department of Astronomy, Türkenschanzstrasse 17, 1180 Vienna, Austria

Abstract. As part of the project Pathways to Habitability (http://path.univie.ac.at/), we study the properties of the stellar winds of low-mass and Sun-like stars, and their influences on the atmospheres of potentially habitable planets. For this purpose, we combine mapping of stellar magnetic fields with magnetohydrodynamic wind models.

Keywords. stars: winds, outflows, stars: mass loss, stars: magnetic fields, MHD,

---

1. Introduction

The Earth is embedded in the extended solar atmosphere we call the solar wind. Such winds are known to emanate from other low-mass and Sun-like stars, including young solar analogues. Very little is currently known about how these winds influence the atmospheres of young habitable planets, largely because such winds are well constrained by neither observations nor theory.

The first physical model for the solar wind was developed by Parker (1958), who modelled the solar wind as a simple 1D isothermal pressure driven wind. Later models of the solar wind produced used a polytropic equation of state, where the thermal pressure is given by $p \propto \rho^{\gamma}$, where γ is the polytropic index (e.g. Totten *et al.* 1995). When $\gamma < 5/3$, these models implicitly heat the winds as they expand, but do not contain any description of the physics responsible for this heating. Despite a huge amount of progress observationally and theoretically, the fundamental mechanisms that drive the solar wind have not yet been determined.

Although there have not been any direct detections of winds from low-mass or Sun-like stars, there are indirect methods of measuring these winds, most notably using measurements of Lyα absorption (Wood 2004). Attempts to directly detect thermal Bremsstrahlung radiation in radio from these winds have so far led to non-detections (Gaidos 2000), putting important upper limits on the wind strengths. As part of the project Pathways to Habitability, we attempt to detect the radio emission from these winds using VLT and ALMA observations.

2. Wind Models

We attempt to constrain the properties of stellar winds using magnetohydrodynamical modelling (e.g. see Fig. 1). We implicitly heat the winds by assuming a polytropic equation of state, as described above. Similar models have been used successfully in previous studies (e.g. Keppens & Goedbloed 1999; Matt & Pudritz 2008; Vidotto *et al.* 2009). Such models contain several free parameters, including the temperature and density at the base of the wind, and the polytropic index, γ. Unfortunately, the free parameters are not well constrained and the results of these models (e.g. wind speeds and mass loss rates) are highly sensitive to these free parameters.

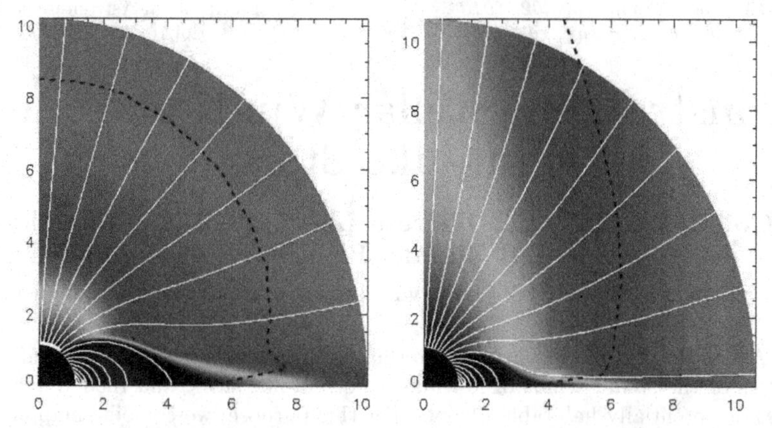

Figure 1. Velocity contour plots showing the expansion of 2D axisymmetric winds with a dipole magnetic field. These simulations were run using the MHD code Nirvana (Ziegler 2005) for the cases of no stellar rotation (left) and a stellar rotation period of one day (right). The dashed black lines show the Alfvén surfaces and the white lines show the structure of the magnetic field. The black equatorial regions show regions of very low velocity, where the stellar plasma is contained within closed magnetic field lines.

3. Magnetic Field Observations

Based on the Zeeman Doppler Imaging (ZDI) technique and the inversion of high-quality Stokes polarization data, we study the origin, strength, and distribution of magnetic fields on young active stars, including T Tauri stars, as a function of age, mass and spectral type. For modelling, we use a state-of-the-art ZDI code from Piskunov & Kochukhov (2002), which has been extended for self-consistent temperature and magnetic mapping of cool active stars (Kochukhov & Piskunov 2009) including treatment of molecular opacities. To obtain complete sets of circular and, for stars bright enough, linear Stokes parameter observations, we have applied and will apply for first-class spectropolarimetric data obtained with frontier instrumentation such as HARPSpol on the ESO 3.6m telescope, ESPaDOnS@CFHT (Hawaii), and NARVAL@TBL (France).

References

Gaidos, E. J., Güdel, M., & Blake, G. A. 2000, *GeoRL*, 27, 501
Keppens, R. & Goedbloed, J. P. 1999, *A&A*, 343, 251
Kochukhov, O. & Piskunov, N. 2009, *ASPC*, 405, 539
Matt, S. & Pudritz, R. E. 2008, *ApJ*, 678, 1109
Parker, E. 1958, *ApJ*, 128, 664
Piskunov, N. & Kochukhov, O. 2002, *A&A*, 381, 736
Totten, T. L., Freeman, J. W., & Arya, S. 1958, *JGR*, 100, 13
Vidotto, A. A., Opher, M., Jatenco-Pereira, V., & Gambosi, T. I. 2009, *ApJ*, 699, 441
Wood, B. 2004, *LRSP*, 1, 2
Ziegler, U. 2005, *A&A*, 435, 385

Magnetic Fields throughout Stellar Evolution
Proceedings IAU Symposium No. 302, 2013
P. Petit, M. Jardine & H. Spruit, eds.

doi:10.1017/S1743921314002191

Bow shocks and winds around HD 189733b

J. Llama, A. A. Vidotto, M. Jardine, K. Wood and R. Fares
SUPA. University of St Andrews. North Haugh. St Andrews. KY16 9SS. UK
email: joe.llama@st-andrews.ac.uk

Abstract. Asymmetries in exoplanet transits are a useful tool for developing our understanding of magnetic activity on both stars and planets outside our Solar System. Near-UV observations of the WASP−12 system have revealed asymmetries in the timing of the transit when compared with the optical light curve. In this proceedings we review a number of reported asymmetries and present work simulating near-UV transits for the hot-Jupiter hosting star HD 189733.

Keywords. stars: activity, coronae, individual (WASP−12, HD 189733), magnetic fields, planetary systems, winds

1. Introduction

An asymmetry has been detected in the *HST* near-UV light curve of the hot-Jupiter WASP-12b when compared to the optical data. The near-UV light curve shows the transit beginning before the optical transit, but finishing at the same time (Fossati *et al.* 2010). Vidotto *et al.* (2010) proposed this early-ingress is caused by the presence of a magnetosphere around the planet. The interaction between the Solar wind and each of the Solar system planets results in the formation of a shock surrounding the planet; however, for magnetised planets, the shock is located further from the planet due to the presence of the planetary magnetosphere. For planets that are located far from their host star, such as Earth and Jupiter, the shock will form directly between the planet and the star, a so-called "dayside shock". For close-in planets, such as hot-Juipiter's, however, an "ahead shock" will form which will occult the stellar disc before the planet which may cause an early-ingress in the transit light curve. If the density of shocked material is high enough then this may show as the presence of additional absorption in the light curve. Llama *et al.* (2011) modelled the near-UV light curve of WASP-12b from Fossati *et al.* (2010) and were able to fit the observations with a simple shock model.

In this proceedings we summarize the findings of Llama *et al.* (2013) where we model a planet and bow shock transiting over a simulated star. We use magnetic maps of the bright K-dwarf HD 189733 (which due to its proximity and relative brightness has been extensively studied) to simulate the stellar wind conditions around the planet transiting planet HD 189733b. From these simulations we are able to prescribe the geometry and density of the shock that should form and simulate near-UV light curves (Llama *et al.* 2013).

2. The Model

We couple numerical simulations of stellar winds with magnetic imaging of HD 189733 to predict the stellar wind conditions around the planet HD 189733b in order to investigate how the geometry of the resultant shock may vary as the planet orbits around the star. We use magnetic surface maps of HD 189733 from June 2007 and also a year later in July 2008 (Fares *et al.* 2010). The magnetic surface maps are used as one of the boundary conditions in our stellar wind simulation. We use BATS-R-US, a three-dimensional

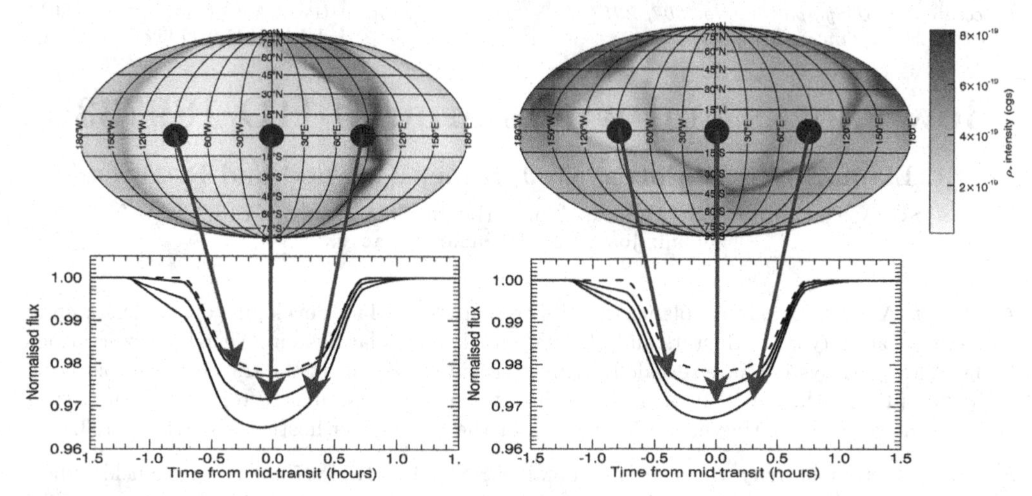

Figure 1. Results of our simulations: June 2007 (left panel) and July 2008 (right panel). Top panel shows the stellar wind density at the orbital distance of the planet. The transit depth and timing is clearly dependent on the local stellar wind conditions for near-UV transits. We find the transits to be very varied, with certain wind conditions causing the transit to be almost indistinguishable from the optical transit, whilst in very dense regions of the wind the transit depth is much deeper and also begins much sooner due to the presence of denser, shocked material.

mangetohydrodynamic (MHD) numerical code developed at The University of Michigan (Powell *et al.* 1999). The output from the wind simulation is a full three-dimensional grid that allows us to determine the local conditions experienced by the transiting planet HD 189733b. The formation of a bow shock is a direct consequence of the relative velocities between the interaction of the planetary magnetic field and the stellar wind occurring at supersonic speeds. We use the shock model of Wilkin *et al.* (1996) coupled with the output of the stellar wind simulations to model the shock around the planet. We are then able to simulate the near-UV light curve as the planet transits over the stellar disc.

3. Results

Figure 1 shows our results. The top image shows the stellar wind density at the orbit of HD 189733b for June 2007 (left) and July 2008 (right). The bottom panel shows three resultant near-UV transits (solid lines) and the optical model for comparison (dashed). The results clearly show that the depth and timing of the transit is dependent on the density of the stellar wind and therefore multiple observations may reveal different transit shapes and help us understand the variability of the stellar wind.

References

Fares, R., Donati, J., Moutou, C., *et al.* 2010, *MNRAS* 406, 409

Fossati, L., Haswell, C., Froning, C, *et al.* 2010, *ApJ* 714, L222

Llama, J., Wood, K., Jardine, M., Vidotto, A. A. *et al.*, 2011, *MNRAS* 416, L41

Llama, J., A. A. Vidotto, Jardine, M., Wood, K., Fares, R., Gombosi, T. I. *et al.*, 2011, *ArXiv e-prints* 1309:2938

Powell, K. G., Roe, P. L., Linde, T. J. *et al.* 1999, *Journal of Computational Physics*, 154, 284

Vidotto, A. A., Jardine, M., & Helling, C. 2010, *ApJ*, 722, L168

Magnetic Fields throughout Stellar Evolution
Proceedings IAU Symposium No. 302, 2013
P. Petit, M. Jardine & H. Spruit, eds.

doi:10.1017/S1743921314002208

Stellar Magnetism and starspots: the implications for exoplanets

Conrad Vilela[1,*], John Southworth[1] and Carlos del Burgo[2]

[1]Astrophysics Group, Keele University,
Keele, Staffordshire, ST5 5BG
[2]Instituto Nacional de Astrofísica, Óptica y Electrónica,
Luis Enrique Erro 1, Sta. Ma. Tonantzintla, Puebla, Mexico
*email: c.vilela@keele.ac.uk

Abstract. Stellar variability induced by starspots can hamper the detection of exoplanets and bias planet property estimations. These features can also be used to study star-planet interactions as well as inferring properties from the underlying stellar dynamo. However, typical techniques, such as ZDI, are not possible for most host-stars. We present a robust method based on spot modelling to map the surface of active star allowing us to statistically study the effects and interactions of stellar magnetism with transiting exoplanets. The method is applied to the active Kepler-9 star where we find small evidence for a possible interaction between planet and stellar magnetosphere which leads to a 2:1 resonance between spot rotation and orbital period.

Keywords. Stellar variability, stellar activity, stellar Magnetism, magnetic fields, starspots, rotation, exoplanet, star-planet interaction

---

1. Introduction

Cool stars (F-M dwarfs), mainly solar-like stars, have partially convective interiors – a radiative core and a convective envelope. The turbulent cyclonic motion of the plasma in the convective layer generates an underlying magnetic dynamo which drives a variety of stellar features. The magnetic fields produced penetrate the stellar atmosphere from photosphere to corona forming starspots, plages, loops, flares, etc (Berdyugina 2005). These features can be observed with a variety of techniques. The more popular are Zeeman-Doppler imaging (ZDI) and spectropolarimetry which provide information on the topology of the magnetic field (Donati & Semel 1990; Phan-Bao *et al.* 2009). However, these techniques become ill-suited for faint stars and slow rotators. With the increasing number of solar-like host-stars discovered by the *K*epler satellite we need to look at different ways of studying the stellar magnetic topology and its effects on planetary companions.

One can use activity proxies, such as the Ca II H&K line inversion (S_{HK} or $\log R'_{HK}$) which trace chromospheric activity (Wilson 1978, Noyes *et al.* 1984). Using this activity proxy Knutson *et al.* (2010) found that exoplanets with no temperature inversion are more likely to orbit active stars as these produce higher UV-fluxes which photo dissociate molecular absorbers (e.g., TiO and VO). A more common and accessible technique, based on photometry, is the use of rotational modulated light curves caused by the presence of starspots. These surface inhomogeneities are formed by local magnetic fields suppressing the vertical heat transport in the convective layer. They are cooler (1000–3000 K) than their surrounding photosphere and are carried through the stellar surface by stellar rotation.

In the exoplanet community starspots are considered a source of noise as they can hamper the detection and characterisation of the companion planet. If present during transit they can bias the determination of the planet radius and density (Czesla *et al.* 2009). The transit shape is also affected, flattening the trough and widening the transit, blurring the transit time and duration (Oshagh *et al.* 2013). On the other hand starspots can also be used as a characterisation tool revealing properties of the system, such as differential rotation rates, rotation-age relations or star-planet interactions. Sanchis-Ojeda & Winn (2011) used occulted spots to provide the obliquity and misalignment of the WASP-4 system. Close-in planets can become circularised where orbital periods are synchronised with stellar rotation due to the interaction with the magnetosphere (Laine *et al.* 2008). Furthermore, this can induce changes in the magnetic topology and tidal interactions which can produce a spin-up of the stellar rotation (Brown *et al.* 2011).

The aim is to produce a robust methodology to model rotational modulated light curves and statistically analyse the effects and interactions of stellar magnetism and exoplanet companions. The following sections describe the method used to map the surface of active stars and the preliminary results when applied to a discovered *Kepler* system – Kepler-9.

Table 1. Definition of spot parameters, description and fitting constraints

Parameters	Description	Constraints
U	stellar unspotted flux	min – max amplitude
i_{axis}	stellar rotational axis inclination	$0° - 90°$
μ	linear limb darkening coefficient	fixed
κ	spot contrast	$0 - 1$
θ	latitude of spot centre	$-90° - 90°$
ψ^a	longitude of spot centre	fixed
γ	spot angular radius	$0° - 90°$
E	spot epoch	min – max timestamp
P^b_{rot}	spot rotational period	$P_{rot}/2 - 2P_{rot}$

Notes:
[a] Spot position is better determined by E (See Croll *et al.* (2006)).
[b] The initial values are obtained from SIGSPEC and left to vary.

2. Methodology

Spot modelling techniques are based on reproducing the rotational modulated light curve into a physical map of the starspot coverage. The main disadvantage of these techniques is the high degeneracy and non-uniqueness between the model parameters. Various approaches have been used to solve this, such as a continuos spot distribution or a fixed number of spots (Lanza *et al.* 2007). However, these approaches either limit the level of activity observed or cannot precisely extract spot properties. In practice, one would like to find a unique un-biased solution to the problem without any loss of information. To achieve this the method presented is divided into two steps, a frequency analysis and spot modelling.

Frequency analysis. Spot P_{rot} can be easily obtained from the stars light curve by analysing the signal frequencies from the rotational modulations. To achieve this we implement a statistical technique based on the Lomb-Scargle periodogram, Significance Spectrum (SIGSPEC) (Reegen 2007). The frequencies computed by SIGSPEC are constrained to the range 0.6 – 0.016 *cycles/day* ($P_{rot} \approx 1.6 - 60$ *days*). The upper limit ensures that SIGSPEC does not detect frequencies which might correspond to background noise, as Solar-like stars tend to be slow rotators. The lower limit ensure that the detected frequency is representative of a full observed cycle, i.e., the same spot has to be in view at least twice within the light curve. This avoid the detection of frequencies belonging to spot or magnetic cycles and long trends. As an additional functionality SIGSPEC can also determine the corresponding phase of the frequency, constraining the spots E.

Spot modelling. To model the rotational modulation we use the 4-term limb darkening spot model from Kipping (2012). The model has 7 parameters described in table 1, of which ψ is calculated using the spots' E. To reduce computational time and avoid introducing degeneracies the models are computed without spot evolution, migration or differential rotation. However, in general these properties can be estimated from the spots themselves. Although P_{rot}, E and to a certain degree θ are constrained, κ and γ are highly correlated and typical optimisation techniques will not provide a global unique solution. To overcome this degeneracy the light curve is fitted using the MultiNest code based on Bayesian statistics and elliptical decomposition (Feroz *et al.* 2013). Given the Bayesian nature of MultiNest we can also obtain the marginalising integral or model evidence used for model selection through Bayes factor (B). As such we iterate through all frequencies obtained from SigSpec. At each iteration a spot is added to the model until the log difference in B (K) between consecutive iterations is less than 2, showing no strong evidence for the current model (Jeffreys 1961).

3. Results from Kepler-9

Kepler-9 is a G type stars with 3 transiting planets observed by *Kepler* for ~ 3 years. Holman *et al.* (2010) reported this star to have photometric variations slightly larger than the Sun and an inversion of the Ca II H&K line core indicative of moderate activity. The similarities between Kepler-9 and the Sun make this a perfect target for the analysis. We only use the first 7 quarters (Q1-Q7) of the *Kepler* data fitting each individual quarter.

We are able to reproduce the Kepler-9 light curve seen in Fig. 1. The residuals show a periodic signal which could be signs of more spots present than those fitted by the models. However, the small value of K (< 0) indicates that fitting more spots, will neither provide a better fit or more information. Hence, this periodicity is interpreted as a combination of spot evolution and migration. Fig. 1 also shows a change in both amplitude and shape of the modulation with time, more precisely between quarters, reproduced by the model (solid line in Fig. 1). This is interpreted as a change in spot number from quarter to quarter indicative of a possible spot cycle (> 90 days) similar to that seen in the Sun. As the models do not include spot evolution, migration or differential rotation, the apparent spot cycle further supports the idea that the periodicity in the residuals is due to spot evolution and migration.

The resulting best fit spot model shows that the vast majority of the rotational modulation is produced by large near-polar spots ($\theta \approx -85° to - 75°$, $\gamma \approx 50° - 90°$) which could correlate with open magnetic field lines. We infer from this that the large near-polar spots are the projection of coronal holes on the photosphere. However, we cannot resolve the stellar surface and, thus, unable to distinguish between a large spot and a group of smaller spots. The alternative explanation is that the large spots ($\gamma > 50°$) correspond to spot groups. If these spots are evolving it is very likely due to smaller spots disappearing within the spot group. This would mean that the large spot would be tracing active regions with complex coronal structures.

Lastly the best fit spot model shows spot P_{rot} in a possible near 2:1 resonance with the planet's orbital period. We find that spots close to the pole have P_{rot} (8 – 12 days) near half the orbital period of planet b (~9.6 days). Spots near the equator show P_{rot} (15 – 22 days) that are closer to half the orbital period of planet c (~19.4 days). This could infer a possible star-planet interaction which can be interpreted as different evolutionary paths for both planets. However, we are unable to identify if the interaction is magnetic or tidal in origin. Furthermore, both planets are close enough to the host-star ($a \approx 0.1 - 0.2au$) to have an effect on the magnetic topology of the star.

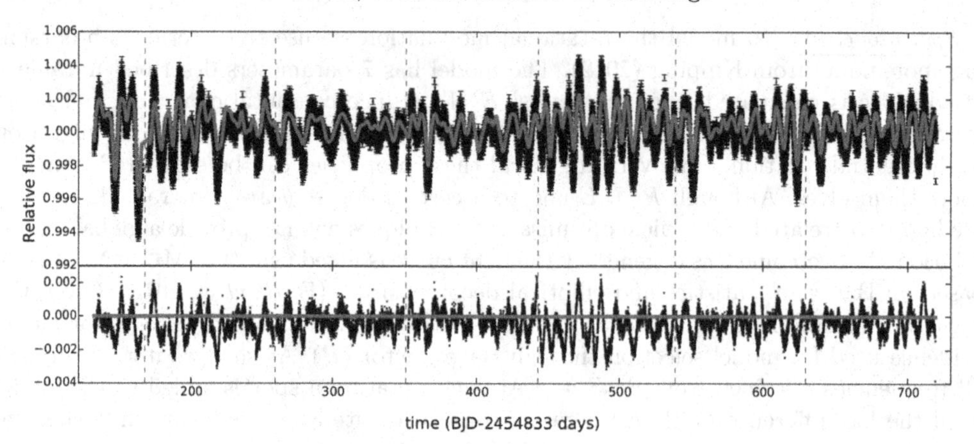

Figure 1. *Top*: Kepler-9 combined light curve (Q1-Q7) with the transits from both planets removed. Black points represent the PDC-MAP data from *Kepler* and the solid line is the best spot model fit for each of the quarters. The vertical dashed lines separate each quarter. *Bottom*: Residual from each quarter from the best spot model fit

References

Berdyugina, S. V. 2005, *Liv. Rev. Solar Phys.*, 2, 8
Donati, J. F. & Semel, M. 1990, *Solar Phys.*, 128, 227
Phan-Bao, N. Lim, J., *et al.* 2009, *ApJ*, 704, 1721
Wilson, O. C. 1978, *ApJ*, 226, 379
Noyes, R. W. Hartmann, L. W., *et al.* 1984, *ApJ*, 279, 763
Czesla, S., Huber, K. F., Wolter, U., Schröter, S., & Schmitt, J. H. M. M. 2009, *A&A*, 505, 1277
Oshagh, M., Santos, N. C., Boisse, I., *et al.* 2013, *A&A*, 556, A19
Kipping, D. 2012 *MNRAS*, 427, 2487
Sanchis-Ojeda, R. & Winn, J. N. 2011, *ApJ*, 743, 61
Brown, D. J. A. Collier Cameron, A., *et al.* 2011, *MNRAS*, 415, 605
Laine, R. O., Lin, D. N. C., & Dong, S. 2008, *ApJ*, 685, 521
Lanza, A. F., Bonomo, A. S., & Rodonò, M. 2007, *A&A*, 464, 741
Croll, B. Walker, G. A. H., *et al.* 2006, *ApJ*, 648, 607
Reegen, P. 2007, *A&A*, 467, 1353
Feroz, F., Hobson, M. P., Cameron, E., & Pettitt, A. N. 2013, *arXiv1306.2144*
Holman, M. J. Fabrycky, D. C., *et al.* 2010, *Science*, 330, 51
Jeffreys, H. 1961, *The Theory of Probability*, Clarendon Press, Oxoford, 3rd edition

Magnetic Fields throughout Stellar Evolution
Proceedings IAU Symposium No. 302, 2013
P. Petit, M. Jardine & H. Spruit, eds.

doi:10.1017/S174392131400221X

On the effects of stellar winds on exoplanetary magnetospheres

V. See[1*], M. Jardine[1], A. A. Vidotto[1], P. Petit[2,3], S. C. Marsden[4] and S. V. Jeffers[5]

[1]SUPA, School of Physics and Astronomy, University of St Andrews, North Haugh, KY16 9SS, St Andrews, UK
*email: wcvs@st-andrews.ac.uk

[2]Université de Toulouse, UPS-OMP, Institut de Recherche en Astrophysique et Planétologie, Toulouse, France

[3]CNRS, Institut de Recherche en Astrophysique et Planétologie, 14 Avenue Edouard Belin, F-31400 Toulouse, France

[4]Computational Engineering and Science Research Centre, University of Southern Queensland, Toowoomba, 4350, Australia

[5]Universität Göttingen, Institut für Astrophysik, Friedrich-Hund-Platz 1, 37077 Göttingen, Germany

Abstract. The habitable zone is the range of orbital distances from a host star in which an exoplanet would have a surface temperature suitable for maintaining liquid water. This makes the orbital distance of exoplanets an important variable when searching for extra-solar Earth analogues. However, the orbital distance is not the only important factor determining whether an exoplanet is potentially suitable for life. The ability of an exoplanet to retain an atmosphere is also vital since it helps regulate surface temperatures. One mechanism by which a planetary atmosphere can be lost is erosion due to a strong stellar wind from the host star. The presence of a magnetosphere can help to shield a planetary atmosphere from this process. Using a simple stellar wind model, we present the impact that stellar winds might have on magnetospheric sizes of exoplanets. This is done with the aim of further constraining the parameter space in which we look for extra-solar Earth analogues.

Keywords. stars: activity, chromospheres, Magnetic fields, mass loss, planetary systems, winds

---

1. Introduction

The first exoplanets were found nearly two decades ago. Since then, the attention has switched from simply looking for exoplanets to characterising them and their potential to host life. As far as we know, all life requires liquid water to survive. This means that any potentially habitable exoplanet will require an Earth-like surface temperature. Since it is unfeasable to measure the surface temperature of every exoplanet we find, we need an observable signature instead to act as a proxy – the habitable zone (Kasting *et al.* 1993). The habitable zone is the set of orbital radii in which it is thought liquid water could exist on the surface of a planet. It is calculated by considering the amount of flux incident on the planet from its host star. Stars which are more massive, and hence more luminous, will have habitable zones which lie further out. This is the most commenly used measure of exoplanetary habitability and quite often the only measure used.

However, it is important to consider other aspects which can affect the potential habitability of a planet. One such factor is the presence of an atmosphere. Indeed, the calculation of the habitable zone assumes this since atmospheres regulate surface temperatures.

If the host star has sufficiently strong stellar winds, the atmosphere can be eroded away. In the Earth's case, this has not happened due to our magnetosphere which diverts the wind around the Earth, shielding us from its erosive effects.

In general, it is the size of the magnetosphere which determines whether it can adequately protect the planet. This is in turn determined by the pressure balance of the system. Three external pressures – the stellar magnetic pressure, the ram pressure of the wind and the ambient thermal plasma pressure – balance the magnetic pressure associated with the planetary magnetosphere. It turns out that the thermal plasma pressure is negligable compared to the other two external pressures, Additionally, the stellar magnetic pressure falls off, with distance from the star, much quicker than the ram pressure meaning the ram pressure dominates for solar type stars. The size of the magnetosphere is therefore determined solely by pressure balance between the ram pressure and the planetary magnetic pressure. It should be noted that the stellar magnetic pressure cannot be ignored for M dwarfs, where the habitable zone lies much closer to the star (Vidotto *et al.* 2013).

This raises two potential problems. Firstly, it is not entirely clear what size of magnetosphere constitutes "adequate" protection. We will assume that exoplanet require at least an Earth-sized magnetosphere since we are searching for Earth-analogues. Secondly, in the absesnce of any direct exoplanetary magnetic field measuresments to date, we need some other observational signiture that might indicate the presence of a sufficiently sized magnetosphere which will be the focus of this work. It should be noted that methods to indirectly detect magnetospheres on planets outside of our solar system have been proposed (Vidotto *et al.* 2010, Llama *et al.* 2011, Llama *et al.* 2013).

2. Model

For this work, we shall use a sample of stars collected by the Bcool collaboration. This is an international collection of scientists studying magnetic activity in cool stars. The stars have effective temperatres between 5000–6000K and masses between 0.5-1.5$M_{\odot}$ and for each star, the Bcool collaboration has measured the chormospheric activity. It is unknown whether these stars are planet hosting but for this study we will assume they are. We place a hypothetical Earth-like exoplanet, with all the characteristics of Earth, around each star and calculate the ram pressure exerted on the planetary magentosphere to determine its size.

To calculate the ram pressure, we use two wind models. The first is based on the wind model of Parker (1958) which is steady, isotropic and isothermal. Solving the momentum equation gives a velocity profile of the wind as a fuction of the distance. Then, using an empirically determined relation from (Mamajek *et al.* 2008), we find X-ray luminosities from the stellar chromospheric activity. We use this as a proxy for the emission measure from which the density at the stellar surface, or equivalently the base of the wind, can be determined. Combining the base density and velocity profile, we can derive a density profile for the wind by assuming mass conservation. Finally a ram pressure profile can be found by combining the veloctiy and density profiles. For this model, five inputs are required - stellar mass, radius, luminosity, chromospheric activity and wind temperature. The first four are observationally determined in the sample but the final one input is observationally unconstrained. We set the wind temperature to 1.3MK so that the the Sun/Earth system is calibrated correctly within our model.

The second wind model is the model of mass loss presented by Cranmer & Saar (2011). This is a more sophisticated model which we use as a check for the more simplistic model outlined above. It differs from the Parker type wind in that both thermal and wave

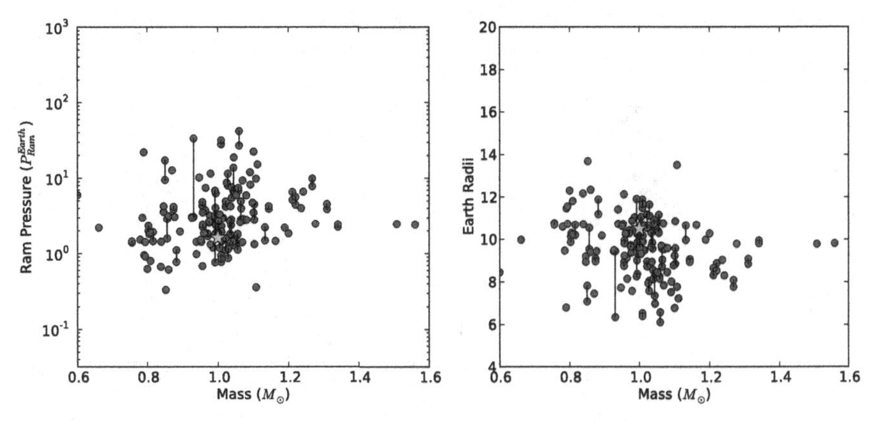

Figure 1. Each hypothetical exoplanet is placed in the habitable zone of its host star. The ram pressure, in units of ram pressure exerted on Earth from the Sun, is calculated using the Parker model of wind (left). The corresponding magnetospheric size for each exoplanet, in units of Earth radii, is also calculated (right). The Earth/Sun case is plotted using the star symbol.

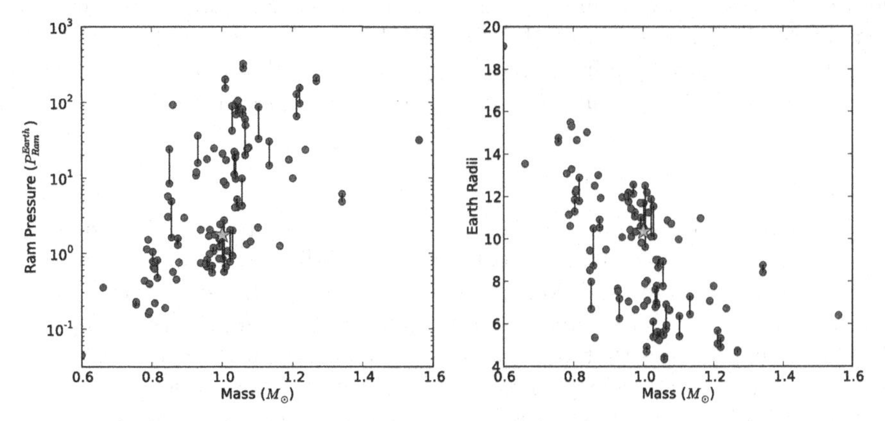

Figure 2. Ram pressures and magnetospheric size are calculated as in Fig 1 but with the Cranmer & Saar model of mass loss.

driving mechanisms are considered when calculating the total mass loss rate from the star. This model also requires five inputs – stellar mass, radius, luminosity, metallicity and rotation period which can all be observationally contrained.

3. Results

To begin with, we placed each hypothetical planet into the habitable zone of its host star. Figure 1 shows the ram pressure exerted on each planet and the corresponding magnetosphere size in Earth radii. A star symbol is used to denote the Earth/Sun system. By placing the planets within the habitable zone, we have forced them to have an Earth-like surface temperature. The figure clearly shows that only a fraction of the planets have at least an Earth-sized magnetosphere or bigger (those with magnetosphere's bigger than roughly $11R_E$). Figure 2 shows the same results but under the Cranmer & Saar model of mass loss. The results of the two models broadly agree with each other with some slight scatter in the results.

Next, we approach the problem from the opposite direction. Figure 3 shows the orbital distance each planet would need to orbit at in order to maintain an Earth-sized magnetosphere for both of the wind models. In this scenario, we have enforced the Earth-sized

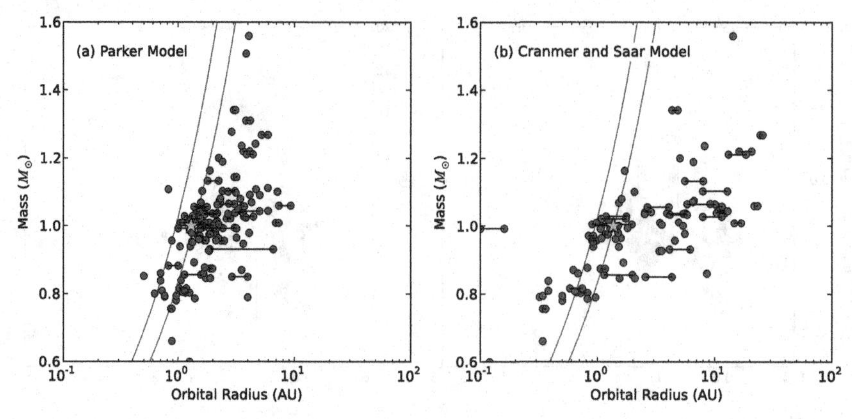

Figure 3. The distance each planet would need to orbit at in order to maintain an Earth-sized magnetosphere for both wind models. The habitable zone has been outlined in green. All symbols are as Fig 1.

magnetosphere condition and find out which of the planets can fulfull the Earth-like surface temperature condition. Again, the two models broadly agree with each other. Also, these results show that only a fraction of the planets can fulfill both of the requirements we outlined before, in agreement with the results from Figures 1 and 2.

4. Implications

The results presented suggest that only a fraction of exoplanets around solar-types can simultaneously fulfill the requirements of an Earth-like surface temperature and an Earth-sized magnetosphere. Additionally, they show that it's the stars with weaker winds which are more likely to host planets able to fulfill both requirements. Observationally, the Parker model indicates that we should look for planets around stars with low chromospheric activity to find such planets. The Cranmer & Saar model show indiates we should look around stars with longer rotation periods which, again, corresponds to lower activity. All of these reulsts point to the importance of characterising the host star when assessing exoplanetary habitabilty rather than focusing only on the orbital distance of the planet.

References

Cranmer, S. R. & Saar, S. H. 2011, *ApJ*, 741, 54

Kasting, J. F., Whitmire, D. P., & Reynolds, R. T. 1993, *Icarus*, 101, 108

Llama, J., Wood, K., Jardine, M., Vidotto, A. A., Helling, C., Fossati, L., & Haswell, C. A. 2011, *MNRAS*, 416, L41

Llama, J., Vidotto, A. A., Jardine, M., Wood, K., Fares, R., & Gombosi, T. I. 2013, arXiv:1309.2938 [astro-ph.EP]

Mamajek, E. E. & Hillenbrand, L. A. 2008, *ApJ*, 687, 1264

Parker, E. N. 1958, *ApJ*, 128, 664

Vidotto, A. A., Jardine, M., & Helling, C. 2010, *ApJL*, 722, L168

Vidotto, A. A., Jardine, M., Morin, J., Donati, J. F., Lang, P., & Russell, A. J. B. 2013, *A&A*, 557, A67

Magnetic Fields throughout Stellar Evolution
Proceedings IAU Symposium No. 302, 2013
P. Petit, M. Jardine & H. Spruit, eds.

doi:10.1017/S1743921314002221

The nature and origin of magnetic fields in early-type stars

Jonathan Braithwaite
Argelander Institut für Astronomie, Auf dem Hügel 71, 53121 Bonn, Germany
email: jonathan@astro.uni-bonn.de

Abstract. I review our current knowledge of magnetic fields in stars more massive than around $1.5M_{\odot}$, in particular their nature and origin. This includes the strong magnetic fields found in a subset of the population and the fossil field theory invoked to explain them; the subgauss fields detected in Vega and Sirius and their possible origin; and what we can infer about magnetic activity in massive stars and how it might be linked to subsurface convection.

Keywords. stars: activity, stars: magnetic fields, stars: early-type, stars: chemically peculiar

1. Introduction

Interest in stellar magnetism has increased in part because of the realisation that magnetic fields, together with rotation, are amongst the most important missing pieces in our understanding of how stars form, evolve and die. In star formation and various modes of stellar death such as γ-ray bursts magnetic fields are known to be crucial, and they also have subtler but important effects at other epochs in a star's life.

Here, I review magnetic fields which we can observe directly with the Zeeman effect, as well as those which have other observational signatures at the stellar surface. [For an introduction to the Zeeman effect in stars and observational techniques to measure it, see Landstreet 2011.] This article does not include discussion of magnetic fields in stellar interiors which have no immediate effect at the surface, for instance small-scale fields which may be important in mixing and stellar evolution.

The focus here is early-type main-sequence (MS) stars. I begin by reviewing observations of the subset of early-type stars which display strong magnetic fields and discussing the theory of the fossil fields they are believed to contain. I then look at the rest of the population, first the intermediate-mass stars ($\lesssim 7M_{\odot}$) where subgauss fields have recently been detected and then the more massive stars. Finally, I discuss possible origins of magnetic fields and some open questions. The relevant observations and theories discussed here are summarised in Table 1.

Table 1. Summary of current knowledge of magnetic fields in early-type stars.

	A and late B	**O and early B**
Magnetic subset ($\lesssim 10\%$)	$B \sim 200$ G to 30 kG steady, large-scale Chemical peculiarities (Ap/Bp) *Fossil field*	$B \sim 200$ G to 10 kG steady, large-scale *Fossil field*
Rest of population	Subgauss fields detected in two stars, probably present in all stars? *Failed fossil field*	No direct detections Indications of magnetic activity *Subsurface convection dynamo*

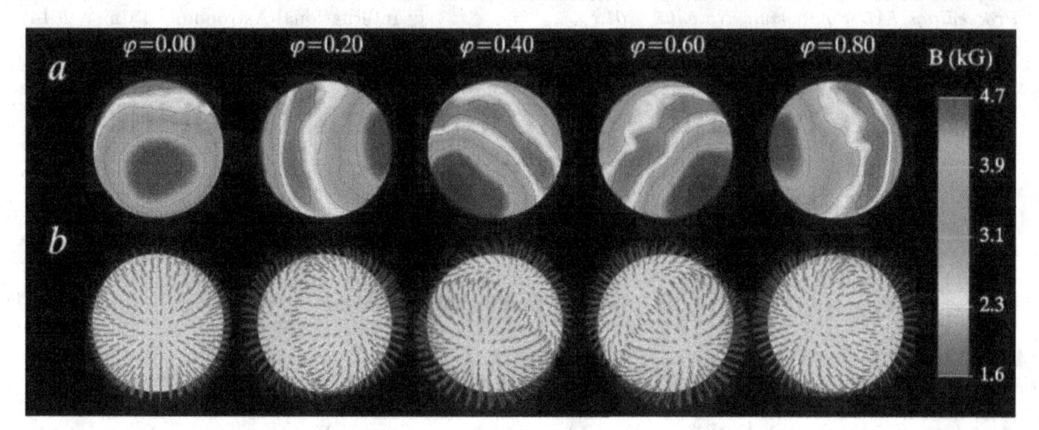

Figure 1. The magnetic field of α^2 CVn viewed at five rotational phases. The top row shows field strengths and the bottow row the field direction. Clearly, α^2 CVn has an approximately dipolar field which is inclined to the rotation axis. From Kochukhov *et al.* 2002.

2. The magnetic subset of intermediate-mass stars: the Ap/Bp stars

t is in these stars that the first magnetic fields were detected outside of the solar system (Babcock 1947), and not surprisingly it is these stars of which we have the best understanding in terms of magnetic properties. A subset of intermediate-mass stars are found to have strong magnetic fields which – in contrast to the fields in convective stars – are dominated by structure on large scales, and are not observed to vary in time. They range in strength from 200 G up to at least 30 kG. There is a real bimodality: whilst the chemically-peculiar Ap and Bp stars all have fields of at least about 200 G, no other intermediate-mass stars have fields above about a few gauss (Aurière *et al.* 2007). The magnetic fraction of the population depends on mass, starting at around 1% at $1.5M_\odot$ and rising to perhaps 10 or 20% at $3M_\odot$, where it seems to level off (Power *et al.* 2007).

The magnetic fields display a range of geometries. A large fraction have fields to which a simple dipole is a good approximation, many have something which can be well described as an 'offset dipole' or a dipole plus quadrupole or similar, and some have more complex fields which cannot be described adequately in terms of low-order spherical harmonics. An example of a star with a very simple field is shown in fig. 1.

As Cowling (1945) pointed out, the Ohmic dissipation timescale in the radiative core of the Sun is long, around 10^{10} yrs; consequently any magnetic field there finding itself in magnetohydrodynamic (MHD) equilibrium can in principle remain there for the Sun's entire lifetime. The same is true of more massive stars; according the the fossil field theory, magnetic stars with radiative envelopes contain just such an equilibrium.

There is not really any other plausible explanation for the strong, large-scale fields observed. For a long time it was discussed how the convective core may host a strong dynamo, the field produced there rising to the surface. However, whilst dynamo activity in the core is almost inevitable (e.g. Browning *et al.* 2004) it seems impossible to get the field to the surface on a sensible timescale (MacGregor & Cassinelli 2003) and with the right geometry. In any case, similar fields to those seen in main-sequence stars have now been found in pre-MS stars without convective cores. Alecian *et al.* (2013) present the results from a survey of 70 Herbig Ae-Be stars, finding that the magnetic fraction is comparable to that amongst the main-sequence stars which these stars will become.

Historically the main challenge to the fossil field theory was to demonstrate that a stable MHD equilibrium can actually exist inside a star. With purely analytic methods it is relatively straightforward, given a star with a particular structure, to construct an equilibrium. Convincingly verifying its stability has proved impossible, however, although it has been possible to demonstrate that certain field configurations are *unstable*, for instance all axisymmetric fields which are either purely poloidal or purely toroidal (Wright 1973, Tayler 1973). The postulation of Wright (1973) and others that a stable configuration must therefore contain both components in a twisted-torus arrangement was confirmed by the discovery of exactly such configurations, using numerical methods (Braithwaite & Spruit 2004).

Essentially, the method consists in evolving the MHD equations in a star containing initially some arbitrary field, as might be left over for instance by a convective dynamo or merger event (see section 6). Braithwaite & Spruit (2004) and Braithwaite & Nordlund (2006) modelled a simplified radiative star: a self-gravitating ball of gas with an ideal gas equation of state, a polytropic index $n = 3$ and ratio of specific heats $\gamma = 5/3$ embedded in an atmosphere with low electrical conductivity. Over few Alfvén timescales, the field organises itself into an equilibrium, after which it continues to evolve only on the very long Ohmic timescale.

There is apparently a large range of equilibria available, and the particular equilibrium which appears depends on the initial conditions. In particular, it seems that the radial energy distribution of the inital field is important. Note that, during the process of relaxation to equilibrium, the gas in a radiative star/zone is restricted by gravity to move around on spherical shells. Consequently it is impossible to transport flux in the radial direction so the total unsigned flux through any spherical shell $\oint |\mathbf{B} \cdot \mathrm{d}\mathbf{S}|$ can only fall. Therefore an initial field which is buried in the interior of a radiative star or zone evolves into a similarly buried equilibrium.

It turns out that if the initial field is somewhat stronger in the interior than near the surface, an approximately axisymmetric equilibrium evolves with both toroidal and poloidal components in a twisted-torus configuration, illustrated in fig. 2 (upper panels). This corresponds qualitatively to equilibria suggested by Prendergast (1956) and Wright (1973). If, on the other hand, the initial field has a flatter radial energy distribution and significant flux connects through the surface of the star, a more complex, non-axisymmetric equilibrium forms – the lower panels of fig. 2. It seems that both axisymmetric and non-axisymmetric equilibria do form in nature: see figs. 1 and 4.

The geometries of these various equilibria have one feature in particular in common, that they can be thought of in terms of twisted flux tubes (illustrated in fig. 3). The simple axisymmetric equilibria can be thought of as a single twisted tube wrapped in a circle (a 'twisted torus') and the more complex equilibria as one or more twisted flux tubes meandering around the star in apparently random patterns. In the equilibria found thus far, the meandering is done at roughly constant radius, a little below the surface. Equilibria where the flux tubes do not lie at constant radius seem possible but it also seems plausible that they are difficult to reach from realistic initial conditions, especially in view of the restriction of motion to spherical shells.

The properties of these non-axisymmetric and axisymmetric equilibria were explored further in Braithwaite 2008 & 2009 respectively.

3. The magnetic subset of high-mass stars

For a multitude of reasons, the detection of magnetic fields in massive stars is more difficult than in lower-mass stars, but advances in technology and methods have improved the situation; see for Henrichs 2012 for a recent review. Thanks to recent surveys (e.g. Wade *et al.* 2013), we now know that around 10% of the population host large-scale fields. The magnetic stars have fields of 200 G - 10 kG, a similar range to the intermediate-mass

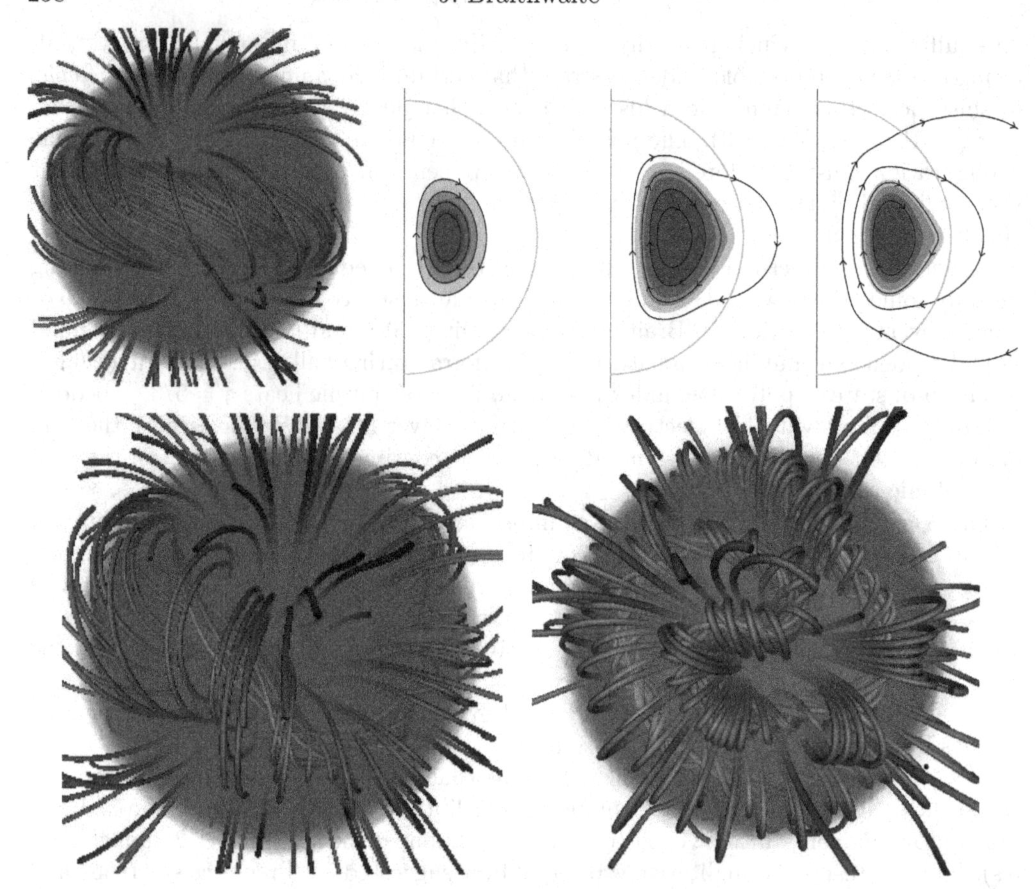

Figure 2. Equilibria found in simulations. *Top left:* an axisymmetric twisted-torus field, consisting of a single twisted flux tube lying in a circle. *Top right:* cross sections of three such approximately axisymmetric equilibria, produced with different radial energy distributions in the initial conditions: on the left, the energy is most centrally concentrated; on the right, a flatter radial energy distribution. Blue shading represents the toroidal component of the field (multiplied by the cylindrical radius) and black lines are poloidal field. *Lower panels:* Non-axisymmetric equilibria: on the left, a somewhat flatter radial energy distribution than the equilibrium at the top right; on the right, a totally flat distribution. The latter corresponds qualitatively to those observed on stars such as τ Sco (see fig. 4). Figures from Braithwaite 2008 & 2009.

stars. Also just like the A stars, their fields have a variety of geometries: whilst some are approximately dipolar, others have a more complicated geometry – see figure 4 for an example. In several other stars similar magnetic fields have been found, dubbed the 'τ Sco clones'.

It is tempting to conclude, therefore, that the phenomenon is simply a continuation of that seen in intermediate-mass stars, and indeed there are no particular theoretical reasons to think otherwise. The historical division between intermediate- and high-mass stars is probably due to difficulty observing the Zeeman effect in hotter stars, and that hotter magnetic stars do not develop chemical peculiarities. There may however be subtle differences due to the greater size of the convective core in hotter stars. If a fossil field is expelled somehow from the core, it might be difficult to maintain a perfect dipole shape at the surface and one expects something more complex; the surface field geometry for instance of τ Sco (figure 4) is consistent with a flat radial energy distribution with little or no magnetic field in the core (see section 2).

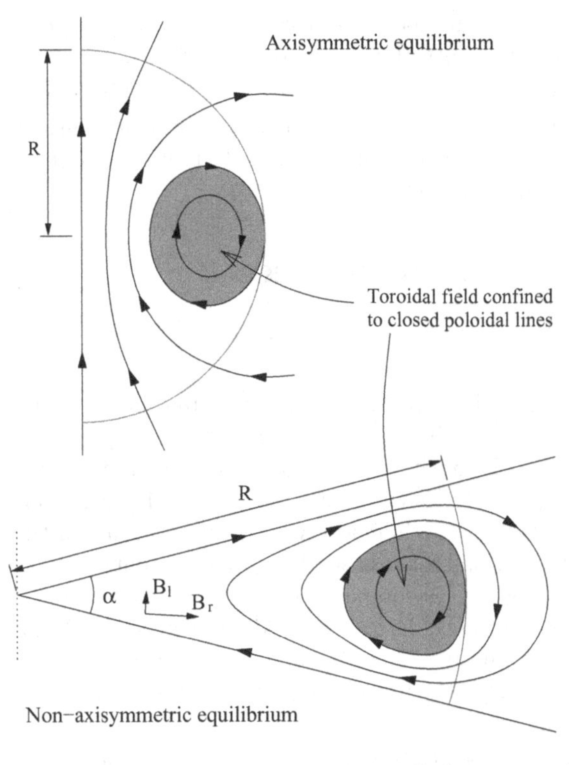

Figure 3. Cross-sections of a twisted flux tube below the surface of a star. Above, the axisymmetric case where the flux tube lies in a circle around the magnetic equator; below, the non-axisymmetric case where the flux tube is narrower and meanders around the star in some apparently random fashion. The stellar surface is shown in green and poloidal field lines (in black) are marked with arrows. The toroidal field (direction into/out of the paper, red shaded area) is confined to the poloidal lines which are closed within the star. Toroidal field outside this area would unwind rather like a twisted elastic band not held at the ends. Figure from Braithwaite 2008.

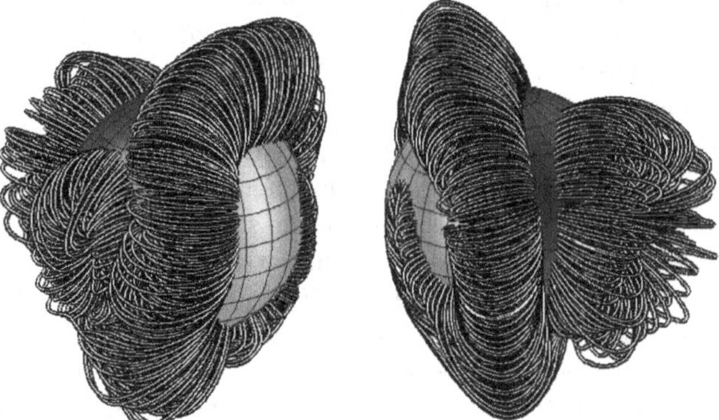

Figure 4. The field geometry on the main-sequence B0 star τ Sco, at two rotational phases, using Zeeman-Doppler imaging. The paths of arched magnetic field lines are presumably associated with twisted flux tubes just below the surface. From Donati *et al.* 2006.

4. The rest of the intermediate-mass population

As far as the 'non-magnetic' intermediate-mass stars are concerned, there has recently been an exciting discovery of magnetic fields in Vega and Sirius, the two brightest A stars in the sky. Zeeman polarimetric observations have revealed weak magnetic fields: in Vega a field of 0.6 ± 0.3 G and in Sirius 0.2 ± 0.1G (Lignières *et al.* 2009, Petit *et al.* 2011). The field geometry is poorly constrained, except that the field should be structured on reasonably large length scales, as cancellation effects would prevent detection of a very small-scale field. Also unconstrained is the existence or otherwise of time variability although Petit *et al.* note that in Vega 'no significant variability in the field structure is observed over a time span of one year'. Given that we have detections in both stars observed, it seems very likely that the rest of the 'non-magnetic' population also have magnetic fields of this kind. There are also theoretical grounds to expect this.

The only model of these fields we have so far is the so-called failed fossil hypothesis (Braithwaite & Cantiello 2013). To understand this model, it is necessary first to consider the formation of the fossil fields described in the previous section. It is informative to compare sizes of terms from the momentum equation

$$\begin{array}{ccccccccc} \frac{d\mathbf{u}}{dt} & = & -\frac{1}{\rho}\nabla P & + & \mathbf{g}_{\rm eff} & + & \frac{1}{4\pi\rho}\nabla\times\mathbf{B}\times\mathbf{B} & - & 2\mathbf{\Omega}\times\mathbf{u} \\ U^2 & & c_{\rm s}^2 & & v_{\rm ff}^2 & & v_{\rm A}^2 & & R\Omega U \end{array} \tag{4.1}$$

where U, $c_{\rm s}$, $v_{\rm ff}$, $v_{\rm A}$, R and Ω are the typical flow speed, sound speed, free-fall velocity, Alfvén speed, length scale (comparable to the stellar radius), and angular frequency of the star's rotation. In the radial direction, obviously the pressure gradient and gravity are in almost perfect balance (note that $\mathbf{g}_{\rm eff}$ includes the centrifugal force). In directions normal to gravity, i.e. on spherical shells, the magnetic field gives rise to motions. In the slowly rotating case where $R\Omega \ll v_{\rm A}$, the Coriolis force is small and the Lorentz force is balanced by inertia; consequently the flow speed U is comparable to $v_{\rm A}$, and the magnetic field evolves on the timescale $R/U \sim \tau_{\rm A}$ where $\tau_{\rm A} \equiv R/v_{\rm A}$ is the Alfvén timescale. As an equilibrium is approached, inertia dies away and its role in balancing the Lorentz force is taken over by the pressure gradient and gravity, which is made possible by non-spherical adjustments to the pressure and density fields.

In a quickly-rotating star where $R\Omega \gg v_{\rm A}$ (accounting for almost all stars), the inertia term is small and the Lorentz force is balanced instead (on spherical shells) by the Coriolis force. This means that the flow speed $U \sim v_{\rm A}^2/R\Omega$ and the field now evolves on the timescale $\tau_{\rm A}^2\Omega$ instead of $\tau_{\rm A}$.

According to the failed fossil theory the magnetic field, instead of having reached an equilibrium long ago (as in the strongly magnetic stars), is still evolving towards equilibrium. Quantitatively, equating the age of the star to the evolution timescale $\tau_{\rm A}^2\Omega$, one can work backwards to find the field strength. In the cases of Vega and Sirius this gives 15 and 5 gauss respectively, the difference coming mainly from Vega's faster rotation. In other words, a field of 15 G evolves dynamically in Vega on a timescale of 400 Myr; it is impossible for a strong field left over from e.g. a pre-MS convective dynamo to decay below that strength in the time available. This somewhat surprising result has an analogy in terrestrial weather systems, where inertia is rarely important and pressure differences would be equalised on the sound-crossing time if the Earth were not rotating.

It is easy to reconcile these predicted field strengths with observed field strengths of 0.6 and 0.2 G: the observations will underestimate the strength of a smaller-scale field, and one naturally expects the surface field to be weaker than the predicted volume-average. Finally since we expect that all of these stars hosted a convective dynamo during the pre-MS, this theory predicts that *all* intermediate-mass stars without strong fossil fields should have magnetic fields of this kind. Finally, note that strongly magnetised stars will still have reached their equilibria relatively quickly: for a field strength of 3 kG and a rotation period of 10 days, $\tau_{\rm A}^2\Omega \sim 2000$ yr.

Note that the failed fossil theory does not take account of various processes which may be going on during the magnetic field's attempt to reach an equilibrium, such as meridional circulation and associated differential rotation; these process may tend to increase the field strength. Further work is necessary to study these effects.

5. The rest of the high-mass population

There have been no direct Zeeman detections of magnetic fields in the rest of the high-mass population, and given that the detection limits are generally somewhat lower than the fields detected so far, it is probably the case that massive stars have the same magnetic bimodality as intermediate-mass stars. All high-mass stars, however, display a wealth of activity and variability not seen in the intermediate-mass stars: line profile variability, discrete absorption components, wind clumping, solar-like oscillations, red noise, photometric variability and X-ray emission (see e.g. Oskinova *et al.* 2012 for a review of these phenomena). Undoubtedly, these are at least partly caused by the strong radiation-driven winds and by the line-deshadowing instability which these winds are subject to, which can produce shocks. However, it is difficult to explain many of these phenomena without also invoking some kind of magnetic activity at the surface.

In principle, massive stars could contain failed fossil fields as described above. Indeed, the younger ages give even stronger fields, although an important difference may be that massive stars never pass through a pre-MS convective phase and begin the MS already with very weak fields. In any case, the presence of convection in layers just below the surface probably makes this irrelevant, as these are expected to produce fields which are stronger and have more interesting observational consequences.

As described by Cantiello *et al.* (2009), massive stars contain two or three thin convective layers close to the surface. These arise because of bumps in the opacity at certain temperatures, which in turn are caused by the ionisation of iron and helium. They are likely to host dynamo activity. Cantiello & Braithwaite (2011) proposed that a magnetic field thus produced can easily rise buoyantly to the surface, thanks to the low density and consequently very short thermal timescale in the overlying layer, which allows heat to diffuse rapidly into magnetic features. This is illustrated in fig. 5. Note that magnetic pressure causes the photosphere to be lower inside magnetic features than in the surroundings. Consequently, magnetic spots on massive stars look bright. This contrasts with spots on convective stars, where the magnetic field has the additional effect of inhibiting upwards heat transfer.

The deepest of these layers – that associated with iron ionisation – is energetically the most interesting. Assuming (a) an equipartition dynamo with $B^2/8\pi \sim \rho u^2/2$ and (b) that $B \propto \rho^{2/3}$ during the buoyant rise of magnetic features through the radiative layer, corresponding to isotropic expansion, field strengths of approximately 5 to 300 G are predicted – as depicted in fig. 6 (for solar metallicity). The field strength depends on the

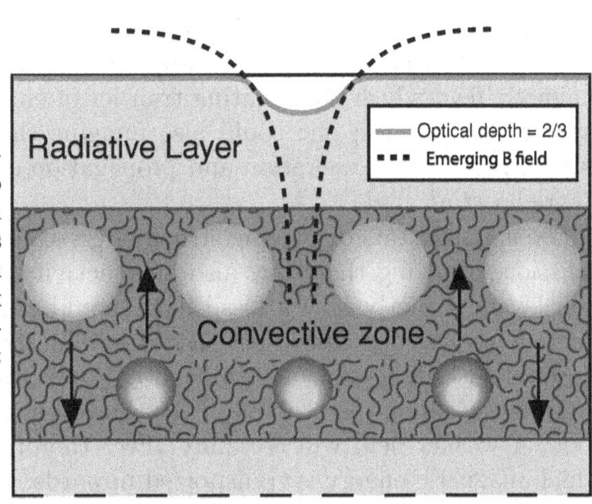

Figure 5. Schematic of the magnetic field generated by dynamo action in a subsurface convection zone. Note that magnetic features should appear as bright spots on the surface, rather than as dark spots as in stars with convective envelopes. From Cantiello & Braithwaite 2011.

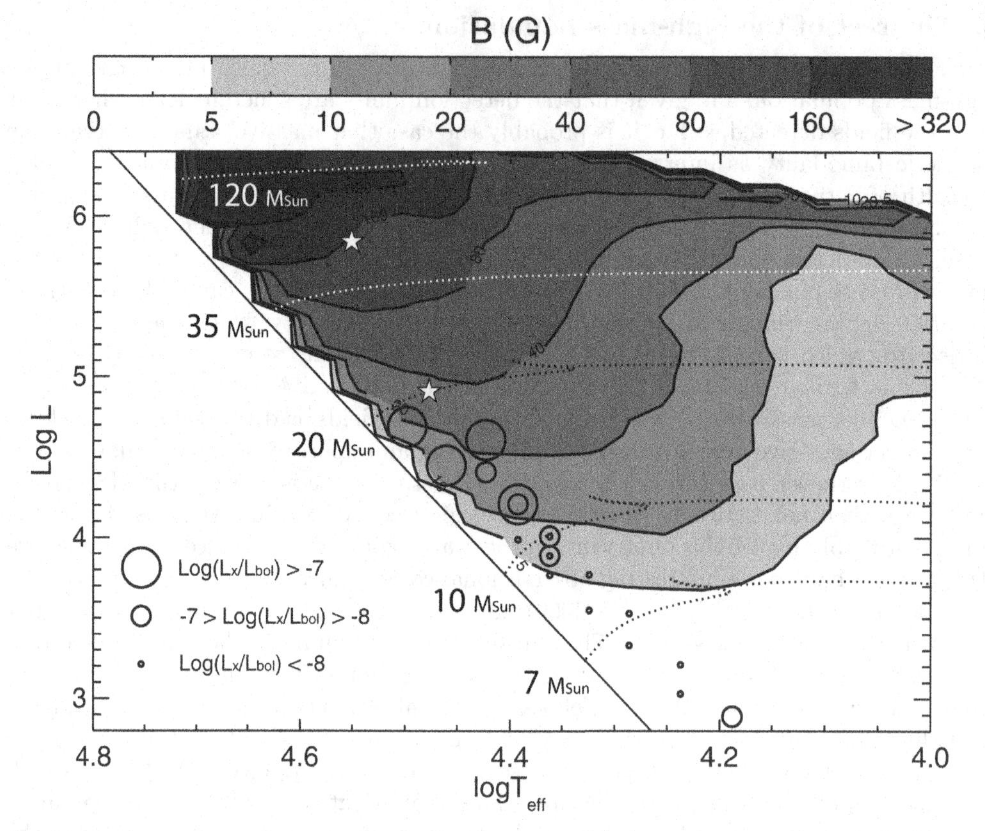

Figure 6. The field strengths predicted at the surfaces of massive stars. This assumes an equipartition dynamo in the convective layer and a $B \propto \rho^{2/3}$ dependence as the field rises through the overlying radiative layer. Also shown are the X-ray luminosities measured in a number of stars, which is at least suggestive of a connection between subsurface convection and X-ray emission. From Cantiello & Braithwaite 2011.

mass and age of the star: higher fields in more massive stars and towards the end of the main sequence. These fields are expected to dissipate energy above the stellar surface and could give rise to, or at least play some role in, the various observational phenomena which are absent in the intermediate-mass stars. Indeed, if the X-ray luminosities of various main-sequence stars are plotted on the HR diagram (fig. 6) a connection with subsurface convection does seem apparent. Of course, we cannot be certain that it is a magnetic field which is mediating transfer of energy and variability from the convective layer to the surface; one could also imagine that internal gravity waves are involved. Simulations of the generation and propagation of such gravity waves were presented by Cantiello *et al.* (2011).

The massive stars displaying strong large-scale fields also display variability and X-ray emission, implying that they also have activity originating in these subsurface convection layers. Exactly how a fossil field should co-exist with a convective dynamo is not clear. A very strong fossil field should inhibit the convection entirely; the field strength required though should be significantly above that corresponding to equipartition with the convection, and is probably not achieved in any star, at least for the iron-ionisation zone. A weaker field will certainly affect the operation of the dynamo and the way in which magnetic energy is transported upwards.

6. Discussion

With dynamo-generated fields, only one theory is required to explain their presence and nature. With fields which are a remnant from an earlier epoch, the questions of their nature and origin are separate. Above, I discussed the nature of fossil and failed fossil fields; I now speculate as to their origins: why do some stars host strong fossil fields, and why do most lack them? Any satisfying formation theory should also explain the range in fossil field strengths of more than two orders of magnitude and the variety of field geometries.

Since there is also a magnetic subset of HAeBe stars, fossil fields can be traced at least as far back as the pre-MS, if not further. The traditional hypothesis has been that variation in magnetic properties on the main sequence can be traced right back to the molecular cloud out of which the star was born. Indeed, the ISM does display a variation in magnetisation, albeit with much less than the range observed in stars. To explain the bimodality we could use some mechanism to destroy fields below a certain threshold; it is interesting to note that the lower cut-off of around 200 G corresponds to the strength at which the magnetic field is in equipartition with the gas pressure at the photosphere. This may or may not be a coincidence.

Another popular hypothesis has been that of mergers. A merger should produce such energy in differential rotation that it is sufficient to convert a fraction $\lesssim 10^{-5}$ of that energy into the magnetic energy of the eventual fossil equilibrium. Ferrario *et al.* (2009) point out that the merger product should be radiative and remain so until the main-sequence, in order to retain the field produced; as described in Braithwaite 2012, a convective star quickly loses any memory of its magnetic past. This might explain why the magnetic fraction increases towards greater masses (Power *et al.* 2007), since more massive pre-MS merger products are more likely to be and remain radiative.

An important clue as to the origin of fossil fields comes from binarity: the binary fraction amongst the magnetic stars has been found to be lower than in the non-magnetic stars (Abt & Snowden 1973). There is apparently a complete lack of Ap stars in binaries with periods of less than about 3 days, except for one known example (HD 200405) with a period of 1.6 days. Obviously, a lack of magnetic stars in close binaries is what one would expect if they are the result of mergers. In the ISM variation hypothesis, it would also not be surprising: a more strongly magnetised cloud core loses more angular momentum and is less likely to break up and form a binary. Puzzling though would still be the binaries with one Ap and one normal A star.

Whatever the process to produce fossil fields, the issue of helicity generation is crucial. Magnetic helicity, a scalar quantity defined as $H \equiv (1/8\pi) \int \mathbf{A} \cdot \mathbf{B} \, \mathrm{d}V$ where $\mathbf{A}$ is the vector potential ($\mathbf{B} = \boldsymbol{\nabla} \times \mathbf{A}$), is what ultimately determines the strength of the fossil equilibrium. This is because it is approximately conserved during relaxation to equilibrium, so that the equilibrium magnetic energy, which one can express in terms of a mean magnetic field strength $\bar{B}$ as $E_{\mathrm{eq}} = (4\pi/3)R^3\bar{B}^2/8\pi$, is given by $E_{\mathrm{eq}} = \zeta H/R$ where R is the stellar radius and ζ is a factor of order unity (either positive or negative) which depends on the geometry of the equilibrium. According to this picture, stars with fossil fields were given a lot of helicity from some process, so that their pre-determined equilibrium field strengths were high. A star with a subgauss field, on the other hand, contains very little helicity and has a correspondingly low pre-determined equilibrium field strength, so low that the equilibrium is never actually reached within the lifetime of the star, owing to the increase of the field evolution timescale $\tau_{\mathrm{A}}^2\Omega$ as the field becomes weaker. Since we expect at least the intermediate-mass stars to host a convective dynamo during their pre-MS, we are left to conclude that this dynamo generates little helicity.

Thinking about negative or positive helicity as a net left-handed or right-handed twist, it seems that the pre-MS dynamo does not produce the required symmetry-breaking for a large non-zero helicity.

Finally, a comment regarding rotation. Whilst 'non-magnetic' A stars generally have rotation periods of a few hours to a day, most Ap stars have periods between one and ten days, and some have periods much greater (see e.g. Abt & Morrell 1995). The slowest periods measured are of order decades and in several cases there are only lower limits. One can speculate that the slow rotation of magnetic stars is a consequence of their magnetism rather than the other way around, especially if the star becomes magnetic while there is still circumstellar material onto which excess angular momentum can be offloaded. Although star-disc interaction is poorly understood, it seems logical that a strong stellar magnetic field leads to a large disc truncation radius, resulting in the spindown of the star until the co-rotation radius becomes comparable. Any theory might struggle though to explain rotation periods of order a century. However, the slow rotation of Ap stars holds only in a very broad sense; rapidly rotating examples like CU Virginis (0.5 d) exist as well. Any given explanation for the slow average rotation may well miss the most important clue: the astonishingly large range in rotation periods, of more than four orders of magnitude.

References

Abt, H. A. & Morrell, N. I. 1995, *ApJS* 99, 135

Abt, H. A. & Snowden, M. S. 1973, *ApJS* 25, 137

Alecian, E., Wade, G. A., Catala, C., *et al.* 2013, *MNRAS* 429, 1027

Aurière, M., Wade, G. A., Silvester, J., *et al.* 2007, *A&A* 475, 1053

Babcock, H. W. 1947, *ApJ* 105, 105

Braithwaite, J. 2008, *MNRAS* 386, 1947

Braithwaite, J. 2009, *MNRAS* 397, 763

Braithwaite, J. 2012, *MNRAS* 422, 619

Braithwaite, J. & Cantiello, M. 2013, *MNRAS* 428, 2789

Braithwaite, J. & Nordlund, Å. 2006, *A&A* 450, 1077

Braithwaite, J. & Spruit, H. C. 2004, *Nature* 431, 819

Browning, M. K., Brun, A. S., & Toomre, J. 2004, *ApJ* 601, 512

Cantiello, M., Langer, N., Brott, I., *et al.* 2009, *A&A* 499, 279

Cantiello, M., Braithwaite, J., Brandenburg, A., *et al.* 2011, *IAU Symposium* 272, 32

Cantiello, M. & Braithwaite, J. 2011, *A&A* 534, A140

Cowling, T. G. 1945, *MNRAS* 105, 166

Donati, J.-F., Howarth, I. D., Jardine, M. M., *et al.* 2006, *MNRAS* , 370, 629

Ferrario, L., Pringle, J. E., Tout, C. A., & Wickramasinghe, D. T. 2009, *MNRAS* 400, L71

Henrichs, H. F. 2012, *Publications de l'Observatoire Astronomique de Beograd* 91, 13

Kochukhov, O., Piskunov, N., Ilyin, I., Ilyina, S., & Tuominen, I. 2002, *A&A* 389, 420

Landstreet, J. D. 2011, *Astronomical Society of the Pacific Conference Series* 449, 249

Lignières, F., Petit, P., Böhm, T., & Aurière, M. 2009, *A&A* 500, L41

MacGregor, K. B. & Cassinelli, J. P. 2003, *ApJ* 586, 480

Oskinova, L., Hamann, W.-R., Todt, H., & Sander, A. 2012, *Proceedings of a Scientific Meeting in Honor of Anthony F. J. Moffat* 465, 172

Petit, P., Lignières, F., Aurière, M., *et al.* 2011, *A&A* 532, L13

Prendergast, K. H. 1956, *ApJ* 123, 498

Tayler, R. J. 1973, *MNRAS* 161, 365

Power, J., Wade, G. A., Hanes, D. A., Aurier, M., & Silvester, J. 2007, *Physics of Magnetic Stars*, 89

Wade, G. A., Grunhut, J., Alecian, E., *et al.* 2013, arXiv:1310.3965

Wright, G. A. E. 1973, *MNRAS* 162, 339

Magnetic Fields throughout Stellar Evolution
Proceedings IAU Symposium No. 302, 2013
P. Petit, M. Jardine & H. Spruit, eds.

doi:10.1017/S1743921314002233

The magnetic characteristics of Galactic OB stars from the MiMeS survey of magnetism in massive stars

G. A. Wade[1], J. Grunhut[2], E. Alecian[3,4], C. Neiner[4], M. Aurière[5], D. A. Bohlender[6], A. David-Uraz[1], C. Folsom[5], H. F. Henrichs[7], O. Kochukhov[8], S. Mathis[9,4], S. Owocki[10], V. Petit[10] and the MiMeS Collaboration

[1]RMC, Canada, [2]ESO, Germany, [3]IPAG, France, [4]LESIA, France, [5]IRAP, France, [6]NRC, Canada, [7]University of Amsterdam, Netherlands, [8]Uppsala University, Sweden, [9]CEA, France,[10]University of Delaware, USA

Abstract. The Magnetism in Massive Stars (MiMeS) project represents the largest systematic survey of stellar magnetism ever undertaken. Based on a sample of over 550 Galactic B and O-type stars, the MiMeS project has derived the basic characteristics of magnetism in hot, massive stars. Herein we report preliminary results.

Keywords. Stars: early-type, Stars: magnetic fields

1. Introduction

Near the main sequence, classical observational tracers of dynamo activity fade and disappear amongst stars of spectral type F, at roughly the conditions predicting the disappearance of energetically-important envelope convection. As an expected consequence, the magnetic fields of hotter stars differ significantly from those of cooler FGKM stars. They are detected in only a small fraction of stars, they are structurally much simpler, and frequently much stronger, than the fields of cool stars (e.g. Donati & Landstreet 2009). They exhibit stability of their large-scale and smaller-scale structures on timescales of decades (e.g. Silvester *et al.* 2013). Most remarkably, their characteristics show no clear correlations with basic stellar properties such as age, mass or rotation.

These puzzling characteristics support a fundamentally different field origin than that of cool stars: that the observed fields are not currently generated by dynamos, but rather that they are *fossil fields*; i.e. remnants of field accumulated or generated during earlier phases of stellar evolution (e.g. Mestel 1999).

The primary aim of the Magnetism in Massive Stars (MiMeS) project is to understand the origin and impact of magnetic fields in hot, massive stars, both from the observational and theoretical perspectives. In this paper we briefly report results from the analysis of the OB stars observed within the MiMeS survey.

2. The MiMeS survey

The MiMeS 'survey component' (SC) was developed to provide critical missing information about the incidence and statistical properties of organized magnetic fields in a large sample of massive stars. Over 4800 high precision (median SNR $\sim$800 per pixel), high resolution ($R \sim 65\,000$) broad-bandpass (364 – 1000 nm) circularly polarized (Stokes V) spectra were acquired for approximately 550 OB stars ranging in spectral type from

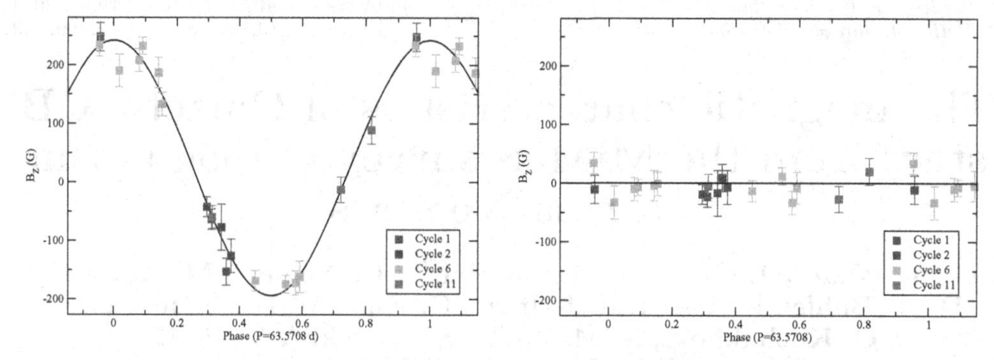

Figure 1. Longitudinal magnetic field of the magnetic O9 subgiant star HD 57682, measured from Stokes V (left) and N (right) spectra acquired with ESPaDOnS and folded according to the rotational period of 63.6 d. Different colours indicate different epochs of observation from December 2008 to December 2010. HD 57682 was discovered to be magnetic within the context of the survey (Grunhut *et al.* 2009); the coherence of the data achieved through continued monitoring firmly establishes the existence and characteristics of the magnetic field, and illustrates the accuracy and precision of MiMeS magnetometry.

B9 to O5, in V magnitude from 0.1 to 13.6, and in luminosity class from V to Ia. Data were acquired using the ESPaDOnS (CFHT), Narval (TBL) and HARPSpol (ESO3.6) spectropolarimeters. Observations were obtained both in the context of competitively-allocated Large Programs (PIs Alecian (ESO), Neiner (TBL), Wade (CFHT)) and PI programs. Reduced, continuum-normalized Stokes I, V and diagnostic null (N) spectra were primarily analyzed using Least-Squares Deconvolution (in particular, the iLSD implementation of Kochukhov *et al.* 2010). Magnetic diagnosis was performed in several ways. First, standard χ^2 analysis (e.g. Donati *et al.* 1997) was performed to quantitatively assess the detection of any signal in the LSD mean V or N profile. Secondly, the longitudinal field was inferred from measurement of the first-order moment of the V and N profiles. Finally, the Bayesian statistics-based method of Petit & Wade (2012) was applied to evaluate the odds ratio and the implied probability distribution of the surface magnetic field strength (again, using both the V and N profiles), under the assumption of a dipolar surface field configuration.

3. Data quality and quality control

The precision of magnetic diagnoses carried out using high resolution Stokes V spectra is a function of signal-to-noise ratio, spectral type and line width. For each target observed within the LPs, the exposure time was normally computed so as to reach a particular sensitivity threshold (0.1, 0.3, 0.5, 1.0 kG estimated surface dipole field) accounting for its particular apparent magnitude, spectral type and line width. Typically, the threshold identified for each star was that which allowed a total exposure time shorter than 2 hours. PI data added to the survey were generally of similar or better quality, due to the nature of many of the PI observing program goals.

In addition to the survey, a major undertaking of the MiMeS project is the detailed analysis of previously known and newly-detected magnetic massive stars as part of the so-called 'targeted component' (TC). Datasets obtained for these stars allow frequent assessment of the reliability of the measurements, which is key to ensuring that the results of the SC are trustworthy. MiMeS quality control focuses on the reproducibility of our measurements and verification of their associated uncertainties (both from the Zeeman signatures and corresponding longitudinal field measurements). This is accomplished in

3 basic ways. First, we confirm that the Stokes V profile and longitudinal field are reproduced within the formal uncertainties in observations of individual magnetic targets acquired at similar rotational phases. Second, we employ the N spectrum to test the instrumental systems for spurious contributions to the polarization. Finally, we examine the statistics of the observed SC sample with no detectable Zeeman signatures in a variety of ways, to ensure that the distribution can be ascribed to noise consistent with the expected observational uncertainties.

4. Results

Of the $\sim$ 550 stars observed within the context of the MiMeS project, approximately 65 show evidence for magnetic fields. Of the detected targets, about 30 were firmly identified as magnetic stars prior to the survey (i.e. these represent the "targeted component"). The statistics that follow are computed ignoring these 30 stars, i.e. based only on the 'blind' MiMeS survey results for $\sim$ 525 stars.

The bulk incidence (i.e. the total number of previously-unknown magnetic stars in the sample relative to the total sample) is $7 \pm 1\%$ (n.b. all incidence uncertainties are computed from counting statistics). Of the approximately 430 B-type stars in the sample previously unknown to host magnetic fields, 32 are found to be magnetic for an incidence of $7 \pm 1\%$. Of the approximately 90 O-type stars in the sample previously unknown to host magnetic fields, 6 are found to be magnetic for an incidence of $7 \pm 3\%$. The incidence as a function of spectral type is illustrated in Fig. 2.

The detected magnetic stars (e.g. Grunhut *et al.* 2009, Alecian *et al.* 2011, Wade *et al.* 2012, Briquet *et al.* 2013) exhibit periodically variable longitudinal fields (with periods in the range of 0.5d to many years) corresponding to organized magnetic fields with important dipole components. The polar strengths of the dipoles range from several hundred G up to 20 kG. The general magnetic characteristics of detected B-type stars and O-type stars are very similar.

Thus we conclude that the incidences and characteristics of large-scale magnetic fields in B and O type stars are indistinguishable, and qualitatively identical to those of intermediate mass stars with spectral types $\sim$F0 to A0 on the main sequence. **The MiMeS survey therefore establishes that the basic physical characteristics of magnetism in stellar radiative zones remains unchanged across more than 1.5 decades of stellar mass, from spectral types F0 ($\sim 1.5\ M_\odot$) to O4 ($\sim 50\ M_\odot$).**

4.1. *Subsample results*

Open clusters: Through the HARPSpol LP, we have acquired magnetic observations of the complete populations of OB stars in 7 Galactic open clusters. In four of the observed clusters, no magnetic stars are detected (although we note that each of these clusters contains only 7 or 8 OB stars). The remaining clusters show incidences of 10-20%, based on samples of 10-60 stars. This can be contrasted with the incidence of $\sim$ 30% inferred by Petit *et al.* (2008) in the ONC. The inferred magnetic incidence vs. cluster age is illustrated in Fig. 2 (right panel).

Of?p stars: We have observed all known Galactic stars of the Of?p class, and detected or confirmed the presence of magnetic field in every star of the sample (see Wade *et al.* 2012 and references therein). We therefore conclude that the **Of?p stars represent a magnetic class, and that their peculiar spectral properties are likely a consequence of the interaction of their winds and magnetic fields** (e.g. Sundqvist *et al.* 2012).

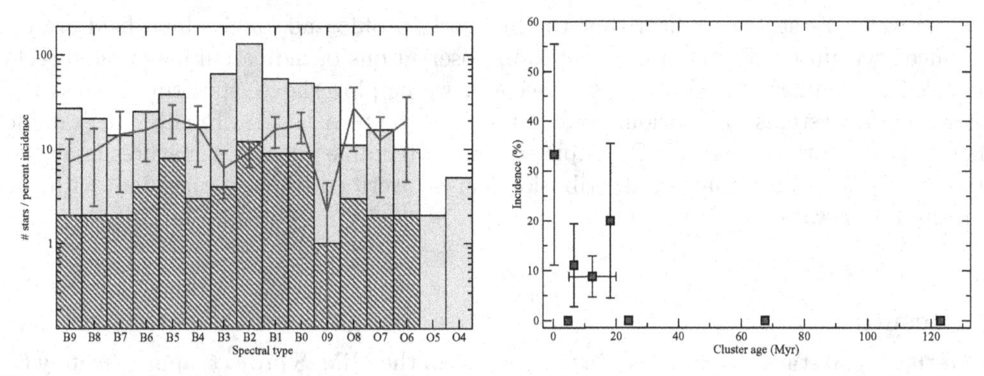

Figure 2. *Left* - Magnetic incidence (fraction of magnetic stars as a function of all stars in the survey sample) versus spectral type. Error bars are inferred from counting statistics. No significant variation of incidence with spectral type is inferred. *Right* - Magnetic incidence versus age for a sample of open clusters observed in the context of the ESO HARPSpol LP, and including results for the ONC by Petit *et al.* (2008).

Pulsating β Cep/SPB stars: We have observed 104 pulsating β Cep and SPB stars. We infer a magnetic incidence in this subsample fully consistent with the larger sample (e.g. Shultz *et al.* 2012), indicating that pulsating β **Cep and SPB stars are not preferentially magnetic.**
Classical Be stars: We have observed a sample of 98 classical Be stars, and failed to obtain any detections despite a magnetic sensitivity similar to that of the larger sample. Based on the magnetic incidence measured for the non-Be B-type stars, we would have expected to detect 10 ± 2 magnetic stars amongst the Be stars. Note that magnetic fields have been detected in other emission-line stars, e.g. Herbig Be stars. Hence we conclude that the lack of detected magnetic classical Be stars is a significant result, indicating that **decretion Keplerian Be discs are not of magnetic origin** (Neiner *et al.* 2012).

5. Conclusion

The MiMeS project provides a broad and robust survey of magnetism in bright Galactic stars with spectral types B9-O4. It establishes that the basic physics of magnetism in stellar radiative zones remains unchanged across 1.5 decades of stellar mass, from main sequence spectral type ~F0 to O4. Papers reviewing the overall project, summarizing and analyzing the magnetic field results, and investigating the physical properties and systematics of various subsamples are currently in preparation, and will be submitted late 2013 through mid 2014.

References

E. Alecian, O. Kochukhov, C. Neiner, G. A. Wade *et al.*, 2011, *A&A*, 536, 6
M. Briquet, C. Neiner, B. Leroy, & P. I. Pápics, 2013, *A&A*, 557, 16
J.-F. Donati, M. Semel, B. D. Carter, D. E. Rees, *et al.*, 1997, *MNRAS*, 291,658
J. Grunhut, G. A. Wade, W. Marcolino, V. Petit *et al.*, 2009, *MNRAS*, 400, 94
O. Kochukhov, V. Makaganiuk, & N. Piskunov, 2010, *A&A*, 524, 5
L. Mestel 1999, in *Magnetic Fields Across the HR Diagram*, ASP Conf. Proc. Vol. 248, 3
C. Neiner, J. Grunhut, V. Petit, A. ud-Doula, A *et al.*, 2012, *MNRAS*, 426, 2738
V. Petit, G. A. Wade, L. Drissen, T. Montmerle, & E. Alecian, 2008, *MNRAS*, 387, 23

V. Petit & G. A. Wade, 2012, *MNRAS*,, 420, 73
J. Silvester, O. Kochukhov & G. A. Wade, 2013, *MNRAS*, submitted
M. Shultz, G. A. Wade, J. Grunhut, S. Bagnulo *et al.*, ApJ 750, 2
J. Sundqvist, A. ud-Doula, S. Owocki, R. Townsend, *et al.*, 2012, *MNRAS*, 423, 21
G. A. Wade, J. Maíz Apellániz, F. Martins, V. Petit, *et al.*, 2012, *MNRAS*, 425, 1278

Magnetic Fields throughout Stellar Evolution
Proceedings IAU Symposium No. 302, 2013
P. Petit, M. Jardine & H. Spruit, eds.

doi:10.1017/S1743921314002245

Magnetic fields of OB stars

A. F. Kholtygin[1], S. Hubrig[2], N. A. Drake[1,3], N. Sudnik[1] and V. Dushin[1]

[1]Saint Petersburg State University, Universitetski pr. 28, Petrodvorets, 198504, Saint Petersburg, Russia, email: afx-afx@mail.ru

[2]Leibniz-Institut für Astrophysik, Potsdam, Germany, email: shubrig@aip.de

[3]Observatório Nacional/MCTI, Rua General José Cristino 77, 20921-400, Rio de Janeiro, Brazil, email: drake@on.br

Abstract. We studied the statistical properties of the magnetic fields of OB stars based on the recent measurements. As the statistically significant characteristic of the magnetic field we use the *rms* magnetic field of the star $\mathcal{B}$. The distribution functions $f(\mathcal{B})$ of magnetic fields of OB stars are evaluated. The function $f(\mathcal{B})$ has a power-law dependence on the $\mathcal{B}$ with an index of about 2-3 and a fast drop below $\mathcal{B} = 100 - 300$ G. We proposed that the compact regions with strong local magnetic fields can contribute to the global magnetic field of O stars.

Keywords. stars: magnetic fields – stars: early-type – stars: spots

---

1. Introduction

At the present time magnetic fields of more than 1000 stars have been detected (Bychkov *et al.* 2009). In order to improve our understanding the nature of the stellar magnetic fields, we investigate the sample of OB stars with measured magnetic fields. As a statistical measure of the magnetic field value we use the root-mean-square (*rms*) magnetic field

$$\mathcal{B} = \sqrt{\frac{1}{n}\sum_{i=1}^{n}(B_l^i)^2}\,, \tag{1.1}$$

where we summarize all measured values of effective magnetic fields B_l^i. Here n is a number of observations. Kholtygin *et al.* (2010) showed that the *rms* field $\mathcal{B}$ depends weakly on the random values of the stellar rotational phase ϕ during the observations.

2. Results

Our sample of the magnetic fields for OB stars consists of the data presented in the catalogue by Bychkov *et al.* (2009) and new data including our recent measurements (Hubrig *et al.* 2011, 2013). Complete list of the references is given in our Stellar Magnetic Fields (SMF) project page (`http://smf.astro.spbu.ru/`). The mean values of magnetic fields, $B_{\rm mean}$, calculated for stars of different spectral types with measured magnetic fields are presented in Fig 1 (left panel).

We analyse the magnetic field distribution function (MFDF) $f(\mathcal{B})$ for all O and B magnetic stars, which can be determined via the following relation: $N(\mathcal{B},\mathcal{B}+\Delta\mathcal{B}) \approx N f(\mathcal{B})\Delta\mathcal{B}$. Here $N(\mathcal{B},\mathcal{B}+\Delta\mathcal{B})$ is the number of stars in the interval of the *rms* magnetic fields $(\mathcal{B},\mathcal{B}+\Delta\mathcal{B})$ and N is the total number of stars with measured field.

The function $f(\mathcal{B})$ for $\mathcal{B} \geqslant \mathcal{B}^{\rm th}$ can be fitted with a power law:

$$f(\mathcal{B}) = A_0\,(\mathcal{B}/\mathcal{B}_0)^{\gamma}\;, \tag{2.1}$$

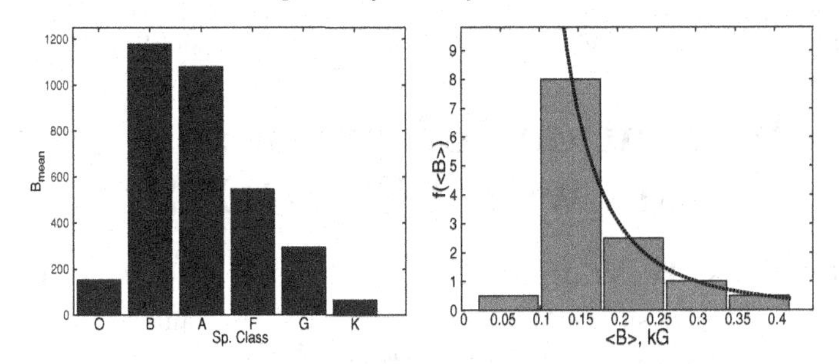

Figure 1. Left panel: Mean magnetic fields for stars of different spectral classes. **Right panel:** Magnetic field distribution function for O stars.

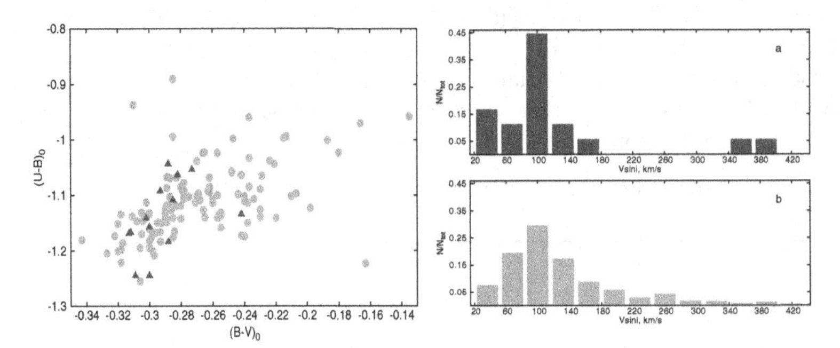

Figure 2. Left panel: Two color diagram for O stars with measured magnetic fields (blue triangles) and with no magnetic fields (green circles). **Right panel:** Distribution of the rotation velocities for magnetic O stars (top) and non-magnetic stars (bottom).

where $\mathcal{B}^{\rm th}$ is a threshold value of the *rms* field, $\mathcal{B}_0 = 1\,$kG. MFDF for O stars is shown in Fig 1 (right panel). The value of $\mathcal{B}^{\rm th}$ is 100 G for O stars and 300 G for B stars. Parameters of the $f(\mathcal{B})$ function are the following: $A_0 = 0.035$, $\gamma = -2.78$ for O stars and $A_0 = 0.34$, $\gamma = -2.09$ for B stars. We can conclude that the distribution of the *rms* mean magnetic fields of OB stars can be described by the power law for $\mathcal{B} > \mathcal{B}^{\rm th}$. The lower value of $\mathcal{B}^{\rm th} = 100\,$G for O stars can be connected with the contribution of the local magnetic fields from the magnetic loops in the stellar photosphere to the global magnetic field (Henrichs & Sudnik 2013). A reason of the sharp decrease of MFDF for $\mathcal{B} < \mathcal{B}^{\rm th}$ remains unknown.

To investigate the difference between magnetic and non-magnetic stars we plot the $(U-B)_0 - (B-V)_0$ diagram for magnetic and non-magnetic O stars in Fig 2 (left panel, blue triangles and green circles respectively). The distribution of the projected rotation velocities for magnetic and non-magnetic O stars are given in Fig 2 (right panel). Analysing the figure we can conclude that there is only a weak difference between magnetic and non-magnetic O stars. The similar conclusion is valid for magnetic B stars.

Acknowledgements. N. A. D. acknowledges support of the PCI/MCTI, Brazil, under the Project 302350/2013-6.

References

Bychkov, V. D., Bychkova, L. V., & Madej, J. 2009, *MNRAS*, 394, 1338

Henrichs, H. & Sudnik, N. P. 2013, *in Massive Stars: From α to Ω, held 10-14 June 2013 in Rhodes, Greece*, (http://a2omega-conference.net, id.71)

Hubrig, S., Schöller, M., Kharchenko, N. V. *et al.* 2011, *A&A*, 528, A151

Hubrig, S., Schöller, M., Ilyin, I. *et al.* 2013, *A&A*, 551, A33

Kholtygin, A. F., Fabrika, S. N., Drake, N. A. *et al.* 2010, *Astron. Lett.*, 36, 370

Magnetic Fields throughout Stellar Evolution
Proceedings IAU Symposium No. 302, 2013
P. Petit, M. Jardine & H. Spruit, eds.

doi:10.1017/S1743921314002257

New observations of chemically peculiar stars with ESPaDOnS†

V. Khalack, B. Yameogo, C. Thibeault and F. LeBlanc

Université de Moncton, Moncton, N.-B., Canada
email: khalakv@umoncton.ca

Abstract. We present the first results of the estimation of gravity and effective temperature for some poorly studied chemically peculiar stars that were recently observed with the spectropolarimeter ESPaDOnS at CFHT. We have analyzed the spectra of HD71030, HD95608 and HD116235 to determine their radial velocity, $v \sin i$ and the average abundance of several chemical species. We have also analyzed our results to verify for possible vertical abundance stratification of iron and chromium in these stars.

Keywords. Stars: chemically peculiar, stars: individual: (HD71030, HD95608, HD116235)

1. Introduction

Accumulation or depletion of chemical elements at certain optical depths brought about by atomic diffusion can modify the structure of stellar atmospheres and it is therefore important to gauge the intensity of such stratification. High resolution (R = 65000) Stokes IV spectra of several CP stars with $v \sin i$ < 35 km s^{-1} were obtained recently with ESPaDOnS in the spectral domain from 3700 Å to 10000 Å. Low rotational velocities of HD71030, HD95608 and HD116235 result in comparatively narrow and unblended line profiles, which are suitable for abundance analysis. They also help produce a hydrodynamically stable atmosphere necessary for the diffusion process to take place. To determine the effective temperature and gravity of these stars (see Table 1), the profiles of some Balmer and He I lines were fitted with the help of FITSB2 code (Napiwotzki *et al.* 2004) in the frame of synthetic fluxes calculated for different values of T_{eff}, $\log g$ and metallicity using the stellar atmosphere code PHOENIX (Hauschildt *et al.* 1997).

2. Spectral analysis and results

Several spectra have been obtained for HD95608 and HD116235 during different nights. We have not detected any significant variability of line profiles and have combined the spectra to obtain one cumulative spectrum for each star taking into account the Doppler shift of the data due to the Earth's orbital motion. In the case of HD71030, we have used a single spectrum, provided by ESPaDOnS, which has sufficient spectral resolution for abundance analysis.

The preliminary results of the abundance analysis are presented in Table 1 for HD71030, HD95608 and HD116235. The line profile simulation is performed using the ZEEMAN2 spectrum synthesis code (Landstreet 1988) and LTE stellar atmosphere model calculated with PHOENIX (Hauschildt *et al.* 1997) for the given values of effective temperature and

† Based on observations obtained at the Canada-France-Hawaii Telescope (CFHT) which is operated by the National Research Council of Canada, the Institut National des Sciences de l'Univers of the Centre National de la Recherche Scientique of France, and the University of Hawaii.

Table 1. Parameters of stellar atmosphere and abundance for studied CP stars.

Parameters	HD71030	HD95608	HD116235
T_{eff}, K	6780 ± 200	9200 ± 200	8900 ± 200
$\log g$	4.0 ± 0.1	4.2 ± 0.2	4.3 ± 0.1
$v \sin i$, km s^{-1}	9 ± 2.0	17.2 ± 2.0	20.2 ± 2.0
v_r, km s^{-1}	38.1 ± 1.0	-10.4 ± 1.0	-10.3 ± 1.0
$\log(FeI/N_{tot})$	-4.35 ± 0.34(148)	-3.95 ± 0.35(139)	-3.86 ± 0.39 (126)
$\log(FeII/N_{tot})$	-4.43 ± 0.34 (23)	-3.82 ± 0.33 (32)	-3.11 ± 0.60 (18)
$\log(CrI/N_{tot})$	-6.11 ± 0.45 (10)	-4.66 ± 1.20 (13)	-5.06 ± 0.89 (3)
$\log(CrII/N_{tot})$	-6.27 ± 0.09 (3)	-5.26 ± 0.59 (14)	-5.14 ± 0.61 (15)
$\log(NiI/N_{tot})$	-5.77 ± 0.26 (40)	-3.63 ± 0.39 (11)	-4.92 ± 0.35 (10)
$\log(TiII/N_{tot})$	-7.32 ± 0.05 (2)	-6.49 ± 0.77 (35)	-6.19 ± 0.83 (6)

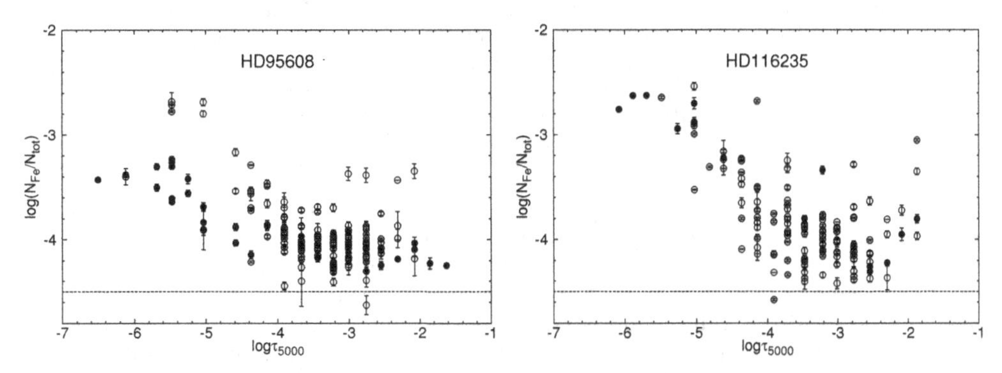

Figure 1. Variation of Fe abundance with optical depth in HD95608 and HD116235. The results for Fe I and Fe II lines are represented by open and filled circles, respectively. Each point provides the element's abundance determined from the analysis of a single line profile taking into account that its core is formed mainly at the given optical depth τ_{5000}. The dashed line represents the solar abundance of iron.

gravity. For each element, we have selected a sample of unblended lines, which are clearly visible in the spectra. The element's abundance, radial velocity and $v \sin i$ were fitted using an automatic minimization routine (Khalack & Wade 2006). For each ion presented in Table 1, the number of analyzed lines is specified in brackets.

For HD71030, HD95608 and HD116235, we have studied vertical stratification of iron and chromium in their atmospheres, taking into account a number of Fe I, Fe II, Cr I and Cr II lines visible in their spectra. The results for Fe I and Fe II lines are in good accordance amongst themselves for each of these three stars. Iron appears to be uniformly distributed in HD71030 and shows some signatures of abundance stratification with optical depth in HD95608 and HD116235 (see Fig. 1). The chromium abundance seems to be constant relative to optical depth in HD71030, but the large scatter of derived chromium abundance does not allow us to determine if chromium is vertically stratified in HD95608 and HD116235.

References

Hauschildt, P. H., Shore, S. N., Schwarz, G. J., Baron, E., Starrfield, S., & Allard, F. 1997, *ApJ*, 490, 803

Khalack, V. & Wade, G. 2006, *A&A*, 450, 1157

Landstreet, J. D. 1988, *ApJ*, 326, 967

Napiwotzki R., Yungelson L., Nelemans G., *et al.* 2004, in: R. W. Hilditch, H. Hensberge & K. Pavlovski (eds.), *ASP-CS* (ASP. San Francisco), 318, 402

Magnetic Fields throughout Stellar Evolution
Proceedings IAU Symposium No. 302, 2013
P. Petit, M. Jardine & H. Spruit, eds.

doi:10.1017/S1743921314002269

The analysis of Li I 6708A line through the rotational period of HD166473 taking into account Paschen-Back magnetic splitting

A. V. Shavrina[1], V. Khalack[2], Y. Glagolevskij[3], D. Lyashko[4], J. Landstreet[5,6], F. Leone[7] and M. Giarrusso[7]

[1]Main Astronomical Observatory, Kyiv, Ukraine, email: shavrina@mao.kiev.ua
[2]Université de Moncton, Moncton, N.-B., Canada, email: khalakv@umoncton.ca
[3]Special Astrophysical Observatory, Nizhnij Arkhyz, Russia
[4]Tavrian National University, Simferopol, Ukraine
[5]University of Western Ontario, London, Canada
[6]Armagh Observatory, Armagh, Northern Ireland - United Kingdom
[7]Università di Catania, Catania, Italy

Abstract. The analysis of Li I 6708 Å line was performed for 6 rotational phases distributed over the whole rotational period (~9.5 years) of HD166473. The magnetic field model was constructed based on the polarimetric measurements from Mathys *et al.* (2007). For each observed phase the modulus of the magnetic field was also estimated from simulation of the Fe II 6147 Å, 6149 Å and Pr III 6706.7 Å line profiles taking into account Zeeman magnetic splitting. The lithium abundance in each phase was obtained from fitting the observed Li I 6708 Å profile with the synthetic one calculated assuming Paschen-Back splitting and estimated magnetic field characteristics from Pr III 6706.7 Å line profile.

Keywords. Stars: chemically peculiar, stars: individual: HD166473

1. Introduction

High-resolution spectra of the strongly magnetic roAp star HD166473 taken for eight rotational phases spread from 0.09 to 0.97 ($P_{rot} = 3513^d.64$) were analyzed to study the variability of the Li I 6708 Å line profile. The line profiles were analyzed by the method of synthetic spectra using a Kurucz model stellar atmosphere with $T_{eff} = 7750$K and $\log g = 4.0$ (Shavrina *et al.* 2006). The magnetic splitting of Fe II 6147 Å, 6149 Å and Pr III 6706 Å lines has been calculated taking into account the Zeeman effect, while for Li I 6708 Å line we have employed the Paschen-Back effect (Khalack & Landstreet 2012, Stift *et al.* 2008). The magnetic field measurements of Mathys *et al.* (2007) have been used to reconstruct a magnetic field configuration employing the method described by Gerth & Glagolevskij (2003). This reconstruction results in the inclination angle of rotational axis to the line of sight $i = 15°$ and the angle $\beta = 75°$.

2. Abundance analysis

The abundances of Li and Pr obtained from the best fit of observed line profiles are shown in the Table 1. Magnetic splitting of Li I line due to Paschen-Back effect is calculated using magnetic field parameters obtained from the modeling of the Pr III line profile. In this procedure we have employed an idea that Li and REE lines are formed near the magnetic poles (see Shavrina *et al.* 2001). We have also estimated the abundances of Ce II, Nd II and Sm II whose lines contribute to the Li blend (see Table 1).

Table 1. Abundance of chemical species at different rotational phases of HD166473.

Phase	0.095	0.26	0.39	0.58	0.64	0.69	0.94	0.00[1]	solar[2]
$\log(N_{LiI}/N_H)$	−8.20		−8.42	−8.23	−8.20	−8.24	−8.23		−10.95
$^6Li/^7Li$	0.0		0.5	0.5	0.5	0.5	0.0		0.03
$\log(N_{CeII}/N_H)$	−7.78		−7.73	−7.60	−7.63	−7.64	−7.78	−7.55	−10.42
$\log(N_{PrIII}/N_H)$	−7.76		−7.86	−7.82	−7.80	−7.80	−7.76	−7.60	−11.28
$\log(N_{NdII}/N_H)$	−8.00		−8.10	−8.22	−8.10	−8.10	−8.30	−7.97	−10.58
$\log(N_{SmII}/N_H)$	−8.45		−8.05	−7.68	−7.68	−7.75	−8.65	−8.25	−11.04
$\log(N_{FeII}/N_H)$	−4.37	−4.42	−4.45	−4.35	−4.35	−4.30	−4.45	−4.31	−4.50

Notes:
[1]Results of Gelbmann *et al.* (2000), [2]solar data are taken from Grevesse *et al.* (2010)

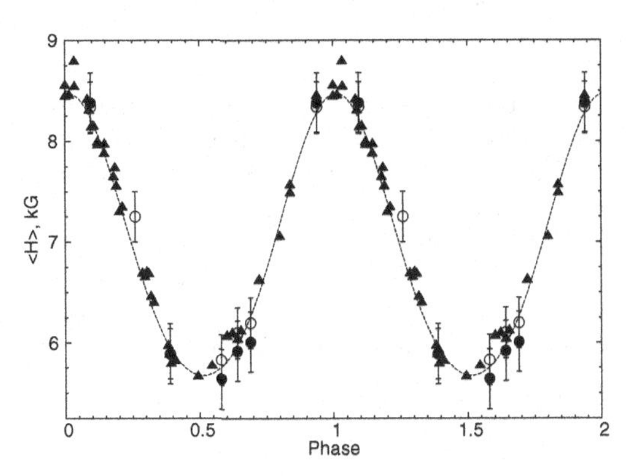

Figure 1. Variation of the mean magnetic field modulus (triangles) with rotational phase for HD166473 (Mathys *et al.* (2007)) and its approximation (dashed curve) in the framework of Gerth & Glagolevskij (2003) model. The magnetic field intensity estimated from the analysis of Pr III 6706 Å line (filled circles) and Fe II 6147 Å, 6149 Å lines (open circles) accords rather well with the Mathys' data almost for all phases.

One can note that:

- all phases show higher than "cosmic" (−8.7 dex) abundance of lithium;
- some differences in the abundance of Li I, the isotopic ratio $^6Li/^7Li$, and REE abundances for different phases, and rather different values of the magnetic field strength obtained from the Pr III 6706.7 Å profile, can be explained by the different location of lithium and REE spots, and by their different stratification with optical depth;
- the value of $^6Li/^7Li$ ratio can be consider as a parameter(measure) of model Li line profile disagreement with observed one;
- additional spectra of this very interesting star (high resolution and high S/N) are required to thoroughly cover the whole rotational period.

References

Gelbmann, M., Ryabchikova, T., Weiss, W. W., Piskunov, N., Kupka, F., & Mathys, G. 2000, *A&A*, 356, 200

Gerth, E. & Glagolevskij, Y. 2003, *Bulletin of SAO*, 56, 25

Grevesse, N., Asplund, M., Sauval, A. J., & Scott, P. 2010, *Ap&SS*, 328, 179

Khalack, V. & Landstreet, J. 2012, *MNRAS*, 427, 569

Mathys, G., Kurtz, D. W., & Elkin, V. G. 2007, *MNRAS*, 380, 181

Shavrina, A. V., Polosukhina, N. S., Khan, S., *et al.* 2006, *Astron. Rep.*, 50, N6, 500

Shavrina, A. V., Polosukhina, N. S., Zverko, J., Khalack, V., Tsymbal, V. V., & Žižňovský, J. 2001, *A&A*, 372, 571

Stift, M. J., Leone, F., & Landi Degl'Innocenti, E. 2008, *MNRAS* 385, 1813

Magnetic Fields throughout Stellar Evolution
Proceedings IAU Symposium No. 302, 2013
P. Petit, M. Jardine & H. Spruit, eds.

doi:10.1017/S1743921314002270

Bp stars in Orion OB1 association

Iosif I. Romanyuk[1] **and Ilya A. Yakunin**[2]

[1]Special Astrophysical Observatory of Russian Academy of Sciences,
369167, Nizhny Arkhyz, Russia
email: roman@sao.ru

[2]Special Astrophysical Observatory of Russian Academy of Sciences,
369167, Nizhny Arkhyz, Russia
email: elias@sao.ru

Abstract. A total of 85 CP stars of various types are identified among 814 members of the Orion OB1 association. We selected 59 Bp stars, which account for 13.4% of the total number of B type stars in the association. The fraction of peculiar B type stars in the association is found to be twice higher than that of peculiar A type stars.

Magnetic field are found in 22 stars, 17 of them are objects with anomalous helium lines. No significant differences are found between the field strengths in the Bp type stars of the association and Bp type field stars. We identified 17 binaries, which make up 20% of the total number of peculiar stars studied which is the standard ratio for CP stars.

Keywords. stars:chemically peculiar-open clusters and associations:individual:Orion OB1

1. Introduction

The Orion constellation hosts one of the most popular groups of early-type stars in the solar neighborhood - the Orion OB1 association. Blaau (1964) identified four regions inside the association - subgroups: a (corresponds to the northen part), b (the Orion's Belt), c (the region located south of Orion's Belt), and d (the very compact region located in the central part of the association) that slightly differ in age and stellar composition.

Most of the objects in the Orion OB1 association are normal hot main sequence stars; however, the association also includes pre-MS objects, like HAEBE stars, T Tau-type stars, and various anomalous (peculiar) stars. Chemically peculiar (CP) stars differ from normal stars by their anomalous chemical composition which shows up in enhanced or weakened intensity of lines of certain elements in the stellar spectrum.

Renson and Manfroid (2009) published the most detailed catalog of CP stars, which includes more than 8200 objects. Over the past quarter-century many new observations of CP stars have been performed. Our aim is to thoroughly analyze massive chemically peculiar and magnetic stars in the Ori OB1 association using all available data. For a review and analysis of the main studies on this subject, see Romanyuk & Yakunin (2012) and Romanyuk *et al.*, (2013).

2. CP stars in the association

Groups of hot stars in the Ori OB1 association have repeatedly attracted the researcher's attention. Here we consider only the issues related to chemically peculiar stars and magnetic field of these objects.

Borra & Landstreet (1979) discovered very strong magnetic fields in a group of B-type stars with enhanced helium lines in young clusters in Orion. Klochkova (1985) performed spectroscopic observations of 24 CP stars to determine the distance moduli and ages of

Table 1. Age of subgroups and number of normal and CP stars

subgroup	age, log t	all stars	CP stars	fraction, %
Ori OB1a	7.05	311	24	7.7
Ori OB1b	6.23	139	21	15.1
Ori OB1c	6.66	350	37	10.6
Ori OB1d	6.0	14	3	21.4

Table 2. Number of CP stars in different subroups

Peculiarity type	total	Ori OB1a	Ori OB1b	Ori OB1c	Ori OB1d
Am	23	6	4	13	0
He-strong	7	1	3	1	2
He-weak	27	7	8	12	0
Si, Si+	19	6	4	8	0
Other	9	3	1	4	1

subroups. Brown *et al.* (1994) reported the results of photometric observations in the Walraven system for 814 stars, for all identified of suspected association members. They determined the effective temperatures, surface gravities, luminosities and masses for all 814 stars. They also detemined the distance moduli and showed that the near and far edges of clouds in the Orion OB1 association are located at the distances of about 320 and 500 pc, respectively. We decided to identify chemically peculiar stars among the 814 objects of this list. We consider a star to be peculiar if it is appeares in the catalog by Renson and Manfroid (2009).

We selected 85 CP stars in the direction of the Ori OB1 association. We list these objects in paper by Romanyuk *et al.*, (2013). Most of them (59 objects) are Bp stars, however, we also found 23 Am and 3 Ap stars. We performed magnetic field observations using the 6m Russian telescope.

The age of subgroups, number of normal and CP stars in each subgroup and fraction of CP stars are presented in Table 1.

We determine distances for most of the stars closer than 250 pc from their Hipparcos parallaxes and we estimate the distances to more distant objects from their temperatures and luminosities. Proper motions have been measured for all CP stars.

The sample of peculiar stars is offset relative to the entire sample both in terms of temperature and luminosity. The fraction of hot stars is greater among peculiar stars. The effective-temperature distribution for the entire sample and CP stars peak at $\log T_e = 3.95$ and $\log T_e = 4.15$ (for details see Romanyuk *et al.*, (2013). The distribution of CP stars in different subroups is presented in Table 2.

The catalog of CP stars by Renson and Manfroid (2009) includes 23 Am stars in the directions of the association. This is surprising given that low-mass Am stars should not have yet evolved enough to settle onto the main sequence. We therefore decided to verify whether the Am stars in question are foreground objects and not members of the association. Parallaxes are available for 14 of the 23 Am stars and they support conclusively the above hypothesis - these objects are located closer than 300 pc. The distances of remaining nine stars can be determined only from analysis of their tempearures and luminosities. The result of analysis is the same: the distances to Am stars are closer than 300 pc.

Occurence frequency of CP stars .

The fraction of CP stars can be seen to be smallest (7.7%) in the oldest subgroup (a) of the association. It is twice higher (15.1%) in the substantially younger subgroup (b). The fraction of peculiar stars is even higher in the youngest subgroup (d), however, it

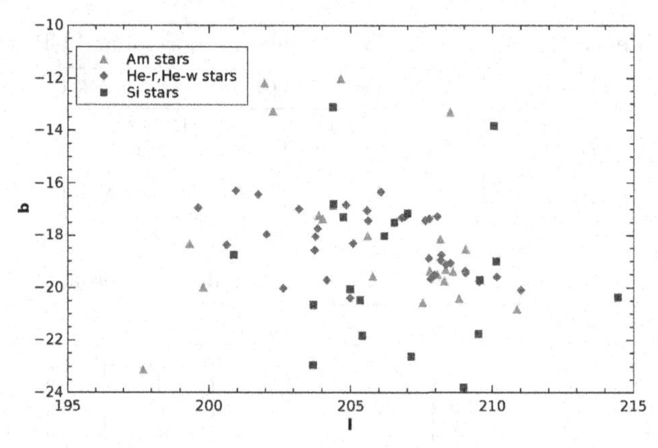

Figure 1. Spatial distribution of CP stars in the Ori OB1 assocation.

contains too few (only 14) objects to allow any statistical conclusions. The lists of Brown *et al.*(1994) contains 661 stars in the two older groups (a) and (c): 81.2% of the total number of stars in the association. We adopt this fraction as standard. Bp stars are much less concentrated in old subroups. Thus the fraction of He-weak stars in the old subgroups (a) and (c) is equal to 70%; that of Si-stars, to 52.6%, and that of He-strong stars, to 28.6%. The discrepancies are quite substantial and significant. The predominance of stars with enhanced helium lines is especially conspicuous in the young subgroups (b) and (d). It is interesting that the fraction of He-strong stars in subgroups (b) and (d), whose ages do not exceed 2 Myr, is three times higher than that of He-weak stars in the same subgroups.

Spatial distribution of CP stars .

Spatial distribution of CP stars in the Orion OB1 association is demonstrated on Fig. 1. The group of Am stars is located separately from the other objects and closer to us. It can be concluded that Am stars are foreground objects which are merely seen projected against the association. Stars with anomalous helium lines belongs to cluster and mostly concentrate in subgroups (b) and (c) −20 objects, whereas subgroup (a) contains only 8 objects. Silicon stars concentrate mostly in subgroups (a) and (c) −14 out of 18 objects.

Binary CP stars.

We identified 17 physical binaries among 85 CP stars of the association, which makes up a typical fraction of 20%. These stars are distributed by peculiarity type as follows: He-strong −5 (out of 7). He-weak −4 (out of 27), SiSi+ −2 (out of 10), Am −1 (out of 23), and other types −5 (out of 9).

The largest fraction of binaries is found among He-strong stars. The two stars with no companion found are HD 36982 and HD 37776. Weak Am stars are very poorly studied, and abnormally low number of binaries may be result of absence of radial velocity measurements. He-weak and SiSi+ stars show the fraction of binaries lower than normal.

3. Magnetic fields

We found 22 magnetic stars in the association of which 21 are Bp stars. Table 3 lists the stars with reliably detected magnetic field.

Eight of the magnetic stars are binary (36.4%). The overwhelming majority (17 out of 22, or 77%) are objects with anomalous helium lines. The fraction of magnetic stars in the inner subgroup (b) is twice higher than in the outer subroups (a) and (c).

Table 3. Magnetic stars in the Ori OB1 association

Star	Sp Pec	B_e extrema	Star	Sp Pec	B_e extrema
HD 35008	B8 Si	−340	HD 36668	B7 He-wk, Si	−2200/+2000
HD 35298	B6 He-wk	−3000/+3000	HD 36916	B8 He-wk, Si	−1100/0
HD 35456	B7 He-wk	−400/+1080	HD 36955	A2 CrEu	−1300/−410
HD 35502	B6 SrCrSi	−2250/−180	HD 37017	B2 He-strong	−2300/−300
HD 35730	B7 He-wk	−450/+250	HD 37058	B2 He-wk, Sr	−1200/+1200
HD 36313	B8 He-wk	−1500/−1100	HD 37140	B8 SiSr	−1050/+400
HD 36429	B6 He-wk	−840/+160	HD 37479	B2 He-strong	−1600/+3500
HD 36485	B2 He-strong	−3700/+3000	HD 37642	B9 He-wk, Si	−3000/+3000
HD 36526	B8 He-wk,Si	−3500/+3400	HD 37687	B7 He-wk	−600/+500
HD 36540	B7 He-wk	−900/+1030	HD 37776	B2 He-strong	−2000/+2000
HD 36629	B3 He-wk	−1300/+1100	HD 290665	B9 SrCrEu	−1600/+5000

We see no significant differences between magnetic stars of the Ori OB1 association and Bp stars in general in terms of magnetic field strength. However, despite poor statistics, He-strong stars can be seen to posses, on the whole, a factor of 1.5 −2 stronger fields than He-wk stars.

4. Conclusion

We thus identified 85 CP stars in the direction toward the young Ori OB1 association. Our CP stars are distributed by peculiarity types as follows: 23 Am stars, 3 Ap stars and 59 Bp stars. The fraction of peculiar B-type stars in the association is twice higher than the corresponding fraction of peculiar A-type stars. The association includes 22 magnetic stars; 21 of them are Bp stars, and only one is an Ap star. We suggest that when the stars were born in the Orion OB1 association the magnetic fields formed mostly in the objects that later developed helium rather than silicon anomalies.

References

Blaau A., 1964, *ARAA*, 2, 236.
Romanyuk I. I. & Yakunin I. A., 2012, *Astrophys Bull.*, 67, 177.
Romanyuk I. I. *et al.*, 2013, *Astrophys Bull.*, 68, 300.
Borra E. F. & Landstreet J. D., 1979, *A&A*, 228, 809.
Klochkova V. G., 1985, *PAZ*, 11,209.
Brown A. G. A. *et al.*, 1994, *A&A*, 289, 101.
Renson P. & Manfroid J., 2009, *A&A*, 498, 961.

Magnetic Fields throughout Stellar Evolution
Proceedings IAU Symposium No. 302, 2013
P. Petit, M. Jardine & H. Spruit, eds.

doi:10.1017/S1743921314002282

"Stellar Prominences" on OB stars to explain wind-line variability

H. F. Henrichs[1] **and N. P. Sudnik**[2]

[1]Astronomical Institute Anton Pannekoek, University of Amsterdam,
Science Park 904, 1098 XH Amsterdam, Netherlands
email: h.f.henrichs@uva.nl

[2]Sobolev Astronomical Institute, Saint Petersburg State University,
Universitetskij pr. 28, Staryj Peterhof, 198504, Saint Petersburg, Russia
email: snata.astro@gmail.com

Abstract. Many O and B stars show unexplained cyclical variability in their winds, i.e. modulation of absorption features on the rotational timescale, but not strictly periodic over longer timescales. For these stars no dipolar magnetic fields have been detected, with upper limits below 300 G. Similar cyclical variability is also found in many optical lines, which are formed at the base of the wind. We propose that these cyclical variations are caused by the presence of multiple, transient, short-lived, corotating magnetic loops, which we call "stellar prominences". We present a simplified model representing these prominences to explain the cyclical optical wind-line variability in the O supergiant λ Cephei. Other supporting evidence for such prominences comes from the recent discovery of photometric variability in a comparable O star, which was explained by the presence of multiple transient bright spots, presumably of magnetic origin as well.

Keywords. stars: early-type, stars: magnetic fields, stars: spots, stars: winds, outflows, stars: atmospheres, stars: rotation, ultraviolet: stars

1. Introduction

Since more than 30 years, spectroscopic UV observations from space have shown that wind variability in massive O and B stars is a wide spread phenomenon. This variability is not strictly periodic, but cyclic (like sunspots) with often a dominant quasi period which scales with the estimated rotation period (days to weeks), or an integer fraction thereof (e.g. Prinja and Howarth (1986), Kaper *et al.* (1996, 1997, 1999), Massa *et al.* (1995), Prinja (1988), Henrichs *et al.* (1988)). The underlying cause or trigger of this variability is, however, unknown. Coordinated ground- and space-based studies show that the major time-variable wind features that are observed in the UV (the so-called DACs (discrete absorption components), must start from very near, or at the stellar surface (e.g. Henrichs *et al.*, 1994, de Jong *et al.*, 2001). The presence of non-radial pulsations or bright magnetic star spots have been suggested as a possible explanation (Henrichs *et al.*, 1994, Cranmer and Owocki, 1996).

Pulsations have been found sofar only for a handful of O stars, mostly by analyzing photospheric spectral line behavior (see Henrichs, 1999), but also from space-based photometry (Walker *et al.*, 2005). Magnetic dipole fields in such stars (except for the well-known class of chemically peculiar stars) have been found since 1998, with the first magnetic O star detection in 1999 (θ^1 Ori C, Donati *et al.*, 2002). Both phenomena are expected, however, to cause strictly periodic variations, and are therefore unlikely the cause of the observed cyclical behavior.

New very promising developments in this field are twofold. First, theoretical studies show that in the sub-surface convective layers in massive stars (Cantiello *et al.*, 2009), magnetic fields can be generated with a short estimated turnover time (Cantiello and Braithwaite, 2011). Second, high-precision space-based photometry of the O giant ξ Per showed rapid variations at the 1 mmag level, incompatible with the observed pulsations, but compatible with the presence of a multitude of corotating bright spots, which live only a few days (Ramiaramanantsoa *et al.* 2013). These spots are suggested to be of magnetic origin as described above, and which are capable of triggering the wind variability in the form of DACs.

In this context, to understand the role of magnetic fields in O and early B stars is a major challenge in massive star research. Here we focus on a simplified model to explain optical wind-line variability in the O supergiant λ Cep, the behavior of which is representative for many other O stars. We conclude with an summarizing overall picture.

2. Cyclical variability in λ Cep O6I(n)fp

The bright runaway star O6I(n)fp star λ Cep ($v \sin i \simeq 214$ km/s, $\log(L/L_{\odot}) \simeq 5.9$, $T_{\rm eff} \simeq 36000$ K, $R \simeq 16R_{\odot}$, $M \simeq 60M_{\odot}$, Markova *et al.* 2004) is a nonradial pulsator ($l = 3, P = 12.3$ h; $l = 5, P = 6.6$ h, de Jong *et al.* 1999), and shows cyclical DACs in the UV resonance lines. Rapid variability have been observed in the He II emission line in 1989 (Fig. 1), as confirmed in later studies at BOAO (Korea) in 2007, used for the analysis below, and in 2012. The dominant period in the UV and optical lines is $\simeq 2$ d. Only redward moving NRP features have been observed, implying an inclination angle greater than, say, 50°. With the adopted radius, the likely rotation period is then $\simeq 4$ d. We also found that the H, He I and other He II lines behave remarkably similarly. This becomes only apparent by considering quotient spectra of subsequent nights (see Fig. 2). We note that we found this covariability in many spectral lines also for other O

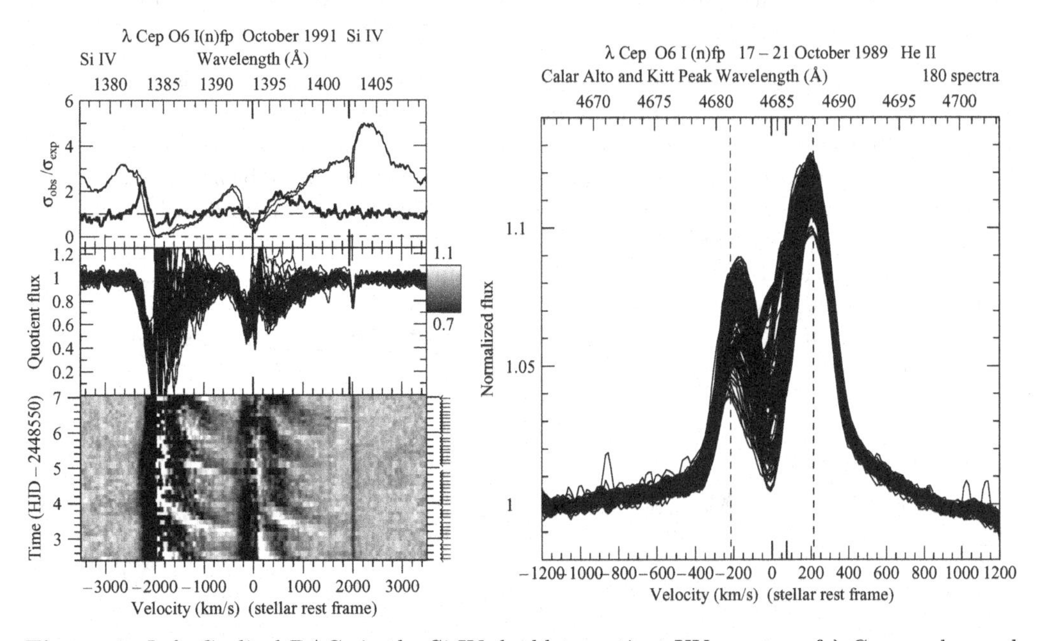

Figure 1. *Left:* Cyclical DACs in the Si IV doublet quotient UV spectra of λ Cep as observed in 1991 (from Kaper *et al.* (1999), Fig. 27). *Right:* Cyclical He II 4686 variability over 4 days as observed in 1989 (Kitt Peak and Calar Alto). Significant changes occur in 15 min.

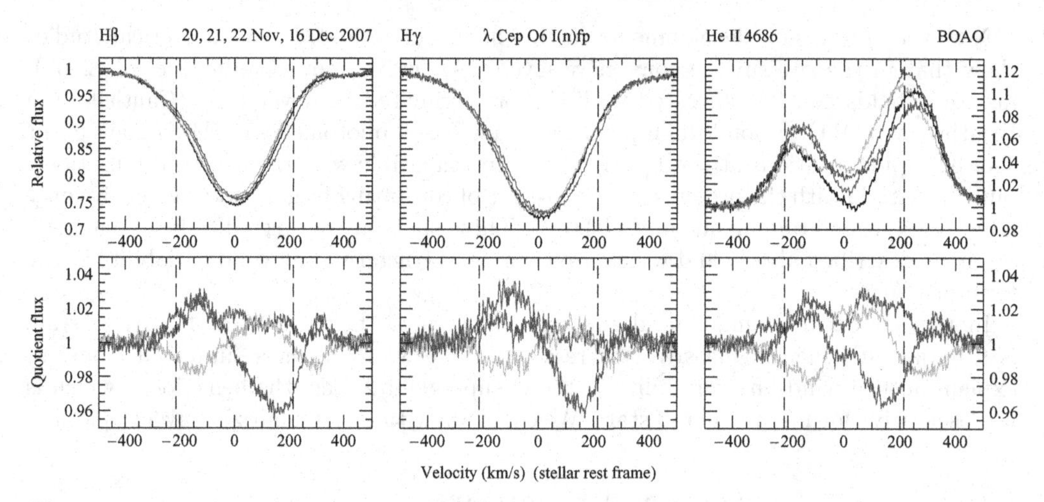

Figure 2. *Top row:* Nightly averages of $H\beta$, $H\gamma$ and He II 4686 lines of the 2007 BOAO dataset. *Bottom row:* Quotient spectra of the same lines of subsequent nights. Note that although the variability of the nightly averages of the $H\beta$ and $H\gamma$ lines are much more subtle than of the He II line, the quotient spectra are qualitatively rather similar. Also note that the variability clearly extends beyond the $|v \sin i|$ range.

stars. This variability extends clearly beyond the $v \sin i$ limits, which implies that the corotating emitting gas sticks out above the stellar surface. This lead us to suggest the term "stellar prominences" to characterize this phenomenon.

3. A simplified model for stellar prominences

As the observed optical profile changes requires emitting gas above the surface, we consider the most simplified model to represent a prominence by a sphere (radius $r \simeq 0.2R_*$, optical depth τ) corotating and touching the surface. The corresponding line profile in velocity space is then given by $F(v) = \exp\left[-A\tau e^{-\frac{1}{2}\left(\frac{v-v_0}{w}\right)^2}\right] \exp\left[(\pi r^2 - A)\tau e^{-\frac{1}{2}\left(\frac{v-v_0}{w}\right)^2}\right]$ in which $0 < A(t) \leqslant \pi r^2$ takes into account the transient and eclipsing geometry, in analogy with exoplanet analysis (including partial eclipses). Each model blob corotates with projected velocity v_0 and has profile width w. The procedure is to adopt a fixed inclination angle (matching $v \sin i$ and R_*) and put a number of blobs around the star to fit the first quotient, which is determined by an assumed rotation period. A least-squares fit gives the best parameters. The remaining quotient spectra should fit correctly, only if the rotation period is correct. This will constrain the rotation period for the adopted stellar parameters. The best fit to the quotients of the He II line (lower right panel of Fig. 2) is shown in Fig. 3. Fitting quotient profiles of other spectral lines, also in other datasets, is work in progress.

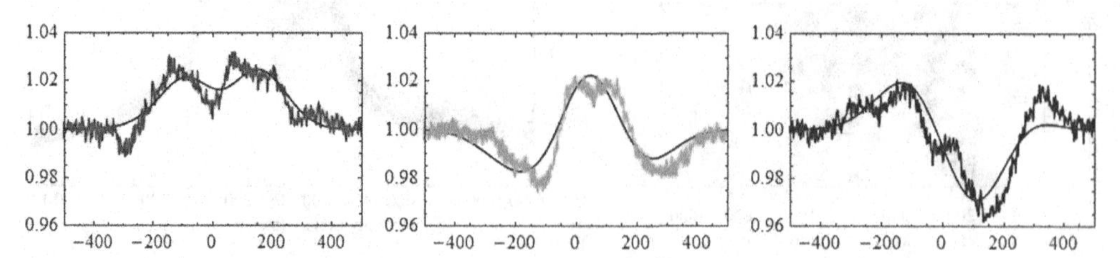

Figure 3. Illustrative model fits to the three He II quotient spectra of the lower right panel of Fig. 2. Six blobs were needed at a 4.07 day rotation period and $i = 55°$. The fit may not be unique, but at least constrains the minimum number of spots and the stellar rotation parameters.

4. Summarizing picture

The overall visualized picture is as follows. The conjectured prominences are located above the corotating magnetic bright spots. Close to the star the enhanced mass flow above a magnetic spot is for simplicity envisaged to be similar to above active regions around sunspots, here labeled as stellar prominences. The mass outflow is stronger above these spots, locating the footpoints of DACs which arise because of the velocity plateau in this outflow, giving enhanced absorption. This part of the wind moves slower then the surrounding wind as it is overloaded because it receives relatively less driving force. The bright spots give rise to small ($\simeq 1$ mmag) photometric variability, as observed in ξ Per, and predicted for λ Cep and other OB stars with DACs. The magnetic spots are bright because of the higher temperature in the deeper layers where the energy is transported by radiation (as opposed to sunspots where the energy transport is by convection). These magnetic fields are the result of the subsurface convection as described by Cantiello and Braithwaite (2011). The lifetime of these generated fields is determined by the relatively short turnover time of this subsurface layer. The maximum strength of the field is estimated by equipartition considerations. The strength of these fields imply the magnetic confinement parameter to be around unity (ud-Doula and Owocki, 2002). The detection, with current instrumentation, of such magnetic field configurations of many magnetic spots distributed over the surface has been studied by Kochukhov and Sudnik (2013). Partial cancellation effects will occur. Such studies, together with observed upper limits, can constrain the number of spots and their distribution.
Future space studies of photometric variability, coordinated with UV and ground-based spectroscopy are needed to test the picture sketched above.

Acknowledgements. We thank Stan Owocki and Marianne Faurobert for constructive and insightful discussions.

References

Cantiello, M., Langer, N., Brott, I., de Koter, A., Shore, S. N., *et al.*, 2009 *A&A*, 499, 279
Cantiello, M. & Braithwaite, J., 2011, *A&A*, 543, 140
Cranmer, S. R. & Owocki, S. A., 1994, *ApJ*, 462, 469
de Jong, J. A., Henrichs, H. F., Schrijvers, C., Gies, D. R., *et al.*, 1999, *A&A*, 345, 172
de Jong, J. A., Henrichs, H. F., Kaper, L., Nichols, J. S., *et al.*, 2001, *A&A*, 368, 601
Donati, J.-F., Babel, J., Harries, T. J., *et al.*, 2002, *MNRAS*, 333, 55
&Henrichs, H. F., 1999, *Lecture Notes in Physics, Berlin Springer Verlag*, 523, 305
Henrichs, H. F., Kaper, L. & Zwarthoed, G. A. A., 1988 *In ESA, Proc. a Decade of UV Astronomy with the IUE Satellite*, Volume 2, 145
Henrichs, H. F., Kaper, L., & Nichols, J. S., 1994, *A&A*, 285, 565
Kaper, L., Henrichs, H. F., Nichols, J. S., *et al.*, 1996, *A&ASS*, 116, 257
Kaper, L., Henrichs, H. F., Fullerton, A. W., *et al.*, 1997, *A&A*, 327, 281
Kaper, L., Henrichs, H. F., Nichols, J. S., & Telting, J. H., 1999, *A&A*, 344, 231
Kochukhov, O. & Sudnik, N., 2013, *A&A*, 554, 93
Markova, N., Puls, J., Repolust, T., & Markov, H., 2004, *A&A*, 413, 693
Massa, D., Fullerton, A. W., Nichols, J. S., *et al.*, 1995, *ApJ*, 452, L53
Prinja R. K., 1988, *MNRAS* 231, 21P
Prinja R. K. & Howarth I. D., 1986, *ApJS* 61, 357
Ramiaramanantsoa, T., Moffat, A. F. J., Chené, A.-N., Desforges, S., Richardson, N. D., Henrichs, H. F., Guenther, D. B., & Kuschnig, R., *et al.*, 2013, *MNRAS* submitted
ud-Doula, A., & Owocki, S. P., 2002, *ApJ*, 576, 413
Walker, G. A. H., Kuschnig, R., Matthews, J. M., *et al.*, 2005, *ApJL*, 623, L145

Magnetic Fields throughout Stellar Evolution
Proceedings IAU Symposium No. 302, 2013
P. Petit, M. Jardine & H. Spruit, eds.

doi:10.1017/S1743921314002294

Partial Paschen-Back splitting of Si II and Si III lines in magnetic CP stars†

Viktor Khalack[1] **and John Landstreet**[2,3]

[1]Université de Moncton, Moncton, N.-B., Canada, email: khalakv@umoncton.ca
[2]University of Western Ontario, London, Canada
[3]Armagh Observatory, Armagh, Northern Ireland – United Kingdom

Abstract. A number of prominent spectral lines in the spectra of magnetic A and B main sequence stars are produced by closely spaced doublets or triplets. Depending on the strength and orientation of magnetic field, the PPB magnetic splitting can result in the Stokes I profiles of a spectral line that differ significantly from those predicted by the theory of Zeeman effect. Such lines should be treated using the theory of the partial Paschen-Back (PPB) effect. To estimate the error introduced by the use of the Zeeman approximation, numerical simulations have been performed for Si II and Si III lines assuming an oblique rotator model. The analysis indicates that for high precision studies of some spectral lines the PPB approach should be used if the field strength at the magnetic poles is $B_{\rm p} >$ 6-10 kG and $V \sin i <$ 15 km s^{-1}. In the case of the Si II line 5041 Å, the difference between the simulated PPB and Zeeman profiles is caused by a significant contribution from a so called "ghost" line. The Stokes I and V profiles of this particular line simulated in the PPB regime provide a significantly better fit to the observed profiles in the spectrum of the magnetic Ap star HD 318107 than the profiles calculated assuming the Zeeman effect.

Keywords. atomic processes – magnetic fields – line: profiles – stars: chemically peculiar – stars: magnetic fields – stars: individual: HD318107

---

1. Introduction

In the analysis and modelling of spectral lines observed in magnetic upper main sequence stars, it is normally assumed that the splitting of the lines is correctly described by the anomalous Zeeman effect. However, in some cases the fine structure splitting of one or both levels involved in a transition is very small, and magnetic fields found in some stars are large enough to produce magnetic splitting comparable in size to this small fine structure splitting. In these cases, the splitting of the line should be calculated taking into account both the fine structure splitting and the magnetic splitting simultaneously. This regime is known as the incomplete or partial Paschen-Back (PPB) effect (Paschen & Back 1921). The partial Paschen-Back regime can occur when observed spectral line profiles are created by closely spaced doublets or triplets.

2. Comparison of PPB and Zeeman profiles

Calculation of the Paschen-Back splitting of spectral lines has been incorporated into the ZEEMAN2 code (Khalack & Landstreet 2012), which allows us to simulate a line

† Based on observations obtained at the Canada-France-Hawaii Telescope (CFHT) which is operated by the National Research Council of Canada, the Institut National des Sciences de l'Univers of the Centre National de la Recherche Scientifique of France, and the University of Hawaii.

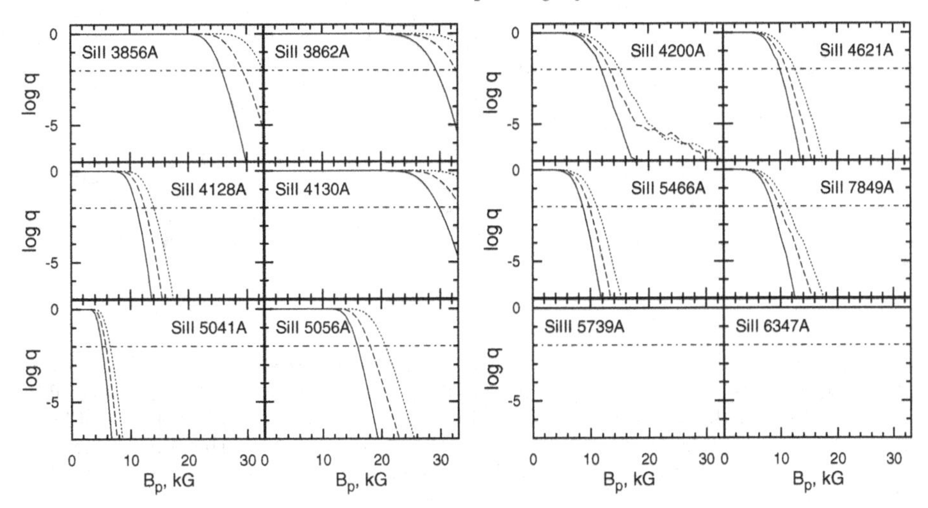

Figure 1. Logarithm of the probability that the noise with $\sigma = 0.04$ (S/N = 250) is masking the difference between the Stokes I calculated assuming PPB and Zeeman splitting, $\log[N_{\rm Si}/N_{\rm H}] = -3.5$ and $V \sin i = 1$ km s^{-1}. The continuous, dashed, and dotted lines correspond to the cases where the axis of magnetic dipole forms an angle $\alpha = 0°$, $45°$, and $90°$ with the line of sight, respectively. The horizontal dash-dotted line corresponds to the probability $q = 0.01$. The bottom of the right-hand panel presents the examples of Si II and Si III lines that show $q \approx 1$ for Stokes I profiles with little or no dependence on the field strength.

profile composed of several spectral lines (blends), some of which are split in the Paschen-Back regime, while the others are split in the Zeeman regime. Initially, this code was created by Landstreet (1988) for the simulation of polarimetric (Stokes $IVQU$) line profiles, and was later modified by Khalack & Wade (2006), who added an automatic minimization of the model parameters using the downhill simplex method. The procedure for calculation of Paschen-Back splitting takes into account the magnetically perturbed energy levels and determines the respective air wavelength and oscillator strength of components, based on the term configurations and the total strength of all lines in the multiplet under consideration.

The simulation of Stokes I profiles of Si II and Si III lines is carried for a star with $T_{\rm eff} = 13000$K, $\log g = 4.0$, zero microturbulent velocity, and an oblique rotator model with a dipolar magnetic field structure, assuming that the field strength at the magnetic pole is B_p and the the axis of magnetic dipole forms an angle $\alpha = 0°$, $45°$ and $90°$ with the line of sight. The observed Stokes I spectra are usually contaminated by observational noise $\sigma \approx \frac{1}{S/N}$. To decide whether the difference between the simulated PPB and Zeeman profiles can be confidently detected above the given noise level we use the *chi square probability function* (Abramowitz & Stegun 1972)

$$q(\chi^2|\nu) = 1 - p(\chi^2|\nu) = \left[\Gamma\left(\frac{\nu}{2}\right)\right]^{-1} \int_{\chi^2/2}^{\infty} t^{\frac{\nu}{2}-1} e^{-t}\, dt \tag{2.1}$$

where ν is the number of resolved elements in the analyzed profile and

$$\chi^2 = \frac{1}{\nu\sigma^2} \sum_{i=1}^{\nu} (I_i^{PB} - I_i^{Ze})^2\,, \tag{2.2}$$

where $I_i^{\rm PB}$ and $I_i^{\rm Ze}$ represent the intensity of the Stokes I profiles at a wavelength point i calculated with the assumption of the PPB and Zeeman splitting, respectively. For $\chi^2 = 35.7$ and $\nu = 20$ the probability that PPB and Zeeman profiles are indistinguishable

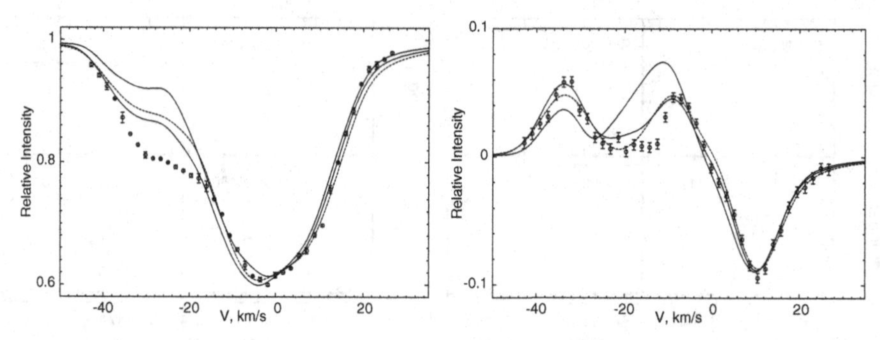

Figure 2. The best fit model for the Stokes I (left) and Stokes V (right) profiles of Si II 5041 Å line in the Zeeman regime (dotted line) and PPB regime (continuous line) in combination with the Fe I and Fe II blends. Open circles define the observed Stokes I and V profiles of Si II 5041 Å line and the dashed curve provides the best fit results for this line assuming PPB splitting without Fe blends. The results for the Zeeman splitting of Si II 5041 Å line without the contribution from the iron lines are not shown here.

Table 1. Approximation of Si II 5041 Å line observed in the spectrum of HD 318107 (phase 0.991) employing the PPB and Zeeman effects (see text for details).

λ, Å	Splitting	$\log[N_{\rm Si}/N_{\rm H}]$	$V\sin(i)$, km s^{-1}	V_r, km s^{-1}	χ^2/ν
5041	PPB	-3.57 ± 0.15	7.9 ± 1.0	-8.4 ± 1.0	16.5
5041	PPB+Fe	-3.80 ± 0.15	7.3 ± 1.5	-8.1 ± 1.0	9.8
5041	Zeeman	-3.74 ± 0.15	7.5 ± 0.8	-8.0 ± 1.0	70.7
5041	Zeeman+Fe	-3.94 ± 0.15	7.5 ± 0.8	-7.7 ± 1.0	40.4

is $q < 0.01$. This confidence level seems to be quite robust to perform an evaluation of the difference between the PPB and Zeeman profiles for individual lines (see Fig. 1).

The Si II 5957 Å, 5979 Å, 6347 Å, 6371 Å and Si III 4552 Å, 4567 Å, 4574 Å lines provide $\log q$ close to zero with almost no dependence on the magnetic field strength for the Stokes I profiles. For the Si II 3856 Å, 3862 Å and 4130 Å lines there is no dependence of $\log q$ on magnetic field strength for $B_{\rm p}$ <20 kG. Meanwhile, for the Si II 4128 Å, 4200 Å, 4621 Å, 5041 Å, 5466 Å and 7849 Å lines, the simulated PPB and Zeeman profiles appear to be different ($q < 0.01$ assuming S/N = 250 and $V\sin i = 1$ km s^{-1}) for a magnetic field strength $B_{\rm p} = $ 5-15 kG.

We have also studied the dependence of the probability (see Eq. 2.1) on the magnetic field strength for higher rotational velocities (1 km s^{-1} $< V\sin i <$ 30 km s^{-1}). The analysis of the Si II 4128 Å, 5041 Å, 5466 Å and 7849 Å lines with S/N = 250 shows that the application of the PPB splitting for the simulation of Stokes I profiles remains important in stars with $V\sin i <$ 15 km s^{-1} when the field strength is $B_{\rm p} >$ 10-15 kG.

From the bottom of the left-hand panel of Fig. 1 we can see that the PPB effect becomes important for simulations of the Stokes I profile of Si II 5041 Å line when $B_p >$5 kG. For this line the difference between the simulated PPB and Zeeman profiles appears due to the so called "ghost" line with $|\Delta J| = 2$ (Khalack & Landstreet 2012). To test the obtained theoretical results, the Stokes I and V profiles of this particular line were analysed for the magnetic Ap star HD318107 ($T_{\rm eff} = 11800$ K, $\log g = 4.2$) using both Zeeman and PPB splitting. The geometry of the magnetic field model used for this star is described by the following parameters: $i = 22°$, $\beta = 65°$, $B_{\rm d} = 25.6$ kG, $B_{\rm q} = -12.8$ kG, $B_{\rm o} = 0.9$ kG that we have adopted from Bailey *et al.* (2011). The weak blending lines Fe I 5040.85 Å, 5040.90 Å and Fe II 5040.76 Å are also taken into account during the simulation, assuming their Zeeman magnetic splitting. From Fig. 2 one can see that the best fit of the observed data is obtained assuming PPB splitting of Si II 5041 Å line and the contribution of the iron blends (see second line in the Tabl. 1). The obtained data are close to the results derived by Bailey *et al.* (2011) from a complex analysis of different spectral lines for this star. The remaining differences between the PPB profiles and the observed spectra may be partially explained in terms of a more complicated actual magnetic field structure and/or horizontal and vertical stratification of the silicon abundance.

3. Summary

If the available polarimetric spectra of the stars with strong magnetic field have S/N ratio higher than 250, the use of PPB splitting during the analysis of spectral lines is necessary to obtain precise results in the framework of an assumed model for the abundance map and the magnetic field structure. In particular, Stokes I profiles of some Si II lines, when calculated with the PPB splitting, differ significantly from those calculated with the Zeeman effect (see Fig. 1). This difference appears due to the different relative intensities and positions of split σ and π-components in the PPB and Zeeman regimes, and due to the "ghost" lines ($|\Delta J| \geqslant 2$) as in the case of Si II 5041 Å line (Khalack & Landstreet 2012). For this line profile a contribution from the "ghost" lines becomes significant for $B_{\rm p} > 5$ kG and an enhanced (by 1 dex) silicon abundance.

References

Abramowitz, M. & Stegun, I. A. 1972, *Handbook of Mathematical Functions*, (10th ed. National Bureau of Standards, Washington), p. 940

Bailey, J. D., Landstreet, J. D., Bagnulo, S. *et al.* 2011, *A&A*, 535, 25

Khalack, V. & Landstreet, J. 2012, *MNRAS*, 427, 569

Khalack, V. & Wade, G. 2006, *A&A*, 450, 1157

Landstreet, J. D. 1988, *ApJ*, 326, 967

Paschen, F. & Back, E. 1921, *Physica*, 1, 261

Magnetic Fields throughout Stellar Evolution
Proceedings IAU Symposium No. 302, 2013
P. Petit, M. Jardine & H. Spruit, eds.

doi:10.1017/S1743921314002300

The Dominion Astrophysical Observatory Magnetic Field Survey (DMFS)

David A. Bohlender and Dmitry Monin

National Research Council Canada, Herzberg Astronomy and Astrophysics Program
5071 W. Saanich Road, Victoria, BC, Canada V9E 2E7
email: `david.bohlender@nrc-cnrc.gc.ca`, `dmitry.monin@nrc-cnrc.gc.ca`

Abstract. In this paper we present a few results from the first three years of an ongoing survey of globally-ordered magnetic fields in relatively faint (down to $V \approx 9$) upper main sequence peculiar stars that we are conducting on the Dominion Astrophysical Observatory (DAO) Plaskett telescope. The DMFS uses the inexpensive DAO polarimeter module, dimaPol, mounted at the Cassegrain focus of the 1.8 m telescope to detect new magnetic stars and determine rotation periods and longitudinal magnetic field curves using medium-resolution ($R \approx 10,000$) circular spectropolarimetry of both the Hβ line and metal lines in an approximately 280 Å wide wavelength region centered on Hβ. By concentrating on the mid-B to A-type peculiar stars, the DMFS provides an extension to the 'Magnetism in Massive Stars' (MIMES) Large Program which concentrated on similar field detections in more massive stars.

Keywords. magnetic fields, instrumentation: polarimeters, stars: magnetic fields

---

1. The DAO Polarimeter Module: dimaPol

The DAO polarimeter module, dimaPol (Monin *et al.* 2012), consists of a fixed achromatic quarter-wave plate, a ferro-electric liquid crystal (FLC) half-wave plate, a simple calcite beam displacer and a mechanical shutter. The quarter-wave plate converts left and right circularly polarized light into linearly polarized light with orthogonal polarization directions. The orientation of the optical axis of the FLC can be rotated by 45° at a rate up to 1 kHz in order to switch the two output beams that exit the beam displacer.

During an exposure, switching of the FLC half-wave plate (at rates between 0.1 to 10 seconds depending on target brightness) is synchronized with charge shuffling on the CCD to reduce instrumental effects. Each observation then results in three spectra on the CCD consisting of distinct left ordinary (LO) and left extraordinary (LE) spectra and a combined right ordinary (RO) and right extraordinary (RE) spectrum.

The magnetic shift produced by the Zeeman effect and observed in a single spectral line is then measured by performing a Fourier cross-correlation of the final LO+LE and RO+RE spectra in a spectral window centered on the spectral line of interest. The magnetic shift in pixels (ΔX) is then translated into a longitudinal magnetic field according to the lines Landé factor (g) and the relation

$$B_l(\mathrm{G}) = \Delta\mathrm{X} \times 0.15/(2 \times 4.67 \times 10^{-13} \lambda^2 \mathrm{g}). \qquad (1.1)$$

For the Hβ line ($g = 1$) this gives B_l (kG) $= 6.8 \times \Delta X$.

2. The DAO Magnetic Field Survey (DMFS)

To date the DMFS consists of multiple observations of more than 125 stars, including well-known magnetic as well as null standards. A number of new magnetic stars have been

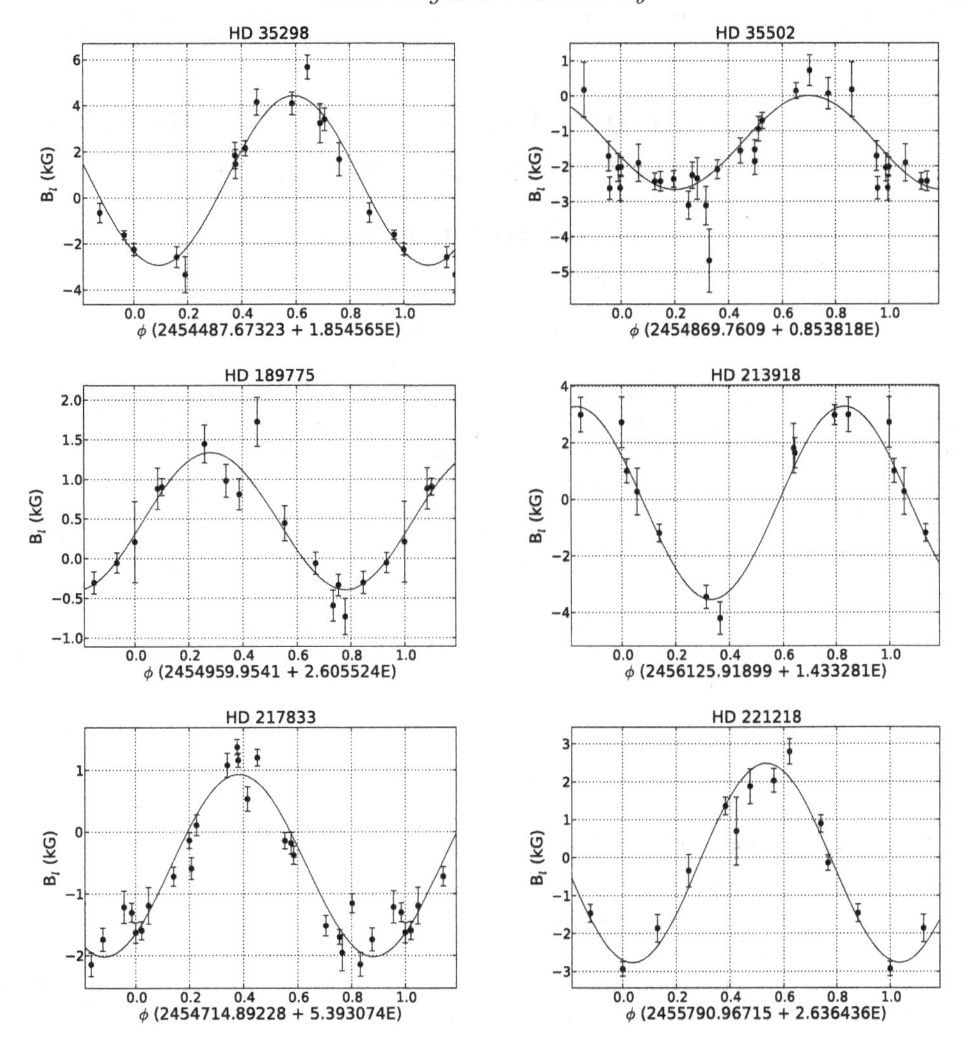

Figure 1. Magnetic field curves for a few of the stars observed in the DMFS. The solid lines through the data points are least-squares sinusoidal fits to the magnetic field variations phased on the periods indicated at the bottom of each plot.

discovered, including several with longitudinal magnetic fields that exceed 2 kG. Objects previously suspected of being magnetic but with a limited number of prior magnetic field observations have also been observed extensively to enable estimates of periods and magnetic field geometries. The unique capabilities of dimaPol also enable us to study the observed (and occasionally very large) differences in stellar longitudinal magnetic fields as measured with the Hβ line and metallic lines. A few magnetic field curves for stars observed in the DMFS are shown in Figure 1. Readers are encouraged to contact the authors to suggest additional program stars to add to the observing program.

References

Monin, D., Bohlender, D., Hardy, T., Saddlemyer, L., & Fletcher, M. 2012, *PASP*, 124, 329

Magnetic Fields throughout Stellar Evolution
Proceedings IAU Symposium No. 302, 2013
P. Petit, M. Jardine & H. Spruit, eds.

doi:10.1017/S1743921314002312

Modeling surface magnetic fields in stars with radiative envelopes

Oleg Kochukhov

Department of Physics and Astronomy, Uppsala University
Box 515, SE-75120 Uppsala, Sweden
email: oleg.kochukhov@physics.uu.se

Abstract. Stars with radiative envelopes, specifically the upper main sequence chemically peculiar (Ap) stars, were among the first objects outside our solar system for which surface magnetic fields have been detected. Currently magnetic Ap stars remains the only class of stars for which high-resolution measurements of both linear and circular polarization in individual spectral lines are feasible. Consequently, these stars provide unique opportunities to study the physics of polarized radiative transfer in stellar atmospheres, to analyze in detail stellar magnetic field topologies and their relation to starspots, and to test different methodologies of stellar magnetic field mapping. Here I present an overview of different approaches to modeling the surface fields in magnetic A- and B-type stars. In particular, I summarize the ongoing efforts to interpret high-resolution full Stokes vector spectra of these stars using magnetic Doppler imaging. These studies reveal an unexpected complexity of the magnetic field geometries in some Ap stars.

Keywords. stars: magnetic fields, stars: early-type, stars: chemically peculiar, polarization

1. Magnetic stars with radiative envelopes

Observational manifestations of magnetic fields in intermediate- and high-mass stars with radiative envelopes differ considerably from the magnetism of solar-type and low-mass stars. As directly observed for the Sun and inferred for many late-type stars, vigorous envelope convection and differential rotation give rise to ubiquitous intermittent magnetic fields, which evolve on relatively short time-scales and generally exhibit complex surface topologies. Although details of the dynamo operation in late-type stars, in particular the relative importance of the convective and tachocline dynamo mechanisms is a matter of debate (Brandenburg 2005) and probably depends on the position in the H-R diagram, it is understood that essentially every cool star is magnetic. Chromospheric and X-ray emission and surface temperature inhomogeneities, which are responsible for characteristic photometric variability, provide an indirect evidence of the surface magnetic fields in cool stars.

In contrast, stars hotter than about mid-F spectral type and more massive than $\sim 1.5 M_{\odot}$ are believed to lack a sizable convective zone near the surface† and therefore are incapable of generating observable magnetic fields through a dynamo mechanism. Nevertheless, about 10% of O, B, and A stars exhibit very strong (up to 30 kG), globally organized (axisymmetric and mostly dipolar-like) magnetic fields that appear to show no intrinsic temporal variability whatsoever. This phenomenon is usually attributed to the so-called fossil stellar magnetism – a hitherto unknown process (possibly related to initial conditions of stellar formation or early stellar mergers) – by which a fraction of

† Very massive stars may generate magnetic fields in the sub-surface Fe convection zone (Cantiello & Braithwaite 2011). However, no signatures of these fields have been discovered so far (Kochukhov & Sudnik 2013).

early-type stars become magnetic early in their evolutionary history. By far the most numerous among the early-type magnetic stars are the A and B magnetic chemically peculiar (Ap/Bp) stars. These stars were the first objects outside our solar system in which the presence of magnetic field was discovered (Babcock 1947). Ap/Bp stars are distinguished by slow rotation (Abt & Morrell 1995) and are easy to recognize spectroscopically by the abnormal line strengths of heavy elements in their absorption spectra. These spectral peculiarities are related to distinctly non-solar surface chemical composition of these stars and non-uniform horizontal (e.g. Kochukhov *et al.* 2004b; Nesvacil *et al.* 2012) and vertical distributions of chemical elements (e.g. Ryabchikova *et al.* 2002; Kochukhov *et al.* 2006). These chemical structures are presumably formed by the magnetically-controlled atomic diffusion (Alecian & Stift 2010) operating in stable atmospheres of these stars.

The chemical spot distributions and magnetic field topologies of Ap stars remain constant (frozen in the atmosphere). Yet, all these stars show a pronounced and strictly periodic (with periods from 0.5 d to many decades) spectroscopic, magnetic and photometric variability due to rotational modulation of the aspect angle at which stellar surface is seen by a distant observer. A subset of cool magnetic Ap stars – rapidly oscillating Ap (roAp) stars – also varies on much shorter time scales ($\sim$10 min) due to the presence of *p*-mode oscillations aligned with the magnetic field (Kurtz & Martinez 2000).

A large field strength and lack of intrinsic variability facilitates detailed studies of the field topologies of individual magnetic Ap stars and statistical analyses of large stellar samples. In this review I outline common methodologies applied to detecting and modeling surface magnetic fields in early-type stars and summarize main observational results. Closely related contributions to this volume include an overview of massive-star magnetism (Wade, Grunhut), a discussion of the stability and interior structure of fossil magnetic fields (Braithwaite), and an assessment of the chemical peculiarities and magnetism of pre-main sequence A and B stars (Folsom).

2. Magnetic field observables for early-type stars

With a few exceptions, investigations of the magnetism of cool stars have to rely on high-resolution spectropolarimetry and to engage in a non-trivial interpretation of the complex polarization signatures inside spectral lines in order to characterize the field topologies (Donati *et al.* 1997). In contrast, a key advantage of the magnetic field studies of early-type stars with stable global fields is availability of a wide selection of magnetic observables that are simple to derive and interpret, but are still suitable for a coarse analysis of the surface magnetic field structure.

The simplest approach to detecting the presence of the field in early-type stars is to perform spectroscopic observation with a Zeeman analyzer equipped with a quarter-wave retarder plate and a beamsplitter. The resulting pair of left- and right-hand circularly polarized spectra will exhibit a shift proportional to the Landé factors of individual spectral lines and to *the mean longitudinal magnetic field* – the line-of-sight field component averaged over the stellar disk. Various versions of this longitudinal field diagnostic technique have been applied by Babcock (1958), Kudryavtsev *et al.* (2006), and Monin *et al.* (2012) to medium-resolution spectra. Landstreet (1980) and Bagnulo *et al.* (2002a) have extended it to, respectively, photopolarimetric and low-resolution spectropolarimetric measurements of polarization in the wings of hydrogen lines.

The mean longitudinal magnetic field represents a particular example of an integral measurement derived from a moment of Stokes V profile (the first moment in this case). Mathys (1989) have generalized the moment technique to other Stokes I and V profile moments. In practice, only *the mean quadratic field* (the second moment of Stokes I) and

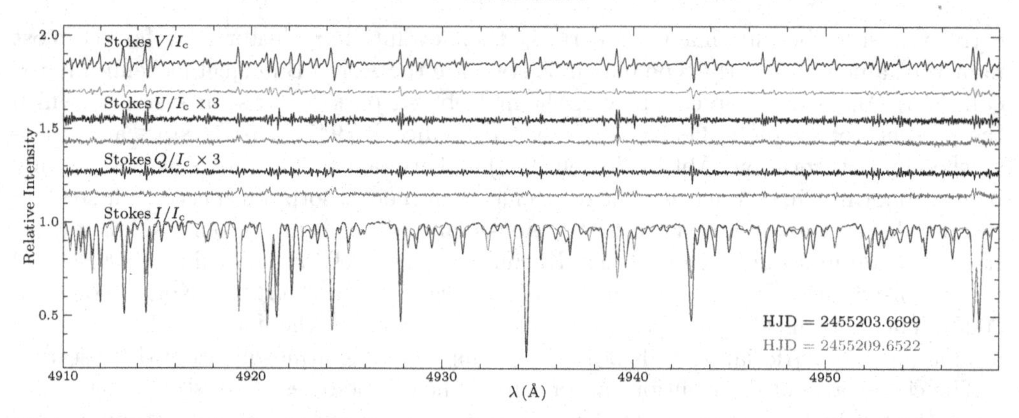

Figure 1. Representative four Stokes parameter Ap-star observations obtained with the HARPSpol spectropolarimeter at the 3.6-m ESO telescope. This figure shows a small segment of the echelle spectra of the cool Ap star HD 24712 at two different rotational phases. The Stokes QUV observations are offset vertically. Clear circular and linear polarization signatures are evident in many individual spectral lines. Detailed analysis of these observations has been published by Rusomarov *et al.* (2013).

crossover (the second moment of Stokes V), in addition to longitudinal field, were systematically studied by Mathys and collaborators using medium-resolution observations of Ap stars (e.g. Mathys & Hubrig 1997).

The three aforementioned magnetic observables can be related to the disk-averaged properties of stellar magnetic field under a number of simplifying and restrictive assumptions (weak lines, weak field, no chemical spots). At the same time, an entirely assumption-free method to diagnose magnetic fields in early-type stars is to measure a separation of the resolved Zeeman-split components in the intensity profiles of magnetically sensitive spectral lines. The resulting *mean field modulus* measurements have been obtained for many Ap stars showing strong magnetic fields in combination with a particularly slow rotation (Mathys *et al.* 1997).

A different approach can be applied to diagnose the transverse magnetic field components which give rise to linear polarization. Cooler Ap stars exhibit measurable *net linear polarization* due to differential saturation of the π and σ components of the strong spectral lines. The resulting net Q and U signals can be detected with a broad-band photopolarimetric technique and related to disk-averaged characteristics of the transverse magnetic field (Landolfi *et al.* 1993).

With the advent of high-resolution spectropolarimeters at the 2–4 m-class telescopes it became possible to directly record and interpret the circular and linear polarization signatures in individual spectral lines. A multi-line LSD (least-squares deconvolution) technique (Donati *et al.* 1997; Kochukhov *et al.* 2010) is often used in conjunction with such observations to obtain very high signal-to-noise (S/N) ratio mean intensity and polarization profiles. LSD analysis greatly facilitates detection of weak magnetic fields (e.g. Aurière *et al.* 2007) and allows to derive the mean longitudinal field, net linear polarization and other profile moments for direct comparison with historic studies. The first high-resolution full Stokes vector investigations were carried out for Ap stars with now decommissioned MuSiCoS spectropolarimeter (Wade *et al.* 2000; Bagnulo *et al.* 2001). More recent analyses took advantage of ESPaDOnS, NARVAL (Silvester *et al.* 2012) and HARPSpol (Rusomarov *et al.* 2013) instruments. An example of exceptionally high quality ($R > 10^5$, $S/N > 400$, 16 rotation phases, coverage of 3800–6910 Å wavelength region; see Rusomarov *et al.* 2013 for details) HARPSpol Stokes spectra of the roAp star

HD 24712 is illustrated in Fig. 1. These observations represent the highest quality full Stokes vector spectra available for any star other than the Sun while covering a much wider wavelength domain than typical for solar polarization observations.

3. Multipolar modeling of magnetic observables

Fitting the phase curves of one or several magnetic observables constitutes the basic method of constraining the stellar magnetic field parameters. A limited information content of integral observables and their relatively simple sinusoidal variation in most Ap stars justifies describing the stellar magnetic field topology with a small number of free parameters. By far the most common approximation is a simple rigidly rotating dipolar field geometry (Stibbs 1950), characterized by an inclination angle of the stellar rotational axis i, magnetic obliquity β, and a polar field strength $B_{\rm p}$. Observations of the phase variation of the mean longitudinal magnetic field alone allows one to constrain $B_{\rm p}$ and β, provided that i is known and not too close to $90^{\rm o}$. In the latter case longitudinal field measurements constrain only the product $B_{\rm p} \sin\beta$.

Occasional deviations of the longitudinal field curves from the sinusoidal shape expected for a dipolar field and the requirement to fit simultaneous measurements of the longitudinal and mean surface fields led to the development of more complex field geometry models, described with additional free parameters. Different low-order multipolar field parameterizations have been considered in the literature. This included a dipolar field offset along its axis (e.g. Preston 1971), an arbitrary offset dipole (Townsend *et al.* 2005), an axisymmetric combination of the aligned dipole, quadrupole, and octupole components (Landstreet 1988), a general non-axisymmetric quadrupolar field (Bagnulo *et al.* 1996), and a potential field geometry formed by a superposition of an arbitrary number of point-like magnetic sources (Gerth *et al.* 1999).

The choice between these different multipolar parameterizations is typically subjective. Stellar observations themselves frequently do not allow one to make a clear-cut distinction between multipolar models established in the framework of different parameterizations, even when several magnetic observables are available for a given star. Indeed, it was demonstrated that the same set of observed magnetic curves can be successfully interpreted with very different actual surface magnetic field distributions, depending on which multipolar parameterization is used (Kochukhov 2006).

Nevertheless, systematic applications of multipolar modeling to a large number of stars allowed to reach interesting conclusions. Using centered dipole fits to the longitudinal field curves, Aurière *et al.* (2007) established the existence of a lower field limit of $B_{\rm p} \approx 300$ G for Ap stars. This threshold of global fossil field strength is likely to be of fundamental importance for understanding the magnetism of intermediate-mass stars (see Lignières, this volume). Among other notable findings one can mention the work by Landstreet & Mathys (2000), who demonstrated that magnetic field axis tends to be more aligned with the stellar rotation axis for Ap stars with long (> 25 d) rotation periods. Bagnulo *et al.* (2002b) confirmed this result using a different multipolar field parameterization. They also found a certain dependence of the relative orientation of the dipolar and quadrupolar components on the stellar rotation rate. Both studies fitted the observed curves of the mean field modulus, longitudinal field, crossover, and quadratic field. Despite the overall statistical agreement, in many individual cases the surface field maps resulting from the application of Landstreet's and Bagnulo's parameterizations appear very different for the same stars. Furthermore, some of the observables are poorly reproduced by either multipolar model, which can be ascribed to the presence of more complex field structures, an unaccounted influence of chemical abundance spots or to shortcomings of the basic assumptions of the moment technique or to a combination of all these effects.

Some applications of the multipolar fitting procedure have incorporated detailed polarized radiative transfer (PRT) modeling of the Zeeman-split Stokes I profiles into solving for the surface field geometry (Landstreet 1988; Bailey *et al.* 2011). This approach enables an independent validation of the magnetic field topology and makes possible to deduce a schematic horizontal distribution of chemical spots in addition to studying the field geometry. However, a feedback of chemical spots on the magnetic observables is not taken into account by these studies.

The most sophisticated and well-constrained non-axisymmetric multipolar models were developed by Bagnulo *et al.* (2000, 2001) for the Ap stars β CrB and 53 Cam using all integral magnetic observables available from the Stokes IV spectra together with the broad-band linear polarization measurements. However, even such detailed models do not guarantee a satisfactory description of the same Stokes parameter spectra from which the magnetic observables are obtained. As found by Bagnulo *et al.* (2001), the multipolar models derived from magnetic observables provide a rough qualitative reproduction of the phase variation of the Stokes V profiles but sometimes fail entirely in matching the Stokes QU signatures observed in individual metal lines. This problem points to a significant limitation of the multipolar models: a successful fit of the phase curves of all magnetic observables is often non-unique and is generally insufficient to guarantee an adequate description of the high-resolution polarization spectra. On the other hand, discrepancies between the model predictions and observations in partially successful multipolar fits cannot be easily quantified in terms of deviations from the best-fit magnetic field geometry model.

4. Interpretation of Stokes parameter spectra

Modeling of high-resolution observations of polarization signatures in individual spectral lines or in mean line profiles represents the ultimate method of extracting information about stellar magnetic field topologies. The wide-spread usage of the LSD processing of high-resolution polarization spectra stimulated development of various Stokes V profile fitting methodologies (Alecian *et al.* 2008; Grunhut *et al.* 2012; Petit & Wade 2012). These studies usually deal with weak-field early-type stars without prominent chemical spots (e.g. magnetic massive or Herbig Ae/Be stars, but not typical Ap stars). The observed LSD profiles are approximated with a dipolar field topology, using a simplified analytical treatment of the polarized line formation. Eventual variations caused by chemical spots or other surface features are not considered. So far, this modeling approach has been applied to a few stars, but it has a potential of providing constraints on magnetic field and other stellar parameters (inclination, $v \sin i$) beyond what can be obtained from the longitudinal field curves (Shultz *et al.* 2012; Kochukhov *et al.* 2013).

A more rigorous approach to the problem of finding the stellar surface magnetic field geometry from spectropolarimetric observations is to perform a full magnetic inversion known as Magnetic (Zeeman) Doppler imaging (MDI). In the MDI methodology developed by Piskunov & Kochukhov (2002) and Kochukhov & Piskunov (2002) the time-series observations in two or four Stokes parameters are interpreted with detailed PRT calculations, taking surface chemical inhomogeneities into account. Simultaneous reconstruction of the magnetic field topology and chemical spot distributions is carried out by solving a regularized inverse problem. Regularization limits the range of possible solutions and is needed to stabilize the iterative optimization process and to exclude small-scale surface structures not justified by the data. Different versions of regularization have been applied for magnetic mapping of early-type stars. Piskunov & Kochukhov (2002) needed only the local Tikhonov regularization (imposing a correlation between neighboring surface pixels) to achieve a reliable reconstruction of an arbitrary magnetic field map from full Stokes vector spectra. However, a more restrictive multipolar regularization (Piskunov & Kochukhov 2002) or a spherical harmonic field expansion (Donati *et al.* 2006) is required to reconstruct a low-order multipolar field in the case when only Stokes I and V observations are available.

Magnetic imaging of Ap stars was recently coupled to a calculation of the atmospheric models that take into account horizontal variations of the atmospheric structure due to chemical spots (Kochukhov *et al.* 2012). However, while the self-consistency between spots and atmospheric models is critical for magnetic mapping of cool stars (Rosén & Kochukhov 2012) and may be needed for mapping He inhomogeneities in He-rich stars, it is generally unnecessary for treating metal spots in Ap stars.

It should be emphasized that, in contrast to temperature or chemical spot imaging from intensity spectra, the MDI with polarization data is not limited to rapid rotators. Polarization is strongly modulated by the stellar rotation even for magnetic stars with negligible $v \sin i$. Numerical experiments and studies of real stars demonstrated that this modulation is sufficient for recovering the field structure at least at the largest spatial scales (e.g. Kochukhov & Piskunov 2002; Donati *et al.* 2006).

Several studies (Kochukhov *et al.* 2002; Lüftinger *et al.* 2010; Rivinius *et al.* 2013) applied MDI to high-quality time-series circular polarization spectra of several early-type magnetic stars. These Stokes V analyses did not find any major deviations from dipolar field topologies. At the same time, they found numerous examples of chemical spot maps showing diverse and complex distributions of chemical elements, often not correlating in any meaningful way to the underlying simple magnetic field geometry. These results are difficult to explain in the framework of atomic diffusion theory because the latter expects a very similar behavior for different elements and a definite correlation between the spots and magnetic field (Alecian & Stift 2010).

A couple of other studies have attempted to examine the surface magnetic field structure in B-type stars with fields deviating significantly from dipolar geometry. A study of the He-peculiar star HD 37776 (Kochukhov *et al.* 2011) has simultaneously interpreted a longitudinal field curve and moderate-resolution Stokes V spectra. This analysis inferred a decisively non-axisymmetric, complex and strong (up to 30 kG locally) magnetic field, but ruled out a record $\sim$100 kG quadrupolar field proposed for this star by previous longitudinal field curve fits. An MDI study of the early B-type star τ Sco (Donati *et al.* 2006) revealed the presence of weak complex magnetic field configuration, which exhibits no appreciable temporal variation (Donati & Landstreet 2009). These two studies have proven that stable complex fields can exist in early-type stars and tend to be found in the most massive objects. Despite these impressive MDI results, it should be kept in mind that the Stokes IV inversions are intrinsically non-unique and their outcome is highly sensitive to additional constraints adopted to stabilize inversions. Details of the magnetic field maps of HD 37776 and τ Sco are likely to change if different regularizations or different forms of spherical harmonic expansion are adopted for magnetic imaging.

Numerical tests of MDI inversions (Donati *et al.* 1997; Kochukhov & Piskunov 2002) have concluded that reconstruction of stellar magnetic field topologies from the full Stokes vector data should be considerably more reliable and resistant to cross-talk and non-uniqueness problems in comparison to the Stokes IV imaging. In particular, a four Stokes parameter inversion is able to recover the field structure without imposing any *a priori* constraints on the global field geometry. The first Stokes $IQUV$ MDI studies exploiting this possibility were carried out for the Ap stars 53 Cam (Kochukhov *et al.* 2004a) and

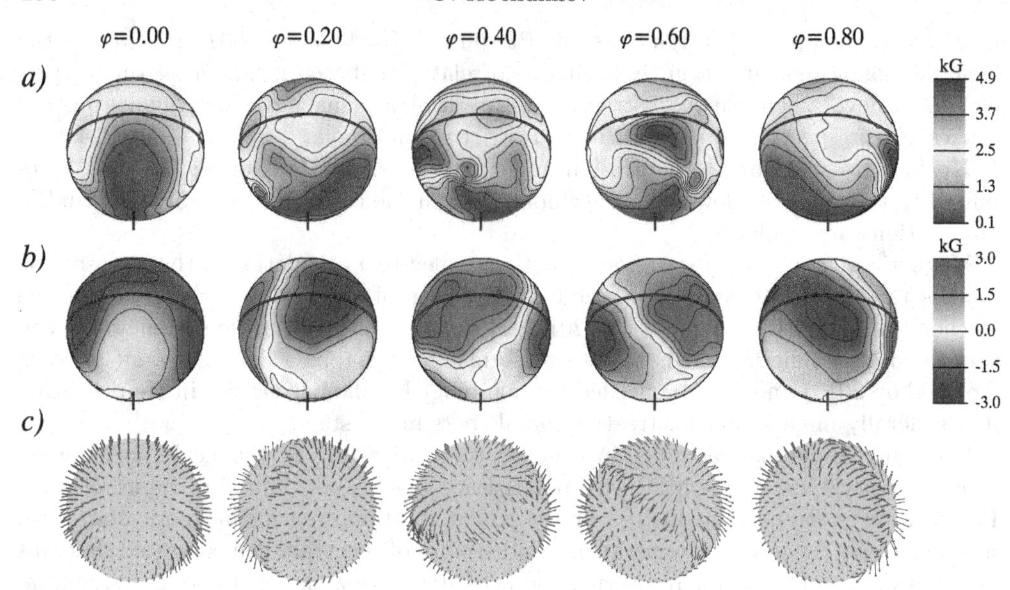

Figure 2. Magnetic field geometry for the prototypical magnetic Ap star α^2 CVn reconstructed by Silvester *et al.* (2013, submitted) using MDI. These magnetic inversions were based on a set of high-resolution four Stokes parameter spectra described by Silvester *et al.* (2012). The star is shown at five different rotation phases and an inclination angle $i = 120°$. The spherical maps show a) surface distribution of the magnetic field strength, b) distribution of the radial magnetic field component, and c) vector plot of magnetic field. The field complexity is evident, especially in the field strength map.

α^2 CVn (Kochukhov & Wade 2010) using the MuSiCoS spectra collected by Wade *et al.* (2000). Both studies succeeded in reproducing the phase variation of the circular and linear polarization signatures in metal lines with the magnetic maps containing small-scale deviations from the dominant dipolar-like field component. Interestingly, it was the inclusion of Stokes QU profiles in the magnetic inversions that allowed to ascertain the presence of complex fields. The deviations from dipolar field configurations occur on much smaller spatial scales than can be described by a quadrupolar field. Thus, the widely adopted dipole+quadrupole expansion may not be particularly useful for interpreting the Stokes $IQUV$ spectra of Ap stars.

The limited resolution, S/N ratio, and wavelength coverage of the MuSiCoS spectra allowed us to model the Stokes $IQUV$ profiles of only 2–3 saturated metal lines. A new generation of MDI studies is currently underway, taking advantage of the higher-quality Stokes profile data available from ESPaDOnS, NARVAL, and HARPSpol spectropolarimeters. In particular, Silvester *et al.* (submitted) have reassessed the magnetic field topology of α^2 CVn using new observations and extending the PRT MDI modeling to a large number of weak and strong Fe and Cr lines. The resulting magnetic maps (Fig. 2) show some dependence of the mapping results on the spectral line choice but generally demonstrate a very good agreement with the magnetic topology found by Kochukhov & Wade (2010) from observations obtained about 10 years earlier. Thus, the small-scale magnetic features discovered in Ap stars by MDI studies do not exhibit any temporal evolution. The new four Stokes parameter observations of α^2 CVn also demonstrate very clearly the necessity of going beyond a low-order multipolar field model and the role of Stokes QU spectra in recognizing this field complexity. As illustrated by Fig. 3, an attempt to reproduce the observations of α^2 CVn with either a pure dipole

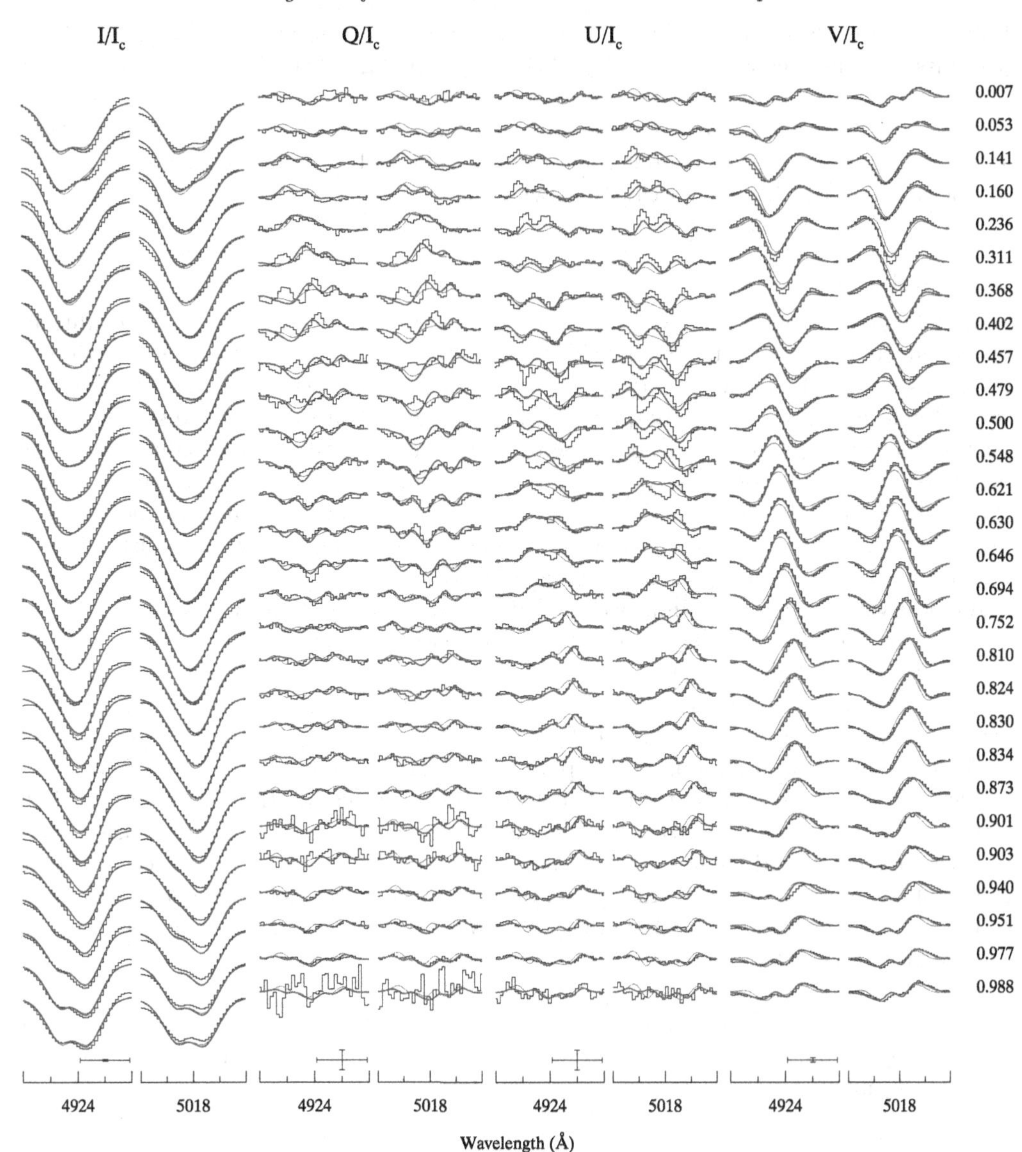

Figure 3. Attempt to reproduce the four Stokes parameter observations (histogram) of the two strong Fe II lines in the spectrum of Ap star α^2 CVn with a dipolar field model (red thin line) and a non-axisymmetric dipole plus quadrupole magnetic configuration (blue thick line). It is evident that none of the models fits the QU observations at all rotational phases, thus requiring a more structured field geometry for this star (see Fig. 2). However, the presence of this small-scale field cannot be recognized from the Stokes IV data alone.

or dipole+quadrupole geometries fails for Stokes QU while providing a reasonable fit to Stokes IV.

Despite an improved sensitivity to complex fields, not all four Stokes parameter magnetic inversions point to local deviations from dipolar field topologies. The ongoing HARPSpol study of the cool Ap star HD 24712 does not reveal any significant non-dipolar field component (Rusomarov *et al.* 2013 and in this volume). The preliminary conclusion of this work is that the previous MDI analysis of this star carried out by

Lüftinger *et al.* (2010) using the Stokes I and V data and assuming a dipolar field did not miss any significant aspects of the field topology. HD 24712 is much cooler, hence less massive and/or older than 53 Cam and α^2 CVn, raising an intriguing possibility of the mass and/or age dependence of the degree of magnetic field complexity in early-type stars. The presence of very complex non-dipolar fields only in relatively massive B-type magnetic stars (HD 37776, τ Sco) agrees with this trend.

5. Conclusions

Modeling of magnetic fields in early-type stars with radiative envelopes has traditionally assumed low-order multipolar field configurations and focused on interpretation of the phase curves of the mean longitudinal magnetic field and other integral magnetic observables derived from moderate-quality circular polarization spectra. The studies based on this methodology have reached several important statistical conclusions about the nature of magnetic fields in Ap and related stars. This includes the discovery of a lower threshold of the surface magnetic field strength, demonstration of an alignment of the magnetic and rotational axes in stars with long rotation periods, and confirmation of long-term stability stability of fossil magnetic fields.

As observational material improves and high-resolution spectra in several Stokes parameters become more widely available, the focus of magnetic field modeling studies gradually shifts to direct interpretation of the polarization signatures in spectral line profiles. The most powerful version of this methodology – magnetic Doppler imaging inversions based on detailed calculation of polarized spectra – has been applied to a handful of Ap stars observed in all four Stokes parameters. These studies revealed significant local deviations from a dominant dipolar field topology, suggesting that the magnetic field structure of early-type stars with fossil fields is more complex than thought before and that the degree of field complexity increases with stellar mass. With the exception of a couple of massive stars with distinctly non-dipolar fields, these small-scale field structures could be recognized and fully characterized only using spectropolarimetric observations in all four Stokes parameters.

Acknowledgements. The author is a Royal Swedish Academy of Sciences Research Fellow, supported by the grants from Knut and Alice Wallenberg Foundation and Swedish Research Council.

References

Abt, H. A. & Morrell, N. I. 1995, *ApJS*, 99, 135
Alecian, E., Catala, C., Wade, G. A., *et al.* 2008, *MNRAS*, 385, 391
Alecian, G. & Stift, M. J. 2010, *A&A*, 516, A53
Aurière, M., Wade, G. A., Silvester, J., *et al.* 2007, *A&A*, 475, 1053
Babcock, H. W. 1947, *ApJ*, 105, 105
Babcock, H. W. 1958, *ApJS*, 3, 141
Bagnulo, S., Landi degl'Innocenti, M., & Landi degl'Innocenti, E. 1996, *A&A*, 308, 115
Bagnulo, S., Landolfi, M., Mathys, G., & Landi Degl'Innocenti, M. 2000, *A&A*, 358, 929
Bagnulo, S., Wade, G. A., Donati, J.-F., *et al.* 2001, *A&A*, 369, 889
Bagnulo, S., Szeifert, T., Wade, G. A., Landstreet, J. D., & Mathys, G. 2002a, *A&A*, 389, 191
Bagnulo, S., Landi Degl'Innocenti, M., Landolfi, M., & Mathys, G. 2002b, *A&A*, 394, 1023
Bailey, J. D., Landstreet, J. D., Bagnulo, S., *et al.* 2011, *A&A*, 535, A25
Brandenburg, A. 2005, *ApJ*, 625, 539
Cantiello, M. & Braithwaite, J. 2011, *A&A*, 534, A140

Donati, J.-F., Howarth, I. D., Jardine, M. M., *et al.* 2006, *MNRAS*, 370, 629
Donati, J.-F. & Landstreet, J. D. 2009, *ARA&A*, 47, 333
Donati, J.-F., Semel, M., Carter, B. D., Rees, D. E., & Collier Cameron, A. 1997, *MNRAS*, 291, 658
Gerth, E., Glagolevskij, Y. V., Hildebrandt, G., Lehmann, H., & Scholz, G. 1999, *A&A*, 351, 133
Grunhut, J. H., Wade, G. A., Sundqvist, J. O., *et al.* 2012, *MNRAS*, 426, 2208
Kochukhov, O. & Piskunov, N. 2002, *A&A*, 388, 868
Kochukhov, O., Piskunov, N., Ilyin, I., Ilyina, S., & Tuominen, I. 2002, *A&A*, 389, 420
Kochukhov, O., Bagnulo, S., Wade, G. A., *et al.* 2004a, *A&A*, 414, 613
Kochukhov, O., Drake, N. A., Piskunov, N., & de la Reza, R. 2004b, *A&A*, 424, 935
Kochukhov, O. 2006, *A&A*, 454, 321
Kochukhov, O., Tsymbal, V., Ryabchikova, T., Makaganyk, V., & Bagnulo, S. 2006, *A&A*, 460, 831
Kochukhov, O. & Wade, G. A. 2010, *A&A*, 513, A13
Kochukhov, O., Makaganiuk, V., & Piskunov, N. 2010, *A&A*, 524, A5
Kochukhov, O., Lundin, A., Romanyuk, I., & Kudryavtsev, D. 2011, *A&A*, 726, 24
Kochukhov, O., Wade, G. A., & Shulyak, D. 2012, *A&A*, 421, 3004
Kochukhov, O. & Sudnik, N. 2013, *A&A*, 554, A93
Kochukhov, O., Makaganiuk, V., Piskunov, N., *et al.* 2013, *A&A*, 554, A61 Kochukhov, O. & Sudnik, N. 2013, *A&A*, 554, A93
Kudryavtsev, D. O., Romanyuk, I. I., Elkin, V. G., & Paunzen, E. 2006, *MNRAS*, 372, 1804
Kurtz, D. W. & Martinez, P. 2000, *Baltic Astronomy*, 9, 253
Landolfi, M., Landi degl'Innocenti, E., Landi degl'Innocenti, M., & Leroy, J. L. 1993, *A&A*, 272, 285
Landstreet, J. D. 1980, *AJ*, 85, 611
Landstreet, J. D. 1988, *ApJ*, 326, 967
Landstreet, J. D. & Mathys, G. 2000, *A&A*, 359, 213
Lüftinger, T., Kochukhov, O., Ryabchikova, T., *et al.* 2010, *A&A*, 509, A71
Mathys, G. 1989, Fundamentals of Cosmic Physics, 13, 143
Mathys, G. & Hubrig, S. 1997, *A&AS*, 124, 475
Mathys, G., Hubrig, S., Landstreet, J. D., Lanz, T., & Manfroid, J. 1997, *A&AS*, 123, 353
Monin, D., Bohlender, D., Hardy, T., Saddlemyer, L., & Fletcher, M. 2012, *PASP*, 124, 329
Nesvacil, N., Lüftinger, T., Shulyak, D., *et al.* 2012, *A&A*, 537, A151
Petit, V. & Wade, G. A. 2012, *MNRAS*, 420, 773
Piskunov, N. & Kochukhov, O. 2002, *A&A*, 381, 736
Preston, G. W. 1971, *ApJ*, 164, 309
Rivinius, T., Townsend, R. H. D., Kochukhov, O., *et al.* 2013, *MNRAS*, 429, 177
Rosén, L. & Kochukhov, O. 2012, *A&A*, 548, A8
Rusomarov, N., Kochukhov, O., Piskunov, N., *et al.* 2013, *A&A*, 558, A8
Ryabchikova, T., Piskunov, N., Kochukhov, O., *et al.* 2002, *A&A*, 384, 545
Shultz, M., Wade, G. A., Grunhut, J., *et al.* 2012, *ApJ*, 750, 2
Silvester, J., Wade, G. A., Kochukhov, O., *et al.* 2012, *MNRAS*, 426, 1003
Stibbs, D. W. N. 1950, *MNRAS*, 110, 395
Townsend, R. H. D., Owocki, S. P., & Groote, D. 2005, *ApJ*, 630, L81
Wade, G. A., Donati, J.-F., Landstreet, J. D., & Shorlin, S. L. S. 2000, *MNRAS*, 313, 823

Magnetic Fields throughout Stellar Evolution
Proceedings IAU Symposium No. 302, 2013
P. Petit, M. Jardine & H. Spruit, eds.

doi:10.1017/S1743921314002324

New Experiments with Zeeman Doppler Mapping

Alex J. Martin[1,2], **Stefano Bagnulo**[1] **and Martin J. Stift**[3]

[1]Armagh Observatory, College Hill, Armagh,
BT61 9DG, United Kingdom
email: ajm@arm.ac.uk

[2]Keele University, Keele, Staffordshire,
ST5 5BG, United Kingdom

[3]Universittssternwarte Wien, Türkenschanzstr. 17,
A-1180 Wien, Austria

Abstract. We present recent experiments using a Levenberg-Marquardt algorithm and the polarised radiative transfer code COSSAM to produce a new ZDM code. Currently the code is able to recover the magnetic parameters of model stars with either a decentred dipole morphology or a morphology consisting of a centred dipole and a quadrupole, while simultaneously calculating multiple chemical abundances (including a basic stratification model). The ZDM code has been tested using both synthetic spectra and real, well studied stars. Additional features are currently being added such as a multipole morphology of arbitrary order and more sophisticated chemical stratification models.

Keywords. SNR, crosstalk, Ap Stars, ZDM, polarimetry

1. Introduction

With spectro-polarimetry instruments now on many of the large telescopes around the world, the study of magnetic fields in stars has taken a large step forward. Using Zeeman-Doppler Mapping it is potentially possible to determining the magnetic configuration of stars and the distribution of the chemical elements over the stellar surface. Several different codes are already available, and our motivation to develop a new one are: 1) to have a higher arbitrariness (by leaving the star's geometric orientation with respect to the observer as a parameter to be recovered); 2) to further explore the advantages of Ada computer programming language; 3) to expand the investigations on the reliability and uniqueness of the results. The latter point is of a special interest for us, because some recent stellar modelling results seem to contradict theoretical results from diffusion theory. Some experiments to recover the distribution of the chemical elements over the stellar surface have already been performed in the non magnetic case and will be presented elsewhere (Stift *et al.*, in prep). Here we present our very first tests to check the impact of spectral resolution, noise, and instrument artifacts on the inversion of the magnetic configuration.

2. Simulating the Observations

To perform our tests, we have considered a homogeneous atmosphere and we have parametrised the magnetic morphology by adopting a multipolar expansion as described, e.g., in Landolfi *et al.* (1998). For each observing set of 10 rotation phases we have considered six spectral lines of two different elements (Fe and Cr). We have considered: SNR = 1000, 300, 100, 30 and 10, over a spectral bin of 0.01Å; Spectal resolutions = 200000, 50000, 25000, 10000, 5000, 2500, 1000; and Crosstalks = 0%, 5%, 10% and 15%.

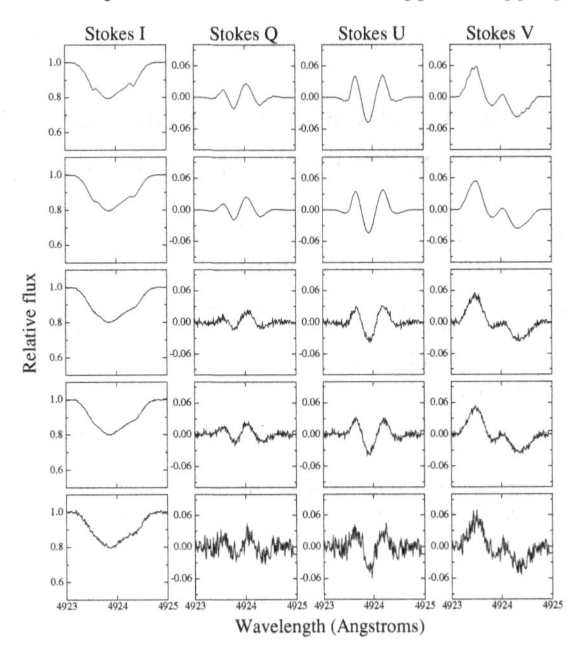

Figure 1. Signal degradation of Stokes $IQUV$ in percentage units with SNR on a spectral bin of 0.01Å. From top to bottom: 1. No Instrumental effects; 2. Spectral resolution: 60000; 3. Spectral resolution: 30000, SNR: 300; 4. Spectral resolution: 30000, SNR: 300, crosstalk 10%; 5. Spectral resolution: 30000, SNR: 100, crosstalk 10%. With the magnetic configuration not being recovered past, spectral resolution: 60000, SNR: 100, crosstalk 10%

3. Results and Conclusions

With no noise or crosstalk the inversion was successful even for the lowest values of spectral resolution, however, below a spectral resolution of 5000 the algorithm did not find the exact model values but, was reasonably close. With no crosstalk and perfect spectral resolution the algorithm recovered the model parameters down to a SNR of 100. In the case of no noise and perfect spectral resolution, for the majority of configurations the algorithm worked down to a crosstalk of 10%. However, we found that crosstalk can have a large impact when the magnetic field is weak or when the configuration leads to a majority Stokes V signal. If SNR and spectral resolution are both considered, with a spectral resolution of 60000 the algorithm is successful down to a SNR of 100 in all but the lowest field strengths. With the addition of crosstalk at 10%, the algorithm will successfully invert for a spectral resolution of 60000 and a SNR between 1000 and 300. Our results show that for a homogeneous atmosphere, if the magnetic configuration can be described by a low order multipolar expansion, and if all Stokes $IQUV$ are available and measured with a commonly available high-resolution spectropolarimeter, then the results are sufficiently robust. Our tests will continue by adding abundance spots, element stratification, and more complex magnetic configurations.

References

Amari, S., Hoppe, P., & Zinner, E., Lewis R. S. 1995, *Meteoritics*, 30, 490

Anders, E. & Zinner, E. 1993, *Meteoritics*, 28, 490

Bernatowicz, T. J., Messenger, S., Pravdivtseva, O., Swan, P., & Walker, R. M. 2003, *Geochim. Cosmochim. Acta*, 67, 4679

Magnetic Fields throughout Stellar Evolution
Proceedings IAU Symposium No. 302, 2013
P. Petit, M. Jardine & H. Spruit, eds.

doi:10.1017/S1743921314002336

Combining magnetic and seismic studies to constrain processes in massive stars

Coralie Neiner[1], Pieter Degroote[2,1], Blanche Coste[1], Maryline Briquet[3] and Stéphane Mathis[4,1]

[1]LESIA, UMR 8109 du CNRS, Observatoire de Paris, UPMC, Univ. Paris Diderot, 5 place Jules Janssen, 92195 Meudon Cedex, France
email: coralie.neiner@obspm.fr

[2]Instituut voor Sterrenkunde, Celestijnenlaan 200D, B-3001 Heverlee, Belgium

[3]Institut d'Astrophysique et de Géophysique, Université de Liège, Allée du 6 Août 17, Bât B5c, 4000 Liège, Belgium

[4]Laboratoire AIM Paris-Saclay, CEA/DSM-CNRS-Université Paris Diderot; IRFU /SAp, Centre de Saclay, 91191 Gif-sur-Yvette Cedex, France

Abstract. The presence of pulsations influences the local parameters at the surface of massive stars and thus it modifies the Zeeman magnetic signatures. Therefore it makes the characterisation of a magnetic field in pulsating stars more difficult and the characterisation of pulsations is thus required for the study of magnetic massive stars. Conversely, the presence of a magnetic field can inhibit differential rotation and mixing in massive stars and thus provides important constraints for seismic modelling based on pulsation studies. As a consequence, it is necessary to combine spectropolarimetric and seismic studies for all massive classical pulsators. Below we show examples of such combined studies and the interplay between physical processes.

Keywords. stars: early-type, stars: magnetic fields, stars: oscillations (including pulsations)

---

1. Modelling oblique magnetic dipoles with pulsations

β Cep is a magnetic pulsating star. It hosts a radial pulsation mode as well as non-radial pulsations of lower amplitude. The line profiles in its spectrum therefore show mainly variations with the pulsation periods in addition to the Zeeman broadening due to the magnetic field. Measurements of the magnetic field in Stokes V clearly show both variations due to the pulsations and the Zeeman signatures of its field. Therefore it is mandatory to take pulsations into account when trying to determine the magnetic field configuration and strength from the Stokes profiles.

For the first time we thus modelled the intensity and Stokes V profiles of a magnetic massive star, β Cep, taking into account its pulsations with the new Phoebe 2.0 code (see Fig. 1), and we compared the results with a standard oblique dipole model without pulsations. Without pulsations we find $i = 89°$, $\beta = 51°$, $B_{pol} = 389$ G, while with pulsations our preliminary results are $i = 70°$, $\beta = 50°$, $B_{pol} = 276$ G. In particular, the resulting field strength seems significantly lower when taking pulsations into account.

2. Taking magnetism into account in seismic studies

The impact of a fossil magnetic field on rotation and mixing can be estimated following two theoretical criteria: (1) the Spruit criterion (Spruit 1999): Above a critical strength, the magnetic field freezes differential rotation and mixing, and the field stays oblique. Otherwise the structure adjusts to a symmetric configuration by rotational smoothing;

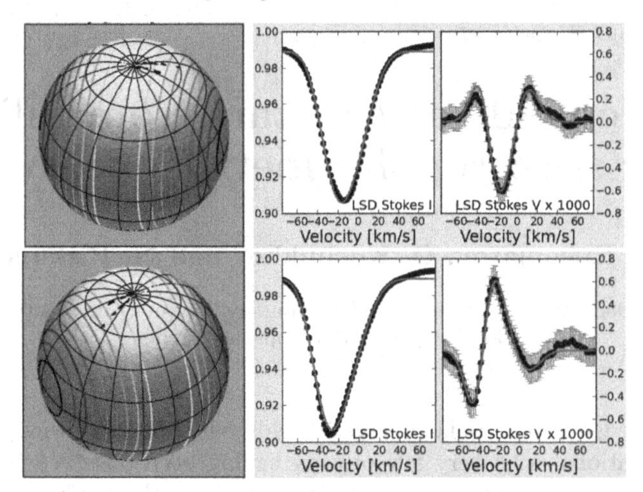

Figure 1. Examples of the modelled surface (left), LSD Stokes I (middle) and V (right) profiles of β Cep fitted with pulsations and magnetic field, at two different phases.

(2) the Zahn criterion (Zahn 2011, Mathis & Zahn 2005): The Lorentz force removes differential rotation along poloidal field lines above a certain field strength and thus removes mixing.

For the magnetic pulsating B star V2052 Oph, the Spruit and Zahn critical field strengths are $B_{crit} = 40$ G and 70 G, respectively. In this star, the measured polar field strength is $B_{pol} = 400$ G (Neiner *et al.* 2012). Therefore we expect no mixing in V2052 Oph. Indeed, a seismic model of the pulsations of V2052 Oph shows no overshoot in this star (Briquet *et al.* 2012). This example shows how a magnetic field study can provide constraints for seismic modelling.

A magnetic field also produces splitting of the pulsation modes and a modification of the amplitude of the pulsation modes. The split multiplet depends both on the strength of the field and on its obliquity (see Shibahashi & Aerts 2000). No magnetic splitting has been identified so far in massive stars. However, the observation with CoRoT of regular splittings in the hybrid B pulsator HD 43317 (Papics *et al.* 2012) and the recent discovery of a magnetic field in this star (Briquet *et al.* 2013) make it an ideal candidate.

3. Conclusions

It is crucial to take pulsations into account when modelling the magnetic field strength and configuration in pulsating massive stars. Moreover, knowing this magnetic configuration provides important constraints on seismic modelling, in particular it constraints the mixing, differential rotation and identification of the modes.

References

Briquet, M., Neiner, C., Aerts, C. *et al.* 2012, *MNRAS*, 427, 483
Briquet, M., Neiner, C., Leroy, B. *et al.* 2013, *A&A*, 557L, 16
Mathis, S. & Zahn. J.-P. 2005, *A&A*, 440, 653
Neiner, C., Alecian, E., Briquet, M. *et al.* 2012, *A&A*, 537A, 148
Papics, P., Briquet, M., Baglin, A. *et al.* 2012, *A&A*, 542A, 55
Shibahashi, H. & Aerts, C. 2000, *ApJ*, 531L, 143
Spruit, H. 1999, *A&A*, 349, 189
Zahn, J.-P. 2011, *IAUS*, 272, 14

Magnetic Fields throughout Stellar Evolution
Proceedings IAU Symposium No. 302, 2013
P. Petit, M. Jardine & H. Spruit, eds.

doi:10.1017/S1743921314002348

Magnetic fields of Ap stars from full Stokes vector spectropolarimetric observations

N. Rusomarov, O. Kochukhov and N. Piskunov

Department of Physics and Astronomy, Uppsala University,
Box 516, SE-75120 Uppsala, Sweden

Current knowledge about stellar magnetic fields relies almost entirely on circular polarization observations, with very few objects having been observed in all four Stokes parameters. We are investigating a sample of Ap stars in all four Stokes parameters using the HARPSpol instrument at the 3.6-m ESO telescope. In the context of this project we recently observed the magnetic Ap star HD 24712 (DO Eri, HR 1217). The resulting spectra have dense phase coverage, resolving power $> 10^5$, and S/N ratio of 300–600. These are the highest quality full Stokes observations obtained for any star other than the Sun. Furthermore, we have achieved good phase coverage for HD 125248 and HD 119419. Typical four Stokes parameters HARPSpol spectra are shown in Fig. 1. An analysis of the full Stokes vector spectropolarimetric data set of HD 24712 has been published in Rusomarov *et al.* (2013).

Here we present preliminary results from the magnetic Doppler imaging analysis of HD 24712. We derived chemical distribution abundance maps and magnetic field maps (Fig. 2) using five FeI lines, three NdIII lines and one NaI line for the case of a dipolar field geometry. Our preliminary results show that dipole field geometry reproduces observed Stokes profiles for our selected lines very well (Fig. 3).

This analysis is the first step towards obtaining detailed 3-D maps of magnetic fields and abundance structures for HD 24712 and other Ap stars that we currently observe with HARPSpol. We plan to study magnetic field and chemical spots in these stars, reconstruct 3D maps for the first time and analyze other stars.

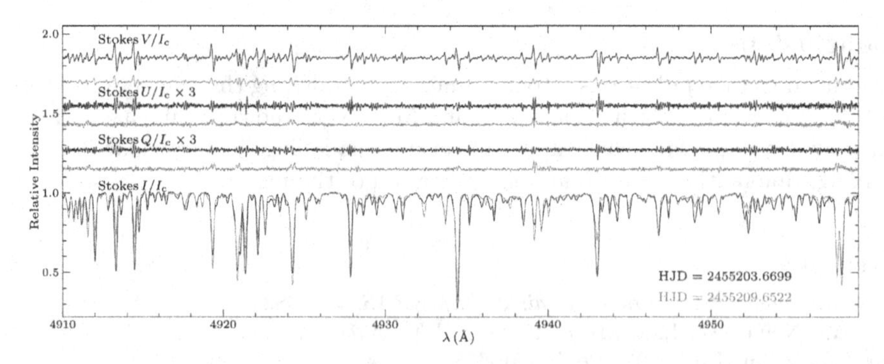

Figure 1. HARPSpol four Stokes parameter spectra of HD 24712. Spectra plotted with black lines were obtained at magnetic maximum; spectra plotted with red lines were obtained around the magnetic minimum. Most lines exhibit strong intensity variations with phase, and show complex strong linear and circular polarization signatures.

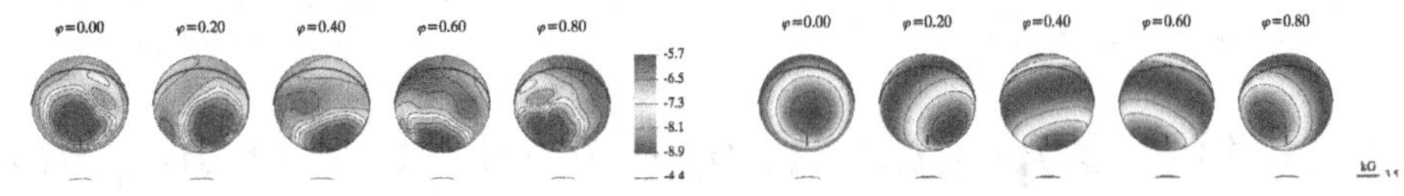

Figure 2. Abundance distribution maps of Nd, Fe and Na on the surface of HD24712, and maps showing a distribution of the magnetic field strength a), radial component b), and field orientation c).

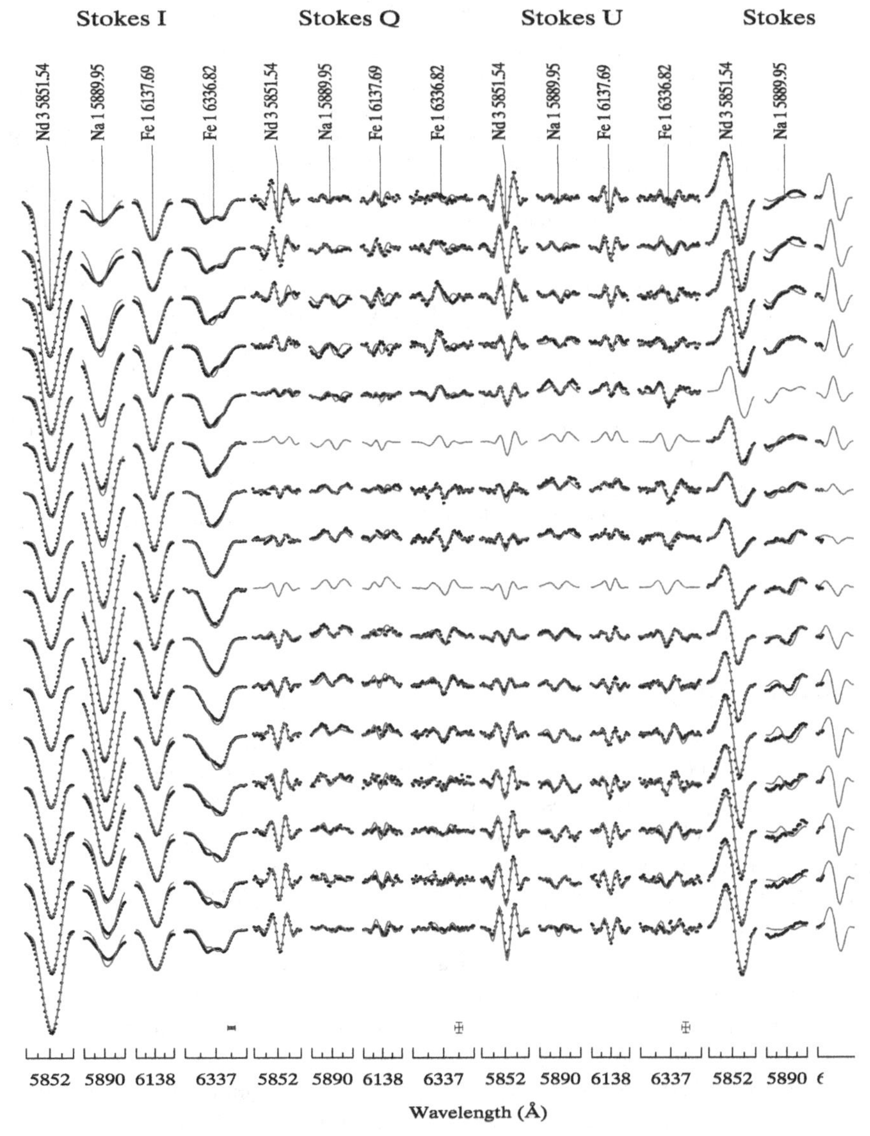

Figure 3. Comparison of the observed (dots) and calculated (lines) Stokes profiles for HD 24712. Phase values increase downwards.

References

Rusomarov, N., Kochukov, O., Piskunov, N., *et al.* 2013, *A&A*, 558, A8

Magnetic Fields throughout Stellar Evolution
Proceedings IAU Symposium No. 302, 2013
P. Petit, M. Jardine & H. Spruit, eds.

doi:10.1017/S174392131400235X

Magnetic Doppler Imaging of He-strong star HD 184927

I. Yakunin[1], G. Wade[2], D. Bohlender[3], O. Kochukhov[4], V. Tsymbal[5] and MiMeS Collaborators

[1]Special Astrophysical Observatory, Nizhniy Arkhyz, Russia 369167
[2]Department of Physics, Royal Military College of Canada, P.O. Box 17000, Station Forces, Kingston, Ontario, Canada, K7K 7B4
[3]National Research Council of Canada, Herzberg Institute of Astrophysics, 5071 West Saanich Road, Victoria, BC, V9E 2E7, Canada
[4]Department of Physics and Astronomy, Uppsala University, SE-751 20, Uppsala, Sweden
[5]Tavrian National University, Vernadskiys Avenue 4, Simferopol, Crimea, 95007, Ukraine

Abstract. We have employed an extensive new timeseries of Stokes I and V spectra obtained with the ESPaDOnS spectropolarimeter at the 3.6-m Canada-France-Hawaii Telescope to investigate the physical parameters, chemical abundance distributions and magnetic field topology of the slowly-rotating He-strong star HD 184927. We infer a rotation period of $9^d.53071 \pm 0.00120$ from $H\alpha$, $H\beta$, LSD magnetic measurements and EWs of helium lines. We used an extensive NLTE TLUSTY grid along with the SYNSPEC code to model the observed spectra and find a new value of luminosity. In this poster we present the derived physical parameters of the star and the results of Magnetic Doppler Imaging analysis of the Stokes I and V profiles. Wide wings of helium lines can be described only under the assumption of the presence of a large, very helium-rich spot.

Keywords. stars:magnetic fields, stars:chemically peculiar

---

1. Observational data

Stokes V spectra of HD 184927 were obtained with the ESPaDOnS spectopolarimeter at CFHT between 2008 August 20 and 2012 June 27. The resolution of all the spectra is R = 65000. The S/N ratio varies, but is typically is about 520. Spectra were reduced using the Upena pipeline feeding the LIBRE-ESPRIT code. Then we applied the LSD procedure to obtain mean Stokes I and V profiles and improve the S/N ratio of our measurements (see Donati J.-F. *et al.*, 1997, MNRAS, 291, 658 for details). Resulting S/N ratio of LSD profiles varies from 1000 to 1700.

We also used medium resolution data obtained at the DAO 1.8-meter Plaskett telescope with the dimaPol spectropolarimeter. The S/N in Stokes V varies between 250 and 450 at 4970 Å. Typically 12-18 10-minute sub-exposures were taken and then combined to produce a single measurement.

To check consistency with previously published data, we included magnetic field measurements from Wade *et al.* (Wade G. *et al.*, 1997, A&A, 320, 172) in our dataset.

2. Period refinement and rotation

To determine rotational period of HD 184927 we used magnetic field data obtained with SAO Zeeman analyzer and UWO photoelectric polarimeter (see Wade *et al.* 1997) and added dimaPol measurements of $H\beta$ and He 4922 line. $H\beta$ measurements (UWO + dimaPol) only give possible periods of 9.531072 and 9.522540, with the latter barely within the 1-sigma error bar of the 1997 period estimate. The best ephemeris for this set is $JD = 2455706.843 + 9.53071 \pm 0.00120d$.

In order to estimate projected rotational velocity we used grid of NLTE TLUSTY models with fixed Teff, log g and microturbulence and combination of different [Si/H] and vsini parameters to calculate several synthetic spectra for nine SiII-SiIII lines. The code computed equivalent widths for each case and compared it to the observed equivalent widths. For these calculations we adopted projected rotational velocity $v \sin i = 9 \pm 2$ km/s.

3. Magnetic field

The longitudinal magnetic field and null measurements from the ESPaDOnS spectra were computed from each LSD Stokes V and diagnostic null profile, using an integration range from -200 to 200 km/s. The computed longitudinal field varies from -200 to 950 G. We used the Oblique Rotator Model to model the magnetic curve assuming that the field can be described by a centred dipole. The best-fit model has the following parameters: beta = 70, Bd = 4500, i = 26. To explore the consistency of magnetic field measured from different elements we built LSD masks containing lines of Si, O, N, Fe and He. The longitudinal magnetic field was also measured from the Balmer lines Halpha and Hbeta.

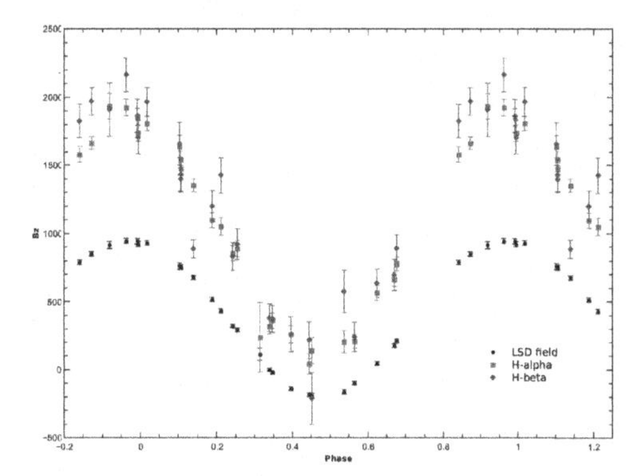

Figure 1. Longitudinal magnetic field from different spectral lines in comparison with LSD field

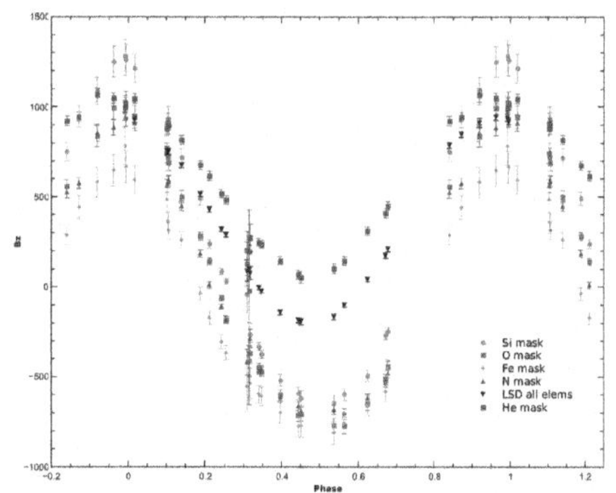

Figure 2. Longitudinal magnetic field from different spectral lines in comparison with LSD field

As one can clearly see, magnetic field from Si, O, N and Fe masks changes its sign while the field from He is not. At the same time the magnetic field of helium lines strongly correlates with the field measured from hydrogen Halpha and Hbeta lines. We find that the lines can be reproduced by a spot with 60° radius and helium abundance N(He)/N(H) = 2.

4. Magnetic Doppler Imaging

Inversions were carried out with a modified Inversl3 code (Kochukhov O. *et al.*, 2013, A&A, 550, A84). The grid of LLmodels atmospheres (Shulyak D. *et al.*, 2004, A&A, 428, 993) was computed for Teff = 22000, log g = 4.0 and a range of He abundance. For this grid we computed NLTE departure coefficients with TLUSTY. These model atmospheres and NLTE departure coefficients were then used in the MDI code to calculate local He I 6678 Åline profiles. Magnetic field was parameterized with a spherical harmonic expansion, similar to the tau Sco analysis by Donati *et al.* (Donati J.-F. *et al.*, 2006, MNRAS, 370, 629).

We performed a dipole+quadrupole fitting, allowing full freedom (toroidal field, independent poloidal radial and horizontal fields) of the spherical harmonic expansion. We found that the field is mostly poloidal, with comparable contributions of the dipole and quadrupole terms. Total fraction of poloidal component is 81.5% and toroidal component is 18.5%.

Magnetic Fields throughout Stellar Evolution
Proceedings IAU Symposium No. 302, 2013
P. Petit, M. Jardine & H. Spruit, eds.

doi:10.1017/S1743921314002361

roAp stars: surface lithium abundance distribution and magnetic field configuration

N. Polosukhina[1], **D. Shulyak**[2], **A. Shavrina**[3], **D. Lyashko**[4], **N. A. Drake**[5,6], **Yu. Glagolevski**[7], **D. Kudryavtsev**[7] **and M. Smirnova**[1]

[1]Crimean Astrophysical Observatory, Nauchnyi, Ukraine, email: polo@crao.crimea.ua
[2]Georg-August University, Göttingen, Germany, email: denis.shulyak@gmail.com
[3]Main Astronomical Observatory of NAS of Ukraine, Kyiv, Ukraine, email: shavrina@mao.kiev.ua
[4]Taurida National V. I. Vernadsky University, Simferopol, Crimean Autonomous Republic, Ukraine, email: dlyashko@gmail.com
[5]Sobolev Astronomical Institute, St. Petersburg State University, St. Petersburg, Russia
[6]Observatório Nacional/MCTI, Rio de Janeiro, Brazil, email: drake@on.br
[7]Special Astrophysical Observatory of RAS, Nizhnii Arhyz, Russia, email: dkudr@sao.ru

Abstract. High-resolution spectra obtained with the 6m BTA telescope, Russia, and with HARPS and VLT/UVES telescopes at ESO, Chile, were used for Doppler Imaging analysis of two roAp stars, HD 12098 and HD 60435, showing strong and variable Li resonance line in their spectra. We found that Li has highly inhomogeneous distribution on the surfaces of these stars. We compared our results with previously obtained Doppler Imaging mapping of two CP2 stars, HD 83368 and HD 3980, and discuss the correlation between the position of the high Li-abundance spots and magnetic field.

Keywords. stars: abundances – stars: chemically peculiar – stars: magnetic field – stars: individual: HD 3980, HD 12098, HD 60435, HD 83368

---

1. Introduction

Spectral observations of a number of magnetic chemically peculiar stars allowed us to discover several Ap stars with abnormally high lithium abundance. However, the behaviour of the Li I 6708 Å line in these stars is still puzzling (Polosukhina *et al.* 1999, 2004, 2012).

In this paper we present the first results of the Doppler Imaging of two rapidly oscillating (roAp) cool magnetic chemically peculiar stars, HD 12098 and HD 60435, showing a remarkable rotational modulation of the Li I line. The high-resolution spectra of the northern star HD 12098 were obtained during 2007 - 2013 with the 6m BTA telescope of SAO, Russia, whereas the HARPS spectra of HD 60435 were downloaded from the ESO Science Archive Facility under request number drake.62954.

2. First results and conclusions

For the investigation of the surface distribution of lithium on the surfaces of HD 12098 and HD 60435, we used the Doppler Imaging (DI) code INVERS12 (Piskunov & Rice 1993; Kochukhov *et al.* 2004). We modelled the resonance Li I doublet at λ 6708 Å taking into account its variable blending with the nearby line of Pr III 6706.7 Å.

Our earlier DI analysis of two other roAp stars, HD 83368 (Kochukhov *et al.* 2004) and HD 3980 (Nesvacil *et al.* 2012), showed that Li is strongly concentrated in the areas of the magnetic poles and depleted in the regions around the magnetic equator.

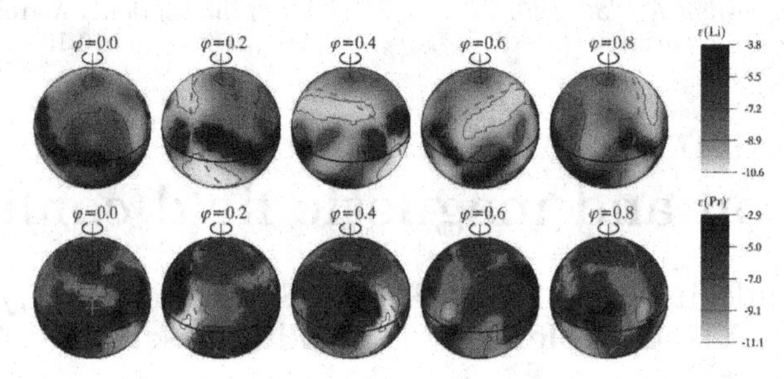

Figure 1. Distribution of Li and Pr on the surface of HD 12098. The star is shown at five rotation phases. The abundance scales are in $\log N(\mathrm{el})/N(\mathrm{H})$ (darker areas correspond to higher element abundance). The dashed line, plus sign and small circle correspond to the magnetic equator, positive and negative poles respectively. The following stellar parameters were used: $T_{\mathrm{eff}} = 7800$ K, $\log\ g = 4.3$, $v \sin i = 10 \pm 2$ km s^{-1}, $B_{\mathrm{p}} = 6.5$ kG, $i \cong 55°$, $\beta = 65°$ (Shavrina *et al.* 2008), $P_{\mathrm{rot}} = 5.460^{\mathrm{d}} \pm 0.001$ (Ryabchikova *et al.* 2005).

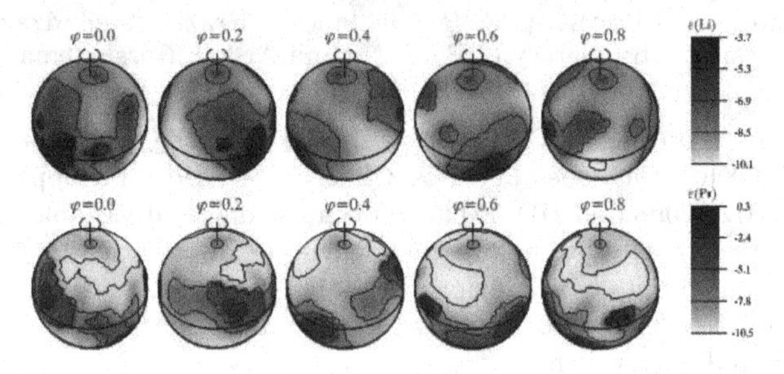

Figure 2. The same as in Fig. 1, but for the star HD 60435. The following stellar parameters were used: $T_{\mathrm{eff}} = 8250$ K, $\log\ g = 4.5$, $v \sin i = 11 \pm 2$ km s^{-1}, $H_{\mathrm{s}} = 3 \pm 1$ kG, $i \cong 47°(133°)$ (Shavrina *et al.* 2001), $P_{\mathrm{rot}} = 7.6793^{\mathrm{d}} \pm 0.0006$ (Kurtz *et al.* 1990)

Preliminary DI of HD 12098 showed that Li has highly inhomogeneous distribution on the surface of this star. However, no such correlation has been found between the magnetic field structure and the Li abundance. For the star HD 60435 the picture is uncertain because of the lack of magnetic field model for this star.

Analysis of the Li I line behaviour in magnetic CP stars can give new challenges to the theoretical calculations of diffusion in the presence of strong magnetic fields. New observations of HD 12098 permitting to achieve the better phase coverage are still needed.

Acknowledgements. N. A. D. acknowledges support of the PCI/MCTI, Brazil, under the Project 302350/2013-6.

References

Kochukhov, O., Drake, N. A., Piskunov, N., & de la Reza, R. 2004, *A&A*, 424, 935
Kurtz, D. W., van Wyk, F., Marang, F. 1990 *MNRAS*, 243, 289
Nesvacil, N., Lüftinger, T., Shulyak, D. *et al.* 2012, *A&A*, 537, A151
Piskunov, N. E. & Rice, J. B. 1993, *PASP*, 105, 1415
Polosukhina, N., Kurtz, D., Hack, M. *et al.* 1999, *A&A*, 351, 283
Polosukhina, N., Shavrina, A., Drake, N. A. *et al.* 2004, *A-Star Puzzle*, Proc. IAU Symp. 224 , Cambridge University Press, 2004, 665
Polosukhina, N., Shavrina, A., Drake, N. A., & Smirnova, M. 2012, *MSAIS*, 22, 129
Ryabchikova, T., Wade, G. A., Aurière, M. *et al.* 2005, *A&A*, 429, 55
Shavrina, A. V., Polosukhina, N. S., Zverko, J. *et al.* 2001, *A&A*, 372, 571
Shavrina, A. V., Polosukhina, N. S., Drake, N. A., & Kudryavtsev, D. O. 2008, *Astrophysics*, 51, 517

Magnetic Fields throughout Stellar Evolution
Proceedings IAU Symposium No. 302, 2013
P. Petit, M. Jardine & H. Spruit, eds.

doi:10.1017/S1743921314002373

Roadmap on the theoretical work of BinaMIcS

Stéphane Mathis[1,2], Coralie Neiner[2], Evelyne Alecian[2,3], Gregg Wade[4] and the BinaMIcS collaboration

[1]Laboratoire AIM Paris-Saclay, CEA/DSM-CNRS-Université Paris Diderot; IRFU /SAp, Centre de Saclay, 91191 Gif-sur-Yvette Cedex, France
email: stephane.mathis@cea.fr

[2]LESIA, UMR 8109 du CNRS, Observatoire de Paris, UPMC, Univ. Paris Diderot, 5 place Jules Janssen, 92195 Meudon Cedex, France

[3]UJF-Grenoble 1/CNRS-INSU, IPAG, UMR 5274, F-38041, Grenoble, France

[4]Dept. of Physics, Royal Military College of Canada, Kingston, K7K 7B4, Canada

Abstract. We review the different theoretical challenges concerning magnetism in interacting binary or multiple stars that will be studied in the BinaMIcS (Binarity and Magnetic Interactions in various classes of Stars) project during the corresponding spectropolarimetric Large Programs at CFHT and TBL. We describe how completely new and innovative topics will be studied with BinaMIcS such as the complex interactions between tidal flows and stellar magnetic fields, the MHD star-star interactions, and the role of stellar magnetism in stellar formation and vice versa. This will strongly modify our vision of the evolution of interacting binary and multiple stars.

Keywords. stars: magnetic fields, stars: binaries (including multiple): close

1. The BinaMIcS project

The BinaMIcS project has been awarded “Large Program” status with two high resolution spectropolarimeters: ESPaDOnS at the CFHT (Hawaii; PIs Alecian/Wade) for 604 hours over 4 years from 2013 to 2016, and Narval at TBL (France; PI Neiner) for 128 hours over 2 years from 2013 to 2014 (renewable for 2 more years). This large amount of time is being used to acquire an immense database of sensitive measurements of polarized and unpolarized spectra of spectroscopic binary (SB2) stars. The program includes 3 components: the detailed study of some known magnetic cool binaries, the detailed study of the few known magnetic hot close binaries, and a survey of a large number of hot binaries to search for magnetic fields. The detailed studies will allow us to obtain magnetic maps and test various models, while the survey part of the project will allow us to obtain statistical results. This database will be combined with new and archival complementary data (e.g. optical photometry, UV and X-ray spectroscopy) as well as theoretical studies, numerical simulations, and modeling (described hereafter), and applied to address the 4 main scientific objectives of the BinaMIcS project: i) what is the impact of magnetic fields during stellar formation, and vice versa; ii) how do tidally-induced internal flows impact fossil and dynamo fields; iii) how do magnetospheric star-star interactions modify stellar activity; iv) what is the magnetic impact on angular momentum exchanges and mass transfers.

2. Roadmap on the theoretical work of BinaMIcS

Magnetic fields are a crucial ingredient in a star's evolution, influencing its formation, the structure of its atmosphere and interior, as well as controlling the interaction with its environment. For binary stars, magnetism is even more significant, as magnetic fields in binary systems will be strongly affected by, and may also strongly affect, the transfer of energy, mass and angular momentum between the components. Therefore, the interplay between stellar magnetic fields and binarity has to be investigated in detail, from both the observational and theoretical point-of-view. First, the incidence and characteristics of magnetic fields are key parameters for understanding the physics of binaries. In higher-mass stars (above 1.5 $M_\odot$) the incidence of magnetic stars in binary systems provides a basic constraint on the detailed origin of the magnetic field, assumed to be a fossil remnant, and whether such strong magnetic fields suppress binary formation or are a result of mergers. Next, in low-mass stars, tidal interactions are expected to induce large-scale 3D shear and/or helical flows in stellar interiors that can significantly perturb the stellar dynamo. Similar flows may also influence the fossil magnetic fields of higher-mass stars. Finally, magnetically driven winds/outflows in cool and hot close binary systems have long been suspected to be responsible for their orbital evolution, while magnetospheric interactions have been proposed to enhance stellar activity. However, the crucial observational constraints required to test these hypotheses are, at present, nearly nonexistent, and dedicated theoretical studies are mandatory to bring the studies of binary (and multiple) stars to a new level of understanding. Within BinaMIcS, we will therefore study theoretically the complex interactions between tides and magnetic fields, i.e.: i) how magnetic fields modify tidal flows described by Zahn 1977, Ogilvie & Lin 2007, Le Bars *et al.* 2010, and Remus *et al.* 2012 and the associated torques applied on each component; ii) how flows driven by tides (as well as precession and libration) can modify dynamo mechanisms (e.g. Le Bars *et al.* 2011) because of the angular momentum they transport (Ω-effect) and of their helicity (α-effect), and the stability of fossil fields; iii) how such external mechanical forcings can compete with convective driving as a function of the mass ratio and of the separation of the components; iv) how do the combined action of tides and magnetic fields modify the orbital dynamics of binary stars and the evolution of their components (e.g. de Mink *et al.* 2013, Mathis & Remus 2013, Song *et al.* 2013). Moreover, we will study the complex MHD interactions between the components of binary stars, i.e.: i) the interactions between stellar wind emitted by each component and the related torques (e.g. Strugarek *et al.* 2012); ii) the magnetospheric interactions and the associated applied torques, helicity exchanges, and modifications of the magnetic activity of the components (e.g. Lanza 2012 and the contribution by S. Gregory in this proceeding). Results will also be applied to the study of star-planet interactions.

References

de Mink, S. E., Langer, N., Izzard, R. G., Sana, H., & de Koter, A. 2013, *ApJ*, 764, article id. 166

Lanza, A. F. 2012, *A&A*, 544, id. A23

Le Bars, M., Lacaze, L., Le Dizès, S., Le Gal, P., & Rieutord, M. 2010, *PEPI*, 178, 48

Le Bars, M, Wieczorek, M. A., Karatekin, Ö, Cebron, D., & Laneuville, M. 2011, *Nature*, 479, 215

Mathis, S. & Remus, F. 2013, *LNP*, 857, 111

Ogilvie, G. & Lin, D. N. C. 2007, *ApJ*, 661, 1180

Remus, F., Mathis, S., & Zahn, J.-P. 2012, *A&A*, 544, id. A132

Song, H. F., *et al.* 2013, *A&A*, 556, id. A100

Strugarek, A., Brun, A.-S., & Matt, S. 2012, *Proceedings of the Annual meeting of the SF2A*, 419

Zahn, J.-P. 1977, *A&A*, 57, 383

Magnetic Fields throughout Stellar Evolution
Proceedings IAU Symposium No. 302, 2013
P. Petit, M. Jardine & H. Spruit, eds.

doi:10.1017/S1743921314002385

Candidate Ap stars in close binary systems

C. P. Folsom[1,2], G. A. Wade[3], K. Likuski[3], O. Kochukhov[4], E. Alecian[5], D. Shulyak[6] and N. M. Johnson[3]

[1]Institut de Recherche en Astrophysique et Planétologie, Toulouse, France
email: colin.folsom@irap.omp.eu
[2]Armagh Observatory, Armagh, Northern Ireland
[3]Department of Phyics, Royal Military College of Canada, Kingston, Canada
[4]Department of Astronomy and Space Physics, Uppsala University, Uppsala, Sweeden
[5]Observatoire de Paris, Meudon, France
[6]Institute of Astrophysics, Georg-August-University, Göttingen, Germany

Abstract. Short period binary systems containing magnetic Ap stars are anomalously rare. This apparent anomaly may provide insight into the origin of the magnetic fields in theses stars. As an early investigation of this, we observed three close binary systems that have been proposed to host Ap stars. Two of these systems (HD 22128 and HD 56495) we find contain Am stars, but not Ap stars. However, for one system (HD 98088) we find the primary is indeed an Ap star, while the secondary is an Am star. Additionally, the Ap star is tidally locked to the secondary, and the predominately dipolar magnetic field of the Ap star is roughly aligned with the secondary. Further investigations of HD 98088 are planned by the BinaMIcS collaboration.

1. Introduction

Magnetic Ap stars in short period binary systems are very rare. Whereas the incidence of other chemically peculiar A stars in close binary systems is at least as large as in single stars, the incidence of Ap stars in close binaries is much lower. This observation may provide insight into the origin of magnetism in A-type stars, and is one of the avenues being pursued by the new Binarity and Magnetic Interactions in various classes of Stars (BinaMIcS) collaboration. As an initial step in this project, we studied three close binary systems which have been suggested to contain Ap stars (HD 22128, HD 56495, & HD 98088), in order to asses the presence of magnetic fields and study the atmospheric chemistry of the components. High resolution spectropolarimetric observations of these stars were obtained with the MuSiCoS instrument at the Observatoire du Pic du Midi.

2. HD 98088

HD 98088 is a SB2 (P = 5.905 days, e = 0.184; Carrier *et al.* 2002), which was identified as chemically peculiar by Abt (1953), and magnetic by Babcock (1958). Carrier *et al.* (2002) studied the system's orbital parameters, but there are no modern magnetic or chemical abundance studies. We present new results from Folsom *et al.* (2013a).

We applied spectral disentangling to the set of Stokes I observations of HD 98088. The disentangled spectra were used for abundance analyses of the two components, by fitting them with synthetic spectra computed with the ZEEMAN code. In the primary, we find strong overabundances of Fe-peak elements and rare earths, and roughly solar abundances for lighter elements, indicating an Ap star. In the secondary, we find overabundances of Fe-peak elements and underabundances of Ca and Sc, indicating an Am star.

In order to assess stellar magnetic fields, we performed Least Squares Deconvolution (LSD) on our observations, producing 'mean' line profiles. In Stokes V, we find clear

Zeeman signatures in the primary's lines, and no signal in the secondary's lines. For the primary, longitudinal magnetic fields were measured from the LSD profiles. We measured the rotation period of the primary from the magnetic variability, and the result agrees well with the orbital period, implying that the system is tidally locked. We find the magnetic field of the primary is predominately dipolar, with a polar strength of 3850 ± 450 G.

Comparing the magnetic and orbital geometries, we find that one magnetic pole of the primary always points roughly towards the secondary. The dipole axis appears to be $15 \pm 5°$ out of the orbital plane, and thus the alignment may not be perfect, but it is very suggestive.

3. HD 22128

HD 22128 is a SB2 (P = 5.086 days, e ∼0; Carrier *et al.* 2002). The primary was proposed as an Ap star by Olsen (1979) and by Abt *et al.* (1979). Here we present new results from Folsom *et al.* (2013b).

An abundance analysis was performed by fitting synthetic SB2 spectra computed with ZEEMAN to the observations. In both stars, we find overabundances of Fe-peak elements and rare earths, and underabundances of Ca and Sc, indicating both components are Am stars. We extracted LSD profiles from our observations, and find no detection of a magnetic signature in any Stokes V profile. Measuring longitudinal magnetic fields from these profiles we find no detection, with uncertainties of ±50 G in the primary, and ±90 G in the secondary.

We conclude that HD 22128 is a close binary containing two very similar Am stars, but neither star is a magnetic Ap star.

4. HD 56495

HD 56495 is a SB2 (P = 27.38 days, e = 0.165; Carrier *et al.* 2002). The primary was proposed as an Ap star based on a marginal magnetic detection by Babcock (1958). We present new results from Folsom *et al.* (2013b).

We performed an abundance analysis, fitting the SB2 spectrum with ZEEMAN. For the primary, we find clear overabundances of Fe-peak elements and underabundances of Ca and Sc, indicating an Am star. For the secondary we find abundances consistent with solar. We computed LSD profiles for our observations, and there is no detection of a magnetic signature in the Stokes V profiles. Measuring longitudinal magnetic fields from the unblended profiles we detect no magnetic field, with uncertainties of ±80 G in the primary, and ±100 G in the secondary.

We conclude that the primary is an Am star and the secondary a normal F star. The results for HD 22128 and HD 56495 suggest that Ap stars in close binaries may be even more rare than previously thought.

References

Abt, H. A. 1953, *PASP* 65, 274

Abt, H. A., Brodzik, D., & Schaefer, B. 1979, *PASP* 91, 176

Babcock, H. W. 1958, *ApJS* 3, 141

Carrier, F., North, P., Udry, S., & Babel, J. 2002, *A&A* 394, 151

Folsom, C. P., Likuski, K., Wade, G. A., Kochukhov, O., Alecian, E., & Shulyak, D. 2013a, *MNRAS* 431, 1513

Folsom, C. P., Wade, G. A, & Johnson, N. M. 2013b, *MNRAS* 433, 3336

Olsen, E. H. 1979, *A&AS* 37, 367

Magnetic Fields throughout Stellar Evolution
Proceedings IAU Symposium No. 302, 2013
P. Petit, M. Jardine & H. Spruit, eds.

doi:10.1017/S1743921314002397

The unusual binary HD 83058 in the region of the Scorpius-Centaurus OB association

M. A. Pogodin[1], N. A. Drake[2,3], E. G. Jilinski[1,3,4] and C. B. Pereira[3]

[1]Pulkovo Observatory of Russian Academy of Sciences, Pulkovskoe shosse 65/1, 196135, Saint Petersburg, Russia email: pogodin@gao.spb.ru

[2]Saint Petersburg State University, Universitetski pr. 28, Petrodvorets, 198504, Saint Petersburg, Russia, email: drake@on.br

[3]Observatório Nacional/MCTI, Rua General José Cristino 77, 20921-400, Rio de Janeiro, Brazil, email: claudio@on.br

[4]Instituto de Física, Universidade do Estado do Rio de Janeiro (UERJ), Rua São Francisco Xavier 524, Maracanã, 200550-900, Rio de Janeiro, Brazil, email: jilinski@on.br

Abstract. We present the results of high-resolution spectroscopy of the binary system HD 83058 situated in the region of the Sco-Cen OB association. On the base of the radial-curve solution we have determined the elements of the orbit and determined the period $P = 2.365102$ days. We have disentangled the spectra of the two components of the system and derived the basic parameters of both components. We have shown that moving features in the Si III line profiles seen in the spectra of the primary can be interpreted in the frame of the assumption of the rotation of local spot-like inhomogeneities on the stellar surface. We have also found that the lines in the spectrum of the secondary show another type of variability.

Keywords. Line:profiles – (Stars:) binaries – Stars: fundamental parameters – Stars: early-type – Stars: spots – Stars: individual (HD 83058)

---

1. Observations and determination of the system parameters

The southern early-type star HD 83058 is situated in the region of the Sco-Cen OB association. Early it has been revealed that: a) the object is a binary system with the orbital period about 2.3 days, b) two components of the system show different types of spectral variability (Telting *et al.* 2006, Jilinski *et al.* 2010). The aim of our study is to determine the orbital elements of the system and the fundamental parameters of its components as well as to investigate the spectral variability of both components of the system.

Seventeen high-resolution spectra ($R = 48\,000$) were obtained in 2007 – 2009 using the FEROS spectrograph at the 2.2m telescope of ESO at La Silla, Chile. We applied the standard method of fitting the observed phase diagram of radial velocities $V\mathrm{r}$ for each components of the binary by the theoretical curve for orbital motion. The least-square method was used for calculations. As a result, we obtained the following values of orbital elements:

$e = (1.7 \pm 0.1)^{-7}$; $P = 2.365102 \pm 0.000022$ days; $\mathrm{MJD}_0 = 54000.8770 \pm 0.0096$ days for the moment when radial velocities of the primary (A) and the secondary (B) components are equal: $V\mathrm{r(A)} = V\mathrm{r(B)} = \gamma$, after what the component A begins to move away from the observer; $K(\mathrm{B}) = 137.1 \pm 0.1\ \mathrm{km\,s^{-1}}$, $K(\mathrm{A}) = 81.6 \pm 0.1\ \mathrm{km\,s^{-1}}$; $\gamma = 12.1 \pm 0.1\ \mathrm{km\,s^{-1}}$.

We determined the basic parameters of the A and B components using the method of constructing the combined synthetic spectrum of the system. For each of the components A and B, we applied the LTE model spectra calculated with the code SYNTH+ROTATE

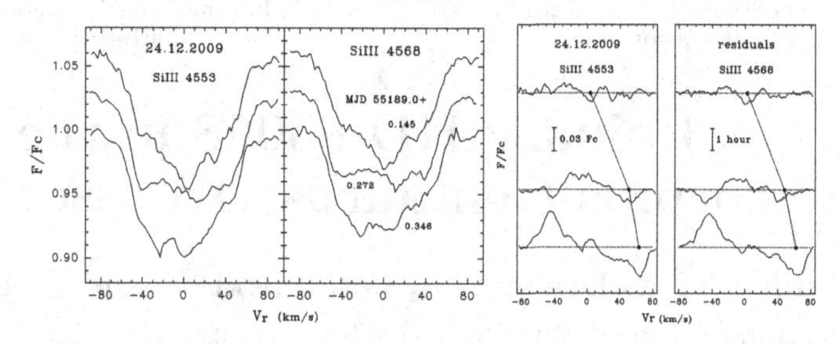

Figure 1. Typical line profiles variations observed in the component A.

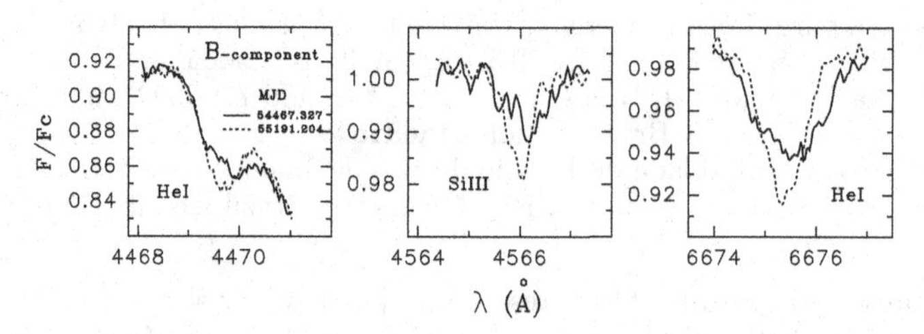

Figure 2. Profiles of some lines in the spectrum of the component B.

(Piskunov 1992) and non-LTE spectra obtained with the code TLUSTY (Hubeny & Lanz 1992). As a result, the following parameters have been obtained:

Component A: $T_{\rm eff} = 25700 \pm 400$ K, $\log g = 4.27 \pm 0.05$, $\xi_{\rm turb} = 5.5 \pm 0.5$ km s^{-1}, $v \sin i = 57.4 \pm 1.0$ km s^{-1};

Component B: $T_{\rm eff} = 19200 \pm 600$ K, $\log g = 4.03 \pm 0.20$, $\xi_{\rm turb} = 5.5 \pm 0.5$ km s^{-1}, $v \sin i = 27.0 \pm 1.4$ km s^{-1};

The ratio of the stellar radii is $R_B/R_A = 0.66 \pm 0.03$.

2. Spectral variability of the components A and B

The spectral variability of the component A manifests itself in a form of moving local features on the Si III line profiles. An example of these variations for the date 24.12.2009 is shown in Fig. 1. We tested the assumption that these features are the result of rotation of spot-like inhomogeneities on the stellar surface, probably of magnetic origin. Analyzing positional change of separate features placed at different latitudes φ, we have obtained that during three dates 24, 25 and 26.12.2009 the features, rotating with approximately the same period $P = 1.10 \pm 0.04$ days were observed at the latitudes $\varphi = 0°$, 41°, and 28°, respectively.

ss In contrast to the component A, the line profiles in the spectrum of the component B demonstrate another type of variability. We have found that in some dates the profiles of all lines become wider, less deep, and slightly red-shifted. Some examples are shown in Fig. 2.

Acknowledgements. N. A. D. acknowledges support of the PCI/MCTI, Brazil, under the Project 302350/2013-6.

References

Hubeny, I. & Lanz, T. 1992, *A&A*, 262, 501

Jilinski, E., Ortega, V. G., Drake, N. A., & de la Reza, R. 2010, *ApJ*, 721, 469

Piskunov, N. E. 1992, *Stellar Magnetism*, 92

Telting, J. H., Schrijvers, C., Ilyin, I. V., *et al.* 2006, *A&A*, 452, 945

Magnetic Fields throughout Stellar Evolution
Proceedings IAU Symposium No. 302, 2013
P. Petit, M. Jardine & H. Spruit, eds.

doi:10.1017/S1743921314002403

Binary and multiple magnetic Ap/Bp stars

Denis Rastegaev[1], Yuri Balega, Vladimir Dyachenko, Alexander Maksimov and Evgenij Malogolovets

Special Astrophysical Observatory,
Nizhnij Arkhyz, Zelenchukskiy region, Karachai-Cherkessian Republic, Russia 369167
[1]email: `leda@sao.ru`

Abstract. We present the results of speckle interferometric observations of 273 magnetic stars most of which are Ap/Bp type. All observations were made at the 6-m telescope of the Special Astrophysical Observatory of the Russian Academy of Sciences. We resolved 58 binary and 5 triple stars into individual components. Almost half of these stars were astrometrically resolved for the first time. The fraction of speckle interferometric binaries/multiples in the sample of stars with confirmed magnetic fields is 23%. We expect that the total fraction of binaries/multiples in the sample with account for spectroscopic short-period systems and wide common proper motion pairs can be twice higher. The detected speckle components have a prominent peak in the ρ distribution that corresponds to the closest resolved pairs. Full version of present paper is available in electronic form at http://arxiv.org/abs/1308.3168.

Keywords. magnetic stars, binary and multiple stars, speckle interferometry

---

1. Sample

A sample of objects for observations was based on the Catalog of Magnetic Stars (Romanyuk & Kudryavtsev 2008). It contains a list of 355 chemically peculiar objects (mostly Ap/Bp) with detected global magnetic fields. We added 17 new magnetic stars discovered after the publication of the catalog. Therefore the total number of stars in the sample is 372. For the majority of stars in the list (322 objects) only the value of the longitudinal component of the field Be is known. For 48 stars the surface fields are determined from the splitting of Zeeman components. The vast majority of the sample objects are brighter than 10^m in the V-band. The stars are uniformly distributed on the celestial sphere, although a relatively small number (about 20%) of objects belong to open clusters of different ages. The BTA can capture only 273 objects from our sample with declinations $\delta > -30°$.

2. Observations

The speckle interferometric observations of 273 magnetic CP stars were carried out at the BTA in 2009–2012. They were performed with the speckle interferometer engineered at the SAO RAS (Maksimov *et al.* 2009). We used the PhotonMAX512 camera based on an internal electron multiplying CCD97 (EMCCD) produced by Princeton Instruments with a 512×512 pixel array. The limiting magnitude of our speckle interferometer is $\approx 15^m$ in the V-band depending on seeing conditions. Basically we employed two filters: 550/20 and 800/100 nm (central wavelength/bandwidth). We took 2000 short exposure images in each filter for almost all observed objects. High quantum efficiency and linearity of the detector permits the maximum magnitude difference between the components to reach up to 5-6^m depending on angular separation and weather conditions (Fig. 1 of Rategaev *et al.* 2013). The minimum angular separation between the components is determined by

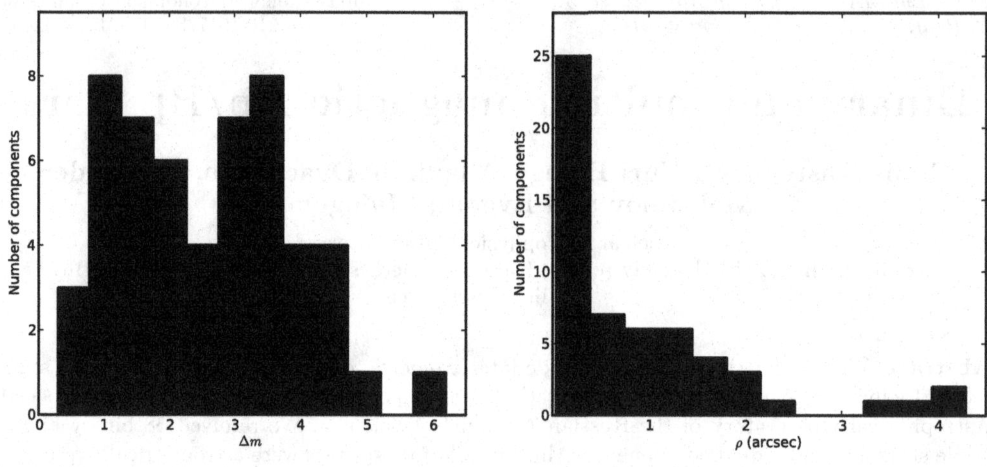

Figure 1.

the diffraction limit of the 6 m telescope. It is $0.022''$ and $0.033''$ for 550/20 and 800/100 nm filters, respectively. The size of the detector's field $4.4 \times 4.4''$ allowed secondary components to be discovered at angular separations as large as $3''$ from the primary star. The accuracy of our speckle interferogram processing method may be as good as 0.02^m, $0.001''$, and $0.1°$ for the component magnitude difference, angular separation, and position angle, respectively.

3. Results

For 63 stars in our sample, we observed speckle interferometric companions. Among the resolved systems 58 are binaries and 5 are triples. Twenty nine companions were resolved astrometrically for the first time. The fraction of speckle interferometric binaries/multiples in the sample of 273 stars with confirmed magnetic fields is 23%. Magnitude difference and angular separation distributions for resolved pairs are shown in Fig. 1. To plot these histograms we used 56 measurements of ρ and 53 Δm. We want to draw attention to the unusual profile of the ρ distribution. Speckle interferometric components of magnetic stars tend to be located close to the primary star. Half of the resolved stars have companions with $\rho < 0.32''$. This result is not a selection effect because close interferometric components are harder to detect than the wide pairs. The distribution of Δm for 28 resolved speckle companions with $\rho < 0.32''$ resembles that on the left half of Fig. 1. Reconstructed images of six systems resolved for the first time on BTA are presented in Fig. 2 of Rategaev *et al.* (2013). The Tab. 1 of Rategaev *et al.* (2013) is a list of all the stars that have speckle components. The systems resolved astrometrically for the first time are marked in bold.

4. Conclusion

According to our research the fraction of speckle interferometric binary and multiple systems in the sample of 273 CP stars with confirmed magnetic fields makes up 23% without account for undetected companions. Generally the speckle interferometric components have orbital periods larger than spectroscopic and smaller than common proper motion pairs. We expect that the total fraction of binaries/multiples in the sample with account for spectroscopic short-period systems and wide common proper motion pairs can be twice higher. The detected speckle components have a prominent peak in the

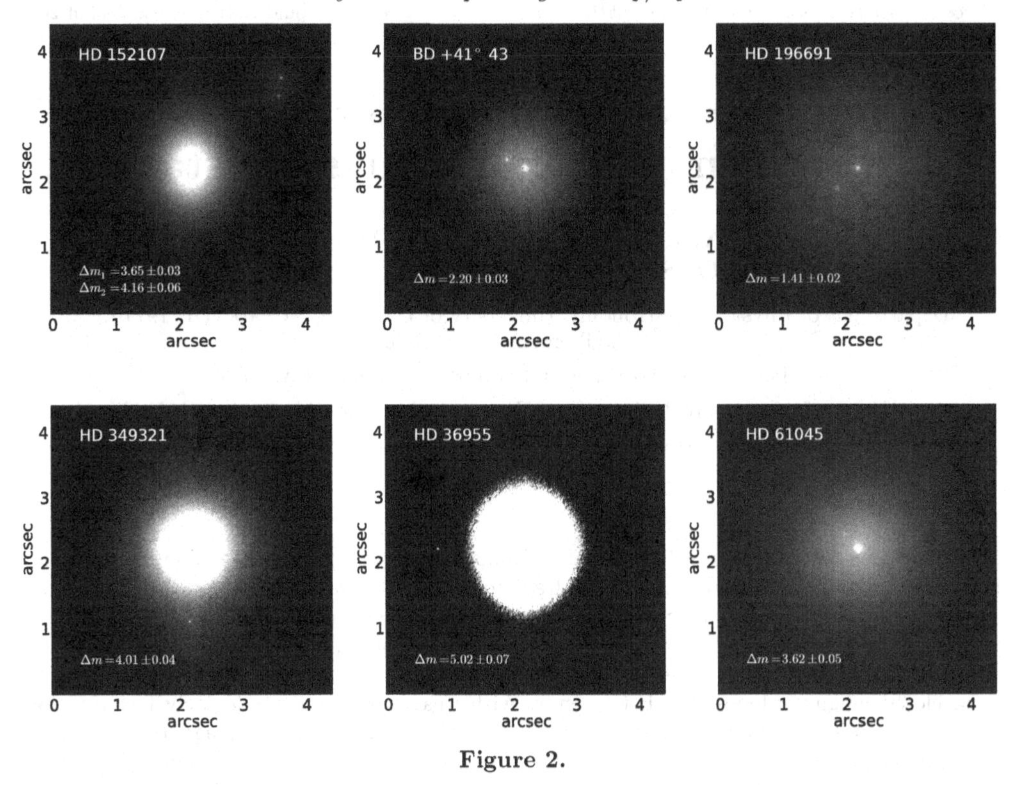

Figure 2.

ρ distribution that corresponds to the closest resolved pairs. More detailed and refined results of the presented study will be published soon.

Acknowledgements

This work was supported by Federal Target Program "Scientific and scientific-pedagogical personnel of innovative Russia" for 2009–2013 years (N 8704) and grant of the President of the Russian Federation for the state support of young Russian PhD scientists (MK-1001.2012.2).

References

Balega, Y. Y., Dyachenko V. V., Maksimov A. F., Malogolovets E. V., Rastegaev D. A. & Romanyuk, I. I. 2012, *AstBu*, 67, 44

Maksimov A. F., Balega Y. Y., Dyachenko V. V., Malogolovets E. V., Rastegaev D. A. & Semernikov E. A. 2009, *AstBu*, 64, 296

Rastegaev D. A., Balega Y. Y., Dyachenko V. V., Maksimov A. F. & Malogolovets E. V. 2013, *arXiv*, arXiv:1308.3168

Romanyuk I. I. & Kudryavtsev D. O. 2008, *AstBu*, 63, 139

Magnetic Fields throughout Stellar Evolution
Proceedings IAU Symposium No. 302, 2013
P. Petit, M. Jardine & H. Spruit, eds.

doi:10.1017/S1743921314002415

Wind channeling, magnetospheres, and spindown of magnetic massive stars

S. P. Owocki[1], A. ud-Doula[2], R. H. D. Townsend[3], V. Petit[1], J. O. Sundqvist[4] and D. H. Cohen[5]

[1]Department of Physics & Astronomy, University of Delaware, Newark, DE 19716 USA
email: owocki@udel.edu

[2]Penn State Worthington Scranton, Dunmore, PA, USA

[3]Dept. of Astronomy, University of Wisconsin-Madison, Madison, WI, USA

[4]Universitaetssternwarte Muenchen, Scheinerstr. 1, 81679 Muenchen, Germany

[5]Dept. of Physics & Astronomy, Swarthmore College, Swarthmore, PA, USA

Abstract. A subpopulation (~10%) of hot, luminous, massive stars have been revealed through spectropolarimetry to harbor strong (hundreds to tens of thousand Gauss), steady, large-scale (often significantly dipolar) magnetic fields. This review focuses on the role of such fields in channeling and trapping the radiatively driven wind of massive stars, including both in the strongly perturbed outflow from open field regions, and the wind-fed "magnetospheres" that develop from closed magnetic loops. For B-type stars with weak winds and moderately fast rotation, one finds "centrifugal magnetospheres", in which rotational support allows magnetically trapped wind to accumulate to a large density, with quite distinctive observational signatures, e.g. in Balmer line emission. In contrast, more luminous O-type stars have generally been spun down by magnetic braking from angular momentum loss in their much stronger winds. The lack of centrifugal support means their closed loops form a "dynamical magnetosphere", with trapped material falling back to the star on a dynamical timescale; nonetheless, the much stronger wind feeding leads to a circumstellar density that is still high enough to give substantial Balmer emission. Overall, this review describes MHD simulations and semi-analytic dynamical methods for modeling the magnetospheres, the magnetically channeled wind outflows, and the associated spin-down of these magnetic massive stars.

Keywords. Stars – early-type, Stars – magnetic fields, Stars – mass loss, Stars – X-rays, Stars – Rotation

---

1. Introduction

Massive, luminous, hot stars lack the hydrogen recombination convection zone that induces the magnetic dynamo cycle of cooler, solar-type stars. Nonetheless, modern spectropolarimetry has revealed that about 10% of O, B and A-type stars harbor large-scale, organized (often predominantly dipolar) magnetic fields ranging in dipolar strength from a few hundred to tens of thousand Gauss. (See contribution by G. Wade in these proceedings). Petit *et al.* (2013) recently compiled an exhaustive list of 64 confirmed magnetic OB stars with $T_{\rm eff} \gtrsim 16\,$kK, along with their physical, rotational and magnetic properties; see figure **??** below.

The review here summarizes efforts to develop dynamical models for the effects of such large-scale surface fields on the radiatively driven mass outflow from such OB stars. The focus is on the properties and observational signatures (e.g. in X-ray and Balmer line emission) of the resulting wind-fed *magnetospheres* in closed loop regions, and on the

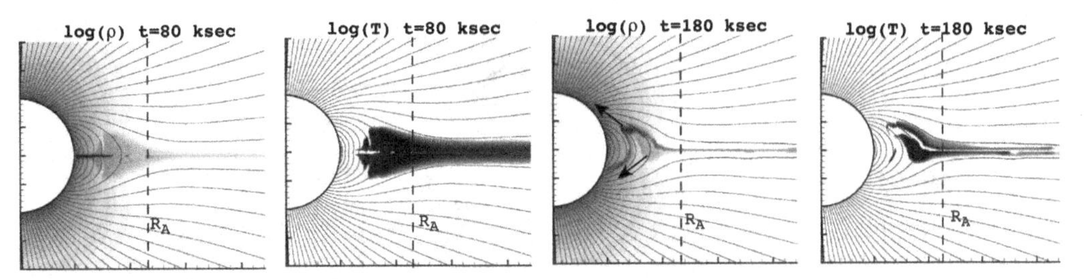

Figure 1. 2D MHD simulation for magnetic channeling and confinement of radiatively driven stellar wind from a non-rotating hot-star with $\eta_* = 15$ (and so $R_A \approx 2.3R_*$), showing the logarithm of density ρ and temperature T in a meridional plane. At a time 80 ksecs after the initial condition, the magnetic field has channeled wind material into a compressed, hot region about the magnetic equator, much as envisioned in the Magnetically Confined Wind Shock (MCWS) paradigm of Babel & Montmerle (1997a,b). But by a time of 180 ksecs, the cooled equatorial material is falling back toward the star along field lines, in a complex 'snake' pattern. The darkest areas of the temperature plots represent gas at $T \sim 10^7$ K, hot enough to produce relatively hard X-ray emission of a few keV. The model reproduces quite well the observed X-ray properties of θ^1 Ori C (Gagné *et al.* 2005).

stellar rotation *spindown* that results from the angular momentum loss associated with magnetically torqued wind outflow from open field regions.

2. MHD of Wind Outflows from Magnetic Hot Stars

2.1. *Wind magnetic confinement parameter and Alfvén radius*

MHD simulation studies (e.g. ud-Doula & Owocki 2002; ud-Doula *et al.* 2008) show that the overall net effect of a large-scale, dipole magnetic field in diverting such a hot-star wind can be well characterized by a single *wind magnetic confinement parameter* and its associated *Alfvén radius*,

$$\eta_* \equiv \frac{B_{eq}^2\, R_*^2}{\dot{M}\, v_\infty} \quad ; \quad \frac{R_A}{R_*} \approx 0.3 + (\eta_* + 0.25)^{1/4} \ , \tag{2.1}$$

where $B_{\rm eq} = B_{\rm p}/2$ is the field strength at the magnetic equatorial surface radius R_*, and $\dot{M}$ and v_∞ are the fiducial mass-loss rate and terminal speed that the star *would have* in the *absence* of any magnetic field. This confinement parameter sets the scaling for the ratio of the magnetic to wind kinetic energy density. For a dipole field, the r^{-6} radial decline of magnetic energy density is much steeper than the r^{-2} decline of the wind's mass and energy density; this means the wind always dominates beyond the Alfvén radius, which scales as $R_{\rm A} \sim \eta_*^{1/4}$ in the limit $\eta_* \gg 1$ of strong confinement.

As shown in figure 1, magnetic loops extending above $R_{\rm A}$ are drawn open by the wind, while those with an apex below $R_{\rm A}$ remain closed. Indeed, the trapping of wind upflow from opposite footpoints of closed magnetic loops leads to strong collisions that form X-ray emitting, *magnetically confined wind shocks* (MCWS; Babel & Montmerle 1997a,b). The post-shock temperatures $T \approx 20$ MK are sufficient to produce the moderately hard ($\sim 2\,$keV) X-rays observed in the prototypical magnetic O-star θ^1 Ori C (Gagné *et al.* 2005). As illustrated by the downward arrows in the density plot at a simulation time $t = 180\,$ksec, once this material cools back to near the stellar effective temperature, the high-density trapped material falls back onto the star over a dynamical timescale.

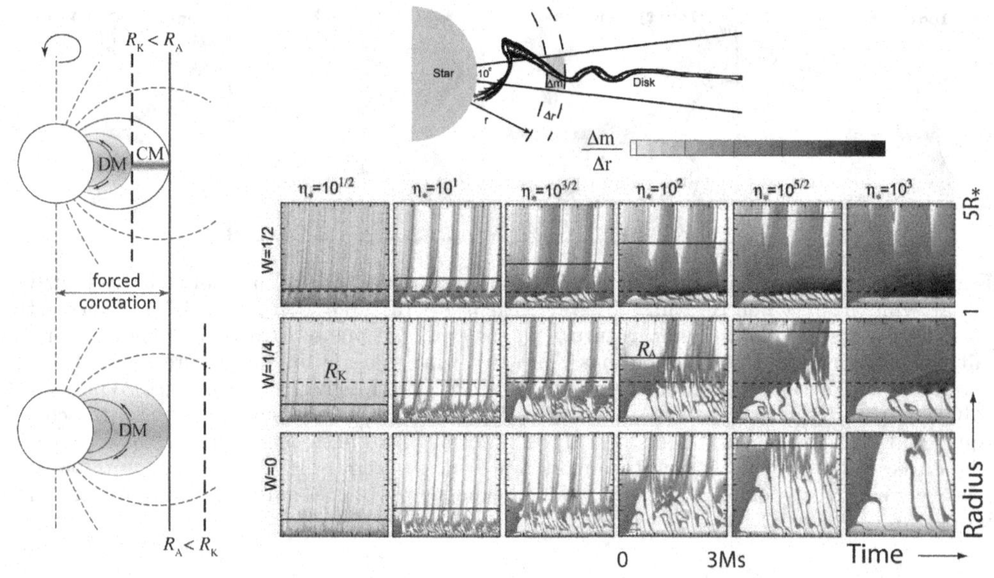

Figure 2. *Left:* Sketch of the regimes for a dynamical vs. centrifugal magnetosphere (DM vs. CM). The lower panel illustrates the case of a slowly rotating star with Kepler radius beyond the Alfvén radius ($R_{\rm K} > R_{\rm A}$); the lack of centrifugal support means that trapped material falls back to the star on a dynamical timescale, forming a DM, with shading illustrating the time-averaged distribution of density. The upper panel is for more rapid rotation with $R_{\rm K} < R_{\rm A}$, leading then to a region between these radii where a net outward centrifugal force against gravity is balanced by the magnetic tension of closed loops; this allows material to build up to the much higher density of CM. *Right, Upper:* Contour plot for density at arbitrary snapshot of an isothermal 2D MHD simulation with magnetic confinement parameter $\eta_* = 100$ and critical rotation factor $W = 1/2$. The overlay illustrates the definition of radial mass distribution, $\Delta m/\Delta r$, within 10° of the equator. *Right, Lower:* Densityplots for log of $\Delta m/\Delta r$, plotted versus radius (1-5 R_*) and time (0-3 Msec), for a mosaic of 2D MHD models with a wide range of magnetic confinement parameters η_*, and 3 orbital rotation fractions W. The horizontal solid lines indicate the Alfvén radius $R_{\rm A}$ (solid) and the horizontal dashed lines show Kepler radius $R_{\rm K}$ (dashed).

2.2. *Orbital rotation fraction and Kepler co-rotation radius*

The dynamical effects of rotation can be analogously parameterized (ud-Doula *et al.* 2008) in terms of the *orbital rotation fraction*, and its associated *Kepler corotation radius*,

$$W \equiv \frac{V_{\rm rot}}{V_{\rm orb}} = \frac{V_{\rm rot}}{\sqrt{GM_*/R_*}} \quad ; \quad R_{\rm K} = W^{-2/3}\, R_* \qquad (2.2)$$

which depend on the ratio of the star's equatorial rotation speed to the speed to reach orbit near the equatorial surface radius R_*. Insofar as the field within the Alfvén radius is strong enough to maintain *rigid-body rotation*, the Kepler corotation radius $R_{\rm K}$ identifies where the centrifugal force for rigid-body body rotation exactly balances the gravity in the equatorial plane. If $R_{\rm A} < R_{\rm K}$, then material trapped in closed loops will again eventually fall back to the surface, forming a *dynamical magnetosphere* (DM). But if $R_{\rm A} > R_{\rm K}$, then wind material located between $R_{\rm K}$ and $R_{\rm A}$ can remain in static equilibrium, forming a *centrifugal magnetosphere* (CM) that is supported against gravity by the magnetically enforced rotation. As illustrated in the upper left schematic in figure 2, the much longer confinement time allows material in this CM region to build up to a much higher density than in a DM region.

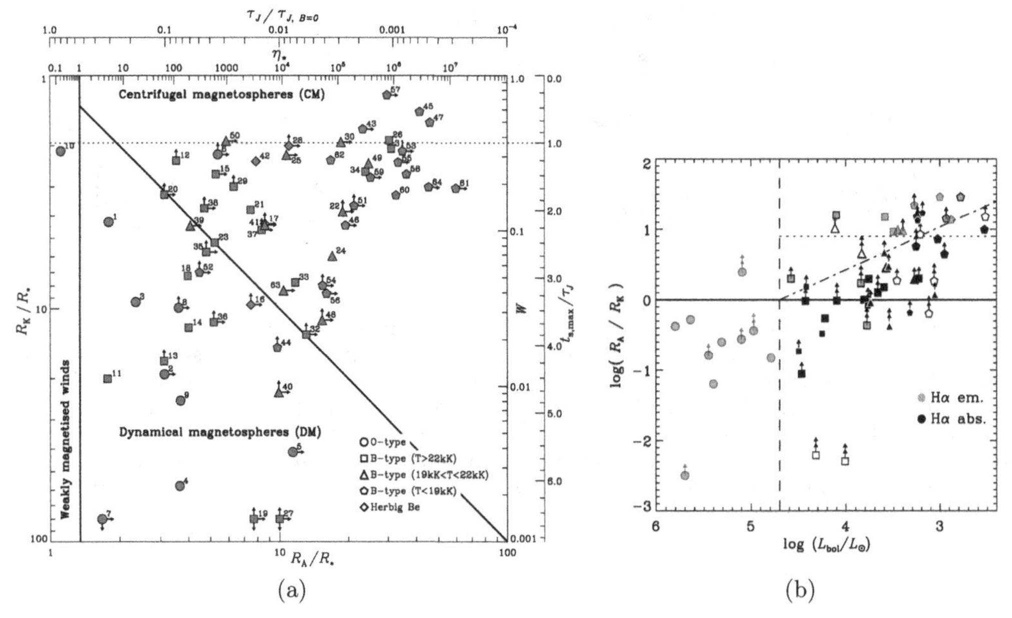

(a) (b)

Figure 3. (a.) Classification of 64 observationally confirmed magnetic massive stars in terms of magnetic confinement vs. rotation fraction, characterized here by a log-log plot of Kepler radius R_K increasing downward vs. Alfvén radius R_A increasing to the right. The labeled ID numbers are sorted in order of decreasing T_{eff}, with stellar identities given in Table 1 of Petit *et al.* (2013). The solid lines separate the magnetosphere domains of weakly magnetized winds (with $\eta_* < 1$), dynamical magnetospheres (DM) with $R_A < R_K$, and centrifugal magnetospheres (CM) with $R_A > R_K$. The additional upper and right axes give respectively the corresponding spindown timescale τ_J, and maximum spindown age $t_{s,\max}$, as defined in §3.4. Rapidly rotating stars above the horizontal dotted line have a maximum spindown age $t_{s,\max}$ that is less than one spindown time τ_J. (b.) Location of magnetic massive stars in a log-log plot of R_A/R_K vs. stellar luminosity. The symbol shadings mark the presence (pink or shaded) or absence (black) of magnetospheric Hα emission, with empty symbols when no Hα information is available. The vertical dashed line represents the luminosity transition between O-type and B-type main sequence stars. The horizontal dotted line and the diagonal dot-dashed line show division of the CM domain according to potential magnetospheric leakage mechanisms.

For full 2D MHD simulations in the axisymmetric case of a rotation-axis aligned dipole, the mosaic of color plots in figure 2 shows the time vs. height variation of the equatorial mass distribution $\Delta m/\Delta r$ for various combinations of rotation fraction W and wind confinement η_* that respectively increase upward and to the right. This illustrates vividly the DM infall for material trapped below R_K and R_A, vs. the dense accumulation of a CM from confined material near and above R_K, but below R_A.

3. Comparison with Observations of Confirmed Magnetic Hot-stars

For the 64 observationally confirmed magnetic hot-stars ($T_{\text{eff}} \gtrsim 16\,$kK) compiled by Petit *et al.* (2013), figure ?? plots positions in a log-log plane of R_K vs. R_A. The vertical solid line representing $\eta_* = 1$ separates the domain of non-magnetized or weakly magnetized winds to left, from the domain of stellar magnetospheres to the right. The diagonal line representing $R_K = R_A$ divides the domain of centrifugal magnetospheres (CM) to the upper right from that for dynamical magnetospheres (DM) to the lower left. Let us now consider how these distinctions in magnetospheric properties organize their observational characteristics.

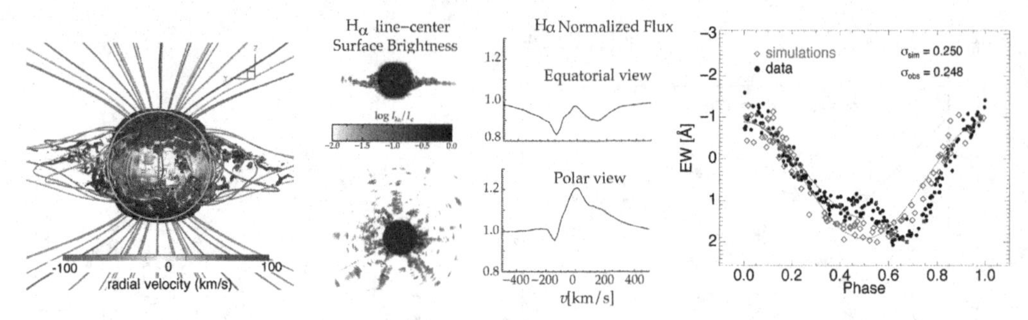

Figure 4. 3D MHD model of the dynamical magnetosphere for the young, slowly rotating (15.4-day period) O7V star θ^1 Ori C (ud-Doula *et al.* 2012). The left panel shows a snapshot of wind structure drawn as isodensity surface, colored to show radial component of velocity. The middle panels shows the predicted equatorial and polar views of Hα line-center surface brightness, along with corresponding line-flux profiles. The right panel compares the observed rotational modulation of the Hα equivalent width (black) with 3D model predictions (red) assuming a pure-dipole surface field tilted by $\beta = 45°$ to the rotation axis, as viewed from the inferred observer inclination of $i = 45°$.

3.1. *Balmer-α line emission from DM and CM*

Figure 3(b) plots these observed magnetic stars in a diagram comparing the *ratio* R_A/R_K vs. stellar luminosity, with now the symbol coded to mark the presence (light shading) or absence (black) of magnetospheric Hα emission. The horizontal solid line marks the transition between the CM domain above and the DM domain below, while the vertical dashed line marks the divide between O- and B-type main sequence stars. Note that *all* O-stars show emission, with all but one (Plaskett's star, which has likely been spun-up by mass exchange from its close binary companion; Grunhut *et al.* 2013) located among the slow rotators with a DM. By contrast, most B-type stars only show emission if they are well above the $R_A/R_K = 1$ horizontal line, implying a relatively fast rotation and strong confinement that leads to a CM.

The basic explanation for this dichotomy is straightforward. The stronger winds driven by the higher luminosity O-stars can accumulate even within a relatively short dynamical timescale to a sufficient density to give the strong emission in a DM, while the weaker winds of lower luminosity B-stars require the longer confinement and buildup of a CM to reach densities for such emission. This general picture is confirmed by the detailed dynamical models of DM and CM emission that motivated this empirical classification.

For the slowly rotating O-stars HD 191612 and θ^1 Ori C (here with respective ID numbers 4 and 3), both 2D and 3D MHD simulations (Sundqvist *et al.* 2012; ud-Doula *et al.* 2012) of the wind-fed DM reproduce quite well the rotational variation of Hα emission. For the 3D simulations of θ^1 Ori C, figure 4 shows how wind material trapped in closed loops over the magnetic equator (left panel) leads to circumstellar emission that is strongest during rotational phases corresponding to pole-on views (middle panel). For a pure dipole with the inferred magnetic tilt $\beta = 45°$, an observer with the inferred inclination $i = 45°$ has perspectives that vary from magnetic pole to equator, leading in the 3D model to the rotational phase variations in Hα equivalent width shown in the right panel (shaded circles). This matches quite well both the modulation and random fluctuation of the observed equivalent width (black dots), though accounting for the asymmetry about minimum will require future, more detailed models that include a secondary, higher-order (non-dipole) component of the inferred surface field.

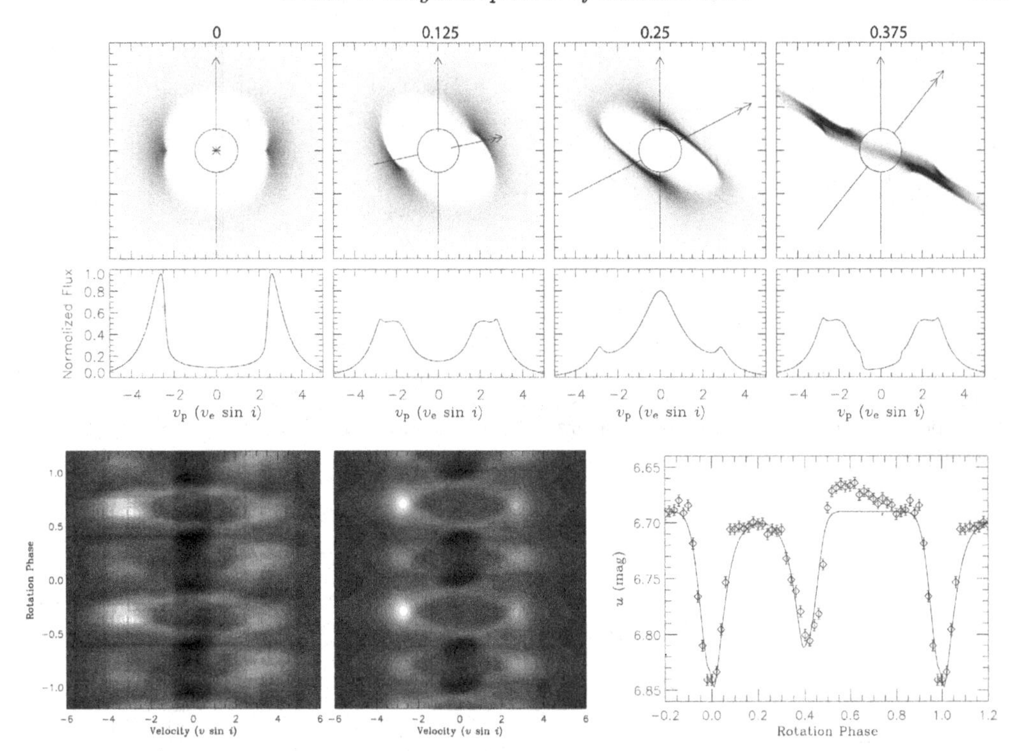

Figure 5. Observational signatures of the CM in the moderately fast rotator (1.2-day period) B2V star σ Ori E, compared with results from the RRM model (Townsend and Owocki 2005). The top row shows surface maps of Hα emission and resulting emission line profiles at the marked rotational phases. The lower-row density plots are associated dynamic Hα spectra, showing the variations relative to the photospheric profile over two rotation periods of $\sim 1.2d$; white indicates emission, and black absorption. The left panel is based on echelle observations of the star, while the right panel is the prediction from the RRM model. The lower-right line plot shows the Strömgren u-band light curve of σ Ori E, revealing the eclipse-like dimmings that occur when its two magnetospheric clouds transit in front of the star. The solid line indicates the predictions of an early RRM model.

3.2. *The Rigidly Rotating Magnetosphere (RRM) model*

In modeling the CM of more rapidly rotating, strongly magnetic B-stars like σ Ori E, a key challenge stems from the fact that their wind magnetic confinement parameters are generally of order $\eta_* \sim 10^6$ or more, far beyond the maximum $\eta_* \approx 10^3$ achieved with direct MHD simulations, which are limited by the Courant stability criterion. As an alternative for this *strong-field limit*, Townsend & Owocki (2005) developed a *Rigidly Rotating Magnetosphere* (RRM) model that uses a semi-analytical prescription for the 3D magnetospheric plasma distribution, based on the form and minima of the total gravitational-plus-centrifugal potential along each separate field line. Townsend *et al.* (2005) applied this RRM model to synthesize the emission from material trapped in the associated CM of σ Ori E. Figure 5 compares the predicted variation of the dynamic emission spectrum over the 1.2 day rotational period with that obtained from echelle observations of the star. The agreement is again very good, providing strong general support for this RRM model for Hα emission from the CM of σ Ori E.

The basic RRM concept has been further developed in a successor *Rigid Field Hydrodynamics* (RFHD) model (Townsend *et al.* 2007), wherein the time-dependent flow along each individual field line is simulated using a 1D hydro code. By piecing together

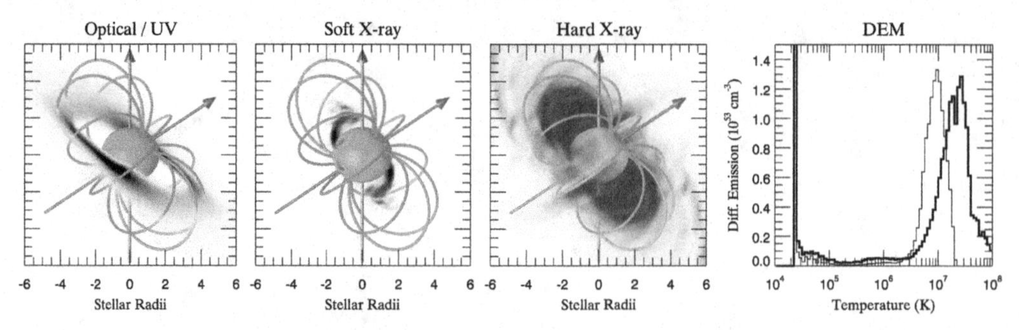

Figure 6. Snapshots from an RFHD model of σ Ori E, showing the spatial distribution of magnetospheric emission measure in three different temperature bins: optical ($T < 10^6$ K), soft X-ray (10^6 K $< T < 10^7$ K) and hard X-ray ($T > 10^7$ K). The plot on the right shows the corresponding differential emission measure, for models with (thin) and without (thick) thermal conduction.

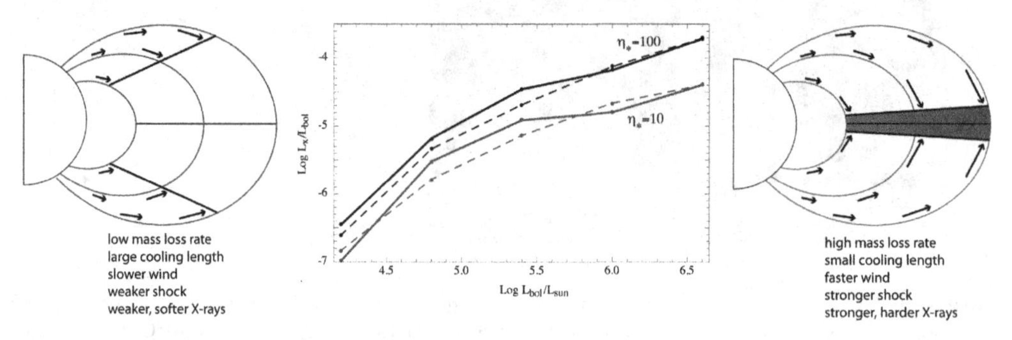

Figure 7. Scaling of X-ray luminosities L_x (for energies $E_x > 0.3$ keV) with stellar bolometric luminosity L_{bol} (center), for X-ray simulations with (solid) and without (dashed) inverse Compton cooling (from ud-Doula *et al.*, in prep). The left and right schematics illustrate the effect of "shock retreat" in reducing the strength and hardness of X-rays in lower luminosity stars with lower mass loss rate and thus less efficient radiative cooling.

independent simulations of many different field lines (typically, several *thousand*!), a 3D picture of a star's magnetosphere can be constructed at modest computational cost, leading in turn to predictions for not only Hα but also for X-ray emission (and other wind-related observables) of magnetospheres in the strong-field limit, as shown figure 6. A powerful aspect of both the RRM and RFHD models is that, within the strong field limit, they are in principle applicable to arbitrary field topologies, not just the oblique dipole configurations considered so far. Thus, for example, they could be used to model the magnetosphere of HD 37776, which harbors high-order multipoles (Kochukhov *et al.* 2011).

3.3. *MHD scalings for X-ray luminosity from MCWS*

A recent focus of MHD simulation parameter studies has been to derive predicted scaling relations for the X-ray luminosity L_x that results from magnetically confined wind shocks (MCWS). The central panel of figure 7 plots the ratio L_x/L_{bol} vs. the bolometric luminosity L_{bol} for models with magnetic confinement $\eta_* = 10$ and 100. The dashed curves assume the post-shock cooling is purely by radiative emission, while the solid curves account also for the effects of inverse Compton cooling. For the most luminous stars, L_x scales in proportion to the wind mass loss rate, which for line-driven winds follows $\dot{M} \sim L_{bol}^{1.6}$; but at lower L_{bol}, the lower $\dot{M}$ means the radiative cooling becomes inefficient. As illustrated in the left vs. right schematic panels, the larger cooling layer forces a "shock retreat" back to lower, slower wind outflow, leading to weaker shocks, and so lower, softer X-ray emission. But overall the X-ray luminosity from this MWCS mechanism is predicted to be well above the canonical $L_x \sim 10^{-7} L_{bol}$ inferred for non-magnetic O-type stars (Chlebowski *et al.* 1989). Work is underway to compare these theoretical predictions with scalings of observed X-ray properties of magnetic massive stars. (See the proceedings contribution by V. Petit.)

3.4. *Magnetic wind braking, spindown time, and spindown age*

Let us now turn to the issue of rotational spindown from magnetic wind braking. In this regard, the case of σ Ori E provides a key testbed, because extended photometric monitoring of the timing of magnetospheric clouds transiting in front of the star (Townsend *et al.* 2008; see also lower right panel of figure 5) has allowed a *direct* measurement of the change in rotation period, yielding a spindown time of 1.34 Myr (Townsend *et al.* 2010). This is remarkably close to the spindown time *predicted* previously by ud-Doula *et al.* (2009), based on the same 2D MHD aligned-dipole parameter study used for figure 2.

This MHD study showed that the angular momentum carried out by a magnetically torqued stellar wind follows the same simple, split-monopole scaling law derived for the Sun by Weber and Davis (1967), $\dot{J} = \frac{2}{3}\dot{M}\,\Omega\,R_{\rm A}^2$ – with, however, the Alfvén radius $R_{\rm A}$ now given by the *dipole* scaling $R_{\rm A} \sim \eta_*^{1/4}$, instead the oft-quoted, stronger scaling ($R_{\rm A} \sim \eta_*^{1/2}$) for a split monopole. This leads to an associated general formula for the rotational braking timescale,

$$\tau_{\rm J} \equiv \frac{I\Omega}{\dot{J}} = \frac{3}{2} f \tau_{\rm mass} \left(\frac{R_*}{R_{\rm A}}\right)^2 \approx 0.15 \frac{\tau_{\rm mass}}{\sqrt{\eta_*}} . \tag{3.1}$$

Here $\tau_{\rm mass} \equiv M/\dot{M}$ is the stellar mass loss timescale, and $f \approx 0.1$ is a dimensionless measure of the star's moment of inertia $I \equiv f M R_*^2$.

If we assume for simplicity a fixed radius R_* and moment of inertia factor $f \approx 0.1$, as well as a constant angular momentum loss rate $\dot{J}$, then the stellar rotation period P will simply increase exponentially with age t from its initial value, $P(t) = P_o e^{t/\tau_{\rm J}}$. This can be used to define a star's spindown *age*, t_s, in terms of the spindown time $\tau_{\rm J}$, and its inferred present-day critical rotation fraction $W = P_{\rm orb}/P$ relative to its initial rotation fraction W_o, $t_{\rm s} = \tau_{\rm J}(\ln W_o - \ln W)$. Taking the initial rotation to be critical, $W_o = 1$, yields a simple upper limit to the spindown age,

$$t_{\rm s,max} = \tau_{\rm J}\ \ln(1/W) . \tag{3.2}$$

If the initial rotation is subcritical, $W_o < 1$, then the actual spindown age is shorter by a time $\Delta t_{\rm s} = \tau_{\rm J} \ln W_o$.

In figure 3(a) the upper axis gives the spindown timescale $\tau_{\rm J}$ (normalized by the value in a non-magnetized wind), while the right axis gives the maximum spindown age $t_{\rm s,max}$ (normalized by the spindown time). Stars above the horizontal dotted line have a maximum spindown age that is *less* than a single spindown time. Together with the $R_{\rm A}/R_{\rm K}$ vs. luminosity plot in figure 3(b), we can identify some important features and trends:

- All the most rapidly rotating stars are cooler B-type with weak winds, and thus weak braking, despite their strong field. The two most extreme examples (ID 45 and 47) may be very close to critical rotation, and so provide a potential link to Be stars, which have *not* been found to have strong ordered fields, but for which rapid rotation is linked to decretion into an orbiting Keplerian disk.
- The only rapidly rotating O-star is Plaskett's star (ID 6), which has likely been spun up by mass exchange with its close binary companion (Grunhut *et al.* 2013). Many O-

stars have very long rotation period, e.g. 538 days for the field star HD 191612 (ID 4), suggesting substantial main-sequence spindown by wind magnetic braking, with a spin-down age comparable to its estimated main-sequence age.

• In contrast, the young Orion cluster star θ^1 Ori C (ID 3) has a moderately slow (15.4-day period) rotation, but is generally thought to be about 1 Myr old (Hillenbrand 1997; Scandariato *et al.* 2012), much less than its maximum spindown age $t_{\rm s,max} \approx 3\tau_{\rm J} \approx 10$ Myr. Thus its ZAMS rotation was likely already quite slow, suggesting significant *pre-*main-sequence braking, e.g. by PMS disk-locking, or through a PMS jet and/or wind. (See review by Pudritz in these proceedings.)

To reinforce the last point, the recent survey of Herbig Ae/Be stars by Alecian *et al.* (2013a,b) concludes that magnetic HeAeBe stars have a slower rotation than those without a detected field. Among their sample of non-magnetic stars they further find that those with lower mass evolve toward the ZAMS with a constant angular momentum, whereas higher mass ($> 5M_\odot$) stars show evidence of angular momentum loss during their PMS evolution, most likely as a result of their stronger, radiatively driven mass loss.

4. Future Outlook

The above shows there has been substantial progress in our efforts to understand the physical and observational properties of massive-star magnetospheres. But there are still important gaps in this understanding and key limitations to the physical realism of the models developed. The following lists some specific areas for future work:

• *3D MHD of Non-Axisymmetric Cases:* Thus far all MHD simulations, whether run in 2D or 3D, have been restricted to cases with an underlying axial symmetry, assuming a purely dipole field either without dynamically significant rotation, or with rotation that is taken to be aligned with the magnetic dipole axis. Fully 3D simulations are needed for both the many stars with an oblique dipole, as well as cases with more complex, higher-order multi-pole fields.

• *Spindown from oblique dipoles or higher-order multipoles:* An important application of these 3D MHD models will be to analyze the angular momentum loss from oblique dipole fields, as well as from higher-order fields. This will allow determination of generalized spindown scalings for complex fields, and provide the basis for interpreting anticipated future direct measurements of magnetic braking in stars with tilted-dipole or higher multi-pole fields.

• *Non-Ideal MHD and magnetospheric leakage:* In MHD simulations of slowly rotating magnetic stars with a DM, the dynamical infall of material back to the star balances the mass feeding from the stellar wind, yielding an overall mass and density that is in quite good agreement with absorption and emission diagnostics. By contrast, in CM simulations the much longer confinement and mass buildup is limited only by eventual centrifugal breakout of regions beyond the Kepler radius (Townsend & Owocki 2005), and this now leads to an overall predicted CM mass and density that significantly exceeds values inferred by observational diagnostics. To understand better the mass budget of CM's, it will be necessary to investigate additional plasma leakage mechanisms, such as the field line interchange transport that is thought to be key to mass balance of planetary magnetospheres (Kivelson & Southwood 2005). In addition to comparison with emission diagnostics of individual stars, this should aim to derive general scaling laws that can explain the trends for Balmer emission seen in Figure 3(b), particularly the boundary between Hα emission and absorption in B-type stars.

- *Rapid rotation and gravity darkening:* To model the rapidly rotating magnetic B-stars with $W = V_{rot}/V_{orb} > 1/2$, there is a need to generalize the lower boundary condition for both MHD and RFHD models to account for stellar oblateness, while also including the effect of gravity darkening for the wind radiative driving. This will also allow a link to Be stars, to constrain upper limits on the dynamical role of (undetected) magnetic fields in their quite distinctively Keplerian (vs. rigid-body) decretion disks. This will also provide a basis for applying such MHD models to PMS disks of HeAeBe stars.

Acknowledgments

This work was carried out with partial support by NASA ATP Grants NNX11AC40G and NNX12AC72G, respectively to University of Delaware and University of Wisconsin. D. H. C. acknowledges support from NASA ADAP grant NNX11AD26G and NASA *Chandra* grant AR2-13001A to Swarthmore College. We thank M. Gagné and G. Wade for many helpful discussions.

References

Alecian, E., Wade, G. A., Catala, C., *et al.* 2013a, *MNRAS*, 429, 1001
Alecian, E., Wade, G. A., Catala, C., *et al.* 2013b, *MNRAS*, 429, 1027
Babel J., & Montmerle T., 1997a, *ApJ*, 485, L29
Babel J. & Montmerle T., 1997b, *A&A*, 323, 121
Chlebowski T., Harnden Jr. F. R., & Sciortino S., 1989, *ApJ*, 341, 427
Gagné M., Oksala M. E., Cohen D. H., Tonnesen S. K., ud-Doula A., Owocki S. P., Townsend R. H. D., & MacFarlane J. J., 2005, *ApJ*, 628, 986
Grunhut, J. H., Wade, G. A., Leutenegger, M., *et al.* 2013, *MNRAS*, 428, 1686
Hillenbrand, L. A. 1997, *AJ*, 113, 1733
Kivelson, M. G. & Southwood, D. J., 2005, *Journal of Geophysical Research (Space Physics)*, **110**, 12209
Kochukhov, O., Lundin, A., Romanyuk, I., & Kudryavtsev, D. 2011, *ApJ*, 726, 24
Owocki S. P., Sundqvist J. O., Cohen D. H., & Gayley K. G., 2013, *MNRAS*, 429, 3379
Petit V., Owocki S. P., Wade G. A., Cohen D. H., Sundqvist J. O., Gagné M., Maíz Apellániz J., Oksala M. E., Bohlender D. A., Rivinius T., Henrichs H. F., Alecian E., Townsend R. H. D., ud-Doula A., & MiMeS Collaboration 2013, *MNRAS*, 429, 398
Sundqvist, J. O., ud-Doula, A., Owocki, S. P., Townsend, R. H. D., Howarth, I. D., & Wade, G. A., 2012, *MNRAS*, **423**, L21
Townsend, R. H. D., 2008, *MNRAS*, **389**, 559
Townsend, R. H. D., Oksala, M. E., Cohen, D. H., Owocki, S. P., & ud-Doula, A., 2010, *ApJ*, **714**, L318
Townsend, R. H. D. & Owocki, S. P., 2005, *MNRAS*, **357**, 251
Townsend, R. H. D., Owocki, S. P., & Groote, D., 2005, *ApJ*, **630**, L81
Townsend, R. H. D., Owocki, S. P., & ud-Doula, A., 2007, *MNRAS*, **382**, 139
ud-Doula A. & Owocki S. P., 2002, *ApJ*, 576, 413
ud-Doula A., Owocki S. P., & Townsend R. H. D., 2008, *MNRAS*, 385, 97
ud-Doula A., Owocki S. P., & Townsend R. H. D., 2009, *MNRAS*, 392, 1022
ud-Doula A., Sundqvist J. O., Owocki S. P., Petit V., & Townsend R. H. D., 2013, *MNRAS*, 428, 2723

Magnetic Fields throughout Stellar Evolution
Proceedings IAU Symposium No. 302, 2013
P. Petit, M. Jardine & H. Spruit, eds.

doi:10.1017/S1743921314002427

X-rays from magnetic massive OB stars

V. Petit[1], D. H. Cohen[2], Y. Nazé[3], M. Gagné[4], R. H. D. Townsend[5], M. A. Leutenegger[6], A. ud-Doula[7], S. P. Owocki[1] and G. A. Wade[8]

[1]Dept. of Physics & Astronomy, University of Delaware, Newark, DE, USA
email: vpetit@udel.edu

[2]Dept. of Physics & Astronomy, Swarthmore College, Swarthmore, PA, USA

[3]GAPHE Dépt. AGO, Université de Liège, Liège, Belgium

[4]Dept. of Geology & Astronomy, West Chester University, West Chester, PA, USA

[5]Dept. of Astronomy, University of Wisconsin-Madison, Madison, WI, USA

[6]Laboratory for High Energy Astrophysics, NASA/GSFC, Greenbelt, MD, USA

[7]Penn State Worthington Scranton, Dunmore, PA, USA

[8]Dept. of Physics, Royal Military College of Canada, Kingston, ON, Canada

Abstract. The magnetic activity of solar-type and low-mass stars is a well known source of coronal X-ray emission. At the other end of the main sequence, X-rays emission is instead associated with the powerful, radiatively driven winds of massive stars. Indeed, the intrinsically unstable line-driving mechanism of OB star winds gives rise to shock-heated, soft emission (~0.5 keV) distributed throughout the wind. Recently, the latest generation of spectropolarimetric instrumentation has uncovered a population of massive OB-stars hosting strong, organized magnetic fields. The magnetic characteristics of these stars are similar to the apparently fossil magnetic fields of the chemically peculiar ApBp stars. Magnetic channeling of these OB stars' strong winds leads to the formation of large-scale shock-heated magnetospheres, which can modify UV resonance lines, create complex distributions of cooled Halpha emitting material, and radiate hard (~2-5 keV) X-rays. This presentation summarizes our coordinated observational and modelling efforts to characterize the manifestation of these magnetospheres in the X-ray domain, providing an important contrast between the emission originating in shocks associated with the large-scale fossil fields of massive stars, and the X-rays associated with the activity of complex, dynamo-generated fields in lower-mass stars.

Keywords. Stars – early-type, Stars – magnetic fields

---

1. Introduction

The connection between the coronal X-ray emission of Sun-like, low-mass stars and their magnetic activity is well established by the strong correlation between their X-ray luminosity and the size of their convection zone, as well as their rotation rates (e.g. Wright *et al.* 2011). These parameters are the main ingredients powering their magnetic dynamos. Most massive OB stars are also X-ray bright (Berghoefer *et al.* 1997), although this emission is not traditionally associated with magnetism for two principal reasons.

First, the internal structure of main sequence stars undergoes major changes with increasing mass, transitioning from a radiative core and convective envelope to a convective core and radiative envelope. The best studied population of magnetic stars massive enough to have radiative envelopes are the so-called chemically peculiar ApBp stars. In contrast to the low-mass cool stars, the ApBp stars have strong, large scale, mostly dipolar magnetic fields, and represent only a sub-population ($\sim 10\%$) of all the A-type and late-B type stars (Power 2007). Recent efforts to characterize the magnetic properties of

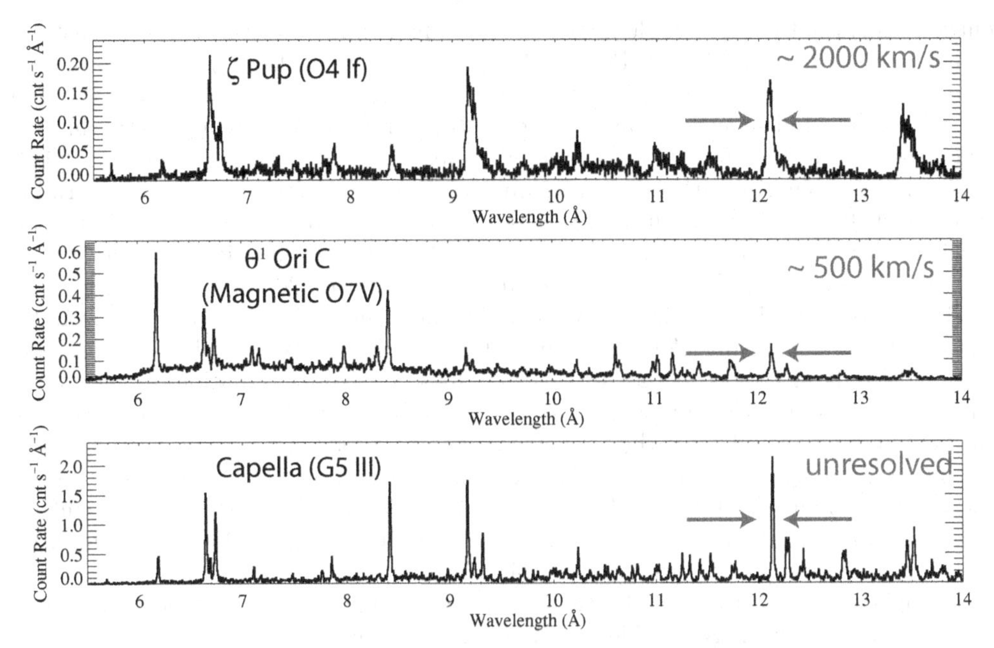

Figure 1. *Chandra* X-ray observatory High Energy Transmission Grating (HETG) spectra of the non-magnetic O-type type star ζ Pup (top), the magnetic O-type star θ^1 Ori C (middle), and the coronal emission from the G-type star Capella (bottom).

more massive OB stars, such as the large survey performed by the Magnetism in Massive Stars Project (Wade *et al.* 2011), have shown that all OBA-type stars share similar magnetic characteristic. This suggests that the fields of these stars are not being contemporaneously generated but are instead a remnant from a previous phase, or event, that occurred earlier during stellar evolution or even formation (e.g. Wade *et al.* and Grunhut *et al.* these proceedings). Moreover, since such large-scale magnetism is not a ubiquitous feature for stars with radiative envelopes, it cannot explain that most massive stars have X-ray emission.

Second, even if shallow sub-surface convection layers associated with opacity peaks of iron (Cantiello *et al.* 2009) could excite stochastic pulsations and perhaps provide mechanical heating at the surface, massive stars still would not have hot coronae. Indeed, their radiatively-driven winds are orders of magnitude denser than solar-type stars. As the wind densities are not low enough for a thermal runaway to occur, their winds stay cool, close to the surface effective temperature (Drew 1989).

Therefore, even OB stars that possess large-scale magnetic fields do not emit X-rays through the same coronal process as low-mass stars. Instead, X-ray production for massive stars is generally related to their strong winds, more specifically from wind material that has been shock-heated by various mechanisms, as reviewed in the following section.

2. X-ray emission processes for OB stars

Most massive OB stars emit relatively soft, stable X-rays (0.5 keV) when compared with low-mass stars (a few keV). They are very bright, $L_X \sim 10^{31}$-10^{33} erg s^{-1}, but this X-ray emission only represents $\sim 10^{-7}$ of their bolometric luminosity (Nazé *et al.* 2011).

This reflects the fact that the most universal shock mechanism, which heats up a few percent of the wind material, is caused by velocity variations in the supersonic wind outflow, intrinsic to the unstable nature of the radiative line-driving (Owocki, Castor & Rybicki 1988, Feldmeier, Puls & Pauldrach 1997). The low shock-jump velocities of these embedded wind shocks (EWS) hence create soft emission that is distributed throughout the whole wind in such a way to produce temporally nearly steady X-ray flux and Doppler-broadened (>1500 km/s) emission lines (e.g. Nazé *et al.* 2013, Cohen *et al.* 2010).

Massive star binaries with collision between their mutual winds at near their terminal speed (~2000 km/s) also can in some cases lead to strong emission ($L_X \sim 10^{31}$-10^{34}) of much harder X-rays (up to 5-10 keV) that will often display drastic variability tied to the orbital period (e.g. Corcoran *et al.* 2010, de Becker *et al.* 2006).

Finally, for stars hosting magnetic fields at their surface, X-rays can be produced through the Magnetically Confined Wind Shock model (MCWS; Babel & Montmerle 1997, ud-Doula & Owocki 2002). In this paradigm, the star's radiation-driven wind is channeled by a large-scale dipole magnetic field such that material is forced to flow along the field lines and collide near the tops of closed loops, producing a shock-heated volume of plasma. Although such shocks are nearly head-on, the magnetic field can only confine the wind up to an Alfvén radius $R_A \approx (B_{eq} R_\star)^{1/2}/(\dot{M} v_\infty)^{1/4}$ and the winds might not have been fully accelerated before reaching the top of the highest loop. Therefore, the X-rays are expected to be softer than for binaries, but still more luminous and energetic than single non-magnetic stars. Furthermore, if the magnetic axis is tilted with respect to the rotational axis, it is possible to observe modulation of the X-ray emission over the rotational period, such as seen for the magnetic O-type star θ^1 Ori C (Gagné *et al.* 2005).

3. X-rays from magnetic OB stars

X-ray observations are valuable to our understanding of magnetic massive stars, as they trace the hot gas and provide constraints on the kinematics and shocks in these magnetospheres.

For example, Fig. 1 compares the X-ray high-resolution spectra of the non-magnetic O-type star ζ Pup, the magnetic O-type star θ^1 Ori C, and the coronal emission from the G-type star Capella, obtained by the High Energy Transmission Gratings (HETG) aboard the *Chandra* X-ray Observatory. The emission lines of ζ Pup are significantly Doppler-broaden, consistent with the expectation of EWS. As can been seen from a comparison with the unresolved coronal lines of Capella, the emission lines of θ^1 Ori C are much narrower, although still resolved, indicative of the low, but non-zero post-shock velocities expected for MCWS.

The X-ray emission from the population of magnetic OB stars as a whole can also guide the models of magnetospheres by examining trends in X-ray luminosity (and eventually X-ray temperature) as a function of stellar/magnetic parameters, such as those predicted by the parameter study presented by Owocki *et al.* (these proceedings, also ud-Doula *et al.* in prep). Such studies could help explain the wide-ranging X-ray properties of magnetic stars that do not all conform at first glance to the simple MCWS paradigm (e.g. Petit 2011, Oskinova *et al.* 2011, Nazé *et al.* 2010).

For example, Fig. 2 presents all available X-ray efficiencies $\log(L_X/L_{bol})$ of hot magnetic OB stars, from the compilation by Petit *et al.* (2013). For stars without centrifugal support of the magnetically trapped material (dynamical magnetosphere), a trend of increased X-ray luminosity with mass-loss rate (which is a function of stellar bolometric luminosity) is expected, as illustrated by the left panel of Fig. 2. For stars with fast enough rotation such that the Kepler radius (R_K; material forced in co-rotation above this radius is centrifugally supported against gravity) is closer to the surface than the Alfvén radius (right panel of Fig. 2), the same trend with luminosity is observed. But

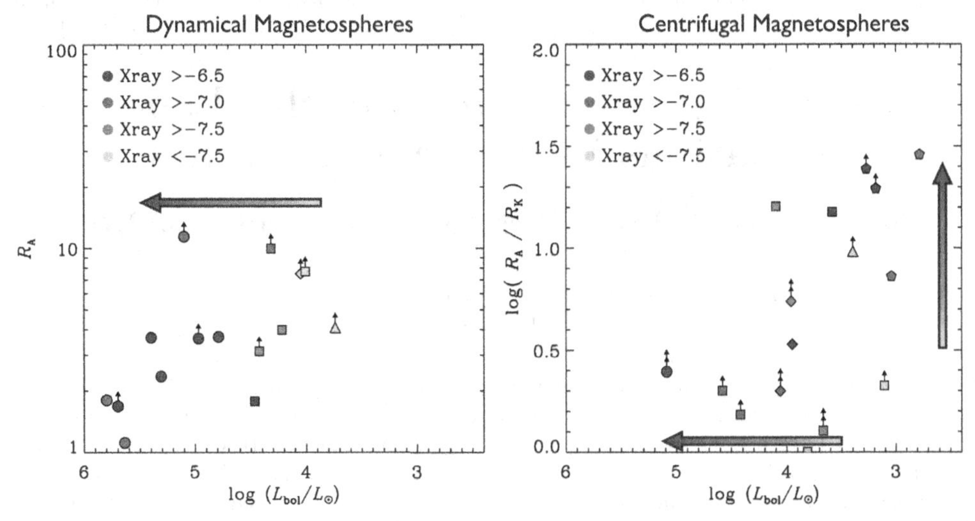

Figure 2. *Left:* X-ray efficiency ($\log(L_X/L_{bol})$; colour scale) of hot magnetic OB stars with dynamical magnetospheres as a function of their position in a diagram of Alfvén radius (R_A) versus luminosity (as a proxy for mass-loss rate). *Right:* X-ray efficiency of stars with centrifugal magnetospheres as a function of the position in a diagram of the ratio of the Alfvén radius to the Kepler co-rotation radius R_K (as a proxy for the size of the supported magnetospheric region) versus the bolometric luminosity. The expected qualitative trends are illustrated with arrows.

there also seems to be a trend of enhanced X-ray luminosity with the size of the centrifugally supported region (vertical axis), which could be explained by an enhancement of the shock-jump velocities from centrifugal acceleration.

References

Babel, J. & Montmerle, T. 1997 *A&A*, 323, 121
Berghoefer, T. W., Schmitt, J. H. M. M., Danner, R., & Cassinelli, J. P. 1997, *A&A*, 322, 167
Cantiello, M. *et al.* 2009, *A&A*, 499, 279
Cohen, D. H., Leutenegger, M. A., Wollman, E. E., Zsargó, J., Hillier, D. J., Townsend, R. H. D., & Owocki, S. P. 2010, *MNRAS*, 405, 2391
Corcoran, M. F., Hamaguchi, K., Pittard, J. M., Russell, C. M. P., Owocki, S. P., Parkin, E. R., & Okazaki, A. 2010, *ApJ*, 725, 1528
De Becker *et al.* 2006, *MNRAS*, 371, 1280
Drew, J. E. 1989 *ApJS*, 71, 267
Feldmeier, A., Puls, J., & Pauldrach, A. W. A. 1997, *A&A*, 322, 878
Gagné, M., *et al.* 2005, *ApJ*, 628, 986
Nazé, Y., Oskinova, L. M., & Gosset, E. 2013, *ApJ*, 763, 143
Nazé, Y. *et al.* 2011, *ApJS* 194, 7
Nazé, Y., ud-Doula, A., Spano, M., Rauw, G., De Becker, M., Walborn, N. R. 2010 *520*, A59
Oskinova, L. M., Todt, H., Ignace, R., Brown, J. C., Cassinelli, J. P., Hamann, W.-R. 2011 *MNRAS*, 416, 1456
Owocki, S. P., Castor, J. I., & Rybicki, G. B. 1988, *ApJ*, 335, 914
Petit, V. *et al.* 2013, *MNRAS*, 429, 398
Petit, V. 2011 Ph.D. Thesis, Université Laval (Canada)
Power, J. 2007, Ph.D. Thesis, Queen's University (Canada)
ud-Doula, A. & Owocki, S. P. 2002, *ApJ*, 576, 413
Wade, G. A. *et al.* 2011 in IAU 272, 118
Wright, N. J., Drake, J. J., Mamajek, E. E., & Henry, G. W. 2011, *ApJ*, 743, 48

Magnetic Fields throughout Stellar Evolution
Proceedings IAU Symposium No. 302, 2013
P. Petit, M. Jardine & H. Spruit, eds.

doi:10.1017/S1743921314002439

Investigating the origin of cyclical spectral variations in hot, massive stars

Alexandre David-Uraz[1], Gregg A. Wade[2], Véronique Petit[3] and Asif ud-Doula[4]

[1]Queen's University, Canada, [2]RMC, Canada, [3]University of Delaware, USA, [4]Penn State University, USA

Abstract. OB stars are known to exhibit various types of wind variability, as detected in their ultraviolet spectra, amongst which are the ubiquitous discrete absorption components (DACs). These features have been associated with large-scale azimuthal structures extending from the base of the wind to its outer regions: corotating interaction regions (CIRs). There are several competing hypotheses as to which physical processes may perturb the star's surface and generate CIRs, including magnetic fields and non radial pulsations (NRPs), the subjects of this paper with a particular emphasis on the former. Although large-scale magnetic fields are ruled out, magnetic spots deserve further investigation, both on the observational and theoretical fronts.

Keywords. Stars: winds, outflows – stars: massive – stars: magnetic fields

1. Introduction

Despite their small numbers, massive stars are known to play an important role in galactic ecology, enriching the interstellar medium with heavy elements processed in their cores, shaping their environment via their fast, dense winds as well as by their spectacular deaths as supernovae. However, their impressive outflows do not only influence their surroundings, but also significantly affect the evolution of massive stars themselves.

While important strides have been taken towards understanding these radiatively-driven, supersonic winds (Castor *et al.* 1975), there still remains a number of pressing questions. For instance, for over 20 years, the community has struggled to explain the presence of cyclical variations in wind-sensitive, UV resonance lines. The most common manifestation of this type of variability appears as "discrete absorption components" (DACs, e.g. Kaper *et al.* 1996), which migrate from zero velocity to approximately terminal velocity over a relatively well-defined timescale, and are thus thought to be linked to structures extending from the base of the photosphere all the way to the outer regions of the wind.

These DACs also possess other revealing properties. Indeed, Prinja (1988) showed an apparent correlation between the projected rotational velocities of stars and the periods of their DACs, suggesting that these variations are rotationally modulated. Furthermore, single observations of over 200 O stars almost all show the presence of a narrow absorption component (NAC) at more or less the terminal velocity (Howarth & Prinja 1989). These narrow features are believed to be snapshots of DACs, leading to the conclusion that the DAC phenomenon is ubiquitous among hot, massive stars.

Cranmer & Owocki (1996) provided a model to understand how these structures are generated. By introducing ad hoc photospheric perturbations on the star (bright spots), they were able to reproduce corotating interaction regions (CIRs), which lead to DAC-like features in synthetic UV spectra. This view has not changed dramatically since and is still considered to be the canonical way of producing DACs. However, the nature of these perturbations is unknown. Because of the ubiquitous quality of DACs, gaining insight on the physical process at their origin should provide meaningful information about all OB stars.

2. Potential physical causes

There are two main probable causes of DACs identified in the literature: magnetic fields and non radial pulsations. Furthermore, we will distinguish between large-scale and small-scale magnetic fields. The following subsections explore each possibility in further detail.

2.1. *Non radial pulsations*

Non radial pulsations are known to exist in a number of well-studied DAC stars (de Jong *et al.* 1999), but it has been difficult to draw any links between both phenomena. Observationally, these pulsations are relatively easy to detect. Theoretically, there are some inconsistencies. Pulsations generally create a pattern of alternating bright and dark regions on the surface of a star. However, Cranmer & Owocki (1996) found that only bright spots reproduce the expected pattern. Additionally, NRP periods are typically on a timescale of a few hours, while DACs have periods of the order of a few days. Nevertheless, it has been postulated that a superposition of modes could create variations consistent with DAC recurrence timescales (de Jong *et al.* 1999). An in-depth analysis of the merits and likelihood of this hypothesis goes beyond the scope of this paper, which focuses primarily on the magnetic properties of DAC stars; therefore, such a discussion will be left for a later study.

2.2. *Large-scale magnetic fields*

Magnetic fields in massive stars are rare and are usually organized and dipolar in nature (Wade & the MiMeS Collaboration 2010). Consequently, they are believed to be of fossil origin. Magnetism has been proposed a number of times as a possible cause for DACs (e.g. Kaper & Henrichs 1994) and given the fact that DACs generally come in pairs (Kaper *et al.* 1996), dipolar fields seem like a reasonable origin, albeit a challenging one. Indeed, if the small rate of detection of magnetic fields in OB stars (7 %, Wade *et al.*, these proceedings) does not already raise red flags, the apparent inconsistency of their stable configurations with the cyclical (rather than periodic) nature of DAC recurrence is troublesome.

Nonetheless, David-Uraz *et al.*, *in prep.* have studied the magnetic properties of 14 OB stars known or believed to have DACs. Using high-resolution spectropolarimetry in the Stokes I and V parameters from ESPaDOnS (Canada-France-Hawaii Telescope) and NARVAL (Télescope Bernard Lyot), they have applied a multi-line signal-enhancing technique (Least-Squares Deconvolution, or LSD, Donati *et al.* 1997) as well as nightly-averaging to obtain high signal to noise ratio V profiles, which were then used to try to detect Zeeman signatures. Assuming the oblique rotator model (Stibbs 1950), two magnetic diagnostics are measured from the data. Longitudinal magnetic field measurements are performed using the first-order moment of the V profiles (Wade *et al.* 2000), and dipolar field strengths are computed by fitting each profile and performing a Bayesian inference-based modeling, as described by Petit & Wade (2012).

The results are unequivocal: no dipolar magnetic fields are detected. All derived values are consistent with a null magnetic field. Furthermore, extremely tight constraints are obtained (the best constraints on both nightly longitudinal field error bars, 4 G, and dipolar field strength, 23 G, are obtained for the sharp-lined O dwarf 10 Lac). Two

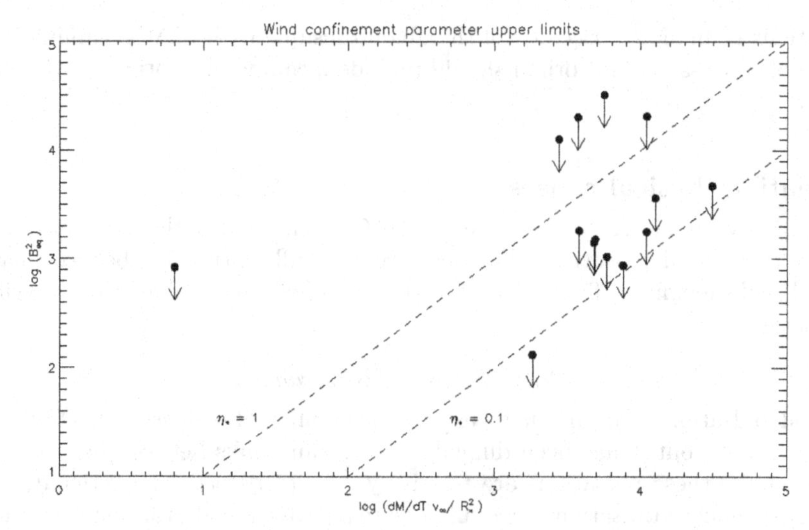

Figure 1. Upper limits on the wind confinement parameter value for 14 well-known DAC stars. The horizontal axis corresponds loosely to the wind kinetic energy density, while the vertical axis is proportional to the magnetic energy density. The dashed lines correspond to wind confinement parameter values of 1 and 0.1, which are considered as important thresholds (see section 2).

interaction mechanisms between the magnetic field and the wind are considered: magnetic wind confinement, and magnetically-induced surface brightness enhancements. In the first case, the magnetic field entraps material by restricting the outflow when it is perpendicular to field lines (i.e. at the equator). Even when the material is not completely confined, this type of mechanism dynamically influences the outflow. This interaction was described by ud-Doula & Owocki (2002), who define the "wind-confinement parameter" essentially as the ratio of the magnetic energy density and the wind kinetic energy density: $\eta_* = ({B_{eq}}^2 {R_*}^2)/(\dot{M} v_\infty)$, where B_{eq} corresponds to the strength of the magnetic field at the equator (which equals half of the dipolar field strength), R_* is the stellar radius, $\dot{M}$ is the mass-loss rate and v_∞ is the terminal velocity of the wind. They showed that above a threshold of $\eta_* = 1$, the wind is magnetically confined, whereas it is dynamically influenced all the way down to $\eta_* = 0.1$. Figure 1 shows the upper limits derived from the Bayesian analysis. Less than half (5) of the points have upper limits above $\eta_* = 1$, slightly more than half (8) are between both thresholds (with a few of them quite close to the lower one), and one point (10 Lac) is under $\eta_* = 0.1$. Clearly, most of these stars do not have magnetically confined winds, therfore a large portion of their winds could not be dynamically influenced by globally-organized dipolar magnetic fields, should they exist.

For the bright spot mechanism, the idea is that the magnetic pressure within a flux tube would decrease the gas pressure (as compared to a point on the surface with no magnetic flux), therefore creating "wells" which probe deeper into the star, which result in hot spots (given the temperature gradient in hot star atmospheres). The luminous flux enhancement can be written as follows: $F'/F = 1 + (3\kappa B^2)/(32\pi g)$, where κ is the mean Rosseland opacity, B is the magnetic field strength and g is the gravitational acceleration. Finally, taking typical values for O dwarfs ($\kappa \sim 1$ and $\log g = 4.0$), it is estimated that a 400 G magnetic field is required to produce a 50% brightness enhancement (this corresponds to the value used by Cranmer & Owocki 1996), which is much greater than all the dipolar field strength upper limits obtained for the sample. In conclusion, combined with the tight constraints obtained observationally, both mechanisms fail to create the required conditions to generate DACs.

2.3. *Small-scale magnetic fields*

On the other hand, the case for magnetism in general is not settled. Indeed, Cantiello & Braithwaite (2011) argue that the iron opacity bump at 150 kK leads to a sub-surface convection zone in the most massive stars, resulting in potentially measurable small-scale surface magnetic fields (or spots). Unfortunately, the characteristics of these fields, as inferred from theory, are not currently very well constrained, leading to significant modeling challenges. While these spots would be very hard to detect (especially if they are distributed randomly, Kochukhov & Sudnik 2013), they present an attractive alternative to other field configurations. Indeed, spots on the Sun for instance vary with time; transient spots seem consistent with the cyclical recurrence of DACs. Furthermore, this fits into a picture where DACs are modulated by rotation.

3. Future work

As mentioned earlier, it has been clearly demonstrated that dipolar fields are not responsible for the DAC phenomenon. Therefore, to further test the magnetic hypothesis, more attention now has to be given to small-scale fields. On the observational side, deeper magnetic measurements can be performed on a limited number of well-suited stars (bright, massive, low projected rotational velocity), and then analyzed using a Bayesian inference technique similar to that used for dipolar fields (only, in this case, using a parametrized spot model under a certain number of assumptions). High-precision photometry has also been suggested as a means of detecting CIRs (e.g. Chené *et al.* 2011, David-Uraz *et al.* 2012, etc.) and deserves to be looked into.

Further numerical simulations should be conducted to show whether magnetic spots can actually produce DAC-like signatures, and if so, what the relevant parameter space is. Non radial pulsations should also be studied more systematically. Finally, if all else fails, it might be necessary to think outside the box, for instance, challenging the idea that DACs are rotationally modulated, and trying to find another mechanism to create them. In the end, that might just be what it takes to solve this long-standing problem.

References

Cantiello, M. & Braithwaite, J. 2011, *A&A*, 534, A140
Castor, J. I., Abbott, D. C., & Klein, R. I. 1975, *ApJ*, 195, 157
Chené, A. N., Moffat, A. F. J., Cameron, C., Fahed, R., *et al.* 2011, *ApJ*, 735, 34
Cranmer, S. R. & Owocki, S. P. 1996, *ApJ*, 462, 469
David-Uraz, A., Moffat, A. F. J., Chené, A. N., Rowe, J. F., *et al.* 2012, *MNRAS*, 426, 1720
de Jong, J. A., Henrichs, H. F., Schrijvers, C., Gies, D. R., *et al.* 1999, *A&A*, 345, 172
Donati, J. F., Semel, M., Carter, B. D., Rees, D. E., & Cameron, A. C. 1997, *MNRAS*, 291, 658
Howarth, I. D. & Prinja, R. K. 1989, *ApJS*, 69, 527
Kaper, L. & Henrichs, H. F. 1994, *Ap&SS*, 221, 115
Kaper, L., Henrichs, H. F., Nichols, J. S., Snoek, L. C., *et al.* 1996, *A&AS*, 116, 257
Kochukhov, O. & Sudnik, N. 2013, *A&A*, 554, A93
Petit, V. & Wade, G. A. 2012, *MNRAS*, 420, 773
Prinja, R. K. 1988, *MNRAS*, 231, 21
Stibbs, D. W. N. 1950, *MNRAS*, 110, 395
Wade, G. A., Donati, J. F., Landstreet, J. D., & Shorlin, S. L. S. 2000, *MNRAS*, 313, 851
Wade, G. A., the MiMeS Collaboration 2010, *arXiv*, 1012, 2925

Magnetic Fields throughout Stellar Evolution
Proceedings IAU Symposium No. 302, 2013
P. Petit, M. Jardine & H. Spruit, eds.

doi:10.1017/S1743921314002440

The dichotomy between strong and ultra-weak magnetic fields among intermediate-mass stars

François Lignières[1,2], Pascal Petit[1,2], Michel Aurière[1,2], Gregg A. Wade[3] and Torsten Böhm[1,2]

[1]CNRS, Institut de Recherche en Astrophysique et Planétologie
14 avenue Edouard Belin, 31400 Toulouse, France
email: francois.lignieres@irap.omp.eu

[2]Université de Toulouse, UPS-OMP, IRAP
31400 Toulouse, France

[3]Department of Physics, Royal Military College of Canada
PO Box 17000, Station Forces, Kingston, Ontario K7K 7B4, Canada

Abstract. Until recently, the detection of magnetic fields at the surface of intermediate-mass main-sequence stars has been limited to Ap/Bp stars, a class of chemically peculiar stars. This class represents no more than 5-10% of the stars in this mass range. This small fraction is not explained by the fossil field paradigm that describes the Ap/Bp type magnetism as a remnant of an early phase of the star-life. Also, the limitation of the field measurements to a small and special group of stars is obviously a problem to study the effect of the magnetic fields on the stellar evolution of a typical intermediate-mass star.

Thanks to the improved sensitivity of a new generation of spectropolarimeters, a lower bound to the magnetic fields of Ap/Bp stars, a two orders of magnitude desert in the longitudinal magnetic field and a new type of sub-gauss magnetism first discovered on Vega have been identified. These advances provide new clues to understand the origin of intermediate-mass magnetism as well as its influence on stellar evolution. In particular, a scenario has been proposed whereby the magnetic dichotomy between Ap/Bp and Vega-like magnetism originate from the bifurcation between stable and unstable large scale magnetic configurations in differentially rotating stars. In this paper, we review these recent observational findings and discuss this scenario.

Keywords. Stars : magnetic fields, instabilities

1. Introduction

The origin of the stellar magnetic fields and their effects on the star structure and evolution are the two basics questions of stellar magnetism. In all stars that possess a convective envelope, a magnetic field is believed to be generated by dynamo mechanism. It is not always possible to detect it, but its presence at the surface of these solar-type stars makes little doubt. For these stars, we also know that a magnetized wind exerts a breaking torque that strongly affects their angular momentum evolution. This illustrates that although limited our level of understanding of the magnetic fields of solar-type stars enables to answer at least some simple questions about their origin and their impact. This is not the case for hotter stars with radiative envelope where the origin of the observed fields as well as their impact on the star evolution remain largely mysterious. The properties of the observed fields are nevertheless well established. They are large scale mostly dipolar fields with dipole strength ranging from 300 G to 30 kG and are remarkably stable over time (Donati & Landstreet, 2009). These fields are usually called fossil fields as a reference to the hypothesis that describes them as remnant of an early phase of the star-life, either from the collapse of a magnetized cloud or during the convective protostellar phase. But they only are detected in a small 5–10% fraction of the intermediate-mass and massive stars and this fact did not receive a convincing explanation yet. From the point of view of stellar evolution, such a lack of understanding of the field generation processes combined with the absence of direct measurements for the vast majority (90–95 %) of stars are a major obstacle to model the effect of a magnetic field on the structure and evolution of typical intermediate-mass and massive stars.

In this paper, we review recent observational results that shed a new light on the 5–10% problem and suggested new scenarios for the origin of hot star magnetism. These results have been obtained thanks to the high sensitivity of a new generation of high-resolution échelle spectropolarimeters (MuSiCoS, Narval, ESPaDOnS, HARPSpol). In particular, their large wavelength coverage allows to sum up the magnetically polarized signal of many spectral lines and thus to detect weak fields. The number of lines available being an important factor, it is comparatively easier to detect weak fields in cooler stars. For example, Aurière *et al.* (2009) showed that a sub-gauss longitudinal field can be detected on the bright late-type star Pollux averaging only a few high S/N Stokes V of Narval or ESPaDOnS. For the same reason, weak fields are easier to detect in intermediate-mass stars than in more massive stars.

Spectropolarimetric surveys of Ap stars as well as of A-type non-Ap stars have first ruled out the possibility that the 5–10% fraction simply reflects the detection limit of magnetic measurements. Indeed, the magnetic fields of Ap/Bp stars were found to be higher than a lower limit of the order of 100 G for the longitudinal field while no field were detected in A-type non-Ap stars. Then two detections of sub-gauss fields on the bright A stars Vega and Sirius have been obtained. Thus, instead of a single class of magnetic stars, intermediate-mass star magnetism is now characterized by two type of magnetisms, Ap/Bp and Vega-like, separated by a two orders of magnitude magnetic desert between 1G - 100 G. In section 2, we present the evidences for the lower bound to the Ap/Bp magnetic fields and the magnetic desert, then the discovery of a new type of sub-gauss magnetism is described in section 3. The implications for the origin of upper-main-sequence magnetism are discussed in section 4.

2. The lower bound of Ap/Bp magnetic fields and the magnetic desert

From the first Zeeman effect measurement (Babcock 1947) to the early 90s, all magnetic field detections in the upper-main-sequence have been achieved among the Ap/Bp stars, a group of late B, A and early F stars showing strong chemical peculiarities. The vector modulus measured from the line broadening in the intensity spectrum ranged from 2 and 30 kG while measurements of the circular polarization of spectral lines allowed to find weaker fields, down to $\sim$ 300 G dipolar fields.

At that time, the detection limit was too high to study Ap/Bp stars with weak magnetic fields or to put strong constraints on the upper bound of the magnetic fields in A-type non Ap/Bp stars (Landstreet 1982). Thus, although this was suspected, it was not possible to confirm that all Ap/Bp stars were magnetic. Moreover a low-field continuation of the Ap/Bp magnetism among non-Ap/Bp stars could not be ruled out.

This last point is neither excluded from our understanding of the link between the chemical anomalies and magnetic fields in Ap/Bp stars. Michaud *et al.* (1970, 1976) indeed proposed that the magnetic fields of Ap/Bp stars are strong enough to avoid macroscopic mixing in the envelope either through differential rotation, thermal convection or stellar winds. He showed that in the quiescent outer layers microscopic diffusion processes produce strong anomalies in the chemical abundances compatible with those observed at the surfaces of Ap/Bp stars. This model also explains why chemical anomalies are either absent (in normal stars) or reduced (in the slowly rotating Am stars) when no fields are detected. Following this picture, we can conclude that all Ap/Bp stars should possess a magnetic field higher than the minimum field strength required to suppress macroscopic mixing and that, if a magnetic field is present at the surface of A-type non-Ap/Bp stars, its strength should be smaller than this minimum field.

From the mid-90s, deep surveys have been conducted with the spectropolarimeters MuSiCoS and Narval at Telescope Bernard Lyot to explore the weakly magnetic Ap/Bp stars and non Ap/Bp stars. Aurière et al. (2007) selected a sample of 28 Ap/Bp stars for which previous attempts had led to no detection or to unreliable ones. Thanks to the improved sensitivity of these instruments, a magnetic field was found in all the stars of the sample confirming that all Ap/Bp stars are magnetic. Moreover, this survey proved the existence of a lower bound to the magnetic field of Ap/Bp stars. The maximum absolute value reached by the longitudinal field as the star rotates B_L^{max} is higher than ~ 100 G, while the dipolar fields obtained by fitting an oblique rotator model - a magnetic dipole inclined with respect to the rotation axis - is higher than $B_{\mathrm{min}} = 300$ G in all but two stars for which the best model is slightly lower but still compatible with the 300 G value within the error bars.

This result is fully compatible with the idea that a minimum field strength is needed to produce the Ap/Bp chemical anomalies. But one must keep in mind that the observed lower bound of Ap/Bp magnetic fields could be either higher or equal to this minimum field. If it is equal, then one would expect to find a low field extension of the Ap/Bp magnetism among the other intermediate-mass stars. Indeed, it is difficult to imagine that a mechanism only generates magnetic fields in a sub-group of stars and that, in addition, the smallest field generated this way is exactly the minimum field required for Ap/Bp chemical anomalies.

The simplest extension one could think of is a population of dipolar-like fields that continuously extends the distribution of Ap/Bp dipolar fields below 300 G. However, surveys among A and late-B stars conducted with MuSiCoS, Narval and HARPSpol (Shorlin *et al.* 2002, Aurière *et al.* 2010, Makaganiuk et al. 2011) as well as deep spectropolarimetric runs dedicated to specific objects (e.g. Wade *et al.* 2006, Kochukhov *et al.* 2011), have ruled out the existence of such a population. These non-detections instead revealed a gap in longitudinal field between 100 G, the lower bound of Ap/Bp magnetism, and the detection limit of the surveys. With MuSiCoS, Shorlin et al.(2002) reached a detection limit of the order of 50 G for the longitudinal fields, while the Narval survey (Aurière *et al.* 2010) set a 3σ upper limit of the longitudinal field to 10 G for most stars of the sample down to ~ 1 G in some bright low v sini targets like Sirius. In the following, we shall speak of the magnetic desert in longitudinal field to describe this gap.

Its discovery among intermediate-mass stars has important consequences : The first one is that Ap/Bp stars constitute a separated class of stars in what concerns their magnetic properties because no low field continuous extension of this large-scale stable magnetism could be found among intermediate-mass stars. The so-called 10% problem of why only a small fraction of stars appears to be magnetic is therefore not due to an observational bias but corresponds instead to a true physical dichotomy between Ap magnetism and other intermediate-mass stars. The lower bound of Ap/Bp star magnetic fields must thus be regarded as a characteristic property of Ap/Bp magnetism whose

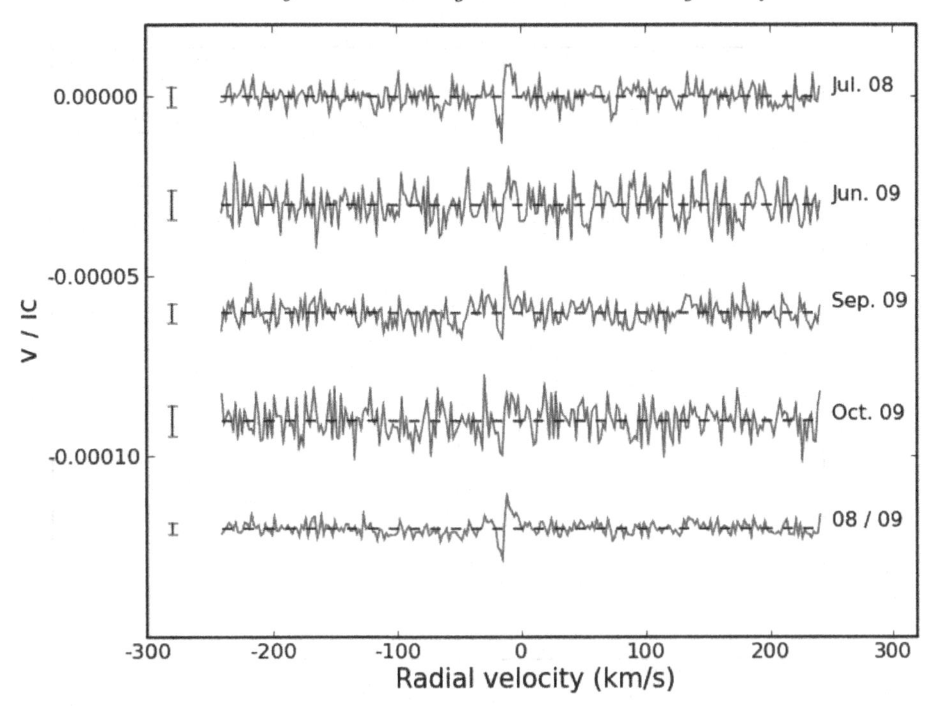

Figure 1. Averaged Stokes V LSD profiles of Vega for 4 different observing runs from July 2008 to October 2009. The red profile is obtained by averaging all 799 spectra taken over this period.

dependency on the star fundamental parameters should provide useful insight into the origin of this magnetism. We shall come back to this point in section 3.

3. Ultra-weak magnetic field

While revealing a magnetic desert among intermediate-mass stars, the spectropolarimetric surveys mentioned in the previous section could only provide an upper bound of the surface magnetic field of A-type non-Ap/Bp stars. Fortunately, a 4-nights Narval run designed to search for pulsations on the A0 star Vega was used as an opportunity to further lower this bound and resulted in the detection of a very small polarimetric signal. The peak Stokes V amplitude divided by the continuum intensity was $V/I_c = 10^{-5}$, for a noise level of $\sigma = 2 \times 10^{-6}$. The corresponding longitudinal magnetic field is -0.6 $\pm$ 0.3 G (Lignières *et al.* 2009). This weak signal called for additional measurements and tests to confirm the presence of a magnetic field. As illustrated on Figure 1, additional observing campaigns conducted with Narval and ESPaDOnS over three years all confirmed this first detection (Petit *et al.* 2010, Alina *et al.* 2012). The possibility of a spurious detection due to the instrument or the reduction process is systematically tested by computing the combination of sub-exposures, the so-called null profiles, for which the signal is expected to cancel. On the other hand, if the signal is of stellar origin, it should be sensitive to the Landé factor of the spectral lines as well as periodically modulated by the star rotation. Figure 2 shows the Stokes V LSD profiles computed for two lines lists having different mean Landé factors. As expected for a Zeeman signature, the amplitude ratio between the two Stokes V profiles is close to the 1.6 ratio of their mean Landé factor. A periodic modulation at 0.678 $\pm$ 0.036 day has also been detected by applying a least squares sine fit period search to each radial velocity bin of the Stokes V LSD profile

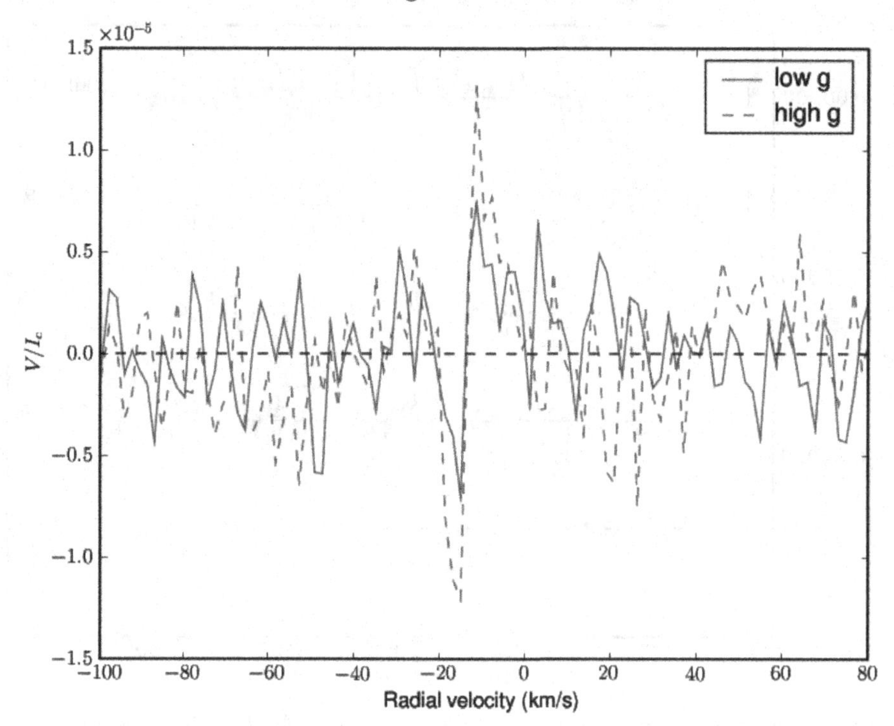

Figure 2. (Colour online) Averaged Stokes V LSD profiles of Vega for two different lines lists. The green and solid profile corresponds to the line list with low Landé factor, the red and dashed profile to high Landé factor. As expected for a Zeeman signal, the amplitude ratio between the two Stokes V profiles is close to the 1.6 ratio between the mean Landé factor of the line lists.

(Alina *et al.* 2012). This value can be compared with two different determinations of the rotation rate that both make use of the gravity darkening effect, that is the effect of the centrifugal force on the local atmospheric parameters. The line shapes fitting by Takeda *et al.* (2008) provided 0.7-0.9 day for the rotation period together with a small 7 degrees inclination angle. By contrast the first model fitting of the interferometric observations gave much higher close-to-break-up periods (0.5–0.6 days) with nevertheless similar inclination angles. In a recent study however, Monnier *et al.* (2012) showed that the error bars of these interferometric determinations were underestimated and a more accurate value now consistent with spectropolarimetric and spectroscopic inferences was obtained thanks to better phase coverages and angular resolutions. Accordingly, the equatorial velocity is 194 ± 6 km s^{-1}, a value only slightly above the average equatorial velocity of normal A0-A1 stars $v_{eq} = 165$ km s^{-1} taken from Royer *et al.* (2007).

Finding the rotation period is key to reconstruct the surface magnetic field through Zeeman-Doppler Imaging. The magnetic field vector appears to be dominated by its radial component whose surface distribution is structured at small lengthscale and characterized by a slightly off-axis polar spot, as seen on Figure 3. This polar spot can already be inferred from the phase-averaged Stokes V concentration in the weakly Doppler-shifted part of the profile.

If we exclude the debated field detections obtained with the low-resolution FORS1/2 instrument at VLT (e.g. Bagnulo *et al.* 2012, Kochukhov 2013), Vega is the first detected magnetic A star which is not an Ap/Bp chemically peculiar star. Its magnetic field distinguishes clearly from Ap/Bp magnetic fields by the strength of the longitudinal component (two orders of magnetic lower than the minimum longitudinal field of Ap/Bp

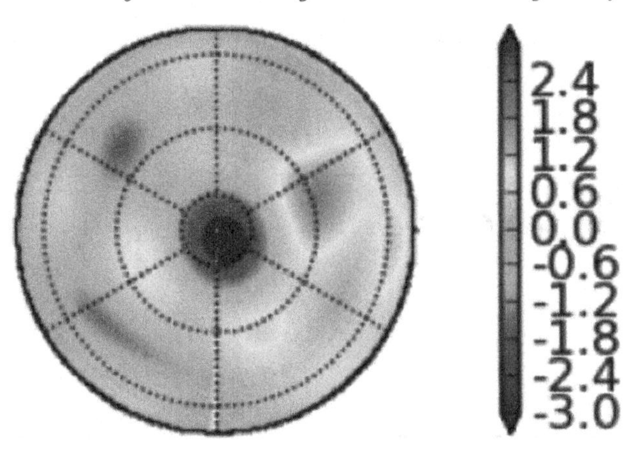

Figure 3. (Colour online) Zeeman-Doppler Imaging of Vega for 2008 July showing the radial component of the magnetic field in polar projection. The color code is expressed in gauss. The spot observed near the pole can be related to the near-zero Doppler velocity feature of the Stokes V profile.

stars taken at its maximum phase) and by the lengthscale of its surface distribution (significantly smaller than for a dipole). It shows that a new - and potentially widespread - class of magnetic stars exists at the low end of the A-type magnetic desert.

The next obvious target is Sirius A, a low v sini bright A1 star with Am-type chemical abundance anomalies. The LSD Stokes V profile obtained by Petit *et al.* (2011) from the average of 442 high S/N ratio spectra clearly shows a polarimetric signal. The V/I_c maximum amplitude is similar to Vega but, contrary to Vega, the V profile is asymmetric about the line center. Asymmetric profiles are not standard although they are observed in some cool stars (Petit *et al.* 2005, Aurière *et al.* 2009) and locally on the Sun surface where they are interpreted as due to the presence of both magnetic field and velocity field radial gradients in the emitting regions (Lopez-Ariste 2002). A model of stellar atmosphere that produces an asymmetric profile similar to Sirius on a global scale is still lacking. Moreover, the quality of the data has not allowed yet to show a significant sensitivity to the Landé factor or a temporal modulation of the polarimetric signal. Thus, by contrast with Vega, more theoretical and observational works are required to confirm the presence of a magnetic field on Sirius. A spurious detection seems however unlikely as the V profile found with ESPaDOnS and Narval has been recently recovered using HARPSpol (see Kochukhov 2013). The rotation of Sirius is not known but as the equatorial velocities of Am stars are always smaller than 120 km s^{-1} (Abt 2009) we know that Sirius rotates more slowly than a typical A0-A1 star.

An ongoing Narval survey investigates the occurrence of Vega-like magnetism among tepid stars. The weakness of the expected signal, the necessity to avoid confusion with hypothetical weak solar-type dynamo field in colder stars, the drastic decrease in the number of lines towards hotter stars and the willingness to avoid a bias towards the slowly rotating Am stars contributed to construct a sample of 10 bright stars restricted to the A0-A2 range (including Vega and Sirius). This survey also intends to investigate how Vega-like magnetism depends on rotation and time. A large occurrence of magnetic field detections would have a direct impact on stellar evolution model by providing the first direct constraint on the value of the magnetic field of a typical intermediate-mass star. Meanwhile, the hypothesis of a widespread magnetism is suggested by the analysis of the Kepler photometry of thousands of A stars. Indeed, according to Balona (2011), a low frequency modulation of the light-curve compatible with a rotational modulation has

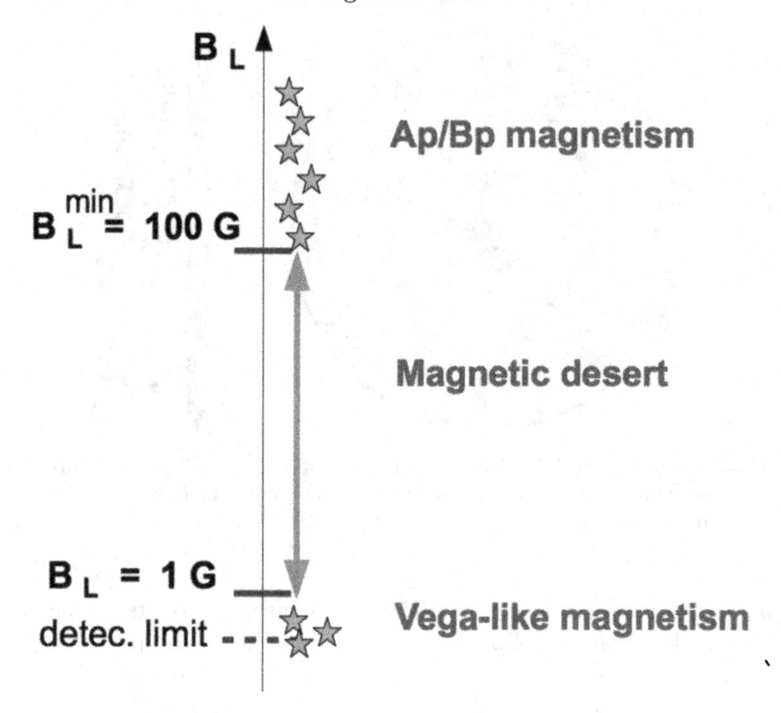

Figure 4. (Colour online) This sketch summarizes the new context of intermediate-mass magnetism set by the recent discoveries of the lower bound of Ap magnetism, the magnetic desert in longitudinal field B_L and the Vega-like magnetism.

been found in 70 % of the A-type Kepler stars, as expected in the presence of starspots or other magnetic corotating features.

4. On the origin of the magnetic dichotomy

The sketch of Fig. 4 illustrates the dichotomy between the Ap/Bp-like and Vega-like magnetic fields of intermediate-mass stars. One can think of two different ways to explain this dichotomy : the first one is to assume that the two types of magnetic fields have different properties because they have been generated by two different processes. Another possibility is that the observed fields have a common origin but during the evolution their magnetic field distribution split into two distinct families of low and high longitudinal fields.

Braithwaite & Cantiello (2013) proposed that Vega-like stars are "failed fossil" magnetic stars meaning that their field, produced during star formation, is still decaying thus not truly fossil. Accordingly, the helicity of the initial field configuration must be low enough to evolve towards a sub-gauss field amplitude at the age of Vega. By contrast, the initial helicity of Ap/Bp-like magnetic stars has to be very high to produce in a relatively short time the observed fossil-like Ap/Bp magnetism. Braithwaite & Cantiello (2013) thus argue that two distinct generation mechanisms must be invoked to explain these very different initial helicities. To account for Ap/Bp stars, they follow Ferrario *et al.* (2009) and Tutukov& Fedorova (2010) who assume that these stars result from the merging of close binaries, their strong fields being produced by a powerful dynamo during the merging phase.

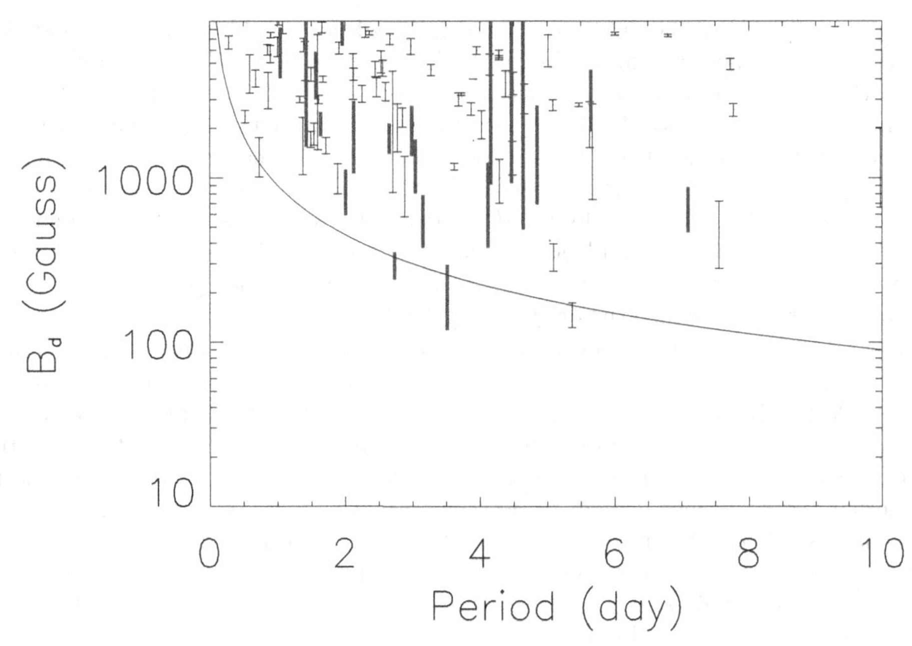

Figure 5. (Colour online) Ap/Bp dipolar field strengths as a function of the rotation rate. The black vertical line segments are dipole strength (with 1σ error bars) derived from the longitudinal field measurements of the Bychkov's catalogue and the inequality : $B_d > 3.3 B_L^{\max}$. The red segments correspond to dipole strength (with 1σ error bars) obtained by fitting an oblique rotator model (taken from Aurière *et al.* 2007). The black continuous line is a $B_d \propto \Omega$ curve.

On the other hand, Aurière et al. (2007) proposed a bifurcation scenario by considering the evolution of a differentially rotating star as a function of its rotation rate and its poloidal field strength. Weak initial poloidal fields cannot prevent their winding-up by differential rotation into strong toroidal fields. The increasingly toroidal configuration is expected to become unstable to an instability, like the Tayler instability, transforming the initially large scale field configuration into a new configuration with opposed signed polarities at the length scale of the instability. On the contrary, if the initial magnetic field is strong enough, Maxwell stresses impose uniform rotation and eventually lead to a stable configuration. Thus, starting from a continuous distribution of poloidal field strength ranging from low to high dipole strengths, this mechanism predicts a sharp decrease of the longitudinal field between the stable large-scale high B_L fields and the unstable configurations where B_L, the surface averaged line-of-sight component, is strongly reduced due to the cancellation of the opposed polarities. In support of this scenario, the observed $\sim$ 300 G lower bound of Ap/Bp magnetic fields (see section 2) happens to be close to an order of magnitude estimate of the critical dipolar field separating the stable and unstable configuration, namely $B_c = (4\pi\rho)^{1/2} r\Omega$, computed for the parameters of a typical Ap star ($\log g = 4, T_{\text{eff}} = 10^4 K, P_{\text{rot}} = 5$ days).

Another important property of B_c is to be proportional to Ω. Thus if this scenario can explain the dichotomy, the lower bound of Ap/Bp magnetic fields should increase with Ω. This can be tested since the spectropolarimetric monitoring of an Ap/Bp star gives access to both the rotation period and the dipole strength. The difficulty is that finding the minimum dipolar fields of Ap/Bp stars for various rotation rate intervals requires a significant sample of Ap/Bp stars for each rotation range. As a first attempt, Aurière & Lignières (2013) compiled data from the Bychkov catalogue (Bychkov *et al.* 2005) although in many cases the data are not good enough for an oblique rotator modelling. Nevertheless with the maximum value of the longitudinal field B_L^{max} available, a lower limit of the dipole strength can be determined from the relation $B_d > 3.3B_L^{max}$ derived from the oblique rotator model. These lower limits are displayed on Fig. 5 with errors bars together with $B_d(\Omega)$ values taken from Aurière *et al.* (2007). The results seem compatible with a $B_{min} \propto \Omega$ law. This data sample is however too limited to obtain an accurate determinations of B_{min}, specially for rapid Ap/Bp rotators. This calls for a comprehensive investigation of the B_{min} dependency on the rotation rates. More generally, we should expect that this lower bound, and its dependency on the stellar parameters, contains important information about the origin of Ap/Bp magnetism.

The physical grounds of the Aurière *et al.* scenario can also be investigated through 2D and 3D numerical simulations. The winding-up of a poloidal field by differential rotation is accessible to 2D axisymmetric simulations while the non axisymmetric instabilities of the wounded-up configuration require full 3D simulations (e.g. Arlt & Rüdiger 2011). Among the open questions is the mechanism that drives differential rotation and whether this mechanism is efficient enough to overcome Maxwell stresses. The seismic analysis of sub-giant stars has shown recently that differential rotation is indeed present in the contracting/expanding radiative interiors of these stars (Deheuvels *et al.* 2012). This points towards pre-main-sequence contraction as a possible source for the differential rotation forcing required in the Aurière *et al.* scenario.

5. Conclusion

While some years ago the fraction of intermediate-mass stars hosting large-scale stable magnetic fields like Ap/Bp stars was unknown, this class of magnetic stars is now known to be clearly separated from the other stars by a two order of magnitude magnetic desert in longitudinal field down to a new type of sub-gauss magnetism. This paradigm shift opens new perspectives for both observational and theoretical studies. Investigating how the lower bound of Ap/Bp magnetic fields depends on the stellar parameters could help to discriminate between the different hypothesis regarding the origin of this magnetism. For example we have seen that the Aurière *et al.* scenario predicts that this lower bound is proportional to the rotation rate. Moreover, the nature and the incidence of Vega-like magnetism among non Ap/Bp stars need to be elucidated although detecting many sub-gauss fields in A stars will be very demanding in observing time. Another important question is whether the dichotomy of intermediate-mass star magnetism is also present among pre-main-sequence intermediate-mass stars and massive stars. This is already suggested by recent spectropolarimetric surveys that detected fossil-like magnetic fields with an incidence similar to Ap/Bp stars (Alecian *et al.* 2013, Wade *et al.* 2013).

Various scenarios have been proposed to account for the magnetic dichotomy. Here we described the bifurcation between stable and unstable magnetic configurations and briefly mentioned the merging scenario for Ap/Bp stars and the failed fossil scenario for Vega-like magnetism (see the contributions by N. Langer and J. Braithwaite in these proceedings for more details). Besides the confrontation with observations, numerical simulations will also help to test these ideas. MHD computations in polytropic stars already enabled to find stable magnetic configurations that might correspond to Ap/Bp stars (Braithwaite 2009). But, except for very strong fields, differential rotation should also play an important role in shaping magnetic configurations in radiative interiors and this effect needs to be fully explored numerically.

Acknowledgements

The authors thank the ANR project IMAGINE and the PNPS (Programme National de Physique Stellaire) of INSU for their financial support.

References

Abt, H. A. 2009, *AJ*, 138, 28
Alecian, E. 2013, arXiv:1310.1725
Alina, D., Petit, P., Lignières, F., *et al.* 2012, *AIP Conference Series*, 1429, 82
Arlt, R. & Rüdiger, G. 2011, *MNRAS*, 412, 107
Aurière, M. & Lignières, F. 2013, *private communication*
Aurière, M., Wade, G. A., Lignières, F., *et al.* 2010, *A&A*, 523, A40
Aurière, M., Wade, G. A., Konstantinova-Antova, R., *et al.* 2009, *A&A*, 504, 231
Aurière, M., Konstantinova-Antova, R., Petit, P., *et al.* 2008, *A&A*, 491, 499
Aurière, M., Wade, G. A., Silvester, J., *et al.* 2007, *A&A*, 475, 1053
Babcock, H. W. 1947, *ApJ*, 105, 105
Bagnulo, S., Landstreet, J. D., Fossati, L., & Kochukhov, O. 2012, *A&A*, 538, A129
Balona, L. A. 2011, *MNRAS*, 415, 1691
Braithwaite, J. & Cantiello, M. 2013, *MNRAS*, 428, 2789
Braithwaite, J. 2009, *MNRAS*, 397, 763
Bychkov, V. D., Bychkova, L. V., & Madej, J. 2005, *A&A*, 430, 1143
Donati, J.-F. & Landstreet, J. D. 2009, *ARA&A*, 47, 333
Deheuvels, S., García, R. A., Chaplin, W. J., *et al.* 2012, textitApJ, 756, 19
Ferrario, L., Pringle, J. E., Tout, C. A., & Wickramasinghe, D. T. 2009, *MNRAS*, 400, L71
Kochukhov, O., Makaganiuk, V., Piskunov, N., *et al.* 2011, *A&A*, 534, L13
Kochukhov, O. 2013, arXiv:1309.6313
Landstreet, J. D. 1982, *ApJ*, 258, 639
Lignières, F., Petit, P., Böhm, T., & Aurière, M. 2009, *A&A*, 500, L41
López Ariste, A. 2002, *ApJ*, 564, 379
Makaganiuk, V., Kochukhov, O., Piskunov, N., *et al.* 2011, *A&A*, 525, A97
Michaud, G., Charland, Y., Vauclair, S., & Vauclair, G. 1976, *ApJ*, 210, 447
Michaud, G. 1970, *ApJ*, 160, 641
Monnier, J. D., Che, X., Zhao, M., *et al.* 2012, *ApJL*, 761, L3
Petit, P., Lignières, F., Aurière, M., *et al.* 2011, *A&A*, 532, L13
Petit, P., Lignières, F., Wade, G. A., *et al.* 2010, *A&A*, 523, A41
Petit, P., Donati, J.-F., Aurière, M., *et al.* 2005, *MNRAS*, 361, 837
Royer, F., Zorec, J., & Gómez, A. E. 2007, *A&A*, 463, 671
Shorlin, S. L. S., Wade, G. A., Donati, J.-F., *et al.* 2002, *A&A*, 392, 637
Takeda, Y., Kawanomoto, S., & Ohishi, N. 2008, *ApJ*, 678, 446
Tutukov, A. V. & Fedorova, A. V. 2010, *Astronomy Reports*, 54, 156
Wade, G. A., Grunhut, J., Alecian, E., *et al.* 2013, arXiv:1310.3965
Wade, G. A., Aurière, M., Bagnulo, S., *et al.* 2006, *A&A*, 451, 293

Magnetic Fields throughout Stellar Evolution
Proceedings IAU Symposium No. 302, 2013
P. Petit, M. Jardine & H. Spruit, eds.

doi:10.1017/S1743921314002452

UVMag: a UV and optical spectropolarimeter for stellar physics

Coralie Neiner[1], Pascal Petit[2], Laurent Parès[2] and the UVMag consortium

[1]LESIA, UMR 8109 du CNRS, Observatoire de Paris, UPMC, Univ. Paris Diderot, 5 place Jules Janssen, 92195 Meudon Cedex, France
email: coralie.neiner@obspm.fr

[2]IRAP, CNRS, UPS-OMP, 14 Avenue Edouard Belin, 31400 Toulouse, France

Abstract. UVMag is a space project currently under R&D study. It consists in a medium-size telescope equipped with a spectropolarimeter to observe in the UV and optical wavelength domains simultaneously. Its first goal is to obtain time series of selected magnetic stars over their rotation period, to study them from their surface to their environment, in particular their wind and magnetospheres. As the star rotates it will be possible to reconstruct 3D maps of the star and its surroundings. The second goal of UVMag is to obtain two observations of a large sample of stars to construct a new database of UV and optical spectropolarimetric measurements.

Keywords. instrumentation: polarimeters, instrumentation: spectrographs, ultraviolet: stars, stars: magnetic fields, stars: winds, outflows, stars: activity, stars: chromospheres

1. Science drivers

Research on stellar magnetism has been progressing very fast from the ground in the last decade, but we are missing information on stellar wind and magnetospheres because there is no UV spectrograph currently available for long fractions of time. To reconstruct the magnetospheres, we need to obtain simultaneous UV and optical spectropolarimetry, continuously over several stellar rotation periods. Of course, the UV domain requires a space mission. Therefore the UVMag consortium proposes to build a dedicated M-size space mission with a telescope of 1.3 m and UV and optical spectropolarimetric capabilities (see http://lesia.obspm.fr/UVMag).

The UV and visible spectropolarimeter will provide a very powerful and unique tool to study most aspects of stellar physics in general and in particular for stellar formation, structure and evolution as well as for stellar environment. For example we plan to study how fossil magnetic fields confines the wind of massive stars and influences wind clumping, how magnetic interactions impact binary stars, how a solar dynamo impacts its planets and how it evolves, how magnetic field, wind and mass-loss influence the late stages of stellar evolution, in which conditions a magnetic dynamo develops, how the angular momentum of stars evolves, how small-scale and large-scale stellar dynamos work and how their cycles influence their environment, what explains the diversity of magnetic properties in M dwarfs, what causes the segregation of tepid stars in two categories: those with sub-Gauss magnetic fields and those with fields above a few hundreds Gauss, what are the timescales over which magnetospheric accretion stops in PMS stars, etc. These questions will be answered by observing all types of stars: massive stars, giants and supergiants, chemically peculiar stars, pre-main sequence stars, cool stars, solar twins, M dwarfs, AGB and post-AGB stars, binaries, etc. Additional possible science includes the study of the ISM, white dwarfs, novae, exoplanets, atomic physics,...

2. UV and optical spectropolarimeter

The spectropolarimeter should ideally cover the full wavelength range from 90 to 1000 nm and at least the most important lines in the domains 117–320 nm and 390-870 nm. Polarisation should be measured at least in Stokes V (circular polarisation) in spectral lines, but the aim is to measure all Stokes QUV parameters (circular and linear polarisation) in the lines and continuum. A high spectral resolution is required, at least 25000 in the UV domain and at least 35000 in the optical, with a goal of 80000 to 100000. The signal-to-noise should be above 100.

Spectroscopy with these specifications in the UV and optical domains is relatively easy to achieve with today's technology and detectors. However, (1) high-resolution spectropolarimetry of stars has never been obtained from space; (2) optical spectropolarimeters available on the ground are large; and (3) it is very important to keep the instrumental polarisation at a low level. Therefore we have started a R&D program to study a space UV+optical spectropolarimeter. Our study is based on existing ground-based spectropolarimeters, such as ESPaDOnS or Narval, and new spectropolarimetric techniques proposed in the literature (e.g. Sparks *et al.* 2012).

3. Observing program

UVMag will observe all types of stars in the magnitude range at least V=3-10. The observing program includes two parts: (1) ~50 stars will be observed over 2 full rotational cycles with high cadence in order to study them in great details and reconstruct 3D maps of their surface and environment. In addition, the solar-like stars among those will be re-observed every year to study their activity cycle; and (2) two spectropolarimetric measurements of ~4000 stars will be obtained to provide information on their magnetic field, wind and environment. This will form a statistical survey and provide input for stellar modelling. The acquisition of the data for these two programs will take 4 years.

4. Conclusions

The UVMag consortium has set the basic requirements for a M-size (1.3 m) space mission to study the magnetospheres and winds of all types of stars. This is the next step to progress on the characterisation and modelling of stellar environments, as well as on important questions regarding stellar formation, structure and evolution. Simultaneous UV and optical spectropolarimetry over long periods of time is indeed the only way to comprehend the full interaction between various physical processes such as the stellar magnetic field and stellar wind. A R&D study is ongoing for the instrument. The M-size mission will be proposed at ESA. A L-size mission (4-8 meter telescope) is also considered with the UVMag UV and optical spectropolarimeter as part of a series of instrument, e.g. on EUVO (Gómez de Castro *et al.*, 2013).

Acknowledgements

The UVMag R&D program is funded by the French space agency CNES.

References

Gómez de Castro, A. I., Appourchaux, T., Barstow, M., Barthelemy, M., Baudin, F., *et al.* 2013, *arXiv*, 1306.3358

Sparks, W., Germer, T. A., MacKenty, J. W., & Snik, F. 2012, *Applied Optics*, 51, 5495

Magnetic Fields throughout Stellar Evolution
Proceedings IAU Symposium No. 302, 2013
P. Petit, M. Jardine & H. Spruit, eds.

doi:10.1017/S1743921314002464

Surface magnetism of cool giant and supergiant stars

Heidi Korhonen[1]

[1]Finnish Centre for Astronomy with ESO (FINCA), University of Turku, Väisäläntie 20, FI-21500 Piikkiö, Finland
email: heidi.h.korhonen@utu.fi

Abstract. The existence of starspots on late-type giant stars in close binary systems, that exhibit rapid rotation due to tidal locking, has been known for more than five decades. Photometric monitoring spanning decades has allowed studying the long-term magnetic activity in these stars revealing complicated activity cycles. The development of observing and analysis techniques that has occurred during the past two decades has also enabled us to study the detailed starspot and magnetic field configurations on these active giants. In the recent years magnetic fields have also been detected on slowly rotating giants and supergiant stars. In this paper I review what is known of the surface magnetism in the cool giant and supergiant stars.

Keywords. stars: activity, stars: late-type, stars: magnetic fields, stars: rotation, stars: spots

---

1. Introduction

A slightly over a century ago Hale (1908) discovered that sunspots are caused by strong magnetic fields in the Sun. Naturally, observing solar-like magnetic fields in other stars is extremely demanding, and their detection had to wait almost a century. Still, stronger fields could be detected already much earlier. The first detections were done in magnetic, chemically peculiar, Ap stars, which have strongest known magnetic fields of non-degenerate stars (see, e.g., Babcock 1947a; Babcock 1947b). Recent years magnetic fields have been discovered in many different classes of stars, from pre-main sequence stars (e.g., Guenther *et al.* 1999) to neutron stars (e.g., Kouveliotou *et al.* 1998), and throughout the main sequence from M-type (e.g., Donati *et al.* 2006; Morin *et al.* 2008) to O-type (e.g., Hubrig *et al.* 2008; Grunhut *et al.* 2009) stars.

In this review the surface magnetism of cool G–M giants and supergiants is discussed. The main emphasis is on the observations of the surface features caused by magnetic fields and the measurement of the magnetic fields themselves, but also the origins of these fields are discussed.

2. Methods for studying stellar activity and magnetism

Obtaining spatial information of the surface of stars, which appear as point sources in our telescopes, is challenging. The two main methods for achieving this are photometry and high resolution spectroscopy through Doppler imaging techniques. Photometry is the easiest and least observing time consuming way of carrying out starspot studies. In many active stars the starspots are so large that they cause brightness variations which can be few tens of percent from the mean light level, thus making them easily observable even from the ground (see, e.g., Chugainov 1966; Montle & Hall 1972). On the other hand, solar-like spots are so small that they are lost in the noise, unless extremely precise observations can be obtained. Currently the by-far best precision and time resolution observations are provided by the Kepler satellite.

Doppler imaging (see, e.g., Vogt *et al.* 1987; Rice *et al.* 1989; Piskunov *et al.* 1990), which provides the best spatial resolution on the stellar surface, is a method that uses high resolution, high signal-to-noise spectroscopic observations at different rotational phases of the star. If the star has a non-uniform surface temperature, i.e., has starspots, the spectral lines show small distortions from the normal Gaussian shape. These distortions move in the line-profile when the position of the starspots on the surface changes, due to the change of line-of-sight velocity caused by the stellar rotation. Surface maps, or Doppler images, are constructed by tracking the movement of these distortions by combining all the observations from different rotational phases and typically comparing them with synthetic model line profiles.

During just the last years a breakthrough using long baseline infrared and optical interferometers has occurred. These facilities now routinely produce aperture synthesis images with milli-arcsecond angular resolution. A variety of targets have been imaged with astonishing result, e.g., bulging stars rotating near their critical limit (Monnier *et al.* 2007) and compact dust disk around a massive young stellar object (Kraus *et al.* 2010). Infrared interferometric imaging has produced amazing results of stellar surfaces and the time is drawing near when even temperature spots on cool stars can be imaged (see Roettenbacher *et al.* 2013), and naturally giant stars are the most fruitful starting point for this due to their large size.

Studying the magnetic fields directly in stars can be done using Zeeman splitting of the spectral lines. This method, which is the easiest one to interpret, requires high spectral resolution and high signal-to-noise ratio to accurately observe the spectral-line profile shapes (see e.g., Johns-Krull & Valenti 1996). Another way to study the magnetic fields themselves is to measure the polarisation of spectral lines (e.g., Donati *et al.* 1997). In astronomy polarisation is usually described using the Stokes vector [I, Q, U, V], where Stokes I gives the total intensity spectrum, Stokes V describes the circular polarisation and Stokes Q and U the linear polarisation. Circular polarisation is mainly sensitive to the line-of-sight magnetic field and the linear one to the perpendicular field, therefore polarisation can be also used to measure the orientation of the field. If spectropolarimetric observations are obtained over the stellar rotation, similarly to what is done in Doppler imaging, the surface magnetic field of the star can mapped using Zeeman-Doppler imaging technique (Semel 1989).

3. Active giants

Cool giant stars are in general very slow rotators. Many of their progenitors were slow rotators when in the main sequence and, even if they were not, the expanding envelope reduces the rotation rate significantly. This makes rotation driven, solar-like, dynamo action very improbable in these stars. Still, there are some cases of giant stars being rapid rotators, and magnetically very active. Reviews of starspots and their properties, also in active giants, are given by Strassmeier (2009) and Berdyugina (2005). The main types of active giants are discussed in the following.

3.1. *RS CVn-type binaries*

The most common type of active giants is the RS Cvn-type binaries, where a G–K giant or sub-giant is partnered with a less massive G–M main sequence star or a sub-giant. These typically tidally locked binaries are still detached systems, and their orbital (and therefore often also rotation periods) are commonly from few days to some tens of days.

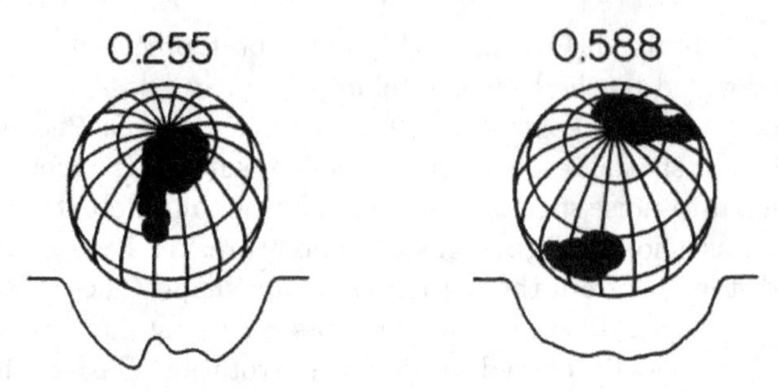

Figure 1. Examples of first Doppler maps of HR1099. (from Vogt & Penrod 1983)

General properties and classification of RS CVn-type binaries are discussed, e.g., by Hall (1976) and Morgan & Eggleton (1979). The first star on which starspots were suggested to be the cause of the brightness variations, AR Lac (Kron 1947), is an RS Cvn-type binary. As is also the first cool star for which Doppler imaging technique was applied, HR1099, shown in Fig. 1 (Vogt & Penrod 1983).

The Doppler imaging of RS CVn binaries often shows very large starspot, like the 12×20 solar radii spot on HD 12545 (Strassmeier 1999), and also very high spot latitudes, or even polar spots (e.g., in KU Peg by Weber & Strassmeier 2001). For a handful of RS CVn-type binaries magnetic field maps have been obtained, among these stars are HR 1099 and II Peg which have maps from multiple epochs. And example of such a maps is shown in Fig. 2. Donati (1999), Donati *et al.* (2003), and Petit *et al.* (2004a) have obtained both spot filling-factor maps and magnetic field maps of HR 1099 on five epochs spanning more than a decade in total. They discovered two distinct azimuthal field regions of opposite polarity around latitudes 30° and 60°. They also conclude that small-scale brightness and magnetic features undergo changes at time scales of 4–6 weeks, whereas the large-scale structures are stable over several years. Recently, Kochukhov *et al.* (2013) published magnetic field and temperature maps of II Peg obtained from seven different epochs spanning in total 12 years. They observe significant field evolution on the time scale of their observations, additionally they did not find a clear correlation between magnetic and temperature features in their maps.

3.2. *FK Comae-type giants*

FK Comae-type giants are a very small group of highly active G–K giants and subgiants (Bopp & Rucinski 1981; Bopp & Stencel 1981). They show activity levels similar to RS CVn binaries. What makes them special is that they are extremely rapidly rotating (FK Com itself has $v \sin i$ around 160km/s), but still they are not part of a binary systems. The most commonly accepted explanation for the rapid rotation is that they are end products of a coalescence of W Uma-type contact binaries (e.g., Bopp & Rucinski 1981; Bopp & Stencel 1981).

Temperature maps of three FK Comae-type stars have been published in the literature. The first one mapped was HD 32918 (YY Men) by Piskunov *et al.* 1990. The map shows mainly equatorial spot configurations. The other two FK Comae stars with temperature maps are FK Com itself (e.g., Korhonen *et al.* 1999; Korhonen *et al.* 2007) and HD 199178 (e.g., Hackman *et al.* 2001). These stars mainly show high latitude spot configurations, but no real polar spots.

Magnetic field of HD 199178 has been mapped using Zeeman Doppler imaging by Petit *et al.* (2004b). Their maps from 2002 and 2003 reveal large regions of azimuthal field, and also changes in the exact field configuration with a time scale of about two weeks. For FK Com no Zeeman Doppler map exists, but its mean longitudinal magnetic field

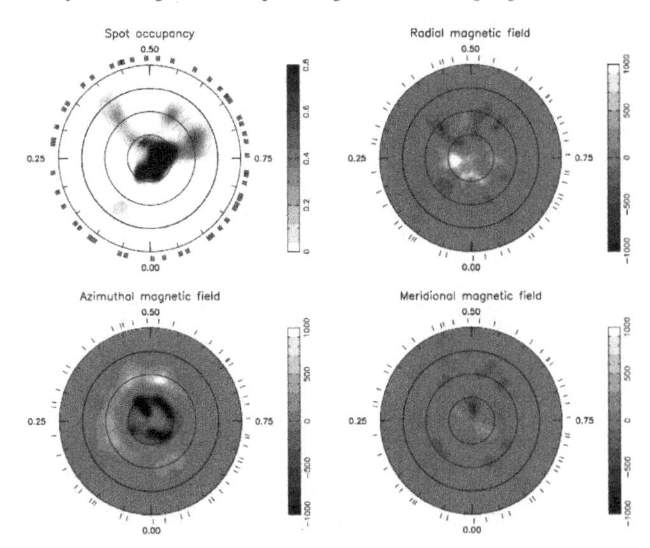

Figure 2. Zeeman Doppler maps of HR 1099. The flattened polar view in the upper-left panel corresponds to a brightness image and the other panels show radial, azimuthal and meridional components of the field in the upper-right, lower-left and lower-right panels, respectively. (from Petit *et al.* 2004a)

has been studied using low resolution spectra (Korhonen *et al.* 2009). These observations indicate that the two high-latitude spots seen in the contemporaneous temperature map could have different polarity.

3.3. *Lithium rich giants*

Some cool giant stars show enhanced Li abundance, and some of these stars are also moderately rapid rotators, and thus magnetically active. Fekel & Balachandran (1993) suggested that during the first dredge-up both angular momentum and Li-rich material could be dredged-up from the stellar interior, creating a rapidly rotating Li-rich giant. Further analysis by Charbonnel & Balachandran (2000) of the Li abundance and rotation in giants did not show clear correlation.

Whatever the cause for the rapid rotation is, it still enhances the dynamo operation in these stars and they become magnetically active. Recently, Kővári *et al.* (2013) published temperature maps of two Li-rich giants, DP CVn and DI Psc. For both stars low latitude spots with relatively small temperature contrasts (600-800 K below the unspotted surface temperature) are recovered. In addition higher latitude spots are recovered, but they either have lower contrast (DP CVn) or smaller extent (DI Psc). Lèbre *et al.* (2009) also detected Stokes V signal with temporal variations on Li-rich giant HD 232862, indicating a presence of magnetic field in this star.

4. Slowly rotating giants

One would not expect to be able to detect magnetic fields on slowly rotating giant stars, because the slow rotation would not enable the dynamo to create strong enough magnetic fields. Still, magnetic fields have been detected, and even mapped, in some slowly rotating giants.

Pollux is a K0 giant with rotation period of 100–500 days. The star has been known for some time to be weakly active (see, e.g., Strassmeier *et al.* 1990a). A weak longitudinal magnetic field of -0.46 ± 0.04 G was detected in Pollux by Aurière *et al.* (2009), see Fig. 2.

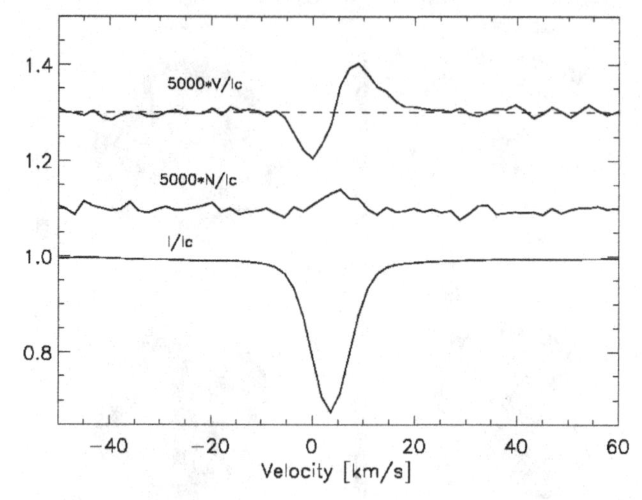

Figure 3. Stokes V, the diagnostic null profile and the Stokes I observation of Pollux. (from Aurière *et al.* 2009)

Another similar case is Arcturus where Sennhauser & Berdyugina (2011) detected a weak longitudinal magnetic field of 0.65±0.26 G and 0.43±0.16 G from two spectra obtained at different times. The origin of the field detected in Pollux and Arcturus is most likely solar-like $\alpha\Omega$-dynamo.

Not only weak fields are detected in slowly rotating giants. One puzzling case is EK Eri, which has rotation period of approximately 300 days (e.g., Strassmeier *et al.* 1990b), and still longitudinal magnetic field as strong as 100 G (Aurière *et al.* 2008). The modelling of Aurière *et al.* (2008) also reveal that the large scale magnetic field is most likely dominated by a poloidal component. They suggest that EK Eri is a descendant of a strongly magnetic Ap star. Similar suggestion has been made to explain the magnetic field measurements of β Cet. Tsvetkova *et al.* (2013) publish two Zeeman Doppler images of β Cet, obtained one year apart and showing very similar magnetic field configurations dominated by a dipole, and very little evolution between the maps. This all implies magnetic Ap star origin of β Cet.

Magnetic fields have also been detected in more evolved M giants. EK Boo is a rapidly rotating (for its class) M5 giant. Konstantinova-Antova *et al.* (2010) detected magnetic field which varied between -0.1 G and -8 G with time. In one of their observations they also detected complex structure in the Stokes V profile, which could indicate dynamo origin of the field. In their sample of nine M giants Konstantinova-Antova *et al.* (2010) also obtained marginal detection of magnetic field in another star, β And.

5. Supergiants

If cool giants are typically slow rotators, then M supergiants are even slower rotators. This makes solar-like dynamo operation in them virtually impossible. In addition, due to the very extended envelope of these stars, also a fossil magnetic field, remnant from a magnetic main sequence progenitor, would be extremely diluted and is not expected.

M supergiants have gigantic convection cells, which have been predicted in simulations (e.g., Freytag *et al.* 2002) and seen in observations of Betelgeuse (e.g., Gilliland & Dupree 1996; Haubois *et al.* 2009). It has also been predicted that local dynamo could operate in the convection cells and create global magnetic field in supergiant stars (Dorch 2004).

The first measurement of the magnetic fields in M supergiants was done for Betelgeuse by Aurière *et al.* (2010). They detected a weak Stokes V signal, and measured longitudinal field of about one Gauss at six different epochs. There was also some evidence of the field strength increasing during the one month that the observations span. The results by Petit *et al.* (2013) indicate that the magnetic elements may be concentrated in the sinking components of the convective flows.

6. Activity cycles

The Sun exhibits well established 11-year spot cycle, and a 22-year magnetic cycle. The Mt. Wilson H&K project (Wilson 1978) has established similar behaviour of the over-all activity in many cool main sequence stars. Unfortunately, magnetic cycles are much more demanding to detect observationally and only some indication have been seen in few stars. The best known example of a magnetic cycle in another star than our Sun is the very short cycle, much shorter than the solar one, detected in the planet hosting F star τ Boo (e.g., Donati *et al.* 2008).

The Mt. Wilson H&K survey provides a unique window to the cyclic activity of 'normal', non-active, stars. The results show that 60% of the 111 lower main sequence stars studied in the project show cyclic activity (Baliunas *et al.* 1998). For the 175 studied (sub)giants the fraction is 40%, and the giants also have a larger fraction of variable activity pattern (not clearly cyclic) than the dwarf stars have (Baliunas *et al.* 1998).

In active giants the activity cycles are easier to study due to the large brightness variations caused by the huge starspots these stars exhibit. Several studies using long-term photometry have been carried out on the cycles in active giants, e.g., by Jetsu *et al.* (1990), Oláh & Strassmeier (2002), and Özdarcan *et al.* (2010). Most studies on active giants reveal complex activity cycles with multiple periodicities. A study by Oláh *et al.* (2009) studied a sample of active stars, among them several giants, using time-frequency analysis. They found that the active stars typically show multiple periodicities, and that the cycle period also changes with time.

In some of the active giants also different kinds of cycles, so-called flip-flop cycles, have been reported. Flip-flops were originally discovered in FK Com by Jetsu *et al.* (1993). In this phenomenon the spot activity concentrates alternatingly on two permanent active longitudes, which are 180° apart from each other. The active longitude remains more active for few years, and then the other active longitude takes over and is more active. The change between the active longitudes has been reported to happen in FK Com every few years (Korhonen *et al.* 2002). The flip-flop phenomenon has been reported also in few other giant stars, e.g., II Peg, σ Gem, EI Eri and HR 7275 (Berdyugina & Tuominen 1998). On the other hand, recent results on the flip-flop phenomenon imply that the timing of the phenomenon is not regular (e.g., Lindborg *et al.* 2011; Hackman *et al.* 2012) and that the change in longitude is often not 180° (e.g., Oláh *et al.* 2006).

7. Surface differential rotation

Differential rotation is one of the main ingredients in the dynamo models, which seek to explain the observed magnetic activity in the Sun and other stars. Therefore, it is crucial for understanding the creation of the observed magnetic field to also investigate the differential rotation.

The theoretical calculations by Kitchatinov & Rüdiger (1999) predict that the differential rotation is larger in the giant stars than in the dwarfs. Barnes *et al.* (2005) measured surface differential rotation for 10 young G2–M2 dwarfs and reported an increase in the magnitude of differential rotation towards earlier spectral types. Similar results have also been obtained by Saar (2011). Küker & Rüdiger (2012) use mean field model to study differential rotation of KIC 8366239, a giant star with rotation period of 70 days. The results are compared to the model of Arcturus (Küker & Rüdiger 2011), which has a rotation period of the order of two years. The model for Arcturus, with its small core, predicts a steep increase of the rotation rate near the center of the star. On the other hand, the faster rotating KIC 8366239, which also is predicted to have a larger core than Arcturus, has similar internal rotation behaviour as the Sun (not the strength, but the behaviour with radius). Surface differential rotation has been measured for a handful of giant stars. The results collected by Marsden *et al.* (2007) indicate that the active (sub)giants show similar strength of surface differential rotation as the young active stars. More measurements of differential rotation on giant stars are needed to answer the question whether the active giants and dwarfs show similar behaviour.

Anti-solar differential rotation, where the polar regions rotate faster than the equator, has been suggested by observations of several active giants, e.g., in HR 1099 by Vogt *et al.* (1999) and in σ Gem by Kővári *et al.* (2007). Kővári *et al.* (2007) studied surface flow patterns on σ Gem from observation spanning 3.6 consecutive stellar rotations, and found evidence for a weak anti-solar differential rotation together with indications of poleward migration of spots. This strong meridional flow hinted at σ Gem would support the hypothesis of Kitchatinov & Rüdiger (2004), which attributes the anti-solar differential rotation to strong meridional circulation. Similar trend is implied by the results of Weber (2007) on several active giants. One should not forget, though, that these meridional flow measurements can be caused by artifacts in maps and have to be confirmed with data from several epochs.

A recent study by Korhonen & Elstner (2011) used snapshots from dynamo simulations to study surface differential rotation obtained using cross-correlation method. The input rotation law could be recovered from the analysis of the model snapshots, but if small-scale fields were included in the models. With using only the large scale dynamo field the input rotation law was not recovered, and usually basically solid body rotation was obtained. This rises the question whether the huge starspot of active giants can actually be created by small scale fields. If they are manifestations of the large-scale dynamo field, then according to the study by Korhonen & Elstner (2011) we would not even expect them to follow the surface differential rotation.

8. Closing remarks

Rapidly rotating G–K giant stars are among the most magnetically active stars known. They show large starspots, choromospheric and coronal activity, frequent flaring and activity cycles. Slowly rotating G–K giants can also have magnetic fields, maybe through weak dynamo action or due to relic fields (being descendants of Ap stars). In some cases magnetic fields have been detected also in M giants, but no clear explanation for the field generation is known. For the faster rotating ones the field could be generated by a solar-like dynamo. Magnetic field has also been detected in the supergiant Betelgeuze, and it has been hypothesised that a local dynamo can operate in its giant convection cells.

This all shows that magnetic fields are present also in evolved stars, and in large fraction of them than earlier thought. Still, the creation of these fields is often not well understood – even the solar dynamo is not fully understood. More observations of magnetism in evolved stars are needed to collect enough clues to unravel its mystery.

Acknowledgments. The author acknowledges the support from an IAU travel grant to participate the conference.

References

Aurière, M., Konstantinova-Antova, R., Petit, P., *et al.* 2008, *A&A*, 491, 499
Aurière, M., Wade, G. A., Konstantinova-Antova, R., *et al.* 2009, *A&A*, 504, 231
Aurière, M., Donati, J.-F., Konstantinova-Antova, R., Perrin, G., Petit, P., & Roudier, T. 2010, *A&A*, 516, L2
Babcock, H. W. 1947a, *ApJ*, 105, 105
Babcock, H. W. 1947b, *PASP*, 59, 112
Baliunas, S. L., Donahue, R. A., Soon, W., & Henry G. W. 1998, *Cool stars, Stellar Systems, and the Sun*, ASP Conference series 154, p. 153
Barnes, J. R., Collier Cameron, A., Donati, J.-F., James, D. J., Marsden, S. C., & Petit, P. 2005, *MNRAS* 357, L1
Berdyugina, S. V. & Tuominen, I. 1998, *A&A* (Letters), 336, L25
Berdyugina, S. V. 2005, *Living Reviews in Solar Physics*, 2, 8
Bopp, B. W. & Stencel, R. E. 1981, *ApJ* (Letters), 247, L131
Bopp, B. W. & Rucinski, S. M. 1981, *Fundamental Problems in the Theory of Stellar Evolution*, Proc. IAU Symposium 93 (Reidel Publishing Co.), p. 177
Charbonnel, C. & Balachandran, S. C. 2000, A&A, 359, 563
Chugainov, P. F. 1966, *IBVS*, 172, 1
Donati, J.-F., Semel, M., Carter, B. D., Rees, D. E., & Collier Cameron, A. 1997, *MNRAS*, 291, 658
Donati, J.-F. 1999, *MNRAS*, 302, 457
Donati, J.-F., Cameron, A. C., Semel, M., *et al.* 2003, *MNRAS*, 345, 1145
Donati, J.-F., Forveille, T., Collier Cameron, A., *et al.* 2006, *Science*, 311, 633
Donati, J.-F., Moutou, C., Farès, R., *et al.* 2008, *MNRAS*, 385, 1179
Dorch, S. B. F. 2004, *A&A*, 423, 1101
Fekel, F. C., & Balachandran, S. 1993, *ApJ*, 403, 708
Freytag, B., Steffen, M., & Dorch, B. 2002, *Astron. Nachr.*, 323, 213
Gilliland, R. L. & Dupree, A. K. 1996, *ApJ* (Letters) 463, L29
Grunhut, J. H., Wade, G. A., Marcolino, W. L. F., *et al.* 2009, *MNRAS*, 400, L94
Guenther, E. W., Lehmann, H., Emerson, J. P., & Staude, J. 1999, *A&A*, 341, 768
Hackman, T., Jetsu, L., & Tuominen, I. 2001, *A&A*, 374, 171
Hackman, T., Mantere, M. J., Lindborg, M., *et al.* 2012, *A&A*, 538, A126
Hale, G. E. 1908, *ApJ*, 28, 315
Hall, D. S. 1976, *Multiple Periodic Variable Stars*, Proc. IAU Colloquium 29 (Reidel Publishing Co.), p. 287
Haubois, X., Perrin, G., Lacour, S., *et al.* 2009, *A&A*, 508, 923
Hubrig, S., Schöller, M., Schnerr, R. S., González, J. F., Ignace, R., & Henrichs, H. F. 2008, *A&A*, 490, 793
Jetsu, L., Huovelin, J., Tuominen, I., Vilhu, O., Bopp, B. W., & Piirola, V. 1990, *A&A*, 236, 423
Jetsu, L., Pelt, J., & Tuominen, I. 1993, *A&A*, 278, 449
Johns-Krull, C. M. & Valenti, J. A. 1996, *ApJ* (Letters), 459, L95
Kitchatinov, L. L. & Rüdiger, G. 1999, *A&A* 344, 911
Kitchatinov, L. L. & Rüdiger, G. 2004, *AN*, 325, 496
Kochukhov, O., Mantere, M. J., Hackman, T., & Ilyin, I 2013, *A&A*, 550, A84
Konstantinova-Antova, R., Aurière, M., Charbonnel, C., *et al.* 2010, *A&A*, 524, A57
Korhonen, H., Berdyugina, S. V., Hackman, T., Duemmler, R., Ilyin, I. V., & Tuominen, I. 1999, *A&A*, 346, 101
Korhonen, H., Berdyugina, S. V., & Tuominen, I. 2002, *A&A*, 390, 179
Korhonen, H., Berdyugina, S. V., Hackman, T., Ilyin, I. V., Strassmeier, K. G., & Tuominen, I. 2007, *A&A*, 476, 881

Korhonen, H., Hubrig, S., Berdyugina, S. V., *et al.* 2009, *MNRAS*, 395, 282
Korhonen, H. & Elstner, D. 2011, *A&A* 532, A106
Kouveliotou, C., Dieters, S., Strohmayer, T., *et al.* 1998, *Nature*, 393, 235
Kővári, Zs., Bartus, J., Strassmeier, K. G., *et al.* 2007, *A&A* 474, 165
Kővári, Zs., Korhonen, H., Strassmeier, K. G., Weber, M., Kriskovics, L., & Savanov, I. 2013, *A&A*, 551, A2
Kraus, S., Hofmann, K.-H., Menten, K. M.,, *et al.* 2010, *Nature*, 466, 339
Kron, G. E. 1947, *PASP*, 59, 261
Küker, M., Rüdiger 2011, *AN*, 332, 83
Küker, M., Rüdiger 2012, *AN*, 333, 1028
Lèbre, A., Palacios, A., Do Nascimento, J. D., Jr., *et al.* 2009, *A&A*, 504, 1011
Lindborg, M., Korpi, M. J., Hackman, T., Tuominen, I., Ilyin, I., & Piskunov, N. 2011, *A&A*, 526, A44
Marsden, S. C., Berdyugina, S. V., Donati, J.-F., Eaton, J. A., & Williamson, M. H. 2007, *AN*, 328, 1047
Monnier, J. D., Zhao, M., Pedretti, E., *et al.* 2007, *Science* 317, 342
Montle, R E, Hall, D. S. 1972, *IBVS*, 646, 1
Morgan, J. G. & Eggleton, P. P. 1979, *MNRAS*, 187, 661
Morin, J., Donati, J.-F., Petit, P., *et al.* 2008, *MNRAS*, 390, 567
Oláh, K. & Strassmeier, K. G. 2002, *AN*, 323, 361
Oláh, K., Korhonen, H., Kővári, Zs., Forgács-Dajka, E., & Strassmeier, K. G. 2006, *A&A*, 452,303
Oláh, K., Kolláth, Z., Granzer, T., *et al.* 2009, *A&A*, 501, 703
Özdarcan, O., Evren, S., Strassmeier, K. G., Granzer, T., & Henry, G. W. 2010, *AN*, 331, 794
Petit, P., Donati, J., Wade, G. A., *et al.* 2004a, *MNRAS*, 348, 1175
Petit, P., Donati, J.-F., Oliveira, J. M., *et al.* 2004b, *MNRAS*, 351, 826
Petit, P., Aurière, M., Konstantinova-Antova, R., Morgenthaler, A., Perrin, G., Roudier, T., & Donati, J.-F. 2013, *Lecture Notes in Physics*, 857, 231
Piskunov, N. E., Tuominen, I., & Vilhu, O. 1990, *A&A* 230, 363
Rice, J. B., Wehlau, W. H., & Khokhlova, V. L. 1989, *A&A*, 208, 179
Roettenbacher, R. M., Monnier, J. D., Harmon, R. O., & Korhonen, H. 2013, these proceedings
Saar, S. 2011, in: D. P. Choudhary, & K. G. Strassmeier (eds.), *Physics of Sun and Star Spots*, Proc. IAU Symposium No. 273 (Cambridge University Press), p. 61
Semel, M. 1989, *A&A*, 225, 456
Sennhauser, C. & Berdyugina, S. V. 2011, *A&A*, 529, A100
Strassmeier, K. G., Fekel, F. C., Bopp, B. W., Dempsey, R. C., & Henry, G. W. 1990a, *ApJS*, 72, 191
Strassmeier, K. G., Hall, D. S., Barksdale, W. S., Jusick, A. T., & Henry, G. W. 1990b, *ApJ*, 350, 367
Strassmeier, K. G. 1999, *A&A*, 347, 225
Strassmeier, K. G. 2009, *Astronomy & Astrophysics Review*, 17, 251
Tsvetkova, S., Petit, P., Aurière, M., *et al.* 2013, *A&A*, 556, A43
Vogt, S. S. & Penrod, G. D. 1983, *PASP*, 95, 565
Vogt, S. S., Hatzes, A. P., Misch, A., & Kürster, M. 1999, *ApJS*, 121, 546
Vogt, S. S., Penrod, G. D., & Hatzes, A. P. 1987, *ApJ* 321, 496
Weber, M. & Strassmeier, K. G. 2001, *A&A* 373, 974
Weber, M. 2007, *AN* 328, 1075
Wilson, O. C. 1978, *ApJ*, 226, 379

Magnetic Fields throughout Stellar Evolution
Proceedings IAU Symposium No. 302, 2013
P. Petit, M. Jardine & H. Spruit, eds.

doi:10.1017/S1743921314002476

Pollux: a stable weak dipolar magnetic field but no planet ?

Michel Aurière[1], Renada Konstantinova-Antova[2,1], Olivier Espagnet [1], Pascal Petit [1], Thierry Roudier [1], Corinne Charbonnel [3,1], Jean-François Donati [1] and Gregg A. Wade [4]

[1]IRAP,Université de Toulouse & CNRS, Toulouse, France
email: michel.auriere@irap.omp.eu

[2]Institute of Astronomy and NAO, Bulgarian Academy of Sciences, Sofia, Bulgaria

[3]Geneva Observatory, University of Geneva, Versoix, Switzerland

[4]Department of Physics, Royal Military College of Canada, Kingston, Ontario, Canada

Abstract. Pollux is considered as an archetype of a giant star hosting a planet: its radial velocity (RV) presents sinusoidal variations with a period of about 590 d, which have been stable for more than 25 years. Using ESPaDOnS and Narval we have detected a weak (sub-gauss) magnetic field at the surface of Pollux and followed up its variations with Narval during 4.25 years, i.e. more than for two periods of the RV variations. The longitudinal magnetic field is found to vary with a sinusoidal behaviour with a period close to that of the RV variations and with a small shift in phase. We then performed a Zeeman Doppler imaging (ZDI) investigation from the Stokes V and Stokes I least-squares deconvolution (LSD) profiles. A rotational period is determined, which is consistent with the period of variations of the RV. The magnetic topology is found to be mainly poloidal and this component almost purely dipolar. The mean strength of the surface magnetic field is about 0.7 G. As an alternative to the scenario in which Pollux hosts a close-in exoplanet, we suggest that the magnetic dipole of Pollux can be associated with two temperature and macroturbulent velocity spots which could be sufficient to produce the RV variations. We finally investigate the scenarii of the origin of the magnetic field which could explain the observed properties of Pollux.

Keywords. Stars:individual:Pollux, Stars:late type, Stars: magnetic field

---

1. Introduction

Pollux (β Geminorum, HD 62509) is a well studied K0III giant neighbour of the sun. It is considered as an archetype of a giant star hosting a planet since its presents periodic sinusoidal radial velocity (RV) variations which have been stable during more than 25 years (Hatzes *et al.* 2006). Now, Aurière *et al.* (2009) discovered a weak magnetic field at the surface of Pollux whose variations could be correlated with the RV ones. We have therefore performed a Zeeman survey of Pollux and collected spectropolarimetric data during 4.25 years i.e. more than for two periods of the RV variations.

2. A spectropolarimetric survey of Pollux with Narval and ESPaDOnS

From 2007 September to 2012 February, using first ESPaDOnS (Donati *et al.* 2006) at CFHT in a snapshot program, then Narval (its twin) at the TBL in a systematic survey, we observed Pollux on 41 dates and got 266 Stokes V series. The observational properties of the instruments and reduction techniques are the same as described by Aurière *et al.* (2009). To obtain a high-precision diagnosis of the spectral line circular polarization, LSD

(Donati *et al.* 1997) was applied to each reduced Stokes I and V spectrum. From these mean Stokes profiles we computed the surface-averaged longitudinal magnetic field B_ℓ in G, using the first-order moment method adapted to LSD profiles (Donati *et al.* 1997). The RV of Pollux was measured from the averaged LSD Stokes I using a gaussian fit. Figure 1 shows the variations of RV (left plot) and B_ℓ (right plot) during the 2007-2012 seasons of our survey: The B_ℓ variations appear to follow those of RV.

3. Modeling the magnetic field of Pollux

Variations of the longitudinal magnetic field .

Figure 1 shows the fit of the variations of RV and B_ℓ with a sinusoid of P = 589.64 d as derived by Hatzes *et al.* (2006). With this period and adjusting the amplitude of the B_ℓ variations to 0.2 G, the sinusoids obtained by least squares fitting have a phase shift of 146 d, i.e. about 25 % of the period.

Zeeman Doppler imaging (ZDI) of Pollux.

To use the whole Zeeman information included in our Stokes V data, and to independently infer the rotational period P_{rot} and the magnetic topology of Pollux, we have used the ZDI method in the version of Donati *et al.* (2006). We limited the number of spherical harmonics to $l < 5$ since increasing the threshold did not change significantly the results. We have followed the approach of Petit *et al.*(2002) for determining the P_{rot} of Pollux, and we find a value of about 587 d as a favoured period. This is very near the RV variation period and both are possible P_{rot} with respect to our error bars. Since the RV period has been found stable during more than 25 years, we consider that it is the real P_{rot} and will use it hereafter in this work. With these parameters, our prefered ZDI model is the following: the inclination angle i is 60° and the poloidal component contains 71 % of the reconstructed magnetic energy. The dipole component corresponds to about 99 % of this poloidal magnetic energy. The mean magnetic field B_{mean} is 0.7 G, and the angle between the rotation and magnetic dipole axis β is about 20°. In these conditions and within the solid rotation hypothesis, the $v \sin i$ of Pollux would be 0.7 km s^{-1}.

4. Origin of the radial velocity variations of Pollux

Since we found that the P_{rot} of Pollux is equal to the period of the RV variations, one has to consider if the weak magnetic field is able to induce these RV variations.

The planet hypothesis

If a planet exists, orbiting around Pollux and being responsible for the RV variations,

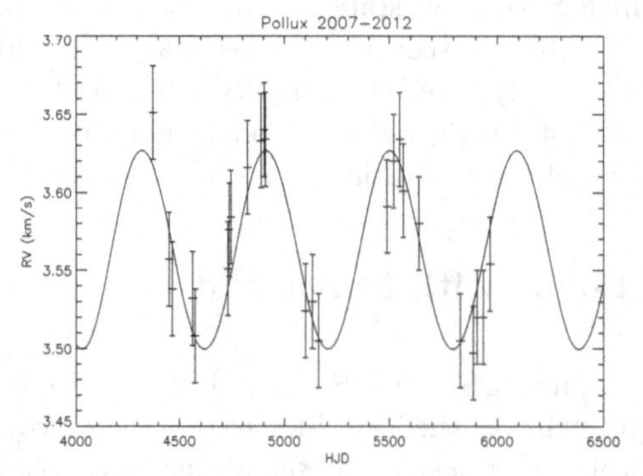

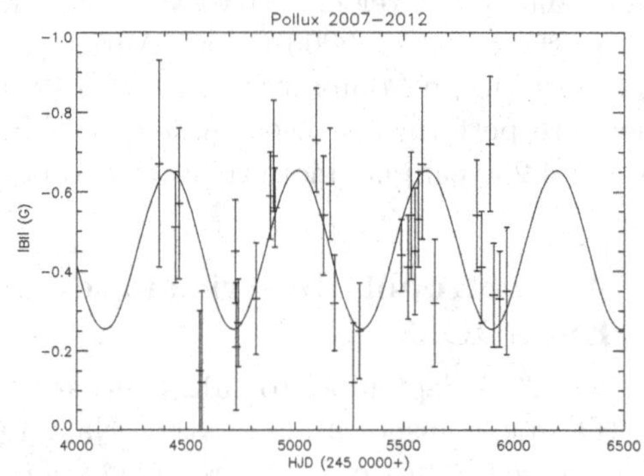

Figure 1. Variations of RV (left plot) and of B_l (right plot) with HJD (245 0000+) in 2007-2012. A sinusoid with P= 589.64 d is fitted for each parameter.

the planet revolution is synchronized with the rotation of the star. From Hatzes *et al.* (2006), the mass of the planet is in the range 3-13 $M_{Jupiter}$ and the semi major axis is a = 1.64 AU. In these conditions we are very far from the case occuring with "hot jupiter" class objects (e.g. Fares *et al.* 2013), and the expected tidal effects are too weak for enabling the synchronization (Pont 2009).

The magnetic and macroturbulence spot hypothesis

A dipole configuration as inferred for Pollux by our ZDI study can be regarded as corresponding to two magnetic spots. Using the SOAP online package (Boisse *et al.* 2012) we have simulated the effects produced by the magnetic configuration of Pollux: we could generate the observed RV variations but photometric variations of larger amplitude than those detected in the Hipparcos data were predicted. Now, Hatzes & Cochran (2000) and Lee *et al.* (2012), proposed the effect of macroturbulent velocity spots linked with a surface magnetic field to explain the RV variations of Polaris and α Per respectively. RV variations up to 100 m s^{-1} were expected for these stars without inducing photometric variations. Consequently, we consider that the magnetic properties of Pollux can make the hosted-planet hypothesis unecessary.

5. Origin of the magnetic field

Pollux is the first giant with a sub-G mean surface magnetic field for which P_{rot} is determined. This P_{rot} is also the longest determined up to now for a giant. If the hypothesis of a planet orbiting around Pollux is not retained, the very long stability of the dipolar magnetic field (more than 25 years or 15 P_{rot}) has to be taken into account to infer the origin of the magnetic field. The high Rossby number *Ro* of 2-3 inferred for Pollux using this P_{rot} and the convective turnover time τ_{conv} derived using the evolutionary models with rotation of Lagarde *et al.* (2012) and Charbonnel *et al.* (this symposium and in preparation) suggest that an α-ω dynamo alone cannot be the origin of the observed magnetic field. Some other possibilities are:

- The high *Ro* and the stability would suggest Pollux being an Ap star descendant. However using B_{mean} = 0.7 G and taking into account the magnetic flux conservation hypothesis we find that an Ap star progenitor of Pollux would have a magnetic strength of about 14 G, i.e. well below the 100-300 G lower limit found for Ap star dipole strength by Aurière *et al.* (2007).

- An extreme case of dynamo when rotation is not involved is a local dynamo as inferred in the case of Betelgeuse (Aurière *et al.* 2010). However, in the case of Pollux

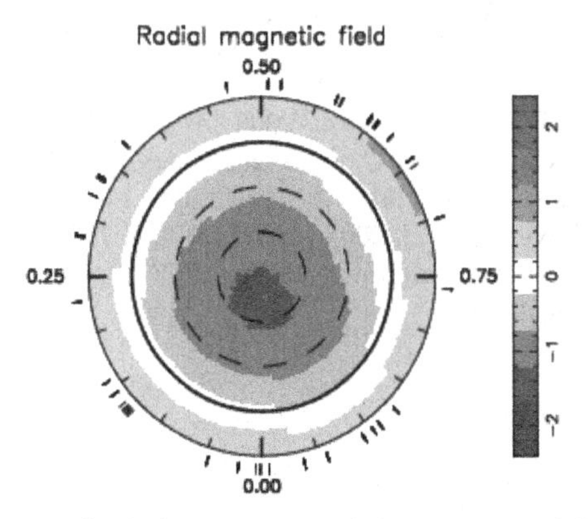

Figure 2. ZDI map of Pollux (radial component of the magnetic field) presented in flattened polar projection down to latitudes of −30°. The magnetic field strength is expressed in G.

the expected number of convective cells is much greater than inferred for Betelgeuse. Furthermore, in the case of Betelgeuse the magnetic field is observed to vary intrinsically on timescales of weeks to months.

- An intermediate case of dynamo that could operate during the red giant branch phase because of the deepening of the convective envelope, is the distributed dynamo. This type of dynamo is expected to occur in M dwarfs, and a long term stability has been reported in the case of the fast rotating and strongly magnetic star V374 Peg (Morin *et al.* 2008). One could suggest that a weak dynamo regime could be stable as well.

6. Conclusion

We have monitored the magnetic field of Pollux during 4.25 years. Our ZDI investigation shows that the P_{rot} is equal to the RV variation period and that the Pollux surface magnetic field is mainly poloidal and dipolar. We then show that photometric and macroturbulent-velocity spots associated to the magnetic poles could explain the RV variations and could then make the hosted-planet hypothesis unnecessary. In this hypothesis, to explain the long term stability of the RV variations, the magnetic field of Pollux should be stable during tens of years. We suggest that a weak-regime dynamo (of interface-type?) could be the origin of the magnetic field of Pollux.

In this way Pollux could be the representative of a class of weakly magnetic G K giants, recently discovered (Aurière *et al.* in preparation, Konstantinova-Antova *et al.*, this symposium). These stars would include bright giants like Alphard and Arcturus, and some of them, like Aldebaran and ϵ Tau also present stable RV variations and are considered to host planets. Though we cannot exclude completely that neither Pollux hosts a planet in addition to its magnetic field, nor that it is a peculiar very stable magnetic star, our investigation suggests that Pollux could be the archetype of a class of weakly magnetic G K giants.

References

Aurière, M, Wade, G. A., Silvester, J., Lignières *et al.* 2007, *A&A*, 475, 1053
Aurière, M., Wade, G. A., Konstantinova-Antova, R. *et al.* 2009, *A&A*, 504, 231
Aurière, M., Donati, J.-F., Konstantinova-Antova, R. *et al.* 2010, *A&A*, 516, L2
Boisse, I., Bonfils, X., & Santos, N.-C. 2012, *A&A*, 545, 109
Donati, J.-F., Semel, M., Carter, B. D. *et al.* 1997, *MNRAS*, 291, 658
Donati, J.-F., Catala, C., Landstreet, J., & Petit P. 2006, in: Casini R., Lites B., eds, *Solar Polarization Workshop n4* ASPC series, 358, 362
Fares, R., Moutou, C., Donati, J.-F. *et al.* 2013, *MNRAS*, 435, 1451
Hatzes, A. P. & Cochran, W. D. 2000, *AJ*, 120, 979
Hatzes, A. P., Cochran, W. D., Endl, M. *et al.* 2006, *A&A*, 457, 335
Lagarde, N., Decressin, T., Charbonnel, C. *et al.* 2012, *A&A* 542, 62
Lee, B.-C., Han, I., Park, M.-G. *et al.* 2012, *A&A* 543, 37
Morin, J., Donati, J.-F., Forveille, *et al.* 2008, *MNRAS*, 384, 77
Petit, P., Donati, J-F., Collier, & Cameron A. 2002, *MNRAS*, 334, 374
Pont, F. 2009, *MNRAS*, 396, 1789

Magnetic Fields throughout Stellar Evolution
Proceedings IAU Symposium No. 302, 2013
P. Petit, M. Jardine & H. Spruit, eds.

doi:10.1017/S1743921314002488

On dynamo action in the giant star Pollux : first results

Ana Palacios[1,2] and Allan Sacha Brun[2]

[1]LUPM, Université Montpellier II,
Place Eugène Bataillon cc -0072, F-34095, Montpellier cedex 5, France
email: ana.palacios@univ-montp2.fr

[2]Laboratoire AIM Paris-Saclay, CEA/Irfu Université Paris-Diderot CNRS/INSU, 91191 Gif-sur-Yvette, France
email:sacha.brun@cea.fr

Abstract. We present preliminary results of a 3D MHD simulation of the convective envelope of the giant star Pollux for which the rotation period and the magnetic field intensity have been measured from spectroscopic and spectropolarimetric observations. This giant is one of the first single giants with a detected magnetic field, and the one with the weakest field so far. Our aim is to understand the development and the action of the dynamo in its extended convective envelope.

---

1. Context and numerical setup

The detection of magnetic fields in the atmosphere of red giant stars using spectropolarimetry (see Aurière and Konstantinova-Antova in this volume) brings the opportunity to get an insight on the secular evolution of magnetic fields in solar-type and intermediate-mass stars. Spectropolarimetry allows to estimate the magnetic fields intensity, the doppler imaging of magnetic red giants being a delicate matter in part due to the usually long rotation period that characterizes these objects. The star Pollux is one of the first giants with a detected magnetic field, and its intensity of about 1 G, is the weakest detected so far. Here, we present preliminary results of a fully 3-D dynamo computation carried out with the Anelastic Spherical Harmonic (ASH) version 2.0 code (Featherstone *et al.* 2013, Brun *et al.* 2004), for the entire convective envelope (75% of the total radius, strong stratification) of a 2.5 $M_\odot$ with R = 9 $R_\odot$ and L = 40 $L_\odot$, well representative of the star Pollux (see Aurière *et al.* 2009). The hydrodynamical progenitor was computed with $\Omega = 2.63\ 10^{-7}$ rad.s^{-1}, corresponding to a v sini ≈ 1.3 km.s^{-1}, in good agreement with published values. Using a magnetic Prandtl number of the simulation of $P_m = 3$, we have next initialized the magnetic field with a multipole l = 3, m = 2 seed field, with an initial total magnetic energy ME less than 10^{-3} of the total kinetic energy.

2. Dynamo onset, internal transport and magnetic properties

Our simulation is still running and as of today the dynamo is in the linear regime with an exponential growth of ME having to enter yet the saturated regime. At this point of the simulation, the total magnetic energy only corresponds to 0.1% of the total kinetic energy. The ME is dominated by non-axisymmetric contributions of the toroidal and poloidal fields. The purely axisymmetric components TME and PME are growing but very weak. The radial energy flux balance has reached an equilibrium. The radial energy transport by the Poynting flux is negligible (at a level of 10^{-5}) at this stage of the simulation. This situation should change once a saturation of the dynamo is reached.

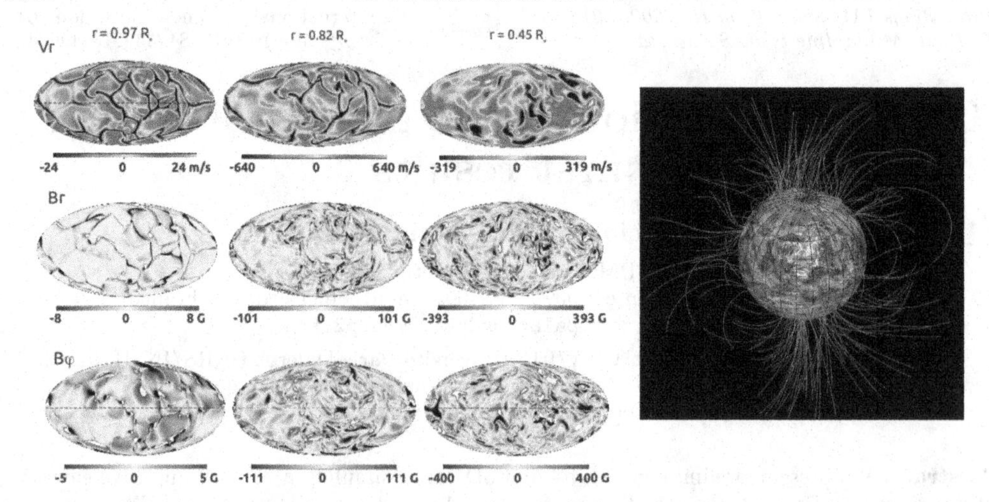

Figure 1. Renderings of the simulation after evolving it for 320 days. Left Fluctuations of the radial velocity (convective patterns upper row), the radial magnetic field (middle row) and the toroidal magnetic field (lower row) at three different depths. Right 3D rendering of the magnetic field and of the extrapolated magnetic field. The lines represent magnetic field lines and the surface is a spherical cut in the convective zone. The colours blue and orange represent negative and positive values of B_r.

As also found in Brun & Palacios (2009) for a more evolved red giant, the enthalpy flux dominates the energy transport. When converted to luminosity it reaches 160% of the total stellar luminosity. This flux arises to compensate the strong inward (negative) kinetic energy flux found in the simulation.

Fig. 1 (left panel) shows the convective patterns achieved consisting of warm large upflows surrounded by a network of cool thin downflows. The differential rotation profile achieved in the simulation is retrograde and cylindrical in the bulk of the domain except near the rotation axis. The rotational state is similar to that achieved in the hydrodynamical progenitor when averaged over the same period of time, indicating that the Lorentz force is too weak for the magnetic field to affect the angular velocity profile at this stage of the simulation.

The radial magnetic field follows the descending flows, while the toroidal field develops within that network. The magnetic activity is larger at the bottom of the convective envelope (see scales Fig.1 left panel). The intensity of the flows is larger at larger depths, with a magnetic energy that can be locally intense but remains globally weak at this stage of the simulation. The global amplitude of the magnetic field is ≈ 9 Gauss. A 3D rendering of the simulation together with the extrapolated magnetic field is also shown in Fig. 1 (right panel). While the magnetic seed is a multipole and the simulation is not mature yet, we clearly see a dipolar configuration appearing (strong l=1 mode).

References

Aurière, M., Wade G., Konstantinova-Antova, R. *et al.* 2009, *A&A*, 504,231
Brun, A. S. & Palacios, A. 2009, *ApJ*, 702, 1078
Brun, A. S., Miesch, M. S., & Toomre, J. 2004, *ApJ*, 614, 1073
Featherstone *et al.* 2014, *in preparation*

Magnetic Fields throughout Stellar Evolution
Proceedings IAU Symposium No. 302, 2013
P. Petit, M. Jardine & H. Spruit, eds.

doi:10.1017/S174392131400249X

The Hertzsprung-gap giant 31 Comae in 2013: Magnetic field and activity indicators

Ana P. Borisova[1], Renada Konstantinova-Antova[1,2], Michel Aurière[2], Pascal Petit[2] and Corinne Charbonnel[3,2]

[1]Institute of Astronomy and NAO, Bulgarian Academy of Sciences
72 Tsarigradsko shosse blvd., BG-1784, Sofia, Bulgaria
email: aborisova@astro.bas.bg

[2]Institut de Recherche en Astrophysique et Plantologie, CNRS, Universitè de Toulouse, France,
[3] Universitè de Genève, Switzerland

Abstract. We have observed the giant star 31 Comae in April and May 2013 with the spectropolarimeter Narval at Pic du Midi Observatory, France. 31 Comae is a single, rapidly rotating giant with rotational period ~ 6.8 d and *vsini* ~ 67 km/s. We present measurements and discuss variability of the longitudinal magnetic field (Bl), spectral activity indicators H_α, CaII H&K, Ca II IR triplet and evolutionary status. Our future aim is to perform a Zeeman-Doppler imaging study for the star.

Keywords. Stars: activity, Stars:individual:31 Comae,Stars:magnetic fields

1. Introduction

31 Comae (HD 111812) is a single *G0III* (Gray *et al.* 2001), rapidly rotating giant with $vsini \sim 67$ km/s, $Teff \sim 5660$ K and $M = 2.6M_\odot$ (Strassmeier *et al.* 2010). The star is variable with a very low light curve amplitude and rotational modulation with a period of ~ 6.8 d. The star displays chromospheric and coronal activity with CaII H&K line emission, super-rotationally broadened coronal and transition-region lines, and X-ray emission of Lx=6.325 $* 10^{30}$ erg s^{-1} (Gondoin 2005). The magnetic field of 31 Comae is interesting to be investigated because of its position in the Hertzsprung-gap region and because of its possible membership of the Coma-Berenices cluster, (Bounatiro 1993).

2. Observations, Results and Conclusions

Observations and Data Processing: Ten Narval spectra, with resolution power of 65 000 and wavelength range from 370 to 1050 nm have been obtained. Libre Esprit (Donati *et al.* 1997) software for automatic extraction of spectra and Least-squares Deconvolution technique (LSD, Donati *et al.* 1997) were used for computing the mean Stokes V and I photospheric profiles. Mean longitudinal field Bl was estimated by the use of the first order moment method (Donati *et al.* 1997, Rees & Semel 1979 Wade *et al.* 2000).

Results: We have detected Zeeman signatures in Stokes V LSD profiles and calculated the corresponding surface Bl of 31 Comae, with values up to 9.5 G and $\sigma_{Bl} < 5.1$ G, (Fig. 1). Very broad CaII H&K absorption profile with a weak chromospheric emission core and S_*index* variations from 0.37 to 0.42 are observed. H_α and CaII IRT are partially filled-in by emission. Activity indicators display moderate variations in the observed period, most pronounced in H_α, (Fig. 2). Variations of Bl do not follow activity indicators changes.

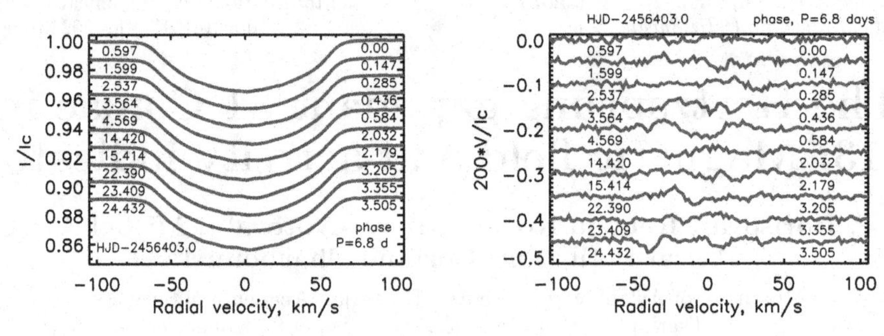

Figure 1. LSD Stokes I (left panel) and Stokes V, (multiplied by 200) photospheric line profiles.

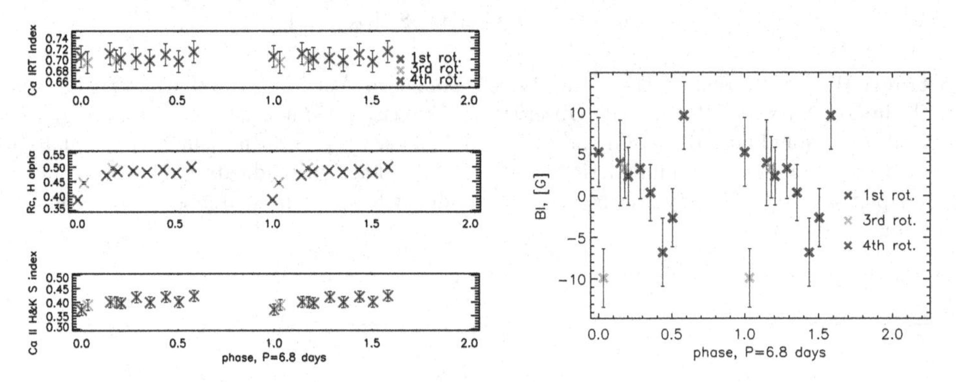

Figure 2. Variations of activity indicators (left panel) and Bl with rotational phase.

Conclusions: Stokes V LSD profiles show composite and variable behaviour thus we might propose a complex structure of its magnetic field. Fast rotation of the star is similar to FK Comae type stars, but Bl is weaker, compared to FK Comae (60 to 272G) (Korhonen *et al.* 2009). The star is also in a different activity level with emission components in Hα and CaII H&K lines not so strong as in FK Comae (Korhonen *et al.* 2009, Strassmeier *et al.* 1990).

Acknowledgements . We are thankful to the TBL team for providing service observations with Narval spectropolarimeter. Observations were funded under the project BG051PO001-3.3.06-0047 financed by the EU, ESF and Republic of Bulgaria. A. B. acknowledge Bulgarian NSF contract DMU 03-87, partial financial support of the TBL, France and the project BG051PO001-3.3.06-0047 for attending the conference.

References

Bounatiro, L. 1993, *A&AS*, 100, 53
Donati, J.-F. *et al.* 1997, *MNRAS*, 291, 658
Gondoin, P. 2005, *A&A*, 444, 531
Gray, R. O. *et al.* 2001, *AJ*, 121, 2148
Korhonen *et al.* 2009, *MNRAS*,395,282
Rees, D. E. & Semel, M. D. 1979, *A&A*, 74, 1
Strassmeier, K. G. *et al.* 1990, *ApJS*, 72, 191
Strassmeier, K. G. *et al.* 2010, *A&A*, 520, A52
Wade, G. *et al.* 2000, *MNRAS*, 313, 823

Magnetic Fields throughout Stellar Evolution
Proceedings IAU Symposium No. 302, 2013
P. Petit, M. Jardine & H. Spruit, eds.

doi:10.1017/S1743921314002506

Magnetic Field Structure and Activity of the He-burning Giant 37 Comae

S. Tsvetkova[1], P. Petit[2], R. Konstantinova-Antova[1,2], M. Aurière[2], G. A. Wade[3], C. Charbonnel[4,2] and N. A. Drake[5,6]

[1]Institute of Astronomy and NAO, Bulgarian Academy of Sciences, Bulgaria
email: stsvetkova@astro.bas.bg

[2]IRAP, UMR 5277, CNRS and Université de Toulouse, France

[3]Department of Physics, Royal Military College of Canada, Ontario, Canada

[4]Geneva Observatory, University of Geneva, Switzerland

[5]Sobolev Astronomical Institute, St. Petersburg State University, Russia

[6]Observatório Nacional/MCTI, Rio de Janeiro, Brazil

Abstract. We present the first magnetic map of the late-type giant 37 Com. The Least Squares Deconvolution (LSD) method and Zeeman Doppler Imaging (ZDI) inversion technique were applied. The chromospheric activity indicators Hα, S-index, Ca II IRT and the radial velocity were also measured. The evolutionary status of the star has been studied on the basis of state-of-the-art stellar evolutionary models and chemical abundance analysis. 37 Com appears to be in the core Helium-burning phase.

Keywords. stars: magnetic fields – stars: abundances – stars: individual: 37 Com

1. Introduction

37 Com is the primary star of a wide triple system (Tokovinin 2008), but the synchronisation effect plays no role for its fast rotation and activity. Its significant photometric and Ca II H&K emission variabilities were presented by Strassmeier *et al.* (1997; 1999) and de Medeiros *et al.* (1999) and interpreted as signatures of magnetic activity.

Observational data for 37 Com were obtained with two twin fiber-fed echelle spectropolarimeters – Narval (2m TBL at Pic du Midi Observatory, France) and ESPaDOnS (3.6m CFHT). We have collected 11 Stokes V spectra for 37 Com in the period January 2010 – July 2010. The Least Squares Deconvolution (LSD) multi-line technique was applied and the surface-averaged longitudinal magnetic field B_l was computed using the first-order moment method (Donati el al. 1997; Wade *et al.* 2000). The Zeeman Doppler Imaging (ZDI) tomographic technique was employed for mapping the large-scale magnetic field of the star (Donati *et al.* 2006).

2. Results

There are significant variations of B_l in the interval from -2.5 G to 6.5 G with at least one sign reversal during the observational period (Fig. 1 left). Also, radial velocity, S-index and line activity indicators Hα and Ca II IRT (854.2 nm) show significant variations, and clear correlations with each other as well as the longitudinal field.

The ZDI mapping (Fig. 1 center and right) reveals that the large-scale magnetic field has a dominant poloidal component, which contains about 88% of the reconstructed magnetic energy. The star has a differential rotation with the following parameters: Ω_{eq} =

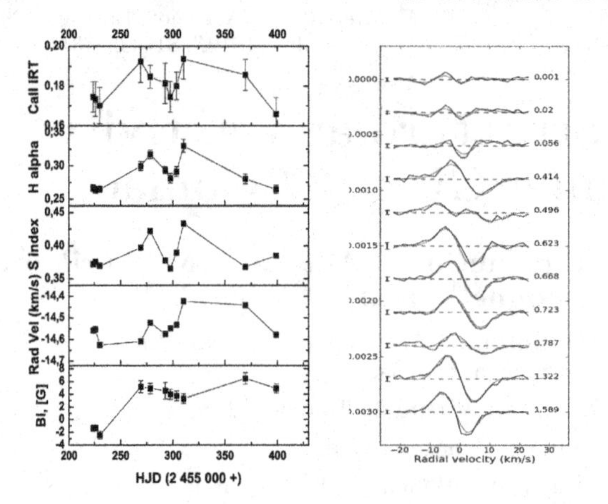
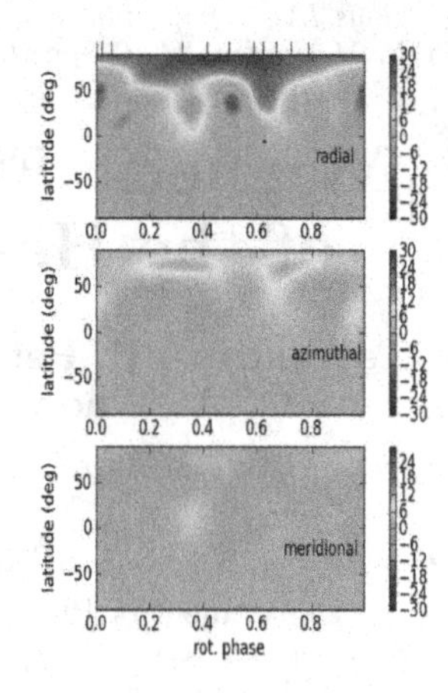

Figure 1. Left: From bottom to top – simultaneous time variaions of B_l with radial velocity (RV), S-index, Hα and Ca II IRT (854.2 nm). **Center:** Normalized Stokes V profiles – observed profiles (black); synthetic fit (red); zero level (dashes lines). The error bars are on the left of each profile. **Right:** The magnetic map of 37 Com. The magnetic field strength is in gauss. The vertical ticks on top of the radial map show the phases when there are observations.

0.06 rad/d (the rotation rate at the equator) and $\Delta\Omega = 0.01$ rad/d (the difference in the rotation rate between the polar region and the equator) (Petit *et al.* 2002).

37 Com shows simpler surface magnetic structure than the fast rotators V390 Aur (Konstantinova-Antova *et al.* 2012) and HD 232862 (Aurière *et al.* in prep.) and shows more complex structure than the slow rotators EK Eri (Aurière *et al.* 2011) and β Ceti (Tsvetkova *et al.* 2013), which are suspected of being descendants of Ap-stars.

The location of 37 Com on the Hertzsprung-Russell diagram was determined on the basis of state-of-the-art stellar evolution models (Charbonnel & Lagarde 2010) and the mass is found to be 5.25 $M_{\odot}$, in a good agreement with the literature. Synthetic spectra in the region containing ^{12}CN and ^{13}CN molecular lines were calculated and compared to our spectra in order to infer the ^{12}C/^{13}C ratio. The best fit was achieved for ^{12}C/^{13}C = 4.0. From these results, it appears that 37 Com is in the core Helium-burning phase.

Acknowledgements

STs thanks the contract BG05PO001-3.3.06-0047. G. A. W. is supported by an NSERC grant. N. A. D. thanks support of PCI/MCTI, Brazil, under the Project 302350/2013-6.

References

Aurière, M., Konstantinova-Antova, R., Petit, P. *et al.* 2011, *A&A*, 534, 139
Charbonnel, C. & Lagarde, N. 2010, *A&A*, 522, 10
de Medeiros, J. R., Konstantinova-Antova, R. K., & da Silva, J. R. P. 1999, *A&A*, 347, 550
Donati, J.-F., Semel, M., Carter, B. D. *et al.* 1997, *MNRAS*, 291, 658
Donati, J.-F., Howarth, I. D., Jardine, M. M. *et al.* 2006, *MNRAS*, 370, 629
Konstantinova-Antova, R., Aurière, M., Petit, P. *et al.* 2012, *A&A*, 541, 44
Petit, P., Donati, J.-F., & Collier Cameron, A. 2002, *MNRAS*, 334, 374
Strassmeier, K. G., Boyd, L. J., Epand, D. H., & Granzer, Th. 1997, *PASP*, 109, 697
Strassmeier, K. G., Serkowitsch, E., & Granzer, Th. 1999, *A&AS*, 140, 29
Tokovinin, A. 2008, *MNRAS*, 389, 925
Tsvetkova, S., Petit, P., Aurière, M. *et al.* 2013, *A&A*, 556, 43
Wade, G. A., Donati, J.-F., Landstreet, J. D., & Shorlin, S. L. S. 2000, *MNRAS*, 313, 823

Magnetic Fields throughout Stellar Evolution
Proceedings IAU Symposium No. 302, 2013
P. Petit, M. Jardine & H. Spruit, eds.

doi:10.1017/S1743921314002518

Strong variable linear polarization in the cool active star II Peg

Lisa Rosén[1], Oleg Kochukhov[1] and Gregg A. Wade[2]

[1]Deptartment of Physics & Astronomy, Uppsala University,
Box 516, SE-75120 Uppsala, Sweden
email: `lisa.rosen@physics.uu.se`
email: `oleg.kochukhov@physics.uu.se`

[2]Department of Physics, Royal Military College of Canada,
PO Box 17000, Station Forces, Kingston, Ontario K7K 7B4, Canada
email: `Gregg.Wade@rmc.ca`

Abstract. Magnetic fields of cool active stars are currently studied polarimetrically using only circular polarization observations. This provides limited information about the magnetic field geometry since circular polarization is only sensitive to the line-of-sight component of the magnetic field. Reconstructions of the magnetic field topology will therefore not be completely trustworthy when only circular polarization is used. On the other hand, linear polarization is sensitive to the transverse component of the magnetic field. By including linear polarization in the reconstruction the quality of the reconstructed magnetic map is dramatically improved. For that reason, we wanted to identify cool stars for which linear polarization could be detected at a level sufficient for magnetic imaging. Four active RS CVn binaries, II Peg, HR 1099, IM Peg, and σ Gem were observed with the ESPaDOnS spectropolarimeter at the Canada-France-Hawaii Telescope. Mean polarization profiles in all four Stokes parameters were derived using the multi-line technique of least-squares deconvolution (LSD). Not only was linear polarization successfully detected in all four stars in at least one observation, but also, II Peg showed an extraordinarily strong linear polarization signature throughout all observations. This qualifies II Peg as the first promising target for magnetic Doppler imaging in all four Stokes parameters and, at the same time, suggests that other such targets can possibly be identified.

Keywords. stars: magnetic field – stars: late-type – stars: individual: II Peg, HR 1099, IM Peg, σ Gem – polarization

1. Introduction

Stellar magnetic fields can be detected by the appearance of signatures in the polarization spectra. The degree of polarization is proportional to the magnetic field strength and a weak field will hence result in weak polarization signatures. Cool stars generally have relatively weak magnetic fields which sometimes makes it difficult to detect any polarization. Since linear polarization is up to 10 times weaker than circular polarization, it is even more difficult to detect linear polarization. Because of this, the current studies of cool star magnetic fields are performed using circular polarization only.

This is not optimal since circular polarization is only sensitive to the line of sight component of the magnetic field vector. The information that can be extracted from circular polarization alone is hence not sufficient to provide a complete picture of the magnetic field geometry. The lack of information becomes especially problematic when the fields are complex which often is the case for cool stars. One of the major problems is cross talk, especially between radial and meridional field components, see e.g. Donati & Brown (1997).

The solution to the problem would be to include linear polarization since linear polarization is sensitive to the transverse component of the magnetic field vector. This means that additional information about the magnetic field geometry can be extracted and when combined with circular polarization the picture of the magnetic field topology will be a lot more complete. This has been shown in several numerical experiments, see e.g. Kochukhov & Piskunov (2002) and Rosén & Kochukhov (2012). The inclusion of linear polarization will also remove the crosstalk.

In a previous study by Kochukhov *et al.* (2011) they managed to detect linear polarization in two cool stars. This study was important since it proved that detection of linear polarization in cool stars is possible. However, their linear polarization observations did not have a sufficient S/N ratio to be useful for magnetic imaging. The goal of our study was therefore to take this one step further by identifying cool stars for which linear polarization could be detected at a level sufficient for magnetic imaging.

2. Observations

All our spectropolarimetric observations were performed at the Canada-France-Hawaii Telescope (CFHT) with the spectropolarimeter ESPaDOnS. ESPaDOnS has a spectral resolution of about 65 000 and a wavelength coverage of 370-1050 nm for a single exposure.

We focused our initial observations on four RS CVn stars, II Peg, HR 1099, IM Peg and σ Gem as they are among the most magnetically active cool stars.

Full Stokes observations of all four stars were performed in periods during 2012 February to July. During this time, II Peg and IM Peg were observed 3 times each, HR 1099 was only observed once and σ Gem was observed 4 times.

As expected, we could not detect any linear polarization signatures in individual spectral lines. We therefore applied the multi-line technique called least-squares deconvolution (LSD) Donati *et al.* (1997). Assuming all lines are self-similar, they can be combined into one mean profile where each line is assigned an individual weight depending on wavelength, magnetic sensitivity and central depth. This procedure will significantly increase the S/N ratio. In Fig. 1 the set of observations with the highest detection probability for each star is shown.

3. Results

σ Gem turned out to have the weakest linear polarization signatures out of the four stars. Despite this, Stokes Q was securely detected in the observation shown in Fig. 1.

The polarization signatures of IM Peg in Fig. 1 look convincing and the detection probability is high. The other two observations of IM Peg were however not as convincing. It should also be mentioned that each of the three observations of IM Peg were obtained at different rotational phases, indicating a varying activity level across the stellar surface. A hint of a Stokes I profile of the secondary can be seen to the blue side of the primary.

HR 1099 was one of the stars that Kochukhov *et al.* (2011) previously detected linear polarization in. In this study we managed to get a similar detection. It should be noted that previous observations were obtained with HARPSpol which has a spectral resolution of more than 100 000 and that the spectral resolution of ESPaDOnS is significantly lower (65 000). In Fig. 1 the secondary is clearly visible in Stokes I, but also, a polarization signature of the secondary can be seen in Stokes V.

II Peg really stands out since the linear polarization signatures are enormous compared to any of the other stars. The amplitudes of the Stokes QU profiles shown in Fig. 1 are actually among the smallest compared to the other II Peg observations. Since II Peg

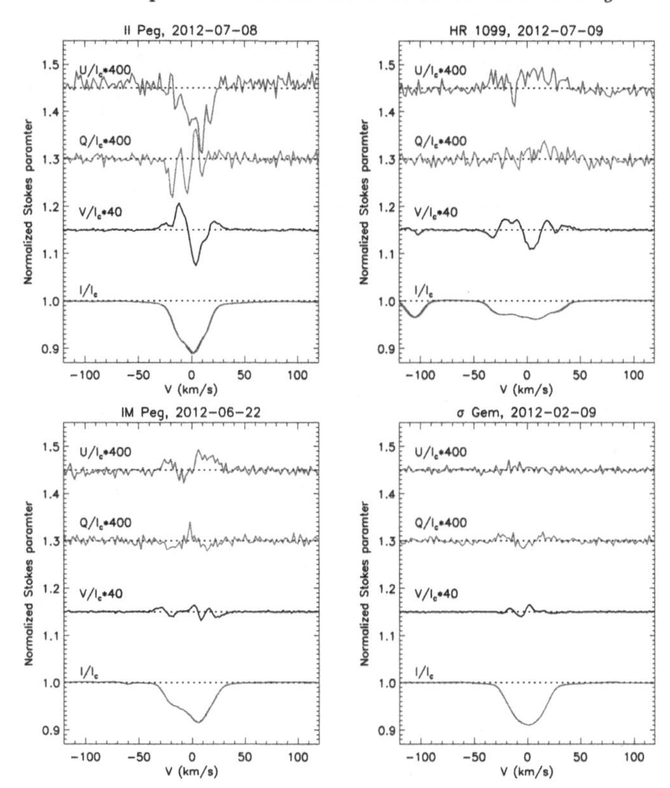

Figure 1. Representative sets of the LSD Stokes $IQUV$ profiles for II Peg, HR 1099, IM Peg, and σ Gem. The polarization profiles are magnified and shifted vertically.

proved to be a very promising target we decided to spend the remaining observations on this star.

4. Further observations and analysis of II Peg

We managed to get 7 more observations of II Peg in 2012 September and also an additional 2 in 2012 December/2013 January. The activity proved to be continuously high throughout all the observations. The mean amplitudes of Stokes QU were as large as the Stokes V amplitudes for some of the other stars and the peak amplitude went as high as $8 \cdot 10^{-4}\ I_{\rm c}$. For a more detailed description of these observations, see Rosén *et al.* (2013).

Even though we had 12 complete sets of Stokes $IQUV$ observations which all had a S/N ratio sufficient for magnetic imaging, they were still not optimal for mapping. The reason is the long time period over which these observations were obtained, about 6 months. The field is constantly evolving and profiles obtained at the same rotational phase but at different observing epochs showed different shapes.

New observations were therefore obtained during 2013 June 15 to July 1. Once again the observations were done at CFHT with ESPaDOnS. During this period another 12 complete sets of observations were performed with good phase coverage. All observations can be seen in Fig. 2.

Once again the activity level was continuously high throughout the observing period. Clear distortions due to temperature inhomogeneities can be seen in all Stokes I profiles. The Stokes QUV profiles vary in shape from one phase to the next.

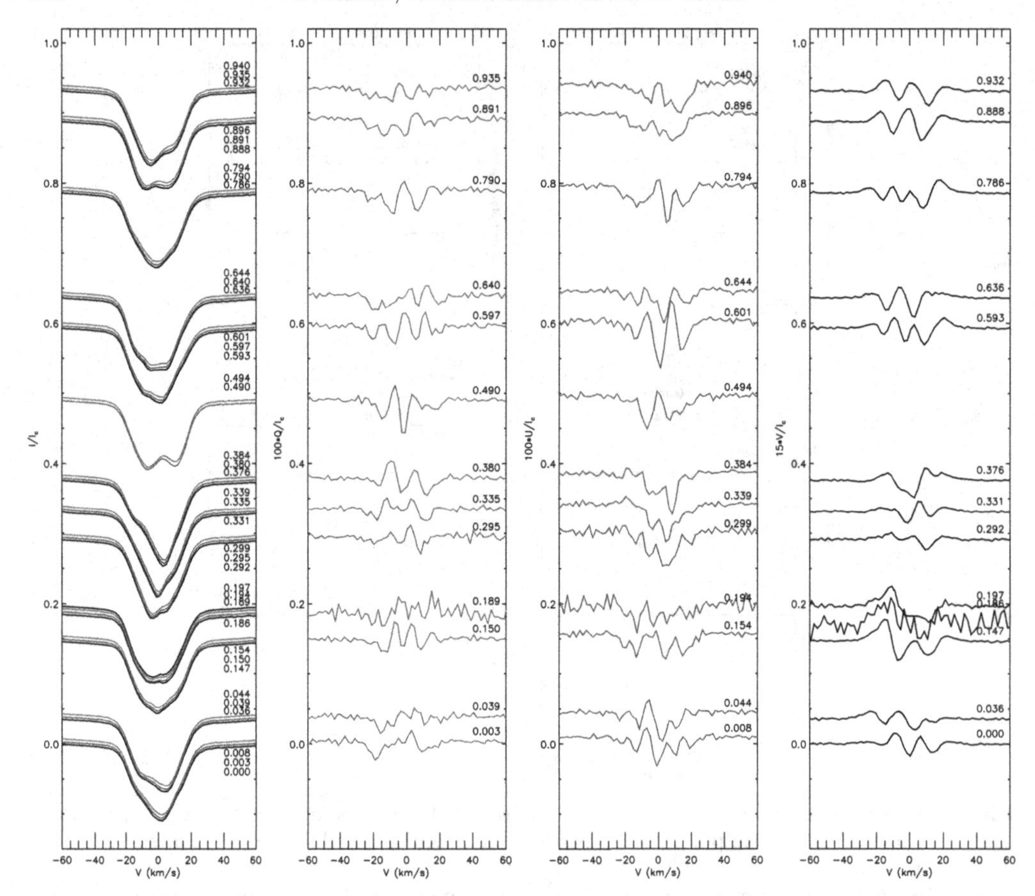

Figure 2. LSD Stokes $IQUV$ profiles of II Peg. Each panel represents a different Stokes parameter. Profiles have been shifted vertically according to the rotational phase and the orbital radial velocity variation has been corrected for. Stokes Q and U (middle panels) have been magnified by a factor of 100 while Stokes V (right panel) has been magnified by a factor of 15.

5. Conclusions

This set of observations is unique since such a set has never previously been obtained for a cool star. It proves that linear polarization can be detected in cool stars at a level suitable for magnetic imaging. It also proves that such observations does not require an ultra-high spectral resolution, but instead can be done at a spectral resolution of about 65 000. The next step will be to reconstruct the magnetic field map using all 4 Stokes parameters self-consistently and simultaneously with temperature spots.

References

Donati, J.-F. & Brown, S. F., 1997, *A&A*, 326, 1135

Donati, J.-F., Semel, M., Carter, B. D., Rees, D. E., & Collier Cameron, A., 1997, *MNRAS*, 291, 658

Kochukhov, O. & Piskunov, N., 2002, *A&A*, 388, 868

Kochukhov, O., Makaganiuk, V., Piskunov, N., Snik, F., Jeffers, S. V., Johns-Krull, C. M., Keller, C. U., Rodenhuis, M., & Valenti, J. A., 2011, *ApJ*, 732, L19

Rosén, L. & Kochukhov, O., 2012, *A&A*, 548, A8

Rosén, L., Kochukhov, O., & Wade G. A., 2013, *MNRAS*

Magnetic Fields throughout Stellar Evolution
Proceedings IAU Symposium No. 302, 2013
P. Petit, M. Jardine & H. Spruit, eds.

doi:10.1017/S174392131400252X

Magnetic fields in single late-type giants in the Solar vicinity: How common is magnetic activity on the giant branches?

Renada Konstantinova-Antova[1,2], Michel Aurière[2], Corinne Charbonnel [3,2], Natalia Drake [4,5], Gregg Wade [6], Svetla Tsvetkova [1], Pascal Petit [2], Klaus-Peter Schröder [7] and Agnes Lèbre [8]

[1]Institute of Astronomy and NAO, BAS
email: renada@astro.bas.bg

[2]IRAP, UMR 5277, CNRS and Universitè Paul Sabatier, Toulouse, France

[3]Geneva Observatory, University of Geneva, Switzerland

[4]Sobolev Astronomical Institute, St. Petersburg State University, St. Petersburg, Russia

[5]Observatorio Nacional/MCTI, Rio de Janeiro, Brazil

[6]Department of Physics, Royal Military College of Canada, Kingston, Ontario, Canada

[7]Departamento de Astronomia, Universidad de Guanajuato, Guanajuato, Mexico

[8]LUPM - UMR 5299 - Université Montpellier II/CNRS, 34095, Montpellier, France

Abstract. We present our first results on a new sample containing all single G,K and M giants down to V = 4 mag in the Solar vicinity, suitable for spectropolarimetric (Stokes V) observations with Narval at TBL, France. For detection and measurement of the magnetic field (MF), the Least Squares Deconvolution (LSD) method was applied (Donati *et al.* 1997) that in the present case enables detection of large-scale MFs even weaker than the solar one (the typical precision of our longitudinal MF measurements is 0.1-0.2 G). The evolutionary status of the stars is determined on the basis of the evolutionary models with rotation (Lagarde *et al.* 2012; Charbonnel *et al.*, in prep.) and fundamental parameters given by Massarotti *et al.* (1998). The stars appear to be in the mass range 1-4 $M_{\odot}$, situated at different evolutionary stages after the Main Sequence (MS), up to the Asymptotic Giant Branch (AGB).

The sample contains 45 stars. Up to now, 29 stars are observed (that is about 64 % of the sample), each observed at least twice. For 2 stars in the Hertzsprung gap, one is definitely Zeeman detected. Only 5 G and K giants, situated mainly at the base of the Red Giant Branch (RGB) and in the He-burning phase are detected. Surprisingly, a lot of stars ascending towards the RGB tip and in early AGB phase are detected (8 of 13 observed stars). For all Zeeman detected stars v sin i is redetermined and appears in the interval 2-3 km/s, but few giants with MF possess larger v sin i.

Keywords. late-type giants, magnetic field, Solar vicinity

1. Introduction

Single late-type giants are an excellent laboratory to study the conditions under which dynamo could operate at different stages of the stellar evolution when significant changes in the structure of intermediate mass stars appear. Different hypotheses were suggested to explain the origin of the magnetic field and activity in giant stars: dynamo operation as a result of planet engulfment (Siess & Livio 1999) or angular momentum dredge-up from the interior (Simon & Drake 1989) for the faster rotators; remnant rotation for the more massive stars; Ap star descendants for slow rotators with a strong MF (Stepien 1993). Our study tries to give an answer what kind of dynamo that operates in RGB and

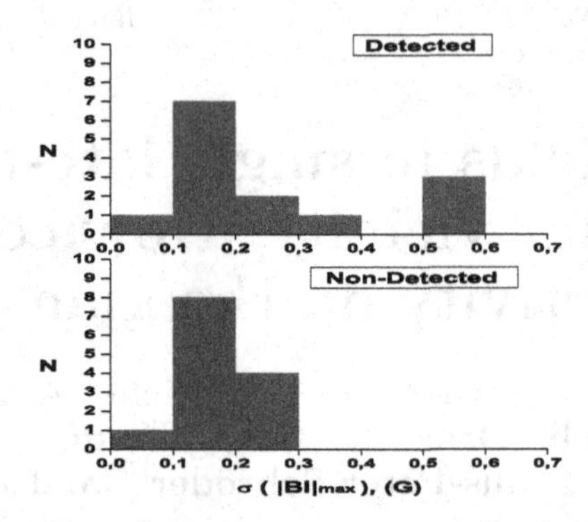

Figure 1. One σ error bars of the B_l measurements for the Zeeman detected and non-detected stars.

AGB stars, what are the properties of the magnetic activity in single giants, how long activity lasts and how common is it.

Our first sample of single G and K giants (more than 50 stars) revealed that about 50 % of these stars possess MF (Aurière *et al.*, in prep.; Konstantinova-Antova *et al.* 2013). The stars are situated mostly in the first dredge-up phase and some are in the He-burning phase. Magnetic field was detected also in stars ascending the RGB and AGB single giants (Konstantinova-Antova *et al.* 2010; Konstantinova-Antova *et al.* 2013). However, this sample was biased because of its selection was made on the basis of fast rotation and/or activity signatures.

2. Selection of the sample. Equipment and Methods.

To understand to what extent late-type giants possess MFs, a new sample entirely independent of activity was selected. Our new sample contains all single G, K and M giants up to $V = 4$ mag in the Solar vicinity, accessible for observations at Pic du Midi with Narval at TBL. These are 45 stars. Their $v \sin i$, log(L) and T_{eff} are specified by Massarotti *et al.* (1998). The stars cover the different evolutionary stages after the MS: Hertzsprung gap (5 stars), the region of the dredge-up phase and He-burning (21 stars), stars ascending the RGB and AGB (19 stars). Their observations enable a precision of Bl measurement of typically 0.2 G, using the LSD method (Donati *et al.* 1997). The stars are observed at least twice with a time interval between the observations of one month and more. For a few stars we also used data obtained in the period 2008 -2010 with Narval at TBL and its twin ESPaDOnS at CFHT.

3. First results

29 of 45 stars are observed up to now (64 % of the sample). 15 of the observed stars (52 %) are Zeemen detected. For the different evolutionary stages: Hertzsprung gap - 2 stars of 5 observed. 40 % of the stars are observed, 50 % of the observed stars detected. Base of the RGB - 14 of 21 observed. 67 % observed, 35 % of the observed stars are detected. Stars ascending towards the RGB tip and AGB stars - 13 of 19 stars observed. 68 % observed, 61 % of the observed stars are detected.

The precision of the B_l determination (Fig. 1) indicates that our aim for a deep magnetic survey is fulfilled. The distribution of error bars confirms that the non-detections are generally not due to poor quality observations.

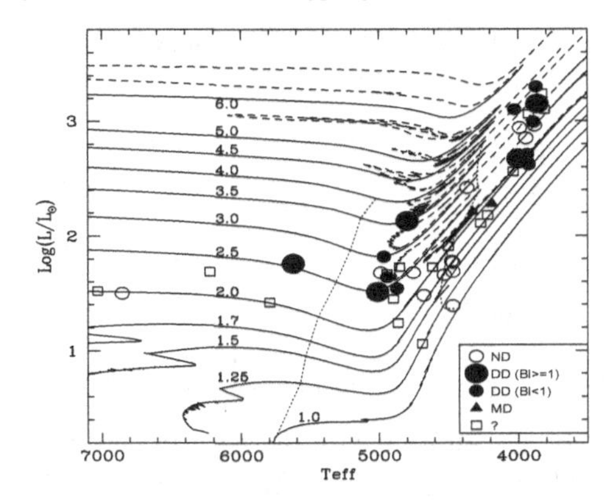

Figure 2. Situation of the studied stars on the HR diagram. Filled circles stand for detections, open ones - for non-detections, triangles - for marginal detections. Squares indicate the stars not observed yet. The dotted lines delimit the first dredge-up phase. Dashed lines designate the He-burning and AGB phases.

The situation of our sample stars on the HRD is presented in Fig. 2. Models with rotation for solar chemical composition (Lagarde *et al.* 2012; Charbonnel *et al.*, in prep.) are used. The detected stars are in the interval 1.7-4 $M_\odot$. There is a tendency for "clumping" at certain evolutionary phases: the first dredge-up and He-burning region, and the region near the upper RGB and early AGB for less massive giants, and the AGB for the more massive ones. Stars with weak and stronger MFs occupy the same regions of the HRD. Only two stars with very weak MF (and marginal detection) appear in the region between the first dredge-up and the tip of the RGB. Also detected and non-detected stars coexist in one and the same regions in the HRD.

According to Gray (2013), it appeared that the v sin i values of Massarotti *et al.* (1998) could be overestimated. We performed a redetermination of v sin i for the detected stars, using the spectrum synthesis method and taking into account the macroturbulence. Most of them have v sin i in the interval 2-3 km/s. The dependence of $|B_l|_{max}$ on v sin i is also presented (Fig. 3). There are some indications that stars with higher v sin i situated in the second "magnetic" region in the HRD possess stronger MF. However, further study is necessary on this topic, including more B_l measurements and determination of the rotation period.

Recent findings by Mosser *et al.* (2012) that core rotation slows down during the RGB phase support the idea that angular momentum dredge-up could be essential for the dynamo operation in most magnetic single giants. Further studies on the processes of angular momentum transfer between the core region and the convective envelope in giants are required.

4. Conclusions

The X-ray study by Schroder *et al.* (1998) gave some hints that all giants in the Solar vicinity could be active at least at the solar level, but the first results of our deep magnetic study do not confirm this expectation (52 % of the stars observed up to now possess MF within the limit of our precision). We study a complete sample of single G,K and M giants in the solar vicinity, selected independently of activity signatures, rotation, etc. The mass range appears to be in the interval 1-4 $M_\odot$. The sample has better coverage of the different evolutionary stages as compared to our previous one. The studied stars appear to have small v sin i, except few M giants. Stars with weak and stronger magnetic

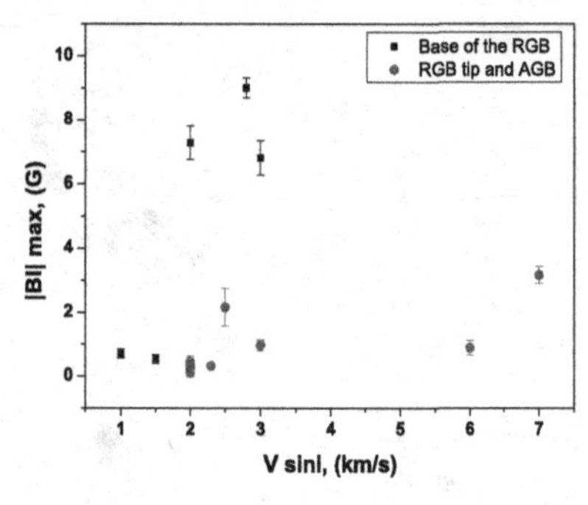

Figure 3. Dependence of $|B_l|_{max}$ on v sin i for the Zeeman detected giants. Squares stand for the region of the base of RGB and He-burning phases, and circles stand for the stars ascending towards the RGB tip and AGB stars.

field coexist in the same place on the HRD. We find no evidence for magnetic giants below 1.7 $M_\odot$. The magnetic G and K giants are at the first dredge-up and He-burning phases. A lot of stars with MF detection are located at the upper part of RGB and in the AGB, contrary to the expectations. There are some indications for a dependence of $|B_l|_{max}$ on v sin i. Is the α - ω dynamo in operation there (predictions by Nordhaus *et al.* 2008)? For our previous sample of 9 giants (for 7 of them MF was detected) we have a good correlation (Konstantinova-Antova *et al.* 2013) to suspect this. Further study of their MF variability, based on the rotation period is required to have an idea about the dynamo type, but for the moment the periods are unknown. Future works on angular momentum transport in the giants interior are also required.

5. Acknowledgements

We thank the TBL and CFHT teams for the service observing. The NARVAL observations in 2008 are granted under OPTICON programs. We also acknowledge NARVAL observations under French PNPS program and CFHT observations under a Canadian program. R. K.-A. is thankful for the possibility to work for two months in 2013 in IRAP, Tarbes as invited researcher. The work on the topic was partly supported under the program RILA/CAMPUS. N. A. D. thanks PCI/MCTI (Brazil) grant under the project 302350/2013-6. C. C. acknowledges support from the Swiss National Foundation. G. A. W. is supported by an NSERC discovery grant. R. K.-A. and S.Ts. are thankful to the TBL for the partial financial support to attend IAUS302.

References

Donati, J.-F., Semel, M., Carter, B. D. *et al.* 1997, *MNRAS*, 291, 658
Gray, D. F. 2013, *AJ*, 146, 29
Konstantinova-Antova, R., Aurière, M., Charbonnel, C. *et al.* 2010, *A&A*, 524, 57
Konstantinova-Antova, R., Aurière, M., Charbonnel, C. *et al.* 2013, *BlgAJ*, 19, 14
Lagarde, N., Decressin, T., Charbonnel, C. *et al.* 2012, *A&A*, 543, 108
Massarotti, A., Latham, D. W., Stefanik, R. P., & Fogel, J. 2008, *AJ*, 135, 209
Mosser, B., Goupil, M. J., Belkacem, K. *et al.* 2012, *A&A*, 548, 10
Nordhaus, J., Busso, M., Wasserburg, G. J., *et al.* 2008, *ApJ*, 684, L29
Schröder, K-P., Hünsch, M., & Schmitt, J. H. M. M. 1998, *A&A*, 335, 591
Siess, L. & Livio, M. 1999, *MNRAS*, 308, 1133
Simon, T. & Drake, S. A. 1989, ApJ *ApJ*, 346, 303
Stepien, K. 1993, *ApJ*, 416, 368

Magnetic Fields throughout Stellar Evolution
Proceedings IAU Symposium No. 302, 2013
P. Petit, M. Jardine & H. Spruit, eds.

doi:10.1017/S1743921314002531

Evolution of magnetic activity in intermediate-mass giants

Philippe Gondoin

European Space Agency, ESTEC,
Postbus 299, 2200 AG, the Netherlands
email: pgondoin@rssd.esa.int

Abstract. The X-ray surface fluxes of intermediate-mass G and K giants are correlated with their rotation periods and Rossby numbers. Empirical relationships are presented that accounts for the X-ray luminosity evolution of single intermediate-mass giants, such as FK Comae-type stars, and of giants in close or long-period binaries, such as RS CVn-type systems, as they evolve off the main sequence towards the top of the red giant branch.

1. Introduction

A major topic of stellar activity is to explain how magnetic phenomena depend on stellar parameters such as rotation and convection. One magnetic field diagnostic for cool stars is coronal X-ray emission. A relation between X-ray luminosity and rotation has been reported for late-type dwarfs but the connection between rotation, convection and magnetic activity is less evident among giants.

Intermediate-mass giants have early-F, A and late B-type progenitors on the main sequence that have no outer convection zones and that are typically rapid rotators (Royer *et al.* 2002). As they evolve off the main sequence, in the shell hydrogen burning stage, they develop thin outer convection zones. The increasing convection zone depth combined with fast rotation is expected to trigger dynamo processes that generate magnetic fields that, by analogy with the Sun, cause the X-ray emission of their outer stellar atmospheres.

2. Rotation-activity relationships and activity evolution on giants

A sample of intermediate-mass G and K giants with 1.5 $M_\odot \leqslant M \leqslant 3.8\ M_\odot$ and with known rotational periods was defined from a sample of single G giants with known rotation periods (Gondoin 2005), and from a list of binaries compiled by Gondoin (2007).

I found evidence that the X-ray surface flux F_X of intermediate-mass G and K giants is correlated with their rotation period P. Confidence in the degree of correlation is not higher when the Rossby number is used in place of the rotation period, but it significantly improves when stellar gravity g is taken into account (Gondoin 2007). The empirical relations are given by:

$$\log(F_X) = -0.73 \times \log(P) + 0.64 \times \log(g/g_\odot) + 7.9 \qquad (2.1)$$

$$\log(F_X) = -0.83 \times \log(Ro) + 0.75 \times \log(g/g_\odot) + 6.6 \qquad (2.2)$$

In order to estimate the X-ray luminosity evolution of single giants and of giants in long-period binary systems, I used empirical models of rotation evolution of single intermediate-mass stars (Gondoin 2005; see Fig.1 left). These rotation evolution models were combined with convection turnover times derived by Gunn *et al.* (1998) to estimate the evolution of the Rossby numbers of intermediate-mass stars during their evo-

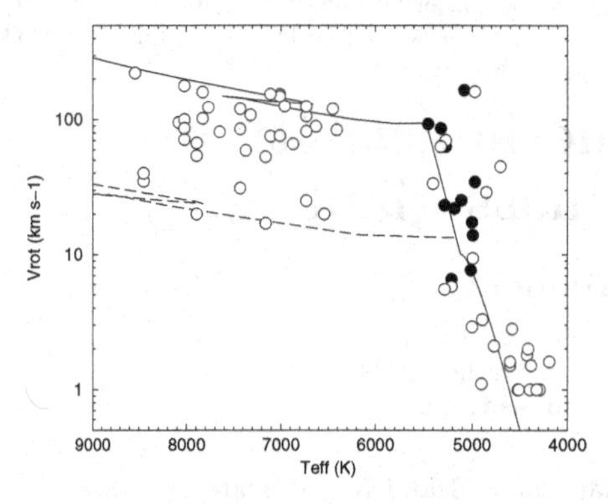

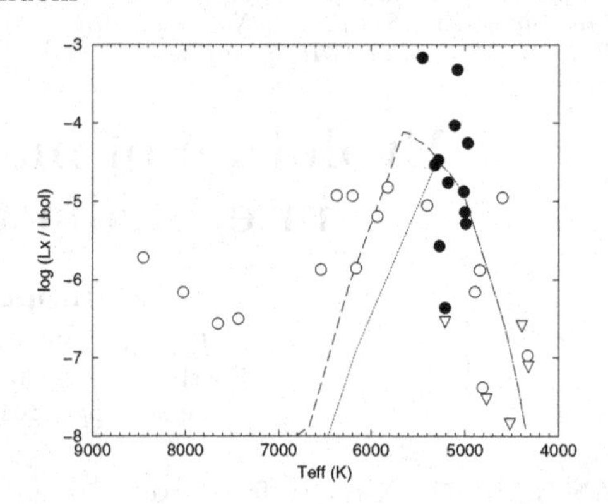

Figure 1. Left: equatorial velocity of a sample of single G giants (black circles) as a function of effective temperature. The open circles are the projected rotational velocities of A,F, G and K single field giants. The solid lines describe rotation evolution models for 2.5 $M_\odot$ (lower curve) and 2.0 $M_\odot$ stars (upper curve) with equatorial velocities of 50 and 300 km s^{-1}, respectively, on the main sequence. Right: X-ray to bolometric luminosity ratio as a function of effective temperature. The lines represent empirical models of L_X/L_{bol} evolution for 2.5 $M_\odot$ (dotted line) and 2.0 $M_\odot$ single stars (dashed line) using rotation evolution models (Gondoin 2005). The triangles represent giants for which only upper limits of the X-ray luminosity are available.

lution off the main sequence. The X-ray fluxes of these giants were then calculated from Eq. 2.2. Evolutionary models computed by Schaller *et al.* (1992) were used to determine the stellar radii and bolometric luminosities of the stars and to infer their X-ray luminosities and X-ray to bolometric luminosity ratios as a function of effective temperature.

The results (see Fig. 1 right) indicate that the X-ray luminosity of single giants, including FK Comae-type stars and giants in long-period binary systems increases by three to four orders of magnitudes between the mid-F and mid-G spectral types due to a deepening of their convection envelope and then decreases sharply as the stars ascend the red giant branch due to a strong rotational braking.

3. Conclusion

I conclude that (i) a relation exists between rotation and X-ray emission among intermediate-mass giants, that do not directly depends on the presence of a companion and that applies to all intermediate-mass giants with either G or K spectral type, (ii) this relation accounts for the large magnetic activity level of single intermediate-mass giants (such as FK Comae-type stars) as they evolve near the bottom of the red giant branch, (iii) gravity is an important parameter in determining the X-ray surface flux of giants, and (iv) a major role played by binarity in the magnetic activity level of intermediate-mass giants in close binaries (such as RS CVn systems) is to provide a mechanism that maintain rapid rotation at a late stage of stellar evolution.

References

Gondoin, P. 2005, *AA*, 444, 531

Gondoin, P. 2007, *AA*, 464, 1101

Gunn, A. G., Mitrou, C. K., & Doyle, J. G. 1998, *MNRAS*, 296, 150

Royer, F., Grenier, S., Baylac, M. O. *et al.* 2002, *AA*, 393, 897

Schaller, G., Schaerer, D., Meynet, G., & Maeder, A. 1992, *AAS*, 96, 269

Magnetic Fields throughout Stellar Evolution
Proceedings IAU Symposium No. 302, 2013
P. Petit, M. Jardine & H. Spruit, eds.

doi:10.1017/S1743921314002543

Surface differential rotation of IL Hya from time-series Doppler images

Zsolt Kővári[1], Levente Kriskovics[1], Katalin Oláh[1] Krisztián Vida[1] János Bartus[1], Klaus G. Strassmeier[2] and Michael Weber[2]

Konkoly Observatory,
[1]Konkoly Thege út 15-17., H-1121, Budapest, Hungary
email: kovari, kriskovics, olah, vida, bartus@konkoly.hu
[2] Leibniz Institute for Astrophysics Potsdam,
An der Sternwarte 16, 14482 Potsdam, Germany
email: kstrassmeier, mweber@aip.de

Abstract. We present a time-series Doppler imaging study of the K-subgiant component in the RS CVn-type binary system IL Hya ($P_{\rm orb}$ = 12.905 d). From re-processing the unique long-term spectroscopic dataset of 70 days taken in 1996/97, we perform a thorough cross-correlation analysis to derive surface differential rotation. As a result we get solar-type differential rotation with a shear value α of 0.05, in agreement with preliminary suggestions from previous attempts. A possible surface pattern of meridional circulation is also detected.

Keywords. stars: activity, stars: imaging, stars: individual (IL Hya), stars: spots, stars: late-type

1. Time-series Doppler images of IL Hya

IL Hya is a double-lined binary star (K0IV + G8V), a typical RS CVn-type system orbiting with a period of 12.905 days. Our time-series spectroscopic dataset were obtained during a 70-night long observing run at NSO in 1996/97. From that we reconstruct 30 time-series Doppler images for two favoured mapping lines (Fe I-6430 and Ca I-6439) using our image reconstruction code TEMPMAP (Rice *et al.* 1989). Adopted astrophysical parameters are listed in Table 1. As samples from the reconstructions, combined (Fe+Ca) maps are shown in Fig. 1, indicating significant changes of the spotted surface over a few rotation cycles.

2. Surface differential rotation and meridional flow

To measure surface DR we employ our method called 'ACCORD' (acronym from Average Cross-CORrelation of consecutive Doppler images), based on averaging cross-correlation function (ccf) maps of subsequent Doppler images. This way the surface differential rotation (hereafter DR) pattern in the ccf-maps could be enhanced, while

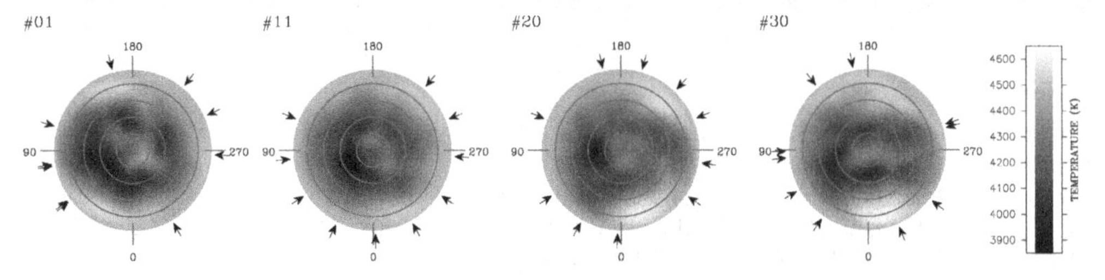

Figure 1. Time evolution of the spotted surface of IL Hya from time-series Doppler imaging.

Table 1. Astrophysical chart of IL Hya based on Weber & Strassmeier (1998)

Spectral type	K0IV (+G8V)
$\log g$	2.5±0.5
$T_{\rm eff}$ [K]	4500±250
$B-V$ [mag]	1.012±0.010
$V-I$ [mag]	0.99±0.01
Distance[a] [pc]	105.9±5.6
$v \sin i$ [km s^{-1}]	26.5±1.0
Inclination [deg]	55±5
$P_{\rm orb}$ [days]	12.905±0.004
Radius[a] [R$_\odot$]	8.1±0.9
Microturbulence [km s^{-1}]	2.0
Macroturbulence [km s^{-1}]	4.0
Chemical abundances	0.9 dex below solar
Mass [M$_\odot$]	≈2.2

[a] based on Hipparcos data

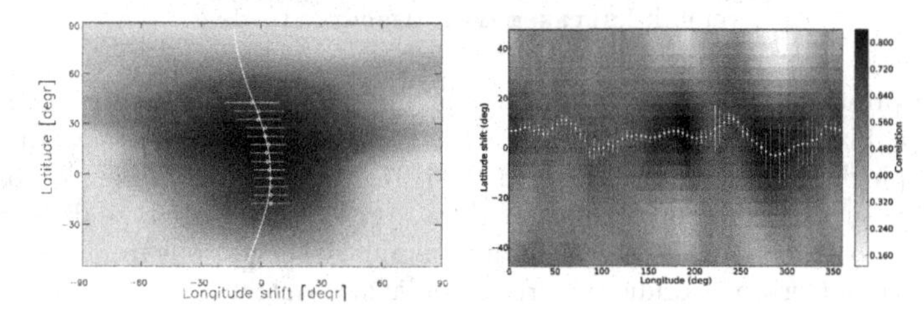

Figure 2. Averaged cross-correlations clearly reveal solar-type DR pattern (left) and common poleward drifting of spots (right).

the unwanted effect of stochastic spot changes are supressed (see Kővári *et al.* 2004, 2007 for details). Applying ACCORD yields solar-type rotation law in the form of $\Omega(\beta) = \Omega_{\rm eq} - \Delta\Omega \sin^2 \beta$ with an equatorial angular velocity $\Omega_{\rm eq}$ of 28.28 ± 0.03 deg/day and $\Delta\Omega = \Omega_{\rm eq} - \Omega_{\rm pole}$ of -1.43 ± 0.15 deg/day, corresponding with a surface shear $\alpha = \Delta\Omega/\Omega_{\rm eq}$ of 0.05 ± 0.01 (see the fitted average ccf-map in the left panel of Fig. 1). This shear is consistent with the value of $\alpha = 0.03 \pm 0.02$ derived by using a different method for a different dataset taken in 1988 (Kővári & Weber 2004). Regarding the reliability of the results read the other paper by Kővári *et al.* in this proceedings.

Latitudinal motion of spots can also be quantified by ACCORD. For this we use only the hemisphere of the visible pole. For a detailed description of the method see Kővári *et al.* (2007). The resulting latitudinal correlation pattern (right panel in Fig. 2) can be converted into an average poleward surface velocity field of 0.5 ± 0.1 km/s, that could be interpreted as the surface pattern of a single-cell meridional circulation.

Acknowledgements

This work has been supported by the Hungarian Science Research Program OTKA K-81421, the Lendület-2009 and Lendület-2012 Young Researchers' Programs of the Hungarian Academy of Sciences and by the HUMAN MB08C 81013 grant of the MAG Zrt.

References

Kővári, Zs., Strassmeier, K. G., Granzer, T., *et al.* 2004, *A&A*, 417, 1047
Kővári, Zs. & Weber, M. 2004, *PADEU* Vol. 14, 221
Kővári, Zs., Bartus, J., Strassmeier, K. G., *et al.* 2007, *A&A*, 474, 165
Rice, J. B., Wehlau, W. H., & Khokhlova, V. L. 1989, *A&A*, 208, 179
Weber, M. & Strassmeier, K. G. 1998, *A&A*, 330, 1029

Magnetic Fields throughout Stellar Evolution
Proceedings IAU Symposium No. 302, 2013
P. Petit, M. Jardine & H. Spruit, eds.

doi:10.1017/S1743921314002555

Magnetic field of the classical Cepheid η Aql: new results

V. Butkovskaya[1], S. Plachinda[1], D. Baklanova[1] and V. Butkovskyi[2]

[1]Crimean Astrophysical Observatory of Taras Shevchenko National University of Kyiv, 98409, Nauchny, Crimea, Ukraine,
email: varya@crao.crimea.ua

[2]Taurida National V. I.Vernadsky University, 95007, Vernadskogo str. 4, Simferopol, Crimea, Ukraine

Abstract. We present the results of the spectropolarimetric study of the classical Cepheid η Aql in 2002, 2004, 2010, and 2012. The longitudinal magnetic field of η Aql was found to be variable with the pulsation cycle of 7.176726 day. The amplitude, phase, and mean value of the field vary from year to year presumably due to stellar rotation or dynamo mechanisms.

Keywords. stars: magnetic fields, stars: oscillations, stars: individual (η Aql)

1. Introduction

Historically the first photoelectric magnetometer observations of η Aql have been performed by Borra *et al.* (1981), and Borra *et al.* (1984). These studies detected no magnetic field on the star with a minimal uncertainty of 7.7 G. The presence of a magnetic field on η Aql was firstly reported by Plachinda (2000), who found that in 1991 the longitudinal component of the magnetic field was variable from −100 to +50 G during pulsation cycle. Wade *et al.* (2002) detected no convincing evidence of a photospheric magnetic field on η Aql during three nights in 2001. They phased their magnetic field values with the pulsation period and noted that around phase 0.8 their results are strongly inconsistent with those of Plachinda (2000). The authors concluded that η Aql is a non-magnetic star, at least at the level of 10 G. More recently, Grunhut *et al.* (2010) detected clear Zeeman signatures in Stokes V for 9 later-type supergiants (including η Aql), confirming the earlier result of Plachinda (2005), who detected the longitudinal magnetic field of two yellow supergiants - ϵ Gem and ϵ Peg. While the presence of a magnetic field on later-type supergiants (including pulsating classical Cepheids) can therefore be considered proven, a doubt on a variability of the magnetic field with the stellar pulsations is shared yet by many astronomers. We present here the new results of the spectropolarimetric observations of η Aql.

2. Observations

Spectropolarimetric observations of η Aql have been performed in the spectral region 6210 - 6270 Å during 60 nights from 2002 to 2012 using the coude spectrograph at the 2.6-m Shajn telescope of the Crimean Astrophysical Observatory (CrAO, Ukraine). The longitudinal magnetic field was calculated using the technique described in detail by Butkovskaya & Plachinda (2007). The technique adopted for spectropolarimetric measurements and magnetic field calculation allows us to exclude spurious magnetic signals produced by variations of spectral line profiles due to stellar pulsations.

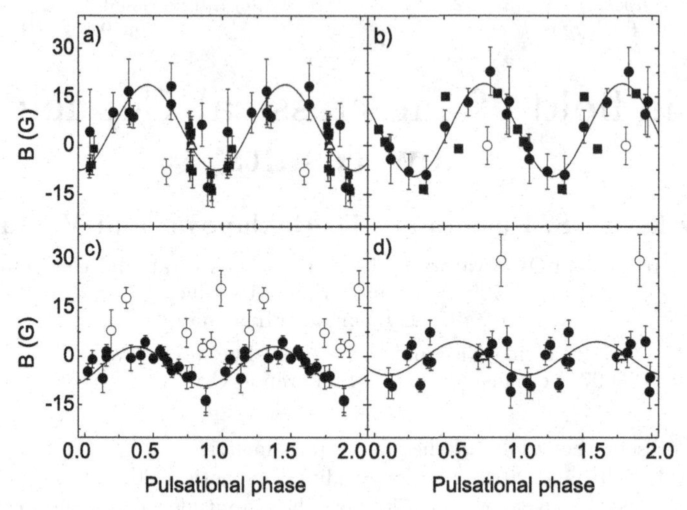

Figure 1. Longitudinal magnetic field of η Aql folded in phase with the 7.176726 day pulsation period: **a)** *closed and open circles*: our data obtained in 2002, *black squares*: data by Wade *et al.* (2002), *open triangles*: data by Grunhut *et al.* (2010); **b)** *closed and open circles*: our data obtained in 2004, *black squares*: data by Borra *et al.* (1981), and Borra *et al.* (1984), (mean $\sigma =$ 13 G); **c)** *closed and open circles*: our data obtained in 2010; **d)** *closed and open circles*: our data obtained in 2012, and *open triangles*: data by Grunhut *et al.* (2010). Full lines represent fitting sinusoids. Open circles represent data that have not been taken into account for the fits. The reason for the deviation from the common curve is unknown (e.g. shock waves or buoyantly rising magnetic flux tubes).

3. Results

The pulsation modulation of the longitudinal magnetic field of η Aql in different years is illustrated in Fig. 1. Our data are supplemented with data of Borra *et al.* (1981), Borra *et al.* (1984), Wade *et al.* (2002), Grunhut *et al.* (2010) All data are folded in phase according to the pulsation ephemeris $JD = 2450100.861 + 7.176726E$ by Kiss & Vinko (2000).

The pulsation modulation of the longitudinal magnetic field of η Aql is not unique: Butkovskaya & Plachinda (2007) discovered variability of the magnetic field of γ Peg (B2 IV) with the radial pulsation period and described possible origins of this variability.

References

Borra, E. F., Fletcher, J. M., & Poeckert, R. 1981, *ApJ*, 247, 569
Borra, E. F., Edwards, G., & Mayor, M. 1984, *ApJ*, 284, 211
Butkovskaya, V. & Plachinda, S. 2007, *A&A*, 469, 1069
Grunhut, J. H., Wade, G. A., Hanes, D. A., & Alecian, E. 2010, *MNRAS*, 408, 2290
Kiss, L. L. & Vinko, J. 2000, *MNRAS*, 314, 420
Plachinda, S. I. 2000, *A&A*, 360, 642
Plachinda, S. I. 2005, *Astrophysics*, 48, 9
Wade, G. A., Chadid, M., Shorlin, S. L. S.., Bagnulo, S., & and Weiss, W. W. 2002, *A&A*, 392, L17–L20

Magnetic Fields throughout Stellar Evolution
Proceedings IAU Symposium No. 302, 2013
P. Petit, M. Jardine & H. Spruit, eds.

doi:10.1017/S1743921314002567

Activity on a Li-rich giant: DI Psc revisited

Levente Kriskovics, Zsolt Kővári, Krisztián Vida and Katalin Oláh

Konkoly Observatory,
Konkoly Thege út 15-17., H-1121, Budapest, Hungary
email: `kriskovics`, `kovari`, `vidakris`, `olah@konkoly.hu`

Abstract. We present a new Doppler imaging study for the Li-rich single K-giant DI Psc. Surface temperature maps are reconstructed for two subsequent rotation cycles. From the time evolution of the spot distribution antisolar-type differential rotation pattern is revealed. We show marks of non-uniform Li-abundance as well. The possible connection between the current evolutionary phase of the star and its magnetic activity is briefly discussed.

Keywords. stars: activity, stars: imaging, stars: individual (DI Psc), stars: spots, stars: late-type

1. Introduction

DI Psc (HD 217352) is a rapidly rotating ($P_{\rm rot}$ = 18.07 days) single K-giant, a new candidate for the small group of Li-rich K-giant stars. The extreme Li-abundance is related to a short evolutionary episode, the helium flash, when different (partly unknown) processes activate Li-production and propagation (Charbonnel & Balachandran 2000). Moreover, the role of rotation and magnetic activity in these processes is also unclear. Surface Li-abundance and the position of DI Psc on the HRD was determined in our recent Doppler imaging study (Kővári *et al.* 2013). In this paper we aim to investigate the time evolution of the surface by Doppler imaging, as well as the Li-distribution on the surface.

2. Doppler imaging results

15 time series spectra of exceptionally high signal-to-noise ratio were taken with NARVAL@TBL in Nov-Dec 2012, covering 40 days (i.e., $\approx 2 \times P_{\rm rot}$). We use the Doppler imaging code `TempMap` by Rice *et al.* (1989) for three lines (Fe I-6430, Ca I-6439 and Li I-6708) to reconstruct the stellar surface in two consecutive rotational cycles. Combined (Fe+Ca+Li) images, as well as their cross-correlation is plotted in Fig. 1. The best fit correlation pattern suggests antisolar-type surface differential rotation with a shear of $\alpha = -0.11 \pm 0.02$. This result should be regarded as a preliminary one, since unexpected rearrangements in spot configuration (e.g., emerging new flux) can disturb seriously the

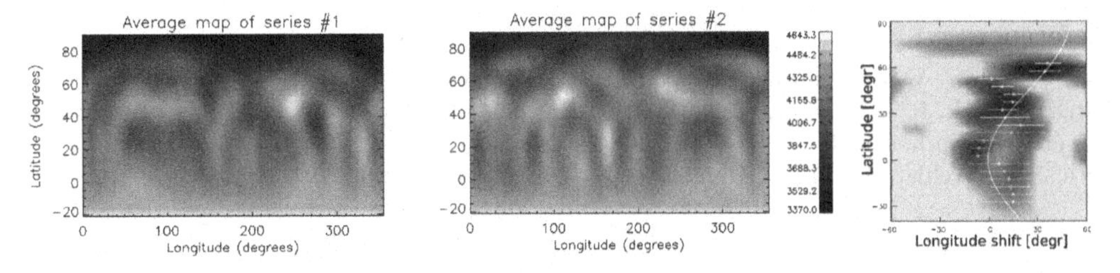

Figure 1. Combined Doppler images for two consecutive rotation periods (left) in Nov-Dec 2012 and the fitted antisolar-type differential rotation on the cross-correlation function map (right).

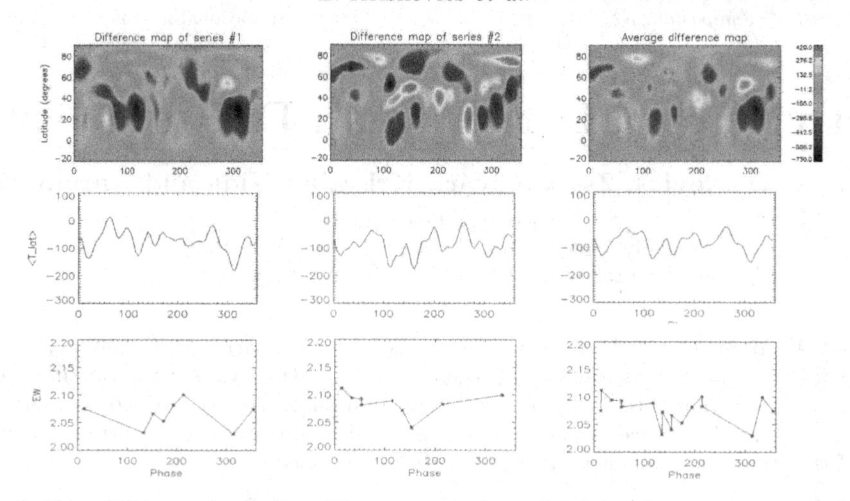

Figure 2. Top: difference maps (i.e., Li-maps are subtracted from the average maps shown in Fig 1 left). Note the similar structure for both rotational cycles. Middle: longitudinal distribution of the mean latitudinal temperature. Bottom: rotational modulation of the Li I-6708 equivalent width.

pure differential rotation pattern, thus yielding false observation. We note, however, that antisolar-type differential rotation was reported also for DP CVn, which is another Li-rich K-giant, the twin of DI Psc (Kővári *et al.* 2013). High Li-abundance might be related to strong meridional circulation (i.e., extra mixing), which, on the other hand, can eventuate antisolar-type differential rotation (cf. Kitchatinov & Rüdiger 2004).

Non-uniform surface Li-abundance would affect the Doppler maps by altering the strength of the local line profiles (the temperature inversion code assumes constant abundance, therefore a higher Li equivalent width is fitted with a lower temperature causing a false cool spot on the map). In order to investigate this behaviour we subtract the Li-maps from the combined (Fe+Ca+Li) ones. The resulting difference maps are plotted in Fig. 2 top, where signs of similar non-uniformity can be marked in both rotation cycles. This is supported by the longitudinal distribution of the mean latitudinal temperature values, as well as by the rotational modulation of the Li I-6708 equivalent width (middle and bottom panels of Fig. 2, respectively). We conclude that DI Psc is an optimal target for studying the connection between surface activity, rotation and surface Li-enrichment.

Acknowledgements

This work has been supported by the Hungarian Science Research Program OTKA K-81421, the Lendület-2009 and Lendület-2012 Young Researchers' Programs of the Hungarian Academy of Sciences and by the HUMAN MB08C 81013 grant of the MAG Zrt.

References

Charbonnel, C. & Balachadran, S. 2000, *A&A*, 309, 563
Kitchatinov, L. L. & Rüdiger, G. 2004, *AN*, 325., 496
Kővári, Zs., Korhonen, H., Strassmeier, K. G., *et al.* 2013, *A&A*, 551., 2
Rice, J. B., Wehlau, W. H., & Khokhlova, V. L. 1989, *A&A*, 208, 179

Magnetic Fields throughout Stellar Evolution
Proceedings IAU Symposium No. 302, 2013
P. Petit, M. Jardine & H. Spruit, eds.

doi:10.1017/S1743921314002579

Search for surface magnetic fields in Mira stars: first results on χ Cyg

Agnès Lèbre[1], Michel Aurière[2], Nicolas Fabas[3], Denis Gillet[4], Fabrice Herpin[5], Pascal Petit[2] and Renada Konstantinova-Antova[6]

[1]Laboratoire Univers et Particules de Montpellier, CNRS & Université de Montpellier, France
email: Agnes.Lebre@univ-montp2.fr

[2]IRAP, CNRS & Université de Toulouse, Observatoire Midi-Pyrénées, Toulouse, France

[3]Instituto de Astrofísica de Canarias & Universidad de la Laguna, Tenerife, Spain

[4]Observatoire de Haute Provence, OSU Pythéas, CNRS & Aix-Marseille Université, France

[5]Laboratoire d'Astronomie de Bordeaux, OASU, CNRS & Université de Bordeaux, France

[6]Institute of Astronomy and NAO, Bulgarian Academy of Sciences, Sofia, Bulgaria

Abstract. So far, surface magnetic fields have never been reported on Mira stars, while observational facilities allowing detection and measurement of weak surface fields through the Zeeman effect have become available. Then, in order to complete the knowledge of the magnetic field and of its influence during the transition from Asymptotic Giant Branch (AGB) to Planetary Nebulae (PN) stages, we have undertaken a search for magnetic fields at the surface of Miras. We present the first spectropolarimetric observations (performed with the Narval instrument at Télescope Bernard Lyot-TBL, Pic du Midi, France) of the S-type Mira star χ Cyg. We have detected a polarimetric signal in the Stokes V spectra and we have established its Zeeman origin. We claim that it is likely to be related to a weak magnetic field present at the photospheric level and in the lower part of the stellar atmosphere. The origin of this magnetic field is discussed in the framework of shock waves periodically propagating throughout the atmosphere of a Mira.

Keywords. Stars: AGB and post-AGB, stars: atmospheres, Magnetic fields, shock waves

1. Introduction

Miras are cool and evolved pulsating stars, belonging to the AGB, the key evolutionary stage of an intermediate mass star before its transition toward the PN stage. These evolved stars are known to be among the main recycling agents of the interstellar medium as they undergo a prodigious mass loss supposedly mainly driven by radiation pressure combined to other factors (magnetism, pulsation) whose respective contributions are not yet fully identified. For Mira stars a radiative shock wave triggered by the pulsation mechanism and propagating periodically throughout the extended stellar atmosphere may help to form and to enrich a circumstellar envelope (hereafter CSE). The morphology of the CSE of an AGB star is known to severely change during its quick transition toward PN (Sahai & Trauger, 1998). The classical or generalized Interacting Stellar Winds model (Kwok, 2000) tries to explain this shaping, but does not fully address the origin of the wind as it fails in reproducing the observed complex structures (such as jets or ansae). Then binarity and/or magnetic fields have often been invoked in order to rule the mass loss geometry and to shape PN's morphology (Soker, 2006; Blackman, 2009). The presence of a weak magnetic field at the surface of Miras is also expected from theoretical considerations : the ejection of massive winds by an AGB could be triggered by a magnetic activity in the degenerated core, leading to a toroidal field configuration of a few 10 G at the stellar surface (Pascoli, 1997).

Observational evidences for magnetic fields around PN and their AGB and post-AGB progenitors have been recently established (Sabin *et al.*, 2007; Vlemmings, 2011). Magnetic fields have been detected and measured within the CSE of AGBs through the polarization from the maser emission of several molecules (OH, H_2O and SiO), located at different distances from the central star. The current status on magnetic field, $B_{//}$ (mean value of the intensity of the magnetic field along the line of sight) throughout the CSE of an AGB star is the following (with one stellar radius $R_* \sim 1$ au):

- $B_{//} \sim$ 5-20 mG, from OH masers, at 1000-10000 au (Rudnitski *et al.*, 2010),
- $B_{//} \sim$ a few 100 mG, from H_2O masers, at a few 100 au (Leal-Ferreira *et al.*, 2013)
- $B_{//} \sim$ a few G (mean value = 3.5 G), from SiO masers, at 5-10 au (Herpin *et al.*, 2006), the innermost detections of magnetic fields within the environment of Miras.

Combining those values (and few detections reported in the very outer regions of the CSE of carbon-rich AGBs obtained from the Zeeman effect in CN line emission, Herpin *et al.*, 2009), Vlemmings (2011) has presented the behavior of magnetic field strength within the environment of AGBs. Throughout the CSE of an AGB, the magnetic field strength presents a clear decrease along r (distance to the central object) favoring a 1/r variation law. Extrapolating this law, for a toroidal field configuration, one can expect a magnetic field strength of the order of few G at a Mira's photosphere. Its detection and characterization are hence possible with a spectropolarimeter like Narval at TBL, as weak fields ($B_{//}$ between 1 to 10 G) have been detected at the surface of cool giants and AGBs with spectral types from K0III to M5III (Aurière *et al.*, 2009; Konstantinova-Antova *et al.*, 2010; Konstantinova-Antova *et al.*, contribution in this volume) and also at the surface of the red supergiant (RSG) Betelgeuse (Aurière *et al.*, 2010).

2. Narval observations of χ Cyg : detection of a Stokes V signature

The S type bright Mira star χ Cyg has a pulsation period of about 408 days. Herpin *et al.*, (2006) have detected a magnetic field (from SiO maser emission) in the inner part of its CSE, reporting a mean value of 5 G at a few stellar radii.

In March 2012, we have performed spectropolarimetric observations of χ Cyg, observed around its maximum light. We have collected 174 Stokes V sequences (circular polarization data) with Narval at TBL. We have performed a Least Squares Deconvolution (LSD) analysis (Donati *et al.*, 1997) onto the combination of these V spectra, using a numeric mask dedicated to a cool and evolved star ($T_{\rm eff}$ = 3 500 K, $\log g$ = 0.5, microturbulence ξ = 2 km.s^{-1} , solar abondances) and involving $\sim$ 14 000 metallic lines (distributed from 380 to 1 000 nm) with atomic parameters and Landé factors taken from the VALD database. In Fig. 1, we report the detection of a statistically significant signal in the averaged LSD Stokes V profile (with a 10^{-5} amplitude level). The Stokes I profile (unpolarized spectrum) is also presented as well as the Null profile (N) diagnosis. The Null profile remains flat within the line, confirming the definite detection in Stokes V. The LSD Stokes I profile presents the typical doubling of metallic lines (Schwarzschild, 1952) due to the presence of a radiative shock wave imprinting complex ballistic motions in the lower part of the stellar atmosphere (Gillet *et al.*, 1983).

3. Physical origin of the detected Stokes V signature

A magnetic origin. We have performed several tests on the combination of the 174 Stokes V spectra, in order to confirm its stellar origin, and we have found that indeed spurious or instrumental effects (crosstalk) do not contribute to its occurence. Moreover, we have also performed LSD analyses using two different numeric masks (both issued from the initial one) and gathering atomic lines with high or low Landé factor values (cut off value = 1.2). In Fig. 2, we present the resulting LSD Stokes V profiles of these analyses, revealing a definite detection when using the numeric mask gathering lines with high Landé factor values. The clear Stokes V signature obtained with this sub mask presents the same structure, amplitude and location than the one obtained when using the initial mask (Fig. 1). This behavior confirms a magnetic Zeeman effect origin for the V signal detected on χ Cyg and it represents the first detection - at the stellar surface of a Mira star - of a weak magnetic field. We have thus computed the longitudinal magnetic field ($B_{//}$) using the first-order moment method (Donati *et al.*, 1997) adapted to the obtained mean LSD I and V profiles : $B_{//} = -0.25 \pm 0.40$ G.

A link with the periodic radiative shock wave. However, the estimation of $B_{//}$ has been made using the complete width of the Stokes I profile, and it could thus be considered as a lower limit. Indeed the Stokes V profile appears to be associated to the blue component of the LSD Stokes I profile (Fig. 1), i.e., to the material which is driven outward by the radiative shock wave. This suggests that the shock may have an impact onto a surface stellar magnetism, likely producing a compressive amplification of a very weak field. Thus we have performed a direct scaling of the Stokes V signal and the blue component of the I profile of χ Cyg, to the classical Zeeman profile obtained in the (non-pulsating) K0III star Pollux (Aurière *et al.*, 2009). Then the Stokes V signal we report for χ Cyg would correspond to the detection of a surface magnetic field with $B_{//}$ of a few Gauss (2-3 G). This result is in good agreement with the measurements performed in the inner part of the CSE of χ Cyg. It also favors a decreasing law throughout the stellar environment in 1/r (associated to a toroidal field configuration). As already reported for the RSG Betelgeuse (Dorch, 2004 ; Petit *et al.*, 2013), we suggest that a local dynamo powered by the convection and/or the atmospheric dynamics may be a likely explanation for the surface magnetism we have detected in χ Cyg, the propagation of the radiative shock wave probably inducing a compressive amplification on a weak surface field.

Further perspectives for this work rely on monitoring Miras around their maximum light and over a large part of their pulsation cycle, so as to precise the relation between

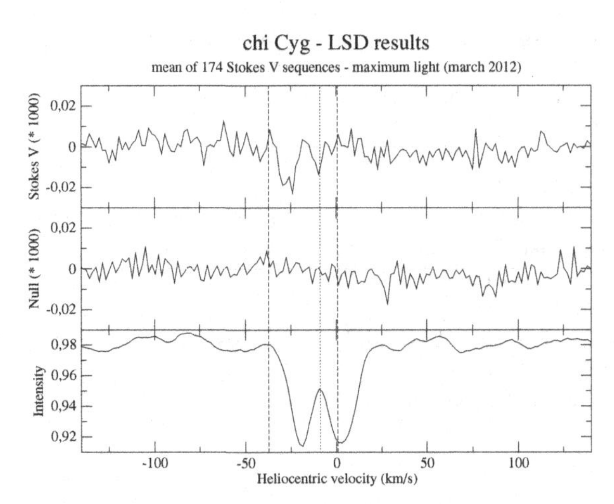

Figure 1. LSD profiles of χ Cyg (in heliocentric velocity) from Narval observations collected in March 2012 (average of 174 V sequences). *Bottom:* Stokes I (unpolarized spectrum). *Middle:* The Null profile (extended by a factor of 1 000). *Top:* Stokes V profile (extended by a factor of 1 000). The vertical dashed lines delimitate the blue line component of the Stokes I profile.

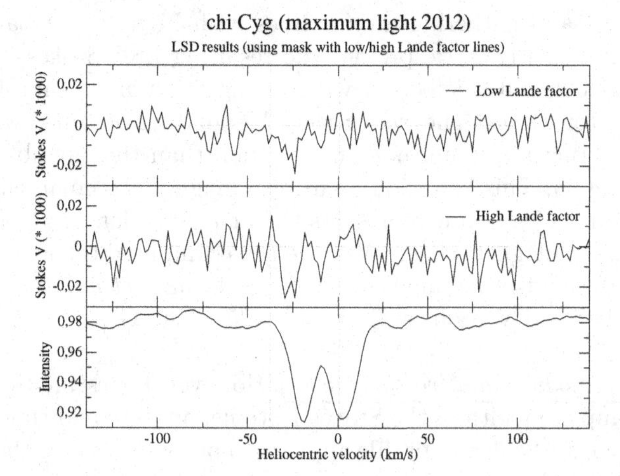

Figure 2. LSD Stokes I and V profiles of χ Cyg (in heliocentric velocity). *Bottom:* LSD Stokes I profile (see Fig. 1). *Middle:* LSD Stokes V profile obtained with a numeric mask gathering atomic lines with high Landé factor values (mean value = 0.9). *Top:* LSD Stokes V profile obtained with a numeric mask gathering atomic lines with low Landé factor values (mean value = 1.5).

the atmospheric shock wave and the stellar magnetism. The sub gauss level detection is required for such observations, and in a next future, SPIRou at CFHT will be very useful for the detection and the investigation of surface magnetic fields in cool and evolved stars.

References

Aurière, M., Wade, G., Konstantinova-Antova, R., *et al.*, 2009, *A&A* 504, 231
Aurière, M., Donati, J. F., & Konstantinova-Antova, R., 2010, *A&A* 516, L2
Blackman, E. G., 2009, in "Cosmic Magnetic Fields", eds. Strassmeier *et al.*, *IAU Symp.* 259, 35
Donati, J. F., Semel, M., Carter, B., Reed, D., & Collier Cameron, A., 1997, *MNRAS* 291, 658
Dorch, S. B. F., 2004, *A&A*, 423, 1101
Gillet, D., Maurice, E., & Baade, D., 1983, *A&A* 128, 384
Herpin, F., Baudry, A., Thum, C., Morris, D., & Wiesemeyer, H., 2006, *A&A* 450; 667
Herpin, F., Baudry, A., Josselin, E., Thum, C., & Wiesemeyer, H., 2009, in "Cosmic Magnetic Fields", eds. Strassmeier *et al.*, *IAU Symp.* 259, 47
Konstantinova-Antova, R., Aurière, M., Charbonnel, C., *et al.*, 2010, *A&A*, 524, A57
Kwok, S., 2000, *ASPC* 199, 9
Leal-Ferreira, M. L., Vlemmings, W. H. T., Kemball, A., & Amiri, N., 2013, *A&A*, 554, A134
Pascoli, G., 1997, *ApJ* 489, 946
Petit, P., Aurière, M. , Konstantinova-Antova, R., Morgenthaler, M., *et al.*, 2013 *LNP* 857, 231
Rudnitski, G. M., Paschenko, M. I., & Colom, P., 2010, Astronomy Reports, Vol. 54, p. 400
Sabin, L., Zijlstra, A. A., & Greaves, J. S., 2007, *MNRAS* 376, 378
Sahai, R. & Trauger, J. T., 1998, *AJ* 116, 1357
Schwarzschild, M., 1952, Transact. IAU VIII, ed. P. Oosterhoff, Cambridge Univ. Press, p 811
Soker, N., 2006, *PASP* 118, 260
Vlemmings, W. H. T., 2011, Proceedings of AsPNe Conf., eds. A. Zijlstra *et al.*, p 91

Magnetic Fields throughout Stellar Evolution
Proceedings IAU Symposium No. 302, 2013
P. Petit, M. Jardine & H. Spruit, eds.

doi:10.1017/S1743921314002580

Magnetic fields around AGB stars and Planetary Nebulae

W. H. T. Vlemmings[1]

[1] Department of Earth and Space Sciences, Chalmers University of Technology, Onsala Space Observatory, SE-439 92 Onsala, Sweden

Abstract. Stars with a mass up to a few solar masses are one of the main contributors to the enrichment of the interstellar medium in dust and heavy elements. However, while significant progress has been made, the process of the mass-loss responsible for this enrichment is still not exactly known and forces beyond radiation pressure might be required. Often, the mass lost in the last phases of the stars life will become a spectacular planetary nebula. The shaping process of often strongly a-spherical PNe is equally elusive. Both binaries and magnetic fields have been suggested to be possible agents although a combination of both might also be a natural explanation.

Here I review the current evidence for magnetic fields around AGB and post-AGB stars pre-Planetary Nebulae and PNe themselves. Magnetic fields appear to be ubiquitous in the envelopes of apparently single stars, challenging current ideas on its origin, although we have found that binary companions could easily be hidden from view. There are also strong indications of magnetically collimated outflows from post-AGB/pre-PNe objects supporting a significant role in shaping the circumstellar envelope.

Keywords. polarization, magnetic fields, polarization, stars-AGB and post-AGB, planetary nebulae

1. Introduction

After the AGB phase, the stellar envelopes undergo a major modification as they evolve to Planetary Nebulae (PNe). The standard assumption is that the initial slow AGB mass loss quickly changes into a fast superwind, generating shocks and accelerating the surrounding envelope (Kwok *et al.* 1978). It is during this phase that the typically spherical circumstellar envelope evolves into a Planetary Nebula. As the majority of PNe are aspherical, an additional mechanism is needed to explain the departure from sphericity. Specifically the discovery that the collimated outflows of the pre-PNe (P-PNe), where such outflows are common, have a momentum that exceeds that which can be supplied by radiation pressure alone (Bujarrabal *et al.* 2001), has led to a revision of the standard theory. The formation mechanism of in particular bipolar PNe is still a matter of fierce debate. Current theories to explain the PNe shapes include binaries, disks, magnetic fields or a combination of these. A promising mechanism could be a binary companion or massive planet that helps maintain a strong magnetic field capable of shaping the outflow (e.g. Nordhaus *et al.* 2007). However, the known fraction of binary systems cannot yet explain the large number of aspherical PNe. Since the shaping mechanism is likely related to the mass loss mechanism responsible for the enrichment of the interstellar medium, a better understanding of this mechanism is crucial.

Here I will review the observational evidence for strong magnetic fields in PNe as well as around their AGB and post-AGB progenitors. I will give an overview of the methods that can be used to study magnetic fields, especially in light of the plethora of new instruments that will be available shortly. Finally, I will discuss a number of questions

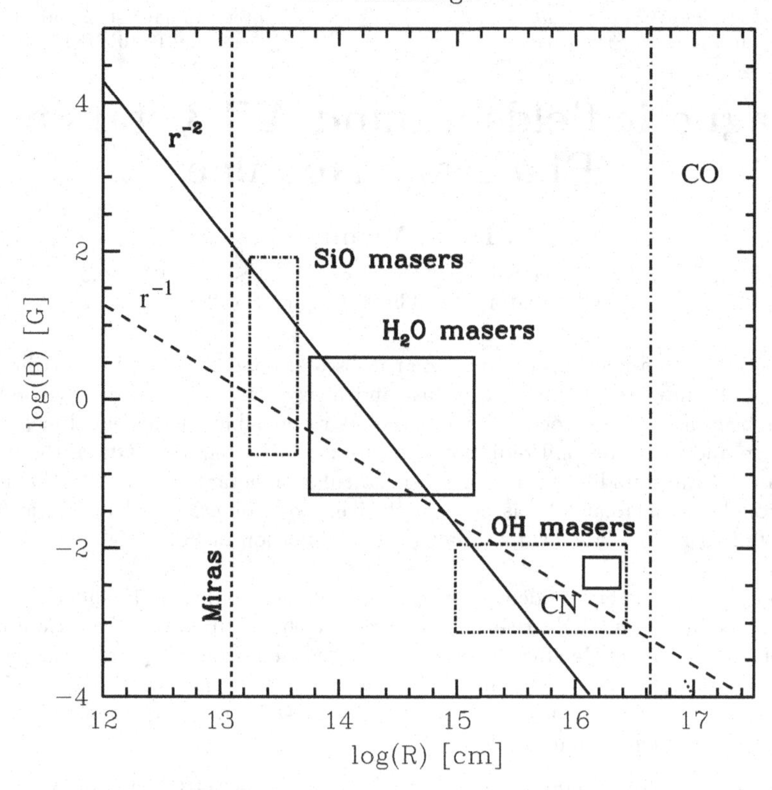

Figure 1. Magnetic field strength vs. radius relation as indicated by current maser polarization observation of a number of Mira stars. The boxes show the range of observed magnetic field strengths derived from the observations of SiO masers (Kemball *et al.* 2009, Herpin *et al.* 2006), H_2O masers (Vlemmings *et al.* 2002, Vlemmings *et al.* 2005, Leal-Ferreira *et al.* 2013), OH masers (e.g. Rudnitski *et al.* 2010) and CN (Herpin *et al.* 2009). The thick solid and dashed lines indicate an r^{-2} solar-type and r^{-1} toroidal magnetic field configuration. The vertical dashed line indicates the stellar surface. ALMA CO polarization observations will uniquely probe the outer edge of the envelope (vertical dashed dotted line).

related this topic that we can expect to be answered with the new instruments in the next few years.

2. Observational Techniques - Polarization

With the exception of observations where the magnetic field strength is estimated assuming forms of energy equilibrium, such as synchrotron observations, the magnetic field strength and structure is typically determined from polarization observations.

2.1. *Circular Polarization*

Circular polarization, generated through Zeeman splitting, can be used to measure the magnetic field strength. It measures the total field strength when the splitting is large and the line-of-sight component of the field when the splitting is small. The predominant source of magnetic field strength information during the late stages of stellar evolution comes from maser circular polarization observations, and particularly the common SiO, H_2O and OH masers. These can show circular polarization fractions ranging from $\sim 0.1\%$ (H_2O) up to $\sim 100\%$ (OH) and are, because of their compactness and strength, excellent sources to be observed with high angular resolution. Unfortunately, the analysis of maser

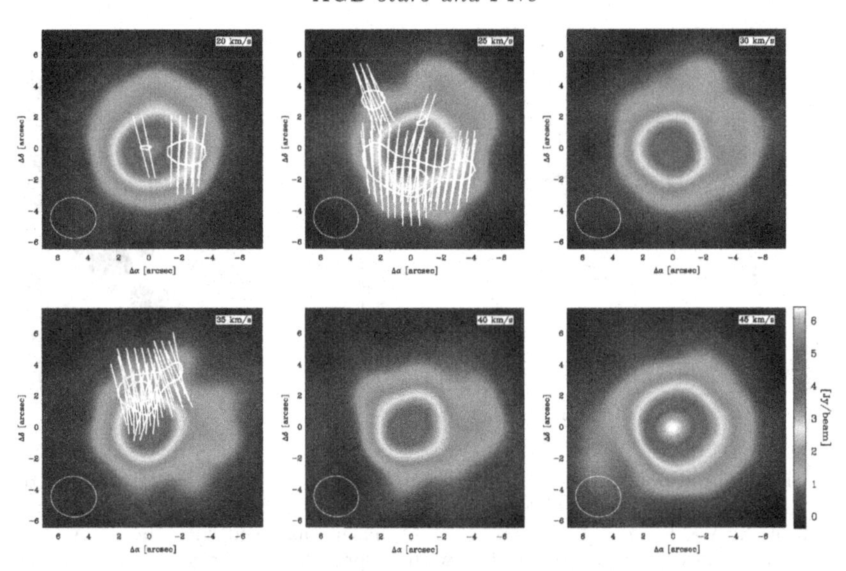

Figure 2. From Vlemmings *et al.* (2012), the polarization of the CO(2-1) line in the circumstellar envelope of IK Tau. The colors are the CO emission and the contours the linearly polarized intensity. The line segments indicate the electric vector polarization angle.

polarization is not straightforward (for a review, see Vlemmings (2012a)), and it has taken a long time before maser observations were acknowledged to provide accurate magnetic field measurements. More recently, the first attempts have been made to detect the Zeeman splitting of non-maser molecular lines, such as CN (Herpin *et al.* 2009). As many of these occur at shorter wavelength in the (sub-)mm regime, the advent of the Atacama Large (sub-)Millimeter Array will further enhance these types of studies.

2.2. *Linear Polarization*

Linear polarization, probing the structure of the plane-of-the-sky component of the magnetic field, can be observed both in the dust (through aligned grains) and molecular lines (through radiation anisotropy - the Goldreich-Kylafis effect). The Goldreich-Kylafis effect on CO has only recently been mapped for the first time in the envelope of evolved stars and will, with ALMA, allow for a systematic study of magnetic fields (Vlemmings *et al.* 2012b). Typical percentages of linear polarization range from up to a few percent (e.g. dust, CO, H_2O masers) to several tens of percent (OH and SiO masers). Again the interpretation of maser polarization depends on a number of intrinsic maser properties, but in specific instances maser linear polarization can even be used to determine the full 3-dimensional field morphology. In addition to the geometry, the linear polarization of most notably dust, can also be used to obtain a value for the strength of the plane-of-the-sky component of the magnetic field. This is done using the Chandrasekhar-Fermi method, which refers to the relation between the turbulence induced scatter of polarization vectors and the magnetic field strength.

3. Current Status - Evolved Star Magnetic Fields

3.1. *AGB Stars*

Most AGB magnetic field measurements come from maser polarization observations (SiO, H_2O and OH). These have revealed a strong magnetic field throughout the circumstellar envelope. In Figure 1, I have indicated the magnetic field strength in the regions of the

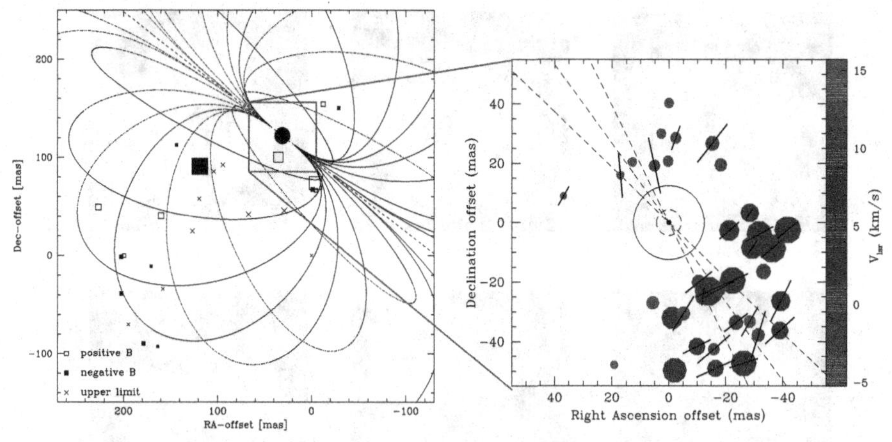

Figure 3. (left) The dipole magnetic field of the supergiant VX Sgr as determined from a fit to the H_2O maser magnetic field observations (Vlemmings *et al.* 2005). (right) Positions and polarization of the VX Sgr $v = 1, J = 5 - 4$ ^{28}SiO masers observed with the SMA (Vlemmings *et al.* 2011). The masers spots are plotted with respect to a weighted mean position of all maser spots. The black vectors are the observed polarization vectors scaled linearly according to polarization fraction. The long dashed inner circle indicates the star and the solid circle indicates the location of the 43 GHz SiO masers. The short dashed circle indicates the minimum radius of the ^{28}SiO masers. The dashed lines indicate the position angle and its uncertainty of the inferred orientation of the dipole magnetic field of VX Sgr observed using H_2O and OH masers (Szymczak *et al.* 2001, Vlemmings *et al.* 2005).

envelope traced by the maser measurements throughout AGB envelopes. While a clear trend with increasing distance from the star is seen, the lack of accurate information on the location of the maser with respect to the central stars makes it difficult to constrain this relation beyond stating that it seems to vary between $B \propto R^{-2}$ (solar-type) and $B \propto R^{-1}$ (toroidal). Future observations of CO polarization might be able to provide further constraints.

As the masers used for these studies are mostly found in oxygen-rich AGB stars, it has to be considered that the sample is biased. However, recent CN Zeeman splitting observations (Herpin *et al.* 2009) seem to indicate that similar strength fields are found around carbon-rich stars.

Beyond determining the magnetic field strength, the large scale structure of the magnetic field is more difficult to determine, predominantly because the maser observations often probe only limited line-of-sights. Even though specifically OH observations seem to indicate a systematic field structure, it has often been suggested that there might not be a large scale component to the field that would be necessary to shape the outflow (Soker 2002). So far the only shape constraints throughout the envelope have been determined for the field around the supergiant star VX Sgr (Fig. 3), where maser observations spanning 3 orders of magnitude in distance are all consistent with a large scale, possibly dipole shaped, magnetic field.

However, the recent observations of the linear polarization of CO (and other molecular lines) caused by the Goldreich-Kylafis effect (Fig. 2, Vlemmings *et al.* 2012b) allows for a much more detailed study of the magnetic field morphology. In the case of IK Tau, the observations indicate a more or less uniform field from close to the star out to a few thousand AU. While the current sensitivity offered by the SMA is not sufficient for a more detailed magnetic field mapping, ALMA will revolutionize this field.

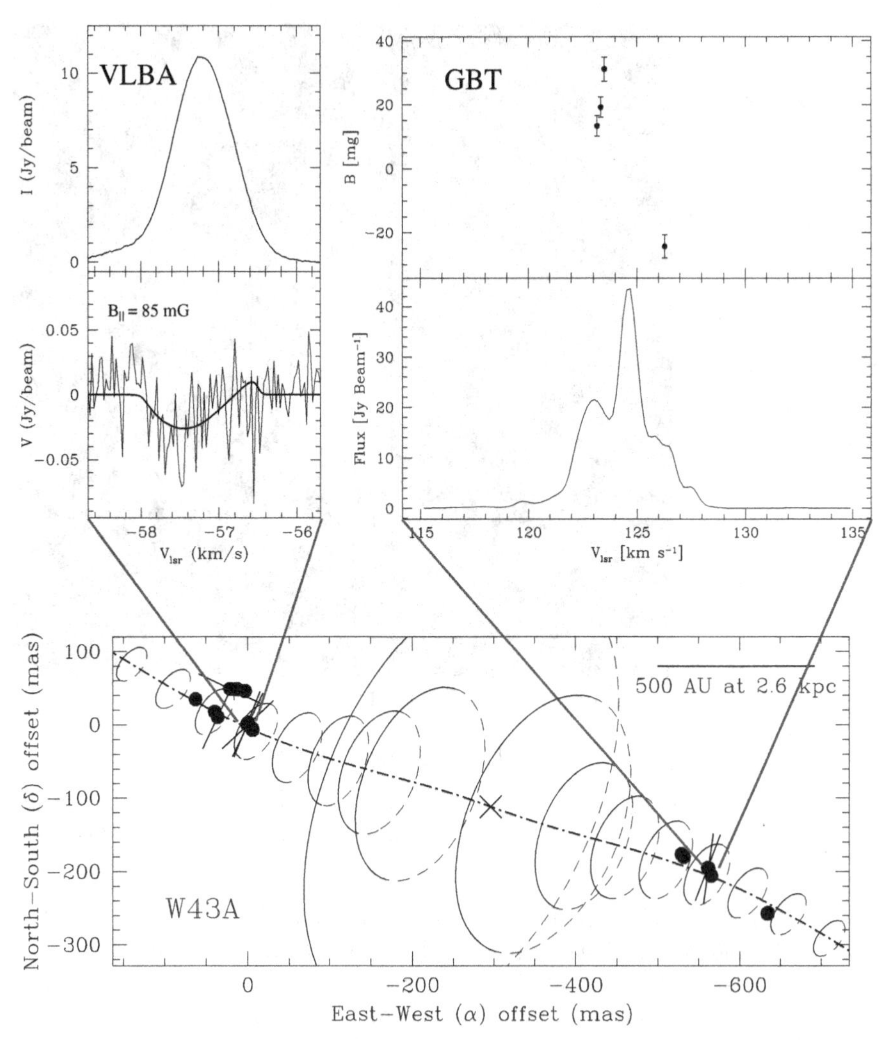

Figure 4. (top left) Total power (I) and V-spectrum for one of the H_2O maser features in red-shifted lobe of the collimated jet of W43A including the best model fit of the V-spectrum corresponding to a magnetic field (Vlemmings *et al.* 2006). (top right) Confirmation of the magnetic field from single-dish GBT observations in the blue-shifted side of the lobe. As expected for being toroidal, the magnetic field reverses sign across the blue-shifted masers (Amiri *et al.* 2010). (bottom) The H_2O masers in the precessing jet (dashed-dotted line) of W43A (indicated by the cross) and the toroidal magnetic field of W43A. The vectors indicate the determined magnetic field direction, perpendicular to the polarization vectors, at the location of the H_2O masers. The ellipses indicate the toroidal field along the jet, scaled with magnetic field strength $\propto r^{-1}$.

Finally, a recent direct measurement of the Zeeman effect on the surface of the Mira variable star χCyg has been reported (Lèbre *et al.*, this proceedings). The apparent field of a few Gauss would be reasonably consistent with the field measured throughout the envelope of other AGB stars.

3.2. *Post-AGB Stars and P-PNe*

Similar to the AGB stars, masers are the main source of magnetic field information of post-AGB and P-PNe, with the majority of observations focused on OH masers. These

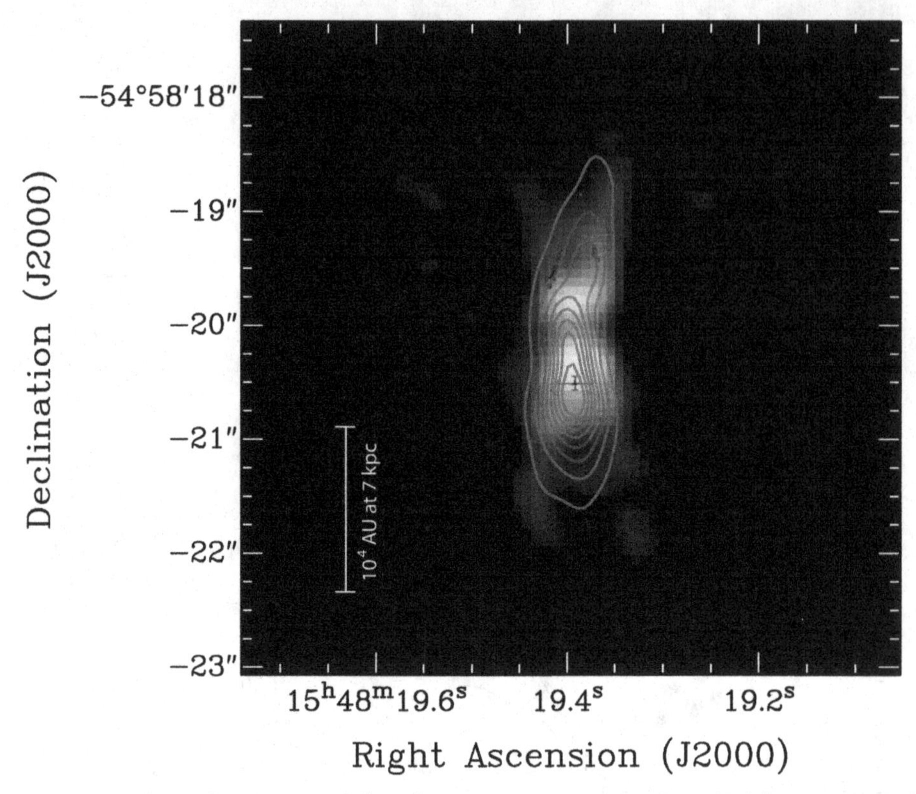

Figure 5. From Pérez-Sánchez *et al.* (2013). The synchrotron jet from IRAS 15445-5449 (contours) on a mid-infrared VLT image. Also indicated are the high-velocity maser features in the Northern lobe of the outflow.

have revealed magnetic field strengths similar to those of AGB stars (few mG) and a clear large scale magnetic field structure (e.g. Bains *et al.* 2003).

The most promising results have come after the detection of the so-called 'water-fountain' sources. These sources exhibit fast and highly collimated H_2O maser jets that often extend beyond even the regular OH maser shell. With the dynamical age of the jet of order 100 years, they potentially are the progenitors of the bipolar (P-)PNe. Although the masers are often too weak for a detection of the magnetic field, observations of the arch-type of the water-fountains, W43A, have revealed a strong toroidal magnetic field that is collimating the jet (Fig. 4 and Vlemmings *et al.* 2006).

Excitingly, the maser results have recently been strengthened by the direct detection of a synchrotron jet from the post-AGB star IRAS 15445-5449 (Pérez-Sánchez et al. 2013). This jet further proves the existence of strong collimating magnetic fields in the late stages of AGB evolution and strengthens the hypothesis that magnetic jets shape the bipolar PNe.

3.3. *Planetary Nebulae*

During the PN phase, masers are rare and weak and until now only the PN K3-35 has had a few mG magnetic field measured in its OH masers (Miranda *et al.* 2001). Fortunately, there are a few other methods of measuring PN magnetic fields. The field orientation in the dust of the nebula can be determined using dust continuum polarization observations and current observations seem to indicate toroidal fields, with the dust alignment likely occurring close to the dust formation zone (Sabin *et al.* 2007 and these proceedings).

Table 1. Energy densities in AGB envelopes

		Photosphere	SiO	H_2O	OH
B	[G]	$\sim 1-10?$	~ 3.5	~ 0.3	~ 0.003
R	[AU]	-	~ 3	~ 25	~ 500
		-	$[2-4]$	$[5-50]$	$[100-10.000]$
$V_{\rm exp}$	[km s^{-1}]	~ 5	~ 5	~ 8	~ 10
$n_{\rm H_2}$	[cm^{-3}]	$\sim 10^{13}$	$\sim 10^{10}$	$\sim 10^{8}$	$\sim 10^{6}$
T	[K]	~ 2500	~ 1300	~ 500	~ 300
$B^2/8\pi$	[dyne cm^{-2}]	$\mathbf{10^{-1.4,+0.6}?}$	$\mathbf{10^{+0.1}}$	$\mathbf{10^{-2.4}}$	$10^{-6.4}$
nKT	[dyne cm^{-2}]	$10^{+0.5}$	$10^{-2.7}$	$10^{-5.2}$	$10^{-7.4}$
$\rho V_{\rm exp}^2$	[dyne cm^{-2}]	$10^{+0.5}$	$10^{-2.5}$	$10^{-4.1}$	$\mathbf{10^{-5.9}}$
V_A	[km s^{-1}]	~ 10	~ 100	~ 300	~ 8

Faraday rotation studies are potentially also able to study the magnetic field in the interaction region between the interstellar medium and the stellar outflow.

In contrast to AGB stars, the central stars of PNe also show atomic lines that can be used to directly probe the magnetic fields on the surface of these stars. Despite earlier claims of the detection of a field of order several kG, recent work shows that the previous detection was spurious and no direct detection has yet been made (Leone *et al.* 2011).

4. Origin of the Magnetic Field

Despite the strong observational evidence for evolved star magnetic fields, the origin of these fields is still unclear. In single stars, differential rotation between the AGB star core and the envelope could potentially result in sufficiently strong magnetic field (Blackman *et al.* 2001). However, as the energy loss due to a rotating magnetic field drag drains the rotation needed to maintain the field within several tens of years, an additional source of energy is needed (e.g. Nordhaus & Blachman 2006). If AGB stars would be able to have a sun-like convective dynamo, magnetically dominated explosions could indeed result from single stars. Alternatively, the energy could be provided by the interaction with a circumstellar disk, although the origin of the disk is then another puzzle.

Another explanation for maintaining a magnetic field is the interaction between a binary companion or potentially a heavy planet, with common-envelope evolution providing paths to both magnetically as well as thermally driven outflows (Nordhaus & Blackman 2006). A companion could be the cause of the precession seen in a number of water-fountain and (P-)PNe jets. However, to date, the majority of the stars with measured magnetic fields do not show any other indication of binarity. However, as shown by our recent ALMA observations of the detached shell carbon AGB star R Scl (Maercker *et al.* 2012), hidden binaries could be fairly commonplace.

5. Effect of the Magnetic Field

Until a more complete sample of magnetically active AGB stars, post-AGB stars and (P-)PNe is known, it is hard to observationally determine the effect of the magnetic field on these late stages of evolution. Starting with the AGB phase, a number of theoretical works have described the potential of magnetic fields in (at least partly) driving the stellar mass-loss through Alfvén waves (e.g. Falcetta-Gonçalves & Jatenco-Pereira 2002), or through the creation of cool spots on the surface above with dust can form easier (Soker 1998). As current models of dust and radiation driven winds are still unable to explain especially the mass-loss of oxygen-rich stars, magnetic fields might provide the missing component of this problem, with tentative evidence already pointing to a relation between the magnetic field strength and mass-loss rate. The recently detected X-ray emission from two AGB stars could potentially be an indication of magnetic stellar activity (Ramstedt *et al.* 2012).

Other theoretical works have focused on the magnetic shaping of the stellar winds (e.g. Chevalier & Luo 1994, García-Segura et al. 2005, Frank & Blackman 2004). But to properly determine the possible effect of the magnetic fields, it is illustrative to study the approximate ratios of the magnetic, thermal and kinematic energies contained in the stellar wind. In Table.1 I list these energies along with the Alfvén velocities and typical temperature, velocity and temperature parameters in the envelope of AGB stars. While many values are quite uncertain, as the masers that are used to probe them can exist in a fairly large range of conditions, it seems that the magnetic energy dominates out to $\sim 50 - 100$ AU in the circumstellar envelope. This would correspond to the so-called 'launch' region of magneto-hydrodynamic (MHD) outflows, which typically extend to no more than $\sim 50R_i$, with R_i the inner-most radial scale of launch engine. A rough constraint on R_i thus seems to be $\sim 1 - 2$ AU, close to the surface of the star.

6. Outlook

While progress in studying the magnetic fields of evolved stars has been significant, a number of crucial questions remain to be answered. Several of these can be addressed with the new and upgraded telescopes in the near future. For example, the upgraded JVLA and eMERLIN will uniquely be able to determine the location of the masers in the envelope with respect to the central star, giving us, together with polarization observations, crucial information on the shape and structure of the magnetic field throughout the envelopes. ALMA will be able to add further probes of magnetic fields with for example high frequency masers (Pérez-Sánchez & Vlemmings 2013) and CO polarization observations, significantly expanding our sample of stars with magnetic field measurements. With the ALMA sensitivity, polarization will be easily detectable even in short observations and thus, even if not the primary goal, polarization calibration should be done. Additionally, ALMA will, as already shown, be able to reveal the hidden binaries and through the study of the circumstellar chemistry investigate peculiarities that can be induced by a magnetically active star. The new low-frequency arrays, including the future Square Kilometer Array, can potentially be used to determine magnetic fields in the interface between the ISM and PNe envelopes through Faraday rotation observations and investigate emission from the interaction between the stellar outflow and magnetic field with a potential planetary companion.

With the advances in the search for binaries and the theories of common-envelope evolution and MHD outflow launching, the new observations will address for example:

- Under what conditions does the magnetic field dominate over e.g. binary interaction when shaping outflows?
- Are magnetic fields as widespread in evolved stars as they seem?
- What is the origin of the AGB magnetic field - can we find the binaries/heavy planets that might be needed?
- Is there a relation between AGB mass-loss and magnetic field strength?

References

Amiri, N., Vlemmings, W., & van Langevelde, H. J. 2011, *A&A*, 532, A149

Bains, I., Gledhill, T. M., Yates, J. A., & Richards, A. M. S. 2003, *MNRAS*, 338, 287

Bujarrabal, V., Castro-Carrizo, A., Alcolea, J., & Sánchez Contreras, C. 2001, *A&A*, 377, 868

Chevalier, R. A. & Luo, D. 1994, *ApJ*, 421, 225

Falceta-Gonçalves, D. & Jatenco-Pereira, V. 2002, *ApJ*, 576, 976

Frank, A. & Blackman, E. G. 2004, *ApJ*, 614, 737

García-Segura, G., López, J. A., & Franco, J. 2005, *ApJ*, 618, 919

Herpin, F., Baudry, A., Thum, C., Morris, D., & Wiesemeyer, H. 2006, *A&A*, 450, 667

Herpin, F., Baudy, A., Josselin, E., Thum, C., & Wiesemeyer, H. 2009, in IAU Symposium, vol. 259 of IAU Symposium, 47

Kemball, A. J., Diamond, P. J., Gonidakis, I., Mitra, M., Yim, K., Pan, K., & Chiang, H. 2009, *ApJ*, 698, 1721

Kwok, S., Purton, C. R., & Fitzgerald, P. M. 1978, *ApJ*, 219, L125

Leal-Ferreira, M. L., Vlemmings, W. H. T., Kemball, A., & Amiri, N. 2013, *A&A*, 554, A134

Leone, F., Martínez González, M. J., Corradi, R. L. M., Privitera, G., & Manso Sainz, R. 2011, *ApJL*, 731, L33

Maercker, M., Mohamed, S., Vlemmings, W. H. T., *et al.* 2012, *Nature*, 490, 232

Miranda, L. F., Gómez, Y., Anglada, G., & Torrelles, J. M. 2001, *Nature*, 414, 284

Nordhaus, J. & Blackman, E. G. 2006, *MNRAS*, 370, 2004

Nordhaus, J., Blackman, E. G., & Frank, A. 2007, *MNRAS*, 376, 599

Pérez-Sánchez, A. F. & Vlemmings, W. H. T. 2013, *A&A*, 551, A15

Pérez-Sánchez, A. F., Vlemmings, W. H. T., Tafoya, D., & Chapman, J. M. 2013, *MNRAS* in press, arXiv:1308.5970

Ramstedt, S., Montez, R., Kastner, J., & Vlemmings, W. H. T. 2012, *A&A*, 543, A147

Rudnitski, G. M., Pashchenko, M. I., & Colom, P. 2010, *Astron. Rep.*, 54, 400

Sabin, L., Zijlstra, A. A., & Greaves, J. S. 2007, *MNRAS*, 376, 378

Soker, N. 1998, *MNRAS*, 299, 1242

Soker, N. 2002, *MNRAS*, 336, 826

Szymczak, M., Cohen, R. J., & Richards, A. M. S. 2001, *A&A*, 371, 1012

Vlemmings, W. H. T. 2007, in IAU Symposium, edited by Booth, R. S. Humphreys, E. M. L. & Vlemmings, W. H. T., vol. 287 of IAU Symposium, 31

Vlemmings, W. H. T., Diamond, P. J., & Imai, H. 2006, *Nature*, 440, 58

Vlemmings, W. H. T., Diamond, P. J., & van Langevelde, H. J. 2002, *A&A*, 394, 589

Vlemmings, W. H. T., Humphreys. E. M. L. & Franco-Hernández, R. 2011, *ApJ*, 728, 149

Vlemmings, W. H. T., van Langevelde, H. J., & Diamond, P. J. 2005, *A&A*, 434, 1029

Vlemmings, W. H. T., Ramstedt, S., Rao, R., & Maercker, M. 2012, *A&A*, 540, L3

Magnetic Fields throughout Stellar Evolution
Proceedings IAU Symposium No. 302, 2013
P. Petit, M. Jardine & H. Spruit, eds.

doi:10.1017/S1743921314002592

Magnetic fields in Proto Planetary Nebulae

L. Sabin[1], Q. Zhang[2], A. A. Zijlstra[3], N. A. Patel[2], R. Vázquez[4], B. A. Zauderer[2], M. E. Contreras[4] and P. F. Guillén[4]

[1]Instituto de Astonomía y Meteorología, Departamento de Física, CUCEI, Universidad de Guadalajara, Av. Vallarta 2602, C. P. 44130, Guadalajara, Jal., Mexico
email: lsabin@astro.iam.udg.mx
[2]Harvard-Smithsonian Center for Astrophysics, 60 Garden Street, Cambridge, MA 02138, USA
[3]Jodrell Bank Centre for Astrophysics, Alan Turing Building, University of Manchester, Manchester, M13 9PL, UK
[4]Instituto de Astronomía, Universidad Nacional Autónoma de México, Apdo. Postal 877, 22800 Ensenada, B. C, Mexico

Abstract. The role of magnetic field in late type stars such as proto-planetary and planetary nebulae (PPNe/PNe), is poorly known from an observational point of view. We present sub-millimetric observations realized with the Submillimeter Array (SMA) which unveil the dust continuum polarization in the envelopes of two well known PPNe: CRL 618 and OH 231.8+4.2. Assuming the current grain alignment theory, we were then able to trace the geometry of the magnetic field.

Keywords. Proto-Planetary nebulae, polarization, dust, magnetic field.

1. Introduction

Proto-planetary nebulae (PPNe) are stellar objects in rapid transition between the asymptotic giant branch (AGB) and the planetary nebula (PN) phases. Contrary to most of their AGB progenitors, PPNe and PNe display more or less pronounced non-spherical morphologies (i.e. elliptical, bipolar, multi-polar or point-symmetric). The evolution from generally spherically symmetric AGB circumstellar envelopes to highly asymmetric morphologies (in PNe for instance) suggests the occurrence of an important and fundamental collimation mechanism. The two most popular models invoked are the binary interaction between a mass-losing AGB star and a close companion and/or magnetic collimation of the wind. We present the results of a study related to the detection of global magnetic fields in two well known PPNe, CRL 618 and OH 231.8+4.2, via the process of thermal dust grain alignment (Sabin *et al.*, submitted).

2. Method

We performed submillimetric observations with the Submillimeter Array (SMA) in polarimetric mode using the compact configuration giving a maximum baseline of 77m. The frequencies range from 330 to 346 GHz (divided into Lower Side and Upper Side Bands). The correlator setup provides a ~0.8 MHz resolution i.e. 0.70 km/s spectral resolution at 345.796 GHz. The data reduction was performed with the software packages MIR and MIRIAD.

3. Results and conclusion

• In the case of CRL 618, the polarization vectors (polarized grains) are well ordered, quasi perpendicularly to the direction of the ionized outflows. As a consequence, we

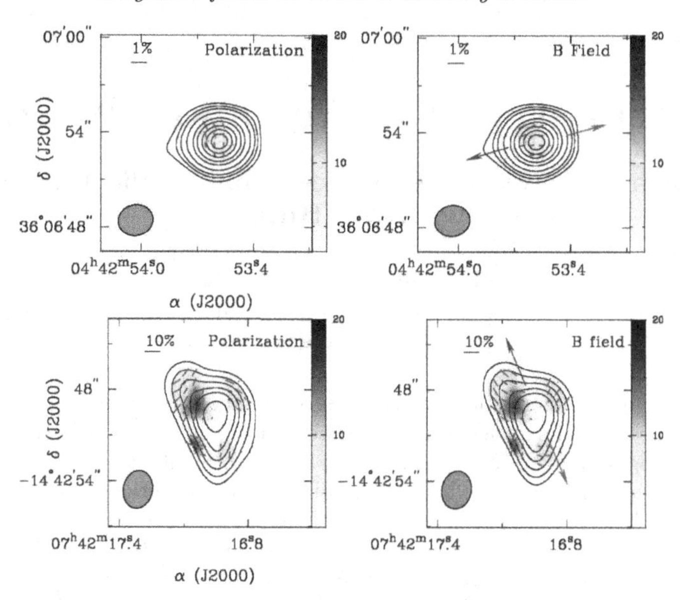

Figure 1. Dust polarization vectors distribution (left panels) and magnetic field distribution (right panels) in CRL 618 (top) and OH231.8+4.2 (bottom). The black contours which indicate the total dust emission are drawn in steps of 0.02 Jy × (3,6,10,20,40,60,90,120,150) for CRL 618 and in steps of 0.02 Jy × (3,6,10,15,20,30,40,50,60) for OH231.8+4.2 . The color images indicate the polarized intensity with its associated scale bar in Jy/Beam on the right. In both cases we also represent the direction of the ionized outflows (arrows).

deduced a clear and well organized poloidal magnetic structure (i.e. aligned with the pairs of ionized outflows).

- Concerning OH 231.8+4.2, the polarization vectors are divided into four groups with different polarization angles which seem to draw the contours of a ring-like structure. The vectors are also well aligned within each region. After rotating these vectors by 90 degrees, we observe an overall X-shaped field structure in OH 231.8+4.2. The field appears particularly patchy and can be linked to a dipole/poloidal configuration. However, some vectors might define the edge of a toroidal field.
- Both PPNe show sign of depolarization i.e. an anti-correlation is seen between the fractional polarization and the total intensity in both objects. The possible causes for this effect have been discussed by Greaves (2002), Lazarian *et al.* (1997) and Cho & Lazarian (2005) for example.
- The O-rich OH 231.8+4.2 shows higher percentage polarization than the C-rich CRL 618. This confirms earlier findings by Sabin *et al.* (2007) that silicate dust shows higher polarization than carbonaceous dust.

The SMA data probed the presence of magnetic fields in the envelopes of two more proto-planetary nebulae. The new findings, which seem to indicate that magnetic fields are common and long lasting, will be helpful for the understanding of the magnetic fields properties in evolved stars.

References

Cho, J. & Lazarian, A. 2005, *ApJ*, 631, 361
Greaves, J. S. 2002, *A&A*, 392, L1
Lazarian, A., Goodman, A. A., & Myers, P. C. 1997, *ApJ*, 490, 273
Sabin, L., Zijlstra, A. A., & Greaves, J. S. 2007, *MNRAS*, 376, 378

Magnetic Fields throughout Stellar Evolution
Proceedings IAU Symposium No. 302, 2013
P. Petit, M. Jardine & H. Spruit, eds.

doi:10.1017/S1743921314002609

Polarimetry of R Aqr and PN M2-9

Silvana G. Navarro[1], Laurence Sabin[1], Julio Ramírez[2] and David Hiriart[2]

[1]Instituto de Astronomía y Meteorología , Universidad de Guadalajara
email: silvana@astro.iam.udg.mx

[2]Instituto de Astronomía, Universidad Nacional Autónoma de México,
email: julio@astroscu.unam.mx

Abstract. The bipolar or more complex morphology observed in planetary nebulae have been explained by two principal hypothesis: by the existence of a companion and an accreting disk or by the effects of magnetic field, (or a combination of both). Symbiotics are binary systems and some of them show morphologies similar to those observed on planetary nebulae. This fact could support the binary hypothesis for PNe. We have therefore performed polarimetric observations of symbiotic systems and some planetary nebulae in order, first to detect linear polarisation with POLIMA at the San Pedro Mártir observatory, and ultimately to prove the existence and physical properties of those disks. We present here the first results of a project dedicated to the analysis of the polarisation observed in evolved objects starting with the PN M2-9 and R Aqr.

Keywords. Polarimetry, planetary nebulae, symbiotic systems

---

1. Introduction

Polarization due to dust scattering is a useful tool to determine the geometry of extended envelopes of Symbiotic systems and proto-planetary nebulae (Scarrott & Scarrott, 1995; Gledhill, 2005). It is also used to trace the circumstellar disk and jets/outflows present in such objects. M2-9 is a widely studied bipolar PN, it shows two coaxial shells and a series of dusty blobs, located simetrically on both lobes. Castro-Carrizo *et al.* 2012 found two ring-shaped structures at the center of this object giving evidence of their binary nature. R Aqr is a symbiotic system formed by a Mira variable (287 days period) and a white dwarf. It has two bipolar and co-axial shells with kinematical ages: 185 and 650 years (Solf & Ulrich, 1985). A high collimated structure (jet) is observed in this object, with bright knots along the jet that evolve in position, brightness and size. UV observations reveal the nature of this jet-like structure (Nichols & Slavin, 2009) They show that the shape of the emission lines is consistent with an ejecta model of a bow shock, supporting the existence of an accretion disk as the origin of the ejected material.

2. Observations and Results

We observed these objects with the polarimeter POLIMA on the 84 cm telescope of San Pedro Mrtir B. C. We used two narrow and two wide filters: Hα (λ_c = 6564 Å, $\delta\lambda$=72 Å, H6819 (6819 Å, 86 Å), Gr (6550 Å, 900 Å) and Gg (4930 Å, 700 Å). The latest are Thuan-Gunn filters. In the case of planetary nebulae M2-9, we determine the polarisation over both lobes, at the regions showed in figure 1a. In this object we detected a considerable polarisation, specially in the regions near to the central object. The polarisation angle does not change notably always near 80 degrees except at the extreme positions E1 and E2 (see table 1). In table 2 we present the polarisation measurements at the center and at the knots observed in R Aqr. The measured polarisation is low and its value depends

Table 1. Polarisation measured in M2-9 over the regions showed in figure 1a

Position	P(%)Hα	Angle	P(%)Gr	Angle	P(%)Gg	Angle
Center	4.6 ± 0.3	80 ± 2	4.3 ± 0.2	63 ± 1.5	3.9 ± 0.3	67 ± 2
1N	8.0 ± 0.5	84 ± 3	6.6 ± 0.5	73 ± 2	6.6 ±0.9	69.6 ± 3
2N	20.6 ± 0.7	72 ± 2	18.9 ± 0.7	72.4 ± 1.4	19.5 ± 1.5	82.2 ± 1.8
3S	11.1 ± 0.5	80 ± 2	16.4 ± 0.6	76.4 ± 1.4	17.2 ± 1.2	105.4 ± 1.7
4S	10.2 ± 0.5	79 ± 2	9.2 ± 0.6	70.3 ± 1.7	8.1 ± 1	75 ± 3
E1	16 ± 2	90 ± 3	13.3 ± 0.4	75.1 ± 1.3	21.6 ± 4	176 ± 4
E2	4 ± 2	43 ± 14	18.3 ± 0.4	80.4 ± 1.3	11.2 ± 2.7	140 ± 6

Table 2. Polarisation measured in R Aqr over the regions showed in figure 1b

Position	P(%)Hα	Angle	P(%)Gr	Angle	P(%)Gg	Angle
Center	1.6 ± 0.2	124 ± 3	0.9 ± 0.2	134 ± 2	1.1 ± 0.2	145 ± 4
Knot 1	4.9 ± 2	86 ± 8	17.4 ± 5.5	36 ± 7	8.1 ±0.9	70 ± 3
Knot 2	5.3 ± 2.9	110 ± 11	30 ± 13	172 ± 9	4.2 ± 1.3	15 ± 6
Knot 3	N/A	N/A	N/O	N/O	32 ± 1	70 ± 1.2
Knot 4	4.5 ± 2.9	71 ± 42	N/O	N/O	7.9 ± 4	160 ± 10
Knot 5	39 ± 21	113 ± 11	N/O	N/O	8.9 ± 10	160 ± 26

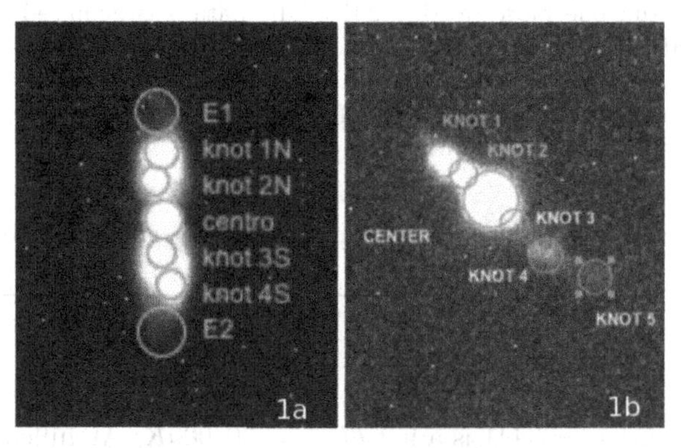

Figure 1. Regions where the polarisation were measured in M2-9 (1a), and in R Aqr (1b).

slightly on the radius of the region used to make the measurements. This is probably due to the low signal to noise in the R Aqr polarisation images.

3. Conclusions

We confirm the polarisation previously detected in M2-9 using the new polarimeter (POLIMA) at San Pedro Mrtir Observatory. In M2-9 we detected variation in the polarisation between the knots nearer to the central object and the farther ones on the jet. We report low polarisation detected on R Aqr. Only in one of the knots (the farther from the center) we detected substantial polarisation. This could imply dust scattered light.

These positive detections allow us to plan new polarimetric and spectro-polarimetric observations at higher resolution to find traces of magnetic field in these objects.

References

Castro-Carrizo, A., Neri, R., Bujabarral, V., *et al.* 2012, *A&A*, 545, A1
Gledhill, T. M. 2005, *MNRAS*, 356, 883
Nichols, J. & Slavin, J. D. 2009, *ApJ*, 699, 902
Scarrott, S. M. & Scarrott, R. M. J. 1995, *MNRAS*, 277, 277
Solf, J. & Ulrich, H. 1985, *A&A*, 148, 274

Magnetic Fields throughout Stellar Evolution
Proceedings IAU Symposium No. 302, 2013
P. Petit, M. Jardine & H. Spruit, eds.

doi:10.1017/S1743921314002610

Measurements of the magnetic field in WD 1658+441

J. Ramírez Vélez[1], D. Hiriart[1], G. Valyavin[2], J.Valdez[1], F. Quiroz[1], B. Martínez[1], S. Plachinda[2] and E. Iñiguez-Garín[1]

[1]Instituto de Astronomía, Universidad Nacional Autónoma de México.
Km. 103 Carretera Tijuana Ensenada, CP 22860, México. email: `julio@astrosen.unam.mx`
[2]Special Astrophysical Observatory, Nizhnij Arkhyz, Karachai-Cherkessian, Russia 369167

Abstract. We present the preliminary results of the measurements of longitudinal magnetic field of the massive white dwarf 1658+441. This star have an hydrogen pure atmosphere (e.g. Dupuis & Chayer, 2003). We have observed the target in a total of 18 hrs during 3 consecutive nights in June 2010 and one more in May 2011. The data was acquired with a prototypical spectropolarimeter at the San Pedro Martir Telescope in Mexico. We have tested the magnetic field measurements with our instrument using the famous Babcock's star obtaining consistent results with previous studies. For our object of study, the WD 1658+441, we have measured variable intensities of the longitudinal magnetic field of B_{long} = 720 kG that oscillates with an amplitude of 130 kG.

Keywords. WD1658+441, HD215441, stellar magnetic fields, spectropolarimetry.

---

1. Introduction

The DA white dwarf 1658+441 is a hot ($T_{eff} = 30,000K$; Archilleos & Wickrasmassinghe 1989), ultra massive star (1.3 $M_{\odot}$), which host an intense variable magnetic field (Liebert *et al.*, 1983).

We have chose this study object because is a suitable target to be observed with the prototypical spectropolarimeter at the 2.1-meters Telescope at the national observatory in "San Pedro Martir", Mexico.

We have performed a minimum of changes to the system Telescope-Spectrograph in order to use it as spectro-polarimeter. Before the slit of the low resolution spectrograph Boller & Chivens, from 150 to 1200 grooves/mm, we have included a polarimetric analyzer module that includes a rotable quarter wave plate and a Savart plate. This configuration allow us to mesure the circular polarized spectra.

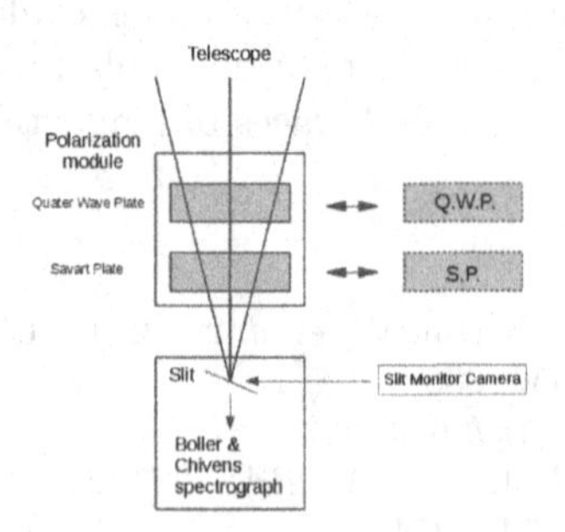

Figure 1. Optical design of the system Telescope - Polarizer Module - Spectrograph

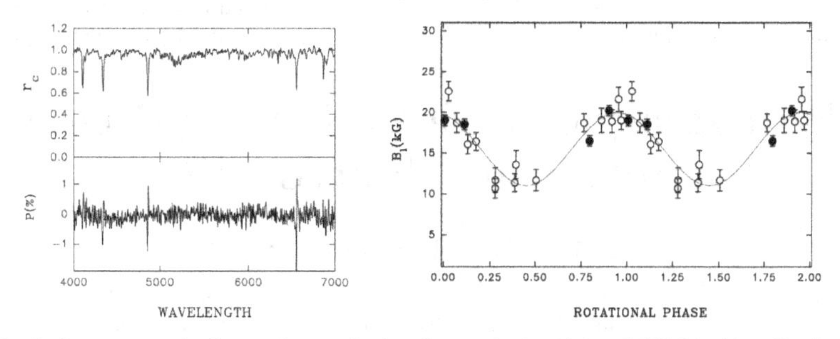

Figure 2. *Left;* spectra in intensity and circular polarization of HD215441. *Right*; magnetic field mesurements in function of the rotational phase of the star (see text for a description).

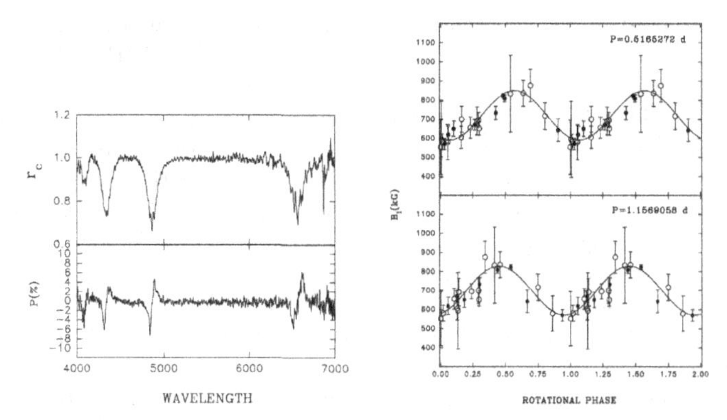

Figure 3. *Left;* spectra in intensity and circular polarization of WD1658+441. *Right*; magnetic field mesurements in function of the rotational phase of the star (see text for a description).

2. Results

In the left panel in Fig. 3, we show an example of the spectra of HD215441 which clearly shows the typical S-shape of the V-Stokes profile. Our measurements of the magnetic field are consistent with the values reported previously. In the right panel, the open circles correspond to Borra & Landstreet (1981), while solid circles correspond to our data.

In Fig. 4, left panel, we show an example of observed spectra of our object of study, the WD1658+441. In the right panel, we have combined our magnetic fields measurements (solid circles) with those of Shtol' *et al.* (1997) (open circles) to analyze the periodicity of the field variations. The power spectra showed two main frequency peaks at 0.5 days and 1.6 days. We interpret these values as the range values for the most probable rotation period of the star. As conclusion, we found that the longitudinal magnetic field intensity of WD1658+441 varies with an amplitud of 130 kG around a value of 720 kG.

Acknowledgements: This study is supported by CONACyT grant number 180817. The authors thank to grant offered by the IAU-Symposium.

References

Achilleos, N. & Wickramasinghe, D. T., 1989 *PASA*, 8, 148–153
Dupuis, J. & Chayer, P., 2003, *ApJ*, 598, 486–491
Borra, A. & Landstreet, J. D., 1981, *ApJS*, 264, 262–272
Liebert, J., Schmidt, G., Green, R., Bychkov, V. & Stolyarov, V., 1997, *AstL*, 23, 48S
Shtol, V., Valyavin, G., Fabrica, S., Stockman, H. & McGraw, J. 1983, *ApJ*, 264, 262–272

Magnetic Fields throughout Stellar Evolution
Proceedings IAU Symposium No. 302, 2013
P. Petit, M. Jardine & H. Spruit, eds.

doi:10.1017/S1743921314002622

Hydromagnetic Equilibria and their Evolution in Neutron Stars

Andreas Reisenegger

Instituto de Astrofísica, Facultad de Física, Pontificia Universidad Católica de Chile
Av. Vicuña Mackenna 4860, 7820436 Macul, Santiago, Chile
email: areisene@astro.puc.cl

Abstract. The strongest known magnetic fields are found in neutron stars. I briefly discuss how they are inferred from observations, as well as the evidence for their time-evolution. I go on to show how these extremely strong fields are actually weak in terms of their effects on the stellar structure. This is also the case for magnetic stars on the upper main sequence and magnetic white dwarfs, which have similar total magnetic fluxes, perhaps pointing to an evolutionary connection. I suggest that a stable hydromagnetic equilibrium (containing a poloidal and a toroidal field component) could be established soon after the birth of the neutron star, aided by the strong compositional stratification of neutron star matter, and this state is slowly eroded by non-ideal magnetohydrodynamic processes such as beta decays and ambipolar diffusion in the core of the star and Hall drift and breaking of the solid in its crust. Over sufficiently long time scales, the fluid in the neutron star core will behave as if it were barotropic, because, depending on temperature and magnetic field strength, beta decays will keep adjusting the composition to the chemical equilibrium state, or ambipolar diffusion will decouple the charged component from the neutrons. Therefore, the still open question regarding stable hydromagnetic equilibria in barotropic fluids will become relevant for the evolution, at least for magnetar fields, which are likely too strong to be stabilized by the solid crust.

Keywords. stars: neutron, stars: magnetic field, stars: magnetars,(stars:) pulsars: general

---

1. *Neutron* stars?

Why *neutron* stars? In the near-vacuum of our laboratories, a free neutron (n) decays spontaneously through the process $n \to p + e + \bar{\nu}_e$ (where p is a proton, e and electron, and $\bar{\nu}_e$ an electron antineutrino), with a half life of only 15 minutes. However, we know neutrons to exist inside atomic nuclei, and compact stars containing a large number of neutrons have been predicted (Baade & Zwicky 1934) and modeled (Oppenheimer & Volkoff 1939) since a long time ago.

The crucial physical ingredient is Pauli's exclusion principle. As a large number of fermions is packed into a small volume, they will fill up all the low-energy quantum states up to the "Fermi energy", which also plays the role of their chemical potential (with a small correction for finite temperatures). Thus, even at zero temperature, the fermions will be moving, providing a "degeneracy pressure" that allows objects such as white dwarfs and neutron stars to resist their very strong gravity. Moreover, if a neutron in the core of a cool, very compact star decays as $n \to p+e+\bar{\nu}_e$, the nearly massless and weakly interacting antineutrino escapes easily, but the proton is held back by gravity, and the electron by the electrostatic potential of the protons, so there will be "Fermi seas" of neutrons, protons, and electrons, with chemical potentials ($\approx$ Fermi energies) μ_n, μ_p, and μ_e. If $\mu_n > \mu_p + \mu_e$, some neutrons (with energies not much below μ_n) will be energetic enough to find unoccupied protons and electron states (above their respective Fermi energies) to decay into, in this process reducing μ_n and increasing μ_p and μ_e. In the opposite case ($\mu_n < \mu_p + \mu_e$), some protons and electrons will be energetic enough to combine by $p + e \to n + \nu_e$, finding an unoccupied neutron state and ejecting an electron neutrino ν_e, thus increasing μ_n and reducing μ_p and μ_e. Both processes will eventually balance in a chemical equilibrium (or "beta equilibrium") state in which $\mu_n = \mu_p + \mu_e$, which implies that a large number of neutrons will coexist with a substantially smaller fraction (few %) of protons and electrons. This fraction is an increasing function of density, meaning that the fluid is *stably stratified*, resisting convective turnover, like water with downward-increasing salinity (Pethick 1991; Reisenegger & Goldreich 1992).

The compulsory presence of protons and electrons, i. e., charged particles, in addition to neutrons in the neutron star core is also crucial for the existence of a magnetic field in these stars, because they allow currents to flow, which act as the source of the field. Since most of their quantum states are occupied, it is difficult to scatter them into a different state. For this reason, the resistivity is low, and currents can flow for a long time without being dissipated, perhaps even more so if, as expected, the protons in much of the core are superconducting (Baym, Pethick, and Pines 1969a,b).

At the very high densities of the "inner core" of a neutron star, neutrons, protons, and electrons could also decay or combine into more exotic particles such as muons, mesons, or hyperons, and it might even be possible for all baryons to dissolve into a degenerate quark-gluon plasma.

On the other hand, in the lower-density outer layers of a neutron star, the state of matter will be closer to those we are accustomed to, with protons and neutrons bound into atomic nuclei that can organize into a solid structure, though also with somewhat unusual properties. On the one hand, it is a strongly compressed solid, in which the Fermi energy of the electrons is much larger than the electrostatic interaction energy between neighboring particles, so its bulk modulus (incompressibility) is much larger than its shear modulus. On the other hand, in the inner crust there are free neutrons coexisting with and moving through the solid, and these neutrons are believed to become superfluid and account for the pulsar glitch phenomenon.

Thus, if we imagine moving inside from the neutron star surface, we expect to encounter quite different states of matter:

- The *outer crust* (densities $\rho \sim 10^6 - 4 \times 10^{11}\,\mathrm{g\,cm^{-3}}$), a solid of heavy nuclei and freely moving electrons.
- The *inner crust* ($\rho \sim 4 \times 10^{11} - 2 \times 10^{14}\,\mathrm{g\,cm^{-3}}$), a solid of even heavier nuclei, freely moving electrons, and freely moving, likely superfluid neutrons.
- The *outer core* ($\rho \sim 2 \times 10^{14} - 10^{15}\,\mathrm{g\,cm^{-3}}$), a liquid composed mostly of neutrons (n), with a relatively small, but increasing fraction (few %) of protons (p), electrons (e), and muons.
- The *inner core* ($\rho \gtrsim 10^{15}\,\mathrm{g\,cm^{-3}}$), in a largely unknown state, likely a liquid containing more exotic particles, such as mesons, hyperons, free quarks, or others.

2. Spin-down, magnetic field, and evidence for its evolution

The electromagnetic radiation received from most neutron stars appears pulsed at a very regular frequency, which slowly decreases in time. This is almost certainly due to the slowing rotation rate Ω of the neutron star, whose radiation is beamed or at least anisotropic. The slow-down is usually modeled (not quite realistically) in terms of a magnetic dipole rotating in vacuum, which loses rotational energy through electromagnetic radiation according to the relation

$$I\Omega\dot{\Omega} \propto -\mu^2\Omega^4, \tag{2.1}$$

where dots indicate time derivatives, I is the moment of inertia, and μ is the magnetic moment of the star. This allows to estimate the spin-down time, $t_s \equiv P/(2\dot{P})$, as a rough estimate of the stellar age (accurate if $\mu =$ constant and the initial rotation rate was much faster than the present one), and the surface magnetic field $B \propto (P\dot{P})^{1/2}$, where $P = 2\pi/\Omega$ is the rotation period.

In nearly all cases, there are no other measurements of the magnetic field strength, and only indirect inferences of its geometry from the pulse profiles. However, the magnetic field clearly plays an important role in neutron star evolution and is present on all known neutron stars. Its magnitude is inferred to be 10^{11-13}G in most objects (the bulk of the so-called "classical pulsars"), as low as 10^{8-9}G in the old, but rapidly spinning "millisecond pulsars", and as high as 10^{14-15}G in the slowly spinning ($P \sim 2 - 12$s), but very energetic "soft gamma repeaters" (SGRs) and "anomalous X-ray pulsars" (AXPs), collectively known as "magnetars". In addition to these, one phenomenologically distinguishes isolated thermal emitters (INSs; $B \sim 10^{13-14}$G), "central compact objects" in supernova remnants (CCOs; $B \sim 10^{10-12}$G), RRATs (intermittent radio pulsars; $B \sim 10^{12-14}$G), and accreting neutron stars (high-mass and low-mass X-ray binaries). A concise overview of these classes of neutron stars, their position on the $P - \dot{P}$ diagram, and their possible connections is given by Kaspi (2010).

There are several lines of evidence suggesting possible evolution of the magnetic field:

• Field decay inferred from the distribution of classical pulsars on the $P - \dot{P}$ diagram: complicated by a number of selection effects, it has been addressed by many authors over the last 35 years, with conflicting results. For a recent analysis, see Faucher-Giguère & Kaspi (2006).

• Very weak (dipole) field of old, recycled pulsars (millisecond pulsars and low-mass X-ray binaries): not yet established whether this is an effect of age (passive magnetic field decay) or induced by accretion (increased resistivity due to heating, magnetic field burial, or motion of superfluid neutron vortices).

• Anomalous braking indices: In very young pulsars, it is possible to measure $\ddot{\Omega}$ and thus construct the "braking index" $n \equiv \Omega\ddot{\Omega}/\dot{\Omega}^2$. Eq. (2.1) with $\mu =$ constant yields $n = 3$, whereas measured values are generally lower, at face value implying an increasing magnetic dipole moment.

• Magnetar energetics: SGRs and AXPs emit copious amounts of high-energy (X and gamma) radiation; in fact, their time-averaged bolometric luminosity exceeds the rotational energy loss given by eq. (2.1). This suggested that their energy source might be the decay of a very strong magnetic field (Thompson & Duncan 1996), later corroborated by the determination of their dipole field as the highest known for any objects (Kouveliotou *et al.* 1998). Note, however, that an even stronger internal field appears to be required to account for the energetics of some of these objects. An interesting, recent discovery has been the detection of quasi-periodic oscillations following SGR flares (Israel *et al.* 2005), which might be magneto-elastic oscillation modes of the neutron star and thus potential probes of its internal magnetic field structure.

3. Strong *and* weak magnetic fields

As already mentioned, the observationally inferred dipole magnetic fields of neutron stars, particularly magnetars ($B \sim 10^{14-15}$G), are the strongest known in the Universe, far exceeding any produced so far on Earth (up to 10^7 G produced in explosions, for very

Table 1. Stars with long-lived magnetic fields

Star type	Upper main sequence	White dwarf	Neutron star
Radius R [km]	$10^{6.5}$	10^4	10^1
Maximum magnetic field B_{max} [G]	$10^{4.5}$	10^9	10^{15}
Maximum magnetic flux $\Phi_{max} \equiv \pi R^2 B_{max}$ [G km^2]	10^{18}	$10^{17.5}$	$10^{17.5}$

short times) or on other stars (up to 10^9G on white dwarfs). An interesting comparison table is given on R. Duncan's web site on magnetars (`http://solomon.as.utexas.edu/magnetar.html`).

On the other hand, neutron stars share with white dwarfs and upper main sequence stars the properties of being mostly or completely non-convecting and having fields appearing to be constant over long time scales and thus likely "frozen in" rather than being rearranged and regenerated by a dynamo process. Table 1 shows that the widely different sizes and observed magnetic field strengths among these three types of stars largely compensate to give quite similar maximum magnetic fluxes $\Phi_{max} \sim 10^{17.5-18}$G km^2 in each type, possibly indicating that the naive hypothesis of flux freezing along the evolution of these stars goes a long way in explaining their magnetic fluxes, despite their very eventful lifes, including core collapse, ejection of a substantial fraction of their mass, differential rotation, and convection.

It is interesting to consider the ratio of gravitational to magnetic energy in these stars,

$$\frac{|E_{grav}|}{E_{mag}} \sim \frac{GM^2/R}{B^2R^3/6} \sim 6\pi^2 G\left(\frac{M}{\Phi}\right)^2 \gtrsim 10^6 \qquad (3.1)$$

which remains constant as the star contracts or expands, as long as it conserves its mass and magnetic flux. The lower bound, based on the numbers in Table 1, shows that all these stars are very highly "supercritical" (in star-formation jargon), so the magnetic forces are much too weak to significantly affect the stellar structure. In this sense, although magnetar fields are the strongest observed in the Universe, they are still very weak in terms of their effect on the stellar structure. Of course, this ignores an eventual additional field component possibly hidden within the star, mentioned in the previous section, to which I will come back below.

4. Axially symmetric, ideal MHD equilibria

For the reasons just exposed, it is almost certainly an excellent approximation to write the physical variables characterizing the stellar fluid as the sum of a non-magnetized "background" plus a much smaller "magnetic perturbation", i. e., density $\rho = \rho_0 + \rho_1$, or pressure $P = P_0 + P_1$, where $|\rho_1|/\rho_0 \sim |P_1|/P_0 \sim B^2/(8\pi P_0) \lesssim 10^{-6}$, according to the estimate of eq. (3.1). In the absence of rotation, the background quantities are spherically symmetric and satisfy the usual hydrostatic equilibrium relation,

$$\frac{dP_0}{dr} + \rho_0 \frac{d\Psi}{dr} = 0, \qquad (4.1)$$

where r is the radial coordinate, and $\Psi(r)$ is the gravitational potential, whose magnetic perturbation I ignore for simplicity ("Cowling approximation"). (In this section, I also ignore the shear forces in the solid crust and possible superconducting components in the neutron star core.) On the other hand, since the magnetic field $\vec{B}(\vec{r})$ cannot be spherically symmetric, the hydromagnetic equilibrium equation for the perturbed quantities is generally a three-component vector equation:

$$\nabla P_1 + \rho_1 \nabla \Psi = \frac{1}{c}\vec{j} \times \vec{B}, \tag{4.2}$$

where c is the speed of light and $\vec{j} = (c/4\pi)\nabla \times \vec{B}$ is the current density.

As explained in § 1, neutron star matter is chemically inhomogeneous, characterized by at least one composition variable Y, such as the ratio of the proton to neutron density, which beta decays adjust to an equilibrium value over very long time scales, but which will be an independent, conserved quantity over dynamical times. If we assume, for now, a single fluid whose composition is frozen in each fluid element, P_1 and ρ_1 above can be considered as independent variables that separately adjust to satisfy the hydromagnetic equilibrium equation (4.2). Of course, two variables are generally not enough to satisfy three scalar equations, so not every magnetic field structure can be realized as a hydromagnetic equilibrium.

The constraint on the magnetic field structure becomes clearest in axial symmetry, in which the magnetic field must take the form

$$\vec{B} = \nabla\alpha(r,\theta) \times \nabla\phi + \beta(r,\theta)\nabla\phi, \tag{4.3}$$

where α and β are (up to this point) arbitrary functions of the spherical coordinates r and θ (but independent of the azimuthal angle ϕ, for which I also used $\nabla\phi = \hat{\phi}/[r\sin\theta]$), and $\nabla \cdot \vec{B} = 0$ is automatically satisfied. In this case, P_1 and ρ_1 must clearly also depend only on r and θ, so the ϕ-component of the left-hand side of eq. (4.2) must be identically zero, imposing the same on the right-hand side:

$$0 = \frac{1}{c}(\vec{j} \times \vec{B})_\phi = \frac{\nabla\beta \times \nabla\alpha}{4\pi r^2 \sin^2\theta}, \tag{4.4}$$

thus the gradients $\nabla\alpha$ and $\nabla\beta$ must be parallel everywhere, and β must be (at least piecewise) a function of α, $\beta(r,\theta) = \beta[\alpha(r,\theta)]$ (Chandrasekhar & Prendergast 1956; Mestel 1956). Once this is imposed, only two non-trivial components of eq. (4.2) remain, and these can generally be satisfied by an appropriate choice of the two independent variables P_1 and ρ_1, as shown for a particular case in Mastrano *et al.* (2011). Thus, no further constraints need to be imposed on the magnetic field to obtain an MHD equilibrium.

Fig. 1 shows what might be an axially symmetric approximation to a realistic magnetic field configuration in a fluid star. The lines shown are the poloidal (meridional) magnetic field lines, i.e., lines of constant α. Outside the star, no substantial currents can be present, and this forces the field to be purely poloidal ($\beta = 0$). Since β is a function of α, we will have $\beta = 0$ everywhere, except on the field lines that close within the star, corresponding to the shaded region on the plot. In this shaded region, both α and β can be non-zero, so the magnetic field lines winds around in a twisted torus, whereas elsewhere $\beta = 0$ but $\alpha \neq 0$, so the field lines are purely poloidal, lying in meridional planes.

Long ago, Tayler (1973) showed that, in a stably stratified star, purely toroidal magnetic fields ($\alpha = 0$, $\beta \neq 0$) are subject to a kink-type instability, in which flux loops slide with respect to each other, almost exactly on surfaces of constant r. Much more recently, Akgün *et al.* (2013) showed that this is true for all toroidal fields, including those confined in a torus, as in Fig. 1, which had not been covered by the conditions imposed by Tayler (1973) or in other previous studies. Similarly, it has long been argued that purely poloidal fields ($\alpha \neq 0$, $\beta = 0$) are also always unstable (Markey & Tayler

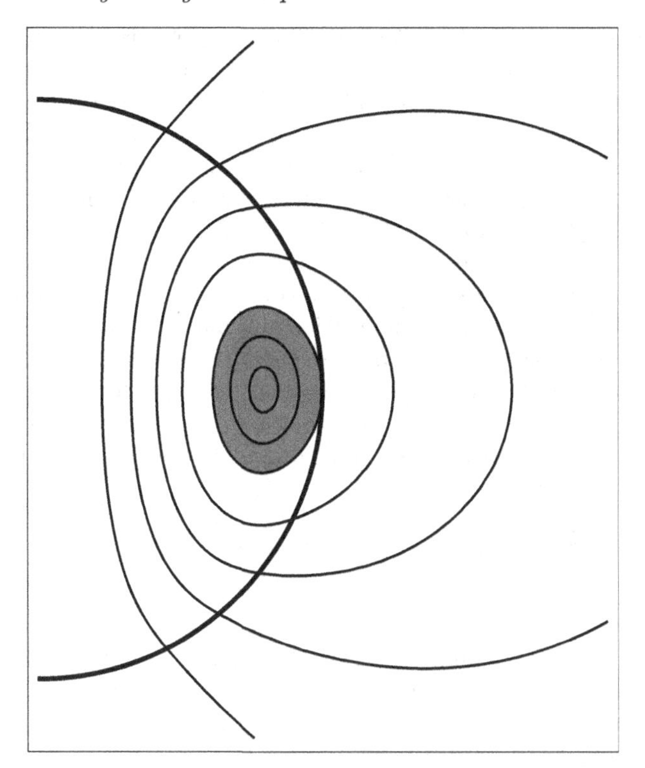

Figure 1. Meridional cut of a star with an axially symmetric magnetic field. The bold curve is the surface of the star, while the thinner curves are poloidal field lines (corresponding to $\alpha =$ constant). The toroidal component of the magnetic field ($\beta \neq 0$) lies only in regions where the poloidal field lines close inside the star (gray region). (Figure prepared by C. Armaza.)

1973; Wright 1973; Flowers & Ruderman 1977; Marchant *et al.* 2011). On the other hand, it was suspected that combined poloidal and toroidal fields might stabilize each other by tying each other together in a magnetic knot as in Fig. 1. This appears to be confirmed by the MHD simulations of Braithwaite and collaborators (Braithwaite & Spruit 2004, 2006; Braithwaite & Nordlund 2006), in which initially complex magnetic fields generically evolve into nearly axially symmetric, twisted-torus configurations like that in Fig. 1, with poloidal and toroidal components of roughly comparable strengths.

The conditions required for the poloidal and toroidal components to stabilize each other were studied numerically by Braithwaite (2009), whereas our group has done a couple of partial, analytical studies (Marchant *et al.* 2011; Akgün *et al.* 2013). Since the energy in the poloidal field component, E_{pol}, can be known or at least estimated from observations (roughly corresponding to E_{mag} in eq. 3.1), it is interesting to write the stability conditions as a (very rough and still not rigorously proven) allowed range for the energy in the hidden, toroidal component, E_{tor}:

$$0.25 \lesssim \frac{E_{tor}}{E_{pol}} \lesssim 0.5 \left[\left(\frac{\Gamma}{\gamma} - 1 \right) \frac{|E_{grav}|}{E_{pol}} \right]^{1/2} . \tag{4.5}$$

The indices γ and Γ characterize, respectively, the equilibrium profile of the star, $\gamma \equiv d\ln P_0/d\ln \rho_0$, and an adiabatic perturbation, which conserves entropy and chemical composition, $\Gamma \equiv (\partial \ln P/\partial \ln \rho)_{ad}$. The Ledoux criterion for stable stratification (stability against convection) requires $\Gamma > \gamma$. In the often assumed barotropic case, $\Gamma = \gamma$, whereas realistic values are $\Gamma/\gamma - 1 \sim 10^{-2}$ for neutron stars (stabilized by a small fraction of chemical impurities, as discussed in § 1) and $\Gamma/\gamma - 1 \sim 1/4$ in the radiative envelopes of upper main sequence stars, which are stabilized by entropy (see Reisenegger 2009 for a more detailed discussion).

Taken at face value, eq. (4.5) implies that, for barotropic stars ($\Gamma = \gamma$), there are no (axially symmetric) stable magnetic fields. However, it is important to note that the stars in Braithwaite's simulations were strongly stratified by entropy, whereas the analysis of Akgün *et al.* (2013) assumed strong stable stratification and made approximations based on this assumption. Thus, strictly speaking, neither of them is applicable to the barotropic case. On the other hand, simulations by Lander & Jones (2012) also suggest that magnetic fields in barotropic stars are generally unstable, and therefore eq. (4.5) might be applicable even in that limit. This issue is being further investigated within our group (Mitchell *et al.*, these Proceedings; Armaza *et al.*, these Proceedings).

It is interesting to rewrite the upper limit on E_{tor} from eq. (4.5) and evaluate it for neutron stars, in the form

$$\frac{E_{tor}}{|E_{grav}|} \lesssim 0.5 \left[\left(\frac{\Gamma}{\gamma} - 1\right) \frac{E_{pol}}{|E_{grav}|}\right]^{1/2} \lesssim 0.5 \times 10^{-4}, \tag{4.6}$$

where eq. (3.1) was used in the second inequality, identifying E_{mag} there with E_{pol} here. This shows that, for realistic poloidal fields, the toroidal component might be substantially stronger, but the total magnetic energy will still be much smaller than $|E_{grav}|$, so even the toroidal field is weak in a dynamical or structural sense.

5. Dissipative processes and field evolution

In the previous section, I have assumed ideal MHD, in the sense that there is a single, conducting fluid interacting with the magnetic field. This is likely a good approximation in the very early stages of the life of a neutron star, in which the relevant time scales are short and the temperature is high. Initially, the gravitational collapse probably leaves a highly convective, differentially rotating proto-neutron star, which eventually settles into a stable MHD equilibrium like those just described, in just a few Alvén times, $t_A \sim R(4\pi\rho)^{1/2}/B \sim (10^{14}\mathrm{G}/B)\mathrm{s}$. Soon afterwards, the temperature decreases enough for the crust to freeze to a solid state, the neutrons of the core and inner crust to become superfluid, and the protons in at least parts of the core to become superconducting.

The crust will thus no longer behave as a fluid. However, the electron currents supporting the magnetic field in the crust will carry along the magnetic flux lines in a process called *Hall drift*, which is non-dissipative but non-linear and has been argued to lead to a Kolmogoroff-like turbulent cascade of energy to small scales (Goldreich & Reisenegger 1992) or at least to the formation of current sheets (Urpin & Shalybkov 1991; Vainshtein *et al.* 2000; Reisenegger *et al.* 2007), which dissipate more quickly than a smooth, large-scale current. On the other hand, "Hall equilibria" have been found, in which the Hall drift does not modify the configuration of the magnetic field (Cumming, Arras, & Zweibel 2004; Gourgouliatos *et al.* 2013 and these Proceedings). If some of these equilibria are stable, which appears to be the case (Marchant *et al.*, in preparation), the magnetic field could evolve into one of them, so the further evolution would be dominated by the much slower Ohmic diffusion, as appears to be observed in many recent simulations (Viganò *et al.* 2013 and references therein; Rea, these Proceedings). The evolution of the magnetic field is likely to generate a Lorentz force that can no longer be balanced by pressure and gravity as in eq. (4.2) and will thus produce shear stresses and strains in the solid, which can break the crust if strong enough (as likely in magnetars), causing the matter and the magnetic field to rearrange. How this occurs and whether this can explain some of the violent events in magnetars is still largely an open question (see Levin & Lyutikov 2012 for a recent discussion).

The core, on the hand, is thought to remain in a fluid state, but here things change as well.

At high temperatures (corresponding to the "strong-coupling" regime in the one-dimensional simulations of Hoyos *et al.* 2008, 2010), the main change is that, over long enough times, neutrons and charged particles can convert into each other through *beta decays*, eventually establishing a chemical equilibrium controlled by only one variable, e.g., the local pressure or density. This means that, in its secular evolution, the fluid will behave as if it were barotropic, with P_1 and ρ_1 in eq. (4.2) proportional to each other, so there is now only one fluid degree of freedom. In this barotropic state, the possible magnetic field structures are much more constrained, and perhaps no purely fluid, stable equilibria exist. If the field is not too strong, the crust might help in supporting a new equilibrium structure in the core, otherwise the field might break the crust and be largely lost from the star.

At lower temperatures (the "weak-coupling" regime of Hoyos *et al.* 2008, 2010), the fluid becomes more and more degenerate, reducing the phase space for interactions and thus the conversion rates between neutrons and charged particles, but also the drag forces between them, so a two-fluid model becomes more applicable. The magnetic field will be coupled only to the charged particles, and it will force them to move relative to the neutrons in a process called *ambipolar diffusion* (Pethick 1991; Goldreich & Reisenegger 1992). If the charged particles are only protons and electrons, whose densities are tied together by the condition of charge neutrality, they will behave as a barotropic fluid, bringing us back to the same situation as in the previous paragraph.

Thus, the evolution of the neutron star magnetic field might unfold as follows (see Fig. 5, as well as a more detailed discussion to be found in Reisenegger 2009). When the neutron star is born, its internal temperature is high, $T \sim 10^{11}$K. Its thermal energy, $E_T \sim 10^{52}(T/10^{11}\mathrm{K})^2$erg, though much smaller than the gravitational binding energy, $|E_{grav}| \sim 10^{54}$erg, is substantially larger than the magnetic energy, $E_{mag} \sim 10^{49}(B/10^{16}\mathrm{G})^2$erg, in the early, relaxed, MHD equilibrium, even for the ultra-strong magnetar fields, $B \sim 10^{14-16}$G. However, neutrino emission cools the star very quickly (much faster than the magnetic field can evolve), until it drops well below the "equipartition" line where $E_T = E_{mag}$, at which point the dissipation of even a small fraction of the magnetic energy can substantially feed back on the thermal evolution, essentially halting the cooling. For a strong magnetic field ($B \gtrsim 10^{16}$G), this will happen in the high-temperature, strong-coupling regime ($T \gtrsim 10^9$K), and the evolution of the magnetic field will be controlled by beta decays, whereas at lower B the low-temperature, weak coupling regime is appropriate, and the evolution occurs through ambipolar diffusion, limited by neutron-charged particle collisions. Depending on field strength, the magnetic field in the crust might reorganize by breaking the latter (violently or causing plastic flow) or through Hall drift, or remain essentially unchanged, providing a fixed boundary condition to the evolution in the core. In any case, the magnetic feedback should leave T essentially constant until the magnetic field has reached a new equilibrium state, compatible with the long-term, barotropic behavior of the liquid core matter. If (as I would conjecture) there are no stable magnetic equilibria in a barotropic, fluid sphere, then these long-lived magnetic equilibria in neutron stars will rely on being stabilized by the solid crust,

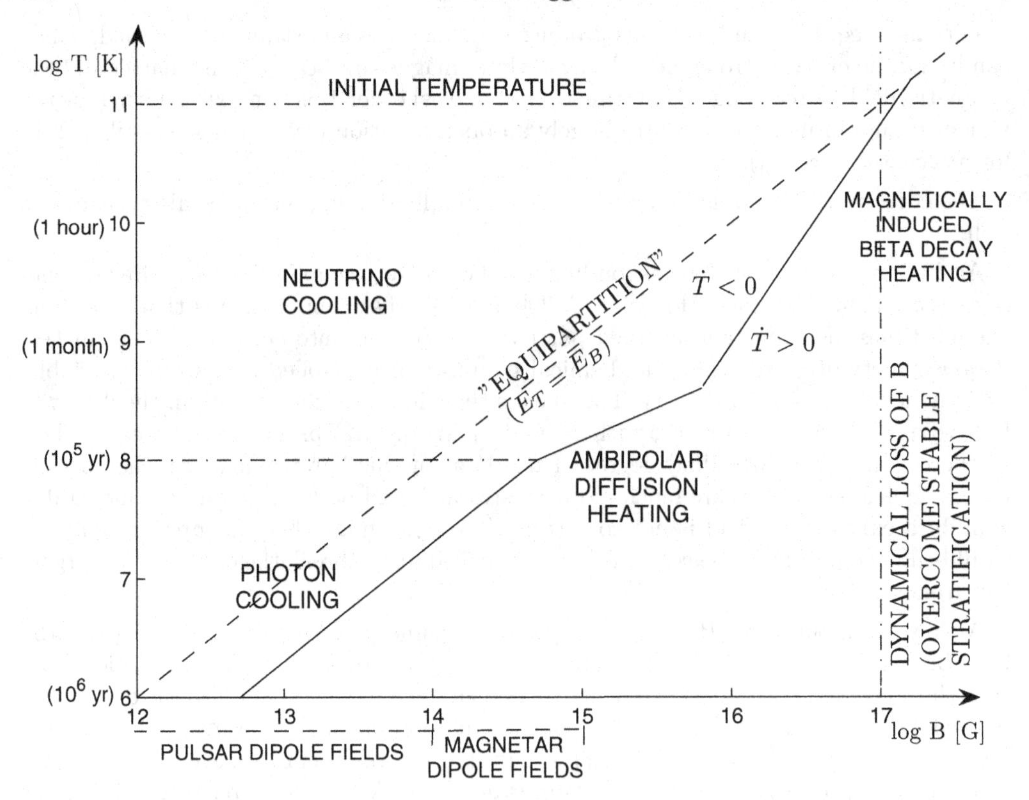

Figure 2. Magnetic field – temperature plane for a non-superfluid neutron star core. The dot–dashed horizontal lines show the initial temperature (just after core collapse), and the transition from neutrino-dominated (modified Urca) to photon-dominated cooling. The dashed diagonal line corresponds to the equality of magnetic and thermal energy. Above and to the left of the solid line, the star cools passively, on the time scales indicated in parenthesis along the vertical axis, without substantial magnetic field decay, so the evolution of the star is essentially a downward vertical line. Once the solid line is reached, magnetic dissipation mechanisms become important and generate heat that stops the cooling until the magnetic field has re-arranged to a new equilibrium state. (Figure prepared by C. Petrovich and first published in Reisenegger 2009.)

and thus the typical field strength must be relatively weak, probably not reaching the magnetar range.

Clearly, the evolution of the magnetic field can be complex and will require numerical simulations to be sorted out in more detail, even in the absence of superfluidity and superconductivity, which I have ignored in the previous discussion. Some aspects of their effects have been considered by other authors, e.g., Glampedakis, Andersson, & Samuelsson (2011).

6. Conclusions

The observed magnetic field strength on the surface of neutron stars appears to be roughly as expected from the flux of their progenitors (massive main sequence stars) and siblings (white dwarfs), although the neutron star birth is accompanied by violent processes that could alter it substantially. Soon after birth, it is likely to reach a stable, ideal-MHD equilibrium, with a poloidal and a toroidal component, which stabilize each other, aided by the compositional stratification of neutron star matter. The subsequent evolution relies on non-ideal-MHD processes such as Hall drift in the solid crust, and beta decays and ambipolar diffusion in the liquid core, all of which will lead to dissipation that temporarily halts the cooling of the neutron star, while the magnetic field re-arranges into a new equilibrium, which probably relies on shear forces in the crust that limit the field strength in this new, long-lived state.

7. Acknowledgments

The author thanks C. Armaza for preparing Fig. 1, C. Petrovich for preparing Fig. 5, and many colleagues and students for enlightening discussions. This work was financially supported by CONICYT International Collaboration Grant DFG-06, FONDECYT Regular Grant 1110213, and the Basal Center for Astrophysics and Associated Technologies (PFB-06/2007).

References

Akgün, T., Reisenegger, A., Mastrano, A., & Marchant, P. 2013, *MNRAS*, 433, 2445
Baade, W. & Zwicky, F. 1934, *PNAS*, 20, 259
Baym, G., Pethick, C., & Pines, D. 1969a, *Nature*, 224, 673
Baym, G., Pethick, C., & Pines, D. 1969b, *Nature*, 224, 674
Braithwaite, J. 2009, *MNRAS*, 397, 763
Braithwaite, J. & Nordlund, Å. 2006, *A&A*, 450, 1077
Braithwaite, J. & Spruit, H. 2004, *Nature*, 431, 819
Braithwaite, J. & Spruit, H. 2006, *A&A*, 450, 1097
Chandrasekhar, S. & Prendergast, K. H. 1956, *PNAS*, 42, 5
Cumming, A., Arras, P., & Zweibel, E. 2004, *ApJ*, 609, 999
Faucher-Giguère, C.-A. & Kaspi, V. M. 2006, *ApJ*, 643, 332
Flowers, E. & Ruderman, M. A. 1977, *ApJ*, 215, 302
Glampedakis, K., Jones, D. I., & Samuelsson, L. 2011, *MNRAS*, 413, 2021
Goldreich, P. & Reisenegger, A. 1992, *ApJ*, 395, 250
Gourgouliatos, K. N., Cumming, A., Reisenegger, A., Armaza, C., Lyutikov, M., & Valdivia, J. A. 2013, *MNRAS*, 434, 2480
Hoyos, J., Reisenegger, A., & Valdivia, J. A. 2008, *A&A*, 487, 789
Hoyos, J. H., Reisenegger, A., & Valdivia, J. A. 2010, *MNRAS*, 408, 1730
Israel, G. L., Belloni, T., Stella, L., *et al.* 2005, *ApJ*, 628, L53
Kaspi, V. M. 2010, *PNAS*, 107, 7147
Kouveliotou, C., Dieters, S., Strohmayer, T., *et al.* 1998, *Nature*, 393, 235
Lander, S. K. & Jones, D. I. 2012, *MNRAS*, 424, 482
Levin, Y. & Lyutikov, M. 2012, *MNRAS*, 427, 1574
Marchant, P., Reisenegger, A., & Akgün, T. 2011, *MNRAS*, 415, 2426
Markey, P. & Tayler, R. J. 1973, *MNRAS*, 163, 77
Mastrano, A., Melatos, A., Reisenegger, A., & Akgün, T. 2011, *MNRAS*, 417, 2288
Mestel, L. 1956, *MNRAS*, 116, 324
Oppenheimer, J. R. & Volkoff, G. M. 1939, *Phys. Rev.* 55, 374
Pethick, C. J. 1992, in *Structure and Evolution of Neutron Stars*, Proceedings of the SENS '90 Conference held in Kyoto, Japan, 1990, eds. D. Pines, R. Tamagaki, & S. Tsuruta, Redwood City, CA: Addison-Wesley, p.115
Reisenegger, A. 2009, *A&A*, 499, 557
Reisenegger, A., Benguria, R., Prieto, J. P., Araya, P. A., & Lai, D. 2007, *A&A*, 472, 233
Reisenegger, A. & Goldreich, P. 1992, *ApJ*, 395, 240
Tayler, R. J. 1973, *MNRAS*, 161, 365

Thompson, C. & Duncan, R. C. 1996, *ApJ*, 473, 322
Urpin, V. A. & Shalybkov, D. A. 1991, *Sov. Phys. JETP*, 73, 703
Vainshtein, S. I., Chitre, S. M., & Olinto, A. V. 2000, *Phys. Rev. E*, 61, 4422
Viganò, D., *et al.* 2013, *MNRAS*, 434, 123
Wright, G. A. E. 1973, *MNRAS*, 162, 339

Magnetic Fields throughout Stellar Evolution
Proceedings IAU Symposium No. 302, 2013
P. Petit, M. Jardine & H. Spruit, eds.

doi:10.1017/S1743921314002634

Hall Effect in Neutron Star Crusts

K. N. Gourgouliatos[1] and A. Cumming[1]

[1]Department of Physics, McGill University, 3600 rue University, Montréal, Québec H3A 2T8, Canada
email: kostasg@physics.mcgill.ca

Abstract. The crust of Neutron Stars can be approximated by a highly conducting solid crystal lattice. The evolution of the magnetic field in the crust is mediated through Hall effect, namely the electric current is carried by the free electrons of the lattice and the magnetic field lines are advected by the electron fluid. Here, we present the results of a time-dependent evolution code which shows the effect Hall drift has in the large-scale evolution of the magnetic field. In particular we link analytical predictions with simulation results. We find that there are two basic evolutionary paths, depending on the initial conditions compared to Hall equilibrium. We also show the effect axial symmetry combined with density gradient have on suppressing turbulent cascade.

Keywords. MHD, stars: magnetic field, stars: neutron

1. Introduction

The magnetic field of neutron stars is anchored in their crust. The crust is an electric conductor, where only electrons have the freedom to move with respect to the static lattice. As most observable properties of neutron stars are directly connected to processes of the magnetosphere and the surface of the neutron star, the understanding of the magnetic field evolution is crucial. Goldreich & Reisenegger (1992) have shown that the evolution of the magnetic field in neutron stars is governed by three processes: Hall drift, ambipolar diffusion and Ohmic dissipation. The relative importance of these processes depends primarily on the intensity of the magnetic field and the electron density, with Hall effect dominating in the crust for magnetic fields $B > 10^{13}$G, while ambipolar diffusion being important in the core where the neutron fraction is larger. The strongly magnetized and with low characteristic ages magnetars have dipole inferred magnetic fields reaching 10^{15}G (Olausen & Kaspi 2013), while older pulsars have weaker fields, suggesting some decay of the field with age.

The evolution of crustal magnetic field due to the Hall effect has been studied analytically (Jones 1988; Vainshtein *et al.* 2000; Reisenegger *et al.* 2007) and numerically (Shalybkov & Urpin 1997; Hollerbach & Rüdiger 2002, 2004; Pons & Geppert 2007; Viganò *et al.* 2012; Kojima & Kisaka 2012). However, there are still some open questions. In particular, it is important to link the analytical work with the results of the simulations, explore a variety of initial conditions and their effect in the structure of the field and investigate the absence of turbulent cascade in crust simulations as opposed to cartesian 3-D box simulations (Biskamp *et al.* 1996; Cho & Lazarian 2009; Wareing & Hollerbach 2009).

Here we present the results of a code we have developed examining the evolution of axially symmetric magnetic fields in neutron star crusts under the influence of the Hall effect and Ohmic dissipation.

2. Hall Effect

An axially symmetric magnetic field can be expressed using two scalar functions Ψ and I, $\boldsymbol{B} = \nabla\Psi \times \nabla\phi + I\nabla\phi$, where ϕ is the azimuthal angle. Since only electrons have the freedom to move, the electric current is given by $\boldsymbol{j} = -n_{\rm e}\mathrm{e}\boldsymbol{v}$, where $n_{\rm e}$ is the electron number density, e is the electron charge and $\boldsymbol{v}$ is the electron velocity. From Ampère's law it is $\boldsymbol{j} = \frac{c}{4\pi}\nabla \times \boldsymbol{B}$, where c is the speed of light, while for a finite conductivity σ the electric field becomes $\boldsymbol{E} = -\frac{1}{c}\boldsymbol{v} \times \boldsymbol{B} + \frac{1}{\sigma}\boldsymbol{j}$. We substitute the electric field in the induction equation and we find

$$\frac{\partial \boldsymbol{B}}{\partial t} = -\frac{c}{4\pi \mathrm{e}}\nabla \times \left(\frac{\nabla \times \boldsymbol{B}}{n_e} \times \boldsymbol{B}\right) - \frac{c^2}{4\pi}\nabla \times \left(\frac{\nabla \times \boldsymbol{B}}{\sigma}\right). \tag{2.1}$$

In spherical polar coordinates (r, θ, ϕ) we define the Grad-Shafranov operator $\Delta^* = \frac{\partial^2}{\partial r^2} + \frac{\sin\theta}{r^2}\frac{\partial}{\partial\theta}\left(\frac{1}{\sin\theta}\frac{\partial}{\partial\theta}\right)$ and $\chi = c/(4\pi \mathrm{e} n_{\rm e} r^2 \sin^2\theta)$ which is related to the effect density gradient and axial symmetry have in the evolution of the magnetic field. The vector equation giving the evolution of the magnetic field reduces to

$$\frac{\partial \Psi}{\partial t} = r^2 \sin^2\theta\, \chi(\nabla\Psi \times \nabla I)\cdot\nabla\phi + \frac{c^2}{4\pi\sigma}\Delta^*\Psi, \tag{2.2}$$

$$\begin{aligned}\frac{\partial I}{\partial t} &= r^2\sin^2\theta[\chi\nabla(\Delta^*\Psi)\times\nabla\Psi + \Delta^*\Psi\nabla\chi\times\nabla\Psi + I\nabla\chi\times\nabla I]\cdot\nabla\phi \\ &+ \frac{c^2}{4\pi\sigma}\left(\Delta^* I + \frac{1}{\sigma}\nabla I \times \nabla\sigma\right).\end{aligned} \tag{2.3}$$

These two equations form a system of non-linear, coupled, differential equations for Ψ and I, which in principle can be solved numerically. Yet, by analytical examination of the above equations we can come to some important conclusions. Assuming high conductivity σ, it is possible to find Hall equilibria solutions satisfying $\frac{\partial\Psi}{\partial t} = \frac{\partial I}{\partial t} = 0$, these solutions have $I = I(\Psi)$ with Ψ satisfying a Grad-Shafranov equation, similar to that of barotropic magnetic equilibria (Cumming *et al.* 2004; Gourgouliatos *et al.* 2013a,b). In the absence of an initial poloidal field ($\Psi = 0$), the toroidal field evolves with the advection of I on surfaces of constant χ obeying Burgers' equation (Reisenegger *et al.* 2007). However, a purely poloidal initial field generates a toroidal field, even if it is not present in the initial state.

3. Simulation - Conclusions

We have developed a finite difference code to study the evolution of the magnetic field because of the Hall effect and Ohmic dissipation. We assume that the electron number density on the surface of the crust is two orders of magnitude smaller than in the crust-core interface and it is $n_{\rm e} \propto (r_* - r)^4$, while the conductivity is $\sigma \propto n_{\rm e}^{2/3}$, the thickness of the crust is $0.2r_*$, where r_* is the radius of the neutron star. The field is confined in the crust of the neutron star without threading the core, while the star is surrounded by a vacuum, therefore, a multipole expansion is used as a surface boundary condition. We have also run cases with constant electron number density $n_{\rm e}$ to allow comparison with previous studies. We have run a variety of initial conditions: purely toroidal field (Figure 1), purely poloidal fields and combinations of poloidal and toroidal fields.

We link the analytical conclusions with the results of our simulations. In particular, a purely toroidal field is advected along lines of constant χ. A Hall equilibrium field does not evolve in a Hall timescale, but in a longer timescale which depends on the Ohmic

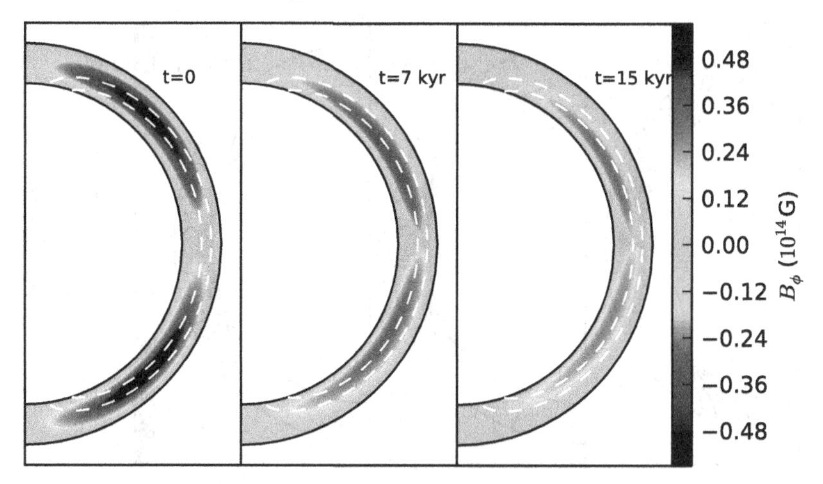

Figure 1. The evolution of a purely toroidal field. I is advected along surfaces of constant χ, drawn with white dashed lines. Within a Hall time-scale shocks form between fields of opposite polarity. For a colour figure please refer to the online version.

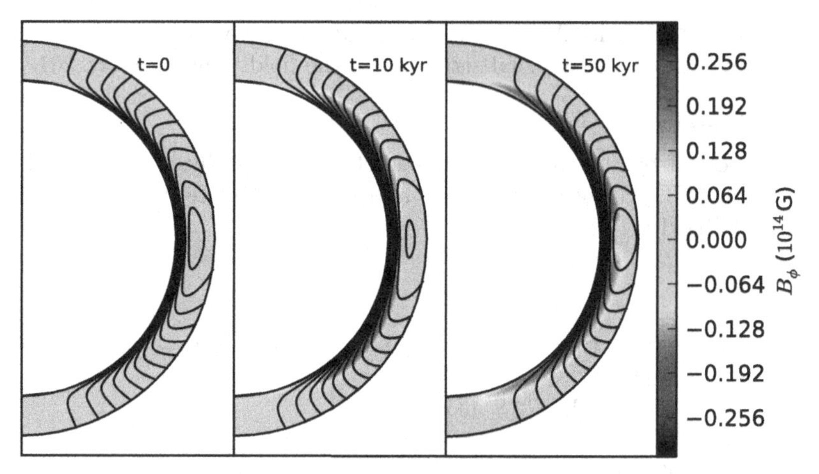

Figure 2. The evolution of poloidal field starting out of Hall equilibrium so that it generates a toroidal field of positive polarity in the northern hemisphere. The field lines are pushed towards the poles initially. For a colour figure please refer to the online version.

timescale. Ohmic dissipation pushes the field out of Hall equilibrium, thus a toroidal field is generated. A purely poloidal field out of Hall equilibrium generates a toroidal field within a Hall timescale. The polarity of the toroidal field generated and subsequent evolution depends on the initial field structure compared to the Hall equilibrium field. If the quadrupolar toroidal field is positive in the northern hemisphere the poloidal field lines are pushed to the poles (Figure 2), otherwise the poloidal field lines are pushed to the equator (Figure 3), leading to distinguishable structures. The toroidal field is subdominant energetically compared to the poloidal, unless the initial conditions impose a strong toroidal field (Ciolfi & Rezzolla 2013). A strong toroidal field twists the poloidal field lines trying to align I with Ψ, while transferring a significant amount of energy to the poloidal field. Assuming a density profile where χ =const. the field generates higher order multipoles, which are suppressed otherwise.

The two evolutionary paths shown in Figures 2 and 3 can be distinguished observationally in terms of the braking indices as they predict different decay rates for the

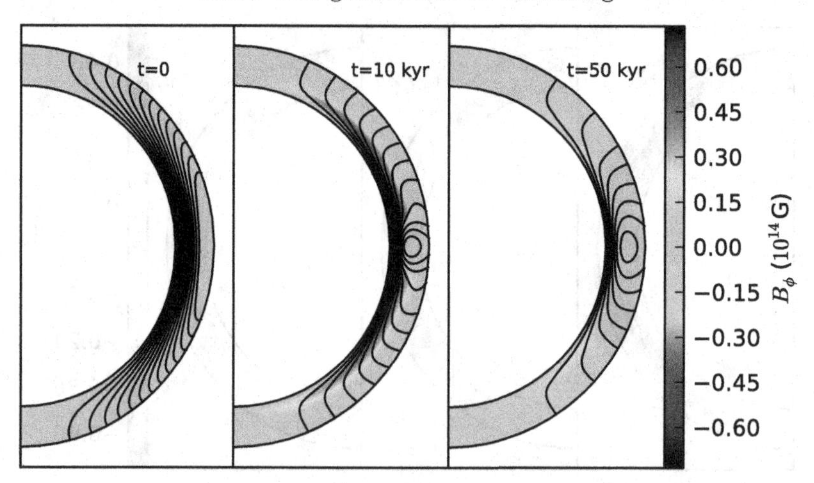

Figure 3. The evolution of poloidal field starting out of Hall equilibrium so that it generates a toroidal field of negative polarity in the northern hemisphere. The field lines are pushed towards the equator. For a colour figure please refer to the online version.

dipole component of the field. Also the surface temperature profile is different, as thermal conductivity depends on the internal structure of the field (Viganò *et al.* 2013).

Acknowledgements

KNG is supported by the Centre de Recherche en Astrophysique du Québec. AC is supported by an NSERC Discovery Grant and is an Associate Member of the CIFAR Cosmology and Gravity Program.

References

Biskamp, D., Schwarz, E., & Drake, J. F. 1996, *Physical Review Letters*, 76, 1264
Cho, J. & Lazarian, A. 2009, *ApJ*, 701, 236
Ciolfi, R. & Rezzolla, L. 2013, *MNRAS*, 435, L43
Cumming, A., Arras, P., & Zweibel, E. 2004, *ApJ*, 609, 999
Goldreich, P. & Reisenegger, A. 1992, *ApJ*, 395, 250
Gourgouliatos, K. N., Cumming, A., Reisenegger, A., *et al.* 2013, *MNRAS*, 434, 2480
Gourgouliatos, K. N., Cumming, A., Lyutikov, M., & Reisenegger, A. 2013, arXiv:1305.0149
Hollerbach, R. & Rüdiger, G. 2002, *MNRAS*, 337, 216
Hollerbach, R. & Rüdiger, G. 2004, *MNRAS*, 347, 1273
Jones, P. B. 1988, *MNRAS*, 233, 875
Kojima, Y. & Kisaka, S. 2012, *MNRAS*, 421, 2722
Olausen, S. A. & Kaspi, V. M. 2013, arXiv:1309.4167
Pons, J. A. & Geppert, U. 2007, *A&A*, 470, 303
Reisenegger, A., Benguria, R., Prieto, J. P., Araya, P. A., & Lai, D. 2007, *A&A*, 472, 233
Shalybkov, D. A. & Urpin, V. A. 1997, *A&A*, 321, 685
Vainshtein, S. I., Chitre, S. M., & Olinto, A. V. 2000, *Phys. Rev. E*, 61, 4422
Viganò, D., Pons, J. A., & Miralles, J. A. 2012, Computer Physics Communications, 183, 2042
Viganò, D., Rea, N., Pons, J. A., *et al.* 2013, *MNRAS*, 434, 123
Wareing, C. J. & Hollerbach, R. 2009, *A&A*, 508, L39

Magnetic Fields throughout Stellar Evolution
Proceedings IAU Symposium No. 302, 2013
P. Petit, M. Jardine & H. Spruit, eds.

doi:10.1017/S1743921314002646

Magnetohydrodynamic equilibria in barotropic stars

C. Armaza[1], A. Reisenegger[1], J. A. Valdivia[2] and P. Marchant[1,3]

[1]Instituto de Astrofísica, Facultad de Física, Pontificia Universidad Católica de Chile,
Av. Vicuña Mackenna 4860, 782-0436 Macul, Santiago, Chile
email: cyarmaza@uc.cl

[2]Departamento de Física, Facultad de Ciencias, Universidad de Chile,
Casilla 653, Santiago, Chile.

[3]Argelander Institut für Astronomie, Universität Bonn,
Auf dem Hügel 71, D-53121, Bonn, Germany

Abstract. Although barotropic matter does not constitute a realistic model for magnetic stars on short timescales, it would be interesting to confirm a recent conjecture that states that magnetized stars with a barotropic equation of state would be dynamically unstable (Reisenegger 2009). In this work we construct a set of barotropic equilibria, which can eventually be tested using a stability criterion. A general description of the ideal MHD equations governing these equilibria is summarized, allowing for both poloidal and toroidal magnetic field components. A new finite-difference numerical code is developed in order to solve the so-called Grad-Shafranov equation describing the equilibrium of these configurations, and some properties of the equilibria obtained are briefly discussed.

Keywords. MHD, stars: magnetic field

---

1. Overview

The persistence of magnetic fields in massive stars and their stellar remnants motivates the study of what physical conditions are involved in sustaining such configurations. In this context, barotropic equations of state, where pressure is a function solely of density, are often assumed to describe the matter within these objects (Yoshida & Eriguchi 2006; Haskell *et al.* 2008; Lander & Jones 2009; Ciolfi *et al.* 2009; Fujisawa *et al.* 2012). Barotropy strongly restricts the range of possible equilibrium configurations and does not strictly represent the realistic stably stratified matter within these objects, which is likely to be an essential ingredient in the stability of magnetic fields in stars on short timescales (Reisenegger 2009).

The question whether magnetic equilibria in barotropic stars can be stable or not remains as an important issue to be answered. Several authors (Lander & Jones 2009; Ciolfi *et al.* 2009; Fujisawa *et al.* 2012; Gourgouliatos *et al.* 2013) have explored the possible axially symmetric equilibria in barotropic stars, generally finding that the fraction of the total magnetic energy corresponding to the toroidal component, $E_{\rm tor}/E_{\rm mag}$, is at most a few %. In addition to not being enough to account for the energy emitted by magnetars, this would be insufficient to stabilize the poloidal component (Braithwaite 2009; Akgün *et al.* 2013), as confirmed by the simulations of Lander & Jones (2012). However, recent simulations (Ciolfi & Rezzolla 2013; Fujisawa & Eriguchi 2013) have shown that higher fractions $E_{\rm tor}/E_{\rm mag}$ are possible, making a more extensive survey of these equilibria relevant. Studying properties of barotropic equilibria could be also relevant considering the scenario in which neutron stars would reach an effectively barotropic state after overcoming stable stratification by means of direct and inverse β–decays and ambipolar diffusion acting on timescales shorter than their lifetime (Hoyos *et al.* 2008; Reisenegger 2009; Reisenegger, these Proceedings; Mitchell *et al.*, these Proceedings).

This ongoing research is focused on obtaining a wide range of numerical barotropic equilibria, paying attention to their main properties. In addition, these results can be considered as a starting point to study in more detail whether magnetic fields in a wide range of barotropic equilibria are stable or not (Mitchell *et al.*, these Proceedings).

2. Barotropic equilibria: the Grad-Shafranov equation

Throughout this work, we take the approach of considering a magnetic star within the context of ideal MHD, that is, a perfectly conducting fluid in dynamical equilibrium described by the Euler equation,

$$\nabla P + \rho\nabla\Phi = \frac{1}{c}\mathbf{J}\times\mathbf{B}, \tag{2.1}$$

where the right-side is the Lorentz force per unit volume. All known stars have a very large fluid pressure P ($P \sim GM^2/R^4$, M being the mass and R the radius), to magnetic pressure $B^2/8\pi$ ratio (B being an estimation of the maximum magnetic field strength), $8\pi P/B^2 \gtrsim 10^6$ (Reisenegger 2009), which suggests that magnetic fields do not play an important role in the structure of these stars, so at first approximation we can consider the star as spherical, with negligible deformations due to magnetic forces. In addition, if axial symmetry is assumed, and spherical coordinates (r,θ,ϕ) are used to describe the model, all scalar quantities are independent of the azimuthal coordinate, and the magnetic field may be expressed as the sum of a *poloidal* (meridional) component, and a *toroidal* (azimuthal) component, each determined by a single scalar function,

$$\mathbf{B} = \mathbf{B}_{\rm pol} + \mathbf{B}_{\rm tor} = \nabla\alpha(r,\theta)\times\nabla\phi + \beta(r,\theta)\nabla\phi, \tag{2.2}$$

which turn out to be constant along their respective field lines (Chandrasekhar & Prendergast 1956). Under this symmetry, the azimuthal component of the magnetic force per unit volume must vanish, which implies a functional relation between these scalar functions, $\beta(r,\theta) = \beta\left(\alpha(r,\theta)\right)$. In this way, both α and β are constant along field lines and, if a vacuum is assumed outside the star, the toroidal field may lie only in regions where the poloidal field lines close within the star. On the other hand, *if a barotropic equation of state, $P = P(\rho)$, is assumed*, the Lorentz force per unit mass must be the gradient of some arbitrary function $\chi(r,\theta)$, which turns out to be a function of α as well, $\chi(r,\theta) = \chi\left(\alpha(r,\theta)\right)$. From this, a non-linear elliptic partial differential equation is found to be the master equation governing barotropic MHD equilibria, the so-called Grad-Shafranov (GS) equation,

$$\frac{\partial^2\alpha}{\partial r^2} + \frac{\sin\theta}{r^2}\frac{\partial}{\partial\theta}\left(\frac{1}{\sin\theta}\frac{\partial\alpha}{\partial\theta}\right) + \beta\beta' + r^2\sin^2\theta\rho\chi' = 0 \tag{2.3}$$

(Grad & Rubin 1958, Shafranov 1966), where primes stand for derivatives with respect to the argument, and both $\beta = \beta(\alpha)$ and $\chi = \chi(\alpha)$ are arbitrary functions, whose form may be chosen depending on the particular magnetic configuration of interest. Under the assumption of weak magnetic field discussed in this section, the density ρ appearing in the GS equation may be replaced by its non-magnetic background counterpart, $\rho = \rho(r)$, such that we solve for the magnetic functions for a *given* density profile, instead of considering the more difficult task of solving self-consistently for the magnetic functions *and* for the fluid quantities, as done, e. g., by Lander & Jones (2012).

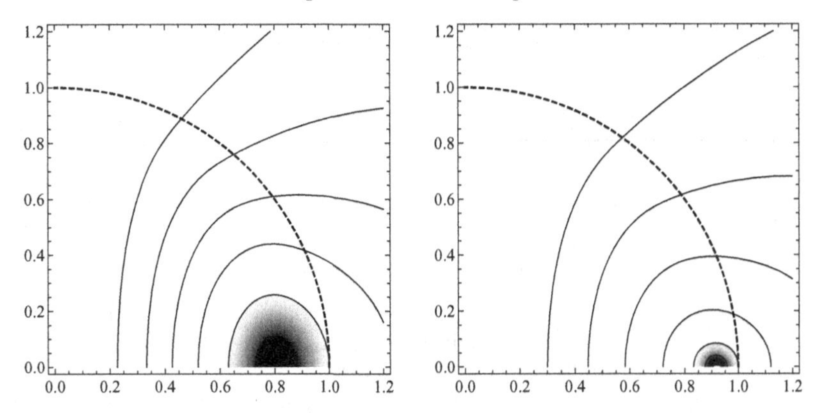

Figure 1. Numerical equilibria found using our code. In both cases, we used $\beta(\alpha)$ as given in Eq. (3.1), together with $\chi(\alpha) = \alpha$. Left: $s = 10$, right: $s = 45$. The dashed line corresponds to the stellar surface. Poloidal field lines are shown in solid black curves while the color map accounts for $(r \sin\theta B_\phi)$.

3. Numerical solutions and discussion

In order to obtain suitable barotropic equilibria, we have developed a finite-difference code to solve numerically the GS equation inside the star, for arbitrary choices of $\beta(\alpha)$, $\chi(\alpha)$ and density profile $\rho(r)$. Outside the star, the function $\alpha(r,\theta)$ may be written as a superposition of multipoles, corresponding to the general solution of the GS equation with both $\beta = 0$ and $\rho = 0$. Solutions found inside the star are matched to the exterior expansion by demanding continuity of the magnetic field components, in order to avoid surface currents. After testing our code, we studied the so-called twisted-torus configuration

$$\beta(\alpha) = \begin{cases} s(\alpha - \alpha_s)^{1.1} & \alpha_s \leqslant \alpha \\ 0 & \alpha < \alpha_s, \end{cases} \tag{3.1}$$

where s is a free parameter accounting for the relative strength between the poloidal and the toroidal component. Here, $\alpha_s \equiv \alpha(R, \pi/2)$ stands for the value of α along the longest poloidal field line closing within the star, R being the stellar radius, so the toroidal field lies in the region $\alpha(r,\theta) \geqslant \alpha_s$ only, as seen in Figure 1. This choice of $\beta(\alpha)$, along with $\chi(\alpha) = \alpha$, allows us to compare with previous works which have taken these same functional forms. For the simple, but reasonably realistic profile $\rho(r) = \rho_c(1 - r^2/R^2)$, we explored the behavior of such configurations by changing the parameter s. Figure 2 shows the magnetic field profile for two different values of s: the larger the value of s, the stronger the toroidal field becomes, although the region where the toroidal field lies shrinks, as shown in Figure 1.

In all cases, the energy stored in the toroidal component is only a few percent of the total magnetic energy, even when the toroidal field strength is comparable to the poloidal one, in agreement with previous works. Furthermore, this ratio seems to be bounded by a certain maximum value as seen in Figure 3 (left panel), above which the ratio would decrease, as already pointed out in the literature (Lander & Jones 2009, Ciolfi *et al.* 2009, Fujisawa *et al.* 2012), although such a behavior for large s and density profile considered here is not reached in our calculations. It is interesting to see, however, that we do reach a maximum in the ratio of toroidal to poloidal flux (Figure 3, right panel).

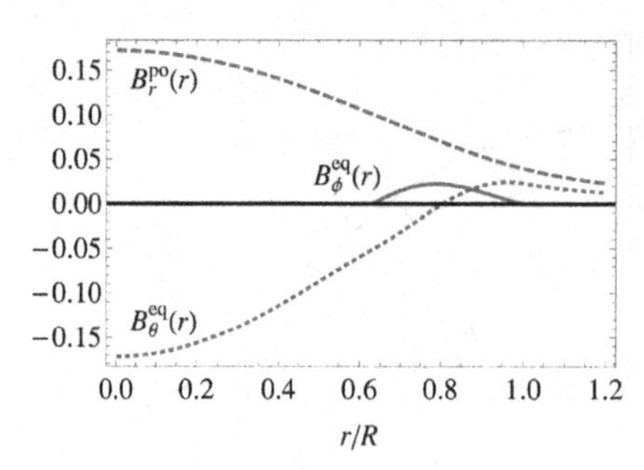

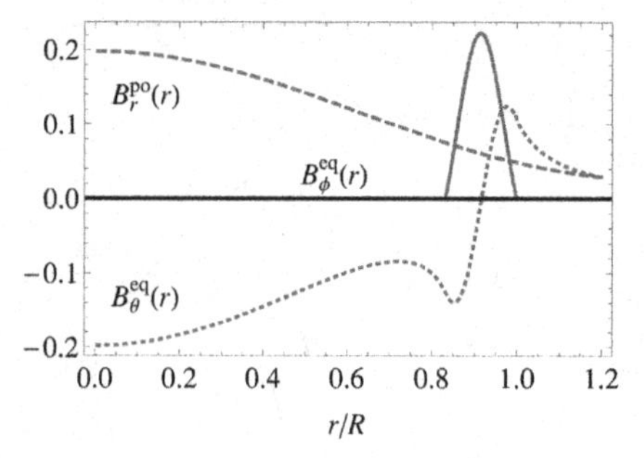

Figure 2. Magnetic field profiles for equilibria in Fig. 1. Left: $s = 10$, with $E_{\rm tor}/E_{\rm mag} \approx 0.5\%$. Right: $s = 45$, with $E_{\rm tor}/E_{\rm mag} \approx 3.5\%$.

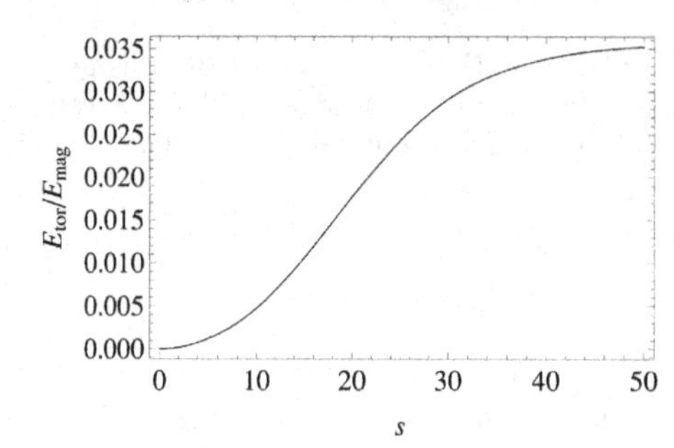

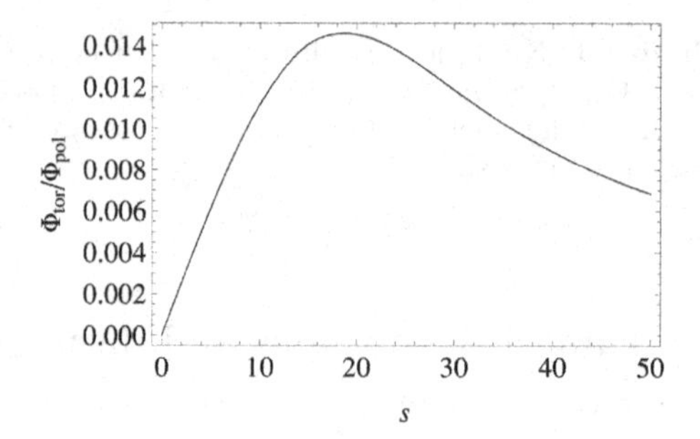

Figure 3. Toroidal-to-total magnetic energy (left) and maximum toroidal-to-maximum poloidal flux (right) ratios as functions of s, using $\rho(r) = \rho_c(1 - r^2/R^2)$ and $\chi' = 1$.

Acknowledgments

All authors are supported by CONICYT International Collaboration Grant DFG-06. CA and PM have been supported by CONICYT Master's Fellowships. AR and JAV are supported by FONDECYT Regular Grants 1110213 and 1110135, respectively. CA, AR and PM are supported by the Basal Center for Astrophysics and Associated Technologies. CA thanks K. N. Gourgouliatos and K. Fujisawa for very stimulating discussions.

References

Akgün, T., Reisenegger, A., Mastrano, A., & Marchant, P. 2013, *MNRAS*, 433, 2445
Braithwaite, J. 2009, *MNRAS*, 397, 763
Chandrasekhar, S. & Prendergast, K. H. 1956, *Proc. Nat. Acad. Sci.*, 42, 5
Ciolfi, R., Ferrari, V., Gualtieri, L., & and Pons, J. A. 2009, *MNRAS*, 397, 913
Ciolfi, R. & Rezzolla, L. 2013, *MNRAS*, 435, L43
Fujisawa, K., Yoshida, S., & Eriguchi, Y. 2012, *MNRAS*, 422, 434
Fujisawa, K. & Eriguchi, Y. 2013, *MNRAS*, 432, 1245
Grad, H. & Rubin, H. 1958, in Proc. *2st Int. Conf. on Peaceful Uses of Atomic Energy*. United Nations, Geneva, 31, 190
Gourgouliatos, K. N., Cumming, A., Reisenegger, A., Armaza, C., Lyutikov, M., & Valdivia, J. A. 2013, *MNRAS*, 434, 2480
Haskell, B., Samuelsson, L., Glampedakis, K., & Andersson, N. 2008, *MNRAS*, 385, 531
Hoyos, J., Reisenegger, A., & Valdivia, J. A. 2008, *A&A*, 487, 789
Lander, S. K. & Jones, D. I. 2009, *MNRAS*, 395, 2162
Lander, S. K. & Jones, D. I. 2012, *MNRAS*, 424, 482
Reisenegger, A. 2009, *A&A*, 499, 557
Shafranov, V. D. 1966, in: *Reviews of Plasma Physics*, (New York: Cons. Bureau), 2, 103
Yoshida, S. & Eriguchi, Y. 2006, *ApJS*, 164, 156

Magnetic Fields throughout Stellar Evolution
Proceedings IAU Symposium No. 302, 2013
P. Petit, M. Jardine & H. Spruit, eds.

doi:10.1017/S1743921314002658

Magnetic field structures inside magnetars with strong toroidal field

Kotaro Fujisawa

Department of Earth Science and Astronomy, Graduate School of Arts and Sciences, University of Tokyo, Komaba, Meguro-ku, Tokyo 153-8902, Japan
email: fujisawa@ea.c.u-tokyo.ac.jp

Abstract. We have analyzed the magnetized equilibrium studies with strong toroidal magnetic fields and found that the negative toroidal current density inside the star is very important for the strong toroidal magnetic fields. The strong toroidal magnetic fields require the strong poloidal current, but the strong poloidal current results in the localized strong toroidal current density in the axisymmetric system. This localized toroidal current changes the magnetic field configuration and makes the size of the toroidal magnetic field region smaller. As a result, the toroidal magnetic field energy can not become large. We need to cancel out the localized toroidal current density in order to obtain the large toroidal fields solutions. We have found and showed that the negative toroidal current cancels out the localized toroidal current density and sustain the large toroidal magnetic field energy inside the star. We can explain the magnetized equilibrium studies with strong toroidal magnetic fields systematically using the negative current density. Physical meaning of the negative current is key to the magnetar interior magnetic fields.

Keywords. Stars: interior - Stars: neutron - Stars: magnetic fields

1. Introduction

Anomalous X-ray Pulsars (AXPs) and Soft Gamma-ray Repeaters (SGRs) are considered as special classes of magnetized neutron stars, magnetars. Magnetars have very strong dipole magnetic fields whose typical values reach about 10^{15} G at their surfaces. Recent observations, however, reported the low magnetic fields SGRs whose dipole fields are much lower than typical values (Rea *et al.* 2010). Their high energy activities would be driven by the internal strong toroidal magnetic fields energy, but we can not observe the internal magnetic fields directly. We need to calculate magnetized stars in equilibrium theoretically to obtain the internal magnetic fields configurations.

Many theoretical studies, however, failed to obtain the magnetized star with strong toroidal magnetic fields (e.g. Ciolfi *et al.* 2009; Lander & Jones 2009; Fujisawa *et al.* 2012 etc.). The typical value of *toroidal* to *total* magnetic filed energy ratio M_t/M is only a few percent. On the other hand, only a few works had obtained the strong toroidal models under special boundary conditions. Glampedakis *et al.* (2012) obtained strong toroidal field models ($M_t/M \sim 25\%$) imposing surface current on the stellar surface. Duez & Mathis (2010) obtained very strong toroidal models ($M_t/M \sim 66\%$) imposing the boundary condition that the magnetic flux on the stellar surfaces should vanish. Therefore, all magnetic field lines in the model are closed and confirmed within the star. These works, however, did not describe the reason why such boundary conditions result in the strong toroidal magnetic fields. In order to answer this question, we have analysed these works carefully and found the negative toroidal current density (opposite flowing toroidal current density) inside the star plays very important role for the strong toroidal magnetic fields (see Fujisawa & Eriguchi 2013).

2. Formulation and arbitrary functions

We assume that the system is stationary and axisymmetry. Then, the magnetic fields are described by two scalar functions Ψ and I as below:

$$\boldsymbol{B} = \frac{1}{r\sin\theta}\nabla\Psi(r,\theta)\times\boldsymbol{e}_\varphi + \frac{I(r,\theta)}{r\sin\theta}\boldsymbol{e}_\varphi, \tag{2.1}$$

where Ψ is a poloidal magnetic flux function and I is a poloidal current flux function. The magnetic flux function Ψ is governed by the equation as

$$\frac{\partial^2\Psi}{\partial r^2} + \frac{\sin\theta}{r^2}\frac{\partial}{\partial\theta}\left(\frac{1}{\sin\theta}\frac{\partial\Psi}{\partial\theta}\right) = -4\pi r\sin\theta\frac{j_\varphi}{c}. \tag{2.2}$$

The source term of this equation is the toroidal current density in the right hand side. The toroidal current density is determined by matter equation. From the axisymmetric condition, I is a conserved quantity along the magnetic field line ($I \equiv I(\Psi)$) in both barotropic and baroclinic systems (see Braithwaite 2009). If the magnetized star is barotropic ($p = p(\rho)$) without differential rotation and meridional flow, the toroidal current density is described using two arbitrary function I and F as

$$\frac{j_\varphi}{c} = \frac{1}{4\pi}\frac{dI(\Psi)}{d\Psi}\frac{I(\Psi)}{r\sin\theta} + \rho r\sin\theta F(\Psi). \tag{2.3}$$

Where F is also a conserved quantity ($F \equiv F(\Psi)$) which is related to the Lorentz force (see Fujisawa *et al.* 2012). If the outside of the star is vacuum, I must vanish in the region. As follow many previous studies, we fix the functional forms of F and I as

$$F(\Psi) = F_0, \quad I(\Psi) = \frac{I_0}{k+1}(\Psi - \Psi_{\rm max})^{k+1}\Theta(\Psi - \Psi_{\rm max}), \tag{2.4}$$

where F_0 and k are constants and $\Psi_{\rm max}$ is a maximum value of Ψ in the vacuum region. We fix $k = 0.1$. I_0 means the local strength of the toroidal magnetic field. Θ is Heaviside-step function. We calculate Eq. (2.2) using Green function relaxation method with arbitrary current sheets. (see Fujisawa & Eriguchi 2013).

3. Results & Discussion

First, we have calculated three magnetized equilibrium states changing the value of I_0. The numerical solutions are displayed in Fig 1. The top panels show the poloidal magnetic field configurations and the strength of the toroidal magnetic fields. The bottom panels show the distributions of the toroidal current density. As the value of I_0 increases, the local strength of the toroidal magnetic field also increases. At the same time, however, the size of the toroidal magnetic field region becomes smaller.

The strong toroidal magnetic fields require the strong poloidal current I (Eq. 2.1), but I is related to the toroidal current density by the first term in Eq. (2.3). As the I increases, the toroidal current density also increases. Since the size of the toroidal magnetic field region is limited by the boundary condition, the toroidal current from the I term is localized within the stellar interior (see Fig. 1). The localized strong toroidal current changes the magnetic fields configurations. The distribution of Ψ also becomes localized and the size of the toroidal field region becomes smaller (Fig. 1). As a result, we fail to obtain the large toroidal magnetic fields solutions. We see similar results as Fig. 1 in the previous studies (see table 2 in Lander & Jones 2009, Fig. 12 in Ciolfi *et al.* 2009, Fig. 4 in Glampedakis *et al.* 2012). Therefore, this behaviour of the toroidal field region is a general feature of axisymmetric magnetized equilibria. We need to cancel out the localized toroidal current from the I term to obtain the strong toroidal magnetic models.

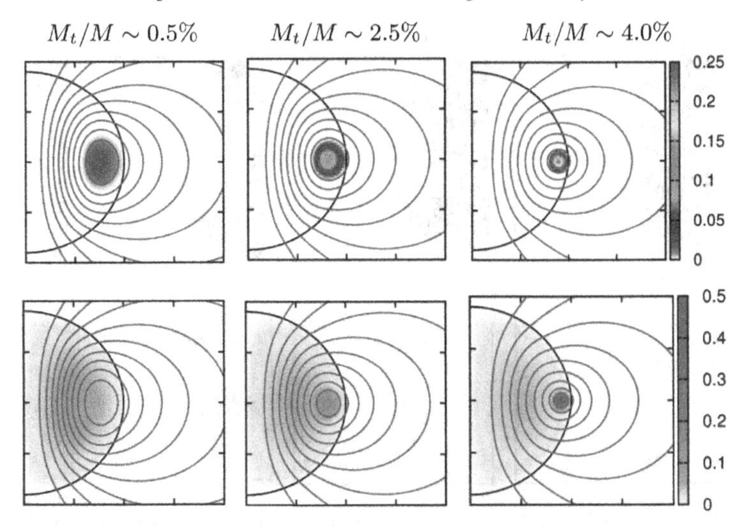

Figure 1. The magnetic field configurations and the strengths of the toroidal magnetic fields (colour maps, top) and the distributions of toroidal current density (color maps, bottom). Each panel shows $I_0 = 5$ (left), $I_0 = 20$ (center) and $I_0 = 40$ (right) solution respectively.

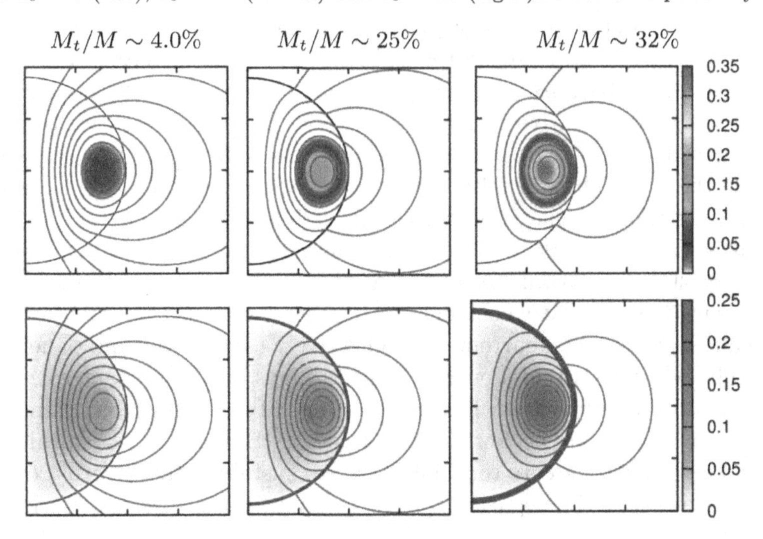

Figure 2. Same as Fig. 1 except for the negative surface current. The width of the surface represents the surface current strength. Each panel shows the solution with weak surface current (left), middle surface current (center) and strong surface current (right).

The surface current can cancel out the toroidal current density from the I term. Next, we have calculated the magnetized equilibrium states with negative surface current (Fig. 2). The negative surface current cancel out the localized toroidal current density and the size of the toroidal magnetic fields does not become small (see from left to right panel). As a result, these models can sustain the strong toroidal magnetic fields energy and the value of the energy ratio reaches $M_t/M \sim 32\%$.

We can explain the magnetized equilibrium states with strong toroidal magnetic fields systematically using the negative toroidal current. We classify the previous studies into three types according to the boundary conditions at the surface, (a): open fields model (e.g. Tomimura & Eriguchi 2005; Lander & Jones 2009; Ciolfi *et al.* 2009), (b): surface current model (Glampedakis *et al.* 2012) and (c): close fields model (Duez & Mathis

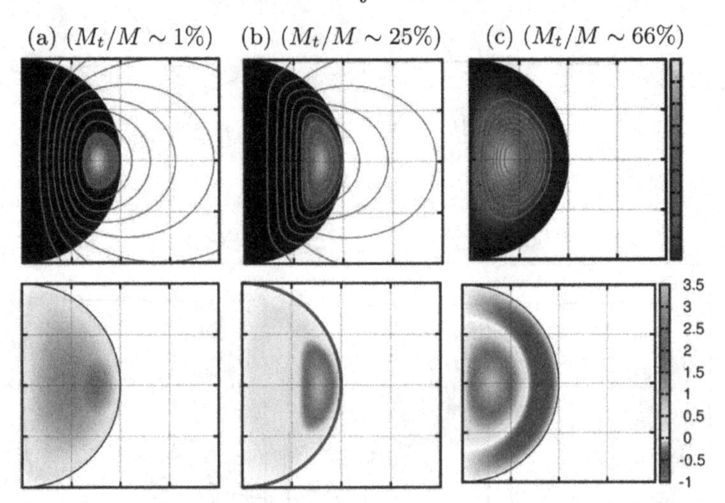

Figure 3. Top: The poloidal magnetic field lines and the toroidal magnetic field strength (color map). Bottom: The toroidal current density distributions (color map) of each model.

2010; Yoshida *et al.* 2012). The panels in Fig. 3 display the magnetic fields configuration (top) and the toroidal current density (bottom) in each model. The typical energy ratios are $M_t/M \sim 1\%$ (a), $\sim 25\%$ (b) and $\sim 66\%$ (c). (b) and (c) have the strong toroidal magnetic fields energy because they have the negative toroidal current density. As seen from Fig. 2, (b) has the negative surface current and (c) has the negative toroidal current density region near the stellar surface, while (a) does not have the negative toroidal current inside the star. These results show the negative toroidal current are required to sustain the strong toroidal magnetic fields inside the star.

Very recently, Ciolfi & Rezzolla (2013) have succeed in obtaining the magnetized equilibria with very strong toroidal magnetic energy ($M_t/M \sim 90\%$). We can also explain their solutions using the negative current density. They changed the functional form of F in order to cancel out toroidal current more effectively (see the bottom panels of Fig. 2 in the paper). The negative toroidal current density plays also important role in this case.

It is important to consider the physical meanings of the negative current. One possibility is magnetic stress on the bottom of the crust (Braithwaite & Spruit 2006). Our results show that the strong core toroidal fields are sustained by the crust of the magnetar. Physical meaning of the negative current is key to the magnetar interior fields.

References

Braithwaite, J., 2009, *MNRAS*, 397, 763
Braithwaite, J. & Spruit, H. C., 2006, *A&A*, 450, 1097
Ciolfi, R., Ferrari, V., Gualtieri, L., & Pons, J. A., 2009, *MNRAS*, 397, 913
Ciolfi, R. & Rezzolla, L., 2013, *MNRAS*, 435, L43
Duez, V. & Mathis, S., 2010, *A&A*, 517, A58
Fujisawa, K. & Eriguchi, Y., 2013, *MNRAS*, 432, 1245
Fujisawa, K., Yoshida, S., & Eriguchi, Y., 2012, *MNRAS*, 422, 434
Glampedakis K., Andersson N., Lander S. K., 2012,*MNRAS*, 420, 1263
Lander, S. K. & Jones, D. I., 2009, *MNRAS*, 395, 2162
Rea, N. *et al.*, 2010, *Science*, 330, 944
Tomimura, Y. & Eriguchi, Y., 2005, *MNRAS*, 359, 1117
Yoshida, S., Kiuchi, K., & Shibata, M., 2012, *Phys. Rev. D*, 86, 044012

Magnetic Fields throughout Stellar Evolution
Proceedings IAU Symposium No. 302, 2013
P. Petit, M. Jardine & H. Spruit, eds.

doi:10.1017/S174392131400266X

Axisymmetric and stationary magnetic field structures in neutron star crusts under various boundary conditions

Kotaro Fujisawa[1] and Shota Kisaka[2]

[1]Department of Earth Science and Astronomy, Graduate School of Arts and Sciences, University of Tokyo, Komaba, Meguro-ku, Tokyo 153-8902, Japan
email: fujisawa@ea.c.u-tokyo.ac.jp

[2] Institute for Cosmic Ray Research, University of Tokyo, 5-1-5 Kashiwa-no-ha, Kashiwa city, Chiba 277-8582, Japan

Abstract. We have calculated many Hall equilibrium states within the neutron star crust under various boundary conditions in order to investigate the influences of the boundary conditions clearly. We have found two important features of these solutions. First, the magnitude of the core magnetic fields affects the *toroidal* to *total* magnetic field energy ratio within the crust (E_t/E). If the core magnetic fields are vanished, the crustal toroidal magnetic fields become weak and the typical energy ratio is only $E_t/E \sim 0.1\%$. If the core magnetic fields are strong, however, the crustal toroidal magnetic fields become strong and the typical ratio reaches $E_t/E \sim 15\%$. Second, the core *toroidal* magnetic fields and the twisted magnetosphere around the star make the size of the crustal toroidal magnetic field regions large. Therefore if the strong core magnetic fields have strong toroidal component, both strength and size of the crustal toroidal magnetic fields become large. These results show that the Hall MHD evolutions would be deeply affected by both inner and outer boundary conditions.

Keywords. Stars: interior – Stars: neutron – Stars: magnetic fields

1. Introduction

The decay of the magnetic field is considered as an important energy source of magnetar activity and neutron star heating, but the Ohmic diffusion cannot be effective physical mechanism for the decay because of the high electrical conductivity inside the neutron star. The Hall effect is considered as a key of the promotion of the magnetic fields decay (e.g. Goldreich & Reisenegger 1992). Recently, Hall MHD numerical simulations (Kojima & Kisaka 2012, Viganò *et al.* 2013) and Hall equilibrium states (Gourgouliatos *et al.* 2013) have been calculated. Since the boundary conditions are very important in both cases, we have calculated the Hall equilibria within the neutron star crust under various boundary conditions in order to investigate the influences of the boundary conditions.

2. Results and Discussion

We have obtained Hall equilibrium states (see Gourgouliatos *et al.* 2013) using Green function method with arbitrary current sheets (see Fujisawa & Eriguchi 2013). For simplicity, we divided the magnetized neutron star interior into three regions in order to include the boundary conditions systematically. Each region is core ($0 \leqslant r \leqslant r_s$), crust ($r_{in} \leqslant r \leqslant r_s$), vacuum or magnetosphere ($r \leqslant r_s$) respectively. Here, r_{in} and r_s denote the core-crust boundary radius and the stellar radius. Numerical results and the ratios E_t/E are displayed in Fig. 1 and Fig. 2.

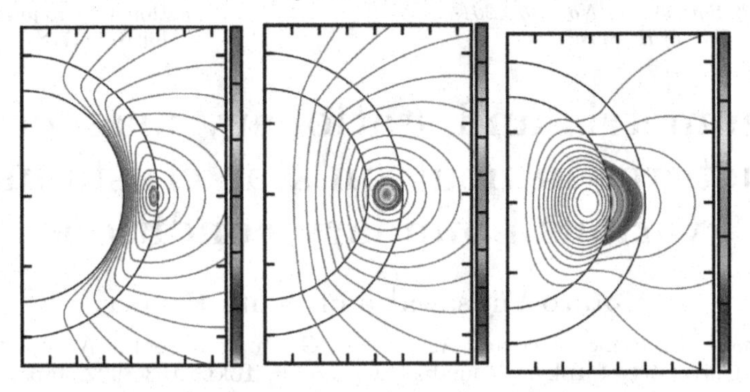

Figure 1. The poloidal magnetic fields configurations and the strength of the crustal toroidal magnetic fields (color maps). A pure crustal magnetic field model (left, $E_t/E \sim 0.5\%$), a crust–core magnetic field model (center, $E_t/E \sim 15\%$) and a very strong core magnetic field model (right, $E_t/E \sim 14\%$) are displayed. The left model has the smallest E_t/E and the center and the right models have larger E_t/E. The right model has the strong core toroidal magnetic field and the size of the toroidal magnetic field region is largest in these models.

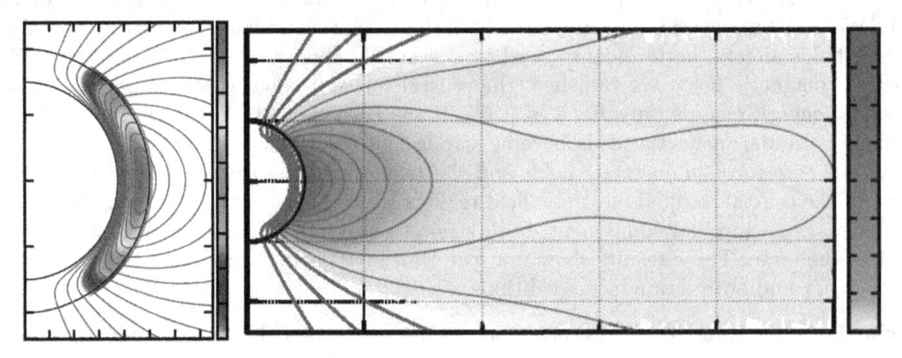

Figure 2. The crustal poloidal fields and the strength of the crustal toroidal fields (left) and the magnetospheric poloidal fields and the strength of the toroidal current density (right). The size of the crustal toroidal magnetic field region is much larger than those of models in Fig. 1.

As we have seen in these figures, the boundary conditions change the Hall equilibria within the crust significantly. These boundary conditions would affect the Hall drifts and evolutions deeply. Since the strong initial toroidal magnetic fields enforce the Hall drift (Kojima & Kisaka 2012) during the Hall MHD evolution and the Hall drift would stop when the magnetic fields reach Hall equilibrium configurations, we can evaluate the efficiency of the Hall drift using the value of E_t/E of our results. If we assume the strong initial toroidal fields, the strong core magnetic fields (right panel in Fig. 1) would weaken the Hall drift because E_t/E is larger. In the pure crustal field case (left panel in Fig. 1), the Hall drift would become more active because E_t/E is smallest in these models.

References

Fujisawa, K. & Eriguchi, Y., 2013, *MNRAS*, 432, 1245

Goldreich, P. & Reisenegger, A., 1992, *ApJ*, 395, 250

Gourgouliatos, K. N., Cumming, A., Reisenegger, A., Armaza, C., Lyutikov, M., & Valdivia, J. A., 2013, *MNRAS*, 434, 2480

Kojima, Y. & Kisaka, S., 2012, *MNRAS*, 421, 2722

Viganò, D., Rea, N., Pons, J. A., Perna, R., Aguilera, D. N., & Miralles, J. A., 2013, *MNRAS*, 434, 123

Magnetic Fields throughout Stellar Evolution
Proceedings IAU Symposium No. 302, 2013
P. Petit, M. Jardine & H. Spruit, eds.

doi:10.1017/S1743921314002671

Magnetars: the explosive character of a small class of strongly magnetized neutron stars

Nanda Rea[1,2]

[1]Institut de Ciències de l'Espai (CSIC–IEEC), Campus UAB, Facultat de Ciències, Torre C5-parell, E-08193 Barcelona, Spain
email: rea@ice.csic.es

[2]Astronomical Institute "Anton Pannekoek", University of Amsterdam, Postbus 94249, NL-1090 GE Amsterdam, the Netherlands

Abstract. I will review our current knowledge on the most magnetic objects in the Universe, a small sample of neutron stars called magnetars. The powerful persistent high energy emission and the flares from these strongly magnetized (10^{15} Gauss) neutron stars are providing crucial information about the physics involved at these extremes conditions, reserving us many unexpected surprises.

Keywords. neutron stars, magnetic fields

1. Introduction

Neutron stars are the debris of the supernova explosion of massive stars, the existence of which was first theoretically predicted around 1930 (Chandrasekhar 1931; Baade & Zwicki 1934) and then observed for the first time more than 30 years later (Hewish *et al.* 1968). They were predicted all along as very dense and degenerate stars holding about 1.4 solar masses in a sphere of 10km radius. We now know many different flavors of these compact objects, and many open questions are still waiting for an answer after decades of studies. The neutron star population is dominated by radio pulsars (thousands of objects), however in the last decades several extreme and puzzling sub-classes of isolated neutron stars were discovered: Anomalous X-ray Pulsars (AXPs), Soft Gamma Repeaters (SGRs; see Mereghetti 2008), Rotating Radio Transients (RRATs; Keane & McLaughlin 2011), X-ray dim Isolated Neutron stars (XINSs; Turolla 2009), and Central Compact Objects (CCOs; Mereghetti 2011). The large amount of different acronyms might already show how diverse is the neutron star class, and on the other hand, how far we are from a unified scenario. These objects are amongst the most intriguing populations in modern high-energy astrophysics and in physics in general. They are precious places to test gravitational and particle physics, relativistic plasma theories, as well as strange quark states of matter and physics of atoms and molecules embedded in extremely high magnetic fields (impossible to be reproduced on Earth). Since their discovery in the late sixties, about 2000 rotational powered pulsars are known to date, thanks to numerous surveys using single dish radio antennas around the world (Parkes, Green Bank, Jodrell Bank, Arecibo), with periods ranging from about 1.5 ms to 12 s (see the ATNF on-line catalog: Manchester *et al.* 2005), and they have magnetic fields ranging between $\sim 10^8 - 10^{15}$ Gauss. The energy reservoir of all those pulsars is well established to be their rapid rotation, having a rotational luminosity $L_{\rm rot} \sim 4\pi^2 I\dot{P}/P^3 \sim 3.9\times10^{46}\dot{P}/P^3$ erg/s. A key ingredient to activate the radio emission is the acceleration of charged particles, which are extracted from the stars surface by an electrical voltage gap ($\Delta V \propto L_{\rm rot}^2$). The voltage gap forms due to the presence of a dipolar magnetic field co-rotating with

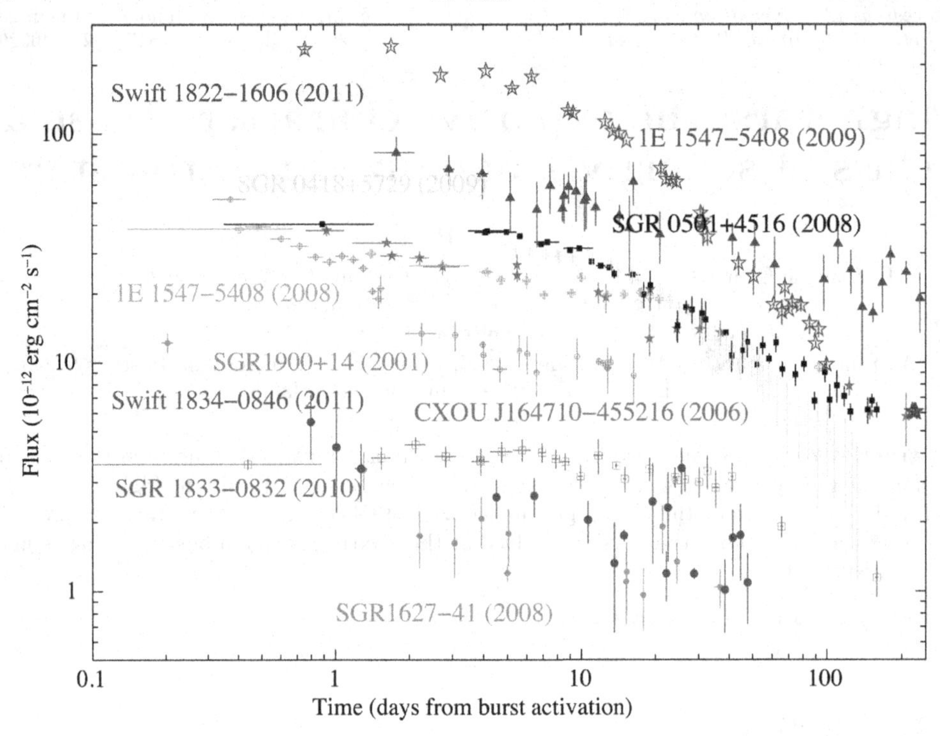

Figure 1. Flux evolution over the first ~ 200 days of all magnetar outbursts (only if observed with imaging instruments, and for which this period span is well monitored). Fluxes are reported in the 1–10 keV energy range, and the reported times are calculated in days from the detection of the first burst in each source. See Rea & Esposito 2011) for the reference for each reported outburst.

the pulsar, and it is believed to extend up to an altitude of $\sim 10^4$ cm with a potential difference $\Delta V > 10^{10}$ statvolts. Primary charges are accelerated by the electric field along the magnetic field lines to relativistic speeds and emit curvature radiation. Curvature photons are then converted into electron-positron pairs and this eventually leads to a pair cascade which is ultimately responsible for the coherent radio emission we observe from radio pulsars. Very energetic pulsars are also observed until the gamma-ray range, most probably in the form of synchrotron photons coming from the acceleration in the so-called outer-gap of the pulsar magnetosphere (Goldreich & Julian 1969; Ruderman & Chen 1975). All isolated pulsar rotational periods are increasing in time. This spin down is quantified by the braking index n, which is defined as: $\dot{\Omega} \propto \Omega^n$ (where $\Omega = 2\pi/P$). With this definition, under the assumption of pure dipole braking, we would expect all pulsars having $n = 3$.

In this review we will report on the state of the art of the study of the strongest magnets in the Universe: the magnetars. However, before presenting these ultra-magnetic objects, it is instructive to indicate how the magnetic field of isolated pulsars is commonly estimated. Assuming that pulsars slow down due to magnetic dipole radiation, the surface dipolar magnetic field at the equator (B_{dip}) can be estimated from the measured pulsar spin period P and its first derivative $\dot{P}$: $B_{dip} \sim 3.2 \times 10^{19}\sqrt{P\dot{P}}$ Gauss (where P is in units of seconds, and we have assumed a neutron star with a radius of 10km, 1.4 solar masses, and being an ortogonal rotator).

The magnetars (comprising AXPs and SGRs; Mereghetti 2008) are a small group of X-ray pulsars (about twenty objects with spin periods between 2–12 s) the emission of

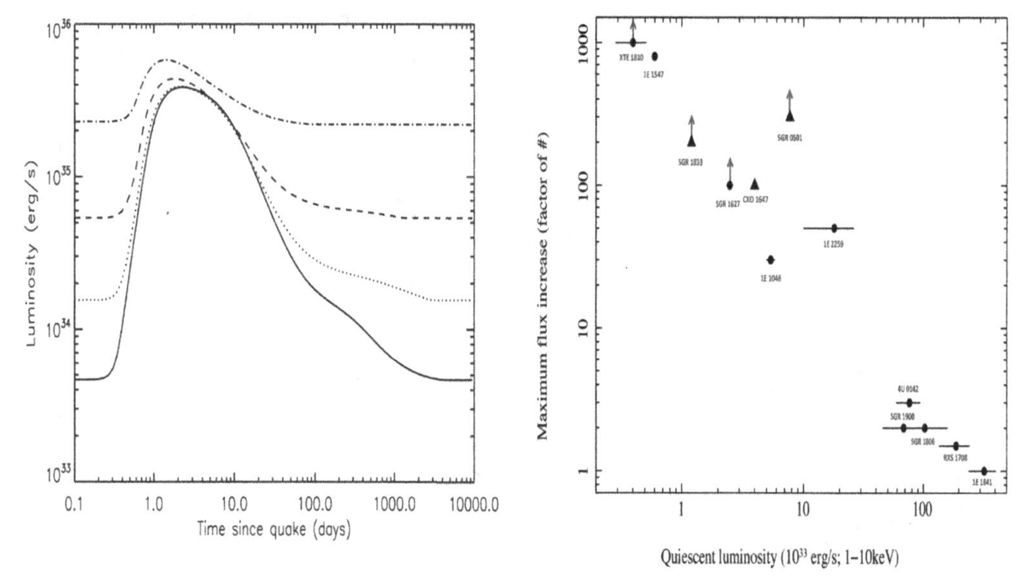

Figure 2. *Left panel*: Luminosity vs. time after energy injection. The models correspond to $E_{\rm oc} = 1.7 \times 10^{41}$ erg, (solid line), 1.7×10^{42} erg (dotted line), 1.7×10^{43} erg (dashed line), and 1.7×10^{44} erg (dash-dotted line). *Right panel*: quiescent luminosity vs. outburst maximum flux increase (all in the 1-10 keV band), for all magnetars showing bursts, glitches or outbursts. See Pons & Rea 2012) for further details.

which is very hardly explained by any of the scenarios for the radio pulsar or the accreting X-ray binary populations. In fact, the very strong X-ray emission of these objects ($L_x \sim 10^{35}$ erg/s) seemed too high and variable to be fed by the rotational energy alone (as in the radio pulsars), and no evidence for a companion star has been found so far in favor of any accretion process (as in the X-ray binary systems). Moreover, roughly assuming them to be magnetic dipole radiator, their inferred magnetic fields appear to be as high as $B_{dip} \sim 10^{14} - 10^{15}$ Gauss. They are then higher than the electron critical magnetic field, $B_Q = m_e^2 c^3/eh \sim 4.4 \times 10^{13}$ G at which an electron gyro-rotating in such magnetic field line gains a cyclotron energy equal to its rest mass. At fields higher than B_Q, QED effects such as vacuum polarization or photon splitting, can take place (see Harding & Lai 2006).

Because of these high B fields, the emission of magnetars was thought to be powered by the decay and the instability of their strong fields (Duncan & Thompson 1992; Thompson & Duncan 1993). This powerful X-ray output is usually well modeled by a thermal emission from the neutron star hot surface (about 3×10^6 Kelvin) reprocessed in a twisted magnetosphere through resonant cyclotron scattering, a process favored only under these extreme magnetic conditions (Thompson, Lyutikov & Kulkarni 2002; Nobili, Turolla, Zane 2008; Rea *et al.* 2008). On top of their persistent X-ray emission, magnetars emit very peculiar flares on short timescales (from fraction to hundreds of seconds) emitting large amount of energy ($10^{40} - 10^{46}$ erg; the most energetic Galactic events after the supernova explosions). They are probably caused by large scale rearrangements of the surface/magnetospheric field, either accompanied or triggered by fracturing of the neutron-star crust, sort of stellar quakes.

Furthermore, magnetars show also large outbursts where their steady emission can be enhanced up to ~1000 times its quiescent level (see Figure 1, and see Rea & Esposito 2011 for recent review on transient magnetars). From the few well monitored events, we

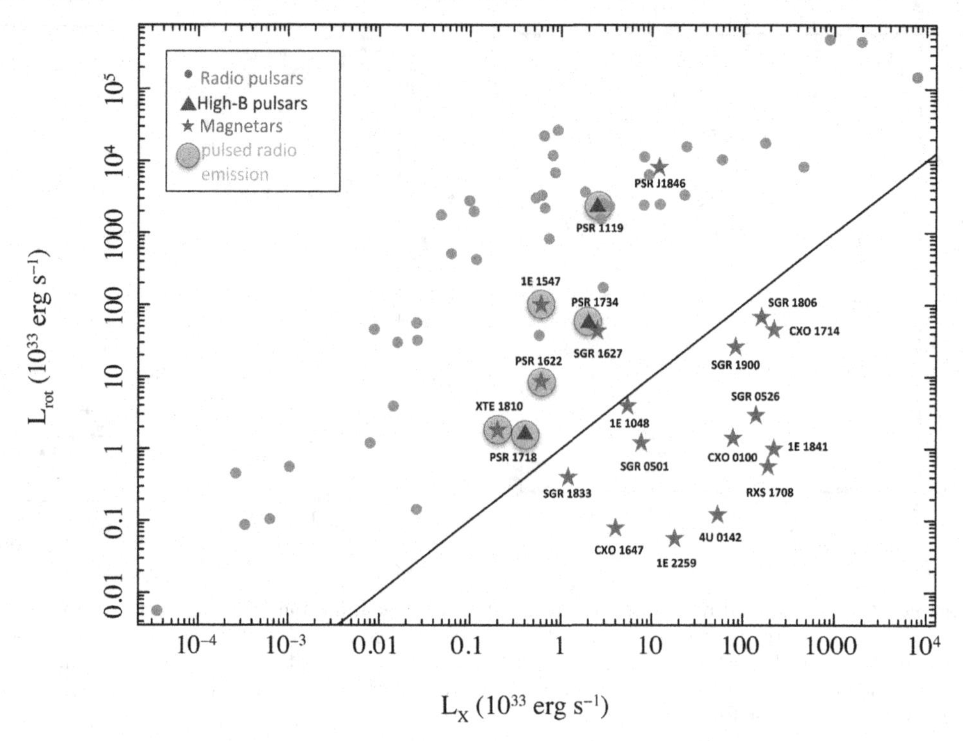

Figure 3. X-ray luminosity versus the spin-down luminosity for all pulsars having a detected X-ray emission (grey filled circles), high-B pulsars (filled triangle), and the magnetars (red stars). Grey shaded circles mark the magnetars and high-B pulsars with detected pulsed radio emission, and the solid line shows $L_x = L_{rot}$. X-ray luminosities are calculated in the 0.5–10 keV energy range, and for variable sources refer to the quiescent emission state (see Rea *et al.* 2012a).

are starting to understand how those outbursts are produced. They are cause by similar crustal fractures as the shorter flares, accompaigned by a strong surface heating, and often by the appearence of additional hot spots on the neutron-star surface. This is what may cause large spectral changes during outbursts, pulse profile variability, and different cooling patterns depending on the outburst. We have recently started to model those outburst decays, and many important physical informations are slowly emerging, i.e. that those month/year-long outbursts saturates at $\sim 10^{36}$ erg/s, due to neutrino cooling processes, and regardless the source quiescent level. This discovery makes magnetar outbursts potential standard candles (Pons & Rea 2012; see Figure 2).

2. Hints to the connection between magnetars and radio pulsars

In the past few years, new discoveries started to shed light on a possible connection between magnetars and the typical radio pulsar population, weakening the strong distinction between these two classes, while pointing to a continuum of magnetar-like emission in the neutron star population. Below we list a few of those key discoveries.

• Magnetars were believed to be radio-quiet sources for a few decades. This was interpreted as the result of a photon splitting process that under magnetic field stronger than the critical electron field (B_Q) is very efficient (Baring &Harding 1998). The discovery in 2004 of transient magnetars, coincided also with the discovery of radio pulsed emission from such sources (Camilo *et al.* 2006; Levin *et al.* 2010). This came as a big surprise, and started the idea of a possible connection between magnetars and the typical radio pulsars, despite the somehow different characteristics of their radio pulsed emission. Furthermore, recently a study of radio magnetars, showed that despite the different radio properties, the radio emission can have the same physical mechanisms as for rotation-powered pulsars (Rea *et al.* 2012a): powered by rotational energy (see also Figure 3), but with different observational properties possibly caused by a different path that a pair cascade might undertake when embedded in a mostly toroidal magnetic field.

• Deep radio surveys discovered a few radio pulsars having dipolar fields larger than the B_Q. Although having magnetic fields in the magnetar range those objects were behaving as normal radio pulsars, and this was interpreted by a different magnetic geometry between the two classes. In 2008 bursting activity, and an X-ray outburst were detected from a high-B pulsar, showing the presence of magnetar-like activity (Gavriil *et al.* 2008; Kumar 2008).

• The extensive follow-up of transient magnetars undergoing an outburst had allowed the most un-expected discoveries. In particular, prompted by detection of typical magnetar-like bursts and a powerful outburst of the persistent emission, a new transient magnetar was discovered in 2009, namely SGR 0418+5729. However, with great surprise after more than 2 years of extensive monitoring, no period derivative was detected, which led to an upper limit on the source surface dipolar field of $B_{dip} < 7.5 \times 10^{12}$ Gauss (Rea *et al.* 2010). For the first time we witnessed a magnetar with a low dipolar magnetic field. This discovery, demonstrated that not only a critical magnetic field ($> B_Q$) was not necessary to have magnetar-like activity, but many apparently normal pulsars can turn out as magnetars at anytime (in fact the discovery of a second low-B magnetar soon followed; Rea *et al.* 2012b).

• Advances in the measure of pulsar breaking indeces showed the existence of objects with indeces n smaller than 3, which would (at least in vacuum) imply an increasing magnetic field with age under the common magnetic breaking picture. In particular this is the case of the high-B pulsar PSR 1734-33 (Espinoza *et al.* 2011), discovered to have $n = 0.9$. This discovery favors models for which the magnetic field is buried into the crust by accretion in the first supernova phases, and start re-emerging during the pulsar life-time (Viganó *et al.* 2012). A similar conclusion has been reached with the discovery of low-B fields in CCOs, whose young age and hot surface temperature are instead pointing to a strong buried magnetic field, despite what is inferred by their period and period derivatives (Halpern *et al.* 2007).

3. Conclusions

We have now understood the importance of certain parameters for the neutron star evolution and diversity: i) the surface dipolar magnetic field strength may not be the only parameter driving their magnetar or radio pulsar nature, ii) magnetars can behave as radio pulsars and vice-versa, possibly powered by a similar mechanism sustained by rotational energy, and iii) an internal strong magnetic field is required to explain the low braking indeces of a few radio pulsars, as well as the emission of the compact central objects, despite the rather low dipolar magnetic field component.

These discoveries show that extremely strong magnetic fields may be extremely common among the pulsar population, rather than an exception. This might imply that supernova explosions should be generally able to produce such strong magnetic fields, hence that most massive stars are either producing fast rotating cores during the explosion to activate the dynamo, or are strongly magnetized themselves (i.e. 1kGauss at least). Furthermore, in this scenario several gamma-ray bursts (not only an irrelevant fraction), might indeed be due to the formation of magnetars, and the gravitational wave background produced by magnetar formation should then be larger than predicted so far.

References

Baade, W & Zwicky, F. 1934 Physical Review 46, 76–77
Baring, M. G. & Harding, A. K., 1998, *ApJ*, 507, L55
Camilo, F., *et al.* 2006, *Nature* 442, 892–895
Chandrasekhar, S. 1931, *ApJ* 74, 81
Duncan, R. C. &Thompson, C 1992, *ApJ* 392, L9–L13
Espinoza, C. M., Lyne, A. G., Kramer, M., Manchester, R. N., & Kaspi, V. M., 2011, *ApJ* 741, L13
Gavriil, F. P., Gonzalez, M. E., Gotthelf, E. V., Kaspi, V. M., Livingstone, M. A., & Woods, P. M. 2008, Science 319, 1802–1805
Goldreich, P. & Julian, W. H., 1969, *ApJ*,157, 869
Harding, A. K. & Lai, D. 2006, *RPPh*, 69, 2631
Hewish, A., Bell, S. J., Pilkington, J. D. H.., Scott, P. F., & Collins, R. A. 1968, *Nature* 217, 709–713
Halpern, J. P.; Gotthelf, E. V.; Camilo, F.; Seward, F. D. 2007, *ApJ* 665 1304
Keane, E. F. & McLaughlin, M. A. 2011, *BASI*, 39, 333
Kumar, H. S. & Safi-Harb, S. 2008, ApJ 678, L43–L46
Levin, L., *et al.* 2010 *ApJ* 721, L33–L37
Manchester, R. N., Hobbs, G. B., Teoh, A., & Hobbs, M. 2005, *AJ* 129, 1993
Mereghetti, S. 2008, *A&AR*, 15 225
Mereghetti, S. 2011, APSS, High-Energy Emission from Pulsars and their Systems, Eds. N. Rea & D. F. Torres, Springer-Verlag Berlin
Nobili, L.; Turolla, R.; Zane, S. 2008, *MNRAS*, 386, 1527
Pons, J. A. & Rea, N. 2011, *ApJ* 750, L6
Rea, N., Zane, S., Lyutikov, M., & Turolla, R. 2007, *Ap&SS* 308, 61
Rea, N.; Zane, S., Turolla, R., Lyutikov, M., & Götz, D. 2008, *ApJ* 686, 1245
Rea, N., Esposito, P., Turolla, R., Israel, G. L., Zane, S., Stella, L., Mereghetti, S., Tiengo, A., Gotz, D., Gogus, E., & Kouveliotou, C. 2010, Science 330, 944
Rea, N. & Esposito, P. 2011, APSS, High-Energy Emission from Pulsars and their Systems, Eds. N. Rea & D. F. Torres, Springer-Verlag Berlin
Rea, N., Pons, J. A., Torres, D. F., & Turolla, R., 2012a, *ApJ*, 748, L12
Rea, N., *et al.* 2012b, *ApJ*, 754, 27
Ruderman, M. A. & Sutherland, P. G.,1975, *ApJ*, 196 , 51
Tendulkhar, S. *et al.* 2012, *ApJ* submitted
Thompson C., Lyutikov M., Kulkarni S. R. 2002, *ApJ* 574, 332
Thompson, C. & Duncan, R. C. 1993, *ApJ* 408, 194
Turolla, R. 2009, *ASSL*, 357, 141
Viganó, D. & Pons, J. A. 2012, *MNRAS*, 425, 2487

Magnetic Fields throughout Stellar Evolution
Proceedings IAU Symposium No. 302, 2013
P. Petit, M. Jardine & H. Spruit, eds.

doi:10.1017/S1743921314002683

NSMAXG: A new magnetic neutron star spectral model in XSPEC

Wynn C. G. Ho[1]

[1]Mathematical Sciences and STAG, University of Southampton, Southampton, SO17 1BJ, UK
email: **wynnho@slac.stanford.edu**

Abstract. The excellent sensitivity of X-ray telescopes, such as Chandra and XMM-Newton, is ideal for the study of cooling neutron stars, which can emit at these energies. In order to exploit the wealth of information contained in the high quality data, a thorough knowledge of the radiative properties of neutron star atmospheres is necessary. A key factor affecting photon emission is magnetic fields, and neutron stars are known to have strong surface magnetic fields. Here I briefly describe our latest work on constructing magnetic ($B \geqslant 10^{10}$ G) atmosphere models of neutron stars and the NSMAXG implementation of these models in XSPEC. Our results allow for more robust extractions of neutron star parameters from observations.

Keywords. radiative transfer, stars: atmospheres, stars: magnetic field, stars: neutron

---

Thermal X-ray radiation has been detected from many radio pulsars and radio-quiet neutron stars. Thermal emission can provide invaluable information on the physical properties and evolution of neutron stars, such as the mass M, radius R, and surface temperature T, which in turn depend on poorly constrained physics of the deep interior, such as the nuclear equation of state and quark and superfluid and superconducting properties at supranuclear densities. In addition, neutron stars are known to possess strong magnetic fields: from $B \approx 10^8 - 10^9$ G in the case of millisecond pulsars to $B \approx 10^{10} - 10^{13}$ G in the case of normal pulsars and even $B \gtrsim 10^{14}$ G in magnetars (see Zavlin 2009; Kaspi 2010; Harding 2013, for reviews).

The observed thermal radiation originates in a thin atmospheric layer (with scale height ~ 1 cm) that covers the stellar surface. Atmosphere properties, such as magnetic field, chemical composition, and radiative opacities, directly determine the characteristics of the observed spectrum. While the surface composition of neutron stars is generally unknown, a great simplification arises due to the efficient gravitational separation of light and heavy elements (Alcock & Illarionov 1980; Hameury *et al.* 1983). A pure hydrogen atmosphere is expected even if a small amount of accretion occurs after neutron star formation; the total mass of hydrogen needed to form an optically thick atmosphere can be less than $\sim 10^{16}$ g. Alternatively, a helium or carbon atmosphere may be possible as a result of nuclear burning on the neutron star surface (Chang & Bildsten 2003; Chang *et al.* 2010). Finally, a heavy-element atmosphere may exist if no accretion takes place or if all the accreted matter is consumed by nuclear reactions.

Steady progress has been made in modeling neutron star atmospheres (see Zavlin 2009, for more detailed discussion and references). Since the neutron star surface emission is thermal in nature, it has been modeled at the lowest approximation with a blackbody spectrum. Early works on more realistic spectra assumed emission from unmagnetized light-element atmospheres, and the resultant spectra exhibit distinctive hardening relative to a blackbody. The inclusion of magnetic fields has many important effects. For example, the presence of a magnetic field causes emission to be anisotropic and polarized (see Mészáros 1992, for review). At $B > e^3 m_e^2 c/\hbar^3 = 2.35 \times 10^9$ G, the binding energy of atoms, molecules, and other bound states increases significantly, and abundances can be appreciable in the atmosphere of neutron stars (see Lai 2001, for review). When field strengths approach and exceed the quantum electrodynamics field $B_{\rm QED} = m_{\rm e}^2 c^3/e\hbar = 4.41 \times 10^{13}$ G, vacuum polarization effects, such as switching between photon polarization modes, become relevant (Lai & Ho 2002; van Adelsberg & Lai 2006), and the atmosphere may even cease to exist as plasma condenses onto the surface (Medin & Lai 2007; Potekhin *et al.* 2012). Most calculations of magnetic neutron star atmospheres focus on a fully ionized hydrogen plasma. Only relatively recently have self-consistent atmosphere models using the latest equation of state and opacity results for strongly magnetized and *partially ionized* hydrogen and mid-Z elements been constructed (Potekhin *et al.* 2004; Mori & Ho 2007).

In previous work, we implemented into XSPEC (Arnaud 1996) our theoretical neutron star magnetic atmosphere X-ray spectra, under the model name NSMAX (Ho *et al.* 2008). These atmosphere spectra are obtained using the partially ionized results of Potekhin *et al.* (2004) and Mori & Ho (2007). Two sets of models are provided: One set with a single surface $\mathbf{B}$ and $T_{\rm eff}$ and a second set which is constructed with $\mathbf{B}$ and $T_{\rm eff}$ varying across the surface according to a magnetic dipole geometry. Magnetic fields and effective temperatures of the models span the range $B = 10^{12} - 3 \times 10^{13}$ G and $\log T_{\rm eff}({\rm K}) \approx 5.5 - 6.7$, respectively. Note that other neutron star atmosphere spectra in XSPEC are either non-magnetic (NSAGRAV: Zavlin *et al.* 1996; NSSPEC: Gänsicke *et al.* 2002; NSATMOS: McClintock *et al.* 2004; Heinke *et al.* 2006) or magnetic but fully ionized hydrogen (NSA: Pavlov *et al.* 1995); the last at only two fields: $B = 10^{12}$ and 10^{13} G. We also note the open source non-magnetic model McPHAC (Haakonsen *et al.* 2012).

We recently implemented into XSPEC a new set of neutron star magnetic atmosphere X-ray spectra, under the model name NSMAXG, which replaces NSMAX. The new model is nearly identical to the old model but with two important differences. The first difference is the inclusion of atmosphere spectra for weaker magnetic fields ($B = 10^{10} - 10^{11}$ G). These spectra are constructed using the method described in Ho *et al.* (2008), and references therein, supplemented by Potekhin & Chabrier (2003) for calculating Gaunt factors and Suleimanov *et al.* (2012) to account for thermal effects (see also Pavlov & Panov 1976; Potekhin 2010; Suleimanov *et al.* 2010). Examples of these spectra are shown in Ho (2013). Note that these weak magnetic field spectra assume a fully ionized hydrogen atmosphere; partially ionized spectra are the subject of current work.

The second important difference between NSMAXG and NSMAX is a change in XSPEC fit parameters. Most XSPEC neutron star atmosphere models (NSA, NSAGRAV, and NSATMOS) use the fit parameters $T_{\rm eff}$, M, R, and either distance d or flux normalization A. The last two are equivalent since $A \propto R^2/d^2$. On the other hand, the fit parameters of NSMAX are $T_{\rm eff}$, A, and gravitational redshift $1+z_{\rm g}$ $[= (1-2GM/c^2R)^{-1/2}]$. There are two reasons for this different choice for NSMAX. The first is that *all* models in XSPEC are calculated assuming that emission arises from the entire visible surface of the neutron star, i.e., $R = R_{\rm NS}$. Thus the same value of R must be used to calculate $z_{\rm g}$. The second reason is that NSMAX is constructed for particular values of surface gravity g $[= (1+z_{\rm g})GM/R^2]$ (similarly NSA is calculated using a single value of g, as well as assuming a fully ionized plasma, which is in contrast to the partially ionized plasma of NSMAX). Thus allowing M and R to vary as fit parameters would not produce consistent results, i.e., the derived values of M and R would not necessarily correspond to the value of g that is used to compute the atmosphere spectrum (see Heinke *et al.* 2006, for comparisons between NSA and NSATMOS). We rectify this second issue by calculating spectra for a range of surface gravities, i.e., $\log g({\rm cm\ s^{-2}}) = 13.6 - 15.4$, thus allowing NSMAXG to use M and R as consistent fit parameters. Note that NSATMOS

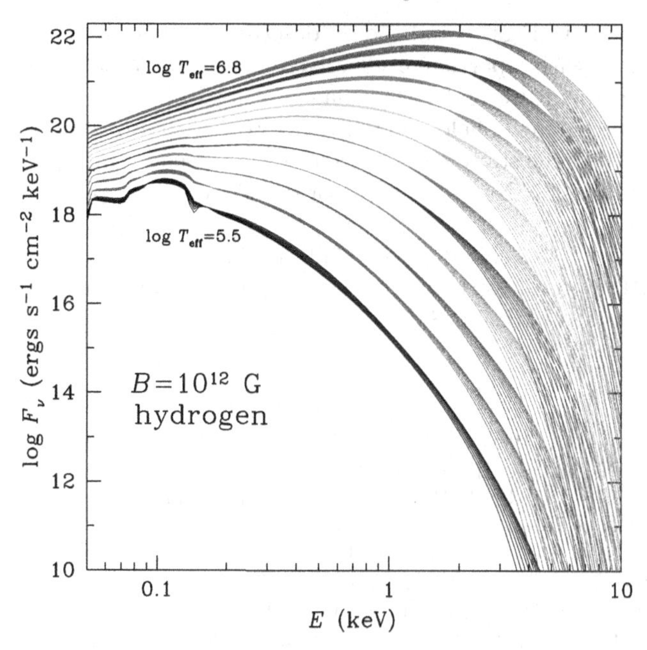

Figure 1. Partially ionized hydrogen atmosphere model spectra for effective temperatures $\log T_{\rm eff} = 5.5 - 6.7$, surface gravities $\log g = 13.6 - 15.4$, and magnetic field $B = 10^{12}$ G.

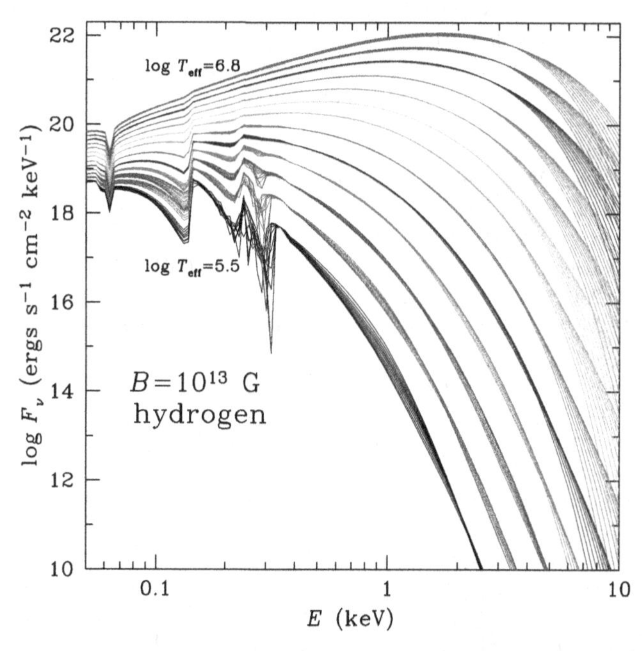

Figure 2. Partially ionized hydrogen atmosphere model spectra for effective temperatures $\log T_{\rm eff} = 5.5 - 6.7$, surface gravities $\log g = 13.6 - 15.4$, and magnetic field $B = 10^{13}$ G.

and NSAGRAV are also calculated for a range of g; however recall that these two are non-magnetic models. Figures 1 and 2 show the resulting NSMAXG spectra for $B = 10^{12}$ and 10^{13} G, respectively. Incidentally, the spectral tables of NSMAX and NSMAXG can easily be made compatible for use with the other XSPEC neutron star fitting routines.

The spectra shown in Figs. 1 and 2 only describe emission from either a local patch of the stellar surface with a particular effective temperature and magnetic field or a star with a uniform temperature and radial magnetic field of uniform strength. By taking into account surface magnetic field and temperature distributions, we can construct more physical models of neutron star emission, which can be used for interpreting and decoding observations (see Ho 2007; Ng *et al.* 2012, for details and examples).

WCGH is grateful to his collaborators, Gilles Chabrier, Kaya Mori, and Alexander Potekhin. WCGH appreciates the use of the computer facilities at KIPAC and acknowledges support from the IAU and STFC in the UK.

References

Alcock, C. & Illarionov, A. 1980, *ApJ*, 235, 534

Arnaud, K. A. 1996, in: G. H. Jacoby & J. Barnes (eds.), *Astronomical Data Analysis Software and Systems V*, ASP Conf. Ser. 101 (San Francisco: ASP), p. 17; see also http://heasarc.gsfc.nasa.gov/docs/xanadu/xspec/

Chang, P. & Bildsten, L. 2003, *ApJ*, 585, 464

Chang, P., Bildsten, L., & Arras, P. 2010, *ApJ*, 723, 719

Gänsicke, B. T., Braje, T. M., & Romani, R. W. 2002, *A&A*, 386, 1001

Haakonsen, C. B., Turner, M. L., Tacik, N. A., & Rutledge, R. E. 2012, *ApJ*, 749, 52

Hameury, J. M., Heyvaerts, J., & Bonazzola, S. 1983, *A&A*, 121, 259

Harding, A. K. 2013, *Front. Phys.*, in press (arXiv:1302.0869)

Heinke, C. O., Rybicki, G. B., Narayan, R., & Grindlay, J. E. 2006, *ApJ*, 644, 1090

Ho, W. C. G. 2007, *MNRAS*, 380, 71

Ho, W. C. G. 2013, in: J. van Leeuwen (ed.), *Neutron Stars and Pulsars: Challenges and Opportunities after 80 years*, Proc. IAU Symposium No. 291 (Cambridge: Cambridge Univ. Press), p. 101

Ho, W. C. G., Potekhin, A. Y., & Chabrier, G. 2008, *ApJS*, 178, 102

Kaspi, V. M. 2010, *Proc. National Academy Sci.*, 16, 7147

Lai, D. 2001, *Rev. Mod. Phys.*, 73, 629

Lai, D. & Ho, W. C. G. 2002, *ApJ*, 566, 373

McClintock, J. E., Narayan, R., & Rybicki, G. B. 2004, *ApJ*, 615, 402

Medin, Z. & Lai, D. 2007, *MNRAS*, 382, 1833

Mészáros, P. 1992, *High-Energy Radiation from Magnetized Neutron Stars* (Chicago: Univ. Chicago Press)

Mori, K & Ho, W. C. G. 2007, *MNRAS*, 377, 905

Ng, C.-Y., et al. 2012, *ApJ*, 761, 65

Pavlov, G. G. & Panov, A. N. 1976, *Sov. Phys. JETP*, 44, 300

Pavlov, G. G., Shibanov, Yu.A., Zavlin, V. E., & Meyer, R. D. 1995, in: M. A. Alpar, Ü. Kiziloğlu, & J. van Paradijs (eds.), *Lives of the Neutron Stars* (Boston: Kluwer), p. 71

Potekhin, A. Y. 2010, *A&A*, 518, A24

Potekhin, A. Y. & Chabrier, G. 2003, *ApJ*, 585, 955

Potekhin, A. Y., Chabrier, G., Lai, D., Ho, W. C. G., & van Adelsberg, M. 2004, *ApJ*, 612, 1034

Potekhin, A. Y., Suleimanov, V. F., van Adelsberg, M., & Werner, K. 2012, *A&A*, 546, A121

Suleimanov, V. F., Pavlov, G. G., & Werner, K. 2010, *ApJ*, 714, 630

Suleimanov, V. F., Pavlov, G. G., & Werner, K. 2012, *ApJ*, 751, 15

van Adelsberg, M. & Lai, D. 2006, *MNRAS*, 373, 1495

Zavlin, V. E. 2009, in: W. Becker (ed.), Ap&SS Lib. Vol. 357, *Neutron Stars and Pulsars* (Berlin: Springer-Verlag), p. 181

Zavlin, V. E., Pavlov, G. G., & Shibanov, Yu.A. 1996, *A&A*, 297, 441

Magnetic Fields throughout Stellar Evolution
Proceedings IAU Symposium No. 302, 2013
P. Petit, M. Jardine & H. Spruit, eds.

doi:10.1017/S1743921314002695

Effects of strong magnetic fields in dense stellar matter

A. Lavagno and F. Lingua

Department of Applied Science and Technology, Politecnico di Torino, I-10129 Torino, Italy
INFN, Sezione di Torino, I-10125 Torino, Italy

Abstract. We study the effects of strong magnetic fields in dense stellar matter within an effective relativistic equation of state with the inclusion of hyperons and $\Delta(1232)$-isobar degrees of freedom. The effects of high magnetic field interactions significantly affect the nuclear equation of state and the macroscopic properties of the star. In this framework we investigate the role of the presence of the Δ-isobars degrees of freedom in structure and in the bulk properties of the compact star.

Keywords. Relativistic equation of state, Δ baryonic resonances, Compact stars

---

Recent observations provided wide evidence that very strong magnetic fields are bound to exist inside and around several neutron stars. Many studies suggest that the magnetic field on the surface of these *magnetar* candidates could rise up to 10^{15} G, reaching even grater values as large as $10^{18}-10^{19}$ G in their cores, see, e.g., Duncan & Thompson (1995), Hurley *et al.* (1999), Mareghetti & Stella (1995) and Chakrabarty, Bandyopadhyay & Pal (1997). Such a strong value of magnetic field may widely influence the property of the nuclear equation of state (EOS) and the bulk properties of the neutron star matter, see, e.g., Cardall, Prakash & Lattimer (2001), Rabhi, Panda & Providencia (2011), Dexheimer, Negreiros, Schramm (2012) and Lopes & Menezes (2012).

In this contribution we are going to study the effect of strong magnetic field in dense stellar star matter in the framework of a relativistic EOS with the inclusion of hyperons and $\Delta(1232)$-isobar degrees of freedom. See Lavagno (2010), Lavagno & Pigato (2012) and Lavagno (2013) for details.

The total Lagrangian density $\mathcal{L}$ can be written as $\mathcal{L} = \sum_b \mathcal{L}_b + \mathcal{L}_m + \sum_l \mathcal{L}_l + \mathcal{L}_M$, where the indexes b and l run over all considered baryons (nucleons, hyperons and Δs) and leptons (electrons and muons) respectively. While m denote the interaction corresponding to the mesons fields (σ, ω, ρ) and the term $\mathcal{L}_M$ corresponds to the lagrangian density of the magnetic field itself. More explicitly, we have

$$\mathcal{L}_b = \overline{\psi_b}\Big[\gamma_\mu\Big(i\partial^\mu - q_b A^\mu - g_{\omega b}\omega^\mu - g_{\rho b} I_{3b}\rho^\mu - (m_b - g_{\sigma b}\sigma)\Big)\Big]\psi_b\,, \tag{0.1}$$

$$\mathcal{L}_l = \overline{\psi_l}\Big[\gamma_\mu\Big(i\partial^\mu - q_l A^\mu - m_l\Big)\Big]\psi_l\,, \tag{0.2}$$

$$\mathcal{L}_m = \frac{1}{2}\Big(\partial_\mu\sigma\partial^\mu\sigma - m_\sigma^2\sigma^2\Big) - \frac{1}{3}bm(g_\sigma\sigma)^3 - \frac{1}{4}c(g_\sigma\sigma)^4 + \frac{1}{2}m_\omega^2\omega_\mu\omega^\mu + \tag{0.3}$$

$$-\frac{1}{4}\Omega_{\mu\nu}\Omega^{\mu\nu} + \frac{1}{2}m_\rho^2\rho_\mu\rho^\mu - \frac{1}{4}\rho_{\mu\nu}\rho^{\mu\nu}\,, \tag{0.4}$$

$$\mathcal{L}_M = -\frac{1}{16\pi}F_{\mu\nu}F^{\mu\nu}\,. \tag{0.5}$$

In this investigation we adopt the so-called GM3 parametrization by Glendenning & Moszkowski (1991), with the same hyperons coupling constants used by Lavagno (2010). The magnetic field contribution is included as an external applied field along the z-axis

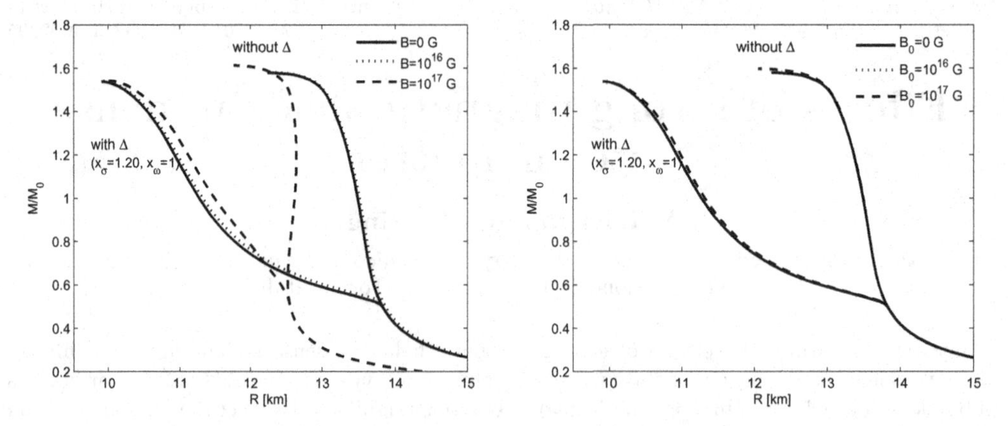

Figure 1. Mass radius relationship with and without Δ-isobars and for different values of uniform (left panel) and density dependent magnetic fields (right panel). A $B_{\rm surf} = 10^{15}$ G is used for different values of the central magnetic field B_0.

through the vector four potential $A^\mu = (0, 0, Bx, 0)$. We are going to study two different configurations of the magnetic field related to its variation with the baryon density. First, we study the EOS with a constant magnetic field strength as a function of baryon density. Second, we introduce the same profile used by Lopes & Menezes (2012) in order to explain the observed difference in strength of magnetic field between the neutron star surface and its core. The general effect of the magnetic field is to introduce a new source of degeneracy for those particles that have an electric charge $q_{b,l} \neq 0$. This degeneracy factor is due to the formation of the so called Landau levels and influences the computation of both energy and particle density.

In Fig. 1, we show the mass-radius relationships without and in presence of the Δ degrees of freedom, with a constant value of magnetic field (left panel) and a density dependent value of B (right panel). The Δ degrees of freedom implies smaller values of radii with respect to the case in which Δs are not taken into account. This effect is totally independent from the value of the magnetic field and the star assumes slightly smaller values of the maximum mass. Moreover, let us observe that the magnetic field contribution seems to be relevant only for field strength greater than 10^{17} G. For this reason, when a density dependent magnetic field is used, only central layers of the star are allowed to assume greater values of B and the mass-radius curves become more similar to the one in absence of magnetic field.

References

Cardall, C. Y., Prakash, M., & Lattimer, J. M. 2001, *Astrophys. J.*, 554, 322
Chakrabarty, S., Bandyopadhyay, D., & Pal, S. 1997, *Phys. Rev. Lett.*, 78, 2898
Dexheimer, V., Negreiros, R., & Schramm S. 2012, *Eur. Phys. J. A*, 48, 189
Duncan, R. & Thompson, C. 1995, *Mon. Not. R. Astron. Soc.*, 275, 255
Glendenning, N. K. & Moszkowski S. A. 1991, *Phys. Rev. Lett.*, 67, 2414
Hurley, K. *et al.* 1999, *Astrophys. J.*, 510, L111
Lavagno, A. 2010, *Phys. Rev. C*, 81, 044909
Lavagno, A. & Pigato, D. 2012, *Phys. Rev. C*, 86, 024917
Lavagno, A. 2013, *Eur. Phys. J. A*, 49, 102
Lopes, L. L. & Menezes D. P. 2012, *Braz. J. Phys.*, 42, 428
Mareghetti, S. & Stella, L. 1995, *Astrophys. J.*, 442, L17
Rabhi, A., Panda, P. K., & Providencia C. 2011, *Phys. Rev. C*, 84, 035803

Magnetic Fields throughout Stellar Evolution
Proceedings IAU Symposium No. 302, 2013
P. Petit, M. Jardine & H. Spruit, eds.

doi:10.1017/S1743921314002701

Search for Stable Magnetohydrodynamic Equilibria in Barotropic Stars.

J. P. Mitchell[1,2], J. Braithwaite[2], N. Langer[2], A. Reisenegger[1] and H. Spruit[3]

[1]Instituto de Astrofísica, Facultad de Física, Pontificia Universidad Católica de Chile, Av. Vicuña Mackenna 4860 7820436 Macul, Santiago - Chile
email: jmitchel@astro.puc.cl

[2]Argelander Institut, University of Bonn, Auf dem Huegel 71, 53121 Bonn - Germany

[3]Max-Planck-Institut für Astrophysik, Karl-Schwarzschild-Str. 1, D-85748 Garching, Germany

Abstract. It is now believed that magnetohydrodynamic equilibria can exist in stably stratified stars due to the seminal works of Braithwaite & Spruit (2004) and Braithwaite & Nordlund (2006). What is still not known is whether magnetohydrodynamic equilibria can exist in a barotropic star, in which stable stratification is not present. It has been conjectured by Reisenegger (2009) that there will likely not exist any magnetohydrodynamical equilibria in barotropic stars. We aim to test this claim by presenting preliminary MHD simulations of barotropic stars using the three dimensional stagger code of Nordlund & Galsgaard (1995).

1. Introduction

The search for magnetohydrodynamic (MHD) equilibria in stars has been an ongoing study for the last few decades. Early works looked at the stability of purely poloidal and purely toroidal magnetic fields in axial symmetry. The configurations in each of these cases were found to be unstable. The instability in the purely toroidal fields was found by Tayler (1973), who showed that the field was prone to de-stabilization due to the kink and and interchange instabilities. In the purely poloidal field case, it was found by Markey & Tayler (1973) and Wright (1973) that the field is unstable in the region around the "neutral line", where the field vanishes. It was then suggested that if MHD equilibria exist, they would consist of a mixed poloidal-toroidal field, as each component would work to stabilize the other. Due to the difficulty of analytically solving the mixed poloidal-toroidal fields, an impasse was reached in the community until the numerical breakthrough of Braithwaite & Spruit (2004) and Braithwaite & Nordlund (2006). The authors used 3-D simulations of a stably stratified star to show that an initially random magnetic field could reach a twisted-torus MHD equilibrium, which was stable on timescales much longer than the Alfvén time.

The aforementioned works of Braithwaite & Spruit (2004) and Braithwaite & Nordlund (2006) had focused on stably stratified stars. Whether or not stable MHD equilibria can exist in a barotropic star is still under debate. It was suggested by Akgun et. al. (2013) that a key ingredient in the stability of the MHD equilibria in stably stratified stars is their stable stratification. This stratification requires very strong forces in the radial direction in order to overcome the stratification. Further conjectures by Reisenegger (2009) have claimed that stable MHD equilibria will not exist in barotropic stars.

On the other hand, there has been extensive work done looking for MHD equilibria in axially symmetric barotropic stars: Tomimura & Eriguchi (2005), Ciolfi *et. al.* (2009),

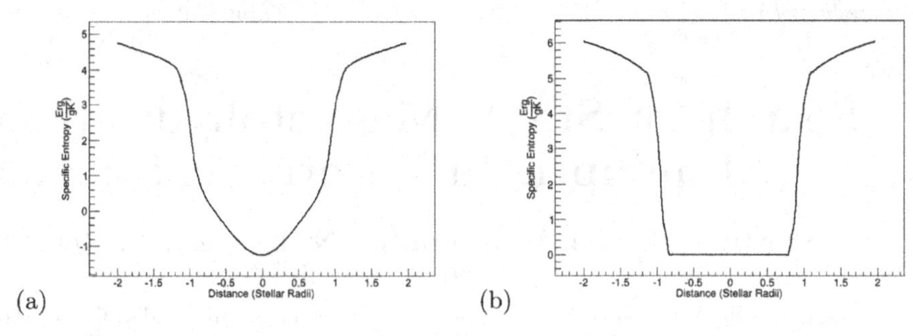

Figure 1. The specific entropy along the x-axis through the computational box for both models. Panel (a) is for the polytrope of polytropic index $n = 3$ the stably stratified star of non-constant specific entropy. Panel (b) is for the polytrope of polytropic index $n = 1.5$. Note the star in this model is barotropic, as the specific entropy is uniform within the star.

Lander & Jones (2009), Reisenegger (2009), Akgün *et. al.* (2013), Armaza *et. al.* (2013). Although these works have found MHD equilibria, the stability of the equilibria have not been investigated and they have all been limited to the axially-symmetric case.

To investigate the possible existence of stable MHD equilibria in barotropic stars, we have evolved initially random magnetic field configurations in barotropic stars to see if stable equilibria can be reached. In Section 2 of the paper we discuss the models used in our simulations. Section 3 contains our preliminary results in the search for stable MHD equilibria in barotropic stars, and Section 4 contains a discussion of our results and of future studies.

2. The Models

To search for stable MHD equilibria in barotropic stars, we use the stagger code of Nordlund & Galsgaard (1995), a three dimensional high order finite-difference MHD code in Cartesian coordinates. The model contains a perfectly conducting spherical star of radius one quarter the size of the grid box embedded in a poorly conducting atmosphere. Two different stars were used in our simulations. The first is an $n = 3$ polytrope, which is stably stratified. This model was used as a comparison to the second star, a polytrope of polytropic index $n = 1.5$, which yields an initially uniform specific entropy, $s = 0$ in the star, and thus a barotropic star. The profiles of the specific entropies of both models can be seen in Fig. 1.

Due to heat diffusion present in the code, the star will not keep a uniform specific entropy, resulting in an entropy gradient and a component of stable stratification. To remedy this, we included a term in the energy equation in the code which forces the specific entropy back to its initial value. The term takes the form :

$$\frac{de}{dt} = ... + \frac{\rho T(s_0 - s)}{\tau_s}, \tag{2.1}$$

where ρ, T, s, and s_0 are the density, temperature, specific entropy, and initial specific entropy respectively, and τ_s is the timescale at which this entropy term forces the star back to its initially barotropic structure. In our models τ_s is a free parameter that was varied in different simulations with the only restriction being that it must be longer than the dynamical timescale of the star, and shorter than the diffusion timescale of the star. With this term included, we are able to force the star to its initial barotropic structure and search for the existence of stable MHD equilibria.

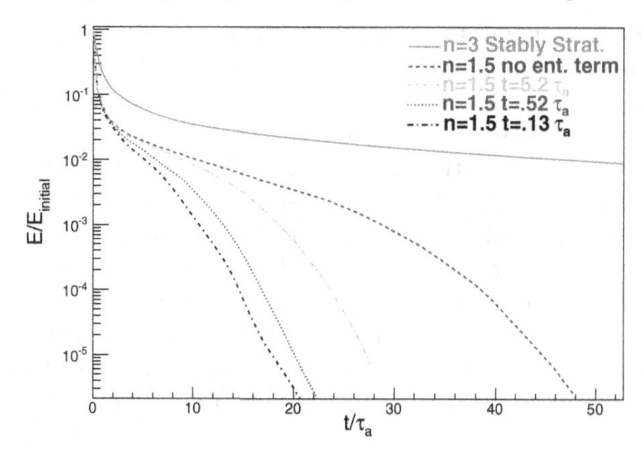

Figure 2. Total magnetic energy relative to initial value versus time for differing models all starting from an initially random magnetic field configuration. The solid gray curve is for a stably stratified model, which reaches a stable MHD equilibrium. All other curves are of a barotropic star with differing values of the τ_s, which are 0.13, 0.52, and 5.2 τ_A for the black dashed-dotted, the red dotted and teal dashed-dotted curves respectively as well as the blue dashed curve which does not include the entropy forcing term. Note that none of the barotropic models reach a stable MHD equilibrium.

3. Results

All our models start out with an initially random magnetic field. A series of simulations were run for the $n = 1.5$ polytrope model with differing values for the entropy timescale, τ_s, presented in Eq. 2.1. We used four different $n = 1.5$ models, one that did not include the entropy term in the energy equation, and three that included the entropy term with timescale values of 0.13, 0.52, and 5.2 τ_A, where τ_A is the Alfvén crossing time of the star defined as $\tau_A = R\sqrt{M/2E}$, where R is the radius of the star, M is the mass of the star and E is the total magnetic energy in the star. To compare to the stably stratified star, we also present one run in which an $n = 3$ polytrope was used. The evolution of the magnetic energy can be found in Fig. 2. The main thing to take away here is that the stably stratified model reaches an equilibrium, as the field energy decays on a very long timescale, while all the barotropic models do not reach a stable equilibrium. This suggests that stable stratification does seem to be important for reaching a stable MHD equilibrium. This is further elucidated by looking at how quickly the magnetic energy decays for different values of the entropy timescale. The smaller values of the entropy timescales keep the stellar structure very close to the initial barotropic structure throughout the evolution, but as this timescale is increased, or the entropy term in Eq. 2.1 is not included for the $n = 1.5$ stars, the decay of the magnetic energy is much slower. This is due to the fact that the structure of these models obtain a slight gradient in their specific entropy profiles and thus a small amount of stratification, which act to slow the decay of the magnetic field slightly.

4. Discussion

We have searched for stable MHD equilibria in barotropic stars with an initially random magnetic configuration. Our results indicate that from the random magnetic configuration used, no stable MHD equilibrium is reached. This does not mean that stable MHD equilibria cannot exist in barotropic stars, but that starting from the particular random field configuration used, such an equilibrium cannot be reached. Further study is needed, different initial magnetic field configurations should be studied, both from random magnetic field configurations, and from the physically motivated equilibria found in the axially-symmetric configurations of Armaza *et. al.* (2013).

We note that the finite time scale τ_s can in fact represent an astrophysically relevant time scale. In entropy-stratified fluids such as the radiative regions of main-sequence stars and white dwarfs, a magnetic flux tube in mechanical equilibrium will have a lower specific entropy than the surrounding gas, but this entropy difference will be slowly erased by heat conduction. In the compositionally stratified neutron star cores, such a flux tube will have a higher fraction of charged particles (relative to neutrons) than its surroundings, but this composition difference can be erased by beta decays or ambipolar diffusion (Reisenegger (2009); Hoyos, Reisenegger & Valdivia (2008); and Reisenegger (2013)). Thus, all these stars behave as stably stratified (non-barotropic) on short time scales, but barotropic on longer time scales, which are controlled by (but substantially longer than) τ_s.

References

Akgün, T. *et al.* 2013, *MNRAS*, 433, 2445
Armaza, C., Reisenegger, A., & Alejandro Valdivia, J. and Marchant, P. 2013, *ArXiv e-prints*, arxiv:1305.0592
Braithwaite, J. & Spruit, H. C. 2004, *Nature*, 431, 819
Braithwaite, J. & Nordlund, A. 2006, *A&A*, 450, 1077
Ciolfi, R. *et al.* 2009, *MNRAS*, 397, 913
Flowers, E. & Ruderman, M. A. 1977, *ApJ*, 215, 302
Hoyos, J., Reisenegger, A., & Valdivia, J. A. 2008, *A&A*, 487, 789
Lander, S. K. & Jones, D. I. 2009, *MNRAS*, 395, 2162
Markey, P. & Tayler, R. J. 1973, *MNRAS*, 163, 77
Nordlund, Å. & Galsgaard, K. 1995, *http://www.astro.ku.dk/ aake/papers/95.ps.gz*
Reisenegger, A. 2009, *A&A*, 499, 557
Reisenegger, A. 2013, *IAU Proceedings*, 302
Tayler, R. J. 1973, *MNRAS*, 161, 365
Tomimura, Y. & Eriguchi, Y. 2005, *MNRAS*, 359, 1117
Wright, G. A. E. 1973, *MNRAS*, 162, 339

Author Index

Aarnio, A. N. – 80
Acreman, D. M. – 80
Alakoz, A. – 38
Alecian, E. – 25, 44, 46, 70, 87, 265, 311, 313
Alencar, S. H. P. – 44, 50
Alleq, N. – 164
Aloy, C. – 64
Aloy, M. A. – 64
Aranguren, S. M. – 64
Argiroffi, C. – 44, 46, 48, 66, 102
Ariste, A. L. – 130, 164
Armaza, C. – 419
Audard, M. – 44, 46
Aurière, M. – 265, 338, 359, 365, 367, 373, 385

Bagnulo, S. – 87, 300
Baklanova, D. – 196, 381
Balega, Y. – 317
Ballot, J. – 222
Banerjee, R. – 10
Bartus, J. – 198, 379
Basri, G. – 216
Bayandina, O. – 38
Bério, P. – 202
Berta-Thompson, Z. K. – 176
Bigot, L. – 202
Bloemen, S. – 222
Bohlender, D. – 306
Bohlender, D. A. – 265, 288
Böhm, T. – 338
Bonito, R. – 48, 66
Borisova, A. P. – 365
Bouvier, J. – 44, 46, 50, 102, 110
Braithwaite, J. – 255, 441
Brandenburg, A. – 134
Briquet, M. – 302
Brown, C. – 138, 148
Brun, A. S. – 114, 363
Butkovskaya, V. – 381
Butkovskyi, V. – 381

Caramazza, M. – 102
Cargile, P. A. – 100
Carter, B. – 138, 148
Castro, M. – 142, 144
Castro-Chacón, J. – 154
Ceillier, T. – 222
Chaboyer, B. – 150
Charbonneau, D. – 176
Charbonnel, C. – 359, 365, 367, 373
Chièze, J.-P. – 66
Chiavassa, A. – 202
Christensen, U. – 166
Christensen, U. R. – 174
Cohen, D. H. – 320, 330
Contreras, M. E. – 398
Coste, B. – 302
Crozet, P. – 164
Cumming, A. – 415

Daiffallah, K. – 126
Damiani, F. – 44, 46
David-Uraz, A. – 265, 334
de Sá, L. – 66
Degroote, P. – 302
del Burgo, C. – 247
Dittmann, J. – 176
do Nascimento Jr., J. D. 144
do Nascimento, J. D. – 142
do Nascimento, J.-D. – 138
Donati, J.-F. – 40, 44, 46, 110, 237, 359
Drake, N. – 373
Drake, N. A. – 270, 309, 315, 367
Duarte, L. – 166, 174
Duarte, T. – 144
Dushin, V. – 270
Dyachenko, V. – 317

Espagnet, O. – 359

Fabas, N. – 385
Fares, R. – 180, 245
Feiden, G. A. – 150
Fichtinger, B. – 243
Folsom, C. – 265
Folsom, C. P. – 87, 110, 313
Fonseca, N. N. J. – 50
Fujisawa, K. – 423, 427

Gagné, M. – 330
García, R. A. – 222
Gastine, T. – 166, 174
Gelly, B. – 164
Giarrusso, M. – 274
Gillet, D. – 385
Glagolevski, Y. – 309
Glagolevskij, Y. – 274
Gondoin, P. – 106, 377
González, M. – 66

Gourgouliatos, K. N. – 415
Gregory, S. – 46
Gregory, S. G. – 40, 44
Grunhut, J. – 265
Grunhut, J. H. – 70
Güdel, M. – 44, 46, 243
Guillén, P. F. – 398
Guinan, E. – 142

Harmon, R. O. – 212
Harries, T. J. – 80
Henrichs, H. F. – 265, 280
Herpin, F. – 385
Herrera, J. – 154
Higa, M. – 200
Hillenbrand, L. A. – 40
Hiriart, D. – 154, 400, 402
Ho, W. C. G. – 435
Holzwarth, V. R. – 44
Hubrig, S. – 270
Huenemoerder, D. – 46
Huenemoerder, D. P. – 44
Hussain, G. – 46
Hussain, G. A. J. – 25, 40, 44

Ibgui, L. – 48, 66
Iñiguez-Garín, E. – 402
Iñiguez-Garín, E. – 154
Irwin, J. – 176

James, D. J. – 100
Jardine, M. – 40, 237, 245, 251
Jeffers, S. – 138
Jeffers, S. V. – 142, 146, 251
Jilinski, E. G. – 315
Jiménez, A. – 222
Johnson, N. M. – 313
Johnstone, C. P. – 243
Juarez, A. J. – 100

Kastner, J. – 46
Kastner, J. H. – 44
Khalack, V. – 272, 274, 284
Kholtygin, A. F. – 270
Kirk, H. – 10
Kisaka, S. – 427
Klassen, M. – 10
Kochukhov, O. – 170, 265, 290, 304, 306, 313, 369
Konstantinova-Antova, R. – 359, 365, 367, 373, 385
Korhonen, H. – 350
Korhonen, H. H. – 212
Kővári, Z. – 198, 379, 383
Kriskovics, L. – 198, 379, 383
Kudryavtsev, D. – 309
Küker, M. – 194
Kurosawa, R. – 54

Lagrange, A.-M. – 202
Landin, N. R. – 112
Landstreet, J. – 274, 284
Landstreet, J. D. – 87
Lang, P. – 237
Langer, N. – 1, 441
Lanz, T. – 66
Lavagno, A. – 439
Lèbre, A. – 373, 385
Le Men, C. – 164
LeBlanc, F. – 272
Leone, F. – 274
Leutenegger, M. A. – 330
Ligi, R. – 202
Lignières, F. – 338
Likuski, K. – 313
Lingua, F. – 439
Littlefair, S. P. – 91
Llama, J. – 245
Lüftinger, T. – 243
Lyashko, D. – 274, 309

Magalhães, V. d. S. – 21
Maggio, A. – 44, 46
Maksimov, A. – 317
Malogolovets, E. – 317
Marchant, P. – 419
Marsden, S. – 138, 146, 148
Marsden, S. C. – 142, 251
Martínez, B. – 402
Martin, A. J. – 300
Mathis, S. – 265, 302, 311
Mathur, S. – 222
Matsakos, T. – 48, 66
Matsuoka, M. – 200
Mayne, N. J. – 40
Mendes, L. T. S. – 112
Metcalfe, T. S. – 222
Micela, G. – 102
Miceli, M. – 48
Mitchell, J. P. – 441
Monin, D. – 288
Monnier, J. D. – 80, 212
Montmerle, T. – 44, 46
Moraux, E. – 102
Morin, J. – 40, 110, 146, 166, 237
Mourard, D. – 202

Nakahira, S. – 200
Nardetto, N. – 202
Navarro, S. G. – 400
Nazé, Y. – 330

Negoro, H. – 200
Neiner, C. – 265, 302, 311, 348
Núñez-Alfonso, J. M. – 154

Oláh, K. – 198, 224, 379, 383
Orlando, S. – 48, 66
Owocki, S. – 265
Owocki, S. P. – 320, 330

Palacios, A. – 363
Parès, L. – 348
Patel, N. A. – 398
Pereira, C. B. – 315
Peres, G. – 48, 66
Pereyra, A. – 21
Perraut, K. – 202
Petit, P. – 110, 138, 142, 146, 251, 338, 348, 359, 365, 367, 373, 385
Petit, V. – 265, 320, 330, 334
Piskunov, N. – 170, 304
Plachinda, S. – 196, 381, 402
Pogodin, M. A. – 315
Polosukhina, N. – 309
Poppenhaeger, K. – 239
Porto de Mello, G. F. – 142
Pudritz, R. E. – 10

Quiroz, F. – 402

Racca, G. – 21
Ramírez, V. J. – 402
Ramírez, J. – 400
Ramirez-Velez, J. – 154
Rastegaev, D. – 317
Rea, N. – 429
Reale, F. – 48, 66
Régulo, C. – 222
Reiners, A. – 146, 156, 166, 170, 216
Reinhold, T. – 216
Reisenegger, A. – 404, 419, 441
Ribas, I. – 142
Rodrigues, C. V. – 21
Roettenbacher, R. M. – 212
Romanova, M. M. – 54
Romanyuk, I. I. – 276
Rosén, L. – 369
Ross, A. J. – 164
Roudier, T. – 359
Rüdiger, G. – 194
Rusomarov, N. – 304
Russell, A. J. B. – 237

Sabin, L. – 398, 400
Sacco, G. G. – 44, 46
Saikia, S. B. – 146
Salabert, D. – 222
Schröder, K.-P. – 373
See, V. – 251
Seemann, U. – 170
Seifried, D. – 10
Semenko, E. A. – 84
Shavrina, A. – 309
Shavrina, A. V. – 274
Shulyak, D. – 170, 309, 313
Smirnova, M. – 309
Southworth, J. – 247
Spagiari, E. – 220
Spruit, H. – 441
Stassun, K. G. – 100
Stehlé, C. – 66
Sthelé, C. – 48
Stift, M. J. – 300
Strassmeier, K. G. – 379
Sudnik, N. – 270
Sudnik, N. P. – 280
Sundqvist, J. O. – 320

Thibeault, C. – 272
Tomida, H. – 200
Townsend, R. H. D. – 320, 330
Tsuboi, Y. – 200
Tsvetkova, S. – 367, 373
Tsymbal, V. – 306

ud-Doula, A. – 320, 330, 334

Valdez, J. – 402
Valdivia, J. A. – 419
Válio, A. – 220
Val'tts, I. – 38
Valyavin, G. – 402
Vaz, L. P. R. – 112
Vázquez, R. – 398
Vida, K. – 198, 224, 379, 383
Vidotto, A. A. – 228, 237, 245, 251
Vilas-Boas, J. W. – 21
Vilela, C. – 247
Vlemmings, W. H. T. – 389

Wade, G. – 306, 311, 373
Wade, G. A. – 44, 87, 265, 313, 330, 334, 338, 359, 367, 369
Waite, I. – 148
Walkowicz, L. M. – 206
Warnecke, J. – 134
Weber, M. – 379
Weisenburger, K. L. – 176
West, A. A. – 176
Wicht, J. – 166

Wolk, S. J. – 239
Wood, K. – 245
Wright, N. J. – 190

Yadav, R. K. – 174
Yakunin, I. – 306
Yakunin, I. A. – 276
Yameogo, B. – 272

Zauderer, B. A. – 398
Zhang, Q. – 398
Zijlstra, A. A. – 398

Printed in the United States
by Baker & Taylor Publisher Services